COMPARATIVE CARDIAC IMAGING

function
flow
anatomy
and quantitation

Edited by

Bruce H. Brundage, MD

Professor of Medicine and
Radiologic Sciences
UCLA School of Medicine

Chief
Division of Cardiology
Harbor-UCLA Medical Center

Director of Research
Saint John's Cardiac Research Center
Torrance, California

AN ASPEN PUBLICATION®
Aspen Publishers, Inc.
Rockville, Maryland
1990

Library of Congress Cataloging-in-Publication Data

Comparative cardiac imaging : function, flow, anatomy, and quantitation / edited by Bruce H. Brundage.
  p. cm.
"An Aspen publication".
Includes bibliographical references.
ISBN: 0-8342-0125-9
1. Heart—Imaging. I. Brundage, Bruce H.
[DNLM: 1. Echocardiography—methods. 2. Heart—radiography. 3. Heart—radionuclide imaging. 4. Magnetic Resonance Imaging—methods. 5. Nuclear Magnetic Resonance—methods. 6. Tomography, X-Ray Computed—methods. WG 141 C737]
RC683.5.I42C66 1990
616.1'20754—dc20
DNLM/DLC
for Library of Congress
89-18591
CIP

Copyright © 1990 by Aspen Publishers, Inc.
All rights reserved.

Aspen Publishers, Inc., grants permission for photocopying for limited personal or internal use. This consent does not extend to other kinds of copying, such as copying for general distribution, for advertising or promotional purposes, for creating new collective works, or for resale. For information, address Aspen Publishers, Inc., Permissions Department, 1600 Research Boulevard, Rockville, Maryland 20850.

The authors have made every effort to ensure the accuracy of the information herein, particularly with regard to drug selection and dose. However, appropriate information sources should be consulted especially for new or unfamiliar drugs or procedures. It is the responsibility of every practitioner to evaluate the appropriateness of a particular opinion in the context of actual clinical situations and with due consideration to new developments. Authors, editors, and the publisher cannot be held responsible for any typographical or other errors found in this book.

Editorial Services: Jane Coyle Garwood

Library of Congress Catalog Card Number: 89-18591
ISBN: 0-8342-0125-9

Printed in the United States of America

1 2 3 4 5

The editor dedicates this book to his teachers, *David Opdyke* and *Melvin Cheitlin*, who taught him the pursuit of excellence, and his colleagues *William Parmley, Kanu Chatterjee, Elias Botvinick,* and *Nelson Schiller,* who were a constant source of stimulating ideas.

Contents

Contributors xv

Preface xix

Part I Techniques

Chapter 1—Echocardiography and Doppler Techniques 1
Steven B. Feinstein, MD, and George A. Williams, MD

    Introduction to Imaging Echocardiography 1
    Ultrasound Imaging in Cardiology 1
    Doppler Techniques in Cardiology 11

Chapter 2—Cardiovascular Nuclear Medicine Techniques 17
Maleah Grover-McKay, MD

    Physics 17
    Instrumentation 17
    Radiopharmaceuticals 23

Chapter 3—Cardiac Computed Tomography: Technical Aspects 29
Douglas P. Boyd, PhD

    Introduction 29
    Principles of Electron Beam, Ultrafast CT Scanning 30
    Ultrafast CT Methods and Applications 31
    Ultrafast CT Performance 33
    Future Technical Developments and Applications 37
    Summary 38

Chapter 4—Magnetic Resonance Imaging: Technical Aspects 41
Ronald M. Peshock, MD, and Donald M. Moore, Jr., MD

    Introduction 41
    Basic Principles of Nuclear Magnetic Resonance 41
    Basic Principles of Imaging 42
    Contrast Mechanisms 43
    Flow Effects 44
    Techniques Relevant to Cardiac MRI 44
    Summary 46

Part II Coronary Artery Disease

Chapter 5—Cardiac Digital Angiography 47
Jonathan Tobis, MD, Walter Henry, MD, and Orhan Nalcioglu, PhD

    Introduction 47
    Capabilities of Digital Processing 49
    Videodensitometry 53
    Assessment of Left Ventricular Function with Digital Imaging 55
    Digital Coronary Angiography 57
    Summary 61

Chapter 6—Coronary Arteriography 65
George T. Kondos, MD

    Introduction 65
    Indications for Coronary Arteriography 65
    Normal Coronary Anatomy 66
    Coronary Angiography Techniques 73
    Radiographic Evaluation of Coronary Anatomy 81
    Complications of Coronary Arteriography 82
    Pitfalls in Interpretation of the Coronary Arteriogram 86
    Limitations of Coronary Angiography 88

Chapter 7—Contrast Media 91
Bruce H. Brundage, MD, Eva V. Chomka, MD, Christopher J. Wolfkiel, PhD, James L. Ferguson, PhD, and Sanjay Deshpande, MD

    Introduction 91
    Chemistry of Contrast Media 91
    Cardiovascular Effects 92
    Hemostatic Effects 96
    Renal Failure 96
    Anaphylactoid Reactions 97
    Cost 98
    Summary 98

Chapter 8—Measurement of Left Ventricular Function from Contrast Angiograms in Patients with Coronary Artery Disease 101
Florence H. Sheehan, MD

    Introduction 101
    Objectives 101
    Methods 101
    Applications 110
    Summary 115

Chapter 9—Digital Angiography for the Assessment of Left Ventricular Function and Myocardial Perfusion in Coronary Artery Disease **119**
Robert A. Vogel, MD

    Introduction 119
    Instrumentation 119
    Subtraction Techniques 119
    Contrast Medium Administration 120
    Ventriculography 121
    Summary 125

Chapter 10—Echocardiographic Evaluation of the Left Ventricle **129**
Nelson B. Schiller, MD

    Physical Principles and Instrumentation in Echocardiography 129
    Evaluation of the Left Ventricle 134
    Evaluation of Coronary Artery Disease and Related Abnormalities of the Left Ventricle 147
    Evaluation of Cardiomyopathies 150
    Functional Evaluation with Stress Testing 156
    Transesophageal Echocardiography for Intraoperative Assessment of Left Ventricular Function 158

Chapter 11—Acute Myocardial Infarction Infarct Avid Scintigraphy **167**
Elias H. Botvinick, MD, and Michael W. Dae, MD

    Introduction 167
    Development of the Method 167
    Specific Clinical Applications 174

Chapter 12—Evaluation of Myocardial Perfusion Using Planar Thallium 201 Imaging at Rest and During Exercise **181**
Sanjiv Kaul, MD

    Background 181
    Thallium 201 Imaging at Rest 185
    Thallium 201 Imaging During Exercise for Detection of Coronary Artery Disease 187
    Exercise Thallium Imaging for Determination of High-Risk Patients and Prognosis 195
    Specific Situations in Which To Use Exercise Thallium Imaging 198
    Dipyridamole Thallium 201 Imaging 200
    Appendix 12-A 207

Chapter 13—Evaluation of Left Ventricular Function in Coronary Artery Disease: Radioisotope Imaging at Rest and During Stress **211**
Richard J. Peterson, MD, and Robert H. Jones, MD

    Introduction 211
    Perspective 211
    Technique 213
    Normal Response to Exercise 214
    Bayesian Concepts 214
    Prevalence 214
    Selection for Catheterization 215
    Prognosis 215
    Prognosis after Acute Myocardial Infarction 219
    Emerging Technologies 219
    Summary 220

Chapter 14—Evaluation of Acute Myocardial Infarction by Computed Tomography ... 223
Bruce H. Brundage, MD, and Eva V. Chomka, MD

Conventional (Slow) Computed Tomography ... 223
Imaging Acute Myocardial Infarction in Humans by Ultrafast Computed Tomography ... 225
Limitations of Other Infarct-Sizing Techniques ... 226

Chapter 15—The Evaluation of Myocardial Blood Flow by Ultrafast Computed Tomography ... 231
Christopher J. Wolfkiel, PhD, Eva V. Chomka, MD, and Bruce H. Brundage, MD

Chapter 16—Evaluation of Left Ventricular Function by Exercise Bicycle Ergometry Ultrafast Computed Tomography ... 239
Eva V. Chomka, MD, and Bruce H. Brundage, MD

Cardiovascular Response to Exercise ... 239
Alternative Methods of Cardiovascular Evaluation ... 240
Ultrafast CT Bicycle Ergometry ... 240
Comments on Exercise Methods ... 248

Chapter 17—Evaluation of Left Ventricular Aneurysm and Thrombus by Ultrafast Computed Tomography ... 251
Eulalia Roig, MD, Eva V. Chomka, MD, and Bruce H. Brundage, MD

Chapter 18—Evaluation of Coronary Artery Disease by Nuclear Magnetic Resonance Imaging ... 257
George E. Wesbey, MD

Physiologic Gating and Acquisition Techniques ... 257
Normal Anatomy, Physiology, and Biophysics ... 258
NMR Angiography in Coronary Artery Disease ... 259
Functional Evaluation of Coronary Artery Disease by NMR ... 260
Canine Studies of Coronary Occlusion by NMR ... 260
Canine Studies of Coronary Occlusion and Reperfusion by NMR ... 261
Canine Studies of Myocardial Ischemia with Gadolinium ... 261
Acute Myocardial Ischemia in Humans ... 263
Clinical Sequelae of Acute Myocardial Ischemia ... 264
Summary of NMR Imaging in Coronary Artery Disease ... 265

Chapter 19—Imaging Techniques for the Evaluation of Coronary Artery Bypass Grafts ... 269
Timothy M. Bateman, MD, and James S. Whiting, PhD

Indirect Evaluation of Bypass Grafts ... 269
Direct Evaluation of Bypass Grafts ... 271
Summary ... 278

Part III Valvular Heart Disease

Chapter 20—Evaluation of Valvular Heart Disease by Contrast Angiography ... 281
Charles R. McKay, MD, and Maleah Grover-McKay, MD

Fluoroscopy and Contrast Cineangiography of the Left Ventricle ... 281
Fluoroscopy and Cineaortography ... 283
Aortic Stenosis ... 284
Aortic Regurgitation ... 285
Mitral Stenosis ... 286
Mitral Regurgitation ... 288
Pulmonic Valve Disease ... 289

Tricuspid Valve Disease 289
Mixed Valve Disease 289
Prosthetic Valve Disease 290
Summary 290

Chapter 21—Echocardiographic Evaluation of Valvular Heart Disease 293
Eric K. Louie, MD

Introduction 293
Ventricular Outflow Tract: Aortic and Pulmonic Valves 294
Ventricular Inflow Tract: Mitral and Tricuspid Valves 300
Contrast and Intraoperative Echocardiography 306

Chapter 22—Doppler Echocardiographic Evaluation of Valvular Heart Disease 311
Eric K. Louie, MD

Blood Flow Velocity Measurement by the Doppler Principle 311
Doppler Echocardiographic Instrumentation 313
Stenosis and Incompetence of the Semilunar Valves 315
Stenosis and Incompetence of the Atrioventricular Valves 321
Doppler Assessment of Prosthetic Heart Valves 325

Chapter 23—The Value of Radioisotope Blood Pool Imaging for Evaluation of Valvular Heart Disease .. 329
Robert O. Bonow, MD

Mitral Stenosis 330
Aortic Stenosis 330
Aortic Regurgitation 331
Mitral Regurgitation 334
Summary 335

Chapter 24—Evaluation of Valvular Heart Disease by Ultrafast Computed Tomography ... 339
Eulalia Roig, MD, Eva V. Chomka, MD, and Bruce H. Brundage, MD

Ultrafast CT 339
Ventricular Function 340
Assessment of Regurgitant Fractions 340
Mitral Valve 340
Aortic Valve 342
Tricuspid Valve 344
Pulmonic Valve 344

Chapter 25—Magnetic Resonance Imaging of Cardiac Valvular Disease 347
Joseph A. Utz, MD

Spin-Echo Magnetic Resonance Imaging in Congenital Valvular Disorders 347
Dynamic Magnetic Resonance Imaging in Acquired Valvular Disease 348
Summary 351

Part IV Primary Myocardial Disease

Chapter 26—Evaluation of Dilated and Hypertrophic Cardiomyopathy by Cardiac Catheterization and Angiography 353
Jeffrey Shanes, MD

Hypertrophic Cardiomyopathy 353
Coronary Anatomy 356
Gradients 356

	Dilated Cardiomyopathy	357
	Summary	359

Chapter 27—Evaluation of Hypertrophic and Dilated Cardiomyopathy by Echocardiography ... 361
Andrew J. Buda, MD

    Hypertrophic Cardiomyopathy ... 361
    Dilated Cardiomyopathy ... 369

Chapter 28—Evaluation of Cardiomyopathy by Radioisotopes ... 375
Robert A. Quaife, MD, and John B. O'Connell, MD

    Dilated Cardiomyopathy ... 375
    Specific Heart Muscle Disease ... 378
    Hypertrophic Cardiomyopathy ... 379
    Restrictive Cardiomyopathy ... 381
    Summary ... 382

Chapter 29—Evaluation of Hypertrophic Cardiomyopathy and Dilated Cardiomyopathy by Ultrafast Computed Tomography ... 385
Eva V. Chomka, MD, and Bruce H. Brundage, MD

    Hypertrophic Cardiomyopathy ... 386
    Restrictive Cardiomyopathy ... 390
    Dilated Cardiomyopathy ... 391

Chapter 30—Evaluation of Cardiomyopathy by Magnetic Resonance Imaging ... 395
Udo Sechtem, MD, and Charles B. Higgins, MD

    Hypertrophic Cardiomyopathy ... 395
    Congestive Cardiomyopathy ... 397
    Doxorubicin Hydrochloride Cardiomyopathy ... 398
    Transplant Rejection ... 398
    Restrictive Cardiomyopathy ... 399
    Summary ... 400

Chapter 31—Ultrasonic Myocardial Tissue Characterization ... 403
Julio E. Pérez, MD, Steve M. Collins, PhD, James G. Miller, PhD, and David J. Skorton, MD

    Basic Concepts of Ultrasonic Tissue Characterization ... 404
    Applications of Ultrasonic Myocardial Tissue Characterization ... 407
    Potential of Ultrasonic Tissue Characterization in Clinical Cardiology ... 413
    Summary ... 414

Chapter 32—Evaluation of Myocardial Metabolism by Nuclear Magnetic Resonance Spectroscopy ... 417
Saul Schaefer, MD

    The NMR Experiment ... 417
    Spectroscopy ... 417
    Spectroscopy of Biologically Important Nuclei ... 421
    Summary ... 426

Part V Pericardial Disease

Chapter 33—Evaluation of Pericardial Disease by Cardiac Catheterization, Fluoroscopy, and Angiocardiography ... 429
Dale C. Wortham, MD, and W. John Nicholas, MD

    Cardiac Tamponade ... 429
    Constrictive Pericarditis ... 432

Effusive-Constrictive Pericarditis	436
Restrictive Cardiomyopathy	437
Congenital Anomalies of the Pericardium	437
Summary	438

Chapter 34—Echocardiography of Pericardial Disease **441**
Ivan A. D'Cruz, MD, FRCP

Echocardiographic (M-Mode and 2-D) Estimation of the Size of Pericardial Effusions	442
Cardiac Tamponade	444
Role of Echocardiography in Pericardial Paracentesis	446
Pericardial Thickening, Sclerosis, and Calcification	446
Pericardial Constriction	447
Absent Pericardium	448
Summary	448

Chapter 35—Computed Tomography in the Diagnosis of Pericardial Disease **451**
William Stanford, MD

Advantages and Disadvantages of CT as Compared with Ultrasound and Magnetic Resonance Imaging	451
Normal Pericardial Anatomy	452
Imaging Techniques and Protocols	452
Congenital Absence of the Pericardium	453
Pericardial Cyst	453
Pericardial Effusion	453
Pericardial Thickening	455
Benign Tumors	456
Malignant Disease	456
Constrictive Pericarditis	456
Associated Lesions	457
Ultrafast CT in Pericardial Disease	457

Chapter 36—Evaluation of Pericardial Disease by Magnetic Resonance Imaging **459**
Madeleine R. Fisher, MD

Anatomy	459
Technique	460
MRI Appearance of the Normal Pericardium	460
Congenital Anomalies	462
MRI Appearance of Diseased Pericardium	462
Summary	466

Part VI Congenital Heart Disease

Chapter 37—Evaluation of Congenital Heart Defects by Echocardiography and Doppler **469**
Elizabeth A. Fisher, MD

Septal Defects/Left-to-Right Shunts	469
Outflow Obstruction/Stenosis	472
Malalignment	475
Great Artery Malposition	476
Univentricular Heart/Ventricular Hypoplasia	476
Summary	478

Chapter 38—Evaluation of Cardiac Anatomy and Function Using Ultrafast Computed Tomography in Patients with Congenital Heart Disease **481**
W. Jay Eldredge, MD

    Introduction 481
    Volume Mode 483
    Flow Mode 484
    Cine Mode 489
    Summary 491

Chapter 39—Magnetic Resonance Imaging of Congenital Heart Disease **493**
Barbara Kersting-Sommerhoff, MD, and Charles B. Higgins, MD

    Imaging Techniques 493
    Normal Anatomy/Situs Determination 494
    Anatomic Evaluation of Congenital Anomalies 496
    Functional Evaluation of Congenital Anomalies with Cine MRI 501
    Summary 502

Part VII Tumors of the Heart

Chapter 40—Imaging of Cardiac Tumors **505**
Mary Jo Bertsch, MD, Eva V. Chomka, MD, and Bruce H. Brundage, MD

    Angiography 506
    Digital Subtraction Angiography 506
    Echocardiography 506
    Nuclear Imaging Techniques 509
    Computed Tomography 510
    Magnetic Resonance Imaging 511
    Summary 511

Part VIII The Great Vessels

Chapter 41—Evaluation of Diseases of the Great Vessels by Angiography and Digital Subtraction Angiography **515**
Bradley G. Langer, MD, and Dimitrios G. Spigos, MD

    Radiological Modalities for Evaluating Great Vessel Disease 515
    Normal Anatomy 518
    Aortic Trauma 519
    Arteriosclerosis and Arteriosclerotic Aneurysms 522
    Aortic Dissection 523
    Infectious and Autoimmune Conditions 524
    Summary 527

Chapter 42—Evaluation of the Great Vessels by Echocardiography and Doppler **531**
Stuart Rich, MD

    Aortic Aneurysms 531
    Aortic Dissection 532
    Coarctation of the Aorta 532
    Aortic Trauma 533
    Patent Ductus Arteriosus 533
    Idiopathic Dilatation of the Pulmonary Artery 533
    Pulmonary Embolism 534
    Estimation of Pulmonary Artery Pressure by Doppler Techniques 534

Chapter 43—Evaluation of Diseases of the Great Vessels by Computed Tomography **537**
Martin J. Lipton, MD, and Heber MacMahon, MD

    Advantages of CT 537
    Technical Considerations and Anatomic Localization of Scan Planes 538
    Contrast-Enhanced Studies of the Great Vessels 539
    Quantitative CT Techniques 541
    High-Speed Cine-CT Scanning 543
    Specific Applications of CT to the Diagnosis of Selected Pulmonary Artery Pathologies and
        Aortic Coarctation 544
    Summary 546

Chapter 44—Evaluation of the Great Vessels by Magnetic Resonance Imaging **549**
Joseph A. Utz, MD

    Vascular Imaging with Magnetic Resonance 549
    Aorta 550
    Pulmonary Arteries 554
    Venae Cavae 555
    Summary 555

Part IX Miscellaneous

Chapter 45—Comparative Safety of Cardiac Imaging Techniques **557**
Christopher J. Wolfkiel, PhD

    Ionizing Radiation Imaging Systems 558
    Nonionizing Radiation Imaging Systems 559
    Summary 560

Index **563**

Contributors

Timothy M. Bateman, MD
Chairman, Advanced Cardiac Imaging
Mid-America Heart Institute of St. Luke's Hospital
Clinical Associate Professor of Medicine
University of Missouri-Kansas City
Cardiovascular Consultants, Inc.
Kansas City, Missouri

Mary Jo Bertsch, MD
Former Cardiology Fellow
University of Illinois at Chicago
Chicago, Illinois
Cardiologist
Macneal Hospital
Berwyn, Illinois

Robert O. Bonow, MD
Senior Investigator
Cardiology Branch
National Heart, Lung, and Blood Institute
National Institutes of Health
Bethesda, Maryland

Elias H. Botvinick, MD
Professor of Medicine and Radiology
University of California Medical Center
San Francisco, California

Douglas P. Boyd, PhD
President
Imatron, Inc.
South San Francisco, California

Bruce H. Brundage, MD
Professor of Medicine and Radiologic Sciences
UCLA School of Medicine
Chief
Division of Cardiology
Harbor-UCLA Medical Center
Director of Research
St. John's Cardiac Research Center
Torrance, California

Andrew J. Buda, MD
Professor of Internal Medicine
Cardiology Division
University of Michigan Medical Center
Ann Arbor, Michigan

Eva V. Chomka, MD
Assistant Professor of Medicine
Cardiology Section
University of Illinois at Chicago
Chicago, Illinois

Steve M. Collins, PhD
Professor of Electrical and Computer Engineering,
   Radiology, and Biomedical Engineering
The University of Iowa
Iowa City, Iowa

Michael W. Dae, MD
Associate Professor of Radiology and Medicine
University of California, San Francisco
San Francisco, California

Sanjay Deshpande, MD
Cardiology Fellow
Cardiology Section
University of Illinois at Chicago
Chicago, Illinois

Ivan A. D'Cruz, MD, FRCP
Professor of Medicine
Medical College of Georgia
Chief, Echocardiography Laboratory
Veterans Administration Medical Center
Augusta, Georgia

W. Jay Eldredge, MD
Director, Section of Cardiac Imaging
Deborah Heart and Lung Center
Browns Mills, New Jersey
Associate Clinical Professor of Pediatrics
Robert Wood Johnson Medical School

Steven B. Feinstein, MD
Assistant Professor
Department of Medicine
Section of Cardiology
University of Chicago
Chicago, Illinois

James L. Ferguson, PhD
Associate Professor of Physiology
Department of Physiology
University of Illinois at Chicago
Chicago, Illinois

Elizabeth A. Fisher, MD
Professor of Clinical Pediatrics
Section of Pediatric Cardiology
Loyola University Medical Center
Maywood, Illinois

Madeleine R. Fisher, MD
Department of Radiology
Good Samaritan Hospital
Los Angeles, California

Maleah Grover-McKay, MD
Department of Internal Medicine
Division of Cardiology
University of Iowa Hospital
Iowa City, Iowa

Walter Henry, MD
Vice Chancellor for Health Sciences and Dean
California College of Medicine
University of California, Irvine
Irvine, California

Charles B. Higgins, MD
Professor of Radiology
Chief, Magnetic Resonance Imaging
Department of Radiology
University of California, San Francisco Medical Center
San Francisco, California

Robert H. Jones, MD
Department of Surgery
Duke Medical Center
Durham, North Carolina

Sanjiv Kaul, MD
Associate Professor of Medicine
Co-Director, Cardiac Imaging
Director, Cardiac Computer Center
University of Virginia School of Medicine
Charlottesville, Virginia

Barbara Kersting-Sommerhoff, MD
Radiologist
Technical University
Department of Radiology
Munich, West Germany

George T. Kondos, MD
Assistant Professor of Medicine
Department of Medicine
Section of Cardiology
Director
Cardiac Catheterization Laboratory
University of Illinois at Chicago
Chicago, Illinois

Bradley G. Langer, MD
Assistant Clinical Professor of Radiology
University of Illinois at Chicago
Attending Radiologist
Cook County Hospital
Chicago, Illinois

Martin J. Lipton, MD
Chairman
Department of Radiology
University of Chicago
Chicago, Illinois

Eric K. Louie, MD
Associate Professor of Medicine
Cardiology Section
Loyola University Medical Center
Maywood, Illinois

Heber MacMahon, MD
Associate Professor
Department of Radiology
University of Chicago
Chicago, Illinois

Charles R. McKay, MD
Department of Internal Medicine
Cardiovascular Division
University of Iowa School of Medicine
Iowa City, Iowa

James G. Miller, PhD
Professor of Physics
Research Professor of Medicine
Washington University School of Medicine
St. Louis, Missouri

Donald M. Moore, Jr., MD
Cardiology Fellow
University of Texas
Southwestern Medical Center at Dallas
Dallas, Texas

Orhan Nalcioglu, PhD
Professor
Radiological Science
California College of Medicine
University of California, Irvine
Irvine, California

W. John Nicholas, MD
Cardiovascular Diseases
Greeley Medical Clinic
Greeley, Colorado

John B. O'Connell, MD
Associate Professor of Medicine
Division of Cardiology
University of Utah Medical Center
Salt Lake City, Utah

Julio E. Pérez, MD
Associate Professor of Medicine
Medical Director, Cardiac Diagnostic Ultrasound
Washington University School of Medicine
St. Louis, Missouri

Ronald M. Peshock, MD
Nuclear Magnetic Resonance Imaging Center
Southwestern Medical Center
University of Texas
Dallas, Texas

Richard J. Peterson, MD
Teaching Scholar in Thoracic and Cardiovascular
  Surgery
Chief Resident—Surgery
Department of Surgery
Duke University Medical Center
Durham, North Carolina

Robert A. Quaife, MD
Cardiology Fellow
University of Utah Medical Center
Salt Lake City, Utah

Stuart Rich, MD
Associate Professor of Medicine
Department of Medicine
Section of Cardiology
University of Illinois College of Medicine at Chicago
Chicago, Illinois

Eulalia Roig, MD
Servicio de Cardiologia
Hospital Clinic
Barcelona, Spain

Saul Schaefer, MD
Assistant Professor of Medicine and Radiology
Division of Cardiology
University of California, San Francisco
San Francisco, California

Nelson B. Schiller, MD
Director of Noninvasive Imaging
Professor of Medicine and Radiology
University of California, San Francisco
Moffitt Hospital
San Francisco, California

Udo Sechtem, MD
Medizinische Klinik III
Universitat Koln
Koln, West Germany

Jeffrey Shanes, MD
Clinical Assistant Professor of Medicine
University of Illinois at Chicago
Chicago, Illinois

Florence H. Sheehan, MD
Research Associate Professor
Cardiovascular Research and Training Center
University of Washington
Seattle, Washington

David J. Skorton, MD
Associate Professor of Medicine and Electrical and Computer Engineering
Department of Internal Medicine
University of Iowa
Iowa City, Iowa

Dimitrios G. Spigos, MD
Chairman
Department of Radiology
Cook County Hospital
Professor of Radiology
University of Illinois at Chicago
Chicago, Illinois

William Stanford, MD
Associate Professor of Radiology
Department of Radiology
The University of Iowa Hospital and Clinics
Iowa City, Iowa

Jonathan Tobis, MD
Director
Cardiac Catheterization Laboratory
Acting Chief of Cardiology
Division of Cardiology
University of California, Irvine
Irvine, California

Joseph A. Utz, MD
Clinical Associate Professor of Radiology
Uniformed Services University of the Health Sciences
Bethesda, Maryland

Robert A. Vogel, MD
Head, Division of Cardiology
Professor of Medicine and Radiology
University of Maryland Hospital and School of Medicine
Baltimore, Maryland

George A. Williams, MD
Chief of Cardiology
John Cochran VA Medical Center
St. Louis, Missouri

George E. Wesbey, MD
Director
Magnetic Resonance Imaging
Department of Radiology
Skripps Memorial Hospital
La Jolla, California

James S. Whiting, PhD
Senior Research Physicist
Cedars-Sinai Medical Center
Assistant Professor of Medicine
UCLA School of Medicine
Los Angeles, California

Christopher J. Wolfkiel, PhD
Research Associate
Cardiology Section
University of Illinois at Chicago
Chicago, Illinois

Dale C. Wortham, MD, LTC, MC
Chief, Cardiology Service
Department of the Army
Walter Reed Army Medical Center
Washington, DC

Preface

Clinical cardiology has witnessed revolutionary change in cardiac imaging over the last 20 years. At the beginning of this period, cardiac catheterization with accompanying angiography was being refined and had already become a powerful diagnostic tool. The development of coronary angiography and, subsequently, preformed catheters totally changed the clinical approach to coronary artery disease. However, clinicians were not fully satisfied with this technique because of its invasive nature. Serious complications and even death could be the unhappy consequences of the procedure. Furthermore, the procedure is very expensive and usually requires hospitalization. Therefore, diagnosticians continued the search for new imaging techniques. Soon echocardiography appeared on the scene and rapidly became a valuable, totally noninvasive tool. It was also portable; the technique could be taken directly to the bedside or employed in a variety of outpatient settings. Its real-time capability and amenability to repetitive longitudinal studies have advanced the understanding of a variety of cardiovascular disorders, particularly those affecting the cardiac valves and left ventricular function. The subsequent refinement of Doppler techniques has further enhanced the value of this imaging method.

Soon after echocardiography arrived on the clinical scene, radioisotope imaging methods came into use as a noninvasive method for evaluating myocardial perfusion. This technique, coupled with treadmill exercise, significantly advanced the understanding and management of coronary artery disease and proved a useful adjunct in the management of a wide variety of other cardiac disorders.

The diagnostician, however, called for better-quality images and recognized the need for better quantification, whether analyzing cardiovascular anatomic features, cardiac performance, or myocardial blood flow. New technologies continue to arrive in response. Advances in computer technology made the processing of vast amounts of imaging data possible, making feasible, for example, the tomographic analysis of cardiac structure and function and the use of digital subtraction angiography in the cardiac catheterization laboratory. Positron emission tomography, employing cyclotron-produced short half-life isotopes, now permits study and evaluation of myocardial metabolism with a refinement never before possible in the clinical environment, providing such important insights as the recognition that chronically noncontractile myocardium can be viable and have return of function after revascularization.

Ultrafast computed tomography is a significant advance over conventional x-ray transmission tomography because it permits motion artifact–free images of the heart; it provides for evaluation of anatomy, function, and myocardial blood flow in a single setting and is minimally invasive. And yet another technique has burst on the scene: nuclear magnetic resonance (or magnetic resonance imaging), a totally noninvasive technique with remarkable spatial resolution and tissue characterization abilities. The clinical use of magnetic resonance spectroscopy is just around the corner and may provide never-before-attainable insights into cardiac metabolism in the clinical environment.

Confronted with this smorgasbord of clinically feasible diagnostic imaging techniques, how does the clinician choose among them? Should all the techniques be employed, to ensure the most accurate diagnosis possible? Can society afford these expensive diagnostic tools? Are there important differences in their safety?

All major imaging techniques—conventional angiography, echocardiography and Doppler, conventional radioisotope techniques, digital subtraction angiography, positron emission tomography, ultrafast computed tomography, and nuclear magnetic resonance—are covered in this book as they relate to the major cardiovascular disorders. Indeed, this is the first book on cardiac imaging to be organized by disease process rather than technique. Discussions, for example, of thallium myocardial scintigraphy, radioisotope blood pool imaging, and ultrafast computed tomography during exercise appear side by side, to assist the reader in contrasting and comparing the relative strengths and weaknesses of techniques that potentially yield overlapping information. Which one is best for the patient? Safest? Least expensive? Each chapter has been written by a recognized expert in the appropriate field. Hopefully, the reader will find the unique disease-oriented presentation useful in maximizing the potential of this amazing diagnostic armamentarium while dealing with the realities and necessities of patient comfort, safety, and cost containment.

Bruce H. Brundage, MD

Part I
Techniques

Chapter 1

Echocardiography and Doppler Techniques

Steven B. Feinstein, MD, and George Williams, MD

INTRODUCTION TO IMAGING ECHOCARDIOGRAPHY

Long before the advent of high technology, ultrasound principles were routinely used for navigation and flight by ocean and airborne mammals. Sonar has been used by the military to detect submarines, which create sound-reflecting microbubbles as they are propelled through the water. In industry and commerce, ultrasound can be used to test the quality of any relatively homogeneous material without destroying the sample; since the reflected sound patterns are predictable, alterations or disruptions (acoustic interface) may indicate a flaw.[1] By recording reflected sound wave impulses and calculating their relative speed through a medium, it is possible to determine the distance, texture, motion, and direction of a target object.

In medicine, the uses of ultrasound are equally diverse and expanding. The concomitant development of electronic miniaturization and high-speed computers, plus the escalating costs of health care worldwide, make ultrasound an ideal imaging device for clinical diagnosis and medical management. Ultrasound presents numerous inherent advantages over other imaging modalities: It uses nonionizing radiation, it's portable, it's easily and rapidly interpreted, it's widespread clinically, it's accepted by patient and physician alike, equipment costs are relatively low, and minimal exam time is required. Ultrasound is used in such diverse specialties as pediatrics, cardiology, radiology, urology, ophthalmology, anesthesiology, cardiac surgery, vascular surgery, nephrology, obstetrics, and gynecology.

Recently, esophageal ultrasound has been used to examine the cardiac structures in outpatients or during surgery. Ultrasound is being used in transrectal prostate imaging, diagnostic radiologic interventions (biopsy, drug instillation), and transvaginal imaging. In addition, newly developed contrast agents permit ultrasound perfusion imaging.[2] Stable, ultrasound-reflective microbubbles that are smaller than red blood cells (4 ± 1 μm) (Molecular Biosystems, Inc.) permit the tracking of blood through the vascular systems of organs visualized by ultrasound techniques (heart, kidney, liver, etc.).

ULTRASOUND IMAGING IN CARDIOLOGY

Specifically in cardiology, ultrasound is assuming an ever-expanding role in the diagnosis and management of heart disease (Figure 1-1). Early in the clinical application of ultrasound techniques, Edler et al.[3] described the use of ultrasound to identify the cardiac structures, via a transducer placed on the chest of patients. Subsequently, M-mode echocardiography, with its rapid data-sampling rate (1000 times per second) and excellent anatomic resolution, became a standard for the noninvasive assessment of cardiac function (Figure 1-2). Feigenbaum[4] recognized early on the utility of this new imaging technique and popularized its use in the United States through seminars,

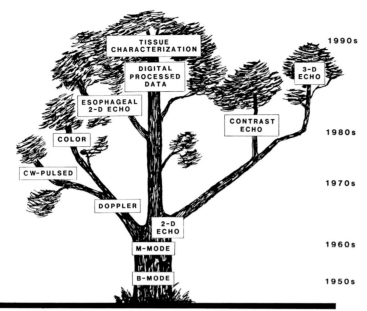

Figure 1-1 Growth of cardiac ultrasound. Schematic representation of the current uses and growth of the ultrasound technique in modern day cardiology. Initially, the single-crystal display (M-mode) evolved into two-dimensional imaging (sector scanning). Doppler, improved esophageal transducers, and a resurgence in contrast echocardiography techniques for myocardial perfusion imaging have been the latest outgrowths of cardiac ultrasound. Future ultrasound devices will be significantly improved in image quality and application (tissue characterization) due to the incorporation of ever-improving electronics into medical imaging systems.

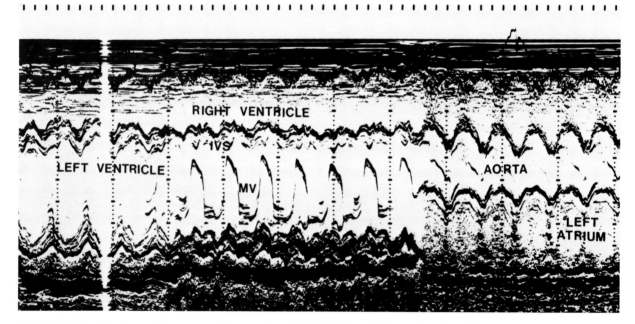

Figure 1-2 Typical M-mode echocardiogram recording, scanning (right to left) from the base of a patient's heart to the apex. Clearly identified by the large specular interfaces are the aortic root, left and right atrial structures, interventricular septum (IVS), and left ventricular cavity. MV is the mitral valve.

manuscripts, and textbooks. The development of sector (two-dimensional) ultrasound imaging by Von Rohn and Thurstone[5] expanded the clinical utility of ultrasound techniques in cardiac disease. Figure 1-3, taken from the Report of the American Society of Echocardiography Committee on Nomenclature and Standards in Two-Dimensional Echocardiography,[6] shows the standardized format devised for recording and displaying two-dimensional echocardiographic images. The traditional image planes include parasternal long- and short-axis views, apical four- and two-chamber views, and the subcostal or subxiphoid views. Recently, Doppler echocardiography (see second part of chapter) has become clinically accepted as a noninvasive means of assessing valvular function and flow.

Ultrasound imaging of the heart, as a means of evaluating ischemic heart disease, is now ready to expand beyond the laboratory and into the operating room, and

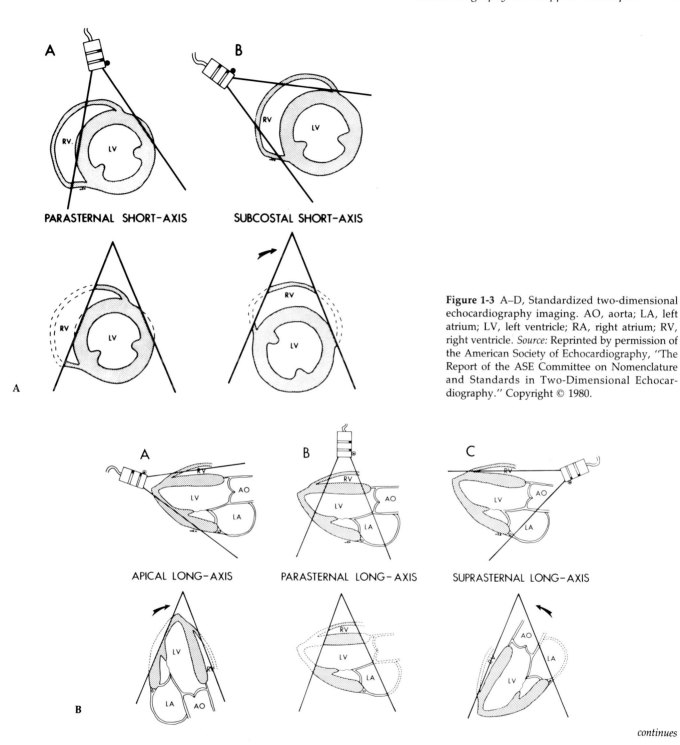

Figure 1-3 A–D, Standardized two-dimensional echocardiography imaging. AO, aorta; LA, left atrium; LV, left ventricle; RA, right atrium; RV, right ventricle. *Source:* Reprinted by permission of the American Society of Echocardiography, "The Report of the ASE Committee on Nomenclature and Standards in Two-Dimensional Echocardiography." Copyright © 1980.

continues

perhaps ultimately into physicians' offices and outpatient facilities. Using echocardiography, direct information can be obtained on the functional status of the myocardium and its perfusion state during cardiac events or surgery.[7] In the future, cardiac ultrasound will likely be used in combination with other modalities (e.g., coronary angiography) for detection of ischemic cardiac disease and for evaluating medical management and interventions (i.e., coronary artery bypass grafts, percutaneous transluminal coronary angioplasty, thrombolysis, etc.).[8–9] The use of ultrasound in transesophageal imaging of the heart has opened new vistas for patient monitoring and improved management of patients undergoing high-risk cardiovascular surgery.[7] The ability to directly visualize the function of the heart during major vascular surgery influences intraoperative management and may ultimately be shown to positively affect operative morbidity and mortality. With the additional use of ultrasound contrast agents during surgery, it is possible to directly assess the com-

Figure 1-3 continued

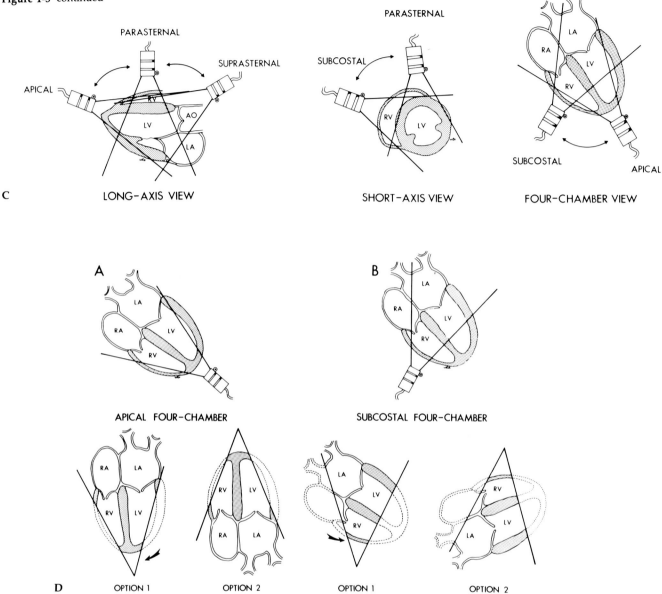

petency of a prosthetic valve and to establish regional perfusion areas following coronary artery bypass graft operations or valve replacement.[10–11]

Another new application of ultrasound is in assessing ischemic coronary disease during exercise two-dimensional echocardiography.[12] The representative characteristics of acute ischemia can be appreciated and displayed using this ultrasound technique. Specifically, during an acute ischemic event, ischemic myocardial tissue fails to thicken, and regional wall motion is reduced. It appears that the duration of the regional wall abnormalities is directly related to the duration of the ischemic period. In the future, ultrasound techniques will be used with any form of myocardial stress test (exercise, supine, bicycle, pharmacologic) to assess the effects of ischemia. It is likely that stress echocardiograms will be used for screening and preoperative assessment of patients with coronary artery disease.[13]

Direct digital acquisition and processing of radio frequency signals will expand the dynamic range of the currently acquired ultrasound signal. The enhanced signal spectrum improves imaged tissue characteristics and facilitates contrast echocardiogram perfusion assessment of myocardial tissue. Volumetric analysis of tissue perfusion and, ultimately, three-dimensional reconstruction programs may eventually be feasible.

Contrast Echocardiography

The development of stable, small microbubbles capable of pulmonary capillary transit has led to a rebirth of interest in contrast echocardiography. Coupled with the recent technological advances made in digital, expanded-range ultrasound systems, intravenous injection of this new transpulmonary contrast agent has made it feasible to image left heart structures.

Interest in contrast echocardiography began in 1968[14] with the first report of the use of a contrast medium (air bubbles) to enhance the ultrasound effect and better define aortic root anatomy. The initial studies utilized manually agitated solutions of saline, 5% dextrose solution or 5% dextrose in water, or indocyanine green to enhance the ultrasound image and provide a dynamic aspect of blood flow imaging to echocardiography. Although the contrast techniques described in 1968 and subsequent years continue to be utilized today in detecting shunts and cavity dimensions in children and adults, overall interest in contrast echocardiography waned, possibly due to the growth of Doppler echocardiography.

However, early in the 1980s a novel use of contrast echocardiography emerged: myocardial perfusion imaging. The pioneering work of Tei et al.[15] and Armstrong et al.[16] used manually prepared microbubbles as nondiffusible intravascular tracers. Because of the external nature of the ultrasound detector system, it was possible to directly image blood flow characteristics within a living organ system without disrupting perfusion of the tissue or organ being studied.

Due to the relatively large size of the manually prepared microbubbles and their short life (seconds to minutes), direct left heart injections of the microbubbles were required. In addition, these microbubbles were generally unable to pass through the capillary system.[17] Thus it was difficult to characterize blood flow based upon dynamic perfusion imaging of the transit time of the manually prepared microbubbles passing through tissue structures. Nevertheless, it was clear that even with the use of manually prepared microbubble contrast agents, contrast echocardiography was an excellent marker of a fixed perfusion deficit, as validated by experimental animal studies using vascular dye, radiolabeled microspheres, and autopsy comparisons. Areas of potential risk of infarction also could be identified using contrast echocardiography.[18–19]

The recent development of sonication techniques[20] for creating contrast agents (see Figures 1-4 to 1-6) has made it possible to produce small, relatively safe and stable microbubbles capable of capillary passage at physiologic transit times. Sonicated microbubbles made possible further identification of perfusion characteristics in the experimental setting, such that it became possible to correlate the degree of coronary stenosis with the loss of indicator (loss of contrast effect within the muscle).[21]

Continued development of sonication techniques led to the clinical use of contrast echocardiography to

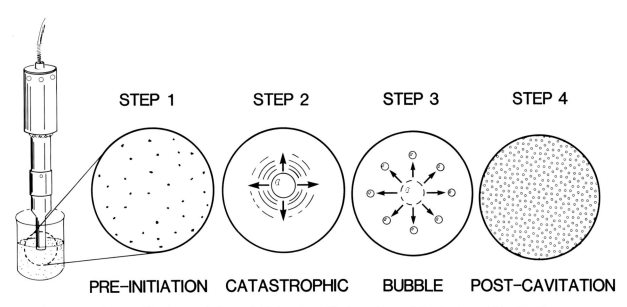

Figure 1-4 Sonication technique. The first step is the pre-initiation phase. The impurities and dissolved gas within a liquid serve as "motes" for the development of resonant bubbles and the cavitation process once energy is applied to the system. In the catastrophic phase a resonant bubble is produced; its collapse leads to microbubbles. The post-cavitation phase is not energy dependent and results from the persistence of post-cavitation microbubbles within a liquid medium. The ultimate size and fate of the post-cavitation microbubbles depends on the physiochemical properties of the medium.

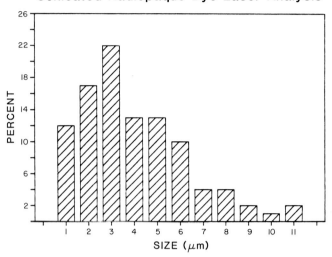

Figure 1-5 A frequency histogram (size in micrometers on abscissa and frequency of occurrence on ordinate) of sonicated Renografin-76, determined by a laser particle counter. The size of a human red blood cell is approximately 8 μm. *Source:* Reprinted with permission from *Journal of the American College of Cardiology* (1988;11:61), Copyright © 1988, American College of Cardiology.

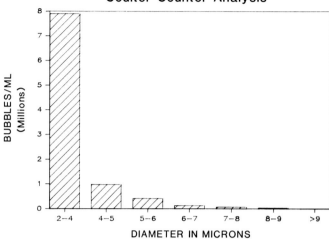

Figure 1-6 A Coulter counter analysis of sonicated albumin microbubbles. Note the smaller mean size and higher concentrations as compared to sonicated Renografin-76 microbubbles. *Source:* Reprinted with permission from *American Heart Journal* (1987; 114[3]:572), Copyright © 1987, CV Mosby Company.

assess myocardial perfusion during cardiac catheterization and cardiac surgery. Thus, for the first time, the coronary anatomy and the regional perfusion characteristics of a patient's myocardial tissue could be at once identified. (See Figure 1-7, showing a coronary angiogram and simultaneous contrast echocardiography perfusion study.) In view of the interest in assessing coronary reserve and preserving myocardial tissue through thrombolytic and other intervention techniques, contrast echocardiography is ideally suited to be "tagged on" to catheterization procedures.[8–9]

Further refinement of the sonication technique led to the development of a stable air-filled albumin microsphere capable of transpulmonary passage,[22] making it possible to image left heart structures (i.e., left ventricle, myocardial tissue, outflow tract, etc.) from a single intravenous injection site. Currently, clinical trials are underway to determine the clinical safety and efficacy of this new contrast agent. However, even as only a research tool, the transpulmonary contrast agent permits serial study of left heart perfusion characteristics under various physiologic and pharmacologic conditions, without invading the animal except for making an intravenous injection, which will ultimately expand our understanding of myocardial perfusion in normal and disease states.

Contrast echocardiography exhibits excellent spatial and temporal resolution in a tomographic format and in a serial fashion. Thus it is now possible to vigorously pursue studies of the microvascular flow characteristics of any organ system that is capable of being imaged with an ultrasound scanner. Early reports have indicated that contrast echocardiography techniques may be used in the assessment of renal perfusion.[23] It is possible that large muscle groups and whole organ systems will some day be imaged, and their perfusion assessed. These applications would be clinically useful in the operating room during bypass graft surgery or at the time of organ transplantation or revascularization.

Although validation and safety issues remain, contrast echocardiography appears to be entering a new and exciting era.

Ultrasound Principles and Digital Ultrasound Imaging

Energy in the form of sound waves surrounds us. A variety of frequencies (pitch) and amplitudes (power) are employed in medical ultrasound for diagnosis and management of disease and normal states. Ultrasound utilizes the detection of sound waves reflected by targets within the path of transmitted sound waves. Echocardiography relies upon the reflectance characteristics created by the variable alteration in sound energy transmitted through differing media. If the media have different acoustic properties (i.e., liquid-air interface or liquid-tissue interface), the transmitted and reflected energy will be altered.

Sound energy is created by applying electrical impulses to piezoelectric crystals. Like all forms of energy, it is propagated through a given medium at varying rates, depending upon the density and homogeneity of the medium. Reflected (backscattered)

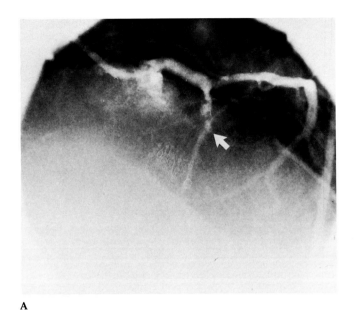
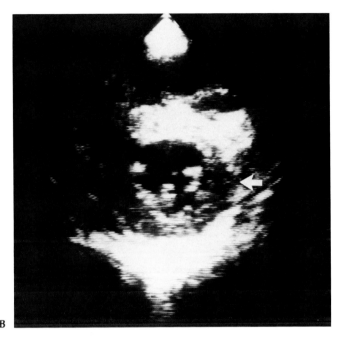
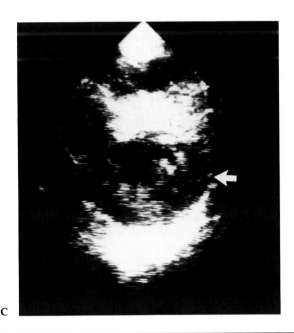

Figure 1-7 A, Photograph of a cineangiographic frame taken from a coronary angiogram, revealing an occluded mid–left anterior descending artery (arrow). B, The cross-sectional two-dimensional echocardiogram at the mitral valve level following injection of 1.5 mL of sonicated Renografin-76 into the left main coronary artery during cardiac catheterization. Note that the anterior and anterolateral myocardial tissue is enhanced from the echocardiography contrast, whereas the lateral and posterior regions are not enhanced. (Arrow points to leading edge of ultrasound contrast effect.) C, A second intracoronary injection of 1.5 mL of sonicated Renografin-76 revealed a similar perfusion defect in the lateral and posterior regions. This two-dimensional echocardiogram was obtained at the mid–papillary muscle level. Thus in this series of photographs it was possible to detect the coronary artery defect and immediately correlate the anatomy and subsequent perfusion defect by combining contrast echocardiography techniques with coronary angiography.

energy is interdependent with the frequency and amplitude of the initiating sound energy source.

The spatial resolution of structures is related to the frequency of the initiating power source. Higher-frequency ultrasound energy (7.5–10 MHz) provides short pulse responses and wavelengths, resulting in improved spatial resolution. However, the transmitted signal power is attenuated proportionally to frequency and distance traveled. Generally, human tissue is attenuated 0.7 dB/MHz per centimeter of depth. Therefore, higher ultrasound frequencies result in greater signal attenuation but enhanced spatial resolution.

Present two-dimensional scanners are classified according to geometry and field of view (linear versus sector). The imaging elements of the transducer design are either mechanically or electronically driven. Traditional linear scanners produce a series of parallel scan lines with a focus at a specific or variable tissue depth. By contrast, sector scanners are designed to interrogate smaller regions of the organ (e.g., regions of myocardium). They use a smaller ultrasound window than linear scanners in order to form the image profile resulting from a group of radians set at specific angles (depending upon the elements within the transducer).

Typically, a sector scanner produces an array of lines originating from the transducer face. The line of firing sequences may be altered, as may the focus of the ultrasound beam. The shape of an acoustic beam is controlled by the number and arrangement of the piezoelectric crystal elements, the transmitted waveform, and the receiver bandwidth characteristics. Gain compensation (for predictably lost signal due to depth) is typically accomplished by depth adjustment.

The reflected ultrasound signal enters the transducer and is processed in the following manner. The reflected signal may contain a radio frequency signal with a dynamic range of approximately 100 to 120 decibels. (The decibel is a measure of sound intensity, consisting of watt per square meter. Sample decibel levels are 120 dB for a thunderclap, 20 dB for a whisper.)

The signal (schematically represented in Figure 1-8) is initially received and processed. Initially, the data are digitized, logarithmically compressed, and enveloped. Due to the front end processing of commercial ultrasound systems, significant signal information is lost. With the advent of microprocessors, miniaturization, and decreasing costs, it is feasible to acquire an enhanced dynamic range of signal. Today, echocardiographic equipment is designed to identify and highlight large specular reflectors (endo- and epicardial surfaces, valve structures, etc.). However, development of ultrasound techniques for tissue characterization and perfusion imaging will rely upon the identification of small intravascular and tissue reflectors (point reflector sources), which will require the full dynamic range of reflected signal.

In order to access the full range of acquired signal, it will be necessary to store the information and call up portions of it for analysis. For example, in tissue characterization studies and myocardial perfusion assessments, it will be necessary to display tissue backscatter variations and contrast enhancement in ranges of 5 to 15 dB above background. This discrete alteration in backscatter contains the critical quantitative information that ultimately will be displayed. To place this in perspective, the ultrasound acoustic interfaces between tissue edges and cavities may exceed 40 to 50 dB. Thus imaging both scatter reflectors (tissue, intravascular microbubbles) and specular reflectors (endocardial and pericardial surfaces) will require a display mechanism that provides a variable dynamic range. Ultimately, the quantitative determination of tissue texture and perfusion assessments will be possible with the development of equipment capable of acquiring and accessing the full "up front" dynamic range of reflected ultrasonic signals.

New Ultrasound Imaging Equipment

Worldwide, investigators have designed and built ultrasound systems that will meet the needs of tissue characterization studies and consequently also improve contrast echocardiography techniques.[24-27] Specifically, the design of a system developed for the Computer Acquisition of Echocardiographic Data (CAED) project, built at the University of Chicago,[28] is shown in Figure 1-9. To accommodate the needs of quantitative perfusion imaging, it was necessary to capture the returning ultrasound signal prior to thresholding and compression effects. This required a high-speed interface (data rates between 1 to 8 megabytes per second) with an average operating rate of 1 megabyte per second channeled through the computer's direct memory access. Basically, the limiting problems in constructing such a device had been two: rapid data rate and volume of data. Typically, the returning signal contains 50,000 data samples per frame, with a data burst rate of 3 megabytes per second. A transducer may be programmed to fire at a rate of 100 times per frame and to sweep 25 frames per second. Previous efforts to acquire portions of the dynamic range of ultrasound data resulted in limiting the recording of data from the scan sweep rate or in applying threshold levels to the data registration, thus reducing the ability to acquire the full dynamic range of data.

The project at the University of Chicago was specifically designed to promote experimental and clinical research in contrast echocardiography. The system needed to be patient-safe and relatively portable for use in the experimental animal laboratory, cardiac catheterization laboratory, and intensive care unit. Additionally, the acquired data had to be available for real-time and instant replay in order to assess the success of the imaging study. Finally, the interface, once constructed, had to be flexible enough to permit adaptation to other commercial scanners and, ultimately, if indicated, to permit direct "raw" radio frequency acquisition of the returning ultrasound signal.

The current operating system includes the use of commercially available semiconductor memory boards for data storage, connected via a VME bus computer

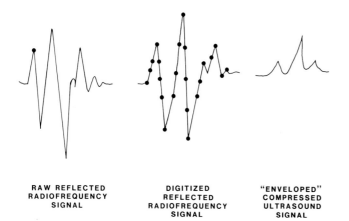

Figure 1-8 Schematic representation of a reflected ultrasound signal. Initially, the "raw" radio frequency signal is received at the level of the transducer. Subsequently, the signal is digitized and ultimately enveloped, resulting in a video signal.

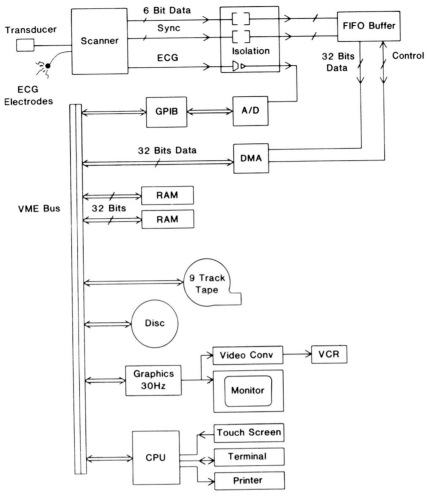

Figure 1-9 Block diagram of the acquisition of echocardiographic data by the CAED computer, designed and built at the University of Chicago. *Source:* Reprinted with permission from *Proceedings of the International Society for Optic Engineering, Visual Communications and Image Processing II* (1987;845:384–395), Copyright © 1987, IEEE Circuits and Systems Society.

system supporting 32-bit data paths serving as the back plane. A microcomputer serves as the input and output coordinator. Ultimately, all data are stored on a nine-track, ½-in magnetic tape. Additional software includes electrocardiogram (ECG) gating with variable timing delays for assessment of perfusion at various intervals within the cardiac cycle. The rapid acquisition of data and the synchronization requirements led to the construction of a buffer system to direct data through the VME bus.

The specific data flow can be traced from the digital output of the ultrasound scanner through the interface and into the buffer. If "raw" radio frequency data are to be acquired, 30 to 40 megabytes per second would be expected. However, if the digitized, enveloped radio frequency signal is recognized (further down the processing scheme), 4 to 7 megabytes per second of data would be expected. The actual system built utilized a 6-bit enveloped and logarithmically-compressed reflected ultrasound signal. The ultrasound data are also acquired after applying time gain compensations.

Once the data are acquired in the buffer, they are downloaded to the data memory access (DMA) boards via the bus. (The DMA boards operate at 8 megabytes per second.) The magnetic tapes containing the digitized ultrasound data are then loaded into a 1600 bpi tape drive system (a component of a modified Data General Systems Eclipse 8800 system). Modified software programs (video intensity pixel analysis, curve fitting, etc.) from other imaging systems have been adapted for use in contrast echocardiogram programs. Researchers analyze the data at an individual physician desk counsel (IPDC) equipped with a track ball, mouse, and keyboard.

REGIONAL PERFUSION AREAS

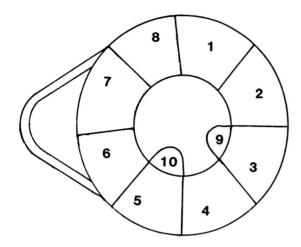

REGIONS

1. Anterior
2. Anterior-lateral
3. Posterior-lateral
4. Posterior
5. Inferior
6. Inferior-septum
7. Mid-septum
8. Anterior-septum
9. Anterior-lateral papillary muscle
10. Posterior-medial papillary muscle

A

Figure 1-10 A, Schematic cross-sectional representation of the myocardium as sectioned into 10 regions. B, The regional perfusion patterns in a patient following an intracoronary injection of 1.5 mL of sonicated Renografin-76. Note the significant alteration in backscatter amplitude (ordinate) in regions 4 and 5, corresponding to the posterior and inferior myocardial regions. These regions are generally perfused from the left circumflex coronary artery system. The regions typically perfused from the left anterior descending artery (8, 1, 2, 3) reveal only a minimal peak (around frame 11 to 12) following the intracoronary injection of sonicated Renografin-76. Regions 6 and 7 are generally perfused by the right coronary artery, and in this patient study, no contrast agent was delivered to the right coronary artery. In this first clinical use of the digital ultrasound system and contrast echocardiography techniques, a 40% mid–left anterior descending coronary artery lesion was identified. Interestingly, the myocardial perfusion pattern, as identified using contrast echocardiography, shows only minimal perfusion effects in the distribution of the left anterior descending artery, whereas the left circumflex region shows a significantly enhanced reflectance pattern, indicating the presence of adequate myocardial perfusion. *Source:* Reprinted with permission from *Journal of the American College of Cardiology* (1988;11:61), Copyright © 1988, American College of Cardiology.

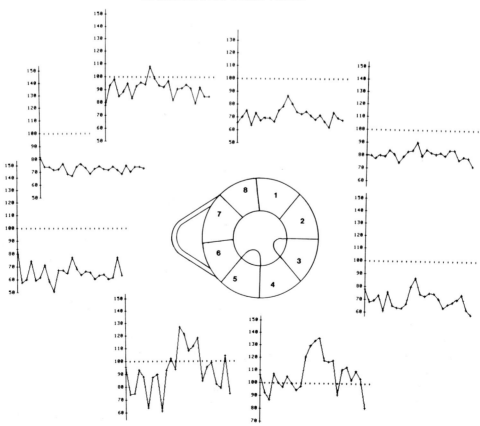

B

The first such digitally acquired data obtained with the system described above from a patient in the cardiac catheterization laboratory are seen in Figure 1-10. The cross-sectional image is divided into eight regions, with each region represented by video intensity backscatter curves. The abscissa is time (cardiac cycle), and the ordinate is the video intensity digitized signal. From this single study it is possible to recognize a significant alteration in reflected signal in regions 4 and 6. These regions correspond to the posterior and inferior regions of the myocardium and, as such, are generally perfused from the circumflex coronary artery.[2] The peak effect can be noted at the 11th to 12th beat following the injection of 1.5 mL of sonicated Renografin-76 into the patient's coronary artery. Note a small increment in signal occurring in regions 8, 1, 2, and 3. These regions correspond to the anteroseptal, anterior, and anterolateral regions of the myocardium. This is the distribution of the left anterior descending artery. Finally, regions 6 and 7 do not demonstrate any significant alteration in backscatter. This is the region supplied by the right coronary artery. No microbubble injection was performed in the right coronary artery for this initial analysis. In this patient no significant coronary artery disease was identified in the circumflex artery at the time of cardiac catheterization. However, the left anterior descending artery demonstrated a 40% stenosis in the mid-artery. The clinical situation that prompted this patient to undergo a cardiac catheterization was chest pain with exertion. Although this single study does not conclusively prove the significance of studying regional perfusion patterns with constrast echocardiography techniques at the time of coronary angiography, it underscores the need for combined functional and anatomic markers to understand and treat ischemic heart disease.

Summary: The Future of Ultrasound Imaging in Cardiology

Recent advances in technology suggest an expanding role for noninvasive techniques in the evaluation and management of ischemic heart disease. Ultrasound techniques are particularly well suited to the challenges of the future in that they are low cost, high resolution, portable, and in extensive clinical use. The concomitant developments of a digitally processed ultrasonic signal, a new transpulmonary ultrasound contrast agent for perfusion assessment, and esophageal ultrasound techniques in the outpatient setting and in the cardiac surgery suite will ensure an exciting future for the assessment of ischemic heart disease and perhaps provide a positive effect on patient management.

DOPPLER TECHNIQUES IN CARDIOLOGY

Imaging echocardiography and Doppler echocardiography use the same sound signal to provide different forms of information about cardiac function. Imaging relies predominantly on information reflected from interfaces between materials of different density, that is, blood and muscle or valves and blood. Doppler echocardiography, on the other hand, relies on backscattered information from red blood cells. These signals are much weaker and are typically not displayed by imaging echocardiography. Doppler techniques use the shift of sound frequency caused by the motion of blood cells to estimate the speed at which the cells are traveling.

The Doppler Principle

The determination of velocity makes use of the Doppler equation, which states that since sound can travel only at a fixed speed in tissue, any sound emitted by a moving object will show a shift in frequency that is proportional to the ratio of the speed of the object to the speed of sound. In its simplest form the Doppler equation is

$$\Delta F = \left[\text{Speed of object (velocity)} \div \text{speed of sound}\right] \times F_0$$

where ΔF is the change in frequency and F_0 is the original frequency. Solving for velocity produces the equation

$$\text{Velocity} = \text{Speed of sound} \times \Delta F \div F_0$$

Echocardiography uses reflected sound, and as a result the apparent frequency shift is doubled. Therefore, for ultrasound the velocity equation must be divided by two, producing the formula

$$\text{Velocity} = \text{Velocity of sound} \times \Delta F \div 2F_0$$

Since the velocity of sound and the generated frequency are constant, it can be seen that a change in velocity produces a proportionate change in frequency shift.

A major determinant of the accuracy of this equation is the angle of the interrogating ultrasound beam to blood flow. As long as the beam of sound is within 20° of the angle of flow, the results will be displayed with approximately 94% accuracy. However, as the angle of intercept between sound and beam increases, the amount of underestimation of velocity becomes proportionately greater; when the beam is 60° away from the interrogated flow, the velocity reported will only be 50% of actual velocity.

Spectral Analysis

While the transducer sends a relatively simple waveform into the tissue, the returning signals are complex, representing variations in blood velocity within the sample volume. The returning signal is processed by the spectral analyzer, which breaks the complex waveform into its component frequencies and measures signal strength at each frequency. Signal strength is a function of the number of red blood cells generating the frequency.

The output of modern Doppler machines is displayed as a spectral tracing: a graph of all the returning velocities on the vertical axis versus time on the horizontal axis (Figure 1-11). Velocities above a baseline represent those moving toward the transducer; those below the baseline represent velocities moving away from the transducer. Signal strength is assigned a gray scale value, becoming progressively darker as signal strength increases. Analysis of the spectral tracing can therefore provide information about all velocities within the sample volume and about the velocity at which the majority of cells are moving.

Doppler Modality

Two major types of Doppler modalities are used in clinical practice, pulsed Doppler and continuous wave Doppler. Each has a specific place in the evaluation of cardiac disease.

Pulsed Doppler

Pulsed Doppler utilizes a single crystal and is similar to imaging echocardiography. A short burst (5 to 7 cycles) of sound is sent out from the transducer. Sampling is done from a specific area within the heart by range gating. Returning signals from the heart are ignored except during the time period representing signals returned from the area of interest. These signals are then gated into the analyzer and the resulting Doppler shift displayed on the spectral tracing. The time gate can be moved closer or farther away from the transducer, allowing sampling of specific areas of flow in the heart.

The major advantage of pulsed Doppler is its ability to localize flow within the heart. However, the peak velocity measured is a function of the sampling rate of the transducer (pulsed repetition frequency, or PRF). To measure velocities accurately, the PRF must be more than twice the frequency shift caused by the moving red blood cells. When the PRF is less than twice the Doppler frequency, a phenomenon called aliasing occurs. A similar effect is seen when watching the wheels of a moving vehicle in moving pictures: The wheels appear to turn normally, then appear to rotate backward, although they are in reality moving faster. The velocity signal obtained by pulsed Doppler appears to "wrap around" from the highest velocity in one direction, and to move more slowly in the opposite direction, although in reality, velocity is increasing. This can be partially compensated for by baseline shifting (Figure 1-12).

In order to localize flow accurately, only one sound sample may be within the heart at any particular time. As sampling is moved farther away from the transducer, the pulsed repetition frequency must be lowered to allow time for the sound packet to travel to the sample volume site and return to the transducer. Thus the maximal measurable velocity falls proportionately as the sample volume is moved deeper into the tissue. A second factor affecting the ability of pulsed Doppler to measure velocities accurately is the carrier frequency. Although higher echocardiographic frequencies improve image resolution, the maximal velocity mea-

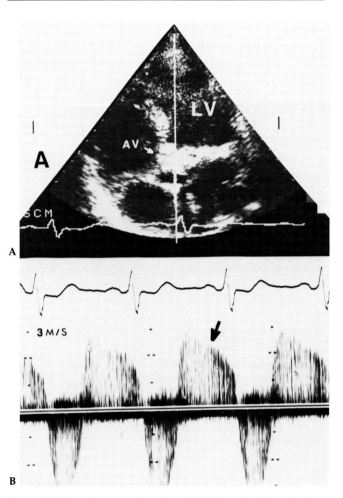

Figure 1-11 Doppler spectral tracing. A, (top) An apical five-chamber view shows a calcified aortic valve (AV) (arrow). LV, left ventricle. The Doppler spectral tracing, B, shows flow both away from the transducer below the baseline and toward the transducer above the baseline (arrow), diagnostic of both aortic stenosis and aortic insufficiency.

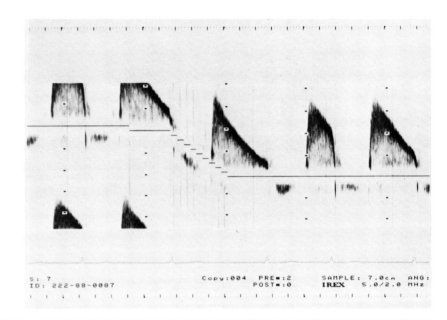

Figure 1-12 Pulsed Doppler tracing showing aliasing (left side of panel). Moving the baseline down allows "unwrapping" of the signal and display of the entire velocity profile. *Source:* Reprinted from *Doppler Echocardiography: The Quantitative Approach* (p 6) by AJ Labovitz and GA Williams with permission of Lea & Febiger, © 1988.

surable is inversely proportional to the original frequency. For example, at a depth of 10 cm away from the transducer, maximal frequency measured with a 2-MHz transducer approaches ± 1 m/sec. A transducer frequency of 3.5 MHz reduces the maximal measurable velocity to ± 0.7 m/sec at the same depth. By shifting the baseline of the Doppler tracing up or down, it is possible to double the maximal measurable velocity. In many cases this allows higher-frequency transducers to measure normal intracardiac velocity. This is especially important in adults, whose normal aortic velocities can approach 1.8 m/sec.

In order to measure high velocities with pulsed Doppler, high PRF Doppler has been utilized. Using this technique, the PRF of the transducer is increased 2 to 4 times. This allows a proportional increase in maximal measurable velocity. Although several sample volumes are present within the heart, they are typically displayed on the screen, and the area of interest can be localized using the two-dimensional image.

The newest clinically useful pulsed Doppler modality is the multigate Doppler. It differs from routine pulsed Doppler in that multiple samples are taken along the line of travel of a single sound burst. Using this technique, a flow map can be developed showing velocity at multiple points. By sampling along multiple B-mode lines and assigning a color value to the received velocities, a color flow map can be superimposed on the anatomic image to produce a "non-invasive angiogram" displaying flow as color within the anatomic structures of the heart.

Continuous Wave Doppler

Continuous wave Doppler requires the use of two separate transducers. One transducer is constantly generating a sound signal, while the second is receiving all reflected sound. The spectrum analyzer receives all signals returning from the heart. Since all signals are analyzed, there is no range (depth) resolution with this modality. There is, in effect, an infinite PRF, and there is no potential limitation to the maximal velocity measured (Figure 1-13).

Continuous wave Doppler can be used for two functions. First, since it is sampling the entire depth of the heart, it is useful for scanning rapidly to identify abnormal signals. Second, it can be used to measure the high velocities associated with abnormalities within the heart. Very high velocities are associated with the pressure gradient caused by flow across a stenotic lesion. By using the Bernoulli equation, these velocities can be used to calculate pressure gradient across an obstructive lesion. The most simple form of the equation is

$$\text{Pressure} = 4V^2$$

where V is the maximal velocity of flow obtained by Doppler. This formula must be modified, however, when the subvalvular velocity is greater than 1 m/sec to read

$$\text{Pressure} = 4(V_2^2 - V_1^2)$$

Flow Patterns

Laminar Flow

When blood is flowing at normal speeds in smooth chambers, all cells move in the same direction. The cells, however, tend to collect along the edges of the stream, forming layers (lamina) of cells and fluid. This type of flow, with cells and fluid moving in the same direction, is called laminar flow.

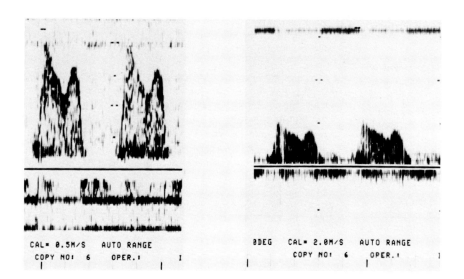

Figure 1-13 Comparison of pulsed (left panel) and continuous (right panel) wave Doppler. The pulsed Doppler tracing displays a well-defined velocity profile. The maximal measurable velocity is 2.5 m/sec. Continuous Doppler displays all velocities and appears more "filled in," and the maximal measurable velocity is 6 m/sec.

Two types of laminar flow occur. During systole, blood in the heart and proximal blood vessels is uniformly accelerating, and all cells and fluid tend to move at the same speed. The result is a flat flow profile of velocities across the vessel. A Doppler signal of a flat flow profile shows a narrow band of signals representing the uniform velocities across the vessel. At the end of systole, blood is no longer being accelerated by the heart but is slowing down (decelerating) as it continues to move forward. The fluid in the center of the vessel, containing fewer cells, tends to move more rapidly and decelerate more slowly than that containing the more densely packed cells along the edge of the vessel. The result is a variety of velocities, rapid in the center of the vessel and progressively slower toward the edges: a parabolic flow profile. The Doppler signal from a parabolic flow profile contains a wide band of velocities.

Nonlaminar Flow

In areas distal to stenotic lesions or at irregularities in vessel walls, the flow becomes more irregular. Areas of stagnation or eddies may occur, resulting in multiple velocities and multiple directions of flow.

Summary: The Special Role of Doppler Techniques

Doppler techniques can provide information not available from imaging echocardiography. Imaging reveals anatomic lesions of calcification, wall motion, and valve deformity. The Doppler examination provides physiologic information about flow characteristics and pressure gradients and diastolic flow orifice measurement. Together the two techniques are useful for noninvasively making comprehensive cardiac diagnoses.

REFERENCES

1. Carlin B. *Ultrasonics: Practical Considerations in the Application of Ultrasonics*. New York: McGraw-Hill Book Co.; 1949:243–264.

2. Feinstein SB, Lang RM, Dick CD, et al. Contrast echocardiography during coronary arteriography in humans: perfusion and anatomic studies. *J Am Coll Cardiol*. 1988;11:59–65.

3. Edler I, Gustafson A, Kaulefous T, et al. Ultrasound cardiography. *Acta Med Scand*. 1961;370(suppl):68–82.

4. Feigenbaum H. *Echocardiography*. Philadelphia: Lea & Febiger; 1972.

5. Von Rohn OT, Thurstone FL. Cardiac imaging using a phased array ultrasound system. *Circulation*. 1976;53:258–262.

6. Henry WL, DeMaria A, Gramiak R, et al. Report of the American Society of Echocardiography Committee on Nomenclature and Standards in Two-dimensional Echocardiography. 1980.

7. Gewertz BL, Kremser PG, Zarins GK, et al. Transesophageal echocardiographic monitoring of myocardial ischemia during vascular surgery. *J Vasc Surg*. 1987;5:607–613.

8. Lang RM, Feinstein SB, Feldman T, et al. Contrast echocardiography for evaluation of myocardial perfusion: effects of coronary angioplasty. *J Am Coll Cardiol*. 1986;8:232–235.

9. Cheirif J, Zoghbi WA, Raizner AE, et al. Assessment of myocardial perfusion in humans by contrast echocardiography, I: evaluation of regional coronary reserve by peak contrast intensity. *J Am Coll Cardiol*. 1988;11:735–743.

10. Goldman ME, Mindich BP. Intraoperative contrast echocardiography to evaluate mitral valve operations. *J Am Coll Cardiol*. 1984;4:1035–1040.

11. Goldman ME, Mindich BP. Intraoperative cardioplegia contrast echocardiography for assessing myocardial perfusion during heart surgery. *J Am Coll Cardiol*. 1985;6:687–694.

12. Armstrong W, O'Donnell J, Dillon J, et al. Complementary value of two-dimensional exercise echocardiography to routine treadmill exercise testing. *Ann Intern Med*. 1986;105:829–835.

13. Ryan T, Vasey CG, Presti LF, et al. Exercise echocardiography: detection of coronary artery disease in patients with normal left ventricular wall motion at rest. *J Am Coll Cardiol*. 1988;11:993–999.

14. Gramiak R, Shah PM. Echocardiography of the aortic root. *Invest Radiol*. 1968;3:356–366.

15. Tei C, Kondo S, Meerbaum S, et al. Correlation of myocardial echo contrast disappearance rate ("washout") and severity of experimental stenosis. *J Am Coll Cardiol.* 1984;3:39–46.

16. Armstrong WF, Mueller T, Kinney E, et al. Assessment of myocardial perfusion abnormalities with contrast enhanced two-dimensional echocardiography. *Circulation.* 1982;66:166–173.

17. Feinstein SB, Shah PM, Bing RJ, et al. Microbubble dynamics visualized in the intact capillary circulation. *J Am Coll Cardiol.* 1984;4:495–500.

18. Kaul S, Dandian NG, Okada RD, et al. Contrast echocardiography in acute myocardial ischemia: in vivo determination of total left ventricular "area of risk." *J Am Coll Cardiol.* 1984;4:1272–1282.

19. Kemper AJ, O'Boyle JE, Cohen CA, et al. Hydrogen peroxide contrast echocardiography quantification in vivo of myocardial risk area during coronary occlusion and of the necrotic area remaining after myocardial reperfusion. *Circulation.* 1984;70:309–319.

20. Feinstein SB, Ten Cate FJ, Zwehl W, et al. Two-dimensional contrast echocardiography, I: in vitro development and quantitative analysis of contrast agents. *J Am Coll Cardiol.* 1984;3:14–20.

21. Ten Cate FJ, Drury JK, Meerbaum S, et al. Myocardial contrast two-dimensional echocardiography: experimental examination at different coronary flow levels. *J Am Coll Cardiol.* 1984;3:1219–1226.

22. Keller MW, Feinstein SB, Watson DD. Successful left ventricular opacification following peripheral venous injection of sonicated contrast agent: an experimental evaluation. *Am Heart J.* 1987;114:570–575.

23. Lang RM, Feinstein SB, Powsner SM, et al. Contrast ultrasonography of the kidney: a new method for in vivo evaluation of renal perfusion. *Circulation.* 1987;75:229–234.

24. Miller JG, Perez JE, Sobel BE. Ultrasound backscatter of myocardial tissue. *Prog Cardiovasc Dis.* 1985;18:85–110.

25. Buda AJ, Delp EJ, Meyer ER, et al. Automatic computer processing of digital 2-dimensional echocardiograms. *J Am Coll Cardiol.* 1983;52:384–389.

26. Rasmussen S, Lovelace DE, Knoebel SB, et al. Echocardiographic detection of ischemic and infarcted myocardium. *J Am Coll Cardiol.* 1984;3:733–743.

27. Monoghan MJ, Quigley PJ, Metcalfe JM, et al. Digital subtraction contrast echocardiography: a new method for the evaluation of regional myocardial perfusion. *Br Heart J.* 1988;59:12–19.

28. Prieto PS, Wood J, Powsner SM, et al. High-speed interface for myocardial sonicated echo contrast studies. Proceedings of the International Society for Optic Engineering, Visual Communications and Image Processing II, IEEE Circuits and Systems Society; 1987;845:384–395.

Chapter 2

Cardiovascular Nuclear Medicine Techniques

Maleah Grover-McKay, MD

The following is a brief description of the technical aspects of cardiovascular nuclear medicine. The reader is encouraged to refer to more thorough discussions presented elsewhere.[1-17]

PHYSICS

Creating images using nuclear medicine techniques is based on detection of radioactive particles emitted during nuclear decay. An atom is composed of a nucleus and its (negatively charged) orbital electrons. The nucleus is composed of the nucleons: neutrons (no charge) and protons (positively charged). Some atoms retain their identity forever (stable), whereas others undergo a spontaneous transformation into another nuclide (radioactive decay). The stability of a nuclide is dependent on the neutron/proton ratio, the pairing of nucleons, and the excess energy required to maintain a nucleon within the nucleus (the nuclear binding energy). An unstable nuclide (termed the "parent") radioactively decays into a more stable nuclide (termed the "daughter") via emission of particles or energy such as gamma rays. Gamma rays are electromagnetic waves with an extremely short wavelength that may be considered packets of energy (photons). Gamma ray energies range from 10 keV to greater than 5 MeV and are unique for each radionuclide. The fractional number of atoms decaying per unit of time (the decay constant) is unique for each radionuclide. The amount of radioactivity is determined by the product of the number of radioactive atoms and the decay constant and is measured in *curies* (Ci) (3.7×10^{10} dps) or *bequerels* (Bq) (1 dps). One mCi equals 37 MBq. The decay constant determines the physical half-life of a radionuclide; the half-life is defined as the time in which the initial radioactivity is reduced by half. After intravenous injection of a radionuclide, the amount of radioactivity remaining in a given patient at a certain time is dependent not only on the physical half-life of a radionuclide but also on the rate of biological elimination, which is variable. The majority of the radioactivity will have decayed after five physical half-lives.

INSTRUMENTATION

General Equipment

The equipment necessary to image radioactive emissions includes a scintillation or gamma camera and a computer. The gamma camera consists of a collimator, a crystal scintillation detector, and the necessary electronic components (Figure 2-1).

Collimator

The collimator is comparable to the lens on a camera; it projects an image of the radioactivity distribution onto the crystal by absorbing all gamma rays except those of a desired trajectory. Collimators are usually made of lead and have holes through which photons

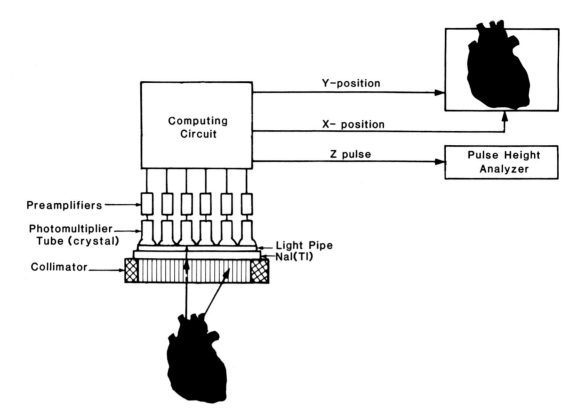

Figure 2-1 Schematic representation of scintillation camera equipped with a parallel hole collimator. Gamma rays from the heart that are perpendicular to the crystal pass through the parallel hole collimator and strike the crystal, producing visible light. The light passes through a light pipe to a photomultiplier tube, which converts the light into current and amplifies the signal. The site of photon production is computed electronically. NaI (Tl), thallium-activated sodium iodide.

pass to strike the detector. Collimators only allow photons traveling in a certain direction to interact with the detector. The part of the collimator between two adjacent holes is termed a septum. If the septa are too thin or the collimator holes are too short, scattered gamma rays may not be stopped by the septa, resulting in image degradation; this is called septal penetration (Figure 2-2). Most imaging in adult nuclear cardiology is performed with parallel hole collimators in which the holes are perpendicular to the face of the detector and parallel to one another. Collimators are available with different spatial resolutions and efficiencies (also called sensitivity), factors that are determined by the collimator material, the size and length of the holes, and the thickness and depth of the collimator and the septa. The higher the efficiency, the lower the resolution, and vice versa. Therefore, choice of a collimator depends on the type of study and the information required.

Crystal Scintillation Detector

When radiation is absorbed in a crystal scintillation detector, scintillations (visible light) are produced. The most commonly used scintillation detector in nuclear medicine is the thallium-activated sodium iodide [NaI (Tl)] crystal (Table 2-1). The thallium creates imperfections in the crystal lattice that trap electrons produced by the interaction of gamma rays with the iodide ions. When one of these electrons is recaptured by an atom, the crystal releases energy in the form of visible light. The intensity of this light is proportional to the gamma ray energy deposited in the crystal. Crystals generally are 10 to 15 in diameter and ¼, ⅜, or ½ in thick. Thicker crystals result in better detection efficiency for higher-energy photons (more gamma rays absorbed) but worse spatial resolution.[18]

Electronic Components

The back side of the crystal interfaces with an array of photomultiplier tubes, which cover the entire crystal. Each photomultiplier tube converts light from the crystal into an electronic pulse and amplifies it.

The total energy absorbed by the detector is proportional to the gamma ray energy. The only energies counted are those that fall within a selected range (called a window) of the photopeak for the radionuclide being imaged. Typically, a symmetrical window of 20% (i.e., photopeak ± 10%) is used. The energy information is used to eliminate scattered photons that have lost energy as they were scattered.

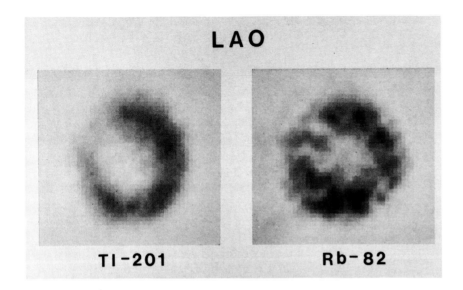

Figure 2-2 Penetration of collimator septa by high-energy rubidium 82 photons. Shown are left anterior oblique (LAO) views of the heart, made using a gamma camera. Thallium 201 was used to obtain the image on the left, and rubidium 82 to obtain that on the right. The 511-keV rubidium 82 photons penetrated the septa despite using a special collimator made of lead plus tungsten (which stops higher energy photons) that had thicker septa than a conventional collimator. Septal penetration results in image degradation.

Once the gamma ray has been converted into light by the detector and into an electronic pulse by the photomultiplier tubes, the location of the gamma ray's origin must be determined. The site of the gamma ray absorption is computed electronically, based on the intensity of light detected by each photomultiplier tube and the location of the photomultiplier tubes that detected the event. This position information is used to intensify a dot on a display screen corresponding to the location of the crystal interaction. Images are formed by taking long exposures of the display screen.

The characteristics of this signal reflect not only the energy of the photon but also the quality of the detector. Therefore, electronic pulses fall into a Gaussian distribution of energies, with the mean centered at the actual photon energy (Figure 2-3). This Gaussian curve is called a photopeak, and its width is an indicator of detector quality. The width of the photopeak at one half its maximum value (also called full width at half maximum, FWHM) is used as a measure of the energy resolution of the detector. This can also be expressed as percent resolution: Percent resolution = (FWHM/photopeak energy) × 100. Typical values for percent resolution are 11% to 14% at 140 keV.

Computer

Image information can also be transferred from the gamma camera to a computer. The computer controls data acquisition, processing (including semiquantitative and quantitative analysis), and image display. Two acquisition modes exist: list and frame.

In list mode, the location of the counts is stored along with time (and sometimes physiologic) markers as consecutive events. The major advantages of list mode are that optimal reformatting of the data can be performed after acquisition and that the time course of radionuclide activity is known. The major disadvantages are that list mode studies require large amounts of

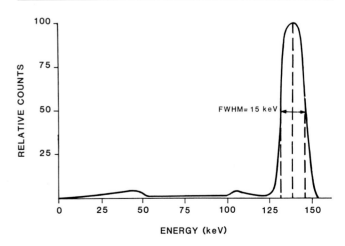

Figure 2-3 Energy spectra from a point source of technetium 99m. The width of this photopeak at half its maximum value, or full width at half maximum (FWHM), is 15 keV. The percent energy resolution equals FWHM/photopeak energy and in this case is (15/140) × 100 = 10.7%.

Table 2-1 Thallium-Activated Sodium Iodide Crystal

Advantages

1. Efficient detectors.
2. Crystal is transparent to light produced.
3. High intrinsic detection efficiency (probability a photon will be absorbed; also called stopping power, sensitivity).
4. Can be made any size and thickness.

Disadvantages

1. Hydroscopic (turn yellow when exposed to water).
2. Fragile.
3. Large crystals are expensive.
4. Energy resolution not as good as with other detectors (e.g., germanium-lithium).

computer memory for storage and additional time for processing.

In frame mode, counts are stored in frames with pixels of a given size and depth. Three types of frame mode studies exist: static, dynamic, and gated. Static images are acquired for a preset time or for a given number of counts. Dynamic studies consist of acquiring sequential static images, usually during short time intervals, in order to examine tracer kinetics. In gated studies an external signal, or gate (e.g., R wave of the electrocardiogram), triggers the computer to acquire data. For electrocardiographically gated studies, the RR interval is divided into a predefined number of frames (usually 16 to 24), each representing a fixed amount of time (typically 30 to 50 milliseconds at rest and 20 to 30 milliseconds during exercise). Thus each portion of the cardiac cycle is represented in a given frame, and the data for many cardiac cycles can be summed, yielding a sufficient number of counts. For gated blood pool studies at rest, approximately 250,000 counts per frame, and during exercise, approximately 150,000 counts per frame are sufficient.

An advantage of having the gamma camera images stored digitally on the computer is that they can be manipulated in a variety of ways. With data processing it is possible to enhance the image quality (e.g., background subtraction, temporal and spatial smoothing), perform quantitative analysis using regions of interest, and obtain data that are not obvious from the images (e.g., thallium washout, phase analysis).

Detection of Abnormalities and Image Quality

The ability to detect abnormalities depends on spatial resolution of the imaging system, distribution of radioactivity between organ and surrounding tissue, object contrast, and statistical fluctuations (noise). Image quality is affected by all of these factors. The better the image quality, the more likely that an abnormality will be detected.

Spatial resolution depends on collimation, distance of the source of the radioactivity to the detector, and photon energy. Optimal photon energy is 140 keV and spatial resolution is worse at lower and higher energies. The object contrast is related to the biodistribution of the radioactivity. The noise is related to the number of detected photons, which is ultimately limited by the radiation burden to the patient. Therefore, radioactive agents with short half-lives are preferable.

As discussed briefly above, proper selection of the collimator, detector, and type of electronic signal processing is extremely important to optimize images. Other factors that may degrade image quality include patient motion and artifacts produced by, for example, the presence of a radionuclide-contaminated foreign object, such as the patient's clothing, in the field of view. It is also important that images be recorded on the proper type of film.

Quantification

Beyond the scope of this chapter are equations describing diffusible and nondiffusible tracers, tracer kinetic modeling, and deconvolution analysis. The latter are mathematical methods that try to correct for bolus dispersion and lack of uniform tracer mixing. However, a brief discussion of the requirements for and problems with quantification is pertinent. For instance, the detector field must be as uniform and the count rate response as linear as possible. The count rate must not exceed the capability of the detector. Also, after a radioactive event is recorded, the detector is unable to image another event for a certain period of time, the dead time. Therefore, at high count rates, true counts are underestimated. Further, attenuation and greater distance from the detector will result in underestimation of the total amount of radioactivity. The size of the object being imaged also affects count recovery (called the partial volume effect[19]). Counts will be underestimated if the object size is less than twice the resolution of the detector.

Planar Imaging

Planar imaging consists of obtaining a two-dimensional image of a three-dimensional radioactivity distribution. The standard views used in cardiac imaging have been anterior, left anterior oblique (best septal view), and left lateral, although right anterior oblique and left posterior oblique have also been used (Figure 2-4).

Tomographic Imaging

Information is lost when three-dimensional radioactivity distribution is compressed into two-dimensional images. This information can be recovered by single photon emission computer tomography (SPECT).

Technique

The internal distribution of radioactivity in any object can be determined if a complete set of projections is available. The gamma camera image is an attenuated projection of radioactivity in the patient. A complete set of projections consists of measurements taken at equal angular increments over 360°. Usually 60 to 120 images are acquired on the computer as the gamma camera revolves around the patient. Images can be acquired either by continuous or discontinuous detector motion.

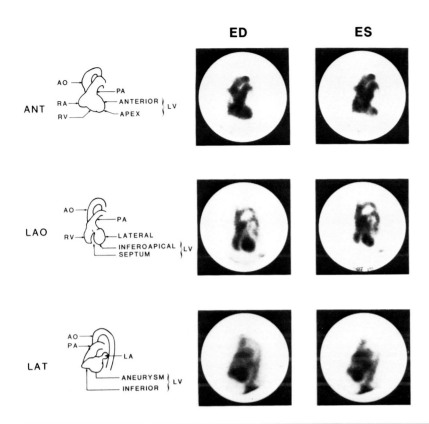

Figure 2-4 Radionuclide angiogram. Shown are the three standard planar views: anterior (ANT, top), left anterior oblique (LAO, middle), and left lateral (LAT, bottom). The anatomic structures are labeled in the schematic drawings on the left. The end-diastolic (ED) images are shown in the middle column and the end-systolic (ES) images on the right. An aneurysm of the proximal inferior wall is seen on the lateral view. AO, aorta; LA, left atrium; LV, left ventricle; PA, pulmonary artery; RA, right atrium; RV, right ventricle.

With continuous acquisition, the imaging angle at a given time is assumed, whereas with discontinuous acquisition the angle is known.

Transaxial images are generated from all the projections by the computer through a process called filtered back projection. Factors that affect image quality are slice thickness (thicker slices yield more counts) and the number of pixels (the larger the number of pixels, the noisier the data).

Because the gamma camera has a large field of view, tomographic images from a large volume (30 cm) are available. Therefore, any oblique-angle image can be obtained by appropriate sampling of the transaxial data. The heart lies at an oblique angle in the thorax. Therefore, cardiac images are routinely reoriented to provide images that are perpendicular to the long axis of the left ventricle (short axis slices) and parallel to the long axis (horizontal and vertical long axis slices) (Figure 2-5).

Data Acquisition

Different imaging protocols have been used for cardiac imaging.[20] Some controversy exists regarding 180° versus 360° of information for thallium 201 cardiac imaging.[21–24] Because thallium 201 emits low-energy x-rays (65 to 80 keV), little useful information is obtained in projections acquired when the gamma camera is posterior and on the right side of the patient. Therefore, thallium 201 SPECT images are routinely acquired over only a 180° arc extending from 45° right anterior oblique to 45° left posterior oblique. This precludes corrections for tissue attenuation and produces some distortions. However, the object contrast in the 180° tomographic images is superior to the 360° images. To decrease distance between the patient and detector, and thus to

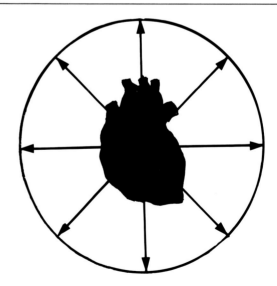

Figure 2-5 Reconstruction of a transverse slice by linear backprojection, represented schematically. Photons emitted from the heart (arrows) create a projection image at each imaging angle. Data from each of the projection images are backprojected onto the transverse slice, thereby creating a composite image containing data from each imaging angle.

improve resolution, elliptical rather than circular acquisition has been employed.

Data Processing

Quantitation of relative myocardial thallium 201 uptake is also possible and may enable more accurate detection of coronary artery disease.[25] Short-axis data are displayed in a bull's-eye image (Figure 2-6).

To evaluate whether the patient has moved up or down during image acquisition, the images can be viewed in cine format. If the patient has moved in relation to the camera, the position of the heart will not be smooth and continuous but will move either up or down at the angle when the motion occurred.

Quality Control

Quality control is important to ensure accuracy of both planar and SPECT images. In addition to the tests that are performed for any planar camera, such as sensitivity, field uniformity, linearity, and resolution, other tests must be performed for SPECT imaging, such as checking the center of rotation and the levelness of the camera head with rotation, volume quantitation, and attenuation correction.[13,26]

Positron Emission Tomography

A positron is an energetic positive electron emitted from certain radionuclides. The positron travels a short distance before combining with an electron in an interaction called annihilation. The masses of the electron and positron are converted by annihilation into electromagnetic energy in the form of two gamma rays, each having an energy of 511 keV. The gamma rays leave the site of this interaction at approximately 180° from each other. This unique feature allows a technique referred to as coincidence detection. Two opposing detectors will record a coincidence event only if both nearly simultaneously detect one of the annihilation photons. This determines the location of the radioactivity as being somewhere along the cylindrical field of view connecting the detectors. Projection information is obtained by a positron camera because a large number of detectors are located symmetrically around the object so that each detector can form multiple coincidences with the detectors across the ring (Figure 2-7). The projection information acquired by the positron camera is reconstructed to form transaxial images by filtered back projection, as described for SPECT.

Positron emission tomography (PET) offers a number of advantages over SPECT. The resolution of the PET system is superior to that achievable with SPECT (4 to 5 mm versus 8 to 15 mm) and the sensitivity is nearly an order of magnitude higher. In addition, correction for tissue attenuation can be more accurately performed, thereby facilitating quantitation of radioactivity. Quantitation of metabolism and blood flow is made possible by (1) the use of an appropriate tracer kinetic model, which requires knowledge of the biochemical and physiological actions of the tracer; (2) determination of the arterial input function via samples from an artery or from a hand vein warmed to dilate capillaries in order to create an "arterialized" vein, or from the images; and

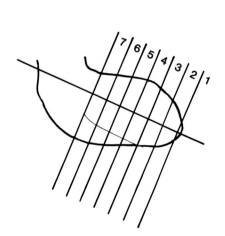
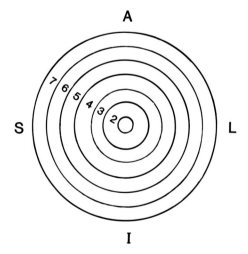

Figure 2-6 Schematic drawing of a bull's-eye image created as follows. Short-axis images of the left ventricle from apex (slice 1) to base (slice 7) are displayed concentrically with the apex at the center. Data for the apex can be obtained either from the short-axis image, as shown here, or from the vertical long-axis image. A, anterior left ventricle; I, inferior left ventricle; L, lateral left ventricle; S, interventricular septum.

Cardiovascular Nuclear Medicine Techniques

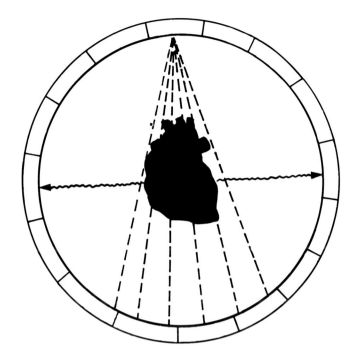

Figure 2-7 Detector fan beam response and coincidence detection of positron emission tomography. Detectors are placed around the object to be imaged. Each detector images the area in its fan beam. Projection images from each detector are backprojected to create the composite image. A true event is recorded when the two gamma rays created during interaction of a positron with an electron arrive nearly simultaneously at detectors 180° apart. This is called coincidence detection.

(3) determination of the tissue tracer concentration, which is measured from the images.[15,27]

Another advantage of PET is that positron-emitting radionuclides exist for physiologically important elements such as oxygen, nitrogen, carbon, and fluorine, and most of them have very short half-lives. While the short half-life is an advantage in terms of radiation burden to the patient and of performing multiple studies, it limits the time available for complicated chemical synthesis and requires an on-site cyclotron to produce the radionuclides. These factors make PET an expensive technique.

RADIOPHARMACEUTICALS

Single Photon–Emitting Radionuclides and Radiopharmaceuticals

Thallium 201

Thallium 201 is produced by a cyclotron (see Table 2-2). Its physical half-life is 73 hours. Patient doses range from 1.5 mCi for planar imaging to 4 mCi for tomographic imaging. The critical organ, defined as the organ receiving the largest absorbed radiation dose, is the kidney, which receives 1.3 rads/mCi. The photons imaged are actually daughter mercury 201–characteristic x-rays; therefore, the imaging photopeak is centered at 80 keV. This low energy is easily attenuated, which creates problems for interpretation of the inferior wall of the heart because of attenuation by the diaphragm and, with women, because of breast tissue. Thallium 201 is most often injected at maximal exercise or following administration of the coronary vasodilator dipyridamole.[28–31] Thallium 201 is a potassium analog and therefore is a tracer of blood flow. Cellular thallium 201 uptake is dependent on its arterial concentration, blood flow, and cellular extraction.[32] The first-pass myocardial thallium 201 extraction, defined as that percentage of thallium 201 reaching the coronary circulation that is extracted on the first transit through the heart, is 85% to 89%.[33–34] The extraction fraction drops slightly with hypoxemia, acidosis, and very high coronary blood flow.

Thallium 201 washout occurs when cellular clearance exceeds uptake. Myocardial thallium washout is related to coronary blood flow; thallium 201 washout is faster in normally perfused myocardium than in ischemic myocardium.[35–39] Other factors that influence thallium 201 washout, and hence image interpretation, include exercise heart rate,[40] the amount of thallium 201 remaining in the arm after injection,[41] and eating a high-carbohydrate meal between the immediate and 4-hour delay images.[42] Regional differences in myocardial thallium 201 uptake and washout enable detection of significant coronary artery disease and prior infarction.[43–45]

Table 2-2 Single Photon–Emitting Radioisotopes

Radionuclide	Half-Life	Imaging Energy (keV)	Administered Dose (mCi)	Absorbed Dose (rads/mCi)	
Thallium 201	73 hours	69–80	1.5–4	Body:	0.24
				Kidney:	1.3
Technetium 99m (Red blood cells)	6 hours	140	20	Body:	0.015
				Blood:	0.06
Indium 111 (Platelets)	28 days	171, 245	0.5	Body:	0.6
				Spleen:	33.5

To assess thallium 201 uptake, immediate images are obtained within 10 to 15 minutes of injection, which is after thallium 201 has cleared from the blood pool and before significant redistribution occurs.[46] To assess thallium 201 washout, delayed images are obtained at 3 to 4 hours after injection. Some areas of decreased perfusion seen at 4 hours may demonstrate improved perfusion (termed delayed redistribution) on subsequent images or after reinjection at rest.[47] Six patterns of thallium distribution can be described on visual analysis of the immediate and delayed images: (1) normal, (2) complete redistribution, (3) incomplete redistribution, (4) delayed redistribution, (5) reverse redistribution, and (6) fixed defect (Figure 2-8). Complete redistribution is consistent with a significant coronary stenosis. Incomplete redistribution at 4 hours may represent prior infarction with peri-infarction ischemia. If further or complete redistribution is seen in this area on subsequent images (i.e., at 8 to 24 hours or after another injection of thallium 201 at rest), a very severe coronary artery stenosis or collateral vessels are often present (Grover-McKay M. Unpublished data). Reverse redistribution, defined as apparent worsening of thallium 201 uptake on the 4-hour images, has been correlated clinically with the presence of nontransmural myocardial infarction and patency of the infarct-related coronary artery.[48–50] A defect that remains fixed on delayed imaging is consistent with prior myocardial infarction.

Technetium 99m

Sodium pertechnetate Tc 99m can be obtained from a generator consisting of the longer-lived parent, molybdenum 99 ($t_{1/2}$, 66 hours). The physical half-life of technetium 99m is 6 hours. Energy of the gamma ray imaged is 140 keV, which is optimal for gamma camera imaging.

Technetium-Labeled Red Blood Cells. Radionuclide angiography, or multigated blood pool studies, is performed using technetium-labeled red blood cells (see Table 2-2). The dose administered to adults is usually 20 mCi. The target organ is the blood, which receives 0.058 rad/mCi. Three methods exist for labeling red blood cells: in vitro, modified in vivo, and in vivo.[51–53] All involve administering stannous pyrophosphate prior to the ^{99m}Tc.

To perform in vivo labeling in adults, 2 to 3 mg of stannous pyrophosphate are injected intravenously 15 to 20 minutes prior to injection of ^{99m}Tc pertechnetate. The ^{99m}Tc binds with high efficiency (>90%) to the hemoglobin molecule (globin portion) in red blood cells in the intravascular space.[52] The in vitro method usually provides the best quality images, although it is

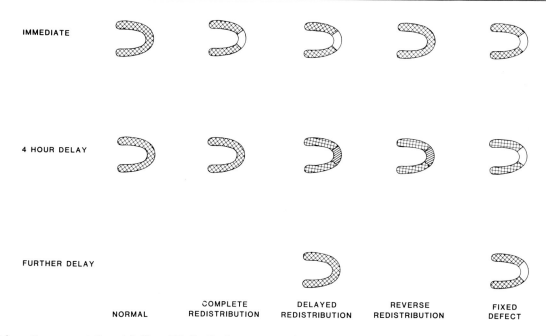

Figure 2-8 Schematic representation of thallium 201 distribution patterns. The immediate images are on the top, the 4-hour delay images in the middle, and the further delay images (i.e., ≥ 8 hours) on the bottom. Normally perfused myocardium is represented by the crosshatched areas, a partial defect by the hatched areas, and a complete defect by the open areas. Complete redistribution occurs when a complete or partial defect on the immediate images subsequently normalizes. Incomplete redistribution occurs when a complete defect becomes partial (e.g., 4-hour delay image compared with immediate image). Delayed redistribution occurs when a complete or partial defect on a 4-hour delay image subsequently normalizes. Reverse redistribution occurs when a normal area subsequently becomes a partial or complete defect. A fixed defect occurs when a partial or complete defect remains partial or complete. (See text for further discussion.)

Table 2-3 Positron-Emitting Radioisotopes

Radionuclide	Half-Life	Administered Dose (mCi)	Maximum Absorbed Dose (rads)	
Blood flow				
Rubidium 82	75 seconds	≤120	Body:	0.15
			Kidneys:	2.16
Oxygen 15 water	2 minutes	30	Body:	0.05
			Lung:	0.28
Nitrogen 13 ammonia	10 minutes	20	Body:	0.02
			Lung:	1.67
Glucose metabolism				
Fluorine 18 deoxyglucose	110 minutes	10	Body:	0.06
			Heart:	1.98
Fatty acid metabolism				
Carbon 11 palmitic acid	20 minutes	20	Body:	0.24
			Liver:	1.10

more time consuming and costly.[54–55] The adequacy of red blood cell labeling can be assessed by the presence of radioactivity in the stomach (parietal cells) and the thyroid, indicating the presence of free ^{99m}Tc, and hence of poor labeling.

Technetium Pyrophosphate. Technetium Tc 99m pyrophosphate has been used to image acute myocardial infarction.[56–57] Myocardial uptake is dependent on several factors, including blood flow, calcium concentration, extent and severity of injury, and time after irreversible injury.[58–60] Pyrophosphate is labeled with technetium by means of a reducing agent such as stannous ion. Technetium Tc 99m pyrophosphate binds to calcium, probably to amorphous calcium phosphate[61] although it may also bind to other sites in irreversibly injured myocardium.[62–63]

Technetium-Labeled Blood Flow Tracers. Two major advantages of technetium-labeled blood flow tracers are availability and the photon energy of 140 keV. Technetium 99m is available from a generator, and its photon energy of 140 keV is less readily attenuated and is within the optimal range for gamma camera imaging. In addition, myocardial function can be simultaneously assessed.[64–65] A class of compounds called isonitriles coupled with technetium 99m is currently under investigation to determine if they can depict myocardial perfusion. Initial results appear comparable to those obtained using thallium 201.[66–67] Currently the most promising agent appears to be technetium 99m methoxyisobutyl isonitrile (MIBI). This tracer binds to an intracellular protein, although it exhibits some redistribution following reperfusion.[68–70] Two injections are required to evaluate myocardial blood flow during exercise and at rest.

Indium 111

Indium 111 has a half-life of 2.8 days and two photopeaks, 171 and 245 keV (see Table 2-2).

Antimyosin Antibody. Antimyosin antibody labeled with ^{111}In is currently undergoing investigation. Besides the ability of antimyosin antibody to detect myocardial infarction,[71–73] it may also be useful for the detection of myocarditis and heart transplant rejection.[74–76]

Platelets. Platelets labeled with ^{111}In have been used to image left ventricular thrombi and to evaluate the thrombogenicity of coronary artery grafts and catheters.[77–84]

Positron-Emitting Radioisotopes and Radiopharmaceuticals

Most of the positron-emitting isotopes (Table 2-3) require a cyclotron for production. One exception is the blood flow tracer rubidium 82 (half-life, 75 seconds), which can be obtained from a strontium 82 generator and therefore can be continuously available.[85–86] The disadvantage of ^{82}Rb is that the average distance the positron travels before it annihilates (positron range) is greater than that for the other positron emitters, resulting in poorer spatial resolution.

Cyclotron-produced positron emitters that have been used for imaging include oxygen 15 (half-life, 2 minutes), nitrogen 13 (half-life, 10 minutes), carbon 11 (half-life, 20 minutes), and fluorine 18 (half-life, 110 minutes). Both ^{15}O-labeled water and ammonia N13 are myocardial blood flow tracers. Carbon 11-labeled compounds have been used to investigate myocardial receptors, carbon palmitate C11 has been used to investigate myocardial fatty acid metabolism, and deoxyglucose fluorine F18 to evaluate exogenous glucose utilization.[17,87–89] Use of both a blood flow tracer and F18 deoxyglucose enables differentiation of normal, ischemic (hence viable), and infarcted myocardium.[90–94]

Special thanks to Dr. Mark Madsen for his critical review of this chapter.

REFERENCES

1. Sorenson JA, Phelps ME. *Physics in Nuclear Medicine*. New York: Grune & Stratton; 1980.

2. Patton JA, Rollo FD. Basic physics of radionuclide imaging. In: Freeman LM, ed. *Freeman and Johnson's Clinical Radionuclide Imaging*. New York: Grune & Stratton; 1984:13–54.

3. McAfee JG, Subramanian G. Radioactive agents for imaging. In: Freeman LM, ed. *Freeman and Johnson's Clinical Radionuclide Imaging*. New York: Grune & Stratton; 1984:55–180.

4. Welch MJ, Kilbourn MR. Positron emitters for imaging. In: Freeman LM, ed. *Freeman and Johnson's Clinical Radionuclide Imaging*. New York: Grune & Stratton; 1984:181–202.

5. Rollo FD, Patton JA. Instrumentation and information portrayal. In: Freeman LM, ed. *Freeman and Johnson's Clinical Radionuclide Imaging*. New York: Grune & Stratton; 1984:203–260.

6. Rabinowitz A, Wexler JP, Blaufox MD. Quantification of the radionuclide image: Theoretical concepts and the role of the computer. In: Freeman LM, ed. *Freeman and Johnson's Clinical Radionuclide Imaging*. New York: Grune & Stratton; 1984:261–314.

7. Goodwin PN. Radiation safety for patients and personnel. In: Freeman LM, ed. *Freeman and Johnson's Clinical Radionuclide Imaging*. New York: Grune & Stratton; 1984:315–328.

8. Ter-Pogossian MM. Positron emission tomography instrumentation. In: Reivich M, Alavi A, eds. *Positron Emission Tomography*. New York: Alan R Liss, Inc; 1985:43–62.

9. Wolf AP, Fowler JS. Positron emitter-labeled radiotracers—chemical considerations. In: Reivich M, Alavi A, eds. *Positron Emission Tomography*. New York: Alan R Liss, Inc; 1985:63–80.

10. Herman GT. Reconstruction algorithms. In: Reivich M, Alavi A, eds. *Positron Emission Tomography*. New York: Alan R Liss, Inc; 1985:103–119.

11. Geltman EM, Bergmann ST, Sobel BE. Cardiac positron emission tomography. In: Reivich M, Alavi A, eds. *Positron Emission Tomography*. New York: Alan R Liss, Inc.; 1985:345–386.

12. Halama JR, Henkin RE. Single photon emission computed tomography (SPECT). In: Freeman LM, ed. *Freeman and Johnson's Clinical Radionuclide Imaging*. New York: Grune & Stratton; 1986:1529–1654.

13. Croft, BY. *Single-Photon Emission Computed Tomography*. Chicago: Year Book Medical Publishers; 1986.

14. Hoffman EJ, Phelps ME. Positron emission tomography: Principles and quantitation. In: Phelps ME, Mazziotta JC, Schelbert HR, eds. *Positron Emission Tomography and Autoradiography*. New York: Raven Press; 1986:237–286.

15. Huang SC, Phelps ME. Principles of tracer kinetic modeling in positron emission tomography and autoradiography. In: Phelps ME, Mazziotta JC, Schelbert HR, eds. *Positron Emission Tomography and Autoradiography*. New York: Raven Press; 1986:287–346.

16. Flower JS, Wolf AP. Positron emitter-labeled compounds: priorities and problems. In: Phelps ME, Mazziotta JC, Schelbert HR, eds. *Positron Emission Tomography and Autoradiography*. New York: Raven Press; 1986:391–450.

17. Schelbert HR, Schwaiger M. PET studies of the heart. In: Phelps ME, Mazziotta JC, Schelbert HR, eds. *Positron Emission Tomography and Autoradiography*. New York: Raven Press; 1986:581–662.

18. Sano R, Tinkel JB, LaVallee CA, Freedman GS. Consequences of crystal thickness reduction on gamma camera resolution and sensitivity. *J Nucl Med.* 1978;19:712–713. Abstract.

19. Hoffman EJ, Huang SC, Phelps ME. Quantitation in positron emission computed tomography, I: Effect of object size. *J Comput Assist Tomogr.* 1979;3:299–308.

20. Greer K, Jaszczak R, Harris C, Coleman RE. Quality control in SPECT. *J Nucl Med Technol.* 1985;13:76–85.

21. Eisner RL, Nowak DJ, Pettigrew R, Fajman WA. Fundamentals of 180° acquisition and reconstruction in SPECT imaging. *J Nucl Med.* 1986;27:1717–1728.

22. Tamaki N, Mukai T, Ishii Y, et al. Comparative study of thallium emission myocardial tomography with 180° and 360° data collection. *J Nucl Med.* 1982;23:661–666.

23. Go RJ, MacIntyre WJ, Houser TS, et al. Clinical evaluation of 360° and 180° data sampling techniques for transaxial SPECT thallium-201 myocardial perfusion imaging. *J Nucl Med.* 1985;26:695–706.

24. Coleman RE, Jaszczak RJ, Cobb FR. Comparison of 180° and 360° data collection in thallium-201 imaging using SPECT. *J Nucl Med.* 1982;23:655–660.

25. Folks R, Banks L, Plankey M, et al. Cardiovascular SPECT. *J Nucl Med Technol.* 1985;13:150–161.

26. Garcia EV, Van Train K, Maddahi J, et al. Quantification of rotational thallium-201 myocardial tomography. *J Nucl Med.* 1985;26:17–26.

27. Weinberg IN, Huang SC, Hoffman EJ, et al. Validation of PET-acquired input functions for cardiac studies. *J Nucl Med.* 1988;29:241–247.

28. Francisco DA, Collins SM, Go RT, Ehrhardt JC, Van Kirk OC, Marcus ML. Tomographic thallium-201 myocardial perfusion scintigrams after maximal coronary artery vasodilation with intravenous dipyridamole. *Circulation.* 1982;66:370–379.

29. Leppo J, Boucher CA, Okada RD, Newell JB, Strauss HW, Pohost GM. Serial thallium-201 myocardial imaging after dipyridamole infusion: Diagnostic utility in detecting coronary stenoses and relationship to regional wall motion. *Circulation.* 1982;66:649–657.

30. Leppo JA, O'Brien J, Rothendler JA, Getchell JD, Lee VW. Dipyridamole–thallium-201 scintigraphy in the prediction of future cardiac events after acute myocardial infarction. *N Engl J Med.* 1984;310:1014–1018.

31. Okada RD, Lim YL, Rothendler J, Boucher CA, Block PC, Pohost GM. Split dose thallium-201 dipyridamole imaging: A new technique for obtaining thallium images before and immediately after an intervention. *J Am Coll Cardiol.* 1983;1:1302–1310.

32. Strauss HW, Harrison K, Langan JK, Lebowitz E, Pitt B. Thallium-201 for myocardial imaging: Relation of thallium-201 to regional myocardial perfusion. *Circulation.* 1975;51:641–645.

33. Weich HF, Strauss HW, Pitt B. The extraction of thallium-201 by the myocardium. *Circulation.* 1977;56:188–191.

34. Grunwald AM, Watson DD, Holzgrefe HH Jr, Irving JF, Beller GA. Myocardial thallium-201 kinetics in normal and ischemic myocardium. *Circulation.* 1981;64:610–618.

35. Beller GA, Watson DD, Ackell P, Pohost GM. Time course of thallium-201 redistribution after transient myocardial ischemia. *Circulation.* 1980;61:791–797.

36. Okada RD, Jacobs ML, Daggett WM, et al. Thallium-201 kinetics in nonischemic canine myocardium. *Circulation.* 1982;65:70–76.

37. Okada RD, Leppo JA, Strauss HW, Boucher CA, Pohost GM. Mechanisms and time course for the disappearance of thallium-201 defects at rest in dogs. *Am J Cardiol.* 1982;49:699–706.

38. Okada RD, Leppo JA, Bouche CA, Pohost GM. The myocardial kinetics of thallium-201 after dipyridamole infusion in normal canine myocardium and in myocardium distal to a stenosis. *J Clin Invest.* 1982;69:199–207.

39. Bergman Sr, Hack SN, Sobel BE. Redistribution of myocardial thallium-201 kinetics in normal and ischemic myocardium. *Circulation.* 1981;64:610–618.

40. Kaul S, Chesler DA, Pohost GM, Strauss HW, Okada RD, Boucher CA. Influence of peak exercise heart rate on normal thallium-201 myocardial clearance. *J Nucl Med.* 1986;27:26–30.

41. Gal R, Port S. Arm vein uptake of thallium-201 during exercise: Incidence and clinical significance. *J Nucl Med.* 1986;27:1353–1357.

42. Wilson RA, Sullivan PJ, Okada RD, et al. The effect of eating on thallium myocardial imaging. *Chest.* 1986;89:195–198.

43. Pohost GM, Zir LM, Moore RH, McKusick KA, Guiney TE, Beller GA. Differentiation of transiently ischemic from infarcted myocardium by serial imaging after a single dose of thallium-201. *Circulation.* 1977;55:294–302.

44. Watson DD, Campbell NP, Read EK, Gibson RS, Teates CD, Beller GA. Spatial and temporal quantitation of plane thallium myocardial images. *J Nucl Med.* 1981;22:577–584.

45. Maddahi J, Garcia EV, Berman DS, Waxman A, Swan HJC, Forrester J. Improved noninvasive assessment of coronary artery disease by quantitative analysis of regional stress myocardial distribution and washout of thallium-201. *Circulation.* 1981;64:924–935.

46. Pohost GM, Alpert NM, Ingwall JS, Strauss HW. Thallium redistribution: Mechanisms and clinical utility. *Semin Nucl Med.* 1980;10:70–93.

47. Gutman J, Berman DS, Freman M, et al. Time to completed redistribution of thallium-201 in exercise myocardial scintigraphy: Relationship to the degree of coronary artery stenosis. *Am Heart J.* 1983;106:989–995.

48. Hecht HS, Hopkins JM, Rose JG, Blumfield DE, Wong M. Reverse redistribution: Worsening of thallium-201 myocardial images from exercise to redistribution. *Radiology.* 1981;140:177–181.

49. Silberstein EB, DeVries DF. Reverse redistribution phenomenon in thallium-201 stress tests: Angiographic correlation and clinical significance. *J Nucl Med.* 1985;26:707–710.

50. Weiss AT, Maddahi J, Lew AS, et al. Reverse distribution of thallium-201: A sign of non-transmural myocardial infarction with patency of the infarct-related coronary artery. *J Am Coll Cardiol.* 1986;7:61–67.

51. Smith TD, Richards P. A simple kit for the preparation of ^{99m}Tc-labeled red blood cells. *J Nucl Med.* 1976;17:126–132.

52. Pavel DG, Zimmer AM, Patterson VN. In vivo labeling of red blood cells with ^{99m}Tc: A new approach to blood pool visualization. *J Nucl Med.* 1977;18:305–308.

53. Callahan RJ, Froelich JW, McKusick KA, Leppo J, Strauss HW. A modified method for the in vivo labeling of red blood cells with Tc-99m: Concise communications. *J Nucl Med.* 1982;23:315–318.

54. Hegge FN, Hamilton GW, Larson SM, Ritchie JL, Richards P. Cardiac chamber imaging: A comparison of red blood cells labeled with Tc-99m in vitro and in vivo. *J Nucl Med.* 1978;19:129–134.

55. Armstrong LK, Ruel JM, Christian PE, Taylor A. Variation in gated blood-pool image quality using in vivo, modified in vivo, and in vitro red blood cell labeling techniques. *J Nucl Med Technol.* 1986;14:63–65.

56. Lyons KP, Olson HG, Aronow WS. Pyrophosphate myocardial imaging. *Semin Nucl Med.* 1980;10:168–177.

57. Willerson JT, Parkey RW, Bonte FJ, Lewis SE, Corbett J, Buja LM. Pathophysiologic considerations and clinicopathological correlates of technetium-99m stannous pyrophosphate myocardial scintigraphy. *Semin Nucl Med.* 1980;10:54–69.

58. Zaret BL, Di Cola VC, Donabedian RK, et al. Dual radionuclide study of myocardial infarction. Relationships between myocardial uptake of potassium-43, technetium-99m stannous pyrophosphate, regional myocardial blood flow and creatine phosphokinase depletion. *Circulation.* 1976;53:422–428.

59. Holman BL, Ehrie M, Lesch M. Correlation of acute myocardial infarct scintigraphy with postmortem studies. *Am J Cardiol.* 1976;37:311–313.

60. Olson HG, Lyons KP, Aronow WS, Brown WI, Greenfield RS. Follow-up technetium-99m stannous pyrophosphate myocardial scintigrams after acute myocardial infarction. *Circulation.* 1977;56:181–187.

61. Buja LM, Tofe AJ, Kulkarni PV, et al. Sites and mechanisms of localization of technetium-99m phosphorus radiopharmaceuticals in acute myocardial infarcts and other tissues. *J Clin Invest.* 1977;60:724–740.

62. Dewanjee MK, Kahn PC. Mechanism of ^{99m}Tc-labeled pyrophosphate and tetracycline in infarcted myocardium. *J Nucl Med.* 1976;17:639–646.

63. Schelbert HR, Ingwall JS, Sybers HD, Ashburn WL. Uptake of infarct imaging agents in reversibly and irreversibly injured myocardium in cultured fetal mouse hearts. *Circ Res.* 1976;39:860–868.

64. Corbett JR, Henderson EB, Akers MJ, et al. Gated tomography with technetium-99m RP-30 in patients with myocardial infarct: Assessment of myocardial perfusion and function. *Circulation.* 1987;76(suppl):217. Abstract.

65. Merz R, Maddahi J, Roy L, et al. Gated Tc-99m MIBI (RP-30) myocardial perfusion study in the detection and localization of myocardial infarction. *Circulation.* 1987;76:IV-217. Abstract.

66. Sia STB, Holman BL. Dynamic myocardial imaging in ischemic heart disease: Use of technetium-99m isonitriles. *Am J Card Imaging.* 1987;1:125–131.

67. Holman BL, Sporn V, Jones AG, et al. Myocardial imaging with technetium-99m CPI: Initial experience in the human. *J Nucl Med.* 1987;28:13–18.

68. Mousa SA, Maina M, Brown BA, Williams SJ. Retention of RP-30 in the heart may be due to binding to a cytosolic protein. *J Nucl Med.* 1987;28:619–620. Abstract.

69. Meerdink DJ, Thurber M, Savage S, Leppo JA. Effect of reperfusion and pacing on the myocardial extraction of thallium-201 and the technetium-labelled isonitrile analog (RP-30). *J Am Coll Cardiol.* 1988;11:33. Abstract.

70. Li Q, Franceschi D, Frank TL, Wagner HN, Becker LC. Overestimation of perfusion in ischemic myocardium by RP-30. *J Am Coll Cardiol.* 1988;11:31. Abstract.

71. Khaw BA, Fallon JT, Beller GA, Haber E. Specificity of localization of myosin specific antibody fragments in experimental myocardial infarction: Histologic, histochemical, autoradiographic and scintigraphic studies. *Circulation.* 1979;80:1527–1531.

72. Khaw BA, Strauss HW, Moore R, et al. Myocardial damage delineated by indium-111 antimyosin Fab and technetium-99m pyrophosphate. *J Nucl Med.* 1987;28:76–82.

73. Khaw BA, Yasuda T, Gold HK, et al. Acute myocardial infarct imaging with indium-111-labeled monoclonal antimyosin Fab. *J Nucl Med.* 1987;1671–1678.

74. Yasuda T, Palacios IF, Dec GW, et al. Indium 111-monoclonal antimyosin antibody imaging in the diagnosis of acute myocarditis. *Circulation.* 1987;76:306–311.

75. Addonizio LJ, Seldin DW, Esser PD, et al. SPECT imaging of cardiac transplant rejection using In-111 antimyosin antibody. *J Nucl Med.* 1986;27:910. Abstract.

76. LaFrance ND, Hall T, Dolher W, et al. In-111 antimyosin monoclonal antibody (AMAb) in detecting rejection of heart transplants. *J Nucl Med.* 1986;27:910–911. Abstract.

77. Datz FL, Taylor AT Jr. Cell labeling: techniques and clinical utility, II: Radiolabeled platelets. In: Freeman LM, ed. *Freeman and Johnson's Clinical Radionuclide Imaging.* New York: Grune & Stratton; 1986:1848–1913.

78. Stratton JR, Ritchie JL, Hamilton GW, Hammermeister KE, Harker LA. Left ventricular thrombi: In vivo detection by indium-111 platelet imaging and two dimensional echocardiography. *Am J Cardiol.* 1981;47:874–881.

79. Stratton JR, Ritchie J. The effects of antithrombotic drugs in patients with left ventricular thrombi: Assessment with indium-111 platelet imaging and two-dimensional echocardiography. *Circulation.* 1984;69:561–568.

80. Ezekowitz MD, Burrow RD, Heath PW, Streitz T, Smith EO, Parker DE. Diagnostic accuracy of indium-111 platelet scintigraphy in identifying left ventricular thrombi. *Am J Cardiol.* 1983;51:1711–1716.

81. Furster V, Kewanjee MK, Kaye MP, Josa M, Metke MP, Chesebro JH. Noninvasive radioisotopic technique for detection of platelet deposition in coronary artery bypass grafts in dogs and its reduction with platelet inhibitors. *Circulation.* 1979;60:1508–1512.

82. Dewanjee MK, Fuster V, Kaye MP, Josa M. Imaging platelet deposition with ^{111}In-labeled platelets in coronary artery bypass grafts in dogs. *Mayo Clin Proc.* 1978;53:327–331.

83. Dewanjee MK. Cardiac and vascular imaging with labeled platelets and leukocytes. *Sem Nucl Med.* 1984;14:154–187.

84. Lipton MJ, Doherty PW, Goodwin DA, Bushberg GT, Prager R, Meares CF. Evaluation of catheter thrombogenicity in vivo with indium-labeled platelets. *Radiology.* 1980;135:191–194.

85. Gennaro GP, Neirinckx RD, Bergner B, et al. A radionuclide generator and infusion system for pharmaceutical quality Rb-82. In: Knapp FF Jr, Butler TA, eds. *Radionuclide Generators: New Systems for Nuclear Medicine Applications.* Washington, DC: Amer Chem Soc, 1984:135–150. Symposium Series no. 241.

86. Grover-McKay M, Schwaiger M, Parodi O, Hoffman EJ, Phelps ME, Schelbert HR. Regional myocardial perfusion evaluated in humans at rest using rubidium-82 with gamma-camera imaging. *Am J Card Imaging.* 1987;1:64–73.

87. Grover M, Schelbert HR. Positron emission computed tomography. In: Buda AJ, Delp EJ, eds. *Digital Cardiac Imaging.* Boston: Martinus Nijhoff Publishers; 1985:240–270.

88. Grover-McKay M, Schelbert HR, Schwaiger M, et al. Identification of impaired metabolic reserve by atrial pacing in patients with significant coronary artery stenosis. *Circulation.* 1986;74:281–292.

89. Maziere M, Comar D, Godot JM, Collard PH, Cepeda C, Naquet R. In vivo characterization of myocardium muscarinic receptors by positron emission tomography. *Life Sci.* 1981;29:2391–2397.

90. Marshall RC, Tillisch JF, Phelps ME, et al. Identification and differentiation of resting myocardial ischemia and infarction in man with positron computed tomography, ^{18}F-labeled fluorodeoxyglucose and N-13 ammonia. *Circulation.* 1983;64:766–778.

91. Schwaiger M, Brunken R, Grover-McKay M, et al. Regional myocardial metabolism in patients with acute myocardial infarction assessed by positron emission tomography. *J Am Coll Cardiol.* 1986;8:800–808.

92. Tillisch J, Brunken R, Marshall R, et al. Reversibility of cardiac wall-motion abnormalities predicted by positron tomography. *N Engl J Med.* 1986;314:884–888.

93. Brunken R, Tillisch J, Schwaiger M, et al. Regional perfusion, glucose metabolism, and wall motion in patients with chronic electrocardiographic Q wave infarctions: Evidence for persistence of viable tissue in some infarct regions by positron emission tomography. *Circulation.* 1986;73:951–963.

94. Brunken R, Schwaiger M, Grover-McKay M, Phelps ME, Tillisch J, Schelbert HR. Positron emission tomography detects tissue metabolic activity in myocardial segments with persistent thallium perfusion defects. *J Am Coll Cardiol.* 1987;10:557–567.

Chapter 3

Cardiac Computed Tomography: Technical Aspects

Douglas P. Boyd, PhD

INTRODUCTION

Computed tomography (CT) has become a preferred method of imaging with x-rays. Because of the cross-sectional format of CT, the problem of superposition present in all other forms of x-ray imaging is completely solved. However, due to the demanding data acquisition and computer reconstruction requirements of CT, it is often more practical to use conventional x-ray techniques, due to the factors of cost, speed, and resolution. Computed tomography replaces x-ray procedures when the benefits of fully three-dimensional imaging outweigh these other considerations. However, there are no inherent limitations to the capabilities of CT that cannot be overcome by advances in technology.

Speed limitations have been the major technical challenge in applying CT scanning to the heart. A research group led by Drs. Wood and Ritman at the Mayo Clinic has explored for several years the potential use of CT in coronary angiography.[1] Their machine, the Dynamic Spatial Reconstructor (DSR), can be thought of as a series of 14 to 28 cinefluoroscopes that rotate about the body. Using computer reconstruction from many simultaneous views of the heart, a three-dimensional cineangiogram is created. The DSR has been valuable for exploring the potential of fully three-dimensional CT in cardiac diagnosis. The DSR concept approaches the problem of high-speed CT from the direction of multiplanar cineangiography using fluoroscopic methods. Thus multiple x-ray tubes and multiple-image intensified video imaging chains are used. By contrast, conventional CT scanners are based on the use of a single x-ray source and a planar array of discrete x-ray detectors that rotate about the body.

The concept of fast cardiac CT can also be approached from the viewpoint of conventional CT. This requires a method of speeding up the gantry rotation of CT scanners from about 1 second for current "fast" machines to about 50 milliseconds, as required to freeze cardiac motion. Figure 3-1 illustrates the types of scanning motion used in conventional CT scanners (whose methods and techniques have been thoroughly described by others[2]). The first two methods, translate-rotate and rotate-rotate, require rotation of the detector array. The rotate-stationary design uses a fixed detector and requires rotation of the source only. Thus the problem of designing a fast cardiac scanner can be reduced to developing a rapidly rotating x-ray source. A number of approaches have been suggested, including multiple rotating x-ray tubes, an array of stationary tubes that are sequentially pulsed, a rotating cathode and fixed anode in a vacuum ring, and the use of a scanning x-ray beam tube.

The scanning electron beam tube concept has emerged as the preferred approach for high-speed cardiac CT.[3] To date, more than 20 machines based on electron beam technology have been installed worldwide. In this chapter I will review the current design and performance characteristics of ultrafast CT, describe the current applications, and discuss the approaches for future advances.

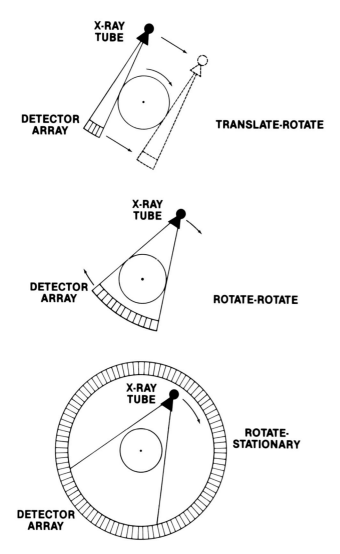

Figure 3-1 Illustration of the types of scanning motion used in the three common types of fan beam CT scanners. Almost all modern CT scanners use either the rotate-rotate geometry or the rotate-stationary configuration.

PRINCIPLES OF ELECTRON BEAM, ULTRAFAST CT SCANNING

A diagram of an ultrafast CT scanner is shown in Figure 3-2. This scanner is designated as the Picker/Imatron FASTRAC in North America and as the Imatron C-100XL overseas. Since the rotating x-ray source is produced by magnetic steering of a high-powered electron beam on a series of four tungsten target rings, there are no mechanical constraints on scanning speed. In addition, since these rings have a large thermal mass and are directly cooled, the heat-loading limitation of conventional rotating anode x-ray tubes is completely removed. Thus this type of system is capable of millisecond scanning speeds and can repeat scans at a high rate to produce high-speed dynamic or cine-tomographic images.

The ultrafast scanner was designed for optimal cardiac imaging applications and therefore is provided with eight 1-cm slices and a basic scan speed of 50 milliseconds. This is provided by sequential scanning of the four target rings and simultaneous acquisition of two slices using a pair of side-by-side stationary detector rings having 432 channels each in 210°. Thus two slices are scanned at a rate of 17 frames per second, allowing for an 8-millisecond interscan delay. A total of eight spaced slices are obtained in 224 milliseconds. In the 50-millisecond multislice mode, 432 views are obtained as the source rotates through 210°.

In a second mode referred to as the high-resolution single-slice mode, the detectors are reconfigured into a single ring of 864 elements. A single target ring is scanned at a scanning rate of 100 milliseconds, giving a total of 864 view samples. In addition, a special radiology collimator is used to provide improved slice collimation and to reduce scattered radiation. This collimator provides for a 6-mm and a 3-mm slice thickness. A 10-mm slice is formed by incrementing the couch during scanning. In addition, a fast raw averaging (FRA) board is used so that 100-millisecond scans can be averaged during scan acquisition, in order to permit averaged scan speeds in a range of 0.1, 0.2, 0.3, . . . 1.2 seconds. This gives the option of higher-dose single-slice scans while preserving the resistance to motion artifact of the basic 100-millisecond scan. Averaged scans are typically used for conventional CT applications outside the chest, in the head and body. The single-slice mode gives a factor of two improvement in spatial resolution, as compared to the multislice 50-millisecond mode.

The transmitted x-ray photons are measured in integrating scintillation crystal–photodiode array detectors, and these signals are digitized by a data acquisition system at a rate of approximately 7 million 16-bit samples per second. The digitized signals are then averaged by the FRA board and temporarily stored in a scan cache memory that currently has a capacity of 80 multislice images or 20 single-slice images. New upgrades are being installed on current ultrafast CT scanners that increase the size of this memory by a factor of two—to 160 images and 40 images, respectively. The scan cache memory is transferred to disk storage at a rate of 0.4 second per multislice image and 1.6 seconds per single-slice image. A more complete description of the hardware and software capabilities of this system is available.[4-5]

The stored scan data are then reconstructed by an asynchronous reconstruction system that begins reconstructing images automatically when new scan data become available. Reconstruction proceeds in parallel with scanning, patient setup, image analysis, and archi-

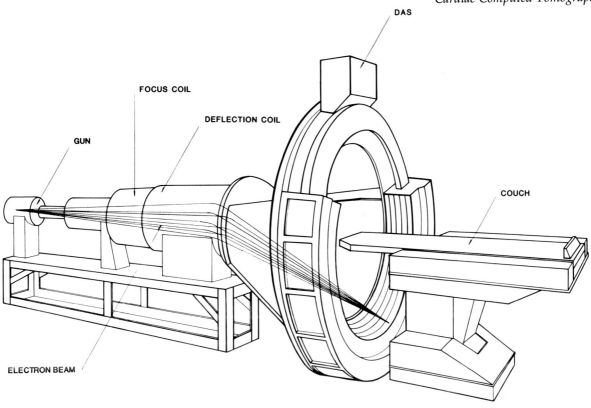

Figure 3-2 Schematic diagram illustrating the principle of electron beam scanning as used in the ultrafast CT scanner. The electron beam is produced in the gun and steered and focused by the coils shown. X-rays are produced on the target rings, sensed in a detector ring, and digitized by the data acquisition system (DAS).

val storage. Thus, typically, several images will be reconstructed prior to completion of a scanning sequence. The reconstruction speed is currently 5 seconds for multislice images and 10 seconds for single-slice images. The images may be viewed and analyzed on either of two independent display consoles. In addition, provision is made for connection to a network of up to 15 additional workstations of various types. The off-line workstations can be low-cost, located remotely, and may be user-programmable.

ULTRAFAST CT METHODS AND APPLICATIONS

A typical hospital installation of the ultrafast scanner is shown in Figure 3-3. Lane[6] has recently described

Figure 3-3 A typical hospital installation of the ultrafast CT scanner. An electrocardiogram-triggering device is placed to the left of the patient couch. A power injector used for intravenous contrast medium injection is seen to the right of the gantry.

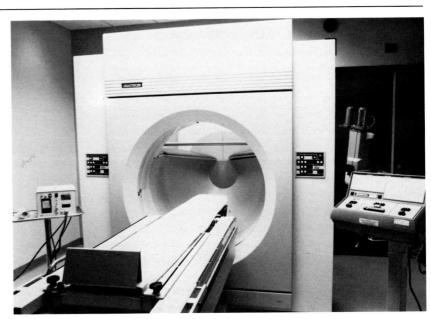

applications of the ultrafast scanner to vascular imaging in general radiology of the body. A number of operational modes intended for specific cardiac applications are provided. These include the flow mode, cine mode, and volume modes. In addition, analysis software to facilitate quantitative measurements and optimal viewing is provided to support the various modes and provide detailed reporting data. These capabilities are described in more detail below.

Flow Mode

The flow mode is designed to sample a given series of slices at defined time intervals during the transit of a bolus of contrast medium through the vessel, chambers, or tissues of interest. Up to eight slices can be sampled in the flow mode. The number of samples per slice is selected so that the total number of slices is either 80 or 160. In cardiac flow studies, each sample is triggered by the electrocardiogram (ECG) signal so that all samples of a given slice are obtained at the same phase of the cardiac cycle. Typically, sampling is performed every other heart beat interval at about 80% of the interval after the R wave. The contrast medium is typically injected as an intravenous bolus of 10 to 50 mL injected over 3 to 5 seconds. Flow data are analyzed by determining the time-density curve in a cursor-defined region of interest and performing a numerical fit on the resulting flow curve.

The simplest flow parameter to measure is the cardiac output. This measurement is available as a byproduct of almost any flow-mode study. The cardiac output is found by measuring the time-density curve in a region of interest inside any chamber of the heart or any great vessel such as the aorta or the pulmonary artery. According to the Stewart-Hamilton equation, the cardiac output is given by the ratio of the amount of contrast medium injected to the area under the time-density curve. These measurements may be made with very small amounts of contrast medium, perhaps as little as 2 to 5 mL.

Blood flow in vessels such as coronary arteries and coronary artery bypass grafts may also be obtained. These measurements are more difficult and involve determining the arrival time of a bolus at several points along the vessel and fitting these data to a straight line to determine the average velocity. An independent estimate of the cross-sectional area of the vessel is made using the brightness-area product method. The vessel flow can then be estimated by the product of velocity and area. In most cases these flow measurements will be performed at rest and under cardiovascular stress.

Tissue blood flow is measured using the time-density curve for a region of interest within the tissue of interest, typically the myocardium. Extraction of myocardial blood flow from the time-density curve fit is dependent on use of a theoretical model describing the tissue blood flow. A number of models have been evaluated. The measurements are typically reliable at normal and low flow rates. At elevated flow rates, such as those obtained during cardiovascular stress, the models become more complex due to the fact that there is overlap between wash-in and washout of the bolus. This is due to the 5- to 10-second widening of the intravenous injected bolus by the time it reaches left heart. Recent data have shown the importance of correcting for changes in myocardial blood volume that occur at elevated flow rates.

Cine Mode for Wall Motion Analysis

In the cine mode the real-time capabilities of ultrafast CT are exploited in order to obtain movies showing the motion of the beating heart. Applications include detection of wall motion abnormalities occurring under stress, precise quantitation of left and right ventricular volumes during the cardiac cycle, and images showing the movement of native and prosthetic valves. The scanning rate is usually 17 frames per second for a pair of simultaneous slices separated by 8 mm. Up to 12 levels may be studied in a single study. This is performed by sequentially scanning pairs of levels on successive heart cycles. For example, six heart cycles will be used for a 12-level study. Since the number of frames per level is limited by the maximum scan cache memory of 80 or 160 levels, cine runs are usually triggered to begin on the R wave, ensuring that several frames describing ventricular contraction from diastole to systole are obtained. For diastolic filling studies, fewer levels can be used in order to cover the entire heart cycle. During pharmacological or exercise-induced stress, fewer samples are needed, due to the elevated heart rate. Typically, 15 frames are used at rest and 8 to 10 during exercise.

Contrast medium injections must be tailored to the timing required by a ventricular motion study. If the right ventricle is to be imaged simultaneously with the left, then a biphasic injection should be used to opacify both simultaneously. The bolus length of each should be long enough to cover the 4 to 8 seconds of the multi-level study. The amount of contrast medium used depends on the precision of wall definition that is required and patient safety and comfort considerations.

In order to quantitate ventricular wall motion, it is necessary to accurately extract the contour describing the endocardial surface and epicardial surfaces from the CT images. This surface is described by a series of contours obtained from each cross-sectional image. The series of contours describes the ventricular volumes at each point during the cardiac cycle.

Several methods are available to determine the myocardial wall contours. The simplest method is to select a density contour level and find the contour that follows this density level. The density value may be selected at the midrange (50%) of the tissue-blood interface or at some other value that is experimentally determined to give accurate results. The epicardial border includes the lateral border of the left ventricle, which involves a tissue-air interface. Here, due to beam hardening, a contour greater than 50% (often 80%) appears to give the best result. An alternative approach to contour definition is to apply the Laplacian mathematics to the image (similar to differentiation) in order to find the points of steepest descent and to use these values to define the contour. This method is provided in automatic edge-finding software.

Volume Imaging for Anatomy

Volume imaging describes a series of modes used to perform high-quality multiple-slice images for anatomic diagnosis. The fastest way to obtain a volume image is to use the eight-slice mode in conjunction with the 50 millisecond multislice capability. This may be either a single or an averaged series of pictures and can be obtained in as little as 224 milliseconds. This mode is used for localization studies or for interactive imaging such as is required by biopsies made by CT-guided needle.

High-resolution volume images are best obtained using the single-slice mode with serial couch incrementation. Low-noise images use 0.8- to 1.2-second averaged images. In the chest 0.1 to 0.4 seconds are usually adequate. When imaging the heart, it is helpful to trigger each serial slice with the ECG, thus ensuring that all slices are in registration with the cardiac cycle. One of the major applications of triggered volume images is to image the coronary arteries. Coronary arteries are well visualized in such images, even without contrast medium, due to the natural contrast of the arterial wall and surrounding fat. Coronary lesions are often easily seen, due to the sensitivity of CT to the small amounts of calcium they contain. The inner lumen of the coronary artery is not differentiated from the wall unless the blood is opacified with contrast medium.

With the rapid serial imaging of ultrafast CT, a series of slices of the coronaries with ECG registration can be obtained in a single breath-holding interval. Thus this stack of images has good registration and may be reformatted and viewed in a variety of oblique planes. This is very helpful in obtaining longitudinal views of coronary arteries. Reformatting into curved, oblique planes may be the optimal way to view such images. Other possible viewing modes include the method of tissue dissolution and reprojection, as demonstrated by the Dynamic Spatial Reconstructor at the Mayo Clinic. These methods are a way of deriving planar images, such as are used in angiography, from the cross-sectional slices of CT.

A third way of presenting three-dimensional data to an observer is shaded surface display. In this method, the surface of structures of interest, such as the left ventricle, is displayed on a monitor as a shaded solid that can be rotated to various views. Shading from a simulated light source aids the eye in determining perspective. These models may be sliced into halves and viewed from the inside. They may also be displayed as a movie, offering the equivalent of a "4-D" movie.

ULTRAFAST CT PERFORMANCE

The most important performance characteristic of a cardiac imaging device is its image quality. Image quality fundamentally determines the capability of the device to resolve details of interest and sets the limits of precision from all quantitative applications. Image quality can be quantitatively described by spatial resolution and image noise (random and structured noise), also referred to as contrast resolution. These fundamental values influence ability to resolve small details, the accuracy of edge measurements, and the precision of density measurements.

Spatial Resolution

The resolution of CT can theoretically match the resolution obtained in other x-ray procedures, but on a practical basis cost considerations tend to limit the maximum resolution that can be designed into a scanner. The many design factors that determine spatial resolution have been previously described.[7] Trends in technology development show that resolution will be improved in the future. The present resolution of the ultrafast CT scanner is indicated in Figure 3-4. The modulation transfer function (MTF) describes the image contrast as a function of spatial frequency. If this frequency is represented by a lead bar pattern, the contrast describes the blurring of the interface between bars. The data in Figure 3-4 were obtained using the method developed by Assimakopoulos et al.[8]

This figure shows that the multislice mode has approximately half the resolution of the single-slice mode. There are two 50-millisecond curves, one for the ring of 432 crystals and the other for the ring of 864 crystals used in the single-slice mode. These crystal readings are summed in pairs for the multislice mode, and the gap between crystals leads to the alteration of the MTF found in images produced by this ring. Thus the upgrade to high-resolution, single-slice scanning resulted in a small degradation of the cardiac images on one detector ring.

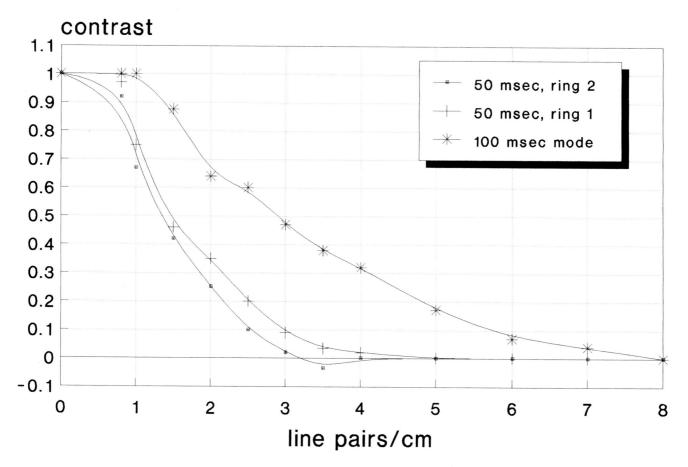

Figure 3-4 Comparison of modulation transfer function (MTF) for each of two slices in the 50-millisecond mode, and for the single-slice, 100-millisecond mode. Ring 2 has somewhat degraded resolution since this ring has double crystals that have slight gaps between them.

Recently the MTF of the single-slice mode has been improved. This series of improvements is shown in Figure 3-5. Most of these improvements were due to software changes that reduce the blurring effect of several interpolation steps that are used during reconstruction. Another major step in resolution is feasible but will be costly due to a great increase in the number of samples that would have to be processed. Current trends indicate an emphasis on obtaining more images quickly, with increasing spatial resolution being less of a priority.

Noise and Artifact

The signal-to-noise ratio in an image consists of random noise attributable to photon Poisson noise and structured noise due to various artifactual sources. Random electronic noise is usually negligible. Poisson noise has a square root relationship to the x-ray intensity and dose used. Using a 600-mA, 130-kV, 50-millisecond scan, the patient dose is approximately 300 mrad and the noise is about 1.2% at the center of a 10-in diameter water phantom. This noise value refers to the standard deviation of the pixel densities and can be lowered by smoothing the image at the expense of spatial resolution. It is currently not practical to improve this value by using higher beam current, due to the melting point limit of the tungsten targets.

Artifactual noise produces streaks, patterns of streaks, and shading variations in density. Streaks generally are caused by inaccurate measurements of position or of x-ray intensity. A variety of calibrations are used to correct as many of these potential inaccuracies as possible. However, drift can occur after calibration and lead to streaks. These drifts are usually tem-

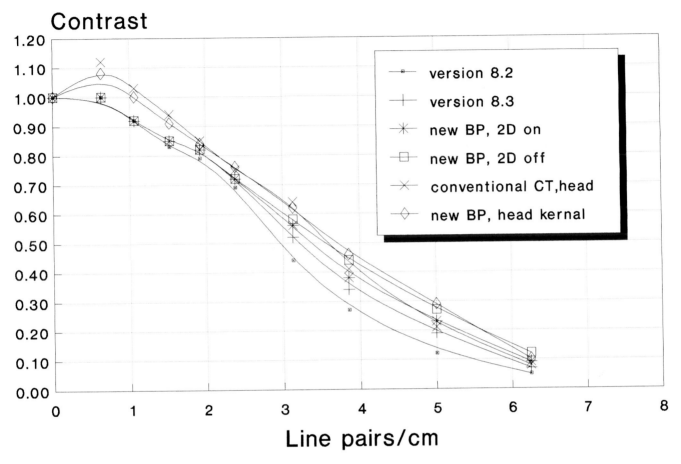

Figure 3-5 Recent improvements in the MTF of the 100-millisecond mode are compared to the MTF of an advanced conventional CT scanner. The improvements shown here were primarily due to improved interpolation schemes used in reconstruction.

perature related but can also be due to stray magnetic fields, such as those produced by a magnetized gurney that has been brought into the scanner room. Streak artifacts are generally minimized by controlling temperature and stray fields and by performing frequent calibrations.

Shading artifacts are caused by inaccuracy in the measurement of the ratio of transmitted x-ray intensity to incident x-ray intensity. The incident x-ray intensity is assumed to be given by an air scan that may have been obtained hours earlier. Thermal expansion of the target system can introduce errors into the assumed incident intensity. The measured transmitted intensity also may have errors due to system nonlinearities or to the presence of scattered radiation. Extensive calibrations are performed to minimize both of these possibilities. Due to the extreme heterogeneity of tissue densities in the chest, this problem can often be a serious limitation to the accuracy of measured CT numbers in the heart, with errors of 3% to 5% not uncommon.

Another type of noise occasionally noticed in images is pattern or structured noise. In the ultrafast scanner this type of noise can usually be traced to nonuniformities in the source intensity, as a function of angle. A typical cause is the presence of spilled contrast medium on the vacuum chamber window. Discontinuities in the tungsten target ring can also cause structured noise. The tungsten rings are made from segments. Under certain conditions, thermal expansion can cause a separation, producing a high-frequency gap in the target ring. This problem has been addressed by using "gap correction" software and by the recent introduction of angled gaps.

Scanning Rates and Limitations

The speed of scanning in an electron beam machine is almost entirely determined by computer restrictions. Recent tests at Imatron have demonstrated the feasibility of 5-millisecond scans, with preservation of an excellent focal spot. However, to take advantage of such speeds it would be necessary to digitize the data and store them at a rate of 130 megabytes per second or faster. Such speeds, although feasible, are not yet economical. Advances in multiplexers, analog-to-digital converters, high-speed digital buses, and fast memory systems will be needed.

At the present time, scan speed is held at either 50 or 100 milliseconds, and current efforts concentrate on increasing the number of these scans that can be obtained in sequence. Using the current 32-megabyte scan cache memory, which can hold 80 256 × 256 images, an eight-level flow study is restricted to 10 time intervals. Likewise, a 12-level cine study can have only eight frames per level, a number that is barely adequate to capture systolic contraction. Diastolic filling studies must use fewer levels. The scan cache memory has recently been doubled by taking advantage of low-cost VME random access memory. Another factor of two in VME memory size is feasible in the future. As the size of the scan cache approaches 128 megabytes, the use of parallel-transfer disk technology becomes attractive. Currently VME disks require 0.4 seconds to store a 50-millisecond image. This number can be reduced by using several disk drives in parallel or several disk heads on a single disk in parallel. When parallel disk technology becomes feasible, the total number of scans that can be stored in sequence will be virtually unlimited.

At the point of unlimited scan cache capacity, other restrictions become important. Heat dissipation in the high-voltage power supply and in the tungsten target system becomes an important issue. Thus higher-capacity power supplies and target systems are actively under development. Recently the cooling rate of the target system was improved by a factor of 30 by adding 20 parallel water-cooling channels and by improving other aspects of the thermal design. With these improvements, several seconds of continuous scanning have become feasible.

One goal of this increase in scan rate and scan capacity is to reduce the amount of contrast medium required to perform a complete cardiac stress examination. Ideally, it will be possible to obtain both a cine and a flow study on a single contrast medium injection. If simultaneous opacification of the right ventricle is required during a left ventricular flow study, a double-bolus injection can be programmed using the power injector. Thus one injection for the resting study and one for the stress study, for a maximum of less than 100 mL, would be a typical requirement.

Accuracy of Time-Density Measurements

The accuracy of time-density measurements is particularly important in determining myocardial perfusion. Since the nearby left ventricle peaks only 1 to 3 seconds earlier than the myocardium and has an absolute enhancement some 10 times greater, density errors can occur in the myocardium. These errors can be of three types. The first, spatial blurring, can result in a portion of the myocardial time-density curve mirroring the left ventricular cavity density. This is because a few percent of the left ventricular cavity enhancement will extend out several millimeters into the endocardium. The second potential source of error is beam hardening. Since x-rays pass through several centimeters of an iodine-filled left ventricle, their average energy changes, which results in an error in measured CT numbers. The third error is attributable to scattered radiation. The measured x-ray intensities may include tens of percent of scattered radiation. The reconstruction program estimates scatter and attempts to subtract it. As iodine fills the left ventricle, the amount and distribution of scattered radiation can change. Both the scattered radiation effect and beam hardening tend to produce broad dark streaks connecting dense objects, such as the spine and the opacified left ventricle, or the sternum and the left ventricle.

Currently these errors, all attributable to left ventricular opacification, limit the accuracy of myocardial perfusion measurements in certain regions, particularly the anterior region. Errors of 50% may be observed in this region; indeed, portions of the time-density curve may have negative values. These problems have been minimized by using large regions of interest for myocardial perfusion analysis, thus averaging out artifactual effects. The use of aortic root injections completely eliminates this source of error and enables myocardial perfusion measurements to be accurately obtained. Software and hardware improvements to improve the accuracy of intravenous myocardial perfusion studies are being actively pursued.

Accuracy of Edge Finding for Wall Motion

Spatial resolution and density resolution combine to set a limit on the accuracy at which a contour can be defined for the endocardial border. Image quality in border definition is important if accurate ventricular volumes are to be obtained. At the present time accuracy studies[9] indicate superior precision for volume determination of both the right and left ventricles. The best results are achieved with good ventricular opacification, using 40 mL or more of the contrast medium. Improvements in image quality—in spatial resolution or density resolution, or both—would allow accurate edge definition with lesser amounts of contrast medium. Since density resolution is fundamentally proportional to patient dose, the best approach would be to improve spatial resolution. An increase by a factor of two in spatial resolution should lead to significantly reduced contrast medium requirements.

FUTURE TECHNICAL DEVELOPMENTS AND APPLICATIONS

High-Resolution Imaging

The next major advance in ultrafast CT will be to improve spatial resolution by a factor of two or more. If the spatial resolution already demonstrated in the single-slice, 100-millisecond mode can be implemented in the multislice, 50-millisecond mode, a dramatic advance in cardiac imaging applications will occur. The technology for implementing such a step is currently available. The number of detector channels must be doubled from 864 to 1728 and the sampling time halved from 40 microseconds to 20 microseconds. This will also require the data acquisition system to process four times as many measurements as now. We had projected that this improvement would become available in approximately 1989.[7] The development of cardiac applications for ultrafast CT has proceeded more slowly than anticipated, so that business development funds for this purpose are not sufficiently available. It now appears that a combination of business funding and government support will enable this upgrade to be introduced by 1991. The development will be partially funded by simply waiting for conventional technology to catch up to the speed requirements needed for high-speed, high-resolution imaging.

Based on the twofold increase in resolution described above, all of the current cardiac applications—flow, function, and anatomic studies—will be greatly improved. In addition, certain new applications will become feasible. With a spatial resolution of 8 lp/cm in 50-millisecond mode and 16 lp/cm in the 100-millisecond mode, CT coronary artery imaging will begin to rival the capability of angiography. Useful images of coronary arteries, demonstrating calcified plaques, have already been a great success.[10–11] With the advent of high-resolution CT, it should also be possible to measure intraluminal narrowing. Thus ultrafast CT will be able to provide a complete coronary artery examination, including screening, anatomic definition of lesions, and measurement of coronary reserve flow under stress.

Very Fast Reconstruction for Real-Time Imaging

A second opportunity for greatly improved performance exists in the reconstruction processor. Currently, 4 seconds of 50-millisecond scans and 10 seconds of 100-millisecond scans are required to reconstruct and display the image. If these times can be reduced to 1 second or less, then real-time fluoroscopic imaging will be feasible. Immediate access to images will provide many major benefits. It will be possible to monitor the examination and to terminate it if a problem is seen to develop. The timing of the scans to capture the peak contrast opacification will be simplified. Currently, the time delay between injection and arrival at the left ventricle must be estimated. With real-time reconstruction, it will be possible to monitor the arrival of contrast medium in the region of interest and manually trigger the study at the optimum point.

Even using present hardware, a fast display is feasible if a restricted reconstruction matrix is specified. For example, if a 128 matrix is used with every other view, the present reconstruction algorithm could be speeded up by approximately four. Thus a preliminary form of real-time reconstruction could be developed with a software development effort. Hardware improvements at full resolution are also expected over the next few years. These improvements will be driven by developments in the semiconductor industry, leading to faster array processors and low-cost VLSI chips that may be used in backprojectors.

Improved Auto Analysis and Report Software, Using Advanced, Low-Cost Workstations

An important part of a diagnostic cardiac examination is the quantitative analysis of the images and extraction of a wide range of parameters. These parameters provide a complete description of the current performance of a patient's heart and may be used to define stages of disease and to follow the course of therapy. Since CT images are already in digital format, it is natural to consider automation of the analysis function. A dedicated computer workstation is the ideal method of implementing a computerized analysis and reporting system for ultrafast CT studies. The ultrafast scanner is equipped with a network facility referred to as the user-programmable offline workstation (UPOW) network. This system allows export of cardiac images from the scanner to a remote workstation for off-line analysis. Nearly all commonly available workstations can be connected to this network. Up to 15 workstations can be accommodated; they may be located at various locations, such as physicians' offices, research offices, and so on.

Software for automated cardiac analysis has been developed by Jon Harman (of Imatron Inc., South San Francisco), using a workstation developed by Virtual Imaging. This workstation uses an IBM-compatible 386 microcomputer that is interfaced to the ultrafast scanner via the UPOW connection described above. The color output of the analysis software is displayed on a 768 × 540 VGA display monitor, and hard copy is printed on 8½ × 11-in paper in color using an inexpensive Hewlett-Packard PaintJet printer. The ventricular analysis program begins by finding the inner and outer borders of the left and right ventricles, using an automated edge finder. Operator guidance is required to indicate the positions of cardiac valves and papillary muscles. Figure 3-6 shows end-diastolic and end-systolic images selected from one level of a cone study following automated edge definition of the left ventricular contours. When the contours have been found for all images at each level, the program then computes regional ejection fractions, wall motion and wall thickening, and the slice volumes and mass. These parameters are then displayed, using polar coordinates to facilitate identification of areas of abnormal movement with the corresponding cross-sectional image. An example of the analysis of systolic and diastolic values for a slice is given in Figure 3-7 (see color insert). Global cardiac parameters such as left ventricular mass, stroke volume, global ejection fraction, and the like are determined by summation of the individual slice values. A complete ventricular work-up, including both the right and left ventricles, is reported on several pages of output.

Further analysis of motion patterns in the left ventricle is facilitated by use of phase imaging and by three-dimensional displays. Phase analysis of the heart involves fitting each pixel to a sine curve of unknown phase. The fitted phase of each pixel is color coded and displayed, to give a phase image in which color represents the relative time of periodic motion. Figure 3-8 (see color insert) is an example of a normal frame from a cine study compared to the derived phase image. Uniform ventricular contraction is represented by a uniform phase for both the right and left ventricles. Lighter areas in this image represent earlier contraction. This method is used to search for sites of cardiac dysrhythmia indicating regions of abnormal electrophysiology.[12]

Since the heart is a three-dimensional structure, it is helpful to view cardiac motion using three-dimensional display techniques. Therefore, Harman has equipped his program with an operator interactive wire mesh display, as illustrated in Figure 3-9 (see color insert). These wire models are displayed as a movie and may be rotated to represent the standard views used in cineangiography. A hard copy of critical frames may be included in the report. The range of computer analysis and presentation that is feasible using a workstation approach is extremely helpful in providing unbiased quantitative information that accurately describes ventricular function.

SUMMARY

Will ultrafast CT be the only examination required for cardiac diagnosis and treatment planning? This is the goal of the physicists and physicians who are working in this field. Certainly there are several other cardiac imaging modalities competing for selection as the preferred imaging approach. Cross-sectional imaging as used in ultrasound and magnetic resonance imaging are the most significant competitors. Cost has been a barrier to attracting more clinical interest in cardiac CT. However, higher cost is compensated by higher throughput and a range of unique, important applications outside the heart. Ultrafast CT is becoming more

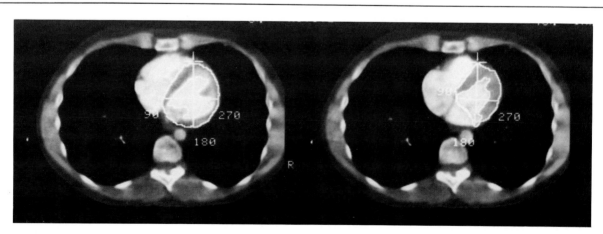

Figure 3-6 Typical auto-traced edges, indicating the contours of the left ventricular borders at systole and diastole.

readily available in hospitals, improving cardiologists' and cardiac radiologists' access to it. Currently such x-ray procedures as cineangiography are the preferred method for imaging coronary arteries. As the power of CT grows due to technological advances, CT will tend to replace conventional x-ray procedures. Recent improvements in ultrafast CT increase the range of cardiac procedures for which CT is the modality of choice. Further improvements in scanning rates and spatial resolution will give further impetus to the growth of cardiac CT applications. Based on the projected technology developments discussed here, CT does have an excellent prospect of becoming the preferred method for complete cardiac diagnosis in the 1990s.

Acknowledgement: The author would like to thank Josh Lack for providing the data in Figures 3-4 and 3-5 and Jon Harman for the data shown in Figures 3-7, 3-8, and 3-9.

REFERENCES

1. Ritman EL, Kinsey JH, Robb RA, Wood E. Physics and technical considerations in the design of the DSR: A high temporal resolution volume scanner. *Am J Radiol.* 1980;134:369.

2. Newton TH, Potts DG, eds. *Radiology of the Skull and Brain.* Vol 5. *Technical Aspects of Computed Tomography.* St Louis: CV Mosby Co; 1981.

3. Boyd DP. Computerized-transmission tomography of the heart using scanning electron beams. In Higgins C, ed. *CT of the Heart: Experimental Evaluation and Clinical Application.* Mt Kisco, NY: Futura; 1983:45–59.

4. Peschmann KR, Napel S, Couch JL, et al. High speed computed tomography, systems and performance. *Appl Optics.* 1985;24: 4052–4060.

5. Ackelsberg SM, Napel S, Gould RG, Boyd DP. Efficient data archive and rapid image analysis for high speed CT. *SPIE J.* 1986; 626:451–457.

6. Lane SD. Power-injected CT contrast opacifies vascular spaces. *Diagn Imaging.* 1988:308–312.

7. Boyd DP, Couch JL, Napel SA, Peschmann KR, Rand RE. Ultrafast cine-CT for cardiac imaging: Where have we been? What lies ahead? *Am J Card Imaging.* 1987;1:175–185.

8. Assimakopoulos PA, Boyd DP, Jaschke W, Lipton MJ. Spatial resolution analysis of computed tomographic images. *Invest Radiol.* 1986;21:260–271.

9. Reiter SJ, Rumberger JA, Feiring AJ, Stanford W, Marcus ML. Precision of measurements of right and left ventricular volume by cine computed tomography. *Circulation.* 1986;74:890–900.

10. Reinmuller R, Lipton MJ. Detection of coronary artery calcification by computed tomography. *Dynamic Cardiovasc Imaging.* 1987;1:139–145.

11. Janowitz WJ, Agatston AS, King D, Smoak KM, Samet P, Viamonte M. High-resolution ultrafast CT of the coronary arteries: New technique for visualizing coronary artery anatomy. Scientific Program of the Radiological Society of North America; November 1988; Chicago; abstract no. 1048.

12. Abbott JA, Scheinman ED, Herre JM, et al. Structural changes in effort induced right ventricular tachycardia. Annual meeting, American Heart Association; November 1988; Washington, DC; abstract no. 1583.

Chapter 4

Magnetic Resonance Imaging: Technical Aspects

Ronald M. Peshock, MD, and Donald M. Moore, Jr, MD

INTRODUCTION

Nuclear magnetic resonance imaging (MRI) is particularly well suited to the evaluation of the heart and vascular system. Advantages of MRI include high inherent contrast between flowing blood and the chamber or vessel wall, good spatial resolution, relative ease of gating to the cardiac cycle, and the ability to obtain images in multiple orientations. In addition, the fact that ionizing radiation and iodinated contrast material are not necessary allows serial evaluations to be performed with little if any risk.

The basic physics of MRI has been the subject of a number of extensive reviews.[1-4] The purpose of this brief review is to introduce the reader to the basic physical principles involved, the multiple contrast mechanisms at play, the flow effects important in cardiovascular imaging, and particular techniques relevant to cardiac imaging. The approach is intended to be intuitive and not rigorous.

BASIC PRINCIPLES OF NUCLEAR MAGNETIC RESONANCE

Simply put, MRI involves exactly what it says, nuclear magnets and resonance. It is dependent upon the fact that the nuclei of certain atoms have a property called magnetic moment. In essence, this property causes the nuclei to behave like tiny bar magnets. When placed in the presence of a magnetic field, these tiny bar magnets tend to line up with the external magnetic field. It is then possible to expose these same tiny bar magnets to a second magnetic field, which is oriented perpendicularly to the main magnetic field. When this additional magnetic field is turned on, the nuclei are tipped away from their original orientation. When this second external magnetic field is then turned off, the nuclei gradually return to equilibrium and their original orientation. In the process of returning to equilibrium, they induce a current that can be detected. It is the energy released in the return of the nuclei to their original condition that is used in magnetic resonance spectroscopy and imaging. (These steps are illustrated in Figure 4-1.)

As can be seen from this brief description of magnetic resonance, this is fundamentally a magnetic phenomenon. Only nuclei that have an odd number of protons and neutrons can have magnetic moment. Examples of nuclei having a magnetic moment are hydrogen, phosphorus, sodium, and carbon. This property arises from the fact that these nuclei have a net electric charge and a property called nuclear spin. A fundamental principle in physics is that a moving electric charge generates a magnetic field. Hence, the combination of net electric charge and nuclear spin produces a magnetic moment. Because of the importance of the nuclear spin, one may see a group of nuclei referred to as "spins" or "nuclear spins." The fact

---

Supported in part by Ischemic SCOR grant HL-17669, Program Project Grant HL-06296, the Moss Heart Fund, and a grant from Diasonics.

■ *Random nuclear moments.*

■ *Uniform magnetic field aligns magnetic moments.*

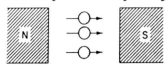

■ *Applied RF burst perturbs magnetic moments.*

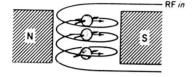

■ *Realignment emits characteristic frequency sensed by RF coil.*

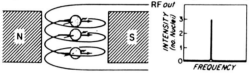

Figure 4-1 Basic steps in nuclear magnetic resonance. See text. RF is radiofrequency. *Source:* Adapted with permission from *American Journal of Cardiology* (1980;46:1278-1283), Copyright © 1980, Cahners Publishing Company, Inc.

the magnetic properties of nuclei are involved led to the term "nuclear magnetic resonance"; note, however, that people frequently refer to it as simply magnetic resonance, particularly in the setting of MRI.

The second fundamental concept of nuclear magnetic resonance is resonance. It turns out that the nuclear spins can only be tipped from their original orientation by energy in certain size packets or "quanta." The energy of these packets is dependent on the local magnetic field and the nucleus (i.e., hydrogen versus phosphorus) being observed. If radio waves containing quanta of the appropriate size are applied, then there is "resonance," and the energy will be absorbed. In a sense, in MRI one takes the nuclei already inside the body and applies a pulse of energy to make them more energetic. (It's as if, in order to do a nuclear medicine study, instead of injecting a radionuclide, one put the patient into a cyclotron to energize the atoms that were already present inside the body.) When the radio wave is turned off, the nuclei that have been magnetically energized within the body will reemit the energy, and this energy is used to form the picture or spectrum.

Resonance is actually a very familiar phenomenon in everyday life. Consider a child on a swing: The child swings back and forth every few seconds with a certain frequency. To make the child swing higher or gain energy, one has to push with the same timing or the same frequency. If one does not, energy is not transferred efficiently into the swing, and the child on the swing does not absorb the energy from the person pushing the swing. The person must push the swing at its resonant frequency for the energy to be absorbed effectively.

Now, as soon as one stops putting energy into a group of nuclei, they begin to lose energy and return to equilibrium. This return to equilibrium is termed relaxation and describes the return of the nuclei to the resting state. This return to equilibrium is also a very familiar idea. After all, if one stops pushing the child on the swing, the motion of the swing gradually declines until the swing comes to rest. We could determine the rate at which the swing loses energy and from that figure determine a relaxation rate or relaxation time for the child on the swing.

In the case of a set of excited nuclei, one describes this return to equilibrium in terms of the T1 relaxation time. This relaxation time describes the loss of energy of the group of excited spins to the surrounding environment. If the spins interact strongly with the environment, then the energy will be lost rapidly and the T1 relaxation time will be short. If they interact inefficiently with the environment, then the energy will be lost slowly and the T1 relaxation time will be long. T1 relaxation times for hydrogen nuclei in water in the body typically range from hundreds to thousands of milliseconds.

Interestingly, a second time constant is required to describe the behavior of a group of excited spins and is termed the T2 relaxation time. This second way of describing the return to equilibrium is slightly less familiar but is still seen in normal day-to-day life. Consider a group of children on a swing set. If all the children are swinging together, they can be described as swinging in unison, or "in phase." As previously discussed, once the parents stop pushing the swings, the swinging children all begin to lose energy, reflecting the loss of energy to the environment, or T1 relaxation. But in addition, something else happens: The swings begin to drift out of phase; they no longer swing together as a group. Without going into detail, in nuclear magnetic resonance the nuclei dephase because of interactions between the individual nuclear spins. For hydrogen nuclei in water in the body, T2 relaxation times are considerably shorter than T1 relaxation times, typically on the order of tens of to a few hundred milliseconds.

BASIC PRINCIPLES OF IMAGING

Magnetic resonance imaging utilizes the principles of both resonance and relaxation in making an image. The typical MRI machine consists of several components: (1) the main magnet, (2) magnetic field gradients, (3) a coil to apply the pulse of energy to tip the nuclei and to

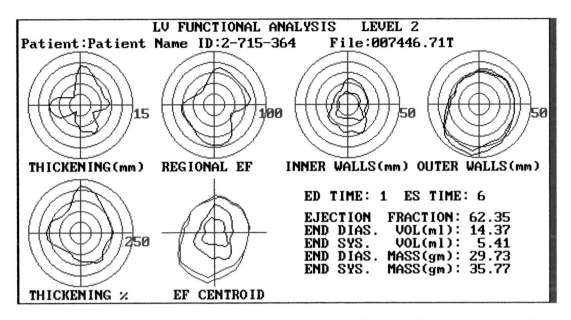

Figure 3-7 The various functional parameters of a single slice have been plotted in polar coordinates to simplify identification of anatomy. This is a single page from a computer-generated report describing left ventricular (LV) function. DIAS., diastolic; ED, end diastole; EF, ejection fraction; ES, end systole; SYS., systolic.

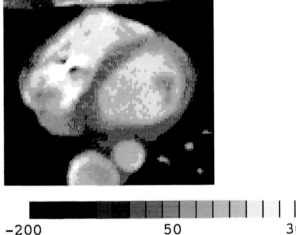

Figure 3-8 Top, A standard gray scale image of one cine frame from a movie of a cross-section of the left ventricle. Bottom, A computed image in which the relative phase of cardiac motion has been mapped into color. By referring to the color scale at the bottom, it is possible to determine the uniformity of ventricular contraction and find points of premature motion.

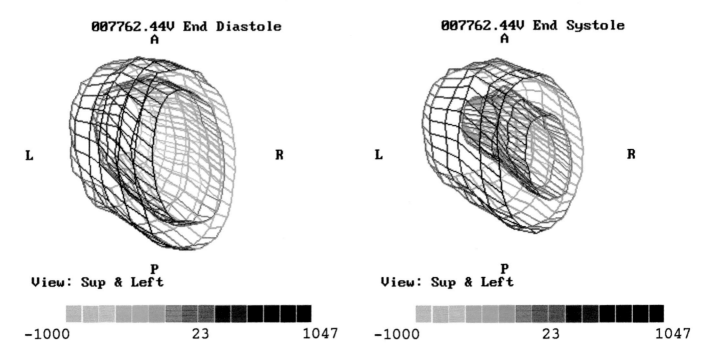

Figure 3-9 These wire mesh models of the epicardial and endocardial surfaces of the left ventricle are useful for showing cardiac motion as a three-dimensional movie. Two frames corresponding to diastole (left) and systole (right) have been selected for inclusion in this patient's hard copy report.

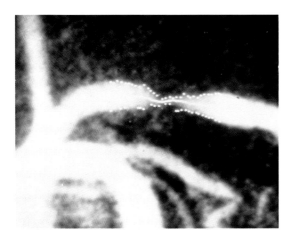

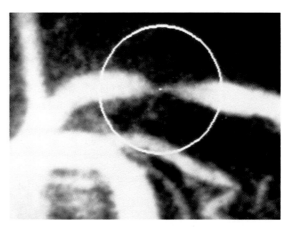

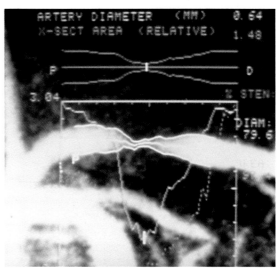

Figure 9-4 An example of automated quantitative coronary arteriography applied to ECG-gated mask mode subtracted coronary arteriograms is shown, following operator identification of the arterial stenosis (top left). Arterial edge boundaries are automatically determined using gradient densitometric criteria (top right). Absolute geometric and relative videodensitometric data are calculated along the arterial segment under assessment and plotted (below right).

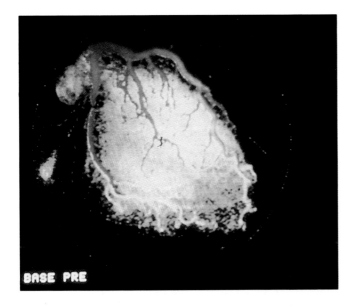
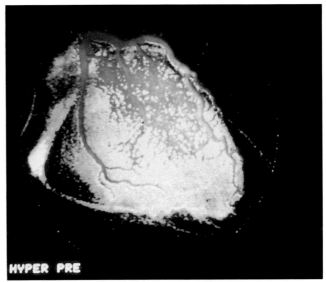
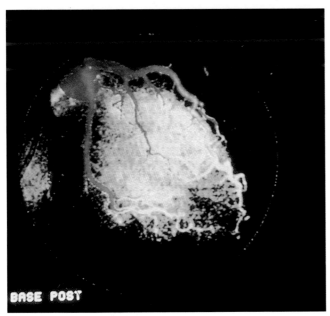
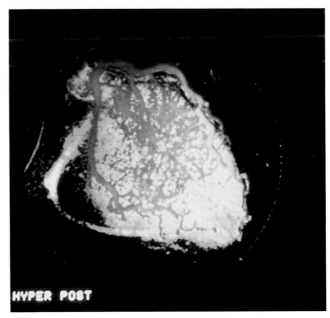

Figure 9-5 Parametric contrast medium appearance pictures obtained prior to (top row) and following (bottom row) left circumflex coronary angioplasty are shown in the right anterior oblique projection. Images were obtained under baseline (left) and hyperemic (right) conditions. Pixels colored red, yellow, and white, appear in the first, second, and third cycles following administration of contrast medium. The absent hyperemic response in the stenotic left circumflex coronary artery distribution prior to angioplasty returns following successful dilation, at which time that distribution has a similar response to the normal left anterior descending coronary artery area. *Source:* Courtesy of J Hodgson, Cleveland, OH.

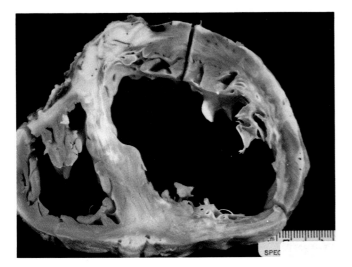

Figure 14-3B The explanted heart cut in a manner similar to that in the scans shown in Figure 14-3A. The area of recent infarction is identifiable by the yellow necrotic tissue. Note the faithful representation of the transmural extent of infarction by computed tomography, nontransmural in the lateral wall and transmural in the posterior wall.

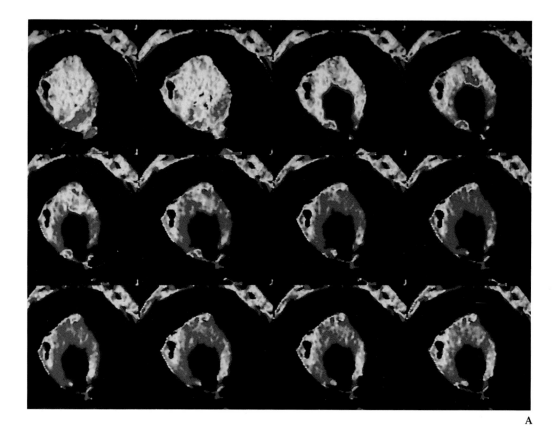

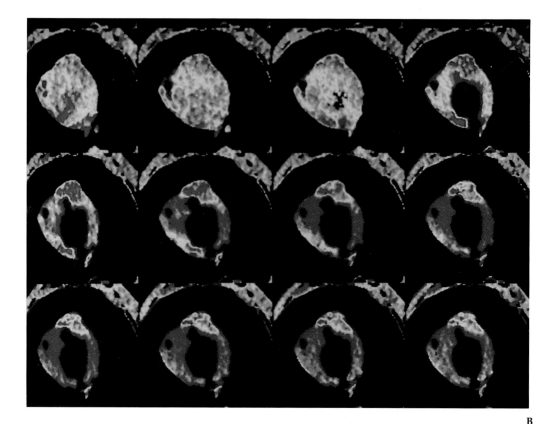

Figure 15-5 Myocardial color perfusion mapping. Two examples (A and B) of color coding myocardial pixels to the change in HU produced by contrast medium in a canine experiment before and after occlusion of the left anterior descending coronary artery.

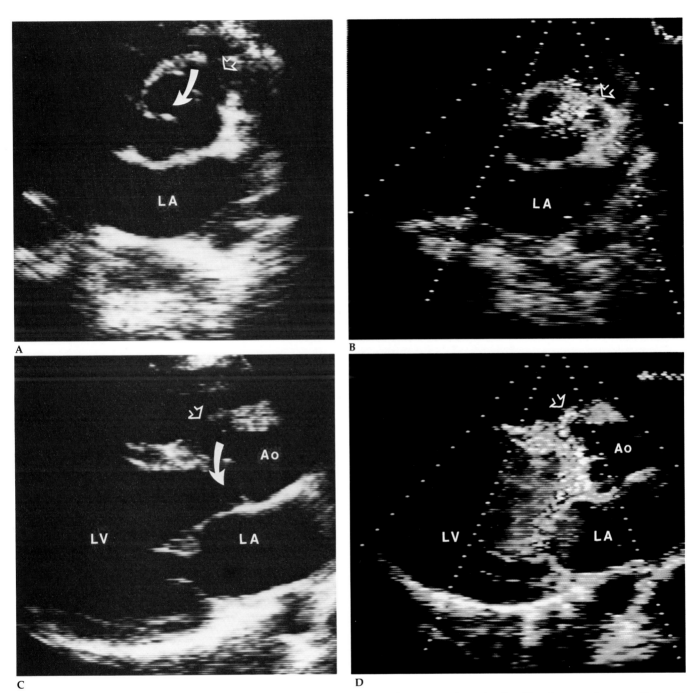

Figure 22-4 Stop-frame views of diastolic two-dimensional echocardiograms and accompanying color flow Doppler maps from a patient with paravalvular aortic regurgitation due to infective endocarditis. A, In the short-axis view at the level of the aorta and left atrium (LA), the channel (open arrow) representing the path of the paravalvular leak permits blood to regurgitate (closed arrow) into the left ventricular outflow tract just proximal to the aortic valve. B, The accompanying short-axis color flow Doppler map overlay shows a speckled blue-green mosaic Doppler signal originating from the paravalvular channel (open arrow) and passing into the left ventricular outflow tract proximal to the aortic valve along the direction of the closed arrow depicted in A. C, In the long-axis view of the left ventricle (LV), left atrium (LA), and aorta (Ao), the channel (open arrow) of the paravalvular leak and the path of its communication (closed arrow) with the left ventricular outflow tract just proximal to the aortic valve are demonstrated. D, The accomanying long-axis color flow Doppler map overlay shows a speckled blue-green mosaic Doppler signal emanating from the channel (open arrow) of the paravalvular leak and proceeding across the left ventricular outflow tract proximal to the aortic valve and anterior to the mitral valve along the direction of the closed arrow depicted in C. Anterior structures are at the tops of the panels.

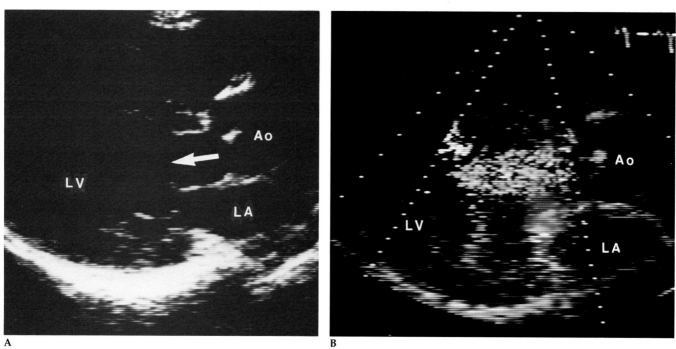

Figure 22-5 Stop-frame views of a diastolic long-axis two-dimensional echocardiogram and accompanying color flow Doppler map from a patient with valvular aortic regurgitation. A, The two-dimensional echocardiogram illustrates the aorta (Ao), left ventricle (LV), and left atrium (LA) along their long axes and the continuity between the anterior aortic root and ventricular septum, in contrast to the anatomic discontinuity noted in Figure 22-4. The path of aortic valvular regurgitation is depicted by the closed arrow. B, The accompanying long-axis color flow Doppler map overlay demonstrates a speckled blue-green mosiac Doppler signal that emanates from the aorta and passes directly to the left ventricle across the aortic valve in the direction of the closed arrow depicted in A. The direction and path of the regurgitant flow are distinctly different from those of the patient with a paravalvular leak illustrated in Figure 22-4.

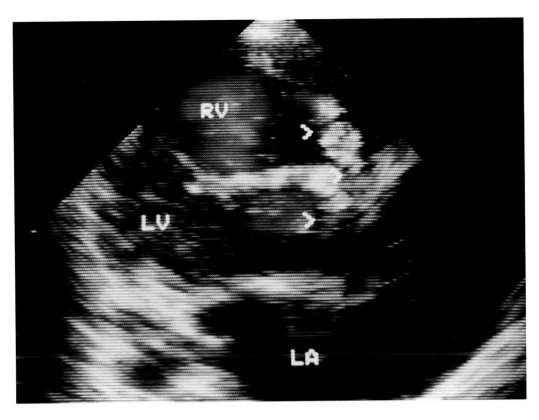

Figure 37-3 Doppler color flow map, parasternal long-axis view. Mosaic jet flow (arrowheads) across the subaortic ventricular septum. The patient has a small residual ventricular septal defect following repair of tetralogy of Fallot. Abbreviations are explained in the legend for Figure 37-2.

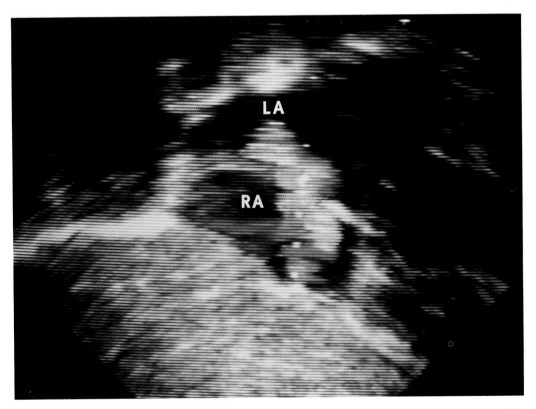

Figure 37-4 Doppler color flow map, same patient as in Figure 37-2. A wide band of red-orange flow (toward the transducer) crosses the midatrial septum. Abbreviations are explained in the legend for Figure 37-2.

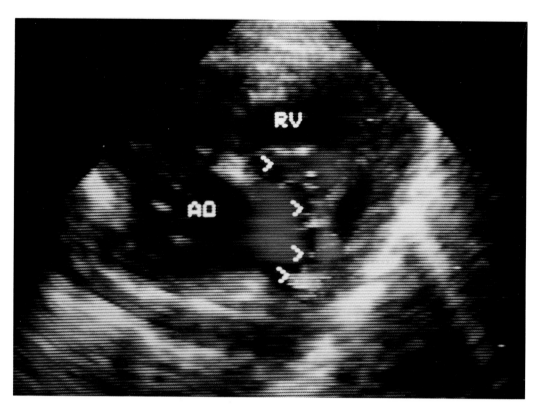

Figure 37-7 Doppler color flow map, parasternal short axis view of great arteries. Color was present in the pulmonary artery root in systole and diastole (shown here). A red-orange jet of diastolic flow (toward the transducer, arrowheads) from a patent ductus arteriosus extends from the descending aorta along the left lateral border of the pulmonary artery root. Antegrade diastolic flow (blue) results from swirling flow in the dilated arterial root. AO, aorta; RV, right ventricle.

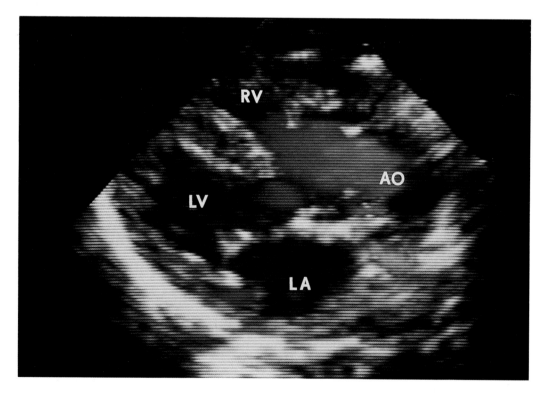

Figure 37-15 Doppler color flow map, parasternal long-axis view. In this systolic frame the right ventricle ejects directly to the aorta. The stream of flow from the right ventricle is oriented away from the transducer and thus is coded blue. The stream of flow from the left ventricle to the aorta is oriented toward the transducer and thus is coded red. AO, aorta; other abbreviations are explained in the legend for Figure 37-2.

receive the energy they emit as they return to equilibrium, and (4) associated electronics and a computer to control the process.

Very briefly, the main magnetic field establishes the initial orientation of the nuclei. Small variations in the magnetic field from point to point are then introduced, using the magnetic field gradients. Since the resonant frequency for a nucleus at a particular location in the body is dependent on the magnetic field at that particular location, it is possible to establish the relationship between points in the body and the frequency. This allows the machine to determine where the signal is coming from in the body. It is the use of magnetic field gradients that allows MRI.

The concept of relaxation then becomes important in that different tissues in the body have different T1 and T2 relaxation times. By applying the pulses of energy in different combinations, referred to as pulse sequences, it is possible to minimize or accentuate the effects of different T1 and T2 relaxation times and flow on the final image. This provides the basis for the intrinsic contrast seen in magnetic resonance images.

CONTRAST MECHANISMS

Now, given that tissues have different T1s and T2s, how is it possible to take advantage of these differences? In a sense, the MRI device is very much like a camera. One takes a series of objects (organs) with different physical characteristics (different T1s and T2s) and then takes a picture using the MRI camera. Just like using a camera, it is possible to set the MRI machine to take a picture that emphasizes particular differences in the objects. In other words, it is possible to set the machine to accentuate differences in T1 or differences in T2 in a particular image. In addition, just like using a camera, it is possible to set the MRI camera to take a very bad picture that obscures differences between tissues.

One of the great strengths of MRI is its flexibility. The types of pulses of energy applied, their timing, and their location in space can be easily altered to accentuate one or more of the characteristics of the object. Moreover, in certain pathologic conditions, the relaxation times T1 and T2 change. If imaging sequences are used that accentuate contrast in the image on the basis of these differences, then it will be possible to distinguish abnormal from normal tissue. If the wrong imaging sequence is used, it may be impossible to identify the abnormal tissue.

Images of an excised heart obtained under different pulsing conditions are shown in Figure 4-2. In the setting of myocardial infarction, both T1 and T2 can increase in the infarct region. In Figure 4-2A, a pulse sequence that emphasizes differences in T1 and mini-

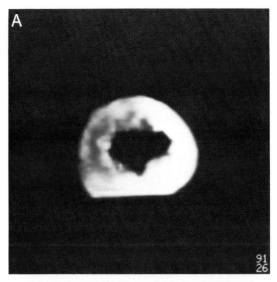

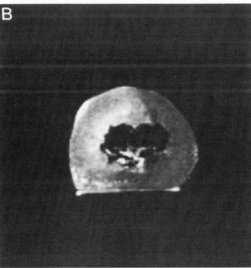

Figure 4-2 Images of an excised heart with a recent myocardial infarction, obtained under different pulsing conditions. A, Image obtained under conditions that accentuate differences in T1 relaxation time, so that the infarct is dark compared to the surrounding normal tissue. B, The same section imaged using a pulse sequence that accentuates differences in T2 relaxation times; the region of infarction is bright because of a prolonged T2 relaxation time.

mizes differences in T2 was used (gradient reversal, short repetition time [TR], short echo time [TE]). In this case, the image becomes darker if the T1 is longer; hence, the infarcted region is dark. In Figure 4-2B, a pulse sequence that emphasizes differences in T2 and minimizes differences in T1 (spin echo, long TR, long TE) was used to image the same heart. Note that by simply changing the settings on the MRI camera (changing the pulse sequence), one can completely reverse the intensities of normal and abnormal tissue. Hence, in interpreting magnetic resonance images, it is always critical to know the pulse sequence that was used to obtain the picture.

FLOW EFFECTS

An important factor in contrast in MRI, particularly for the cardiovascular system, is the fact that MRI is intrinsically sensitive to motion. This sensitivity is what leads to the intrinsic contrast between flowing blood and vessel wall or chamber lumen in cardiovascular MRI. In the case of blood flow, these motion effects can be broken into two basic groups: time-of-flight effects and dephasing effects.[5] Time-of-flight effects derive from the fact that a series of pulses is required to produce a magnetic resonance image. If hydrogen nuclei in the blood "see" certain combinations of these pulses, they will appear white in the magnetic resonance image. If they move out of the imaging plane before they are exposed to all the pulses, then the region will appear dark because no signal will be present. Under other pulsing conditions it is possible to decrease the signal from blood by applying a number of pulses upstream to the final imaging plane. Examples of transit time effects are seen in Figure 4-3A.

Specific types of transit time effects include high-velocity signal loss, in which blood moves out of the imaging plane before it can receive all pulses, and flow-related enhancement.[6] In this latter phenomenon, spins that have not "seen" previous pulses enter the slice, allowing additional signals to appear.

Dephasing effects occur when there is a range of blood flow velocities within a given "voxel," or imaging volume. If there is a small range of velocities within a voxel, then the dephasing effect is small. If there is a wide range in velocities in a voxel, then there is increased dephasing, which leads to cancellation and loss of signal intensity.[7] Dephasing effects are frequently seen in regions of turbulent flow in which there is a wide range or dispersion of blood velocities. The dispersion of velocities in a region is also the basis for the turbulence map used in color flow Doppler echocardiography. Hence, dephasing effects are frequently seen in a magnetic resonance image in regions where turbulence is seen on color flow studies. An example of signal loss due to spin dephasing is seen in Figure 4-3B. Spin dephasing also occurs near the margins of vessels where there is a large gradient or large range of velocities.

TECHNIQUES RELEVANT TO CARDIAC MRI

A variety of imaging techniques are presently employed in cardiac MRI. The most common technique utilized is the standard, gated multislice spin echo. In this approach, imaging pulses are linked to the cardiac cycle using electrocardiogram (ECG) or pulse gating. The use of a multislice technique means that the differ-

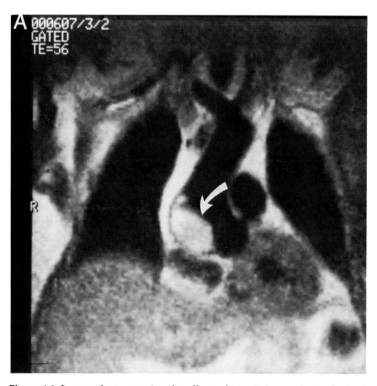
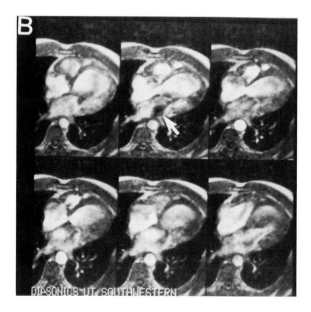

Figure 4-3 Images demonstrating the effects of transit time and spin dephasing. A, Coronal spin echo image of a patient with a proximal aortic dissection. High-velocity signal loss leads to the loss of signal in the ascending aorta. Slow blood flow in the false channel of the dissection (arrow) leads to high signal intensity. B, Multiple frames from a standard gradient reversal cine sequence in which turbulence in a jet of mitral regurgitation leads to spin dephasing and resultant loss of signal in the region of the jet (arrow).

ent slices taken through the heart are acquired at different points in the cardiac cycle. In other words, the images are not simultaneous in time. This is markedly different from all other imaging techniques presently used in cardiology. This technique provides excellent anatomic definition but gives limited information on function, again because the slices are not acquired at the same point in the cardiac cycle.

In a standard spin echo sequence, the TR is linked to the patient's heart rate. Hence, given typical heart rates of 60 to 100 beats per minute, this implies a TR of 600 to 1000 milliseconds. This TR means that T1 effects will be present in the image if the data are acquired with every heart beat. To reduce T1 effects, one typically has to go to every other heart beat, thus lengthening the TR. An advantage of the spin echo approach is that one can easily adjust the TE to increase or decrease the amount of T2 effect on image intensity. Typically, the TE used in MRI of the heart ranges from 30 to 80 milliseconds, with larger values used to increase the degree of T2 effect on the final image intensity.

A technique that provides the high resolution of multislice imaging with multiple cardiac phases is shown in Figure 4-4. The multislice, multiphase technique[8] is obtained by acquiring multiple multislice sequences, with different slices being acquired at different points in the cardiac cycle. By repeating the multislice acquisition, it is possible to obtain all slices at all points in the cardiac cycle. However, to obtain five slices at five points in the cardiac cycle would require five data acquisitions. This means that it would take five times as long as a single standard multislice technique. Such a series of data acquisitions typically takes from 30 to 40 minutes, limiting its clinical applicability.

A second general approach obtains images of a single slice or of several slices at multiple points in the cardiac cycle. This has been termed FLASH (Fast Low Angle Shot), GRASS (Gradient Recall Acquisition of Steady State), or cine MRI.[9] As shown in Figure 4-5, this technique utilizes partial flip angle pulses and gradient reversal sequences to allow framing at rates up to 30 frames per second. In this technique, one sacrifices multiple slices for improving the framing rate or temporal resolution in a single slice. Typically, as one tries to image more slices, the framing rate per slice drops. This type of pulse sequence has been widely applied in the evaluation of cardiac function and valvular regurgitation.

In cine MRI, the TR is typically in the range of 22 to 50 milliseconds, which implies potential framing rates of from 20 to 45 frames per second. Typical TEs are from 7 to 15 milliseconds. Note that this type of sequence will generate images in which differences in T2 have little effect on the final signal intensity while differences in T1 generally have a greater effect.

More recently, it has become possible to obtain an entire magnetic resonance image in a fraction of a second. This technique, termed "echo-planar" imaging, has been described by Mansfield[10] and Pykett and Rzedzian[11-12] and is illustrated in Figure 4-6. In this technique, the magnetic field gradients are changed very rapidly, making it possible to obtain all of the information required to form an image in the space of a single pulse. This technique allows essentially real-time imaging. At present, this high-speed imaging is associated with some loss in spatial resolution (i.e., pixel size is on the order of 3×3 mm for echo-planar, as compared to 1.7×1.7 mm for standard cardiac MRI).

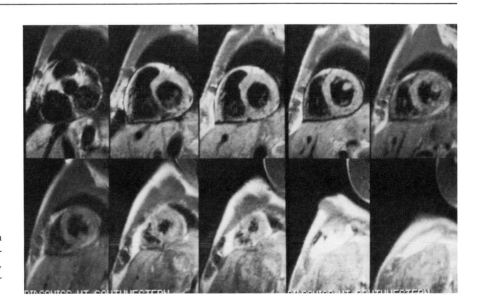

Figure 4-4 Series of images obtained with a multislice, multiphase technique. By permuting the order in which the slices are acquired, one obtains each anatomic level at end-diastole.

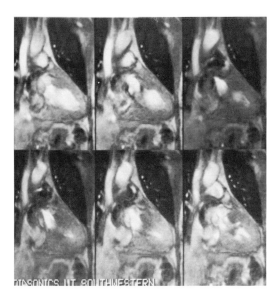

Figure 4-5 Series of images obtained using cine MRI sequence (gradient reversal, partial flip, flow compensated sequence with TR of 50 milliseconds and TE of 12 milliseconds). This type of sequence has been particularly useful in the evaluation of cardiac function.

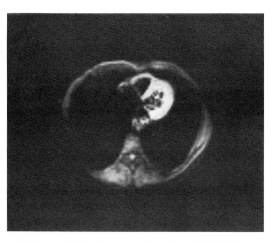

Figure 4-6 Image obtained using echo-planar MRI. This image was acquired in 30 milliseconds, compared to minutes for the previous images.

SUMMARY

Magnetic resonance imaging is fundamentally based on the interaction of the magnetic fields of nuclei in the body and external magnetic fields. Its strengths derive from the sensitivity of these nuclear magnets to changes in their local environment.

REFERENCES

1. Goldman MR, Pohost GM, Ingwall JS, Fossel ET. Nuclear magnetic resonance imaging: potential cardiac applications. *Am J Cardiol.* 1980;46:1278–1283.

2. Young SW. *Magnetic Resonance Imaging: Basic Principles.* New York: Raven Press; 1988.

3. Bradley WG, Newton TH, Crooks LE. Physical principles of nuclear magnetic resonance. In: Newton TH, Potts DG, eds. *Advanced Imaging Techniques.* San Anselmo, Calif.: Clavadel Press; 1983.

4. Pykett IL. NMR imaging in medicine. *Sci Am.* 1982;246:78–88.

5. Dumoulin CL. Flow imaging. In: Budinger TF, Margulis AR, eds. *Medical Magnetic Resonance: A Primer—1988.* Society of Magnetic Resonance in Medicine; 1988.

6. Bradley WG, Waluch V, Lai K, et al. The appearance of rapidly flowing blood on magnetic images. *Am J Radiol.* 1984;143:1167–1174.

7. von Schulthess GK, Higgins CB. Blood flow imaging with MR: spin phase phenomena. *Radiology.* 1985;157:687–695.

8. Crooks LE, Barker B, Chang H, et al. Magnetic resonance imaging strategies for heart studies. *Radiology.* 1984;153:459–465.

9. Frahm J, Haase A, Matthaei D. Rapid NMR imaging of dynamic processes using the FLASH technique. *Magn Reson Med.* 1986; 3:321–327.

10. Doyle M, Turner R, Cawley M, et al. Real-time cardiac imaging of adults at video frame rates by magnetic resonance imaging. *Lancet.* 1986;2(8508):682.

11. Rzedzian RR, Pykett IL. Instant images of the human heart using a new, whole body MR imaging system. *Am J Radiol.* 1987; 149:245–250.

12. Pykett IL, Rzedzian RR. Instant images of the body by magnetic resonance. *Magn Reson Med.* 1987;5:563–571.

Part II

Coronary Artery Disease

Chapter 5
Cardiac Digital Angiography

Jonathan Tobis, MD, Walter Henry, MD, and Orhan Nalcioglu, PhD

INTRODUCTION

The benefit of converting radiographic images from a film-based method to a computerized format is that the digital images can be manipulated to enhance the information within the image with greater facility.[1-2] In addition, conversion to a computerized system facilitates quantitative analysis of the anatomic structures. Although background subtraction was one of the first methods used to enhance images, the versatility of digital imaging resides in the multiple ways that the images can be mathematically manipulated.[3-6] At the present time, digital acquisition of cardiovascular images is best performed in an unsubtracted format to obtain coronary angiograms, whereas the subtraction process is still most useful for enhancing low-dose–contrast left ventriculograms. This review will describe some of the various ways in which digital processing has been applied to cardiovascular imaging. Those aspects of digital angiography that have been found most useful in the daily process of cardiac catheterization and coronary angioplasty will also be summarized.

The method of background subtraction has been available for many years and has been applied to other fields besides radiographic imaging. Astronomers use the subtraction method to distinguish new stars among the thousands of points of information that are present on a photographic plate. To accomplish this, an initial image of a section of the sky is taken. The initial image is then compared to subsequent images. The two images are superimposed, and the stars that are in the same position on both plates cancel each other out; that is, they are subtracted. Any star that does not align with one in the original image can be quickly appreciated.

The same subtraction method is used in medical radiographic imaging. With film-based systems, angiograms can be enhanced by superimposing a radiograph taken before contrast injection upon one obtained following the intra-arterial administration of contrast medium (Figure 5-1). This process is laborious and is only useful for selected cut films. To improve the subtraction process, computers were employed by several pioneer groups led by Heintzen and Mistretta and their respective co-workers.[7-8] Once the images are in a digital format, an image-processing computer can compare the images and provide a subtracted image that accentuates the differences between the initial and all subsequent images. The computer can process the images in real time at 30 frames per second, which permits the application of the subtraction enhancement process to cardiac structures despite their being in constant motion (Figure 5-2).

Although digital angiography was initially developed as a means for obtaining intravenous first-pass angiograms, it has proved to be most beneficial as an adjunct to invasive cardiac catheterization. One reason for this change of emphasis is the computer's unique capacity to manipulate images and obtain functional information about blood flow, iodine contrast medium density, and ventricular performance. In the area of cardiovascular diagnosis, digital angiography has been used to obtain left ventriculograms with a lower dose

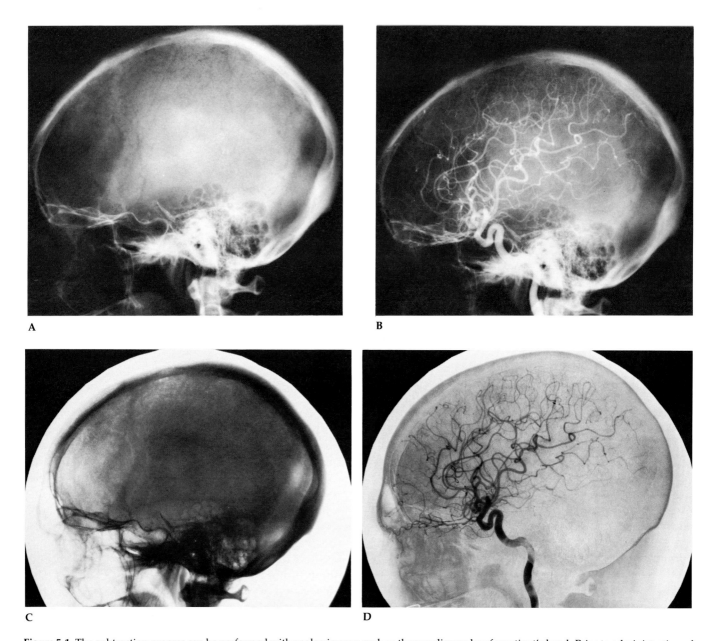

Figure 5-1 The subtraction process can be performed with analog images such as these radiographs of a patient's head. Prior to administration of contrast medium (A), the bones and soft tissues are recorded. Following contrast medium injection (B), there is opacification of the carotid and cerebral arteries, but there is interference in visualization of the smaller arteries because of x-ray absorption by the bones and soft tissues. To enhance the contrast of the fainter iodinated structures, a negative radiographic image of the skull is obtained photographically (C) and is superimposed upon the images with iodine (B) to obtain the "subtracted" image (D).

(12 mL) of iodinated contrast medium than is used during standard film-based angiography.[9–10] This permits multiple left ventriculograms to be performed, which in turn facilitates the performance of interventional studies, such as atrial pacing to assess the functional significance of coronary artery stenosis. During selective coronary angiography, digital acquisition of the images has been shown to yield diagnostic information equal to cineangiography using 35-mm film.[11] In addition, the digital format provides immediate enhancement of the coronary images through computer processing techniques such as edge sharpening, contrast amplification, and fourfold image magnification. Computer software programs are also readily applied to facilitate the quantitation of coronary lesions, either by edge detection or videodensitometric analysis. Computer processing of selective coronary angiograms has also been used to study the relative appearance time of iodinated contrast medium as a measure of coronary flow reserve across stenotic vessels.[12] (See Chapter 9.) A useful application

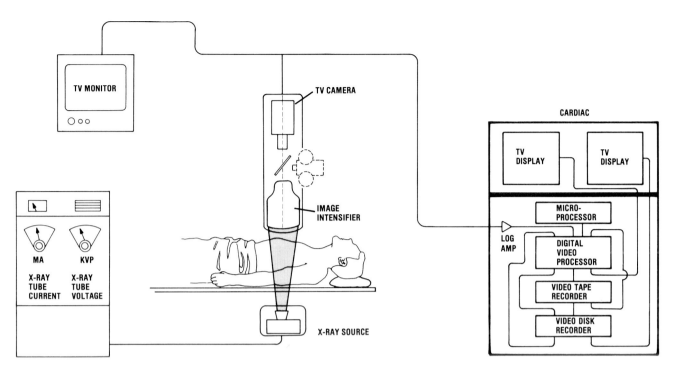

Figure 5-2 The catheterization laboratory that has been equipped for digital acquisition includes the usual x-ray generator, which can be adapted to provide dual energy exposures. The image from the intensifier tube is captured by a television camera and sent to a digital processor, which converts the image into a computerized format. The images are manipulated mathematically by the microprocessor and then stored on videotape, video disk, or digital disk.

---

of digital angiography is digital roadmapping, which provides a continual image of the coronary anatomy to function as a guide during angioplasty.[13]

CAPABILITIES OF DIGITAL PROCESSING

Digitization Process

Computers typically process radiologic images by transforming the continuous black-and-white gradations of an x-ray picture into a matrix in which the shade of gray in each small segment of the image is assigned a number that corresponds to the intensity of x-ray absorption in that image segment (Figure 5-3). Instead of recording the x-ray image on radiographic film, the x-rays that penetrate the body and strike the image intensifier are transformed by a television camera into a television image, as for fluoroscopy. Each television frame is created by a rapidly moving electron beam that traces out 524 horizontal lines from the top to the bottom of the image. This television or video image is described as an analog image because there is a continuous change in electric voltage along each horizontal line, corresponding to the varying light intensity across the image. Digital image–processing computers take the gray scale information within the analog video image and convert it, or digitize it, into a series of binary numbers that can be utilized by the computer for manipulating the image. Each television frame is converted into a matrix of 512 × 512 sections. This process establishes the x-y coordinates of each section (also referred to as a picture element, or pixel) in the new computer image. In a 512 × 512 matrix, there are 262,144 pixels. Each pixel is assigned an integer number that corresponds to the intensity of the electrical signal in each small region of the analog or video image. The scale of integer numbers used depends upon the number of bits of information that are available for storing information in each pixel. For current applications, there are usually eight bits available per pixel. Since the computer uses a binary system, this translates into 2^8 or 256 numbers that can be used to describe the black-and-white intensity of each pixel in the image. In terms of human perception, our visual system can differentiate approximately 15 to 20 shades of gray, so that the digital image with 256 shades of gray appears to us as a continuous or smooth gradation of gray from pure black (digital number of 0) to pure white (digital number of 256). In addition, our perception of the sharpness of the image or spatial resolution depends on the number of boxes or pixels that are used to form the digital image. If only 64 × 64 pixels are used, the image has an obvious checkered pattern, which is referred to as "pixelization" of the image. However, when denser matrices are used, such as the 512 × 512 format, then the pixels become so small that

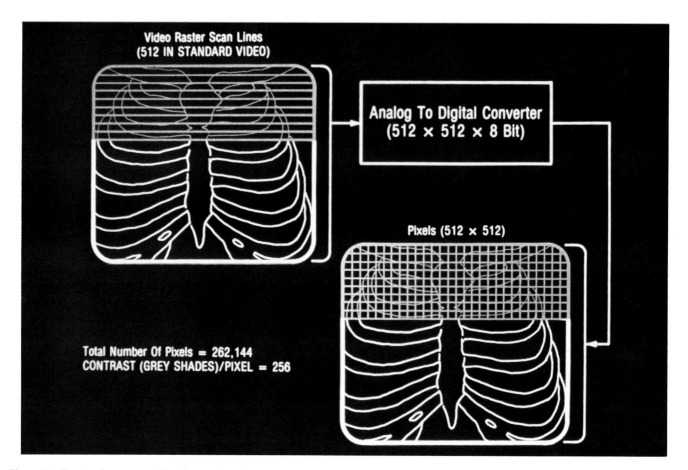

Figure 5-3 During the process of analog-to-digital conversion, the computer takes the television image and assigns to it a spatial matrix that is 512 × 512 spaces. Each one of these spaces, or pixels, has a gray scale differentiation of 256 shades, since each pixel has 8 bits ($2^8 = 256$) of information.

our visual system does not recognize the borders of the individual pixels (unless highly magnified), and we perceive a continuous function of smooth spatial outlines. As the size of the object that is imaged decreases, the need for finer spatial resolution increases. A matrix of 750 × 750 pixels approximates the limit of spatial resolution sharpness that can be obtained with a 7-in image intensifier. Higher spatial resolution is limited by quantum noise within each pixel. In addition, the amount of numbers that have to be handled by the computer limits the speed with which the x-ray images can be processed, so that one must strike an appropriate balance between image resolution and processing speed.

The fact that there are computers available to perform these numerical feats attests to the dramatic improvements in computer capabilities that have occurred over the last 20 years. In 1968, when computers were initially applied in medicine for analyzing the pressure recordings from a cardiac catheterization laboratory, a room full of computers was required to perform 200 analog-to-digital conversions per second.[14] By 1981, computers the size of a large tape recorder were available and could process 8 million analog-to-digital conversions per second. Translated into images, this means that a 512 × 512 pixel matrix can be processed 30 times per second, which is equivalent to television framing rates and usual cineangiographic recording rates.

Mask Mode Subtraction

Once the x-ray image has been transformed into a digital format, enhancement of the image is possible because the computer treats each image as a series of numbers that can be manipulated with such mathematic functions as addition and multiplication or more complicated functions such as logarithmic amplification or Fourier transformation.[15–16] One of the earliest methods of digital image processing was mask mode subtraction, described by Mistretta.[17] During this process, an attempt is made to subtract out extraneous structures in the angiographic image, such as soft tissue and bone densities (Figure 5-4). In order to accomplish this, an initial x-ray image of the area of interest, for example

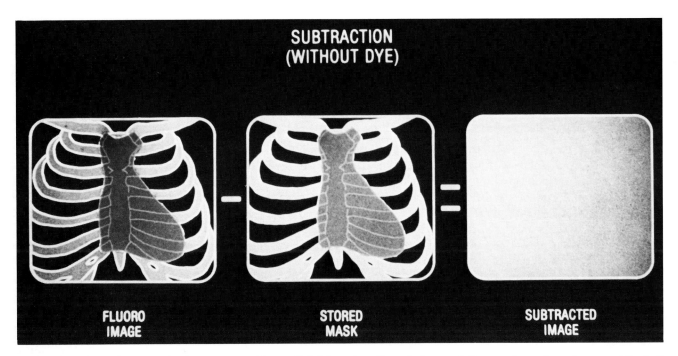

Figure 5-4 Once the analog image has been converted into a digital format, mathematical computations can be performed very rapidly. In the subtraction process demonstrated in this figure, each pixel in the digitized matrix image that is stored in computer memory as the mask is subtracted from incoming video images. If there is no change in the image (as would occur with motion or the injection of contrast medium), then the subtraction process results in a neutral image with all of the x-ray absorption due to soft tissues and bones canceling out.

the neck or thorax, is taken, digitized by the computer, and stored in the memory of the computer for subsequent use. This stored image is referred to as a mask. Another x-ray image of the neck or thorax can then be taken, digitized and compared with the original image in the computer memory. The two images are compared pixel by pixel for the intensity of the gray scale information contained within each corresponding pixel. If the gray scale information in each pixel in the two images is equivalent, the pictures will cancel each other out during the subtraction process. The resulting image will have a uniform gray scale value. In addition, no bones or soft tissue will be visualized, provided there was no motion of the patient between the time when the mask and the subsequent images were acquired. If a diluted amount of iodinated contrast medium is present in the carotid arteries or the heart at the time the second image is obtained, it will be more easily visualized because soft tissues and bone densities will be subtracted out and will no longer obscure the contrast medium–filled artery or ventricle. This process of mask mode subtraction has been effectively used with intravenous injections to visualize the cardiac chambers and the aorta and to screen patients for atherosclerotic involvement of the carotid, renal, or iliac arteries.[18–22] Adequate studies can be obtained in approximately 80% of cases.

One of the major reasons for inadequate studies is misregistration artifact. This occurs when there is movement in the image between the time when the mask is acquired and the subsequent contrast medium–filled images are obtained. The motion of background tissues creates misregistration, and there will be unequal subtraction or cancellation of the digital numbers within certain pixels near the boundaries of bone and soft tissues. The resulting subtracted images have light or dark streaks and decreased visualization of the diluted iodine. In addition, these streaks can be confused with arterial stenoses; thus the recognition of misregistration is important. Fortunately, postprocessing techniques can be used to improve image quality and reduce problems of misregistration artifact. One of the advantages of digital angiography is that, unlike film-based radiography in which the image cannot be altered after development, digital images can be restored from the computer memory and manipulated after the angiogram is acquired. For example, if there is a significant misregistration streak in the subtracted image, then a new mask can be recalled from computer memory and subtracted from the frames with iodine. A series of different masks can be tried until one is found that has a minimum of misregistration. In addition, the computer can move the mask relative to the second image in postprocessing the digital angiogram, to further reduce misregistration artifact. Software programs that use a continuous update of the mask, synchronous to the cardiac cycle, have recently been developed.[23] This process, which is called dynamic mask subtraction, produces images with a minimum of misregistration artifact and thus further enhances iodine contrast.

Dual Energy Subtraction

Another method that can be applied to enhance contrast and reduce motion artifact is a process called dual energy subtraction. In this method, two different x-ray energies are chosen to expose the image. One image is obtained at 60 kVp and a second is obtained at 120 kVp. The difference in soft tissue attenuation at these x-ray energy levels is utilized to diminish tissue densities when the pair of images are subtracted from each other. Compared to mask mode subtraction where soft tissue misregistration can induce artifacts, pairs of dual energy images can be obtained every 30th of a second, so that there is minimal movement of cardiac structures between the image pairs. The reduction in motion artifact during dual energy subtraction permits panning of the image intensifier over the thorax during the study. In addition, respiratory movement does not interfere with the image quality.[24-25] Experimental studies have shown good image quality in thin subjects, but there is about a 70% reduction in signal-to-noise ratio that may interfere with image quality in thicker subjects.

Cardiac Imaging with Digital Angiography

In order to take advantage of the benefits of digital imaging, it was first necessary to compare cardiac images obtained during standard film-based angiography with digitally acquired images, to determine whether ventricular volumes and ejection fraction correlated well between the two imaging techniques. Comparison studies were initially performed with intravenous administration of iodine contrast medium, and the left ventricle was imaged in a first-pass mode after the iodine bolus traveled through the right ventricular and pulmonary phases.[26-32] As distinguished from digital acquisition of peripheral arterial studies, in which the framing rates are 2 to 8 per second, for cardiac ventricular imaging the digital images had to be acquired at 30 frames per second. The image-processing computers available in 1981 did not have the capacity to store images in a digital format at this rapid framing rate. Therefore, initial comparison studies were performed by acquiring the images at 30 frames per second, subtracting a mask image in real time, and reconverting the digital information into an analog format for storage on videotape. Computer technology advanced rapidly, so that by 1983 complete digital storage of 512 × 512 matrix images at 30 frames per second was accomplished. Digital storage made it easier for images to be recalled for postprocessing without degradation of the image by videotape noise.

For the first-pass angiograms, 30 to 40 mL of contrast medium was injected by hand over 2 to 3 seconds through a 6F introducing sheath placed in the femoral vein. The angiograms were obtained with fluoroscopic exposure of 8 mA and 70 to 90 kVp, with 4 mm of aluminum filtration to diminish low-energy x-ray penetration. Standard 35 mm film-based cineangiograms were obtained in the 30° right anterior oblique projection with direct left ventricular injection of 40 mL of contrast medium. Left ventricular volumes were measured by the area-length technique from both the digital and film angiograms. Initial studies demonstrated good correlations in the measured volumes at end-diastole and end-systole and for the calculated ejection fractions. In addition, there were no premature ventricular contractions during the left phase of digital angiograms, whereas 18% of patients had ventricular tachycardia during the standard intraventricular injection of contrast medium with the film-based angiograms.[10]

The spatial resolution of the intravenous angiograms was not as good as that in the direct intraventricular angiograms. However, there were occasional patients with large apical aneurysms who had better visualization of the left ventricle on the digital study because there was improved mixing of the blood pool with the iodine contrast medium following intravenous administration. It is perhaps more appropriate to compare these studies with first-pass radionuclide images. From that perspective, first-pass digital angiograms have 10 times the spatial resolution of radionuclide images and therefore are very accurate for assessing ventricular wall motion at rest and during stressful interventions.[33-36] Later developments in digital angiography permitted pulse mode, cine acquisition with standard radiographic exposure energies, which improved the signal-to-noise ratio of intravenous as well as direct intraventricular studies.

Direct Intraventricular Studies

The initial studies that were performed with digital subtraction angiography used intravenous administration of contrast medium because it was less invasive. In addition, the early expectation was that contrast enhancement through digital processing might provide the long-sought-after means of obtaining intravenous coronary angiograms. Unfortunately, the ability to visualize the coronary anatomy with intravenous injections was not realized even with digital enhancement of the contrast. There are several reasons for this, including the marked dilution of the contrast medium bolus by the time it reached the coronary arteries, the fact that motion artifact and misregistration are more significant when imaging small structures like the coronary arteries, and the effect of overlap of the contrast medium–filled left ventricle during the first-pass method, which diminishes the ability to perceive the coronary arteries adequately.

The next step in the development of digital imaging was to apply the method during invasive arterial and left ventricular studies but to utilize significantly lower doses than those injected during standard film-based angiography. This could offer several important advantages in performing angiograms, including greater safety when studying children or patients who cannot tolerate large volumes of hyperosmolar contrast medium. Various groups of investigators have reported on the accuracy of low-dose digital left ventriculograms compared to standard 40-mL injections of contrast using cine film–based ventriculography.[37–38] Digital angiograms usually require 10 to 15 mL of contrast, diluted 1:1 in water and injected at the rate of 10 mL/sec over 3 seconds. The lower total volume and decreased osmolarity of the digital angiograms provided excellent opacification of the left ventricle, with a significantly lower incidence of contrast medium–induced premature contractions (Figure 5-5). In one study, 6 of 30 (20%) patients had ventricular tachycardia during standard film-based acquisition, and no reliable ventricular volume data could be obtained. In contrast, no patient had sustained premature ventricular contractions during the digital ventriculograms.[38]

Because of the lower iodine dose, three to four digital subtraction left ventriculograms can be obtained with the equivalent contrast medium dose of one film-based angiogram. This permits catheterization studies to be performed with a lower total dose of iodine in diabetic patients and patients with renal or cardiac dysfunction. In routine cases, multiple left ventriculograms can be obtained, thus obviating the need for a biplane laboratory. Alternatively, multiple lower-dose digital angiograms can be used to assess left ventricular function during intervention studies.

VIDEODENSITOMETRY

An angiogram is a two-dimensional projection image or shadowgram of a three-dimensional object. The outline of a left ventriculogram or coronary angiogram only demonstrates the widest boundary in the x-y plane that is filled by iodine contrast medium. Digital angiograms

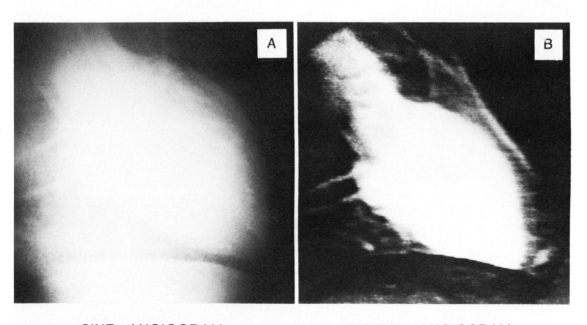

LOW DOSE VENTRICULOGRAPHY

CINE ANGIOGRAM
12 ml OF CONTRAST MEDIUM

DIGITAL ANGIOGRAM
12 ml OF CONTRAST MEDIUM

Figure 5-5 The contrast enhancement that can be achieved with the digital subtraction process is demonstrated in this figure. A shows a left ventriculogram that was obtained on cine film after injecting 12 mL of contrast medium directly in the ventricle. In B, the same image was processed digitally with the soft tissues subtracted in mask mode. There is a fourfold contrast enhancement with the digital process.

also contain information about the depth of the object of interest in the plane perpendicular to the imaging plane. The process that measures density values within an angiogram is called videodensitometry.[39]

Digital acquisition of angiograms facilitates densitometric counting because the digital number assigned to each pixel in the image corresponds to the depth of the iodinated contrast material that the x-ray beam traversed.[40] In order to obtain a linear relationship between digital number and contrast medium depth, the electronic signal from the absorbed x-rays must be logarithmically amplified. Logarithmic amplification is necessary because x-rays are absorbed as an exponential function of the depth of the tissue or contrast material. The formula for x-ray intensity attenuation is

$$I_E = I_O e^{-ux}$$

where I_E is the intensity of x-rays at the image intensifier, I_O is the intensity of the incident x-rays, u is the attenuation coefficient at a particular energy of x-rays, and x is the distance traversed within the tissue or contrast medium–filled object.

This kind of information may be lost when film is used because logarithmic amplification is not performed, and the gray scale of the film is often chosen to saturate the white levels of iodine to improve contrast visualization. The result of saturation is that very white, high-contrast images are obtained on film, but information is lost about relative amounts of iodinated contrast material within the vasculature. Methods have been developed to digitize the cine film image and then perform the quantitative methods that have been used with direct digital acquisition.[41]

Since the image in digital angiography is represented by a matrix of numbers, the information within each pixel can be logarithmically amplified very quickly. The digital numbers within each pixel then correspond to the concentration of iodine in that pixel. By adding up the digital numbers over an area of interest such as the left ventricle, a number is obtained that corresponds to the total volume of the ventricle. For the method to be accurate, there must be equal mixing of the iodine throughout the blood volume of the ventricle, and background densities have to be subtracted.

The theory behind videodensitometry is similar to the photon-counting technique of radionuclide imaging. Therefore, similar software algorithms have been used for performing the densitometric computations. First, an approximate region of interest around the left ventricle is chosen. A background area is also described by the operator, which is used to subtract out background densities from the region of interest that are not coming from the iodinated blood volume. In order to determine absolute ventricular volumes, the concentration of iodine within the ventricle would have to be known. Although this is not feasible at the current time, the digital numbers represent relative ventricular volumes at end-diastole and end-systole. Ejection fraction can be calculated densitometrically because these relative volumes can be used to compute a relative change in volume from end-diastole to end-systole. In a clinical comparison study with 25 patients who had first-pass digital left ventriculograms, the ejection fraction by videodensitometry correlated well with the ejection fraction by the area-length method ($r = 0.88$).[42]

The videodensitometric approach offers unique advantages for assessing right ventricular ejection fraction. One of the difficulties in trying to measure right ventricular volumes by angiography is that the right ventricle does not conform to a simple geometric shape. In addition, the right ventricle has numerous trabeculae and interstices, which make it difficult to use a mathematic formula to derive its volume from a two-dimensional angiogram. The densitometric method can be used to derive the relative volume of any irregular object without using any assumptions about its three-dimensional geometry. Radionuclide imaging of the right ventricle has gained widespread support because the photon-counting technique is based on densitometric principles. However, a major problem of radionuclide imaging is its limited spatial resolution. During intravenous first-pass digital studies, it was noted that the tricuspid valve annulus moves toward the apex and shortens the long axis of the right ventricle by about 20%. Densitometric methods such as radionuclide angiography, which uses only one fixed region of interest to outline the right ventricle at end-diastole, neglect the apical displacement of the tricuspid valve annulus. The result of using one fixed region of interest is that some right atrial counts are included in the end-systolic volume calculation. The increased spatial resolution of digital angiograms permits excellent visualization of the tricuspid annulus and allows identification of the proper boundary of the right ventricle at end-diastole and end-systole. In one study of 19 patients who had first-pass digital right ventriculograms, it was found that a videodensitometric analysis with two separate regions of interest gave a closer correlation with the ejection fraction calculated by the area-length technique than did the method using only one region of interest, whereas the radionuclide data did not correlate very well with either the digital or the area-length method.[43]

Another application of videodensitometry for cardiovascular imaging is in the study of cardiac physiology. The densitometric analysis of an iodine bolus as it passes through the left ventricle yields a density-time curve. The points on this curve represent relative volume in the ventricle. The filling pattern during diastole can be used to derive a relative measure of ventricular filling. The densitometric filling curves have correlated with Doppler velocity patterns of flow across the mitral valve. In addition, left ventricular compliance can be assessed using the density-time curves. If a catheter is placed in the left ventricle during a first-pass digital ventriculogram, then simultaneous pressure and density curves can be generated. The area-length method can be used to scale the density curve at end-diastole and end-systole, corresponding to the maxima and minima of the curve. The resulting data can be used to derive a pressure-volume curve to analyze left ventricular compliance, without the laborious task of calculating ventricular volume frame by frame from a cine film.

Although there are no geometric assumptions in the densitometric approach, as with any methodology there are certain assumptions and potential problems inherent to the technique. With x-ray imaging systems, there are physical limitations that could create non-linear absorption in the system. These include veiling glare from the image intensifier screen, x-ray beam scatter from the patient, and beam hardening since x-rays without a uniform energy are used. A significant amount of physics research in digital imaging is being devoted to understanding these issues of videodensitometry and minimizing their effect.[44-45]

ASSESSMENT OF LEFT VENTRICULAR FUNCTION WITH DIGITAL IMAGING

Digital angiography has been applied to the evaluation of left ventricular function by assessing wall motion contraction patterns at rest and during various stresses employed to induce ischemia. After the digital left ventricular images are obtained, the end-diastolic and end-systolic images can be recalled from computer memory and displayed on the video monitor. The outline of the left ventricle can be defined by the operator using an x-y digitizer pad or light pen.

Since the image is in a digital format, the x-y coordinates of the outline can be readily analyzed by computer programs to assess regional wall motion. In addition, functional images that give a visual representation of the change in wall motion can be displayed. This is a form of analysis called parametric imaging, and it is accomplished by taking the left ventricular image at end-diastole while it is filled with contrast medium and using that image as the stored mask. All subsequent images are then subtracted from the end-diastolic image. The resultant picture describes the wall motion between end-diastole and end-systole, which is equivalent to the displacement of blood producing the stroke volume. If there are akinetic or dyskinetic regions, these are usually highlighted. Although visually appealing, the parametric images have not been as useful as the more quantitative methods, especially when used to define stress-induced wall motion abnormalities.

Bicycle Stress Testing

Functional assessments of coronary stenoses have been performed with stress induced by supine bicycle exercise and with atrial pacing. During the exercise test, first-pass intravenous digital ventriculograms are obtained at rest and immediately following supine bicycle exercise.[46-47] In a manner analogous to first-pass radionuclide exercise stress testing, changes in global ejection fraction and the development of segmental wall motion abnormalities are taken as a measure of the severity of coronary artery stenoses. In one study of 19 patients who underwent supine bicycle exercise stress tests with first-pass digital angiograms, the development of wall motion abnormalities and a fall in ejection fraction were more sensitive than the development of chest pain or ST segment response on the electrocardiogram for identifying patients with coronary disease.[48] Adequate first-pass ventriculograms could not be obtained in 16% of patients because of significant misregistration artifact created by the increased respiratory motion following exercise. Although the experience of Yianakis and co-workers[46] at the Cleveland Clinic and Goldberg and Borer[47] at New York Hospital indicates that this method is clinically useful, others prefer to use pacing in conjunction with cardiac catheterization as the method of inducing myocardial stress.

Atrial Pacing Studies

Pacing the atrium at incremental heart rates is a form of stress that can reproducibly create ischemic conditions in patients with coronary artery disease. This technique has been combined with digital acquisition of left ventriculograms in order to assess the functional significance of coronary artery stenoses. Although the coronary anatomy is defined angiographically during cardiac catheterization, there is often disagreement in interpretation of the severity of a specific lesion.[49] As a result, a stenosis can be underestimated or overestimated depending on the position of the lesion, its internal contour, and the radiographic projection used. A functional assessment of the hemodynamic effect of a coronary lesion provides an independent measure of the severity of the stenosis by allowing one to evaluate the effect that the lesion has on reserve ventricular performance.[50] Digital ventriculograms are very beneficial in performing atrial pacing studies because the low dose of iodine contrast medium permits multiple ventriculograms to be obtained (Figure 5-6).

In a study of 21 patients who were referred for cardiac catheterization to evaluate symptoms of chest pain, digital left ventriculograms were obtained at rest and at the peak heart rate during atrial pacing.[51] The

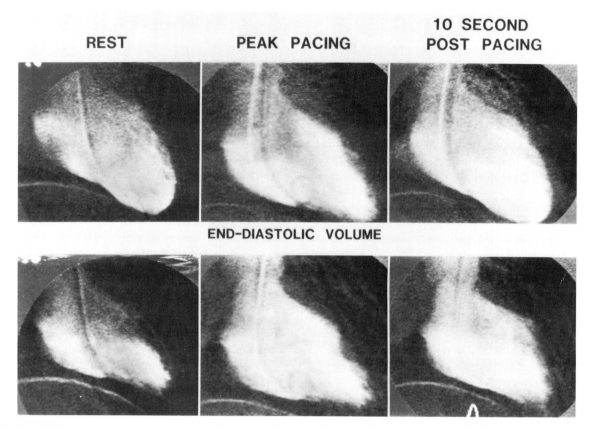

Figure 5-6 The ability to use low doses of contrast medium in performing left ventriculography permits multiple studies to be obtained. These three ventriculograms were obtained with a total of 36 mL of contrast medium. The purpose of the study was to determine the optimal time for assessing the left ventricular wall motion abnormalities induced during an atrial pacing stress test.

pacing protocol increases the atrial rate by 10 beats per minute until the patient develops chest pain or until 85% of the maximum predicted heart rate is obtained. Both the development of ST segment changes on the electrocardiogram (3 of 15 patients) and the occurrence of chest pain (8 of 15 patients) were insensitive markers of ischemia. However, the ejection fraction fell or did not increase in 14 of the 15 patients (93%) with coronary artery disease, whereas it increased in 5 of 6 patients (83%) without coronary artery disease. In addition, segmental wall motion abnormalities that developed during pacing corresponded to myocardial areas supplied by specific stenotic arteries.

In addition to providing information that is useful in clinical decision making, digital angiography can be used during other intervention studies to answer questions of cardiac physiology. One of the concerns about exercise or pacing stress tests is the time at which the analysis of left ventricular function should be performed. Should analysis of left ventricular motion be obtained at the peak heart rate, or shortly thereafter? Is it correct to compare ejection fraction at rest to values obtained at peak heart rates, when preload and afterload are significantly different from baseline values? Low-dose digital ventriculograms are useful to answer these physiologic questions because three to four ventriculograms can be performed with the same iodine contrast medium dose as one standard film-based left ventricular angiogram. In a group of 21 patients, 12-mL digital ventriculograms were obtained at rest, at the peak pacing rate, and within 10 seconds after the pacemaker was discontinued.[52] In a second group of 19 patients, digital ventriculograms were obtained at rest, peak pacing, and at 30 seconds after pacing was stopped. In all 11 patients in the two groups who were without coronary artery disease by angiography, the left ventricular volume decreased at end-diastole and end-systole during atrial pacing and returned to baseline values by 10 seconds after pacing stopped. The ejection fraction increased or did not fall by more than two percentage points in 10 of the 11 patients (91%). In patients with coronary artery disease, the end-diastolic volume decreased as expected during pacing. However, the end-systolic volume did not decrease as much or even increased, thereby causing a decrease in ejection fraction. In the patients with coronary artery disease, the ejection fraction fell by more than two percentage points in 25 of 29 patients (86%). The sensitivity of the atrial pacing study was highest when the ventriculogram was obtained at the peak pacing rate, compared with a sensitivity of 52% during the postpacing studies. This result was found despite the fact that the preload and afterload (as measured by left ventricular end-diastolic pressure and systemic blood pressure) were unchanged from baseline during the 10- or 30-second postpacing studies. In addition, 21 of the 40 patients had treadmill stress tests performed prior to cardiac catheterization. The double product achieved during the treadmill stress test was similar to that achieved during atrial pacing in the patients with coronary artery disease. However, the sensitivity and specificity of the pacing study were superior to that obtained with the treadmill stress test in this population. Thus digital angiography has proved to be very useful in performing intervention studies to understand physiologic questions of ventricular function during ischemia.

DIGITAL CORONARY ANGIOGRAPHY

Digital acquisition following selective injection of contrast medium into the coronary arteries has proved in our experience to be one of the most important benefits of the clinical application of digital angiography. The initial studies with selective coronary artery injections were performed in 1982 with computer systems that did not have a large digital storage capacity. Images were obtained at 30 frames per second, digitized and subtracted in real time from a stored mask, and then reconverted to an analog format for storage on videotape. In order to decrease motion artifact resulting from the movement of the heart itself, a blurred mask was used that summated 16 frames (one half second) of the thorax prior to the injection of contrast material. Continuous fluoroscopic energy levels were used at 20 mA, with 75 to 90 kVp. Because of the analog storage and other variables in the imaging chain, these images had a moderate amount of electronic noise. Nevertheless, a comparison study of 31 patients demonstrated that coronary stenoses were well visualized, and quantitative measurements of the stenoses correlated well with measurements made from cine film angiograms.[53] These initial studies indicated that contrast visualization within the arteries and during the myocardial blush phase was improved about fourfold over cine film.

A videodensitometric analysis was applied to these early digital coronary angiograms in an effort to assess myocardial blood flow as reflected in the distribution of iodine contrast material. Various regions of interest were placed over epicardial coronary arteries or over regions of myocardium perfused by specific coronary branches. The digital angiogram was replayed through the computer to derive density-time curves within the various regions of interest. These perfusion-washout curves yielded information about the time of arrival and the distribution of iodine contrast medium within the myocardium. Whiting and co-workers found a good correlation for assessing myocardial blood flow between digital coronary angiograms and radiolabeled microspheres.[54] In addition, they have shown that quantitative information about the hemodynamic effect of coronary stenoses can be obtained by performing densitometric flow analyses at rest and after the hyperemic response induced by contrast medium. There are several theoretical and practical difficulties in obtaining meaningful myocardial perfusion studies during digital coronary angiography. These include motion of the heart muscle through the region of interest, overlap of blood vessels into the region, and the variable depth of myocardium being analyzed in the z axis of the region of interest. An alternative approach to measuring myocardial blood flow has been performed using cine-computed tomography with intravenous contrast medium injection. With this modality, the digital angiograms are obtained such that the three-dimensional space within the region of interest can be reconstructed to study 1-cm–thick muscle sections.[55] Theoretically, these myocardial perfusion studies should be more accurate than the global, two-dimensional images obtained with digital angiography.

Another approach to understanding the hemodynamic significance of coronary stenoses using selective coronary digital angiography has been developed by Vogel and co-workers.[12] In this method, coronary angiograms are obtained at end-diastole, and sequential end-diastolic images are subtracted from one another to derive a composite image of the time of arrival of the iodine bolus at each diastolic frame, over six cycles. The position of arrival of the bolus peak at each diastole is color coded to improve visualization. Color-mapping studies were performed at rest and within 10 seconds after the hyperemic stimulus of iodine contrast medium. In a nonstenotic artery, the hyperemic response should produce a more rapid peak contrast medium arrival compared to rest. In a stenotic artery that no longer has the autoregulatory tone to permit the expected increase in flow following a hyperemic stimulus, one would expect to see a slower time of arrival of the contrast medium. Vogel has shown that these time-of-arrival maps correlate with the angiographic determination of stenoses and that they can be used before and after transluminal angioplasty as an independent measure of the hemodynamic effects of the angioplasty.[56]

Digital Storage of Coronary Angiograms

By 1983 digital imaging devices were developed that had a digital storage capacity of 80 megabytes. A

512 × 512 image requires approximately one fourth of a megabyte of digital disk space. This meant that approximately 300 images could be stored on the 80-megabyte digital disk. The rate of acquisition of these digital images was also limited to eight frames per second. Despite these temporary technical limitations, the coronary images acquired and stored in a digital format proved to have superior detail compared with analog videotape storage. Since there was no film to develop, digital coronary angiograms were available for review immediately after acquisition (Figure 5-7). The images could be played back in an unsubtracted format or rapidly processed with computer algorithms to enhance contrast with mask mode subtraction, to sharpen coronary boundaries with edge detection algorithms, or to magnify the image fourfold to focus attention on specific lesions (Figure 5-8). The unsubtracted as well as the postprocessed enhanced images could then be downloaded onto ¾-in videotape for replay at conferences. The information could also be stored in a completely digital format on digital magnetic tape. Although the digital archiving was more time consuming, no information was lost because of the introduction of electronic noise. The imaging chain was also used differently compared to the earlier studies with analog storage. The images were obtained in a pulsed radiographic mode using 600 mA at 75 to 85 kVp. Also, the signal from the image intensifier was transformed by a progressive scan television camera instead of the usual interlaced mode. Progressive scanning permitted multiple scrubbing of the input phosphor between images to diminish "ghost" artifacts.

Studies performed with the improved digital acquisition methods demonstrated a very good correlation of quantitative measurements of coronary artery stenoses obtained with either standard 35-mm cine film or digital angiography.[9] In order to improve digital storage capacity, a 475-megabyte digital disk was added to the system in 1984, so that approximately 1800 images could be acquired before it would be necessary to clear the disk. Additionally, a 475-megabyte parallel transfer disk was incorporated into the computer hardware; this disk permitted 512 × 512 digital images to be acquired at 30 frames per second.

Analysis of Coronary Artery Stenoses

One of the major advantages of digital acquisition of coronary angiograms is that the computerized format permits rapid quantitative determination of coronary stenosis. The operator can use computer graphics to outline the boundary of a coronary stenosis in order to quickly calculate the percent diameter narrowing. A second method of measuring stenotic segments uses a videodensitometric approach (Figure 5-9).[57] In the densitometric method, the operator chooses a region of interest over a nonstenotic segment and a second region over the stenosis.[58] Exact boundary detection is not necessary since the computer subtracts background information and determines the relative density of contrast between the two areas. Studies using phantom models suggest that the densitometric technique is independent of eccentric or irregular geometries often

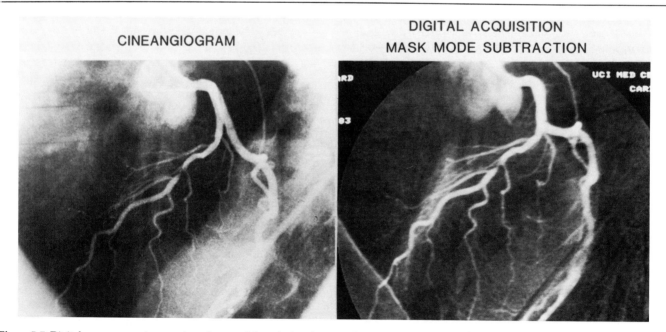

Figure 5-7 Digital coronary angiograms have less spatial resolution than cine film angiograms. However, quantitative analysis of stenoses reveals similar findings for either method of acquisition. In addition, the digitally acquired angiograms are more easily measured by quantitative analysis because they are already in a computerized format.

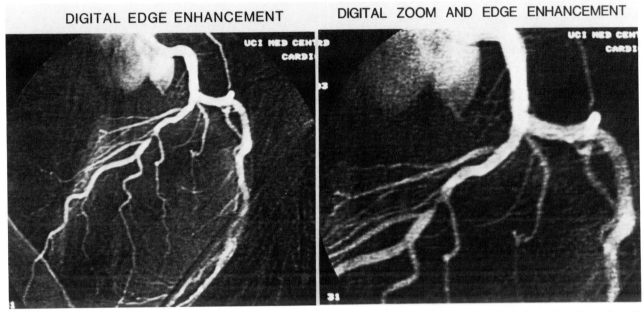

Figure 5-8 The digital coronary angiogram shown in Figure 5-7 has been processed by the computer to enhance the boundary edges. A particular region of interest can be magnified fourfold by the computer for greater ease of performing quantitative analysis of stenoses. In real time, the observer passes a cursor over the image of the artery on the monitor, and the computer makes it appear as if a magnifying glass is being passed over the artery.

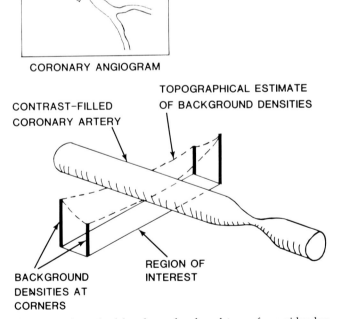

Figure 5-9 A method has been developed to perform videodensitometry from unsubtracted digital images. This method bypasses the need for mask mode subtraction, which can induce errors due to misregistration artifact. In this technique, the densities within a narrow region of interest near the artery are quantified and subtracted from the density within the artery segment to account for x-ray absorption from soft tissues and noniodinated objects. The unsubtracted digital images can thus be obtained while panning the image intensifier over the thorax.

found in atherosclerotic lesions. The densitometric analysis can be used with the edge detection method following calibration for magnification, to derive the absolute minimum lumen diameter in terms of millimeters of a coronary stenosis (Figure 5-10). This can be achieved densitometrically even if the boundary of the lesion is too small to accurately measure with boundary detection methods.

Assessment of the severity of coronary artery narrowing is made primarily by angiographic determination of percentage of diameter narrowing relative to an assumed normal portion of the artery. This determination may be misleading for several reasons. Of greatest importance, pathologic studies have demonstrated that atherosclerosis is a diffuse process, so that even when the angiogram only demonstrates a single stenosis, less than 25% of the length of the arteries will be free of disease and truly normal.[59-60] In addition, stenoses commonly occur at bifurcations, so that it may be difficult to identify a normal proximal portion of the same arterial segment. White et al. reported that measurements of percent diameter narrowing correlate very poorly with a physiologic assessment of stenosis severity as determined by the hyperemic flow response following temporary coronary occlusion during open heart surgery.[61] Legrand and co-workers found a similarly poor correlation between measurements of percent stenosis and coronary blood flow reserve as measured by digital flow maps following a hyperemic stimulus.[62]

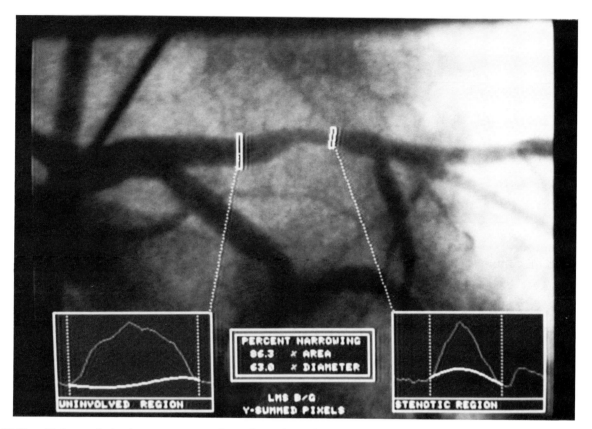

Figure 5-10 Quantitative analysis of a coronary stenosis can be performed very rapidly by the computer. In this image, the computer has determined the percent diameter narrowing and the percent area narrowing using a densitometric analysis of the iodine absorption profiles over the two regions of interest. The computer can also derive the absolute lumen diameter and area in millimeters by calibrating the pixel size with the known dimensions of the catheter in the image.

Digital angiography was useful in determining whether percent stenosis or absolute minimum lumen diameter is more accurate because both methods of quantitative analysis can be compared to a functional assessment of the stenosis by performing atrial pacing stress tests. In a study of 27 patients with coronary artery disease, the effect of atrial pacing on segmental radial wall motion was analyzed and compared with the two methods of measurement of stenosis severity.[63] Whereas percent diameter narrowing was correlated with segmental wall motion with a coefficient of $r = -0.44$, minimum lumen diameter correlated much more closely, $r = 0.78$ ($P<0.05$). The improved correlation suggests that the absolute lumen diameter more accurately predicts the functional impairment induced by pacing because the absolute diameter is more directly related to the resistance to flow than to percent narrowing.

With regard to the amount of narrowing, the data from this study indicated that when an epicardial coronary artery is narrowed to a diameter of less than 1.5 mm, a functional impairment induced by atrial pacing is more likely to occur in the myocardium supplied by the narrowed artery. If the artery is narrowed to less than 1.5 mm in diameter, there is a greater chance of having a positive study both by global ejection fraction and segmental wall motion. Thus a minimum lumen diameter less than 1.5 mm may be useful as a predictor of which patients will be symptomatic during stress.

Digital Roadmapping

Of the many ways in which digital images can be processed, one of the most exciting has been digital roadmapping.[64] Digital roadmapping is a method to assist visualization of the coronary anatomy during angioplasty. Currently, when an angioplasty is performed under fluoroscopy, the operator attempts to have a mental image of where the lesion is located. This image may be based upon cineangiograms that were taken at a previous catheterization. Alternately, videotape replays on another monitor are used. During the angioplasty, iodine contrast medium is injected through the guiding catheter or the distal balloon catheter. However, there is often poor flow through these catheters, and visualization is difficult because of low concentration and the transient nature of the injection.

Digital angiography can be helpful during angioplasty by providing a constant image of the coronary arteries as a guide or roadmap for the angiographer. Prior to loading the guiding catheter with the dilation system, a digital coronary angiogram is obtained, processed by mask mode subtraction, and stored in computer memory. An end-diastolic frame is chosen that demonstrates the lesion, and the image is recalled from memory and displayed on the television monitor above the catheterization table for use as the roadmap. The digital coronary roadmap image is then interlaced with the live fluoroscopic video image. When the balloon dilation catheter and steerable guidewire are advanced under fluoroscopy, the operator sees the catheter and guidewire travel over a superimposed image of the coronary arteries. If a wrong branch is entered, it is easily recognized. The exact placement of the balloon across the stenosis is also aided by the digital roadmap. At the present time, only a single end-diastolic image is used as a roadmap. However, Nissen and Elion have developed computer programs that replay the entire digital angiogram triggered from the patient's electrocardiogram to provide a continuous roadmap that moves in concert with the motion of the heart as the dilation catheter is advanced.[65]

Another way of roadmapping is to display the digital images on a different monitor from the one showing the live fluoroscopic image. The immediate recall and clarity of digital images provide superior image display compared to videotaped images.

Coronary Angiography with Aortic Root Injection

As an alternative to selective coronary artery imaging, the ability to enhance contrast medium with digital processing permits one to obtain images with aortic root injection of contrast medium. Aortic root digital angiograms have been obtained with a total of 20 to 30 mL of contrast material injected during two diastolic phases. The images are processed immediately and are reviewed before selective coronary angiography is performed. This procedure is useful when left main or ostial stenosis is suspected because the early recognition of these lesions with nonselective digital enhancement should help diminish the incidence of left main dissection during selective coronary angiography. The less invasive digital aortic root angiograms also correctly identify proximal and mid left anterior descending, circumflex, and right coronary artery disease in 85% of the cases; however, the degree of accuracy is not as good as with selective cannulation.

An additional use of the digital subtraction aortic root angiogram has been to help locate the origin of saphenous vein bypass graft insertions emanating from the aorta. Once the graft insertion is identified, the digital aortogram can be stored as a digital roadmap. The digital roadmap is then interlaced on the television monitor with the live fluoroscopic image to help position the diagnostic or angioplasty guiding catheter into the ostium of the vein grafts.

SUMMARY

It is clear that the use of computers for cardiovascular imaging has produced dramatic improvements over the last few years. Digital processing can be used to obtain peripheral or ventricular angiograms with less invasive intravenous injection of iodinated contrast material. Alternatively, lower amounts of contrast material can be used during intra-arterial or direct ventricular angiography. The ability to perform low-dose ventriculograms permits multiple observations to be obtained, which is beneficial for assessing cardiac physiology during intervention studies. Selective coronary digital angiography has proved to have several advantages over film-based systems. The digital images are immediately available for review during the catheterization, the images can be manipulated to improve contrast information, and the digital format allows ready access for quantitative analysis either by edge detection or videodensitometric techniques. The ability to immediately recall images and to manipulate the contrast medium allows digital roadmapping to be performed, which assists the clinician in advancing of the guidewire and balloon during coronary angioplasty.

Clinicians have found various applications of digital angiography useful for their laboratories. At the University of California, Irvine, the current preference is to obtain all cardiac catheterization studies in a digital format. This includes the left ventriculogram with direct injection of 12 mL of contrast medium, as well as coronary angiography that is acquired digitally in a nonsubtracted format. The digital coronary images are then postprocessed after the procedure, and the magnified images as well as the quantitative studies of lumen diameter are recorded on a ¾-in videocassette. Each patient study has a separate cassette that records the left ventriculogram in a subtracted mode, followed by the quantitative wall motion analysis graphics. We have found the atrial pacing studies to be very useful in trying to determine whether a specific stenosis is hemodynamically significant and often use this information during our cardiac catheterization conferences to help decide whether an invasive procedure should be performed. These studies are particularly helpful when the patient cannot exercise on a treadmill or if the thallium studies are questionable. During coronary angioplasty procedures, the digital roadmap technique is often beneficial and sometimes invaluable as an aid for guiding the wire and balloon to the proper position.

The application of computer acquisition of radiographic images has been rapid and impressive. The expectation is that future advances in computer technology will permit even more dramatic developments in digital angiography. An intravenous angiographic method of screening for coronary artery disease may yet be achieved.

REFERENCES

1. Tobis J, Nalcioglu O, Henry W. Digital angiography: The implementation of computer technology for cardiovascular imaging. *Prog Cardiovasc Dis.* 1985;28:195–212.

2. Manicini J, Higgins C. Digital subtraction angiography: A review of cardiac applications. *Prog Cardiovasc Dis.* 1985;28:111–141.

3. Mistretta CA, Crummy AB. Diagnosis of cardiovascular disease by digital subtraction angiography. *Science.* 1981;214:761–765.

4. Kruger RA, Mistretta CA, Houk TL, et al. Computerized fluoroscopy techniques for intravenous study of cardiac chamber dynamics. *Invest Radiol.* 1979;14:279–287.

5. Kruger RA, Mistretta CA, Houk TL, et al. Computer fluoroscopy in real time for noninvasive visualization of the cardiovascular system. *Radiology.* 1979;130:49–57.

6. Brennecke R, Brown TK, Bursch J, et al. Digital processing of videoangiocardiographic image series using a minicomputer. *Proc Computer Cardiol.* IEEE Computer Society. 1976:255–260.

7. Heintzen PH, Brennecke R, Bursch JH. Digital cardiovascular radiology. In: Hohne KH, ed. *Digital Image Processing in Medicine.* Berlin: Springer-Verlag; 1981;1–14.

8. Mistretta CA, Ort MG, Cameron JR, et al. Multiple images subtraction technique for enhancing low contrast periodic objects. *Invest Radiol.* 1973;8:43–44.

9. Vas R, Diamond GA, Forrester JS, et al. Computer enhancement of direct and venous-injected left ventricular contrast angiography. *Am Heart J.* 1981;102:719–728.

10. Tobis J, Nalcioglu O, Johnston WD, et al. Left ventricular imaging with digital subtraction angiography using intravenous contrast injection and fluoroscopic exposure levels. *Am Heart J.* 1982;104:20–27.

11. Tobis J, Nalcioglu O, Iseri L, et al. Detection and quantitation of coronary artery stenoses from digital subtraction angiograms compared with 35 mm film cine angiograms. *Am J Cardiol.* 1984;54:489–496.

12. Vogel R, LeFree M, Bates E, et al. Application of digital techniques to selective coronary arteriography: Use of myocardial contrast appearance time to measure coronary flow reserve. *Am Heart J.* 1984;107:153–164.

13. Tobis J, Roeck W, Johnston WD, et al. Digital angiographic coronary roadmapping: a new aid for assisting coronary angioplasty. *Am J Cardiol.* 1985;56:237–241.

14. Henry WL, Crouse L, Stenson RE, et al. Computer analysis of cardiac catheterization data. *Am J Cardiol.* 1968;22:696.

15. Houk TL, Kruger RA, Mistretta CA, et al. Real-time digital K-edge subtraction fluoroscopy. *Invest Radiol.* 1979;14:270–278.

16. Ergun DL, Mistretta CA, Kruger RA, et al. A hybrid computerized fluoroscopy technique for noninvasive cardiovascular imaging. *Radiology.* 1979;132:739–742.

17. Mistretta CA. The use of a general description of the radiological transmission image for categorizing image enhancement procedures. *Optical Eng.* 1974;13:134.

18. Crummy AB, Strother CM, Sackett JF, et al. Computerized fluoroscopy: Digital subtraction for intravenous angiocardiography and arteriography. *Am J Radiol.* 1980;135:1131–1140.

19. Christenson PC, Ovitt TW, Fisher HD, et al. Intravenous angiography using digital video subtraction: Intravenous cervicocerebrovascular angiography. *Am J Radiol.* 1980;135:1145–1152.

20. Meaney TF, Weinstein MA, Buonocore E, et al. Digital subtraction angiography of the human cardiovascular system. *Am J Radiol.* 1980;135:1153–1160.

21. Strother CM, Sackett JF, Crummy AB, et al. Clinical applications of computerized fluoroscopy. *Radiology.* 1980;136:781–783.

22. Moodie DS, Yiannikas J, Gill CC, et al. Intravenous digital subtraction angiography in the evaluation of congenital abnormalities of the aorta and aortic arch. *Am Heart J.* 1982;104:628–634.

23. Elion JL, Fischer PL, Nissen SE. Time stretching strategies for optimal temporal alignment of image sequences of unequal duration. *IEEE Computers in Cardiol.* 1986:619–622.

24. Riederer SJ, Kruger RA, Mistretta CA. Limitations to iodine isolation using a dual beam non–k-edge approach. *Med Phys.* 1981;8:54.

25. Brody WR, Cassel DM, Sommer FG, et al. Dual-energy projection radiography: Initial clinical experience. *Am J Radiol.* 1981;137:201.

26. Higgins CB, Norris SL, Gerber KH, et al. Quantitation of left ventricular dimensions and function by digital video subtraction angiography. *Radiology.* 1982;144:461–469.

27. Norris SL, Slutsky RA, Mancini J, et al. Comparison of digital intravenous ventriculography with direct left ventriculography for quantitation of left ventricular volumes and ejection fractions. *Am J Cardiol.* 1983;51:1399–1403.

28. Nichols AB, Martin EC, Fles TP, et al. Validation of the angiographic accuracy of digital left ventriculography. *Am J Cardiol.* 1983;51:224–230.

29. Goldberg HL, Borer JS, Moses JW, et al. Digital subtraction intravenous left ventricular angiography: Comparison with conventional intraventricular angiography. *J Am Coll Cardiol.* 1983;1:858–862.

30. Kronenberg MW, Price RR, Smith CW, et al. Evaluation of left ventricular performance using digital subtraction angiography. *Am J Cardiol.* 1983;51:837–842.

31. Nissen SE, Booth D, Waters J, et al. Evaluation of left ventricular contractile pattern by intravenous digital subtraction ventriculography: Comparison of cineangiography and assessment of interobserver variability. *Am J Cardiol.* 1983;52:1293–1298.

32. Engels PHC, Ludwig JW, Verhoeven LAJ. Left ventricular evaluation by digital video subtraction angiography. *Radiology.* 1982;144:471–474.

33. Vas R, Diamond GA, Forrester JS, et al. Computer-enhanced digital angiography: correlation of clinical assessment of left ventricular ejection fraction and regional wall motion. *Am Heart J.* 1982;104:732–739.

34. Mancini GBJ, Norris SL, Peterson KL, et al. Quantitative assessment of segmental wall motion abnormalities at rest and after atrial pacing using digital intravenous ventriculography. *J Am Coll Cardiol.* 1983;2:70–76.

35. Johnson RA, Wasserman AG, Leiboff RH, et al. Intravenous digital left ventriculography at rest and with atrial pacing as a screening procedure for coronary artery disease. *J Am Coll Cardiol.* 1983;2:905–910.

36. Mancini GBJ, Peterson KL, Gregoratos G, et al. Effects of atrial pacing on global and regional left ventricular function in coronary heart disease assessed by digital intravenous ventriculography. *Am J Cardiol.* 1984;53:456–461.

37. Sasayama S, Nonogi H, Kawai C, et al. Automated method for left ventricular volume measurement by cineventriculography with minimal doses of contrast medium. *Am J Cardiol.* 1981;48:746–751.

38. Tobis JM, Nalcioglu O, Johnston WD, et al. Correlation of 10-milliliter digital subtraction ventriculograms compared with standard cineangiograms. *Am Heart J.* 1983;105:946–952.

39. Trenholm BG, Winter DA, Mymin D, et al. Computer determination of left ventricular volume using videodensitometry. *Med Biol Eng Comput.* 1972;10:163–173.

40. Bursch JH, Heintzen PH, Simon R. Videodensitometric studies by a new method of quantitating the amount of contrast medium. *Eur J Cardiol.* 1974;1:437–446.

41. Reiber JHC, Gerbrands JJ, Booman F, et al. Objective characterization of coronary obstructions from monoplane cineangiograms and three-dimensional reconstruction of an arterial segment from two orthogonal views. 1982;93–100.

42. Tobis J, Nalcioglu O, Seibert A, et al. Measurement of left ventricular ejection fraction by videodensitometric analysis of digital subtraction angiograms. *Am J Cardiol.* 1983;52:871–875.

43. Johnston WD, Tobis JM, Seibert JA, et al. A video-densitometric method of computing right ventricular ejection fraction from intravenous digital subtraction angiograms. *Circulation.* 1982;66:II–61. Abstract.

44. Nalcioglu O, Roeck WW, Pearce JG, et al. Quantitative fluoroscopy. *IEEE Trans Nucl Sci.* 1981;NS-28:219–223.

45. Nalcioglu O, Seibert JA, Roeck WW. The requirements for and capabilities of x-ray video systems to provide quantitative information. *Proc Soc Photo-optical Instrum Eng.* 1982;318:445–453.

46. Yianakis J, Simpfendorfer C, Detrano R, et al. Stress digital subtraction angiography to assess presence of coronary artery disease in patients without myocardial infarction. *Circulation.* 1983;68:III–41.

47. Goldberg HL, Moses JW, Borer JS, et al. Exercise left ventriculography utilizing intravenous digital angiography. *J Am Coll Cardiol.* 1983;2:1092–1098.

48. Tobis J, Nalcioglu O, Johnston WD, et al. Exercise digital subtraction angiograms in patients with coronary artery disease. *Circulation.* 1982;66(suppl):11–229.

49. Zir LM, Miller SW, Dinsmore RE, et al. Interobserver variability in coronary angiography. *Circulation.* 1976; 52:627–632.

50. Gould KL, Lipscomb K, Hamilton CW. Physiologic basis for assessing critical coronary stenosis. *Am J Cardiol.* 1974;33:87–94.

51. Tobis J, Nalcioglu O, Johnston WD, et al. Digital angiography in assessment of ventricular function and wall motion during pacing in patients with coronary artery disease. *Am J Cardiol.* 1983;51:668–675.

52. Tobis J, Iseri L, Johnston WD, et al. Determination of the optimal timing for performing digital ventriculography during atrial pacing stress tests. *Am J Cardiol.* 1985;56:426–433.

53. Tobis J, Nalcioglu O, Seibert JA, et al. Coronary angiography performed with real time digital subtraction. *Circulation.* 1982;66:II–60(suppl). Abstract.

54. Whiting JS, Drury JK, Pfaff JM, et al. Digital angiographic measurement of radiographic contrast material kinetics for estimation of myocardial perfusion. *Circulation.* 1986;73:789–798.

55. Rumberger JA, Feiring AJ, Lipton MJ, et al. Use of ultrafast computed tomography to quantitate regional myocardial perfusion: a preliminary report. *J Am Coll Cardiol.* 1987;9:59–69.

56. Aueron F, Vogel RA, Bates ER, et al. Comparative effects of percutaneous transluminal coronary angioplasty and coronary artery bypass surgery on chronic coronary flow reserve. *J Am Coll Cardiol.* 1984;3:506.

57. Sander T, Als AV, Paulin S. Cine-densitometric measurement of coronary arterial stenoses. *Cathet Cardiovasc Diagn.* 1979;5:229–245.

58. Tobis J, Nalcioglu O, Johnson W, et al. Videodensitometric determination of minimum coronary artery luminal diameter before and after angioplasty. *Am J Cardiol.* 1987;59:38–44.

59. Waller BF, Roberts WC. Amount of narrowing by atherosclerotic plaque in 44 nonbypassed and 52 bypassed major epicardial coronary arteries in 32 necropsy patients who died within 1 month of aortocoronary bypass grafting. *Am J Cardiol.* 1980;46:956–962.

60. Feldman RI, Nichols WW, Pepine CJ, Conetta DA, Conti CR. The coronary hemodynamics of left main and branch coronary stenoses: The effects of reduction in stenosis diameter, stenosis length, and number of stenoses. *J Thorac Cardiovasc Surg.* 1979;77:377–388.

61. White CW, Creighton BW, Doty DB, et al. Does visual interpretation of the coronary arteriogram predict the physiologic importance of a coronary stenosis? *New Engl J Med.* 1985;310:819–824.

62. Legrand V, Mancini GBJ, Bates ER, Hodgson JMcB, Gross MD, Vogel RA. Comparative study of coronary flow reserve, coronary anatomy and results of radionuclide exercise tests in patients with coronary artery disease. *J Am Coll Cardiol.* 1986;8:1022–1032.

63. Tobis J, Sata D, Nalcioglu N. Correlation of minimum coronary lumen diameter with left ventricular functional impairment induced by atrial pacing. *Am J Cardiol.* In press.

64. Tobis J, Johnston WD, Montelli S, et al. Digital coronary roadmapping as an aid for performing coronary angioplasty. *Am J Cardiol.* 1985;56:237–241.

65. Elion JL, Nissen SE, Fischer PLC, Booth DC. Gated digital roadmapping: a new computer based imaging support system for angioplasty. *J Am Coll Cardiol.* 1987;90:70A. Abstract.

Chapter 6
Coronary Arteriography

George T. Kondos, MD

INTRODUCTION

Ever since its development, the use of coronary arteriography has rapidly grown. It continues to be the only widely available technique to accurately define the angiographic anatomy of the coronary arteries. Its obvious limitation is that it only provides anatomic information; it does not provide important physiological information regarding flow abnormalities. Initially, coronary angiography solely provided diagnostic information. Currently coronary arteriography is used in percutaneous coronary angioplasty and thrombolytic agent administration during myocardial infarction. It is therefore of utmost importance that physicians performing coronary arteriography become familiar with the coronary anatomy and its variations, in order to make appropriate clinical decisions. The ultimate goal of coronary arteriography is the complete visualization of the coronary anatomy and the assessment of the coronary artery collaterals, the type of coronary pathology, and the presence of coronary anomalies. If each of these important factors is not assessed, an incomplete study eventuates, rendering an accurate clinical decision unlikely.

INDICATIONS FOR CORONARY ARTERIOGRAPHY

The indications for coronary arteriography vary from institution to institution and from physician to physician. The indications should be continually assessed and modified to meet the needs of clinicians, as understanding of coronary artery disease advances.

The initial evaluation to establish the diagnosis and prognosis of the patient with presumed ischemic heart disease is noninvasive in nature. When noninvasive testing is equivocal, strongly positive, or cannot be accomplished, coronary arteriography is generally recommended.

The most common indication presently is the evaluation of chest pain. In patients with atypical chest pain and a nondiagnostic treadmill stress test, coronary arteriography is essential in making an accurate diagnosis. In patients with angina pectoris, coronary arteriography is indicated when symptoms are not controlled with medical therapy, the patient is unable to tolerate the side effects of medical therapy, or there is an early positive treadmill study. In general, patients with unstable angina should undergo coronary arteriography to evaluate the need for percutaneous coronary angioplasty or coronary artery bypass grafting. Coronary angiography is indicated in patients suspected of having coronary vasospasm, prior to the administration of ergonovine for provocation of vasospasm. Other indications include the patient with postinfarct angina or postinfarct positive treadmill stress examination. The return of symptoms after coronary angioplasty or coronary revascularization is an additional indication for evaluation. Another very important group of patients is asymptomatic patients with a positive treadmill examination. Coronary arteriography is indicated in this subset, especially if risk factors or any early positive treadmill findings are present. Finally, coronary arteriography continues to be used as a research tool to study the development, progression, and regression of coronary artery disease.

NORMAL CORONARY ANATOMY

For purposes of uniformity I have chosen the Coronary Artery Surgery Study (CASS) system of defining the coronary segments and names of branches (Figure 6-1).[1] The names given to the coronary artery branches and coronary artery segments are highly variable. In the following description of the coronary arteries, the left anterior descending, circumflex, and right coronary arteries will be divided into CASS segments for ease of identification.

Right Coronary Artery

The right coronary artery ostium is generally located in the right half of the right aortic sinus. The proximal portion of the right coronary artery extends from the origin of the vessel to a point halfway to the take-off of the acute marginal branch. The right coronary artery then travels within the right arterioventricular groove. The proximal segment of the right coronary artery is best seen in the 60° left anterior oblique (LAO) projection (Figure 6-2). The first branch of the right coronary artery is the conus branch, which encircles the right ventricular outflow tract and is directed superiorly and anteriorly. In approximately 40% of patients, the conus branch has a separate orifice anterior to the right coronary artery origin. Unrecognized selective cannulation of this vessel may lead to the erroneous conclusion that the right coronary artery has been totally occluded or that a left-dominant system exists (Figure 6-3). In addition, the conus branch may give important collaterals to the left anterior descending. Selective cannulation of the conus branch is indicated when the posterior descending coronary artery is not visualized as arising from the right coronary artery or circumflex vessel. The second branch of the right coronary artery is the sinus node artery. In 60% of patients, the sinus node artery arises from the proximal right coronary artery, 38% from the circumflex and 2% from both the right and circumflex vessels. The right coronary artery courses posteriorly and superiorly toward the spine. In the right anterior oblique (RAO) projection, the sinus node and the conus branches are directed posteriorly and anteriorly, respectively (Figure 6-4). In the LAO projection, the sinus node branch is recognized by its constant division into two rami. The anterior ramus encircles the superior vena cava and supplies the sinus node. The posterior branch supplies the superior and posterior walls of the left atrium. Occasionally the sinus node branch may have a separate orifice from the right coronary artery and this orifice may be inadvertently cannulated. The last branches from the distal segment of the proximal right coronary are various right ventricular branches, which are directed anteriorly and best viewed in the RAO projection. A straight lateral projec-

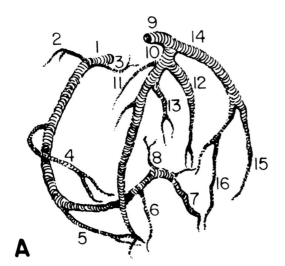
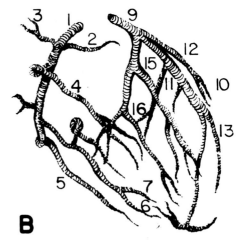

Figure 6-1 Left and right coronary arteries in (A) left anterior oblique (LAO) and (B) right anterior oblique (RAO) views. 1, right coronary artery; 2, conus branch; 3, sinus node artery; 4, right ventricular branch; 5, acute marginal; 6, right posterior descending; 7, right posterolateral branch; 8, atrioventricular nodal artery; 9, left main; 10, left anterior descending; 11, first septal; 12, first diagonal; 13, second diagonal; 14, left circumflex; 15, obtuse marginal; 16, left posterolateral. *Source:* Reprinted from *Cardiology* by W Parmley and K Chatterjee (Eds) with permission of JB Lippincott Company, © 1987.

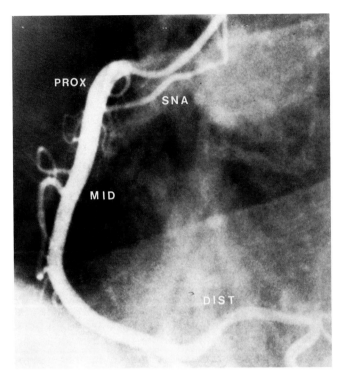

Figure 6-2 Right coronary artery, LAO view. SNA, sinus node artery; PROX, proximal; MID, middle; DIST, distal. *Source:* Reprinted from *Cardiology* by W Parmley and K Chatterjee (Eds) with permission of JB Lippincott Company, © 1987.

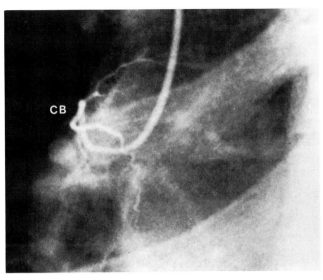

Figure 6-3 LAO view, selective cannulation of the conus branch (CB). *Source:* Reprinted from *Cardiology* by W Parmley and K Chatterjee (Eds) with permission of JB Lippincott Company, © 1987.

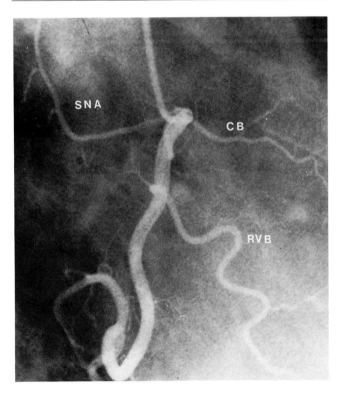

Figure 6-4 Right coronary artery, RAO view. SNA, sinus node artery; CB, conus branch; RVB, right ventricular branch. *Source:* Reprinted from *Cardiology* by W Parmley and K Chatterjee (Eds) with permission of JB Lippincott Company, © 1987.

tion is sometimes needed to evaluate this portion of the right coronary artery if a lesion is suspected and the right ventricular branches obscure or minimize a suspected coronary lesion.

The mid–right coronary artery extends from the end of the proximal right coronary artery to the take-off of the acute marginal branch; it is best seen in the LAO projection. The branches of the mid–right coronary artery include variable numbers of right ventricular branches. The terminal branch of the right coronary artery is the acute marginal branch, which supplies the acute margin of the right ventricle. The acute margin of the right ventricle is formed by the anterior and inferior walls of the right ventricle. On occasion, a right atrial branch may arise opposite the origin of the acute marginal branch and travel posteriorly to supply the right atrium.

The distal right coronary artery begins after the take-off of the acute marginal branch and includes the origin of the right posterior descending coronary artery. The posterolateral segment of the right coronary artery extends from the posterior descending to the take-off of the last right posterolateral branch. To avoid underestimating or missing coronary artery disease, it should be realized that these segments are poorly visualized in standard oblique projections because of foreshortening. Steep cranial RAO or LAO angulation for better visualization is recommended. In the standard LAO projection, the distal right coronary artery appears as an inverted U shape (Figure 6-5). From the first segment of the U arises the posterior descending coronary artery, traveling caudally toward the apex. At the top of the inverted U is the origin of the atrioventricular (AV)

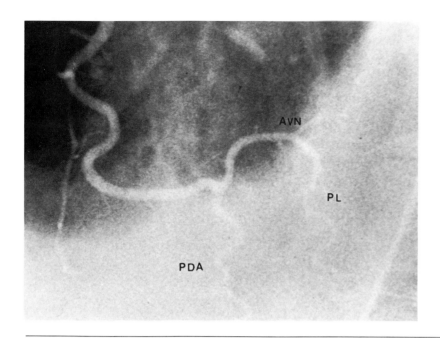

Figure 6-5 Right coronary artery, LAO cranial view. AVN, atrioventricular nodal artery; PDA, posterior descending artery; RPL, right posterolateral artery. *Source:* Reprinted from *Cardiology* by W Parmley and K Chatterjee (Eds) with permission of JB Lippincott Company, © 1987.

nodal artery directed cephalad in the opposite direction of the posterior lateral branches. At the top of the inverted U is located what is termed the crux of the heart, implying the center point of the heart where the atria and ventricles meet. As the right coronary artery continues, the last segment of the U gives off varying numbers of posterolateral branches supplying the posterior wall of the left ventricle. At times, confusion exists about distinguishing the right posterior descending coronary artery from its posterolateral branches. In the RAO projection, small septal perforators are given off from the posterior descending, which are not seen from the posterolateral branches. The take-off of the posterolateral branches distal from the origin of the AV nodal also serves to help distinguish these vessels.

On occasion, dual posterior descending vessels may course in the posterior interventricular sulcus. Another variation in 3% of patients is the circum-marginal posterior descending. This branch arises from the right coronary artery before the take-off of the acute marginal branch. The circum-marginal terminates in the posterior interventricular sulcus as the only vessel in this area, or with a posterior descending vessel.[2]

It is important to recognize the presence of a nondominant right coronary artery so as to avoid mistaking this for a totally occluded vessel. In this instance, the right coronary artery is small and bifurcates early, supplying only the right atrium and right ventricle (Figure 6-6).

Left Coronary Artery

The left main coronary artery generally arises from the left coronary sinus and travels to the left and slightly posteriorly. It has a variable length of 1 to 24 mL. In 2% of patients, the left main is nonexistent, and separate orifices for the left anterior descending and right coronary arteries are present. In 12% of patients, a short left main coronary is present, which may result in selective cannulation of either the left anterior descending or the circumflex vessel. Both of these variations must be recognized to avoid misinterpreting a total occlusion of either major left coronary vessel.

The entire course of the left main coronary artery is best seen in a shallow RAO projection (Figure 6-7). The bifurcation of the left main coronary artery is not best visualized in this projection because the left main coronary artery travels in a horizontal direction. The LAO projection with minimal caudal angulation helps to define the bifurcation area. If this projection is not helpful, a shallow LAO with steep cranial angulation may be useful.

The left main terminates in two branches, the left anterior descending coronary artery, which travels anteriorly in the anterior interventricular sulcus, and the circumflex coronary, which travels posteriorly in the left atrioventricular sulcus (Figure 6-8). Occasionally the left main will give rise to an intermediate vessel that is directed laterally. This trifurcation of the left main is best visualized in either a caudal LAO or an RAO view.

The left anterior descending is best described by dividing it into a proximal, a middle, and a distal segment. The proximal left anterior descending extends from the origin of the left anterior descending to the origin of the first septal perforator. The middle segment extends from the first septal perforator to the take-off of the second diagonal. The distal segment extends from the second diagonal to the termination of the left ante-

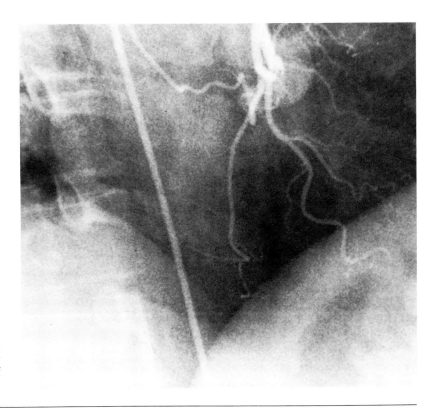

Figure 6-6 Nondominant right coronary artery, RAO view. The two branches supply the right ventricle and right atrium. *Source:* Reprinted from *Cardiology* by W Parmley and K Chatterjee (Eds) with permission of JB Lippincott Company, © 1987.

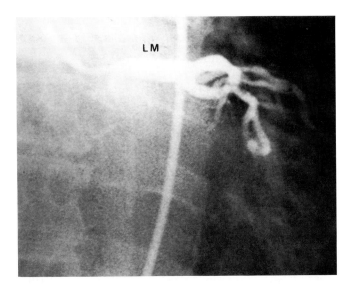

Figure 6-7 Shallow RAO view visualizing the proximal and mid-left main coronary artery (LM). *Source:* Reprinted from *Cardiology* by W Parmley and K Chatterjee (Eds) with permission of JB Lippincott Company, © 1987.

rior descending. The proximal left anterior descending may not be well visualized because of overlap with the intermediate or circumflex vessels. In this instance the following views may be helpful: RAO with steep caudal angulation, LAO with caudal angulation, straight left lateral with slight caudal angulation, or LAO with steep cranial angulation.

Along the course of the left anterior descending, two or more diagonal vessels may arise, supplying the anterolateral wall of the left ventricle. The first diagonal on occasion may be a large vessel and may give rise to septal perforators. In the LAO projection, these vessels are directed downward and to the right, while the left anterior descending courses directly downward. The origins of the diagonals are best seen in a LAO projection with steep cranial angulation. On occasion the origins of the first and second diagonals may best be viewed with an RAO projection with cranial angulation. In the RAO view it may be difficult to distinguish the left anterior descending and the diagonal vessels. In the shallow RAO view, the diagonals are seen to course over the anterolateral surface of the heart, whereas the left anterior descending tavels inside the border to the heart. Another helpful way to distinguish these vessels is the movement toward one another of the diagonal and obtuse marginal vessels and the lack of lateral motion of the left anterior descending during systole.

From one to six septal perforators arise from the left anterior descending along its course. The first septal perforator may be considerable in size and at times be as large as a first diagonal branch. In the LAO cranial projection, the septal vessels travel downward and to the left, as opposed to the left anterior descending, which travels directly downward. In the shallow RAO projection, the septal perforators course at 90° angles to the left anterior descending.

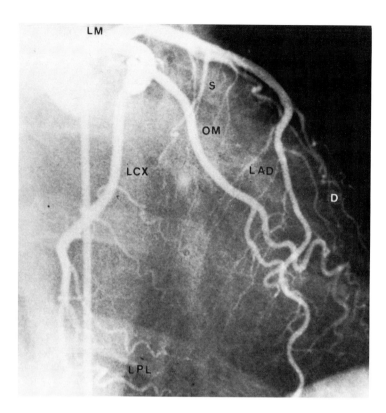

Figure 6-8 Left coronary artery, shallow RAO view. In the shallow RAO projection, the left anterior descending artery travels within the border of the heart, while the two diagonal arteries course over the anterolateral aspect of the heart. LM, left main; LCX, left circumflex; OM, obtuse marginal; LPL, left posterolateral; LAD, left anterior descending; S, septal; D, diagonal. *Source:* Reprinted from *Cardiology* by W Parmley and K. Chatterjee (Eds) with permission of JB Lippincott Company, © 1987.

On occasion dual left anterior descending coronary arteries may exist (Figure 6-9).[3] A dual left anterior descending is seen in less than 1% of patients. In this instance the left anterior descending bifurcates into short and long branches. The short branch supplies the anterior interventricular septum, terminating before the apex of the left ventricle. The long branch may have a variable course, traveling in the anterior interventricular sulcus, then coursing epicardially for a variable distance and re-entering the anterior interventricular sulcus near its distal portion. Alternatively, the long branch may travel entirely intramyocardially or along the surface of the right ventricle. If this anatomic variant is not suspected, the wrong left anterior descending may be bypassed at the time of surgery.

The distal left anterior descending may terminate before, at, or beyond the apex of the left ventricle. The left anterior descending at its termination may also give rise to two left and right recurrent branches.

The circumflex, after arising from the left main, abruptly angulates posteriorly, traveling in the left atrioventricular groove. For purposes of description the circumflex is divided into three segments: proximal, distal, and atrial ventricular.

The proximal circumflex coronary segment is defined from the origin of the circumflex to the origin and includes the first obtuse marginal branch. After the first

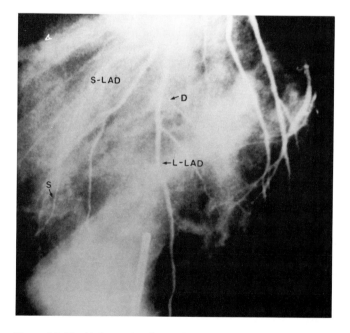

Figure 6-9 Dual left anterior descending, LAO view. L-LAD, long left anterior descending; S-LAD, short left anterior descending; S, septal artery; D, diagonal artery. *Source:* Reprinted from *Cardiology* by W Parmley and K Chatterjee (Eds) with permission of JB Lippincott Company, © 1987.

marginal branch, the circumflex is divided into two portions as it courses in the atrioventricular groove. The first half is termed the distal circumflex, and the second half the atrial ventricular circumflex segment. In 40% of patients, the sinus node artery arises from the circumflex vessel, coursing posteriorly and superiorly.

When the circumflex reaches the obtuse margin of the heart, defined as the border of the inferior and lateral walls of the left ventricle, three obtuse marginal branches are given off, which travel parallel to one another. The first obtuse marginal supplies the lateral wall of the left ventricle. The second obtuse marginal is generally the largest of the marginal vessels and supplies the obtuse margin of the heart. The third obtuse marginal or left posterolateral branch courses toward the lateral wall of the left ventricle inferior to the obtuse margin of the heart. The circumflex coronary artery and its branches are best visualized in the RAO projection.

On occasion a single large obtuse marginal branch may arise from the circumflex, supplying areas of the heart supplied by the second and third obtuse marginal branches. The obtuse marginal branch on occasion may function as the terminal portion of the circumflex vessel.

After the obtuse marginal branches are given off, the circumflex continues in the atrioventricular groove and gives off at least one posterior left ventricular branch. In the case of a dominant-left system, the circumflex gives off the posterior descending coronary artery and the AV nodal artery.

As with the right coronary artery, the distal and atrial ventricular segments of the circumflex coronary artery are best visualized in the LAO cranial projection. At times, a left atrial circumflex branch may arise from the circumflex coronary artery, continuing in the AV groove to supply most of the left atrial wall.

Coronary Artery Dominance

Coronary anatomy has generally been divided into three types of circulations, even though the left ventricle receives most of its blood supply from the left coronary artery.[4] The types of circulations are left-dominant, right-dominant, and codominant or balanced. Coronary dominance by definition indicates which coronary system, left or right, supplies most of the posterior wall of the left ventricle. A left-dominant system (Figure 6-10) is present if the right coronary artery terminates before giving off a posterior descending branch. A right-dominant system is present if the right coronary artery gives off a posterior descending branch and at least one posterolateral branch. A codominant or balanced circulation is present if the terminal branch of the right coronary artery is the pos-

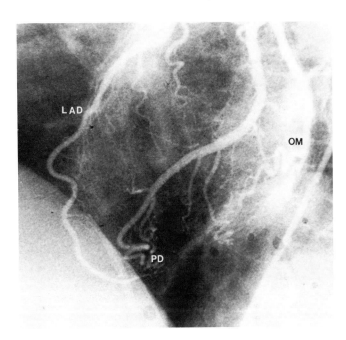

Figure 6-10 Left-dominant circulation, LAO view. LAD, left anterior descending; OM, obtuse marginal; PD, left posterior descending. *Source:* Reprinted from *Cardiology* by W Parmley and K Chatterjee (Eds) with permission of JB Lippincott Company, © 1987.

terior descending and at least one posterolateral branch is given off the circumflex coronary artery. A right-dominant system is present in 84% of patients, 12% have a left-dominant system, and the remaining 4% have a balanced or codominant circulation.

Coronary Artery Collaterals

Coronary artery communications connect all of the major coronary arteries. The development of collateral flow in these connections is dependent upon the development of obstructive coronary artery disease. As the degree of stenosis increases, angiographically visible collaterals develop. In the normal heart, collaterals are generally not angiographically visualized.

Collaterals perform many important functions, among which is the preservation of myocardial function. Collaterals are capable of providing adequate resting myocardial blood flow if supplied by a nondiseased vessel. Ischemia generally results during conditions that require increased myocardial flow. Two general types of collaterals exist. Intercoronary collaterals supply coronary communications between the major coronary vessels. Intracoronary collaterals provide communication between the same coronary artery. Typical instances of coronary collateral circulation are illustrated in Figures 6-11 to 6-13.

To optimally visualize the collateral vessels, the angiographer must record cine images during the first

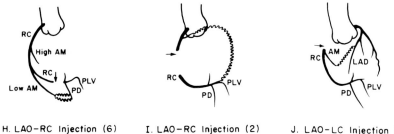

A. RAO-LC Injection (28)　　B. LAO-LC Injection (24)　　C. LAO-LC Injection (17)

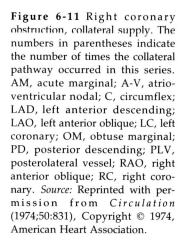

D. RAO-RC Injection (9)　E. LAO-RC Injection (9)　F. RAO-LC Injection (9)　G. LAO-LC Injection (6)

Figure 6-11 Right coronary obstruction, collateral supply. The numbers in parentheses indicate the number of times the collateral pathway occurred in this series. AM, acute marginal; A-V, atrioventricular nodal; C, circumflex; LAD, left anterior descending; LAO, left anterior oblique; LC, left coronary; OM, obtuse marginal; PD, posterior descending; PLV, posterolateral vessel; RAO, right anterior oblique; RC, right coronary. *Source:* Reprinted with permission from *Circulation* (1974;50:831), Copyright © 1974, American Heart Association.

H. LAO-RC Injection (6)　I. LAO-RC Injection (2)　J. LAO-LC Injection (2)

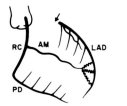 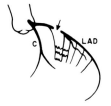 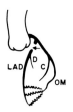

A. RAO-RC Injection (28)　B. RAO-LC Injection (27)　C. LAO-LC Injection (17)

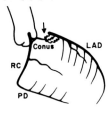 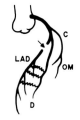 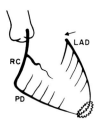

D. RAO-RC Injection (15)　E. LAO-LC Injection (6)　F. RAO-RC Injection (3)

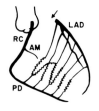

G. RAO-RC Injection (3)

Figure 6-12 Left anterior descending obstruction, collateral supply. The numbers in parentheses indicate the number of times the collateral pathway occurred in this series. AM, acute marginal; C, circumflex; D, diagonal; LAD, left anterior descending; LAO, left anterior oblique; LC, left coronary artery; OM, obtuse marginal; PD, posterior descending; RAO, right anterior oblique; RC, right coronary artery. *Source:* Reprinted with permission from *Circulation* (1972;50:834), Copyright © 1974, American Heart Association.

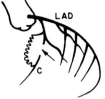 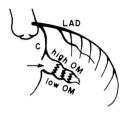 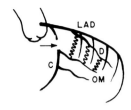

A. RAO-LC Injection (7)   B. RAO-LC Injection (6)   C. RAO-LC Injection (5)

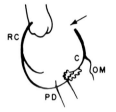 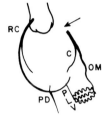

D. LAO-RC Injection (2)   E. LAO-RC Injection (2)

Figure 6-13 Left circumflex obstruction, collateral supply. The numbers in parentheses indicate the number of times the collateral pathway occurred in the series. C, circumflex; D, diagonal; LAD, left anterior descending; LAO, left anterior oblique; LC, left coronary; OM, obtuse marginal; PD, posterior descending; PLV, posterolateral vessel; RAO, right anterior oblique; RC, right coronary. *Source:* Reprinted with permission from *Circulation* (1974;50:835), Copyright © 1974, American Heart Association.

LAO (Figure 6-14) and RAO injections of the left and right coronary arteries, until the coronary veins are well visualized. Visualization of the coronary artery collaterals is of utmost importance because they may provide the only means of assessment of a totally occluded vessel. Coronary collaterals may be better visualized after injection of 300 μg of intracoronary nitroglycerin.

CORONARY ANGIOGRAPHY TECHNIQUES

Successful coronary angiography involves optimal preprocedural preparation. Prior to the procedure the angiographer should evaluate the patient, determine the need for the procedure, and explain potential complications. A definite plan of approach determined by the history and physical exam should be decided prior to the procedure (i.e., femoral or brachial). The patient may take medications prior to the procedure along with small amounts of liquids. Appropriate sedation is administered prior to arrival in the catheterization laboratory. The author recommends meperidine 50 to 70 mg, given intramuscularly. An alternative regimen includes oral administration of diazepam 10 mg PO and diphenhydramine 50 mg prior to the procedure.

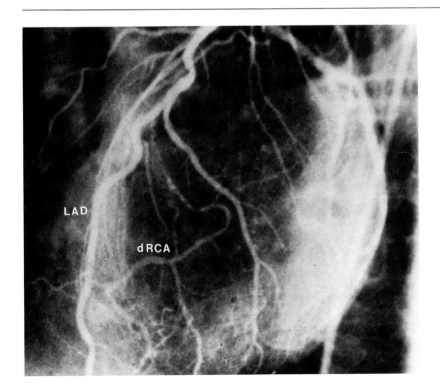

Figure 6-14 Left coronary artery injection. LAO view demonstrating the distal right coronary artery (dRCA) filling by collaterals from the left anterior descending (LAD). *Source:* Reprinted from *Cardiology* by W Parmley and K Chatterjee (Eds) with permission of JB Lippincott Company, © 1987.

Percutaneous Femoral Approach

The angiographer must be familiar with the anatomy of the inguinal area (Figure 6-15). The inguinal ligament extends from the anterior superior iliac spine to the pubic symphysis. The femoral artery is palpated as it courses under the inguinal ligament. The femoral artery should be entered approximately two fingerbreadths below the inguinal crease. Complications may arise if the femoral artery is entered too high because of inability to compress the puncture site; massive blood loss into the pelvis may result, with large hematoma formation. Attempting to enter too low may result in inadvertent cannulation of the superficial femoral artery, with subsequent need for arterial repair. Prior to entrance into the femoral artery, the skin and subcutaneous tissue are infiltrated with lidocaine; intermittent aspiration during anesthesia administration is imperative to avoid inadvertent intravascular administration of the anesthetic agent. After the infiltration is complete, a small stab wound over the femoral artery is made with a scalpel. The stab wound is enlarged with a small Kelly clamp. While palpating the femoral artery with the left hand, a Seldinger needle is advanced at a 30° to 45° angle with respect to the horizontal plane of the patient. When the angiographer feels that the artery has been entered, the stylet is removed. The free flow of blood from the needle signifies that the artery has been entered. If both sides of the artery have been penetrated and the periosteum of the pelvic bone has been hit, the stylet should be removed, local anesthesia should be administered after aspiration, and the needle should be continuously withdrawn until the free flow of blood is observed. An alternative to the Seldinger needle is a hollow core needle without a stylet, which allows one to puncture one side of the artery.

Once free flow of blood has been established, a 0.035-in J-tipped movable or nonmovable core guidewire (Figure 6-16) is inserted through the needle and advanced to the level of the diaphragm. If difficulty is encountered in passing the guidewire, the angiographer must determine the cause of the difficulty by fluoroscopic examination. If the guidewire does not advance beyond the tip of the needle, it is likely that the needle is in a subintimal or extravascular position. The guidewire must then be removed and free backflow of blood re-established.

Once the guidewire is above the level of the diaphragm, the needle is removed while constant pressure is applied over the artery to prevent the formation of a hematoma. An 8F dilator is placed over the guidewire to dilate the subcutaneous tissue and femoral artery. The dilator is removed, the guidewire is wiped to remove clots, and the appropriate angiographic catheter is inserted over the wire and placed in the descending aorta. It is the author's preference to use

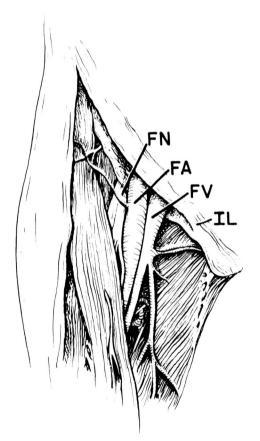

Figure 6-15 Anatomy of the right inguinal area. FN, femoral nerve; FA, femoral artery; FV, femoral vein; IL, inguinal ligament. *Source:* Reprinted from *Cardiology* by W Parmley and K Chatterjee (Eds) with permission of JB Lippincott Company, © 1987.

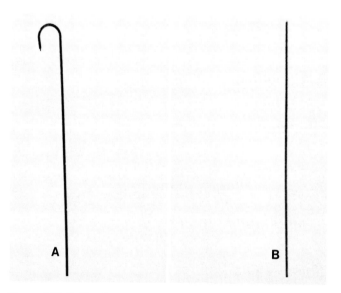

Figure 6-16 Guidewires used in coronary arteriography. A, 0.035-in J-tipped movable core wire; B, 0.035-in straight movable core wire. *Source:* Reprinted from *Cardiology* by W Parmley and K Chatterjee (Eds) with permission of JB Lippincott Company, © 1987.

one of the various sheaths with a hemostatic valve to facilitate catheter exchanges and limit blood loss.

Once the catheter is in the descending aorta, the guidewire is removed and wiped. The catheter is aspirated to remove any clot that may have formed and subsequently flushed. After the catheter has been flushed, 3000 U of heparin is given, and the catheter is advanced around the aortic arch. All catheter flushing should be done in the descending aorta to avoid clot embolization to the central nervous system.

Left Coronary Artery Cannulation

Cannulation of the left coronary artery is usually accomplished with a preformed Judkins catheter. The left Judkins is available in four sizes (3.5 to 6; Figure 6-17). The numbers refer to the length of the secondary curve in centimeters (Figure 6-18). The catheter is advanced retrograde on profile along the aortic root. Once the left coronary is entered, the angiographer must note the catheter tip pressure and waveform to avoid damping or ventricularization. Either of these phenomena may signify a proximal high-grade stenosis with catheter wedging, or the tip of the catheter may be against the vessel wall. Injection of contrast medium with damping or ventricularization may result in coronary dissection. The angiographer must reposition the catheter prior to injection. If repositioning the catheter is unsuccessful, the angiographer may either attempt injections into the sinus of Valsalva to detect high-grade proximal obstructions or use small intracoronary injections with prompt removal of the catheter after each injection.

Patients with dilated aortic roots or high take-offs of the left main may pose some challenging angiographic dilemmas. A 5- or 6-cm Judkins catheter may be employed. In patients with small aortic roots, a 3.5-cm Judkins catheter is used. The position of the catheter tip is useful in helping decide the appropriate size of the Judkins catheter. Upon entering the left main coronary artery, the catheter tip should be in a horizontal position. If the catheter tip is pointing superiorly, a catheter with a larger secondary curve should be used. If the catheter tip is pointed downward, a catheter with a smaller secondary curve should be used. If a Judkins 4-cm catheter is too large, the operator may carefully advance the catheter into left sinus of Valsalva to decrease the primary curve. The operator then withdraws the catheter sharply until the left coronary is engaged. Extreme care must be taken when attempting this maneuver to avoid dissecting the left main coronary artery.

In patients with a short left main, a superior take-off of the left main, or separate origins of the left anterior descending and left circumflex coronary arteries, an Amplatz-type catheter may be useful. These catheters are available in four sizes, each with a larger curve (I to IV; Figure 6-19). The advantage to the Amplatz catheter in these situations is its responsiveness to rotation (torque control).

The Amplatz catheter is advanced on profile along the aortic root into the ascending aorta. The tip of the catheter usually passes the left coronary ostium and lodges in the left sinus of Valsalva. As the catheter is continually advanced, the tip will move superiorly and engage the left main coronary artery. Once the catheter is engaged, slight withdrawal will allow for better catheter positioning. If the Amplatz catheter is advanced further, the catheter tip will disengage from the coronary ostium.

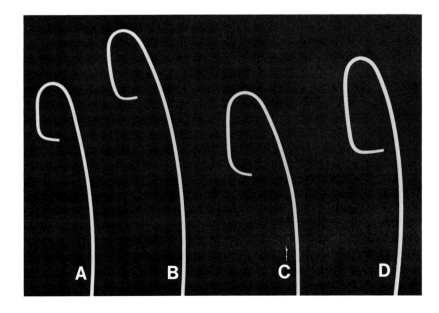

Figure 6-17 Left Judkins coronary catheters. A, 3.5L; B, 4L; C, 5L; D, 6L. *Source:* Reprinted from *Cardiology* by W Parmley and K Chatterjee (Eds) with permission of JB Lippincott Company, © 1987.

Right Coronary Artery Cannulation

The Judkins right coronary catheter is available in four sizes (3 to 6), which reflect the distance in centimeters from the primary curve to the halfway point of the secondary curve (Figure 6-20). The right coronary artery may be cannulated in one of two manners. The catheter is advanced to the level of the aortic valve. The catheter is then rotated in a clockwise manner and pulled back to allow the torque to be applied to the distal catheter tip. Alternatively, the catheter may be positioned 4 cm above the aortic valve and clockwise torque imparted while the catheter is advanced until the right coronary artery is cannulated. Once the right coronary artery is cannulated, the catheter tip pressure and waveform must be checked prior to contrast injection to prevent inadvertent dissection.

Difficulty may be encountered while attempting to cannulate the right coronary artery because of overtorquing of the catheter. Occasionally, selective cannulation of the conus branch occurs because a separate ostium is present. If this happens, the catheter should be removed and the right coronary orifice recannulated. Alternatively, the conus branch may arise superiorly from the right coronary artery. In this circumstance, the conus branch may be selectively cannulated, and again the catheter should be removed and repeat cannulation performed. If these maneuvers fail and the conus branch is repeatedly cannulated, a larger right Judkins catheter with a more inferiorly pointing tip should be used. The same considerations and precautions are taken when damping or ventricularization of the right coronary pressure are observed as with left coronary arteriography.

Bypass Graft and Internal Mammary Artery Cannulation

The key to successful bypass graft evaluation is the angiographer's knowledge of the surgeon's graft placement techniques. In addition, when available, the operative report should be reviewed prior to coronary angiography.

The graft to the right coronary artery is usually attached to the aorta approximately 2 cm above the origin of the native right coronary artery. The grafts to

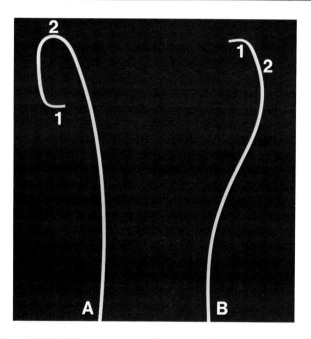

Figure 6-18 Judkins coronary catheters. A, 4L; B, 4R. 1, primary curve; 2, secondary curve. *Source:* Reprinted from *Cardiology* by W Parmley and K Chatterjee (Eds) with permission of JB Lippincott Company, © 1987.

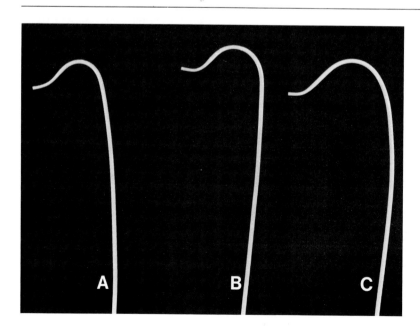

Figure 6-19 Amplatz catheters. A, AL I; B, AL II; C, AL III. *Source:* Reprinted from *Cardiology* by W Parmley and K Chatterjee (Eds) with permission of JB Lippincott Company, © 1987.

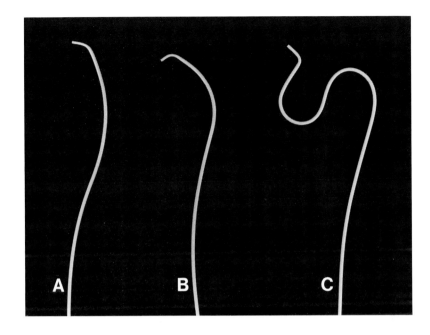

Figure 6-20 Right Judkins coronary catheters. A, 3.5R; B, 4R; C, 5R. *Source:* Reprinted from *Cardiology* by W Parmley and K Chatterjee (Eds) with permission of JB Lippincott Company, © 1987.

the left anterior descending and diagonal branches are attached superiorly and to the left of the native right coronary artery or right coronary graft. The grafts to the obtuse marginals are superior and to the left of the grafts to the left anterior descending or diagonal grafts.

Each of the grafts may be cannulated with either a Judkins right coronary catheter, a right bypass graft catheter, an A2 multipurpose catheter, or a Fromm catheter (Figure 6-20). As each of these catheters is advanced on profile to the ascending aorta, the catheter may inadvertently catch the lip of the bypass graft; care must be taken to avoid damaging the bypass graft.

Internal mammary artery cannulation is accomplished with an internal mammary artery catheter. This catheter is similar to a Judkins right coronary artery catheter, with the exception of a tighter primary curve and a longer tip (Figure 6-21).

The origin of both the right and left internal mammary arteries is from the anterior, inferior surfaces of the subclavian artery. To cannulate the left internal mammary, the catheter is advanced to the origin of the left subclavian, and the catheter tip is rotated anteriorly until the internal mammary artery is cannulated. If difficulty is encountered entering the left subclavian artery, a 0.035-in straight movable core guidewire may be inserted into the catheter and advanced into the subclavian artery. The catheter may then be advanced over the wire. The wire is then removed and the catheter further manipulated until the internal mammary artery is cannulated.

Cannulation of the right internal mammary artery is generally more problematic. The internal mammary catheter is advanced into the aortic arch. The catheter is slowly torqued counterclockwise and advanced. With

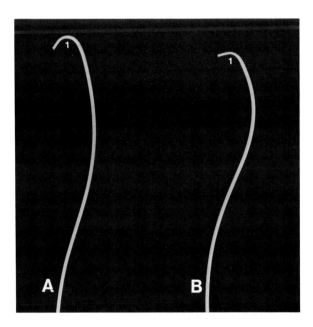

Figure 6-21 Comparison of (A) internal mammary artery catheter and (B) right coronary artery catheter. Note the tighter primary curve (1) of the internal mammary catheter compared to the right coronary catheter. *Source:* Reprinted from *Cardiology* by W Parmley and K Chatterjee (Eds) with permission of JB Lippincott Company, © 1987.

this maneuver the catheter will point superiorly, preferentially entering the innominate and subclavian arteries rather than the ascending aorta. Once in the subclavian, the catheter is manipulated anteriorly and inferiorly until the internal mammary artery is cannulated.

Prior to injecting the internal mammary artery, the distal tip pressure must be monitored. Injection of con-

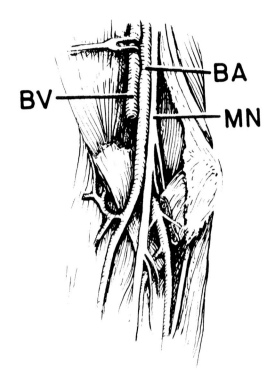

Figure 6-22 Anatomy of the right antecubital fossa. On occasion the vein may lie on top of the artery. Careful dissection is required to separate the vessels. BV, brachial vein; BA, brachial artery; MN, median nerve. *Source:* Reprinted from *Cardiology* by W Parmley and K Chatterjee (Eds) with permission of JB Lippincott Company, © 1987.

trast medium when the pressure is damped may result in dissection of the internal mammary artery. The injection of the internal mammary artery generally results in an unpleasant burning experience in the patient's anterior chest. This results from branches of the internal mammary that supply the anterior chest wall muscles. The pain may be lessened by the use of nonionic contrast agents or the dilution of ionic contrast material 1:1 with 1% lidocaine.

Brachial Artery Approach

Prior to attempting the brachial approach, the angiographer must be familiar with the anatomy of the left and right brachial fossae (Figure 6-22). The brachial artery is palpated two to three fingerbreadths above the brachial crease. Alternatively, it may be palpated at the level of the median epicondyle. The skin and subcutaneous tissue overlying the brachial artery are anesthetized. A horizontal incision is made over the artery. After locating the artery by blunt dissection, umbilical tapes are placed proximally and distally and used for traction and hemostasis. A small transverse arteriotomy is made with an iris scissors, a number 11 blade, or an 18-gauge needle. Five thousand units of heparin are delivered distally using an angiocath attached to a syringe to prevent distal thrombosis. A 7F or 8F 80-cm Sones catheter is introduced into the brachial artery and passed under fluoroscopic guidance to the ascending aorta (Figure 6-23). Upon advancing the catheter, resistance may be met as the catheter travels down the branches of the subclavian or brachial

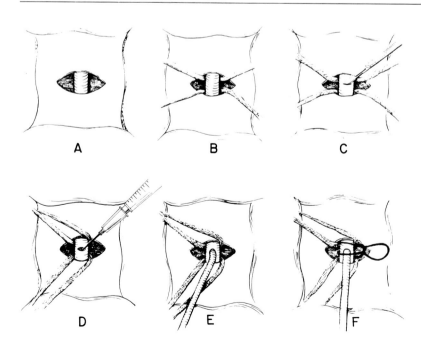

Figure 6-23 Brachial cutdown technique. A, A transverse incision is made in the skin overlying the brachial artery. B, The brachial artery is isolated by two umbilical tapes placed proximally and distally. C, A transverse incision is made in the artery. D, Heparin 5000 U is delivered distally. E, Insertion of Sones catheter through the arteriotomy. F, Rubber band in place over the proximal portion of the brachial artery to aid in hemostasis. *Source:* Reprinted from *Cardiology* by W Parmley and K Chatterjee (Eds) with permission of JB Lippincott Company, © 1987.

arteries. Whenever resistance is met, the catheter should be withdrawn, rotated, and readvanced under fluoroscopic guidance. If difficulty is encountered entering the ascending aorta, a 0.035-in J-tipped guidewire is inserted into the Sones catheter and advanced to the ascending aorta. This technique is particularly helpful in large, tortuous subclavian arteries where the catheter preferentially travels down the descending aorta.

On occasion, spasm is encountered upon insertion of the catheter into the brachial artery, making catheter advancement difficult. When this occurs, the patient generally feels extreme discomfort. The catheter should be removed and reintroduced after the pain is gone and the patient relaxes.

Left Coronary Artery Cannulation

The Sones catheter is advanced with the image intensifier in the 45° LAO projection. In this projection the left and right sinuses of Valsalva are on the left and right of the fluoroscopic image screen, respectively. The catheter is advanced to the left sinus of Valsalva, and a small loop is formed until the catheter tip points superiorly. With gentle, continued advancement of the catheter, the tip moves up the aortic wall until the left coronary ostium is selectively engaged. Occasionally, with dilated aortic roots, selective engagement of the left ostium is impossible, and coronary arteriography may be attempted with the catheter holes adjacent to the coronary ostium. However, contrast medium enhancement of the coronary artery will not be optimal.

Right Coronary Artery Cannulation

In a similar manner as in left coronary artery cannulation, the Sones catheter is advanced to the right sinus of Valsalva. A small loop is formed, and the tip of the catheter is advanced up the aortic wall until the right coronary is selectively engaged. Alternatively, the Sones catheter may be advanced into the left sinus of Valsalva, where a small loop is formed. Counterclockwise rotation is done, sweeping the anterior aortic wall until the right coronary ostium is engaged. At all times care must be used when engaging the right coronary artery with the Sones catheter to prevent deep penetration of the right coronary artery, which may result in dissection.

A number of other catheters, including the right and left Amplatz, A2 multipurpose, or Judkins, may be used via the brachial approach (Figure 6-24).

Bypass Graft and Internal Mammary Artery Cannulation

As described with the Judkins approach, the coronary bypass graft location is predictably positioned. To cannulate the right coronary graft, the Sones catheter is advanced to a level above the orifice of the graft. With further advancement, the catheter will selectively engage the graft orifice. Cannulation of left coronary grafts is similarly accomplished by advancing the catheter to the graft orifice. With slight counterclockwise rotation and retraction of the catheter, the left graft orifices are selectively cannulated. Catheterization of the left internal mammary artery is best accomplished via the left brachial artery. If left internal mammary artery visualization is required, the entire coronary angiogram may be done via the left brachial artery. The catheter is positioned, once in the subclavian, similar to the Judkins cannulation of the internal mammary arteries.

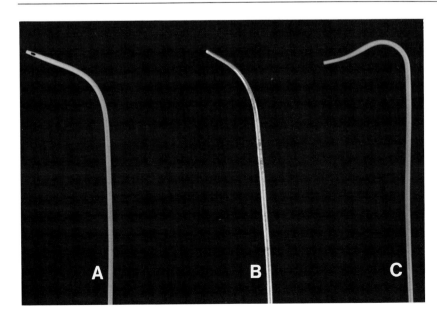

Figure 6-24 Angiographic catheters used during the brachial cutdown approach. A, A2 multipurpose—end-hole, two side-holes; B, Sones catheter—end-hole; C, brachial coronary catheter—end-hold. *Source:* Reprinted from *Cardiology* by W Parmley and K Chatterjee (Eds) with permission of JB Lippincott Company, © 1987.

On completion of coronary arteriography, the catheter is removed and back-bleeding is allowed proximally and distally. If back-bleeding does not occur, a Fogarty catheter is inserted proximally and distally. Once brisk back-bleeding is observed, 3000 U of heparin is given distally via the arteriotomy. Arterial clamps are placed proximally and distally to maintain hemostasis during the repair. The artery is closed with 6-0 Tevdek by a pursestring or transverse interrupted suture closure technique (Figure 6-25). After arterial repair, if the radial artery cannot be palpated, the arteriotomy sutures must be removed, back-bleeding allowed, and a Fogarty thrombectomy catheter utilized. The subcutaneous tissue and skin are approximated using 3-0 silk with an interrupted mattress suturing technique.

Percutaneous Brachial Approach

The percutaneous brachial approach is an alternative to the cutdown technique. The brachial artery is palpated slightly above the elbow crease. The skin and subcutaneous tissue are anesthetized. The skin above the artery is punctured with a number 11 blade. A hollow Cook needle is advanced at a 70° angle with the skin surface. Once the artery is entered, a 0.035-in J-tipped guidewire is advanced through the needle into the brachial artery. The needle is then removed with continued pressure on the brachial artery to prevent hematoma formation. A 7F or 8F sheath is passed over the wire. The Sones catheter is then advanced through the sheath. Once in the brachial artery, 4000 U of heparin is given. When the procedure is finished, the catheter and sheath are removed, and adequate hemostasis is obtained by pressing over the brachial artery with enough pressure allowing the radial pulse to be felt faintly.

Evaluation of Coronary Artery Spasm

The author recommends that patients with a significant chest pain history suggestive of coronary vasospasm with normal or insignificant coronary artery disease undergo provocative ergonovine testing.[5-7] Review of the video recording is important prior to ergonovine administration to rule out any significant obstructive coronary disease.

After routine arteriography, the patient is given incremental doses of ergonovine at 3-minute intervals (0.05 mg, 0.1 mg, 0.15 mg, 0.2 mg). Aortic pressure and a 12-lead electrocardiogram are taken prior to the administration of each dose of ergonovine if the patient has not experienced any chest discomfort. If there are no significant changes, the next dose of ergonovine is given, to a total dose of 0.4 mg of ergonovine. Occasionally, an additional dose of 0.3 mg of ergonovine may be given if there has been no significant change in heart rate or mean aortic pressure. In patients with a strong likelihood of coronary vasospasm, an initial ergonovine dose of 0.025 mg may be prudent.

After the last dose of ergonovine is given or if the patient has experienced his or her typical chest discomfort, coronary arteriography is done. The standard Judkins technique requires catheter exchanges for visualizing each coronary artery. It is the author's practice to attempt to predict the most likely vessel causing the patient's vasospasm and then to start with the appropriate right or left Judkins coronary catheter. The use of the brachial approach obviates catheter exchanges dur-

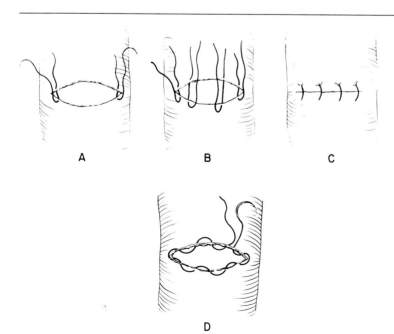

Figure 6-25 Brachial artery closure techniques. Interrupted suture closure: A, Sutures are placed at either end of the arteriotomy; B, Two additional sutures are spaced equally between the first two sutures; C, The sutures are tied. D, Pursestring suture technique. Care must be taken to place the sutures close to the site of the arteriotomy to avoid arterial stenosis. *Source:* Reprinted from *Cardiology* by W Parmley and K Chatterjee (Eds) with permission of JB Lippincott Company, © 1987.

ing coronary arteriography after the administration of ergonovine. If the patient develops chest pain, coronary arteriography should be done before further doses of ergonovine are administered. In the case of documented coronary vasospasm, 300 μg of intracoronary nitroglycerin may be given along with nifedipine 10 mg sublingually to relieve the spasm. Medical therapy must be continued until the vasospasm resolves, and this must be documented arteriographically.

The author recommends that nitrates and calcium channel blockers be stopped at least 24 to 48 hours prior to the administration of ergonovine. If these agents are not discontinued, it is difficult to interpret a negative provocative ergonovine challenge test. Ergonovine should only be administered in the cardiac catheterization laboratory where direct intracoronary administration of nitroglycerin is possible to reverse induced coronary vasospasm (Figure 6-26).

RADIOGRAPHIC EVALUATION OF CORONARY ANATOMY

To adequately visualize the coronary anatomy, the vessels must be visualized in several projections to eliminate foreshortening or overlap. Because of current imaging equipment, coronary arteries may be visualized with the x-ray beam directed perpendicularly to the vessel, avoiding foreshortening. With various degrees of rotation and angulation, vessel overlap may also be avoided.

Currently, the cardiovascular projection used is defined by the position of the image intensifier rather than the direction the x-ray beam is traveling. The degree of angulation in the transverse plane is measured from the midline, whereas the degree of angulation in the sagittal plane is measured from the vertical (Figure 6-27).[8–11]

Multiple projections must be obtained during the angiographic procedure. Additional projections are utilized to delineate areas of overlap and to further define the presence of questionable obstructive disease. The author recommends the following views: (1) left coronary artery: 10° to 20° LAO, 60° LAO, left lateral/10° caudal, 45° LAO/20° cranial, 10° RAO, 60° RAO/20° caudal; (2) right coronary artery: 30° LAO, 30° RAO. Table 6-1 lists some useful angiographic projections. The angiogram must be analyzed in a systematic approach. Each coronary artery and segment must be examined. If the arterial supply to a region of myocardium is not identified, an occluded vessel or anomalous origin of a coronary artery must be suspected.

The left main coronary artery is best visualized in the shallow RAO projection. Other views must be obtained if this vessel is not adequately visualized in this projection or if a question with respect to obstructive dis-

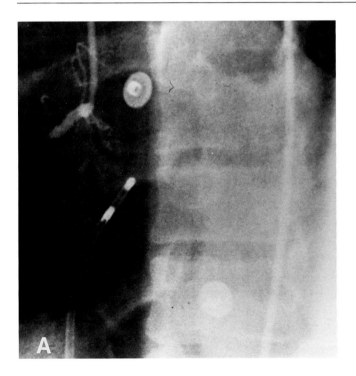
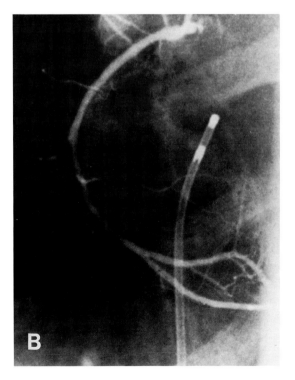

Figure 6-26 A, Right coronary after administration of 0.15 mg of ergonovine. B, Spasm was relieved after 300 μg of intracoronary nitroglycerin was given. No significant obstructive disease is seen. *Source:* Reprinted from *Cardiology* by W Parmley and K Chatterjee (Eds) with permission of JB Lippincott Company, © 1987.

Paulin

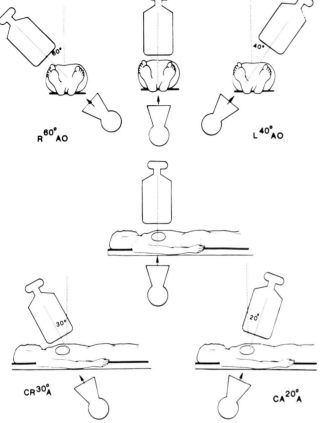

Figure 6-27 Cardiovascular nomenclature. (Top), The image intensifier has been rotated in a transverse plane around the long axis of the body. (Bottom), The image intensifier has been rotated in the sagittal plane. CR³⁰° A, 30° cranial angulation; CA²⁰° A, 20° caudal angulation. *Source:* Reprinted with permission from *Catheterization and Cardiovascular Diagnosis* (1982;7:341), Copyright © 1982, Alan R Liss Inc.

---

ease exists.[12] The distal left main is best seen in a direct anteroposterior projection with 20° to 30° cranial angulation. When the left main travels in a horizontal or cephalad direction, a 45° LAO/20° caudal projection is used to evaluate the distal left main bifurcation. When the left main travels inferiorly, a 45° LAO/20° cranial projection will elongate the left main and provide excellent visualization of the distal bifurcation. In cases of severe left main obstruction, the left main may be visualized by ostial injections. In unusual situations total occlusion of the left main coronary artery may exist.[13–14] If repeated attempts to cannulate the left main are unsuccessful, or if a small stump of the left main is visualized, the right coronary should be injected to look for collateralization to the left coronary system (Figure 6-28).

The proper evaluation of the left anterior descending and left circumflex coronaries depends upon the angle between the two vessels. When the angle is less than 30°, overlap may occur in the RAO projection. The following views are useful in this situation: 60° LAO, 45° LAO/20° cranial, or 45° RAO/20° caudal. If the 45° RAO/20° caudal is not helpful, a 45° RAO/20° cranial should be used. In a horizontal heart, the proximal left anterior descending and bifurcation area may best be evaluated with an LAO caudal projection (Figure 6-29). The entire left anterior descending coronary artery may be best evaluated with a left lateral projection accompanied by 10° to 15° caudal angulation. Occasionally an RAO cranial projection will best visualize the mid–left anterior descending.

The diagonal vessels are best identified in the LAO cranial projections. Misinterpretation may occur in the RAO projection when the left anterior descending is occluded; the diagonal or large septal perforator may resemble the occluded left anterior descending.

If the proximal circumflex is obscured by the left anterior descending or intermediate coronary arteries in the standard projections, an RAO caudal should be used. The proximal circumflex may also be visualized in the left lateral or the 45° LAO/30° cranial projection. The obtuse marginal vessels are best seen with the standard RAO or RAO caudal projections.

Visualization of the right coronary artery may be problematic. The proximal and conduit portions of the right coronary artery are best seen in the standard LAO projections. In this projection, the posterior descending and posterolateral vessels are foreshortened. The standard RAO view visualizes the posterior descending and posterior lateral branches. The cranial RAO and LAO views are helpful in visualizing the take-off of the posterior descending and posterior lateral vessels (Figure 6-30). On occasion a straight left lateral view may be helpful in visualizing the midportion of the right coronary artery. In this view the right ventricular branches do not overlap the mid–right coronary segment.

COMPLICATIONS OF CORONARY ARTERIOGRAPHY

Complications related to coronary arteriography can be classified as major complications, including death, myocardial infarction, and cerebral vascular accidents. Minor complications are related to arterial injury, local neurological complications, and electrophysiologic disturbances.

Major Complications

Table 6-2 summarizes four major literature studies that address complications during cardiac catheterization.[15–18] As a result of these studies, it may be concluded that the more advanced the coronary atherosclerotic disease is, the higher the complication rate. In

Table 6-1 Radiographic Projections of the Coronary Arteries

Coronary Artery	Radiographic Projections	Comments
Left main	10° RAO	Generally the best view for the left main
	45° LAO/20° cranial	Evaluation of the left main bifurcation with an inferior orientation
	45° LAO/20° cranial	Horizontal left main, distal left main, and distal left main bifurcation
	20°–30°/cranial	Distal left main evaluation
Left anterior descending	30° LAO	Evaluation of proximal and mid–left anterior descending
	30° RAO	Evaluation of proximal and mid–left anterior descending
	45° LAO/30° caudal	Evaluation of proximal left anterior descending with a horizontal
	45° LAO/20° cranial	Evaluation of proximal and mid–left anterior descending and origins of diagonals
	Left lateral/10° caudal	Evaluation of proximal, mid, and distal portions of the left anterior descending
	30° RAO/20° caudal	Evaluation of mid and distal left anterior descending
Left circumflex	45° LAO/20° cranial	Evaluation of proximal circumflex
	40° LAO	Evaluation of mid circumflex
	30°–40° RAO	Evaluation of obtuse marginal vessels
	45° LAO/20° cranial	Evaluation of obtuse marginal vessels
	10°–20° RAO/20° caudal	Evaluation of proximal and mid circumflex
	45° LAO/20° cranial	Evaluation of the bifurcation of the right coronary artery and posterior descending branches
	45° RAO/20° cranial	Evaluation of posterior descending and posterolateral branches
Right coronary	30° RAO	Evaluation of the mid–right coronary artery posterior descending and posterolateral branches
	30° LAO	Evaluation of proximal right coronary artery

Source: Reprinted from *Cardiology* by W Parmley and K Chatterjee (Eds) with permission of JB Lippincott Company, © 1987.

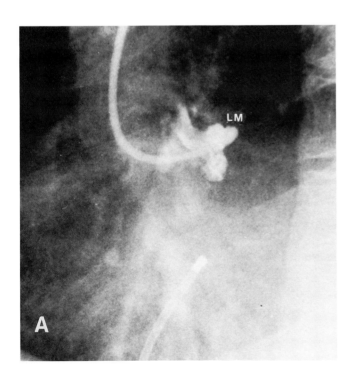 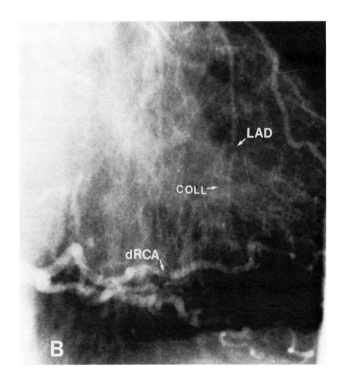

Figure 6-28 A, Total occlusion of the distal left main (LM) coronary artery, LAO projection. B, Right coronary artery injection showing collaterals to the left system. dRCA, distal right coronary artery; COLL, collaterals; LAD, left anterior descending. *Source:* Reprinted from *Cardiology* by W Parmley and K Chatterjee (Eds) with permission of JB Lippincott Company, © 1987.

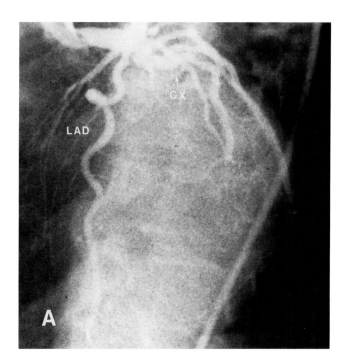
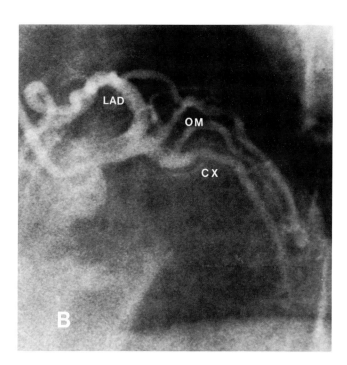

Figure 6-29 Left coronary artery. A, LAO view. B, LAO caudal view. The origins of the left anterior descending (LAD), left circumflex (CX), and obtuse marginals (OM) are clearly seen in the caudal view, without overlap. *Source:* Reprinted from *Cardiology* by W Parmley and K Chatterjee (Eds) with permission of JB Lippincott Company, © 1987.

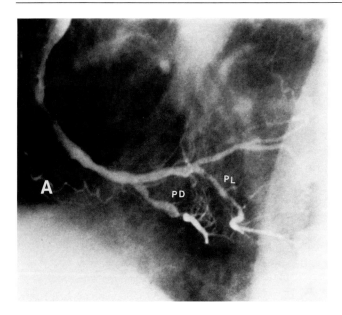
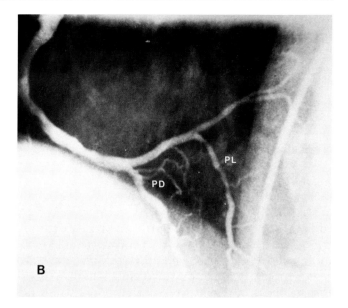

Figure 6-30 A, Right coronary artery, LAO projection. B, Right coronary artery, LAO/cranial projection. With cranial angulation, foreshortening is eliminated; the distal vessels are better visualized. PD, posterior descending; PL, right posterolateral. *Source:* Reprinted from *Cardiology* by W Parmley and K Chatterjee (Eds) with permission of JB Lippincott Company, © 1987.

patients with significant left main coronary artery disease, the mortality rates may range from 0.76% to 0.86%. The age of the patient undergoing cardiac catheterization also influences the complication rate. The incidence of death varies from 1.75% of patients less than 1 year old to 0.25% of patients older than 60 years.

Death from cardiac catheterization can be prevented if high-risk patients are previously identified. These high-risk patients include infants less than 1 year old and patients with advanced coronary atherosclerotic disease, left ventricular dysfunction, and other severe noncardiac medical conditions such as renal insufficiency or severe peripheral or cardiovascular disease.

Table 6-2 Complications of Cardiac Catheterization

Investigator	Years Study Conducted	Number of Patients	Complications	(%)
Braunwald	1966–1968	12,367	Death	(0.44)
			MI*	(0.04)
			VF†	(0.70)
			Arterial	(1.40)
			CNS‡	(0.05)
Adams et al.	1970–1971	46,904	Death	(0.45)
			MI	(0.61)
			VF	(1.28)
			Arterial	(1.62)
			CNS	(0.23)
Davis et al.	1975–1976	7515	Death	(0.20)
			MI	(0.25)
			VF	—
			Arterial	(0.74)
			CNS	—
Kennedy et al.	1978–1979	53,581	Death	(0.14)
			MI	(0.07)
			VF	(0.56)
			Arterial	(0.57)
			CNS	(0.07)

*MI, myocardial infarction; †VF, ventricular fibrillation; ‡CNS, central nervous system.

During the cardiac catheterization procedure, hemodynamic parameters must be monitored continuously. Pharmacological pretreatment is given prior to cardiac catheterization to optimize hemodynamic parameters. In patients suspected of having significant left main disease, the angiographer must take care in cannulating the left main coronary artery, watching the catheter pressure and waveform and using a limited number of projections.

Factors that lead to death may also predispose the patient to developing acute myocardial infarction. To avoid this complication, pretreatment of patients with unstable angina is mandatory. In cases of refractory angina that occurs during cardiac catheterization, aggressive medical intervention is necessary and intra-aortic balloon pumping should be considered. Adequate heparinization during cardiac catheterization has also helped to prevent myocardial infarction. It has been the author's practice not to reverse the anticoagulation for brachial or femoral approaches. In patients with allergies to fish or insulin-dependent diabetics, protamine should not be given. If protamine is given, the patient should be carefully watched for development of an anaphylactoid reaction.[19]

Central nervous system complications are best avoided by using meticulous angiographic technique. Guidewires must be wiped carefully and catheters aspirated and flushed vigorously after guidewire removal.

In patients with advanced cerebral vascular disease, the angiographer must avoid entering the branch vessels of the aortic arch. Likewise in patients with abdominal or thoracic aneurysms, catheter exchanges must be made above the area of the aneurysm to avoid plaque embolization. Also in patients with suspected left ventricular thrombus, care must be taken when advancing the catheter into the left ventricle to avoid dislodging the thrombus. Systemic heparinization has helped to lessen some of these complications; however, the best way to avoid complications, especially of the central nervous system, is by using meticulous catheter technique and reducing procedure duration to a minimum.

Minor Complications

In early studies of cardiac catheterization, the incidence of arterial complications ranges from 3.6% to 6%. In the recent study of Kennedy et al.[17] the vascular complication rate was much lower, occurring in approximately 0.57% of patients.

The vascular complications vary depending upon the route of arterial entry. Thrombosis and trauma of the artery are more common with the brachial approach; this has been especially noted in laboratories where this approach is taken infrequently. The vascular complications include dissection, which may involve the femoral, aortic, subclavian, or coronary arteries. To avoid this complication, careful technique with respect to guidewire and catheter advancement should be used. Catheters and guidewires must never be forced. If difficulties are encountered, free backflow of blood must be established before the catheter and guidewire are manipulated further. Dissection of the coronary arteries is a rare complication. This generally results from vigorous catheter manipulation or injection of contrast when the catheter is wedged against the arterial wall. Dissection more commonly occurs in the right coronary artery, and it occurs more commonly in patients with normal coronary arteries. This complication may eventuate in myocardial infarction. Surgical intervention is usually indicated when coronary dissection occurs. Ostial dissection of the left main coronary artery is best prevented by avoiding manipulation of the catheter as it enters in the orifice of the left main coronary artery.

Thrombosis is best prevented by good catheterization technique. When using the brachial technique, distal anticoagulation is mandatory along with routine Fogarty catheterization of the distal and proximal vessel prior to arterial repair. If the radial pulse is not palpable after repair, the sutures must be removed and the vessel explored.

When using the percutaneous femoral approach, similar vascular complications may occur. Hemorrhage and pseudoaneurysm formation are some of the complications that may result. Hemorrhage generally occurs as a result of early ambulation after cardiac catheterization. It is the author's practice to delay ambulation for at least 6 hours after femoral artery catheterization. In patients who are heavy or in whom difficulty maintaining hemostasis was observed, or when a hemotoma is present, ambulation is allowed only after 12 hours of bed rest.

The development of a pseudoaneurysm, that is, a hematoma communicating with a ruptured artery, generally results from poor compression of the femoral artery after cardiac catheterization. In this case the patient generally complains of pain and local swelling in the leg. This complication may not manifest at times until several weeks after the cardiac catheterization has been completed. Peripheral vascular surgical intervention is always required because of the high incidence of delayed rupture of the pseudoaneurysm.

The development of arteriovenous fistulas following cardiac catheterization is also a well-described complication. This generally results when the vein and artery are overlying each other and, on introduction, the catheter traverses both vessels. Surgical repair of this condition is also mandatory.

When the pulse distal to the entry site is not palpated after the cardiac catheterization procedure has been completed, surgical exploration is mandatory to avoid subsequent ischemic complications and the ultimate loss of the limb. If the pulse is weaker after catheterization, systemic heparinization with careful monitoring for a short period of time may be done. If improvement does not occur, arterial exploration to remove the thrombus or to repair a small intimal flap at the site of entrance into the femoral artery or brachial artery must be done. Complete occlusion by thrombosis of the femoral artery requires urgent surgical exploration and removal of the thrombus.

Local neurological complications may result from brachial or femoral approaches. These complications are a direct result of damage to the median or femoral nerve, generally occurring during anesthetic administration. Other median nerve complications may result during cutdown or during suturing of the arteriotomy when the brachial approach is used. Occasionally large hematomas may compress the brachial or femoral nerves, causing neurological symptoms.

Injury to the median nerve results in a sensation of numbness, tingling, and slight weakness of the hand. These symptoms generally resolve spontaneously in approximately 2 weeks. Alternatively, if the patient is complaining of significant hand pain and weakness, exploration of the area must be done to determine the etiology of the symptoms.

Ventricular fibrillation, ventricular tachycardia, or various bradyarrhythmias have all been noted during cardiac catheterization. The incidence of ventricular fibrillation varies between 0.56% and 1.28%.

Dysrhythmias during coronary arteriography occur from inadvertent entry of the catheter into the left ventricle or after the injection of the coronary arteries. Injection of the right coronary artery more commonly causes major rhythm disturbances. If ventricular fibrillation persists, prompt defibrillation and resuscitative measures are indicated. The development of bradycardia can be prevented by use of a standby ventricular pacemaker. It is the author's practice not to leave the pacing catheter in the right ventricle during arteriography. If a bradyarrhythmia occurs and necessitates the use of a pacemaker, the pacemaker electrode is advanced into the right ventricle and left in position until coronary arteriography is completed.

In summary, the factors important in minimizing complications include experience, constant monitoring of cardiac catheterization technique, and the speed with which the procedure is done. However, it should be remembered that despite meticulous attention to detail, complications will occur. It has been noted that complications occur in populations with advanced arteriosclerotic disease or advanced age. Prompt recognition of complications is mandatory, and appropriate actions to ameliorate them should be taken immediately.

PITFALLS IN INTERPRETATION OF THE CORONARY ARTERIOGRAM

Incomplete Study

An incomplete study eventuates when an inadequate number or type of projections have been taken. The number and type of projections must be tailored to each patient. The angiographic procedure must define the entire coronary arterial tree.

Poor Contrast Medium Injection

Inadequate opacification of the arterial system may result in the appearance of luminal narrowing when the coronary arteries may in fact be normal. The force and rate of injection must be individualized. The force generated should be enough to opacify the entire coronary vessel, ensuring adequate mixing of the contrast agent and blood. It should be recognized that the pressure generated during injection of contrast medium into the coronary arteries should be higher than the aortic diastolic pressure, to ensure the replacement of blood with contrast material in the coronary arteries.

Selective Cannulation of the Coronary Vessels

The unrecognized selective cannulation of coronary vessels may lead the angiographer to the inappropriate conclusion that the unopacified vessel is totally occluded (Figure 6-31). Occasionally, if the conus branch is injected, ventricular fibrillation may result.

On occasion, selective cannulation of coronary vessels is indicated. For example, in cases of total occlusion of the left anterior descending in the presence of normal anterior wall motion and absence of left to left or right coronary to left anterior descending collaterals, the selective injection of the conus branch is indicated to look for collateral supply to the occluded left anterior descending coronary artery. In addition, in case of a short left main coronary artery, selective injections of the left anterior descending and circumflex coronary arteries may be required.

Inadequate X-Ray Penetration of Projections

Attenuation of the x-ray beam and the resultant inadequate penetration generally results from use of newer sagittal projections and attempts to penetrate the liver, especially when the patient does not take an adequate breath to move the diaphragm downward. Poor visualization of the coronaries may result, leading to erroneous or improper diagnoses. These problems may best be avoided by instructing the patient to continually inspire while taking the appropriate cine pictures. In addition, the appropriate x-ray settings must be used for heavier patients.

Catheter-Induced Coronary Spasm

Catheter-induced spasm may occur in any of the coronary vessels. Coronary spasm should be suspected when damping of the catheter waveform results prior to coronary injection, backflow of contrast medium into the aortic sinus is not seen, or there is absence of chest pain during the administration of ergonovine. Unrecognized catheter spasm may lead the angiographer to the erroneous conclusion that significant coronary obstructive disease may be present. When catheter spasm is noted, intracoronary injection of 300 μg of nitroglycerin and repeat arteriography should be done. If catheter spasm persists, cusp injections of contrast material may be helpful to eliminate the possibility of significant obstructive coronary disease.

Myocardial Bridges

The epicardial coronary arteries may course intramyocardially.[20–21] In this situation, because ventricular muscle is overlying the coronary artery for part of its course during systole, an area of obstruction may be seen (Figure 6-32). The key to recognizing myocardial bridging is the absence of evident obstruction during diastole. Generally, myocardial bridging is inconsequential; however, at times it may be associated with ischemic episodes.

Totally Occluded Coronary Artery

The total proximal occlusion of a coronary artery may be difficult to recognize. Large first septal perforators or diagonal vessels may be mistaken for the left anterior descending coronary artery. To avoid missing a totally occluded vessel, the angiographer must be familiar with the typical course of the vessel in the various

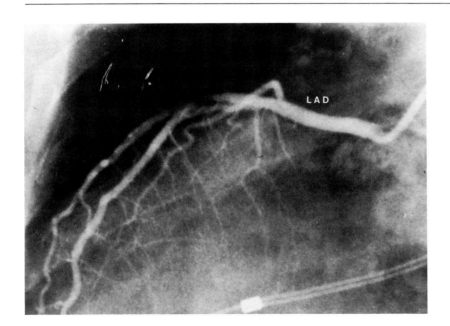

Figure 6-31 Selective cannulation of the left anterior descending (LAD) artery. The circumflex vessel is not visualized. *Source:* Reprinted from *Cardiology* by W Parmley and K Chatterjee (Eds) with permission of JB Lippincott Company, © 1987.

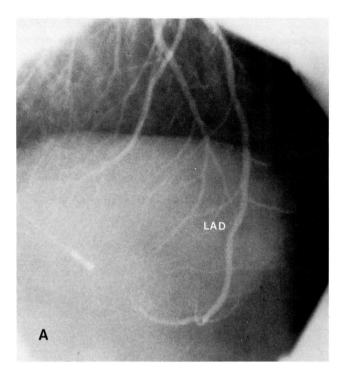
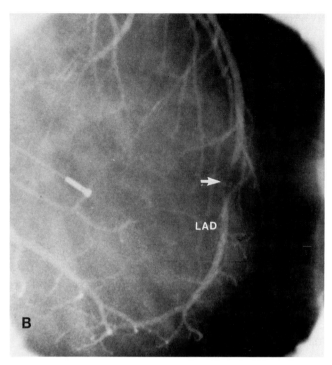

Figure 6-32 Myocardial bridging. A, Normal-appearing left anterior descending (LAD) artery. B, During systole, apparent narrowing of the left anterior descending is indicated by arrow. *Source:* Reprinted from *Cardiology* by W Parmley and K Chatterjee (Eds) with permission of JB Lippincott Company, © 1987.

radiographic projections, and a careful search for collateral vessels must be done (Figure 6-33).

LIMITATIONS OF CORONARY ANGIOGRAPHY

The original purpose of coronary arteriography was to diagnose the presence and severity of coronary artery disease. With the advent of coronary angioplasty, thrombolytic therapy, and new cholesterol-lowering techniques, the accurate measurement of actual coronary stenosis became mandatory to adequately evaluate each of these new therapeutic strategies. Unfortunately, although the coronary arteriogram allows us to understand the anatomy of the coronary vasculature, it gives us no direct understanding of the

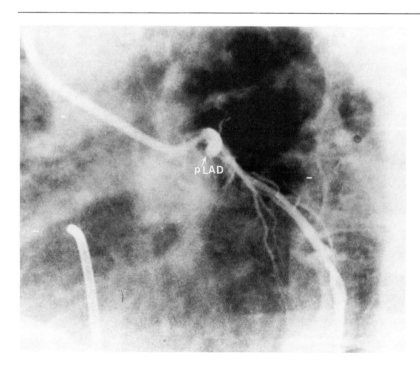

Figure 6-33 Left coronary artery, LAO projection demonstrating total occlusion of the proximal left anterior descending (pLAD) artery. *Source:* Reprinted from *Cardiology* by W Parmley and K Chatterjee (Eds) with permission of JB Lippincott Company, © 1987.

underlying physiology related to the coronary obstruction process.[22]

Degrees of coronary stenosis have been estimated using percent narrowing or absolute cross-sectional luminal area. Each of the techniques has serious limitations, not taking into account numerous characteristics of the coronary stenosis.

There exists a large interobserver and intraobserver variability in the interpretation of the coronary arteriogram. It has been estimated that a 20% to 40% interobserver and a 7% to 10% intraobserver variability exists.[23-24] Because of these observations, more objective means of evaluating coronary artery stenosis are currently being developed.[25-27]

To date, coronary arteriography continues to remain a powerful tool for the assessment and management of patients with coronary artery disease. The limitations and pitfalls of coronary arteriography must be realized if this tool is to be used to its fullest potential.

REFERENCES

1. The Principal Investigators of CASS and Their Associates. The National Heart, Lung and Blood Institute Coronary Artery Surgery Study (CASS). *Circulation.* 1981;63:1-40.

2. McAlpine WA. *Heart and Coronary Arteries.* New York: Springer-Verlag; 1975:203.

3. Spindola-Franco H, Groser R, Solomon N. Dual left anterior descending coronary artery: Angiographic description of important variants and surgical implications. *Am Heart J.* 1983;105:445-455.

4. Fergusson DJG, Kamada RO. Percutaneous entry of the brachial artery for left heart catheterization using a sheath: Further experience. *Cathet Cardiovasc Diagn.* 1986;12:209-211.

5. Maseri A, Severi S, Des Res M, et al. Variant angina: One aspect of a continuous spectrum of vasospastic myocardial ischemia. *Am J Cardiol.* 1978;42:1019-1035.

6. Heupler FA, Proudfit WL, Razqui M, et al. Ergonovine maleate provocative test for coronary arterial spasm. *Am J Cardiol.* 1978;41:631-640.

7. Heupler FA. Syndrome of symptomatic coronary arterial spasm with nearly normal coronary arteriograms. *Am J Cardiol.* 1980;45:873-881.

8. Miller RA, Felix WG, Warkentin DL, et al. Angulated views in coronary arteriography. *Am J Radiol.* 1980;134:407-412.

9. Green CE, Elliot LP, Rogers WJ, et al. The importance of angled right anterior oblique views in improving visualization of the coronary arteries, I: Caudocranial view. *Diagn Radiol.* 1982;142:631-635.

10. Green CE, Elliot LP, Rogers WJ, et al. The importance of angled right anterior oblique views in improving visualization of the coronary arteries, II: Craniocaudal view. *Diagn Radiol.* 1982;142:637-641.

11. Grover M, Slutsky R, Higgins C, et al. Terminology and anatomy of angulated coronary arteriography. *Clin Cardiol.* 1984;7:37-43.

12. Elliot LP, Bream PR, Soto B, et al. Significance of the caudal left anterior oblique view in anayzing the left main coronary artery and its branches. *Diagn Radiol.* 1981;139:39-43.

13. Crosby JK, Mellons HA, Burwell L. Total occlusion of the left main coronary artery. Incidence and management. *J Thorac Cardiovasc Surg.* 1979;77:389-391.

14. Goldberg S, Grossman W, Markis JE, et al. Total occlusion of the left main coronary artery: A clinical hemodynamic and angiographic profile. *Am J Med.* 1978;64:3-7.

15. Adams DF, Fraser DB, Abrams HL, et al. The complications of coronary arteriography. *Circulation.* 1973;48:609-618.

16. Davis K, Kennedy WJ, Kemp HG, et al. Complications of coronary arteriography from the collaborative study of coronary artery surgery (CASS). *Circulation.* 1979;59:1105-1112.

17. Kennedy WJ, Baxley WA, Bunnell IL, et al. Mortality related to cardiac catheterization and angiography. *Cathet Cardiovasc Diagn.* 1982;8:323-340.

18. Braunwald E. Cooperative study deaths related to cardiac catheterization. *Circulation.* 1968;37(5)(suppl 3):17-26.

19. Stewart WJ, McSweeney SM. Increased risk of severe protamine reactions in NPH insulin-dependent diabetics undergoing cardiac catheterization. *Circulation.* 1984;70:788-792.

20. Bloor CM, Lowman RM. Myocardial bridges in coronary angiography. *Am Heart J.* 1963;65:195-199.

21. Geiringer E. The mural coronary artery. *Am Heart J.* 1950;50:359-368.

22. White CW, Creighton BW, Doty DB, et al. Does visual interpretation of the coronary arteriogram predict the physiologic significance of a coronary stenosis? *N Engl J Med.* 1984;310:819-824.

23. Zir LM, Miller SW, Dinsmore RE, et al. Interobserver variability in coronary angiography. *Circulation.* 1976;53:627-632.

24. DeRouen TA, Murray JA, Owen W. Variability in the analysis of coronary arteriograms. *Circulation.* 1977;55:324.

25. Brown BG, Bolson E, Frimer M, et al. Quantitative coronary arteriography: Estimation of dimensions, hemodynamic resistance, and atheroma mass of coronary artery lesions using the arteriogram and digital computation. *Circulation.* 1977;55:329-337.

26. Gould KL, Kelley KO, Bolson EL, et al. Experimental validation of quantitative coronary arteriography for determining pressure-flow characteristics of coronary stenosis. *Circulation.* 1982;66:930.

27. Ledbetter DC, Selzer RH, Gordon RN, et al. Computer quantitation of coronary angiograms. In: Miller HA, Smith EV, Harrison DC, eds. *Noninvasive Cardiovascular Measurements.* Stanford, Calif: Society of Photo-Optical Instrumentation; 1978:167:17.

Chapter 7
Contrast Media

*Bruce H. Brundage, MD, Eva V. Chomka, MD, Christopher J. Wolfkiel, PhD,
James L. Ferguson, PhD, and Sanjay Deshpande, MD*

INTRODUCTION

The density of most tissue, other than fat and bone, is very nearly the same as that of blood. Therefore, the transit of x-rays through myocardium and blood is attenuated to the same degree, and the two are indistinguishable on roentgenograms. This fact means that to define the internal anatomy of the heart and blood vessels, a contrast medium must be employed.

Currently it is estimated that over 8 million radiographic procedures that require the use of contrast medium are performed annually. Contrast agents have been used with diagnostic x-ray very nearly from its inception in 1895. Many compounds were tried, including bismuth, thorium, and strontium, but iodine was soon identified as the ideal agent.[1]

Soon (1922), iodine was combined with oil bases for intrathecal use. However, toxicity due to poor excretion stimulated development of water-soluble iodine compounds that were more rapidly eliminated (1931). Further development of water-soluble iodine contrast media has continued until the present time. Monoionic salts, usually sodium, have largely been replaced by di-ionic salts such as sodium and meglumine. These agents, still in use today, are hyperosmolar compared to blood and extracellular fluid. Currently it is believed that many of the adverse reactions experienced by patients after the intravascular injection of contrast medium are due to the cations present. This belief has led to the development of nonionic contrast media that are less hyperosmolar.[2]

The ideal cardiovascular contrast medium should provide optimum opacification, be water soluble and rapidly excreted, have no adverse effects, be isosmolar with blood and extravascular fluid, have a low viscosity to permit rapid injection through small catheters, and be nonionic, stable, and inert.[3]

CHEMISTRY OF CONTRAST MEDIA

The ionic contrast agents in use today are composed of meglumine, sodium, or a combination of meglumine-sodium salts of tri-iodinated benzoic acid derivatives, diatrizoate or iothalamate. These compounds dissociate into a radiopaque anion with three iodine atoms and two cations, giving a ratio of iodine atoms to osmotic particles of 3 to 2. These compounds are therefore sometimes referred to as ratio −1.5 media.[4] Meglumine salts have lower osmolalities than the sodium salts, so they are the preferred salt, alone or in combination with lesser amounts of the sodium salt. The available nonionic contrast media are iohexol, iopamidol, and metrizamide. They are tri-iodinated benzene derivatives with polyhydroxyalkyl substituents that enhance water solubility without ionic dissociation. They give three iodine atoms per one osmotically active particle and therefore are known as ratio −3 media.

The properties of the common ionic and nonionic preparations are compared in Table 7-1.[5] Ionic contrast media with an iodine content of at least 350 mg/mL

Table 7-1 Physical Properties of Equi-Iodine Concentrations of Selected Low-Osmolality and Ionic Contrast Media

Agent	Concentration (Iodine % w/v)	Iodine (mg/mL)	Osmolality (mOsm/kg)	Viscosity (Centipoise) 25°C	37°C	List Price ($) for 50-mL Vial
Conventional media						
Diatrizoate sodium (various mfr.)	50	300	1522–1550	3.2–3.4	2.4	5.55*
Diatrizoate sodium (8%)—meglumine (52%) (various mfr.)	60	292	1420–1539	5.9–6.2	4.0–5.0	5.83*
Low-osmolality media						
Ioxaglate sodium (19.6%)—meglumine (39.3%) (Hexabrix, Mallinckrodt)	58.9	320	600	15.7	7.5	45.50
Iohexol (Omnipaque, Winthrop-Breon)	64.7	300	709	10.4	6.8	45.90
Iopamidol (Isovue, Squibb)	61	300	616	8.8†	4.7	45.99

* Average of available products.
† Determined at 20°C.

Source: Adapted with permission from "Catalog of Intravascular Contrast Media" by HW Fischer, Radiology (1986;159:561–563), Copyright © 1986, Radiological Society of North America.

have osmolalities of 2049 to 2300 mOsm/kg. Nonionic agents in concentrations of at least 350 mg of iodine per milliliter have osmolalities of 796 to 862 mOsm/kg.[4] The usual half-life of contrast media is 90 to 120 minutes. They are nearly all excreted by glomerular filtration, with less than 1% being eliminated in the feces. There is no significant in vivo deiodination or biotransformation of these agents. They appear to act specifically as contrast media, with little extraneous activity.

CARDIOVASCULAR EFFECTS

The cardiovascular effects of contrast media are somewhat complex.[6–7] The effects are dependent on the site of injection and the contrast agent employed. Sodium salts have the most profound hemodynamic effect, closely followed by meglumine salts. All the nonionic agents produce lesser changes. Low-osmolarity ionic dimers have less effect than the meglumine salts but more than the nonionic agents. The cardiovascular effects can be categorized by the resultant changes in electrophysiology, myocardial contractility, vascular tone, cardiovascular reflexes, and blood volume. These changes are brought about by one or more of the following characteristics of contrast media: hyperosmolality, cation content, and chemical configuration.

Intracoronary injection of contrast medium slows the heart rate by reducing sinoatrial node impulse formation and prolongs the PR interval by slowing atrioventricular (AV) nodal conduction. T wave alterations are most profound when ionic agents are employed.[8] Ventricular fibrillation is the most clinically important consequence of the electrophysiologic effects of contrast media. Although uncommon (less than 1% or 2% in most laboratories), it appears to be even less frequent when nonionic agents are employed.[8–9] Although the reason is not entirely clear, it appears that the electrophysiologic effects of contrast media are most likely due to the hypocalcemia produced by direct binding of calcium ion by the radiopaque anion and other sequestrant agents present. The addition of calcium to contrast media, the elimination of sequestrants like ethylenediamine tetra-acetic acid (EDTA) and the lack of an available anion (nonionic agents) greatly ameliorate the electrophysiologic effects. Hyperosmolality may also play a role in effects on the sinoatrial and AV nodes.[6–7]

When contrast medium is injected into a coronary artery, there is an almost immediate decrease in myocardial contractility. This depression results in a fall in blood pressure and dp/dt max and an increase in left ventricular end-diastolic pressure and end-diastolic and end-systolic dimension.[6–7] This decrease in myocardial contractility is apparently due to the sequestration of calcium ion and can be ameliorated by the addition of calcium to ionic contrast media. Ischemia appears to prolong the depression of myocardial contractility. Depression of myocardial contractility is not seen with nonionic contrast agents. The initial decrease is followed within 8 to 10 seconds by an increase in blood pressure and dp/dt max, possibly as the result of stimulation of the coronary artery mechanoreceptors or direct release of catecholamines from the myocardium.[5] Furthermore, the initial drop in blood pressure stimulates baroreceptors to produce a compensatory rise in blood pressure by increasing heart rate and peripheral vascular tone.

Left ventricular angiography and aortography produce similar electrophysiologic and hemodynamic effects as coronary arteriography, although not as marked.[10] Again, nonionic contrast media have a much diminished effect. Within 1 minute after left ventriculography, there is a substantial rise in left ventricular end-diastolic pressure, which peaks in 3 to 5 minutes and may persist for up to 15 to 20 minutes.[11-12] Although controversy still exists as to the mechanism, this rise is most likely or predominantly due to the increase in circulating blood volume created by the injection of hyperosmolar contrast medium into the circulation. The rise in left ventricular end-diastolic pressure is usually associated with an increase in cardiac output. These hemodynamic responses are largely blunted when nonionic or low osmolar ionic contrast media are used.

Within a few seconds after contrast media come into contact with a blood vessel wall, there is rather marked vasodilation. This phenomenon produces the characteristic flush or sensation of warmth that patients experience. The vasodilation is due to the direct action of contrast media on vascular smooth muscle. Some suggest this is a direct toxic effect.[5] However, the flush is largely ameliorated when nonionic agents are used, suggesting a role for hyperosmolality. This direct effect also explains the increase in coronary flow well known to occur after direct coronary injection (see Chapter 9).

The intravenous injection of large amounts of an ionic contrast medium, as used in digital subtraction angiography and x-ray computed tomography (CT), may also cause substantial alteration in hemodynamics.[13] However, we recently performed a series of experiments in dogs using ionic and nonionic contrast media; the results indicate that the smaller doses of intravenous ionic contrast medium used in ultrafast computed tomography may have relatively little hemodynamic effect.

Methods

Thirty-three conditioned mongrel dogs, weighing 25 to 30 kg, were used for the study. The animals were fasted, except for water, for 24 hours and then anesthetized with 30 to 35 mg/kg of intravenous sodium pentobarbital. The animals were mechanically ventilated, and arterial blood gases were maintained in the physiologic range. The femoral artery and vein were cannulated for administration of contrast medium and drugs, measurement of blood pressure, and blood withdrawal. A right lateral thoracotomy was performed, and the left atrium was cannulated via a right middle lobe pulmonary vein. The aortic and left atrial pressures were measured using a Gould-Statham p23DB strain gauge manometer and recorded on an electrocardiogram (ECG) by an Electronics for Medicine DR 12 physiologic recorder throughout the experiment.

Measurement of Myocardial Blood Flow by Radioactive Microspheres

Myocardial blood flow determinations were made by injecting approximately 3×10^6 microspheres (3M Co.) of the $15u \pm 1u$ size labeled with one of four nuclides (chromium 51, cerium 141, scandium 46, strontium 85) into the left atrium while sampling aortic blood with a Harvard pump via the femoral artery catheter (withdrawal rate, 4.1 mL/min). The microspheres were sonicated for at least 30 minutes before infusion. After removal from the sonicator, the microspheres were mechanically shaken by a Vortex mixer until immediately prior to their infusion. Ten milliliters of saline was infused through the left atrial catheter in 10 seconds to flush the microspheres into the circulation. The withdrawal of the aortic reference sample began before the injection of the microspheres and continued for 90 seconds after the left atrial catheter was flushed. A large number of microspheres were used to ensure that at least 400 microspheres would be trapped in each piece of myocardium where blood flow measurements were made, so as to ensure reproducible measurements.[7] At the conclusion of the experiment, the animals were sacrificed with a 10-mL infusion of a supersaturated solution of KCl, and their hearts were removed and fixed in formalin for 1 week. The atria and right ventricle of each heart were trimmed away, and four adjacent 1-cm slices of the mid–left ventricle were obtained by sectioning the heart perpendicularly to the left ventricular long axis. Each slice was cut into approximately 24 equal pieces weighing 0.5 to 1 g (Figure 7-1). The 96 pieces from the four slices were then counted in a Beckman 9000 gamma deep well scintillation counter to determine radioactivity. Blood flow for each piece was determined by previously reported methods and expressed in milliliters per 100 g of myocardium per minute.[8-9]

$$\text{Flow} = \frac{\text{Tissue counts/tissue weight}}{\text{Total counts injected}} \times \text{cardiac output} \times 100 \ (\text{mL/100g/min})$$

$$\text{Cardiac output} = \frac{\text{Total counts}}{\text{Reference sample counts}} \times 4.1 \ (\text{mL/min})$$

The Effect of Ionic Contrast Medium Infusion on Myocardial Blood Flow after Acute Myocardial Infarction. As part of a study determining the size of acute myocardial infarcts by CT imaging, myocardial blood flow measurements were made in 13 dogs before, during, and 30 minutes after the infusion of 2 mL/kg of iothalamate (400 mg/mL of iodine, 1600 mOsm/L). This ionic contrast agent was infused intravenously over 5 to 6 minutes. In order to minimize autoregulatory effects on coronary flow, 8 mg/kg of chromonar, a potent coronary vasodilator, was given 10 minutes prior to the initial blood flow measurements.[10] For the purposes of

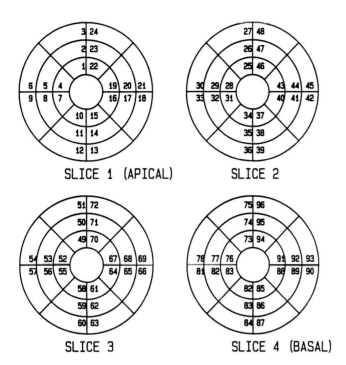

Figure 7-1 The mid–left ventricle was cut into four 1-cm slices perpendicularly to the long axis. Each slice was sectioned into 24 pieces weighing between 0.5 and 1 g. The radioactivity was measured for each piece.

this study, blood flow was averaged for all uninfarcted regions (always more than 60 sections of myocardium).

The Effect of Nonionic Contrast Medium Infusion on Myocardial Blood Flow. In order to determine if there were any differences between the effect of an intravenous infusion of ionic versus nonionic contrast medium on myocardial blood flow, 10 dogs were studied. Six animals were pretreated with chromonar (8 mg/mL) and then received 2 mL/kg of iohexol (370 mg/mL of iodine, approximately 1169 mOsm/L) over 5 to 6 minutes. Another 4 dogs received iohexol in the same manner but were not treated with chromonar. Myocardial blood flow was measured before, during, 20 minutes after, and 30 minutes after contrast infusion.

The Effect of Bolus Injections of Ionic and Nonionic Contrast Media on Myocardial Blood Flow. The effect of small intravenous bolus injections (0.35 mL/kg in 1 to 3 seconds) of ionic and nonionic contrast media on myocardial blood flow was studied in 10 animals. Five received 0.35 mL/kg of meglumine diatrizoate (376 mg/mL of iodine, 1689 mOsm/L), and five received 0.35 mL/kg of iohexol (370 mg/mL of iodine, approximately 1169 mOsm/L). Myocardial blood flow, cardiac output, heart rate, aortic pressure, and left atrial pressure were measured before, during, and after each of three boluses of contrast medium. Ten minutes elapsed between each bolus. This group of animals did not receive the coronary vasodilator chromonar.

Statistical Analysis

Measurements before and after contrast media infusion were analyzed for statistical significance by the paired t test. Repetitive measurement of myocardial blood flow and hemodynamic changes after repeated boluses of contrast medium were tested for significance by one-way analysis of variance. Differences in hemodynamic variables before and after contrast medium were evaluated for significance by the paired t test.

Results

The Effect of Ionic Contrast Medium Infusion on Myocardial Blood Flow after Acute Myocardial Infarction. Thirteen dogs received infusions of 2 mL/kg of iothalamate over a period of 5 to 6 minutes (10 mL/min) in order to visualize the experimentally produced acute myocardial infarction. The animals were pretreated with chromonar. The average myocardial blood flow (mL/100 g/min) for uninfarcted myocardium was 219 mL ± 155 (mean ± SD) before administration of contrast medium and 313 mL ± 213 during the last minute of contrast medium infusion (Figure 7-2). The difference between the two mean flows was significant ($P<0.01$).

The Effect of Nonionic Contrast Medium Infusion on Myocardial Blood Flow. Six dogs were infused with 2 mL/kg of iohexol after pretreatment with chromonar. Myocardial infarction was not produced in these dogs. The myocardial blood flow before infusion was 604 mL ± 305 and was 618 mL ± 256 during the last minute of infusion (Figure 7-3). The same infusion in four dogs not pretreated with chromonar was associated with myocardial blood flows of 126 mL ± 14 before infusion and 156 mL ± 18 during the last minute of infusion

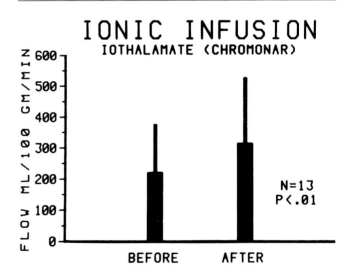

Figure 7-2 Myocardial blood flow increases 44% after the intravenous infusion of 2 mL/kg of iothalamate in spite of pretreatment with chromonar.

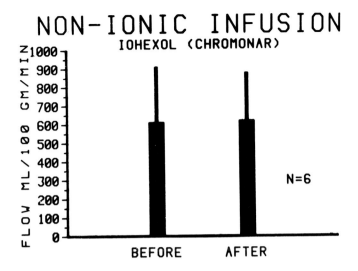

Figure 7-3 Myocardial blood flow does not change after the intravenous infusion of iohexol and pretreatment with chromonar.

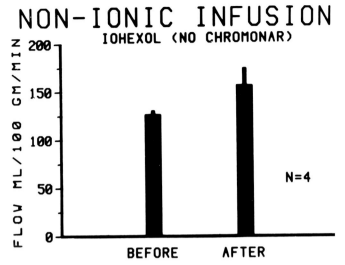

Figure 7-4 Myocardial blood flow does not change significantly after the intravenous infusion of iohexol. The animals were not pretreated with chromonar.

(Figure 7-4). The mean flows after contrast infusion were not significantly different from preinfusion flows in either group.

The Effect of Bolus Injections of Ionic and Nonionic Contrast Media on Myocardial Blood Flow. Myocardial blood flow did not change significantly after three consecutive boluses of 0.35 mL/kg of meglumine diatrizoate in five dogs or iohexol in five dogs (Table 7-2 and Figures 7-5 and 7-6). Furthermore, heart rate, blood pressure, left atrial pressure, and cardiac output did not change significantly in either group of animals (Table 7-2).

The results of this study indicate that large intravenous infusions of ionic contrast medium (2 mL/kg) increase myocardial blood flow in dogs even after pretreatment with a potent coronary vasodilator. This change is presumed to occur because a sufficient concentration of contrast medium reaches the coronary circulation and produces a further increase in coronary flow. However, equal amounts of the nonionic contrast agent iohexol did not significantly change myocardial blood flow with or without pretreatment with chromonar. These findings confirm that nonionic contrast medium has less effect on hemodynamics than ionic contrast medium.

The animals that received the large-volume infusion of ionic contrast medium had undergone coronary artery ligation 2 to 48 hours before the infusion as part

Table 7-2

Meglumine Diatrizoate (N=5)	Myocardial Blood Flow (mL/100 g/min)	Heart Rate (beats/min)		Mean Aortic Pressure (mm Hg)		Mean LA Pressure (mm Hg)	
		Before	After	Before	After	Before	After
Control	142 ± 69	156 ± 20	151 ± 24	138 ± 9	135 ± 10	12 ± 3	13 ± 3
Bolus 1	125 ± 32	147 ± 24	150 ± 25	131 ± 9	135 ± 10	14 ± 3	12 ± 4
Bolus 2	113 ± 35	150 ± 25	150 ± 25	135 ± 11	135 ± 11	14 ± 4	15 ± 5
Bolus 3	125 ± 38	140 ± 18	136 ± 13	135 ± 18	135 ± 17	14 ± 4	15 ± 3
Iohexol n=5							
Control	106 ± 47	152 ± 9	152 ± 9	150 ± 28	152 ± 23	14 ± 4	13 ± 4
Bolus 1	107 ± 44	152 ± 9	152 ± 9	140 ± 15	155 ± 30	13 ± 4	13 ± 4
Bolus 2	123 ± 47	156 ± 8	158 ± 9	147 ± 26	148 ± 23	14 ± 4	12 ± 4
Bolus 3	94 ± 50	156 ± 11	156 ± 11	143 ± 15	152 ± 27	12 ± 5	11 ± 5

LA, left atrial.

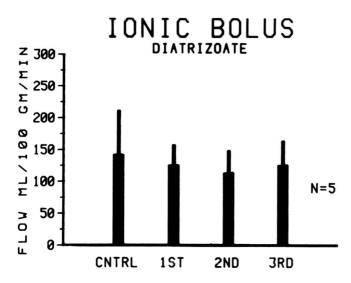

Figure 7-5 Repeated intravenous bolus injections of 0.35 mL/kg of meglumine diatrizoate does not change myocardial blood flow.

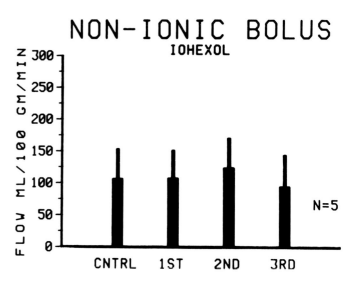

Figure 7-6 Repeated intravenous bolus injections of 0.35 mL/kg of iohexol does not change myocardial blood flow.

of a study of acute myocardial infarction. The infarction could have depressed cardiac function and may explain the lower myocardial blood flows before and during infusion (219 mL and 313 mL) compared to the animals that received a nonionic contrast infusion (604 mL and 618 mL) and did not have coronary ligation (Figures 7-2 and 7-3). The 44% increase in myocardial blood flow in the infarct animals receiving the ionic contrast infusion could be due to an increase in cardiac output produced by an augmentation of circulatory volume due to the contrast medium, rather than a direct vasodilatory effect on the coronary arteries.

The development of ultrafast CT scanners has eliminated the need for large infusions of contrast medium to detect changes in myocardial density. Satisfactory myocardial time-density curves can be produced in humans with intravenous injections of 0.35 mL/kg of contrast medium. This dose of either ionic or nonionic contrast medium did not perturb myocardial blood flow or hemodynamics, even after three repetitions.

In summary, this study in animals indicates that large doses of intravenous ionic contrast medium (2 mL/kg) may increase myocardial blood flow, but smaller doses (0.35 mL/kg) of nonionic or ionic media do not increase myocardial blood flow, even when given repeatedly. Therefore, measurement of regional myocardial blood flow or ventricular function by ultrafast CT should be unaffected by the doses of either type of contrast agent used.

Indeed, the hemodynamic response to contrast media is complex. Several mechanisms play a role, and often in opposite directions and at different times. The predominant changes are due to an initial transient depression in cardiac contractility, followed by a reflex reaction to the initial drop in blood pressure that leads to an increase in heart rate and blood pressure. This is followed in a minute or so by the effects of an increase in circulating blood volume. Many of these hemodynamic changes are minimized or prevented by the use of nonionic contrast media.

HEMOSTATIC EFFECTS

Recent reports have raised concern that nonionic contrast agents may be less anticoagulant than conventional ionic media.[14] Consequently, a symposium was convened to discuss current knowledge. The conclusion of the symposium was that nonionic contrast media are anticoagulant, though less so than ionic agents.[15] Rouleaux formation and red blood cell aggregation are more frequent at the blood-nonionic contrast interface, but the red blood cells are easily dispersed and their aggregation does not seem to be of any clinical consequence. Clot formation in catheters during coronary angiography appears to be related more to technique than to the use of nonionic agents.

RENAL FAILURE

Contrast media are one of the leading causes of hospital-acquired renal insufficiency.[16] Two thirds of cases of contrast-induced nephropathy follow intravenous pyelography, nearly one third occur after angiography, and about 5% after x-ray CT.[17]

Contrast-induced nephropathy can be an asymptomatic, transient laboratory abnormality or result in severe oliguria. The need for dialysis, however, is rare. Renal insufficiency develops in 24 to 48 hours, and if oliguria occurs, it usually lasts 2 to 5 days. During the oliguric phase, fluid and electrolyte imbalance may occur, as well as acidosis and azotemia. Spontaneous diuresis usually occurs within 2 to 8 days, and renal function returns to baseline in 10 to 14 days.[17]

The major pathologic finding in contrast medium–induced nephropathy involves proximal tubule epithelial cells. Vacuolization and patchy tubule cell necrosis are observed.[17] The findings are consistent with those observed in other osmotic nephroses. Contrast medium interaction with renal membranes and precipitation of renal protein in the intratubular lumen, causing obstruction, have also been suggested as mechanisms of the acute renal failure. The true incidence of acute renal failure secondary to contrast medium is difficult to determine because few prospective studies have been done. In many cases the renal insufficiency is clinically silent and is therefore not detected unless renal function is routinely evaluated for several days after exposure to a contrast medium. With ionic agents, the incidence in patients with previously normal renal function is said to be less than 0.15% to 1.5%.[17] This number probably represents the incidence of clinically recognized renal failure after exposure to a contrast agent. A recent prospective study of a nonionic contrast medium found laboratory evidence of renal insufficiency in 8.3% of patients.[18] None of these patients required renal dialysis or were even particularly ill as a consequence.

A number of risk factors for developing acute renal failure from contrast agents have been suggested (Table 7-3).[17] It seems clear that pre-existent renal insufficiency greatly increases the risk for developing acute renal failure after exposure to a contrast medium. With ionic agents, serum creatinine levels greater than 1.6 mg/dL were associated with a 6% incidence of renal insufficiency, and nonionic agents are associated with a sharp rise in the incidence of "laboratory" renal dysfunction when serum creatinine levels exceed 1.2 mg/dL.[17–18] However, the incidence of clinically significant renal failure is still small.

Diabetes is thought to be a particularly important risk factor for contrast-induced nephropathy. In one study 76% of diabetic patients with mild renal insufficiency (serum creatinine, 2 mg/dL) showed deterioration of renal function after receiving contrast medium.[19] However, Davidson et al., using a nonionic agent, could not demonstrate that diabetes was a risk factor.[18]

Probable risk factors are multiple myeloma, dehydration, and a previous history of contrast medium–induced nephropathy. The risk with multiple myeloma is less than 4%; however, the large number of reported cases make a cause-and-effect association likely.[17] Dehydration leads to increased levels of vasopressin, and contrast media induce further release, which possibly predisposes the patient to contrast nephropathy. In any case, adequate hydration prior to contrast medium exposure is important preventive therapy.

Other possible risk factors are large contrast medium dose, age, proteinuria, vascular disease, hyperuricemia, congestive heart failure, and hepatic disease.[17] A recent prospective study evaluating a nonionic agent could not demonstrate independent risk for age, sex, diabetes, heart failure, hyperuricemia, peripheral vascular disease, cerebrovascular disease, hypertension, coronary artery disease, reduced ejection fraction, contrast medium dose, or state of hydration.[18] It may be that many of these possible risk factors are really only important as they affect the state of renal function.

ANAPHYLACTOID REACTIONS

An estimated 1% to 2% of unselected patients have an anaphylactoid response to contrast media. Fortunately, only 1 in 1000 cases is severe, and fatal consequences are reported to be between 1 in 12,000 to 1 in 75,000.

The response has been termed "anaphylactoid" because the clinical manifestations are similar to immunoglobulin E (IgE)–mediated anaphylaxis but without IgE antibodies being involved.[20] Current thinking is that contrast media somehow nonimmunologically stimulate the contact system in endothelial and mast cells so that histamine and bradykinin are released.[20–21] Some investigators believe the hyperosmolality of the ionic contrast agents plays a role. This observation is supported by anecdotal reports of fewer anaphylactoid responses to nonionic agents. Unfortunately, there are no large, controlled, prospective randomized studies comparing ionic and nonionic agents in this regard. A relatively small preliminary report suggests the incidences are not different, being 1% with an ionic medium and 0.8% wth a nonionic medium.[22]

Table 7-3 Risk Factors in the Development of Contrast Medium–Induced Acute Renal Failure

Definite Risk Factors
1. Pre-existing renal insufficiency
2. Diabetes mellitus

Probable Risk Factors
1. Multiple myeloma
2. Dehydration
3. Prior contrast medium–induced acute renal failure

Possible Risk Factors
1. Large contrast medium load
2. Advanced age
3. Proteinuria
4. Vascular disease
5. Hyperuricemia
6. Congestive heart failure
7. Hepatic disease

The anaphylactoid symptoms are primarily related to smooth muscle contraction, vasodilation, and the increased capillary permeability mediated by histamine. Symptoms and signs can be categorized as local, wheal-flare, and systemic; they may be further divided into upper respiratory, lower respiratory, and circulatory.[23] They include urticaria, angioedema, laryngeal edema, bronchospasm, urge to defecate or urinate, rhinitis, conjunctivitis, hypotension, shock, convulsions, cardiac arrest, and death. Other symptoms that are probably not due to histamine and bradykinin release are nausea, vomiting, a metallic taste, cardiac dysrhythmias, angina, pulmonary edema, and fever and chills.[21,23] Most symptoms occur within 5 minutes of administration, 50% within the 1st minute and 66% within 3 minutes.[23–26] Most deaths occur within 1 hour.[26]

A prior history of a contrast medium anaphylactoid event is highly predictive of another such reaction on repeat exposure. The chance of a repeat reaction is between 17% and 60%.[20] The chance for a repeat reaction can be greatly reduced (to 8%) by pretreatment with corticosteroids and diphenhydramine or these drugs plus ephedrine (to 4.1%).[20] To be effective, the corticosteroids must be administered 12 hours before exposure to contrast media.[20–21] The diphenhydramine and ephedrine can be administered 1 hour before the procedure.[20] H_2 blockers do not provide any added protection.[25] One report recommends the routine use of prophylactic corticosteroids with ionic agents as an alternative to expensive nonionic agents.[21] In this study, when compared to placebo, all reactions were reduced by 31% and severe reactions by 62%. The incidence for all reactions requiring therapy was 1.2%, compared to 0.9% for a nonionic medium given without steroid prophylaxis.[26]

A history of a previous allergy, the use of sodium or meglumine salts, and a high contrast medium dose are associated with a higher incidence of reactions.[21] Pretreating with a scratch test or intradermal or intracutaneous injection is of no predictive value.

COST

A major issue has arisen with the introduction of nonionic contrast agents—their cost. The average cost of a nonionic agent is 10 to 20 times that of ionic agents. It is estimated that the universal use of nonionic contrast agents would increase the annual cost of contrast media in the United States from $92 million to $1.15 billion.[27] At the present time nonionic agents should be reserved for special circumstances, such as direct arterial injection to reduce pain, history of previous anaphylactoid reaction, and pre-existent renal insufficiency.

SUMMARY

Contrast media are widely employed in the performance of cardiovascular radiography. The recent development of nonionic agents has improved patient comfort and safety, but unfortunately at a great increase in cost. Therefore, ionic contrast media are still used extensively. Knowledge of the hemodynamic effects, impact on renal function, and potential for anaphylactoid reaction of these agents is mandatory when using them for the diagnosis of cardiovascular disease.

REFERENCES

1. Bordley J III, Harvey M. *Two Centuries of American Medicine: 1776–1976*. Philadelphia: WB Saunders Co, 1976:314–326.

2. Almen TJ. Contrast agent design: Some aspects of low osmolality. *J Theor Biol*. 1969;24:216–226.

3. Gainger RG. Osmolality of intravascular radiological contrast media. *Br J Radiol*. 1980;53:739–746.

4. Swanson D. Conventional or low-osmolality: Picking the right contrast media. *Diagn Imaging*. 1988;10:191–210.

5. Swanson DP, Thrall JH, Petty PC. Evaluation of intravascular low-osmolality contrast agents. *Clin Pharm*. 1986;5:877–891.

6. Higgins CB. Overview of radiovascular effects of contrast media: Comparison of ionic and nonionic media. *Invest Radiol*. 1984;19:5187–5190.

7. Higgins CB. Overview and methods used for the study of the cardiovascular actions of contrast materials. *Invest Radiol*. 1980;15(suppl):S188–S193.

8. Gertz EW, Wisneski JA, Chiu D, et al. Clinical superiority of a new nonionic contrast agent (Iopamidol) for cardiac angiography. *J Am Coll Cardiol*. 1985;5:250–258.

9. Piao ZE, Murdock DK, Hwang MH, et al. Contrast media-induced ventricular fibrillation: A comparison of Hypaque-76, Hexabrix and Omnipaque. *Invest Radiol*. 1988;23:466–470.

10. Vik-Mo H, Rosland GH, Folling M, et al. Hemodynamic and electrocardiographic consequences of high and low osmolality contrast agents on left ventriculography. *Cathet Cardiovasc Diagn*. 1988;14:143–149.

11. Brundage BH, Cheitlin M. Left ventricular angiography as a function test. *Chest*. 1973;64:70–74.

12. Brundage B, Cheitlin M. Ventricular function curves from the cardiac response to angiographic contrast: A sensitive detector of ventricular dysfunction in coronary artery disease. *Am Heart J*. 1974;88:281–288.

13. Higgins CB, Gerber KH, Mattrey RF, et al. Evaluation of the hemodynamic effects of intravenous administration of ionic and nonionic contrast materials. *Radiology*. 1982;142:681–686.

14. Grollman JH Jr, Lui CK, Astone RA, et al. Thromboembolic complications in coronary angiography associated with the use of nonionic contrast media. *Cathet Cardiovasc Diagn*. 1988;14:159–164.

15. Brettman M. Clinical summary and conclusions: ionic versus nonionic contrast agents and their effects on blood components. *Invest Radiol*. 1988;23(suppl):S378–S380.

16. Hou SH, Bushinsky DA, Wish JB, et al. Hospital acquired renal insufficiency: A prospective study. *Am J Med*. 1983;74:243–248.

17. Coggins CH, Fang LST. Acute renal failure associated with antibiotics, anesthetic agents, and radiographic contrast agents. In: Brenner BM, Lazarus JM, eds. *Acute Renal Failure*. Philadelphia: WB Saunders Co; 1983:283–320.

18. Davidson CJ, Heathy M, Morris KG, et al. Cardiovascular and renal toxicity of a nonionic radiographic contrast agent after cardiac catheterization. *Ann Intern Med*. 1989;110:119–124.

19. Harkovan S, Kjillstrand CM. Exacerbation of diabetic renal failure following intravenous pyelography. *Am J Med*. 1977;63:939–946.

20. Greenberger PA, Patterson R, Radin RC. Two pretreatment regimens for high-risk patients receiving radiographic contrast media. *J Allergy Clin Immunol*. 1984;74:540–543.

21. Lasser EC, Berry CC, Talner LB, et al. Pretreatment with corticosteroids to alleviate reactions to intravenous contrast material. *N Engl J Med*. 1987;317:845–849.

22. Davies P, Richardson RE. A randomized long-term study of acute and delayed reactions to nonionic and an ionic medium. *Invest Radiol*. 1988;23(suppl):S209.

23. Hildreath EA. Anaphylactoid reactions to iodinated contrast media. *Hosp Pract*. 1987;22:77–95.

24. Shehadi WH. Death following intravascular administration of contrast media. *Acta Radiol Diagn*. 1985;26:457–461.

25. Greenberger PA, Patterson R, Tapio CM. Prophylaxis against repeated radiocontrast media reactions in 857 cases. Adverse experience with cimetidine and safety of B-adrenergic antagonists. *Arch Int Med*. 1985;145:2197–2200.

26. Shrott KM, Behreuds B, Clauss W, et al. Iohexol in der Ausscheidungsurographie: Ergebnisse des drug-monitoring. *Fortschr Med*. 1986;140:153–156.

27. Latchow RE. A case for universal use of low osmolality contrast media. *Admin Radiol*. 1986;24–28.

Chapter 8

Measurement of Left Ventricular Function from Contrast Angiograms in Patients with Coronary Artery Disease

Florence H. Sheehan, MD

INTRODUCTION

More than 50 years ago, Tennant and Wiggers made the observation that coronary artery occlusion causes left ventricular dysfunction.[1] Since then, the relationship between levels of reduction in coronary blood flow and corresponding decreases in left ventricular function has been demonstrated in experimental animals. More recently, the development of quantitative methods of measurement has enabled an increasingly detailed characterization of global and regional left ventricular function. This chapter will discuss these methods and their applications in ischemic heart disease.

OBJECTIVES

Quantitative analysis of left ventricular angiograms of patients with coronary artery disease (CAD) is useful for the following purposes: (1) to precisely determine the magnitude of dysfunction and the level of compensation, (2) to measure the response to a therapeutic intervention, and (3) to assess the patient's risk for cardiac events and prognosis for survival. A possible fourth application is diagnosis of CAD. However, experimental studies have shown that left ventricular function does not sensitively reflect the severity of coronary artery stenosis since contraction remains normal until coronary blood flow is reduced below a critical threshold. Furthermore, left ventricular dysfunction is not specific for CAD because regional or global abnormalities may also result from cardiomyopathy or valvular disease, nor in CAD patients is it specific for infarction since unstable angina may also depress function.

Left ventricular volume and ejection fraction (EF) are the most frequently measured parameters. However, analysis of regional wall motion provides more detailed information on the functional status of ischemic and nonischemic myocardium. The choice of parameter to be used is largely governed by the purpose for which the assessment is performed. The following section will discuss the methods for measuring the various parameters, including their accuracy and limitations. The applications of these methods to patient evaluation will follow.

METHODS

Calculation of Ventricular Volume

Due to its demonstrated accuracy, volume determination from contrast ventriculography is the gold standard against which other imaging modalities are compared. The method has been validated not only against post-mortem hearts but also in vivo. From the measurement of volume, a number of clinically useful parameters can be derived (Figure 8-1).

The principal considerations for accurate assessment of left ventricular chamber volume are: (1) assumptions

```
UH    -111111              SVCAD , DEMO                                          Male   Cath:  27-OCT-76
DEMO$: SVDEMO.DAT                                                                Report date:   17-JUN-87

Parameter           Value           Value/BSA
  EDV (cc)          145.            72.
  ESV (cc)          54.             27.
  SV (cc)           91.             45.
  C.O. (L/min)      7.02            3.48
  MASS (gm)         0.              0.
  RRI (sec)         0.78
  HR (B/min)        77.4
  EF                0.63
  BSA (m2)          2.02            173 cm  88.0 Kg
Study ID:  (      , 0)  (      , 0)  (      , 0)  (      , 0)  (ALK  , 1)
RAO  CINEfor  Frames   193-147   CFAP = 0.542   CFLT = 0.000

         Frame   #:   EDV: 193        ESV: 172
```

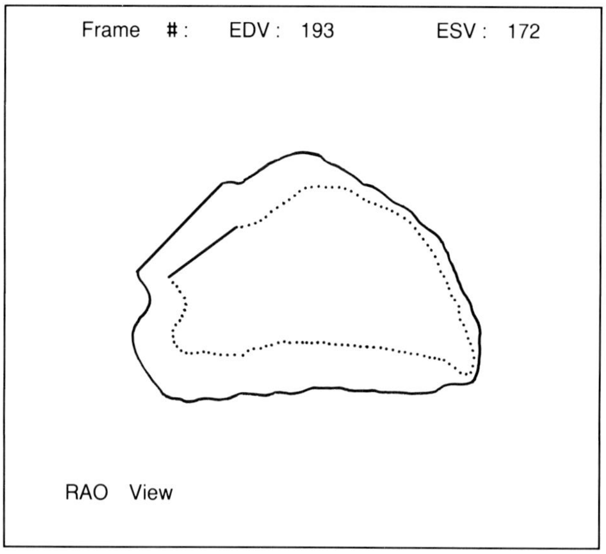

RAO View

Figure 8-1 An example of the report of cardiac volumes generated from the University of Washington Cardiac Catheterization Laboratory. EDV, end-diastolic volume (mL); ESV, end-systolic volume (mL); SV, stroke volume (EDV − ESV [mL]); CO, cardiac output (SV × HR × 10^{-3} [L/min]); RRI, R to R interval on the electrocardiogram (sec); HR, heart rate (bpm); EF, ejection fraction (SV/EDV); BSA, body surface area, calculated from the height and weight [DuBoi D[16a]] (Note: Indices are calculated as volume/BSA); ALK, initial of person who traced the contours; CFAP, correction factor in the anteroposterior (AP) (or right anterior oblique [RAO]) view; CFLT, correction factor in the lateral (or left anterior oblique [LAO]) view if biplane.

---

regarding the geometry of the left ventricle, (2) the variation in contour associated with different projection angles, (3) correction for the volume of the papillary muscles and trabeculae carneae, (4) variability in tracing the endocardial contours, and (5) correction for magnification due to x-ray beam divergence, pincushion distortion, and additional magnification associated with projecting the image for tracing.

Several approaches have been developed for measurement of left ventricular volume. Their accuracy is determined by filming cadaver hearts filled with barium paste and comparing the volume estimated from the projected image with the true volume. Volumes calculated using the area-length method agreed closely with the known volume ($r = 0.995$, standard error of the estimate = 8.2 mL).[2]

In this method, the short axis in each projection is calculated from the measured area and long axis, using the equation for the area of an ellipse:

$$D = \frac{4A}{\pi L}$$

where D is the short axis, A is the area of the projected chamber, and L is the length of the chamber. Volume (V) is then computed as

$$V = \frac{\pi}{6} \times L \times D_a \times D_b$$

where L is the longer measured chamber length of two orthogonal projections and D_a and D_b are the calculated short axes. Although other approaches have been proposed, they are less accurate. The exception is Simpson's rule, which also yields accurate volumes but requires more elaborate computation.

The area-length method was originally developed for anteroposterior (AP) and lateral films. Regression equations have also been derived to calculate volume from biplane ventriculograms in the 30° right anterior oblique (RAO) and 60° left anterior oblique (LAO) projections[3] and other views.[4] Because the calculated length of the short axis is similar in orthogonal views, accurate volume determinations can also be made from single-plane AP or 30° RAO images.[3,5–7] Since the ventriculogram in the 60° projection is often foreshortened, some investigators advocate 15° to 30° cranial angulation in the LAO view, which provides a full-length view of the left ventricle.[8–9]

Calculated left ventricular volumes consistently overestimate true volumes. This has been attributed to the geometric figure used to model the left ventricle or to the volume of the papillary muscles, trabeculae carneae, and chordae tendineae. Therefore, to adjust for this, regression equations have been developed for the standard views (Table 8-1). These equations depend on the population studied as well as on the gold standard used for comparison. Of the equations for the single-plane AP projection, that of Sandler and Dodge[5] was derived from the greatest number of observations made at all phases of the cardiac cycle. Neither of the equations for the 30° RAO projection was based on a large population of subjects; they were validated against different standards but differ by only 5.2 mL/100 mL volume.

The accuracy of volume determination has also been demonstrated in vivo. Angiographic stroke volumes correlate well with stroke volumes computed using Fick or indicator-dilution techniques in patients without valvular regurgitation.[10–12] Correlation coefficients are about 10% lower for in vivo validation studies, but this is due at least in part to the variability of the Fick and indicator-dilution techniques. One question sometimes raised is the accuracy of angiographic volume determinations in patients with abnormally shaped ventricles, such as those with aneurysms. However, the applicability of the area-length method to nonellipsoid chambers is amply illustrated by its accuracy in calculating the volume of the right ventricle[13] and of the left and right atria.[14–15]

Variability in Volume Calculation

Regardless of the geometric reference figure to which the left ventricle is theoretically fit, in practice accuracy in volume determination depends on such factors as the quality of the image and the confidence with which the endocardial contour can be delineated and traced. For example, variability is relatively higher at end-systole when the thickening of trabeculae makes the endocardial border more irregular.

Interobserver variability ranges from 6.6 mL to 20 mL for end-diastolic volume, from 5.7 mL to 10 mL for end-systolic volume, and from 0.04% to 0.05% for ejection fraction.[16–19] Beat-to-beat variability, assessed by a single observer, is similarly low[16] despite the depressant effect of contrast medium. In fact, measurement of cardiac dimensions using epicardial markers in human subjects showed that the end-systolic volume and EF do not change significantly from their preinjection values until seven beats after injection.[20] Even then, the changes are small (Table 8-2). On the other hand, volume and function may change considerably between serial ventriculograms in clinically stable patients, in part reflecting differences in hemodynamic state.[16] For example, both volume and EF decrease in a linear inverse relationship with increases in heart

---

Table 8-1 Regression Equations for Left Ventricular Volume Determination*

Projection	Regression		Number of Observations	Standard Error of Estimate (mL)	Gold Standard	Reference
AP/lateral	V' = 0.928	V − 3.8	54 (9)†	8.2	casts	Dodge et al.[2]
AP	V' = 0.951	V − 3.0	1204 (55)	15.0	biplane	Sandler and Dodge[5]
AP	V' = 1.00	V + 9.6	15‡	24	biplane	Kennedy et al.[6]
AP	V' = 0.788	V + 8.4	16‡	28.8	biplane	Kasser and Kennedy[7]
RAO 30°	V' = 0.81	V + 1.9	15	24	biplane	Kennedy et al.[6]
RAO 30°	V' = 0.938	V − 5.7	11	5.5	casts	Wynne et al.[3]
RAO 30°/LAO 60°	V' = 0.989	V − 8.1	11	8	casts	Wynne et al.[3]

*Key: V, volume computed from the cine image using the area-length method; V', volume after correction by regression equation; RAO, right anterior oblique; LAO, left anterior oblique; AP, anteroposterior.
†Parentheses indicate the number of subjects if multiple images/subject were studied.
‡These studies had 10 patients in common.

Table 8-2 Reproducibility of Volume and Ejection Fraction Determinations*

Source of Variability	EDV (mL)	ESV (mL)	EF (%)	Statistic	Reference
Intraobserver	4.47	4.16	0.03	SEE	Dodge et al.[18]
	4.8	4.7	0.02	SEE	Chaitman et al.[19]
Interobserver	8.3	7.3	0.04	SEE	Dodge et al.[18]
	10.8	5.7	0.04	SEE	Chaitman et al.[19]
	10	6	0.04	SEE	Rogers et al.[17]
	29	10	0.05	MD	Cohn et al.[16]
Beat-to-beat	6	4	0.02	MD	Cohn et al.[16]
Contrast effect† beat #3	↑2 (P=NS)	↑6 (P=NS)	NS	paired t	Hammermeister and Warbasse[122]
	↑6.9	↑5.8	—	paired t	Carleton[123]
Contrast effect beat #7	NS	↓7.3	0.04	paired t	Vine et al.[20]
Projection, AP/lat vs. RAO/LAO	22	12	0.05	SEE	Rogers et al.[17]
Study-to-study, 30 min	19	17	0.04	MD	Cohn et al.[16]
Study-to-study, 90 min	20	14	0.04	MD	Cohn et al.[16]
Study-to-study, 24 hr	20	32	0.13	MD	Cohn et al.[16]

*Key: SEE, standard error of the estimate; MD, mean difference; EDV, end-diastolic volume; ESV, end-systolic volume; EF, ejection fraction; AP/lat, anteroposterior/lateral; RAO/LAO, right anterior oblique/left anterior oblique; NS, not significant.
†Contrast effect noted with successive cardiac cycles after injection.

rate.[21] To minimize study-to-study variability, it is suggested that serial studies be traced by the same observer.

A correction factor to compensate for magnification and pincushion distortion can be calculated by filming a grid of known dimension at the level of the patient's left ventricle.[7] Such a grid may consist, for example, of fine wires embedded in a 10 × 10-cm Plexiglas plate at 1-cm intervals. The location of the patient's heart may be approximated as midchest level. Care must be taken to record the heights of the image intensifier, the table, and the x-ray tube (if movable) at the time of ventriculography so that the equipment can be reset to this configuration after removing the patient in order to film the grid. When the ventriculogram is projected for tracing, the number of grid squares that span the ventricle is counted, and their projected width is measured.[7] The ratio of the number of grid squares to the measured span of grid squares is the correction factor. Because inaccuracies are cubed, the use of a standard correction factor instead of an individually calculated one may cause an error of up to 16% in volume calculation. The higher the magnification mode, the smaller the correction factor. This factor is entered into the equation for volume calculation.[2] When biplane ventriculography is performed, correction factors must be calculated for each view.

Regional Wall Motion Analysis

By its nature, CAD causes regional abnormalities in left ventricular function. However, in addition to the hypokinesia of the ischemic region, there may be hypokinesia in the adjacent normally perfused regions, and distant regions may display either hypo- or hyperkinesia.[22-23] Parameters of global function reflect the net impact of these regional abnormalities.[24] Insight into regional contributions to global function is provided by analysis of wall motion.

Accuracy of Wall Motion Measurement

Unlike the methods for volume calculation, whose accuracy could be validated from post-mortem heart studies, there is no gold standard for wall motion measurement. Attempts have been made to track the motion of markers or landmarks located on the epicardium, midwall, or endocardium.[25-28] However, due to the endocardial infolding that occurs during systole, points on the end-diastolic contrast border do not correspond to points on the end-systolic border.[29] As a result, the various methods model, rather than measure, wall motion.

The lack of a gold standard has had three important sequelae. First, it has fostered the invention of almost as many approaches to quantifying regional function as there are investigators interested in doing it. Second, it has made validation dependent on empiric criteria such as the method's sensitivity and specificity in distinguishing the motion of normal subjects from that of patients with known CAD.[30] Finally, since wall motion cannot be directly measured, it cannot be distinguished angiographically from the translational motion of the heart within the chest. Methods have been proposed, again empiric in nature, to correct for translation by realigning the end-systolic contour relative to the end-diastolic contour. However, the value of these proposed methods remains questionable because they increase variability, introduce artifact, and have no theoretical validity.[31-33] At present there is no accepted method for distinguishing wall motion from the swinging of the heart, although most investigators recognize the need for one.

Comparison of Methods

Since accuracy in measuring wall motion cannot be evaluated when comparing methods for analyzing regional function, selection must be based on other criteria. The principal theoretical consideration is the assumption implied by each method concerning directionality of wall motion. Performance criteria evaluate the method's ability to detect, measure, and express regional abnormality; these encompass the following: (1) reliability or reproducibility, (2) homogeneity of function in normal subjects, (3) ability to focus on the region of interest, and (4) sensitivity and specificity in identifying hypo- and hyperkinesis.

Assumptions Concerning Direction of Motion. Although the literature abounds with methods for performing wall motion analysis, most are based on either a rectangular or a radial coordinate system (Figure 8-2).[34–35] The motion of discrete points around the contour is measured along the hemiaxes or radii and normalized for patient-to-patient differences in heart size by dividing by the end-diastolic dimension. The resulting shortening fraction is a linear analog of the EF. These coordinate systems can also be used to define regions; the change in the area of each region can similarly be expressed as an area ejection fraction.

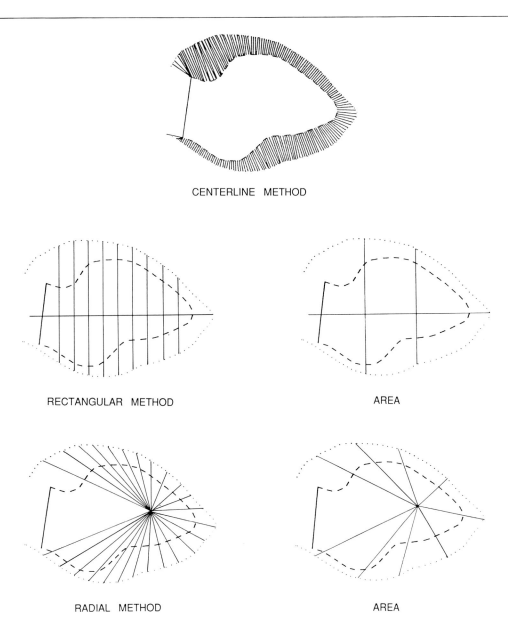

Figure 8-2 The most commonly used methods of wall motion analysis involve construction of a radial or rectangular coordinate system. Motion is calculated either as shortening along the radii or hemiaxes or as the area reduction in the regions they create.

Two coordinate system models have been developed, based on marker tracking. Ingels and co-workers analyzed the motion of metallic markers surgically implanted in the midwall of patients undergoing coronary bypass surgery. They found that a radial coordinate system with origin on the end-systolic long axis, 69% of the distance from the anterior aspect of the aortic valve to the apex, best agreed with marker motion.[35] Slager and co-workers instead analyzed the ventriculograms of normal subjects. They used an automated border recognition system to trace the left ventricular contours frame by frame throughout one cardiac cycle. Irregularities on these endocardial contours that could be identified on successive frames were considered to be landmarks. Their displacements were used to define 20 pathways in a long axis apex coordinate system. In this landmark model, motion is expressed in terms of its contribution at each of the 20 pathways to the volume EF.[28]

The problems with coordinate system methods are several. First, with the exception of the landmark model, they assume that all motion proceeds inward toward a central long axis or origin, which is convenient but of questionable validity.[25,36–37] Second, the coordinate system is usually based on a long axis drawn from the aortic valve plane to the apex. However, since the apex is not an anatomic landmark and is the least reproducibly visualized part of the ventricular contour,[30–31] reliance on its identification may introduce error. Some investigators have constructed a radial coordinate system around a so-called center of mass.[38] Actually, only the center of the cavity is determined from the endocardial contour; to calculate the center of the myocardial mass, the epicardial contour must also be traced. In any case, others have found this approach to be relatively insensitive in detecting wall motion abnormalities.[39–41] Finally, the ideal wall motion program should be applicable to all views of the left ventricle as imaged, using noninvasive techniques as well as angiography, and perhaps also to the right ventricle. However, the coordinate system methods were developed for the AP or RAO projection of the left ventricle. Rectangular coordinates are inappropriate for the rounded contours of the LAO projection of the left ventricle, which is frequently foreshortened, or for the short-axis view seen on two-dimensional echocardiography. A radial approach could be utilized for the LAO, if it does not require identification of the apex. Neither radial nor rectangular coordinates are suitable for the indented and nonellipsoidal contours of the right ventricle.

Some of the constraints inherent in measuring motion along coordinate system grids may be avoided by evenly dividing the end-diastolic and end-systolic contours into the same number of points and measuring motion between corresponding points on the two contours. However, this approach incorporates a new assumption—that the contour shortens homogeneously around the left ventricle and along its length. This assumption is not valid in patients with CAD because ischemia reduces local fiber shortening; the resulting mismatch of points in the end-diastolic and end-systolic contours may give the artifactual appearance of motion to akinetic regions.

The centerline method measures wall motion without reference to a coordinate system or axis (Figure 8-3).[42] Instead, a "centerline" is constructed midway between the end-diastolic and end-systolic contours. Motion is measured along 100 chords drawn perpendicularly to the centerline, that is, in the direction of locally defined vectors. To adjust for patient-to-patient differences in heart size, the measured motion of each chord is divided by the length of the end-diastolic perimeter length, yielding a shortening fraction.

Motion in Normal Subjects. To detect abnormality, it is necessary to define the limits of normal. Therefore, one of the empiric criteria frequently used in evaluating methods of wall motion analysis is the degree of variability observed in normal subjects. The rationale lies not only in the expectation that they will be homogenous but also in the fact that a narrow normal range improves sensitivity for detecting abnormal motion (Figure 8-4).

The normal limits for wall motion vary by location around the left ventricle. This regional variation in normal is a consistent finding independent of the method used to measure motion (Figure 8-3).[27,43–45] As a result, a threshold that identifies hypokinesis in one region may lie within the normal range in another. For example, akinesis represents a highly significant decrease in function along the anterior and inferior walls but lies within two standard deviations of the normal mean at the apex (Figure 8-3C). Therefore, in measuring abnormality in wall motion of a single patient or group, it is necessary to refer to the normal values. This may be unwieldy, however, if motion is measured at a large number of chords. One solution is to convert the shortening fraction calculated at each chord or radius into z-scores by expressing them in units of standard deviations (SD) from the normal mean.[43] This approach was adopted in the centerline method (Figure 8-3D) because it allows comparison of motion measured in different regions of the same heart as well as in different hearts and enables calculation of motion in a region of interest. In addition, the SD units indicate both the severity and the significance of the wall motion abnormality. The sign reflects the direction of abnormality: Positive values signify hyperkinesis and negative values hypokinesis.

Variability. Wall motion measurements are subject to the beat-to-beat, interobserver, intraobserver, and

Left Ventricular Function and Contrast Angiograms   107

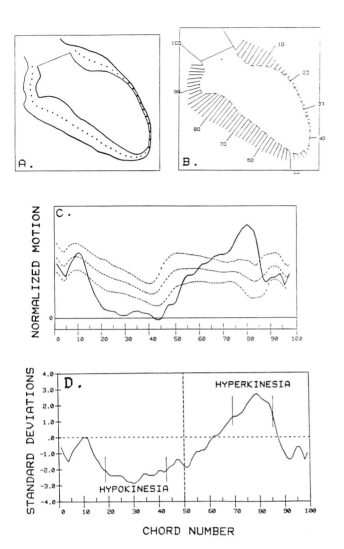

Figure 8-3 Wall motion analysis by the centerline method. A, A centerline is constructed midway between the end-diastolic and end-systolic endocardial contours. B, One hundred chords are drawn perpendicular to and evenly spaced along the centerline. C, Motion at each chord is normalized by the end-diastolic perimeter length to yield a shortening fraction. The patient's wall motion is presented in comparison to the normal mean ± 1 standard deviation (SD). D, The vertical bars delimit the most hypokinetic part of the infarct artery territory and the most hyperkinetic part of the opposite wall. The severity of these abnormalities in chord motion is expressed in units of SDs from the normal mean. The circumferential extent of hypo- or hyperkinesis can be easily calculated as the number of chords with motion below or above some threshold (e.g., ≤ -2 SD or $\geq +1$ SD) and expressed as a percentage of the 100 chords. *Source:* Reprinted with permission from *Circulation* (1985;71:1121–1128), Copyright © 1985, American Heart Association.

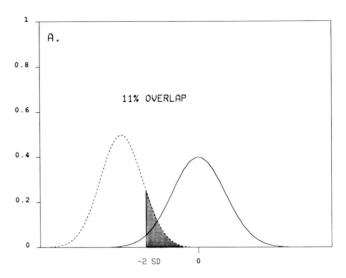

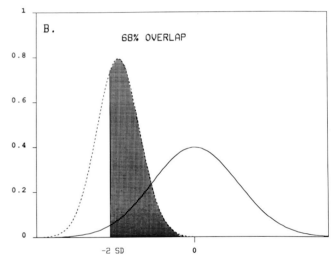

Figure 8-4 Effect of variability on diagnostic accuracy. A, Probability distribution functions for a normal (——) and disease (— —) population. When the variability of the measurement is low and the standard deviation is narrow, there is virtually no overlap between the two groups. B, When variability is greater, as indicated by a higher standard deviation, the overlap between the same two populations increases from 11% in A to 68% here.

study-to-study variability that occur in tracing left ventricular contours for volume calculation. Studies have shown that reproducibility is least affected by beat selection or repetitions by the same observer and is poorest between serial angiograms.[19,31,46–47] In addition, there is point-to-point variability in the motion measured in discrete chords. This will "average out" when calculating motion over a region,[31] which may explain the greater sensitivity of area methods over radial or rectangular chord methods.[43]

These sources of variability affect regional more than global function. For example, normal variability in wall motion measured by the centerline method averages 31%, ranging from 18% at the base to 57% at the apex, compared with 8% for EF in the same normal population.[31] One reason for this is that differences in the translational motion of the heart within the chest increase variability for wall motion but not EF. Although in theory realignment of the contours should adjust for translational motion, in practice the application of empirical methods increased rather than decreased variability.[31] The greater variability in wall motion tends to make it less sensitive than the EF in detecting abnormalities. However, this is more than balanced by the increased sensitivity derived from being able to focus on the region of the left ventricle that is of interest.

Focus on the Region of Interest. In evaluating patients with CAD, it is often desirable to measure wall motion in a particular region of the left ventricle. Since more than one coronary artery may be stenosed, a regional analysis allows one to focus on the desired arterial territory. For example, one might wish to measure response to therapy at the site of acute infarction in a patient with multivessel disease. Such focus is impossible when measuring the circumferential extent of dysfunction using parameters such as the akinetic segment length or the proportion of the left ventricular contour having hypokinesis. These length parameters are primarily suitable for patients with single-vessel disease.

To perform a regional analysis, many investigators partition the left ventricle into four or more regions, using a coordinate system (Figure 8-2) and calculating area EFs. This approach has the advantage of reducing point-to-point variability, as noted above. The disadvantage of measuring area EF is that abnormality will be underestimated whenever the dysfunctional region does not fit into any of the predefined areas, a frequent occurrence. Just as stenoses vary in location along the length of a coronary artery, so wall motion abnormalities resulting from coronary stenoses vary in location around the left ventricular contour. In a study of patients with acute anterior infarction, the location of the most hypokinetic 20% of the left ventricle varied so much that no fixed region could have captured more than 50% of the cases.[42] In response to this problem, the centerline method (1) applies a "sliding window" to the wall perfused by the artery of interest or arterial territory, (2) selects the most abnormally contracting region within that arterial territory, and (3) calculates the mean motion of chords in that region (Figure 8-3D). Each arterial territory is defined simply as the set of chords whose motion is significantly depressed compared to normal in patients with single-vessel disease.[48] In the validation studies performed on the centerline method, the average motion of chords lying in the most hypokinetic 50% of the arterial territory yielded the closest correlation with the severity of coronary artery stenosis.[49–50] Thus this single parameter provides a measure of the severity of dysfunction precisely in the region of interest.

Sensitivity and Specificity. In the absence of a gold standard, methods of wall motion analysis have been compared using empiric criteria. That is, methods that (1) yield a narrow range of normal values, (2) accurately distinguish patients with CAD from normal subjects or patients with severe coronary artery stenosis from those with milder disease, or (3) agree with visual identification of regional abnormalities have been considered more valid. The results obtained from applying these criteria depend on the populations of patients and normal subjects selected for testing (their size, homogeneity, and differences from one another) and on the statistical approach.[39,51–52] Therefore, it is not surprising that comparative studies sometimes disagree on the relative merits of the various methods (Table 8-3). Nevertheless, certain trends can be perceived. Methods based on a rectangular coordinate system yield poorer results than the area, radial (especially that of Ingels et al.[35]), and centerline methods. The landmark model of Slager et al.[37] has not been compared to other methods.

Technical Considerations for the Performance and Analysis of Contrast Ventriculography

Ventriculography

The goal is to obtain images of adequate contrast to allow accurate delineation of the endocardial contours at end-diastole and end-systole. Experience has shown that certain angiographic techniques are helpful in achieving this goal. The catheter should be positioned to maximize opacification of the ventricular chamber while avoiding ectopy. Use of a pigtail catheter helps to minimize ectopy, the most common reason for failure to obtain an analyzable study, and to avoid damaging the endocardium by force of injection. To rapidly fill the ventricle with contrast medium, a power injection of 35 to 50 mL over 2 to 3 seconds is performed. Smaller volumes are adequate if digital subtraction techniques are used to facilitate identification of the endocardial border by removing noncardiac structures such as ribs and pulmonary vessels.[53–54] Ventriculographic images are most commonly recorded on 35-mm cine film in the 30° RAO projection. However, the AP view or biplane studies in orthogonal projections are also used, in the AP/lateral or 30° RAO/60° LAO views.

Tracing the Left Ventricular Contour

Quantitative analysis of ventricular function should be performed on a normal sinus beat. Due to postextrasystolic potentiation, the function measured from postectopic beats is enhanced and therefore nonrepresentative. Selection of the cardiac cycle to be analyzed is greatly facilitated by recording the electrocardiogram during ventriculography, either directly on the cine film (as a "cine trace") or on a strip chart recorder. If a recorder is used, it should incorporate markings indicating the timing of each angiographic frame, or at least when filming began and ended, so that the beat selected from the electrocardiogram can be readily iden-

Table 8-3 Comparison of Sensitivity and/or Specificity of Methods of Wall Motion Analysis in Detecting Regional Dysfunction*

Methods	Reference
Radial$_{ED\ COM}$ > radial$_{ES\ COM,LA}$	Papietro et al.[38]
Radial$_{LA}$ > radial$_{ES\ COM,ED+ES\ COM}$	Karsch et al.[39]
Radial > area > rectangular	Karsch et al.[39]
Area, radial$_{midwall}$ >> radial$_{ED+ES\ COM}$, rectangular in anterior MI	Lorente et al.[52]
Radial$_{ED+ES\ COM}$, radial$_{midwall}$ >> rectangular, area in infarct MI	Lorente et al.[52]
Radial$_{midwall}$ > rectangular > radial$_{ED+ES\ COM}$	Colle et al.[41]
Area > radial$_{LA}$, rectangular	Gelberg et al.[43]
Centerline, rectangular >> radial$_{midwall}$	Sheehan et al.[124]
Centerline = radial$_{midwall}$ = rectangular	Lamberti et al.[125]
Centerline >> radial$_{midwall}$	Ginzton et al.[126]
Radial$_{midwall}$ > radial$_{LA}$, radial$_{ES\ COM}$, rectangular	Ingels et al.[35]

*Key: A>>B,C indicates that method A was significantly better than methods B and C, and that B and C do not differ from each other; A>B indicates method A was better than method B but not compared statistically; ED COM, origin at end-diastolic center of mass; ES COM, origin at end-systolic center of mass; ED + ES COM, origin of each contour at that contour's center of mass; LA, origin at midjoint of long axis; Area, specifies that regional ejection fractions were measured; Radial$_{midwall}$, refers to the method of Ingels et al.[35]; MI, myocardial infarction; Centerline, refers to the centerline method of Sheehan et al.[124]

tified on the ventriculogram. Alternatively, one can count the number of frames between consecutive end-diastoles to determine the regularity of the rhythm during ventriculography, and to select a non-postectopic beat of representative cycle length.

Angiographic contrast agents depress myocardial contractility and cause reactive hyperemia in coronary arteries. This effect can be minimized by (1) delaying ventriculography for several minutes after the preceding angiogram to allow hemodynamics to return to baseline, (2) using a nonionic contrast medium[55] and (3) selecting one of the first seven cardiac cycles after contrast medium injection for quantitative analysis.[20]

End-diastole is selected as the frame following the peak of the R wave on the electrocardiogram (if recorded) at which chamber volume is greatest. End-systole is the frame of minimum volume. Some investigators have suggested that end ejection is a more appropriate end-systole.[56] However, this is difficult to put into practice because the closure of the aortic valve is clearly visible in only 20% to 35% of ventriculograms and the duration of valve closure is finite rather than instantaneous, so that completion of valve closure often occurs several frames after the time of minimum volume.[57] In patients with asynchronous motion, selection of the end-systolic frame may require that the contours be traced from two or three frames and their volumes calculated to find the time of minimum volume.

Careful delineation of the endocardial contour is essential for accurate and reproducible measurement of left ventricular dimension and function. Tracing of the contour is most easily performed in a darkened room using an overhead-mounted projector that displays the image on a table. It is more awkward and fatiguing to trace from the screen of the cine projector or from images projected on the wall. The contours are traced onto paper or transparencies and then entered into a computer using an x-y digitizing tablet. Computer-assisted systems are available for digitizing the contours directly from the projection screen; models that allow portions of the traced contour to be edited or corrected without requiring the entire contour to be redrawn are preferable. This is important because only from ventriculograms of superior quality can the contours be readily identified from the selected frame alone. Generally it is necessary to visually track wall motion while running the projector back and forth, trace a segment of border, and repeat the process.

The standard practice is to exclude the papillary muscles and trabeculae carneae from the chamber. It is useful to keep a record of who traced the ventriculogram, the quality of the ventriculogram or the confidence with which the contour could be identified, and the frame numbers selected for end-diastole and end-systole. The latter can be determined easily by setting the frame counter of the projector to zero at the beginning of the ventriculogram.

Computation

Whereas volume can be computed manually using only a planimeter and hand calculator, the analysis of wall motion is impractical without computer assistance. Fortunately, there are an increasing number of commercial products available, which offer a range of services, relieving the angiographer from the need to hire a programmer to develop analytical software. In the ideal system, once the left ventricular contours are traced and digitized, subsequent calculations are performed on the stored set of x-y coordinates; the user never needs to retrace or redigitize. A menu is provided for the user to select analysis programs. For example, if a biplane ventriculogram is filmed in the 30° RAO/60° LAO projections, one may wish to calculate volume from the

RAO view alone but measure wall motion biplane using method X. The results are printed in report form, presented on the terminal screen for review, and/or stored in a database together with other clinical parameters for that patient. The database of angiographic data can be cross-referenced by patient identification to data from other cardiac diagnostic laboratories. The database is flexible, easy to use, and can be revised to accommodate changing needs. Not only can data on specific patients be retrieved but also more complex queries can be handled. For example, the efficacy of an angioplasty in relieving symptoms and improving exercise tolerance can be determined by comparing pre- versus post-procedure treadmill tests and correlating with factors such as the severity of residual stenosis, the change in the pressure gradient across the stenosis after dilation, or the improvement in regional and global left ventricular function. Statistical programs are available that can accept selected data from the databases, perform a range of analyses, and graph the results.

APPLICATIONS

Normal Values

The normal mean and standard deviation for end-diastolic, end-systolic, and stroke volumes, EF, and wall motion have been obtained using the ventriculograms of patients who underwent diagnostic cardiac catheterization and were found to have normal cardiac anatomy. For volume and EF, the results obtained by different studies are quite similar (Table 8-4). It should be noted that the standards based on angiographic normals may be more lenient than standards obtained by studying normal volunteers or subjects with a low probability of disease, determined by Bayesian analysis. This is because angiographic normals may include patients with abnormal cardiac function on exercise testing.[58]

The normal mean and SD for wall motion measurements vary by the method used as well as by location around the ventricular contour. The motion measured in a given patient needs to be displayed in a format that allows easy identification of abnormalities and assessment of their severity and significance. Figures 8-3C and D and 8-5 illustrate three approaches, as adapted to the centerline method.

Evaluation of the Patient with CAD

The left ventricular EF is the most commonly used parameter of function. However, coronary disease causes regional myocardial ischemia, resulting from inadequate perfusion by one or more stenosed arteries. Global function may be impaired when involvement is severe or extensive, or it may be mantained by means of compensatory mechanisms. Therefore, evaluation of the patient requires examination of regional function and measurement of volume and EF.

Visual Assessment

In standard clinical practice, the ventriculogram is inspected visually, to make a rapid assessment of endocardial motion. This allows detection of asynchronous contraction, an abnormality that may be missed by quantitative analysis as routinely performed on the end-diastolic and end-systolic contours alone. The disadvantage of visual assessment is the lack of quantitation: Although severe dysfunction (i.e., akinesis or dyskinesis) can be identified reliably, attempts to grade levels of hypokinesis have been poorly reproducible.[19]

Correlation between Left Ventricular Function and CAD

Most patients with chronic angina and no prior infarction have a normal EF. Wall motion abnormalities, if present, correlate well in location with those of the stenosed coronary artery.[34] Significant regional hypokinesis, defined as wall motion depressed more than two SD below normal (< -2 SD) as measured by the centerline method, is uncommon in resting studies in the absence of a critical stenosis (Figure 8-6).[50] As demonstrated in experimental studies, the relationship between coronary artery stenosis severity and ventricular function in the ischemic region is nonlinear; function decreases exponentially once blood flow falls below resting levels.[59] The critical threshold for

Table 8-4 Normal Values for Volume in Adults*

Projection	EDVI	ESVI	SVI	EF	N	Reference
AP/lateral	70 ± 20	24 ± 10	45 ± 13	67 ± 8	16	Kennedy et al.[90]
RAO/LAO	72 ± 15	20 ± 8	51 ± 10	72 ± 8	17	Wynne et al.[3]
	78 ± 15	—	50 ± 13	64 ± 7	29	Hammermeister et al.[127]

*Key: EDV, end-diastolic volume; ESV, end-systolic volume; SV, stroke volume; EF, ejection fraction; AP, anteroposterior; RAO/LAO, right anterior oblique/left anterior oblique; I, index—all volumes are normalized by body surface area and expressed in mL/m².

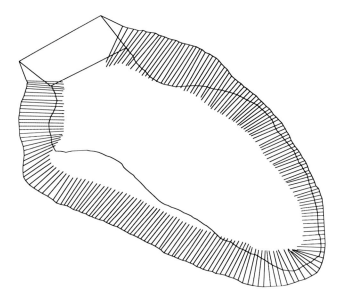

Figure 8-5 Graphic presentation of wall motion abnormality. The patient's motion is indicated by the endocardial contours. For comparison, the lengths of the 100 chords represent normal motion as anticipated from the end-diastolic contour.

unstable angina or subendocardial infarction has been defined as a minimum stenosis diameter $\leq 0.88 \pm 0.14$ mm or $\leq 0.64 \pm 0.08$ mm, respectively,[60] using quantitative coronary analysis.

Myocardial Infarction

In patients with acute myocardial infarction due to thrombosis of the left anterior descending (LAD) or right coronary (RCA) arteries, the severity of hypokinesis in the infarct site averages -2.7 SD in the 30° RAO projection. Comparable hypokinesis is observed with thrombosis of a nondominant circumflex (CFX) artery but is better visualized from a 60° LAO view.[61] The EF may bear little relationship to the severity of hypokinesis in the infarct site, because it is influenced by the size of the infarct[62] and by the function of the surrounding and remote myocardium.[63] The appearance of acute compensatory hyperkinesis in the noninfarcted region may boost the EF to normal levels; conversely, hypokinesis in this region due to previous infarction or severe multivessel disease may lower the EF beyond the level expected from the severity of dysfunction in the acute infarct site (Figure 8-7).[63-64]

Hyperkinesis in noninfarcted myocardium subsides to normal motion within 3 days.[65] In some patients, hyperkinesis is still present on follow-up studies performed at hospital discharge or later.[48] Although the development of regional hypertrophy as a mechanism of chronic hyperkinesis has been demonstrated in experimental animals,[66] chronic hyperkinesis is uncommon clinically unless the infarct artery was reperfused acutely with thrombolytic agents.[67-68] The EF reflects the function of all regions of the left ventricle. In the chronic stage, its correlation with the severity of hypokinesis in the infarct site is considerably higher than during acute infarction ($r = 0.41$ versus $r = 0.79$).[63] The magnitude of compensatory hyperkinesis is inversely related to the extent of hypokinesis in the infarct site; the larger the infarct, the less normal myocardium is left to compensate by hypercontracting.[69] Consequently, dilation occurs in patients with extensive dysfunction as the ventricle attempts to compen-

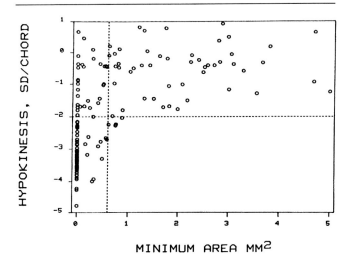

Figure 8-6 Relationship between the severity of hypokinesis and coronary artery stenosis. Severe hypokinesis depressed more than 2 SD below normal is uncommon until the stenosis exceeds the threshold associated with myocardial infarction. *Source:* Reprinted with permission from *American Journal of Cardiology* (1983;52:431–438), Copyright © 1983, Cahners Publishing Company, Inc.

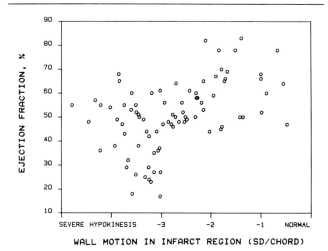

Figure 8-7 The correlation between the EF and the severity of hypokinesis in the infarct region is poor when assessed at the time of acute infarction ($r = 0.47$). This is due to the influence of wall motion abnormalities, either hypo- or hyperkinesis, in the opposite wall of the left ventricle. *Source:* Reprinted with permission from *European Heart Journal* (1985;6 Supplement E:117–125), Copyright © 1985, Academic Press London Ltd.

sate by the Frank-Starling mechanism. The proportion of the endocardial contour displaying hypokinesis has been correlated with elevation in end-diastolic volume and pressure and the development of heart failure.[70-71]

To a limited extent, infarct size can be estimated from the function of the left ventricle. The severity and location of regional hypokinesia and akinesia have been correlated with the degree of fibrosis in the myocardium measured by biopsy or at post-mortem exam.[59,72] In addition, hypokinesia measured by the centerline method has been correlated with (1) total creatine phosphokinase release in patients who achieved reperfusion after thrombolytic therapy,[42] (2) the size of the perfusion defect seen in intracoronary thallium scintigrams,[73] (3) the infarct size measured using nuclear magnetic resonance imaging,[74] and (4) uptake of antimyosin antibody.[75] Despite this apparently impressive array of evidence, it should be noted that infarct size can only be roughly estimated from function measurements because function also declines in the presence of ischemia alone, and experimental studies have demonstrated hypokinesis in normally reperfused myocardium bordering an acutely ischemic area.[76] In patients who undergo thrombolytic therapy or revascularization acutely, the measurement of ventricular function must be delayed by at least 1 week to allow time for the myocardium to recover.[77-78]

Preoperative Evaluation

Measurement of left ventricular function is useful when evaluating patients for surgery, to determine the risk of the procedure, estimate outcome, and assess the prognosis for survival postoperatively.

Coronary Artery Bypass Graft Surgery. Because of their excellent survival on medical therapy, patients with one- or two-vessel disease are not referred for bypass surgery unless they demonstrate persistent symptoms despite medical therapy. In three-vessel disease, however, the benefit of surgery on survival appears to depend on left ventricular function. Those with an EF ≤50% or segmental contraction abnormalities at rest were significantly benefitted by surgery in both the Veterans Administration (VA) and Coronary Artery Surgery Study (CASS) studies,[79-80] even though depressed left ventricular function is an important predictor of increased operative mortality.[81] In patients with three-vessel disease and an EF ≥50%, the trial results were controversial, with surgery improving survival in the European but not the VA or CASS studies.[82] In this group, additional testing to identify inducible ischemia has been suggested. It has been shown angiographically that the stress of supine exercise or pacing can elicit dysfunction in patients with normal function at rest or worsen the resting abnormality.[83] Comparison of rest and exercise function, which is more conveniently done using noninvasive testing, has been shown to enhance sensitivity in detecting ischemia,[84] and in identifying high-risk patients.[85] However, the benefit of surgical therapy in this group has not been determined yet.

Aneurysmectomy. Ventricular aneurysm is a common complication of myocardial infarction that may cause congestive heart failure, angina, or dysrhythmias. Aneurysmectomy may be performed to obtain symptomatic relief, but the operative mortality is significant. Most studies performed to identify patients who are likely to benefit from a resection have found that the EF of the contractile section correlated best with survival, although there is some disagreement on how this should be calculated.[86-87]

Evaluating Response to Intervention

The application of quantitative methods for analysis of left ventricular function is particularly useful in clinical trials. They provide accurate and reproducible measurements, reducing the number of patients who must be enrolled to achieve statistically significant results. Also, left ventricular function correlates well with survival yet is more precise than clinical endpoints such as death, which may be due to other causes.

Preliminary Considerations

The measurement of left ventricular function before therapeutic intervention provides a baseline to ensure that the patient populations are comparable before treatment and allows measurement of the change in function that occurs in response to treatment. However, the level of left ventricular function at follow-up may also be used as the endpoint.

The advantage of measuring follow-up function as the endpoint is that it correlates with prognosis and clinical status and indicates how close the intervention has come to restoring a normal EF and wall motion. The disadvantage is a lack of sensitivity if variability in the patient's baseline function is large relative to the response to treatment. In comparison, measurement of the change in function between baseline and follow-up study is more sensitive to small treatment effects if they are consistent and is independent of the level of dysfunction present acutely. However, there are a few caveats to this approach. First, it is subject to study-to-study variability, which is great enough to make the assessment of changes in individual patients inadvisable.[88] Second, the change in function seen after an intervention may be influenced by the pretreatment status.[89] Finally, some approaches to wall motion analysis are subject to error due to regression toward the mean, which is the statistical likelihood that a region identified by chance as displaying abnormal contraction on the baseline study will be abnormal on restudy.[88] To avoid this problem, the centerline method selects the region of hypokinesis separately for each study, without reference to the region selected in other studies of the same patient and constrained only by the limits of the arterial territory.[48] This approach has the additional benefit of reducing study-to-study variability due to shifts in the assignment of chord numbers to locations around the ventricular contour.

When evaluating response to therapy, there is often a tendency to dichotomize patients according to whether they improved or not. This facilitates certain types of statistical analysis but degrades the informational content of a continuous variable such as ventricular function. Patients' EFs are usually defined as abnormal if they are ≤50% to 55%, which is 2 SD below the mean of a normal population.[90] Change in global function is considered significant if it exceeds the variability of the measurement, for example, if change in EF ≥ 5%. Similar definitions are employed for wall motion, for example, change in hypokinesis ≥ 1 SD by the centerline method.

Response to Thrombolytic Therapy

Quantitative analysis of regional wall motion has been particularly useful for evaluating the efficacy of thrombolytic therapy in salvaging left ventricular function in patients with acute myocardial infarction. One of the earliest observations made was the comparative insensitivity of the EF.

Due to acute compensatory hyperkinesis in the wall opposite the infarct, many patients have a normal EF despite severe hypokinesis in the infarct region. Also, the EF often fails to improve between acute and follow-up studies, or to reflect even significant recovery of contraction in the infarct region of reperfused patients, because hyperkinesis in the noninfarct region usually declines to normal levels.[48,91] Thus measurement of wall motion is more sensitive than the EF in reflecting the effect of thrombolysis on the function of the infarct site.

The application of quantitative methods has increased the amount of useful information gained from clinical studies. Although many studies have reported a beneficial effect of thrombolytic therapy on left ventricular function compared either to placebo-treated[92–94] or to unsuccessfully reperfused patients,[48,91] there have also been studies with negative results.[95–97] Even among the former, there was considerable variability in patient response. To identify better the patients most likely to benefit, the factors that influence recovery of left ventricular function after reperfusion were sought. The most important are the time delay from the onset of symptoms of infarction to treatment and the severity of residual stenosis in the infarct artery after thrombolytic therapy (Figures 8-8 and 8-9).[89] Other factors include the severity of myocardial dysfunction present acutely and whether the infarct artery was totally or only subtotally occluded at the time of initial angiography. Reperfusion may also enhance function in the noninfarct region by promoting chronic compensatory hyperkinesis, in addition to improving function in the infarct region.[67] This probably contributed to the EF increase seen in patients with collaterals to the infarct artery[98] in the National Heart, Lung, and Blood Institute (NHLBI)–sponsored trial of thrombolysis in myocardial infarction (TIMI). Wall motion

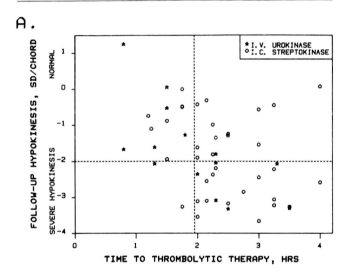

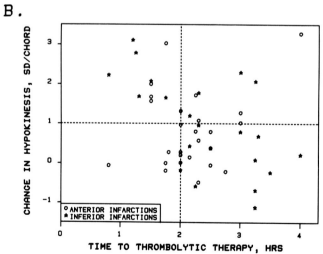

Figure 8-8 Time to treatment is an important determinant of functional recovery after thrombolysis. A, The change in wall motion between acute and follow-up (≥2 weeks later) studies is significantly greater in patients who received intravenous (I.V.) urokinase or intracoronary (I.C.) streptokinase within 2 hours after symptom onset than in those treated later. B, The earlier-treated patients had less residual hypokinesis at follow-up than did patients treated more than 2 hours after onset of infarction ($P<0.025$). *Source:* Reprinted with permission from *Journal of the American College of Cardiology* (1985;6:518–25), Copyright © 1985, American College of Cardiology.

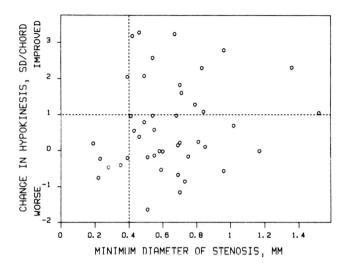

Figure 8-9 The severity of residual stenosis in the infarct artery measured within 1 hour after thrombolytic therapy also influenced recovery of ventricular function in reperfused patients. Wall motion did not improve in the infarct site if the minimum stenosis diameter remained narrowed to ≤0.4 mm. *Source:* Reprinted with permission from *Circulation* (1985;71:1121–1128), Copyright © 1985, American Heart Association.

analysis has assisted in these investigations by providing quantitative measurements pertinent to the infarct site and surrounding regions. From a statistical viewpoint, the use of precise endpoints reduces the number of patients who must be enrolled to demonstrate a significant relationship.

Response to Stimulation

The left ventricular functional response to inotropic stimulation has been used to assess viability and contractile reserve in patients with poor resting function being evaluated for coronary artery bypass. Early studies showed that patients whose preoperative EF or wall motion increased with post-extrasystolic potentiation had improved left ventricular function and a better prognosis following bypass surgery than did nonresponders.[99–100] Similar results were obtained following nitroglycerin or epinephrine infusion.[100,101] Noninvasive techniques have since been developed for this purpose.

The problem is more acute in patients undergoing thrombolytic therapy. This is known to carry a significant restenosis and reinfarction rate, especially if the residual stenosis is severe.[102] Yet the viability of the reperfused myocardium cannot be determined from the level of resting left ventricular function for at least 1 week after reperfusion due to the delay in recovery of myocardial energy metabolism. Some studies have reported that post-extrasystolic potentiation, measured before reperfusion, predicts subsequent recovery in function in both reperfused and nonreperfused patients.[103–104] These findings may be useful for evaluating myocardial viability in patients admitted too late after onset of infarction to qualify for thrombolytic therapy. However, preliminary studies in my laboratory suggest that the time from onset of infarction until treatment is a more powerful predictor.[105]

Response to Coronary Artery Bypass Graft Surgery

There have been many studies of change in left ventricular function following coronary bypass surgery. The results of the earlier studies were often conflicting, with some reporting improved function[106–107] and others finding no change.[108–109] These discrepancies have been attributed to (1) study-to-study variability due to changes in hemodynamic status between the pre- and postoperative angiograms; (2) use of qualitative rather than quantitative methods to analyze ventricular function; (3) timing of the follow-up study too early after surgery, when sympathetic activity may still have been elevated; and (4) failure to distinguish patients with normal preoperative function from those with depressed EF or regional wall motion, since the former cannot be expected to improve.[110] In addition, these investigators advocated biplane ventriculography, although others have found the LAO projection to contribute little additional information for assessing ventricular function.[61,111] When these sources of variability were controlled, bypass surgery was found to improve myocardial performance in patients with depressed left ventricular function preoperatively and patent grafts. Function did not change in patients with normal preoperative function or occluded grafts.[110]

Response to Percutaneous Transluminal Coronary Angioplasty

By the advent of angioplasty, noninvasive techniques of evaluating rest and exercise left ventricular function were widely available. Accordingly, the efficacy of angioplasty has been reported in terms of an improved radionuclide EF response.[112]

Because of the repeated coronary occlusions performed during an angioplasty procedure, there was initial concern about the cumulative effects of intermittent ischemia on ventricular performance. Serruys and co-workers analyzed the time course of changes in regional and global function and found that the effect of coronary occlusion on wall motion was profound but had reversed by the end of the procedure.[113]

Evaluating Prognosis

A large number of studies have been performed to identify variables predictive of poor prognosis in patients with CAD. Such studies have consistently reported the left ventricular EF to be one of the (if not the most) powerful predictors of survival in patients treated medically or surgically.[116] In patients with myocardial infarction, a depressed EF predischarge predicts mortality and need for rehospitalization.[117-118] These findings are supported by studies showing that clinical parameters indicative of poor left ventricular function and the presence of an aneurysm or of extensive wall motion abnormality also identify high-risk patients.[81,119-120]

Data from these and other studies have provided the basis for formulating treatment plans. For example, Epstein et al. have suggested an approach to selecting patients for cardiac catheterization based on the clinical course postinfarction and the presence of inducible ischemia on exercise testing performed with noninvasive imaging techniques.[121]

The ability to identify patients at high risk of morbid events is also useful because such individuals can be singled out for closer observation and possibly more aggressive therapy. More importantly, data on the characteristics distinguishing these patients may assist in the conduct and assessment of clinical trials intended to evaluate interventions.

SUMMARY

The measurement of left ventricular volume, EF, and wall motion provides a powerful tool for evaluating the effect of CAD on left ventricular function and compensatory reserve, the response to therapeutic interventions, and the prognosis of the patient. Although technological advances now facilitate the application of quantitative methods to analysis of left ventricular function, they should be paired with increased rather than slackened quality control in processing the ventriculographic data. Awareness of sources of error, variability, and unproved assumptions in the various methods aid in interpreting the computed results.

REFERENCES

1. Tennant R, Wiggers CJ. Effect of coronary occlusion on myocardial contraction. *Am J Physiol.* 1935;112:351-361.

2. Dodge HT, Sandler H, Ballew DW, Lord JD Jr. The use of biplane angiocardiography for the measurement of left ventricular volume in man. *Am Heart J.* 1960;60:762-776.

3. Wynne J, Green LH, Mann T, Levin D, Grossman W. Estimation of left ventricular volumes in man from biplane cineangiograms filmed in oblique projections. *Am J Cardiol.* 1978;41:726-732.

4. Lange PE, Onnasch D, Farr FL, Heintzen PH. Angiocardiographic left ventricular volume determination: Accuracy, as determined from human casts, and clinical application. *Eur J Cardiol.* 1978;8:449-476.

5. Sandler H, Dodge HT. The use of single plane angiocardiograms for the calculation of left ventricular volume in man. *Am Heart J.* 1968;75:325-334.

6. Kennedy JW, Trenholme SE, Kasser IS. Left ventricular volume and mass from single-plane cineangiocardiogram: A comparison of anteroposterior and right anterior oblique methods. *Am Heart J.* 1970;80:343-352.

7. Kasser IS, Kennedy JW. Measurement of left ventricular volumes in man by single-plane cineangiocardiography. *Invest Radiol.* 1969;4:83-90.

8. Als AV, Paulin S, Aroesty JM. Biplane angiographic volumetry using the right anterior oblique and half-axial left anterior oblique technique. *Radiology.* 1978;126:511-514.

9. Rogers WJ, Smith LR, Bream PR, Elliott LP, Rackley CE, Russell RO. Quantitative axial oblique contrast left ventriculography: Validation of the method by demonstrating improved visualization of regional wall motion and mitral valve function with accurate volume determinations. *Am Heart J.* 1982;103:185-193.

10. Hugenholtz PG, Wagner HR, Sandler H. The in vivo determination of left ventricular volume: comparison of the fiberoptic-indicator dilution and the angiocardiographic methods. *Circulation.* 1968;37:489-508.

11. Arvidsson H. Angiocardiographic determination of left ventricular volume. *Acta Radiol.* 1961;56:321-339.

12. Dodge HT, Hay RE, Sandler H. An angiocardiographic method for determining left ventricular stroke volume in man. *Circ Res.* 1962;11:739-745.

13. Lange PE, Onnasch D, Farr FL, Heintzen PH. Angiocardiographic right ventricular volume determination: Accuracy, as determined from human casts, and clinical application. *Eur J Cardiol.* 1978;8:477-501.

14. Sauter HJ, Dodge HT, Johnston RR, Graham TP. The relationship of left atrial pressure and volume in patients with heart disease. *Am Heart J.* 1964;67:635-642.

15. Graham TP Jr, Atwood GF, Faulkner SL, Nelson JH. Right atrial volume measurements from biplane cineangiocardiography. *Circulation.* 1974;49:709-716.

16. Cohn PF, Levine JA, Bergeron GA, Gorlin R. Reproducibility of the angiographic left ventricular ejection fraction in patients with coronary artery disease. *Am Heart J.* 1974;88:713-720.

17. Rogers WJ, Smith LR, Hood WP Jr, Mantle JA, Rackley CE, Russell RO Jr. Effect of filming projection and interobserver variability on angiographic biplane left ventricular volume determination. *Circulation.* 1979;59:96-104.

18. Dodge HT, Sheehan FH, Stewart DK. Estimation of ventricular volume, fractional ejected volumes, stroke volume, and quantitation of regurgitant flow. In: Just H, Heintzen PH, eds. *Angiocardiography: Current Status and Future Developments.* Berlin: Springer-Verlag; 1986:99-108.

19. Chaitman BR, DeMots H, Bristow JD, Rosch J, Rahimtoola SH. Objective and subjective analysis of left ventricular angiograms. *Circulation.* 1975;52:420-425.

20. Vine DL, Hegg TD, Dodge HT, Stewart DK, Frimer H. Immediate effect of contrast medium injection on left ventricular volumes and ejection fraction: A study using metallic epicardial markers. *Circulation.* 1977;56:379-384.

21. Ricci DR, Orlick AE, Alderman EL, Ingels NB Jr, Daughters GT II, Stinson EB. Influence of heart rate on left ventricular ejection fraction in human beings. *Am J Cardiol.* 1979;44:447-451.

22. Lang T, Corday E, Gold H, et al. Consequences of reperfusion after coronary occlusion: Effects on hemodynamic and regional myocardial metabolic function. *Am J Cardiol.* 1974;33:69-81.

23. Theroux P, Franklin D, Ross J Jr, Kemper WS. Regional myocardial function during acute coronary artery occlusion and its modification by pharmacologic agents in the dog. *Circ Res.* 1974;35:896-907.

24. Feild BJ, Russel O, Dowling JT, Rackley CE. Regional left ventricular performance in the year following myocardial infarction. *Circulation.* 1972;46:679–689.

25. McDonald IG. The shape and movements of the human left ventricle during systole. *Am J Cardiol.* 1970;26:221–230.

26. Kong Y, Morris JJ Jr, McIntosh HD. Assessment of regional myocardial performance from biplane coronary cineangiograms. *Am J Cardiol.* 1971;27:529–537.

27. Ingels NB Jr, Daughters GT II, Stinson EB, Alderman EL. Measurement of midwall myocardial dynamics in intact man by radiography of surgically implanted markers. *Circulation.* 1975;52:859–867.

28. Slager CJ, Hooghoudt TEH, Serruys PW, et al. Quantitative assessment of regional left ventricular motion using endocardial landmarks. *J Am Coll Cardiol.* 1986;7:317–327.

29. Hugenholtz PG, Kaplan E, Hull E. Determination of left ventricular wall thickness by angiocardiography. *Am Heart J.* 1969;78:513–522.

30. Sandor T, Paulin S, Hanlon WB. Left ventricular wall motion analysis using operator-independent contour positioning. *Comput Biomed Res.* 1984;17:129–142.

31. Sheehan FH, Stewart DK, Dodge HT, Mitten S, Bolson EL, Brown BG. Variability in the measurement of regional ventricular wall motion from contrast angiograms. *Circulation.* 1983;68:550–559.

32. Leighton RF, Drobinski G, Fontaine GH, Eugene M, Frank R, Grosgogeat Y. Effects of correcting apical displacement on regional wall motion in severely damaged ventricles. In: *Computers in Cardiology.* Long Beach, Calif: IEEE Computer Society; 1983:173–176.

33. Clayton PD, Jeppson GM, Klausner SC. Should a fixed external reference system be used to analyze left ventricular wall motion? *Circulation.* 1982;65:1518–1521.

34. Herman MV, Heinle RA, Klein MD, Gorlin R. Localized disorders in myocardial contraction: Asynergy and its role in congestive heart failure. *N Engl J Med.* 1967;277:222–232.

35. Ingels NB Jr, Daughters GT II, Stinson EB, Alderman EL. Evaluation of methods for quantitating left ventricular segmental wall motion in man using myocardial markers as a standard. *Circulation.* 1980;61:966–972.

36. Goodyer AVN, Langou RA. The multicentric character of normal left ventricular wall motion. Implications for the evaluation of regional wall motion abnormalities by contrast angiography. *Cathet Cardiovasc Diagn.* 1982;8:225–232.

37. Slager CJ, Hooghoudt TEH, Reiber JHC, Schuurbiers JCH, Booman F, Meester GT. Left ventricular contour segmentation from anatomical landmark trajectories and its application to wall motion analysis. In: *Computers in Cardiology.* Long Beach, Calif: IEEE Computer Society; 1979:347–350.

38. Papietro SE, Smith LR, Hood WP Jr, Russell RO Jr, Rackley CE, Rogers WJ. An optimal method for angiographic definition and quantification of regional left ventricular contraction. In: *Computers in Cardiology.* Long Beach, Calif: IEEE Computer Society; 1978:293–296.

39. Karsch KR, Lamm U, Blanke H, Rentrop KP. Comparison of nineteen quantitative models for assessment of localized left ventricular wall motion abnormalities. *Clin Cardiol.* 1980;3:123–128.

40. Skorton DJ, Collins SM, Kerber RE. Digital image processing and analysis in echocardiography. In: Collins SM, Skorton DJ, eds. *Cardiac Imaging and Image Processing.* New York: McGraw-Hill Book Co; 1986:171–205.

41. Colle JP, LeGoff G, Page A, Besse P. Validite de la methode de stanford dans l'evaluation des dyskinesies segmentaires du ventricule gauche. *Arch Mal Coeur.* 1982;75:395–406.

42. Sheehan FH, Bolson EL, Dodge HT, Mathey DC, Schofer J, Woo HW. Advantages and applications of the centerline method for characterizing regional ventricular function. *Circulation.* 1986;74:293–305.

43. Gelberg HJ, Brundage BH, Glantz S, Parmley WW. Quantitative left ventricular wall motion analysis: A comparison of area, chord and radial methods. *Circulation.* 1979;59:991–1000.

44. Harris LD, Clayton PD, Marshall HW, Warner HR. A technique for the detection of asynergic motion in the left ventricle. *Comput Biomed Res.* 1974;7:380–394.

45. Shapiro E, Marier DL, St John Sutton MG, Gibson DG. Regional non-uniformity of wall dynamics in normal left ventricle. *Br Heart J.* 1981;45:264–270.

46. Clayton PD, Klausner SC, Blair TL, Jeppson GM, Liddle HV. Sources and magnitude of variability in measurements of regional left ventricular function. In: Sigwart U, Heintzen PH, eds. *Ventricular Wall Motion.* Stuttgart: Georg Thieme Verlag; 1984:90–99.

47. Sigel H, Nechwatal W, Stauch M. Quantitative evaluation of regional and global parameters of left ventriculography with different methods in an intra- and interobserver test. *Z Kardiol.* 1981;70:742–747.

48. Sheehan FH, Mathey DG, Schofer J, Krebber HJ, Dodge HT. Effect of interventions in salvaging left ventricular function in acute myocardial infarction: A study of intracoronary streptokinase. *Am J Cardiol.* 1983;52:431–438.

49. Sheehan FH, Dodge HT, Mathey DG, Brown BG, Bolson EL, Mitten S. Application of the centerline method: Analysis of change in regional left ventricular wall motion in serial studies. In: *Computers in Cardiology.* Long Beach, Calif: IEEE Computer Society; 1982:97–100.

50. Sheehan FH, Brown BG, Dodge HT, Bolson EL, Mitten S. Quantitative analysis of the relationship between coronary artery stenosis and regional left ventricular wall motion In: Sigwart U, Heintzen PH, eds. *Ventricular Wall Motion.* Stuttgart: Georg Thieme Verlag; 1984:198–205.

51. Lorente P, Azancot I, Masquet C, et al. Comparison of decision rules in assessing regional wall motion. In: *Computers in Cardiology.* Long Beach, Calif: IEEE Computer Society; 1984:83–88.

52. Lorente P, Azancot I, Masquet C, Babalis D, Duriez M, Slama R. Relationships between single-vessel coronary artery obstructions and wall motion dysfunction analyzed by four computer-based methods. *Int J Cardiol.* 1985;7:361–374.

53. Mancini GBJ, Norris SL, Peterson KL, et al. Quantitative assessment of segmental wall motion abnormalities at rest and after atrial pacing using digital intravenous ventriculography. *Am J Cardiol.* 1983;2:70–76.

54. Vas R, Eigler N, Miyazono C, et al. Digital quantification eliminates intraobserver and interobserver variability in the evaluation of coronary artery stenosis. *Am J Cardiol.* 1985;56:718–723.

55. Higgins CB, Gerber KH, Mattrey RF, Slutsky RA. Evaluation of the hemodynamic effects of intravenous administration of ionic and nonionic contrast materials. *Radiology.* 1982;143:681–686.

56. Marier DL, Gibson DG. Limitations of the two frame method for displaying regional left ventricular wall motion in man. *Br Heart J.* 1980;44:555–559.

57. Sheehan FH, Dodge HT, Mathey DG, Szente A, Woo HW, Bolson EL. Measurement of abnormalities in the timing and extent of motion from frame-by-frame analysis of contrast left ventriculograms. In: *Computers in Cardiology.* Long Beach, Calif: IEEE Computer Society; 1983:35–39.

58. Rozanski A, Diamond GA, Forrester JS, Berman DS, Morris D, Swan HJC. Alternative referent standards for cardiac normality. *Ann Intern Med.* 1984;101:164–171.

59. Schwarz F, Flameng W, Thiedemann K-U, Schaper W, Schlepper M. Effect of coronary stenosis on myocardial function, ultrastructure and aortocoronary bypass graft hemodynamics. *Am J Cardiol.* 1978;42:193–201.

60. McMahon MM, Brown BG, Cukingnan R, et al. Quantitative coronary angiography: Quantitation of the critical stenosis in patients with unstable angina and single-vessel disease without collaterals. *Circulation*. 1979;60:106–113.

61. Sheehan FH, Schofer J, Mathey DG, et al. Measurement of regional wall motion from biplane contrast ventriculograms: A comparison of the 30 degree RAO and 60 degree LAO projections in patients with acute myocardial infarction. *Circulation*. 1986;74:796–804.

62. Miller RR, Olson HG, Vismara LA, Bogren HG, Amsterdam EA, Mason DT. Pump dysfunction after myocardial infarction: Importance of location, extent and pattern of abnormal left ventricular segmental contraction. *Am J Cardiol*. 1976;37:340–344.

63. Sheehan FH, Szente A, Mathey DG, Dodge HT. Assessment of left ventricular function in acute myocardial infarction: The relationship between global ejection fraction and regional wall motion. *Eur Heart J*. 1985;6(Suppl E):117–125.

64. Stamm RB, Gibson RS, Bishop HL, Carabello BA, Beller GA, Martin RP. Echocardiographic detection of infarct-localized asynergy and remote asynergy during acute myocardial infarction: Correlation with the extent of angiographic coronary disease. *Circulation*. 1983;67:233–244.

65. Schmidt WG, Sheehan FH, von Essen R, Vebis R, Effert S. Serial angiographic analysis of ventricular functional recovery after intracoronary streptokinase. *J Am Coll Cardiol*. 1987;9:61A. Abstract.

66. Theroux P, Ross J Jr, Franklin D, Covell JW, Bloor CM, Sasayama S. Regional myocardial function and dimensions early and late after myocardial infarction in the unanesthetized dog. *Circ Res*. 1977;40:158–165.

67. Hooghoudt TEH, Serruys PW, Reiber JHC, Slager CJ, Van den Brand M, Hugenholtz PG. The effect of recanalization of the occluded coronary artery in acute myocardial infarction on left ventricular function. *Eur Heart J*. 1982;3:416–421.

68. Martin GV, Sheehan FH, Stadius M, et al. Intravenous streptokinase for acute myocardial infarction: Effects on global and regional systolic function. *Circulation*. 1988;78:258–266.

69. Sheehan FH, Dodge HT, Bolson EL, Mitten S, Mathey DG. Relationship between hypokinesis and compensatory hyperkinesis and their effect on global left ventricular function. *J Am Coll Cardiol*. 1983;1:734. Abstract.

70. Klein MD, Herman MV, Gorlin R. A hemodynamic study of left ventricular aneurysm. *Circulation*. 1967;35:614–630.

71. Kitamura S, Kay JH, Krohn BG, Magidson O, Dunne EF. Geometric and functional abnormalities of the left ventricle with a chronic localized noncontractile area. *Am J Cardiol*. 1973;31:701–707.

72. Ideker RE, Behar VS, Wagner GS, et al. Evaluation of asynergy as an indicator of myocardial fibrosis. *Circulation*. 1978;57:715–725.

73. Schofer J, Sheehan FH, Spielmann R, Wygant J, Bleifeld W, Mathey DG. Early intracoronary thallium/technetium-pyrophosphate scintigraphy is a reliable method to predict myocardial salvage. *Circulation*. 1986;74(Suppl II):II–273. Abstract.

74. Johns JA, Yasuda T, Gold HK, et al. Estimation of site and localization of myocardial infarction by magnetic resonance imaging. *J Am Coll Cardiol*. 1986;7:196. Abstract.

75. Khaw BA, Gold HK, Yasuda T, et al. Scintigraphic quantification of myocardial necrosis in patients after intravenous injection of myosin-specific antibody. *Circulation*. 1986;74:501–508.

76. Gallagher KP, Gerren RA, Stirling MC, et al. The distribution of functional impairment across the lateral border of acutely ischemic myocardium. *Circ Res*. 1986;58:570–583.

77. Heyndrickx GR, Millard RW, McRitchie RJ, Maroko PR, Vatner SF. Regional myocardial functional and electrophysiological alterations after brief coronary artery occlusion in conscious dogs. *J Clin Invest*. 1975;56:978–985.

78. Lavallee M, Cox D, Patrick TA, Watner SF. Salvage of myocardial function by coronary artery reperfusion 1, 2, and 3 hours after occlusion in conscious dogs. *Circ Res*. 1983;53:235–247.

79. The Veterans Administration Coronary Artery Bypass Surgery Cooperative Study Group. Eleven-year survival in the Veterans Administration randomized trial of coronary bypass surgery for stable angina. *N Engl J Med*. 1984;21:1333–1339.

80. CASS Principal Investigators and Their Associates. Myocardial infarction and mortality in the coronary artery surgery study (CASS randomized trial). *N Engl J Med*. 1984;310:750–758.

81. Kennedy JW, Kaiser GC, Fisher LD, et al. Multivariate discriminant analysis of the clinical and angiographic predictors of operative mortality from the collaborative study in coronary artery surgery (CASS). *J Thorac Cardiovasc Surg*. 1980;80:876–887.

82. European Coronary Surgery Study Group. Long-term results of prospective randomised study of coronary artery bypass surgery in stable angina pectoris. *Lancet*. 1982;2:1172–1180.

83. Sharma B, Taylor S. Localization of left ventricular ischaemia in angina pectoris by cineangiography during exercise. *Br Heart J*. 1975;37:963–970.

84. Borer JS, Kent KM, Bacharach SL, et al. Sensitivity, specificity and predictive accuracy of radionuclide cineangiography during exercise in patients with coronary artery disease. Comparison with exercise electrocardiography. *Circulation*. 1979;60:572–580.

85. Bonow RO, Kent KM, Rosing DR, et al. Exercise-induced ischemia in mildly symptomatic patients with coronary-artery disease and preserved left ventricular function. *N Engl J Med*. 1984;311:1339–1345.

86. Kiefer SK, Flaker GC, Martin RH, Curtis JJ. Clinical improvement after ventricular aneurysm repair: Prediction by angiographic and hemodynamic variables. *J Am Coll Cardiol*. 1983;2:30–37.

87. Kapelanski DP, Al-Sadir J, Lamberti JJ, Agnostopoulos CE. Ventriculographic features predictive of surgical outcome for left ventricular aneurysm. *Circulation*. 1978;58:1167–1174.

88. Jeppson GM, Clayton PD, Blair TJ, Liddle HV, Jensen RL, Klausner SC. Changes in left ventricular wall motion after coronary artery bypass surgery: Signal or noise. *Circulation*. 1981;64:945–951.

89. Sheehan FH, Mathey DG, Schofer J, Dodge HT, Bolson EL. Factors determining recovery of left ventricular function following thrombolysis in acute myocardial infarction. *Circulation*. 1985;71:1121–1128.

90. Kennedy JW, Baxley WA, Figley MM, Dodge HT, Blackmon JR. Quantitative angiography, I: The normal left ventricle in man. *Circulation*. 1966;34:272–278.

91. Stack RS, Phillips HR III, Grierson DS, et al. Functional improvement of jeopardized myocardium following intracoronary streptokinase infusion in acute myocardial infarction. *J Clin Invest*. 1983;72:34–95.

92. The I.S.A.M. Study Group. A prospective trial of intravenous streptokinase in acute myocardial infarction (ISAM): Mortality, morbidity, and infarct size at 21 days. *N Engl J Med*. 1986;314:1465–1471.

93. Anderson JL, Marshall HW, Bray BE, et al. A randomized trial of intracoronary streptokinase in the treatment of acute myocardial infarction. *N Engl J Med*. 1983;308:1312–1318.

94. Serruys PW, Simoons ML, Suryapranata H, et al. Preservation of global and regional left ventricular function after early thrombolysis in acute myocardial infarction. *J Am Coll Cardiol*. 1986;7:729–742.

95. Khaja F, Walton JA, Brymer JF, et al. Intracoronary fibrinolytic therapy in acute myocardial infarction. *N Engl J Med*. 1983;308:1305–1311.

96. Leiboff RH, Katz RJ, Wasserman AG, et al. A randomized, angiographically controlled trial of intracoronary streptokinase in acute myocardial infarction. *Am J Cardiol*. 1984;53:404–407.

97. Rentrop RP, Feit F, Blanke H, et al. Effects of intracoronary streptokinase and intracoronary nitroglycerin infusion on coronary angiographic patterns and mortality in patients with acute myocardial infarction. *N Engl J Med*. 1984;311:1457–1463.

98. Sheehan FH, Braunwald E, Canner P, et al. The effect of intravenous thrombolytic therapy on left ventricular function: A report on tissue plasminogen activator and streptokinase from the thrombolysis in myocardial infarction (TIMI Phase I) trial. *Circulation*. 1987;75:817–829.

99. Hamby RI, Aintablian A, Wisoff BG, Hartstein ML. Response of the left ventricle in coronary artery disease to postextrasystolic potentiation. *Circulation*. 1975;51:428–435.

100. Cohn PF, Gorlin R, Herman MV, et al. Relation between contractile reserve and prognosis in patients with coronary artery disease and a depressed ejection fraction. *Circulation*. 1975;51:414–420.

101. Banka VS, Bodenheimer MM, Shah R, Helfant RH. Intervention ventriculography. Comparative value of nitroglycerin, post-extrasystolic potentiation and nitroglycerin plus post-extrasystolic potentiation. *Circulation*. 1976;53:632–637.

102. Harrison DG, Ferguson DW, Collins SM, et al. Rethrombosis after reperfusion with streptokinase: Importance of geometry of residual lesions. *Circulation*. 1984;69:991–999.

103. Azancot I, Beaufils P, Masquet C, et al. Detection of residual myocardial function in acute transmural infarction using postextrasystolic potentiation. *Circulation*. 1981;64:46–53.

104. Hodgson JM, O'Neill LWW, Laufer N, Bourdillon PDV, Walton JA Jr, Pitt B. Assessment of potentially salvageable myocardium during acute myocardial infarction: use of postextrasystolic potentiation. *Am J Cardiol*. 1984;54:1237–1244.

105. Sheehan FH, Mathey DG, Dodge HT, Schofer J. Postextrasystolic potentiation predicts recovery of LV function in patients reperfused late after infarction. *J Am Coll Cardiol*. 1985;5:495. Abstract.

106. Bourassa MG, Lesperance J, Campeau L, Saltiel J. Fate of left ventricular contraction following aortocoronaryvenous grafts. Early and late postoperative modifications. *Circulation*. 1972;46:724–730.

107. Chatterjee K, Swan HJC, Parmley WW, Sustaita H, Marcus HS, Matloff J. Influence of direct myocardial revascularization on left ventricular asynergy and function in patients with coronary heart disease: With and without previous myocardial infarction. *Circulation*. 1973;47:276–286.

108. Hammermeister KE, Kennedy JW, Hamilton GW, et al. Aortocoronary saphenous-vein bypass: failure of successful grafting to improve resting left ventricular function in chronic angina. *N Engl J Med*. 1974;290:186–192.

109. Shepherd RL, Itscoitz SB, Glancy DL, et al. Deterioration of myocardial function following aorto-coronary bypass operation. *Circulation*. 1974;49:467–475.

110. Wolf NM, Kreulen TH, Bove AA, et al. Left ventricular function following coronary bypass surgery. *Circulation*. 1978;58:63–70.

111. Cohn PF, Gorlin R, Adams DF, Chahine RA, Vokonas PS, Herman MV. Comparison of biplane and single plane left ventriculograms in patients with coronary artery disease. *Am J Cardiol*. 1974;33:1–6.

112. Rosing DR, van Raden MJ, Mincemoyer RM, et al. Exercise electrocardiographic and functional response after percutaneous transluminal coronary angioplasty. *Am J Cardiol*. 1986;53:35C–41C.

113. Serruys PW, Wijns W, Van den Brand M, et al. Left ventricular performance, regional blood flow, wall motion, and lactate metabolism during transluminal angioplasty. *Circulation*. 1984;70:25–36.

114. Topol EJ, O'Neill WW, Langburd AB, et al. A randomized, placebo-controlled trial of intravenous recombinant tissue-type plasminogen activator and emergency coronary angioplasty in patients with acute myocardial infarction. *Circulation*. 1987;75:420–428.

115. O'Neill W, Timmis GC, Bourdillon PD, et al. A prospective randomized clinical trial of intracoronary streptokinase versus coronary angioplasty for acute myocardial infarction. *N Engl J Med*. 1986;314:812–818.

116. Hammermeister KE, DeRouen TA, Dodge HT. Variables predictive of survival in patients with coronary disease. *Circulation*. 1979;59:421–430.

117. Norris RM, Barnaby PF, Brandt PWT, et al. Prognosis after recovery from first acute myocardial infarction: Determinants of reinfarction and sudden death. *Am J Cardiol*. 1984;53:408–413.

118. Multicenter Postinfarction Research Group. Risk stratification and survival after myocardial infarction. *N Engl J Med*. 1983;309:321–336.

119. Marchlinski FE, Buxton AE, Waxman HL, Josephson ME. Identifying patients at risk of sudden death after myocardial infarction: Value of the response to programmed stimulation, degree of ventricular ectopic activity and severity of left ventricular dysfunction. *Am J Cardiol*. 1983;52:1190–1196.

120. Maisel AS, Gilpin E, Hoit B, et al. Survival after hospital discharge in matched populations with inferior or anterior myocardial infarction. *J Am Coll Cardiol*. 1985;6:731–736.

121. Epstein SE, Palmeri ST, Patterson RE. Medical intelligence. Evaluation of patients after acute myocardial infarction. Indications for cardiac catheterization and surgical intervention. *N Engl J Med*. 1982;24:1487–1492.

122. Hammermeister KE, Warbasse JR. Immediate hemodynamic effects of cardiac angiography in man. *Am J Cardiol*. 1973;31:307–314.

123. Carleton RA. Change in left ventricular volume during angiocardiography. *Am J Cardiol*. 1971;27:460–463.

124. Sheehan FH, Bolson EL, Dodge HT, Mitten S. Centerline method—comparison with other methods for measuring regional left ventricular motion. In: Sigwart U, Heintzen PH, eds. *Ventricular Wall Motion*. Stuttgart: Georg Thieme Verlag; 1984:139–149.

125. Lamberti C, Gnudi G, Bombardi F, Sanders WJ. Test of methods for ventricular wall motion analysis. In: *Computers in Cardiology*. Long Beach, Calif: IEEE Computer Society; 1985:467–470.

126. Ginzton LE, Berntzen R, Lobodzinski S, Thigpen T, Laks MM. Computerized quantitative segmental wall motion analysis during exercise: Radial vs. centerline left ventricular segmentation. In: *Computers in Cardiology*. Long Beach, Calif: IEEE Computer Society; 1985:157–160.

127. Hammermeister KE, Brooks RC, Warbasse JR. Rate of change of left ventricular volume in man, I: Validation and peak systolic ejection rate in health and disease. *Circulation*. 1974;49:729–738.

Chapter 9

Digital Angiography for the Assessment of Left Ventricular Function and Myocardial Perfusion in Coronary Artery Disease

Robert A. Vogel, MD

INTRODUCTION

Although intravenous contrast medium–administration cardiac and peripheral angiography and analog image subtraction were implemented decades ago,[1] digital subtraction angiography has been widely applied clinically only during the past few years. The technologic developments pioneered at the universities of Wisconsin and Arizona, and the Mayo Clinic and Kiel in Europe that have made this possible[2-5] have now been embodied in commercial systems that are being tested in numerous medical centers. Digital angiography was introduced as a technique for enabling the visualization of peripheral arterial structures using intravenous contrast medium administration. Soon after introduction, right and left ventricular imaging was reported using the same approach. More recently, digital left ventriculography has been increasingly performed using low-dose direct contrast medium administration, a development that has paralleled trends in the entire digital radiographic field. Unlike peripheral arteriography, cardiac applications have often involved quantitative measurements and parametric imaging that take greater advantage of the numerical nature of the images involved. Parametric (functional) imaging has proved useful for the assessment of myocardial perfusion by enabling the measurement of regional coronary flow reserve. These recent cardiac applications of digital ventriculography and parametric coronary flow imaging are summarized in this chapter.

INSTRUMENTATION

The digital radiographic process uses conversions of fluoroscopic or radiographic images into digital format for subsequent image enhancement and storage. Important technical features of this process include radiographic technique, image matrix size, and framing rate. In general, neither ventriculographic nor parametric flow imaging require high spatial resolution, allowing for use of cineradiographic exposure. Unlike most standard catheterization laboratory cineradiography, which uses floating radiographic parameters that adapt to image intensifier output, all digital radiographic techniques require constant parameters. Digital ventriculography is generally performed in 256 × 256 eight-bit pixel matrix resolution but mandates at least 30-frame-per-second temporal sampling. Parametric flow imaging is generally performed in 512 × 512, eight-bit resolution using electrocardiogram (ECG)-gated acquisition at a framing rate of once per cardiac cycle.

SUBTRACTION TECHNIQUES

All subtraction techniques have the purpose of increasing the visualization of low concentrations of contrast medium by eliminating densities due to overlying bones and soft tissues. To achieve this, subtraction of two images that differ by one or more variables is accomplished either before or after image storage.

These variables include time, x-ray energy, and density.[6] The most commonly used and simplest technique for performing digital enhancement is mask mode subtraction. This method subtracts a single or time-averaged frame (the mask), usually acquired prior to contrast medium administration, from each of several frames acquired following contrast medium administration. This enhances the visualization of the contrast medium bolus, as it is the only element not common to both the mask and each of the several post–contrast medium administration images. Owing to the exponential attenuation of the x-ray beam, logarithmic analog-to-digital conversion must be employed so as to provide equivalent subtraction properties independent of overlying tissue attenuation. Because of nonlinear image intensifier and plumbicon transfer functions, contrast medium detection is always affected to some extent by overlying tissue attenuation.

For ventriculographic imaging, both an average mask (averaged in time over 0.5 to 1 second or over one cardiac cycle) and a set of ECG-gated separate masks have been found adequate, although the latter is theoretically preferable (Figure 9-1).[7] Parametric flow imaging must utilize ECG-gated mask mode subtraction to minimize noncorrespondence of the mask and contrast medium–containing images. As is pointed out by use of the latter technique, cardiac applications differ from peripheral digital radiography owing to the constant, cyclic cardiac motion. Patient motion is the greatest problem for mask mode subtraction, as the mask and contrast medium–containing images differ by from a few seconds for intraventricular injection up to about 10 seconds for intravenous contrast medium injection. Patient motion during this interval causes nonalignment of both the cardiac and overlying thoracic structures, which produces artifact on the subtracted image. In clinical practice, great attention must be paid to ensuring that patients do not breathe, move, or do the Valsalva maneuver during image acquisition.

Time-interval differencing is another time-based subtraction technique, but it differs from the others in that the images being subtracted are temporally separated by a short (30 to 60 milliseconds) interval (Figure 9-2).[8] Subtraction of these post–contrast medium administration frames displays contrast medium bolus transit and wall motion that have occurred within the time difference between the frames. An advantage of this approach is its ability to display functional and phasic information. Dyskinetic ventricular segments and the atrioventricular valve planes are readily evident using this subtraction method. This method is considerably less influenced by patient motion (owing to the short time interval), but only changing structures are visualized. Thus akinetic ventricular segments and much of the ventricles during end-diastole and end-systole are not visualized.

CONTRAST MEDIUM ADMINISTRATION

Although proposed as a means for reducing the invasiveness of standard arteriography, digital radi-

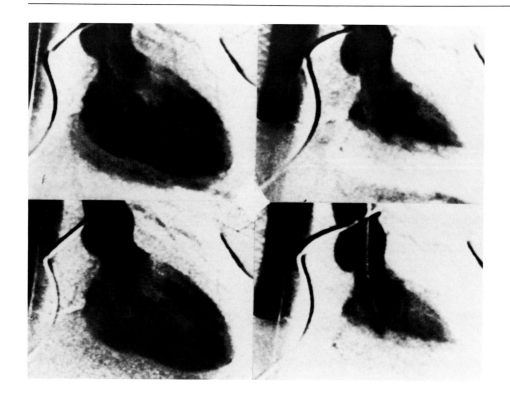

Figure 9-1 End-diastolic (top and bottom, left) and end-systolic (top and bottom, right) right anterior oblique projection digital left ventriculograms are shown using mean mask mode subtraction (top left and right) and ECG-gated mask mode subtraction (bottom left and right). Ten milliliters of contrast medium, diluted to 20 mL, was injected directly into the left ventricle via a pigtail catheter, in both instances.

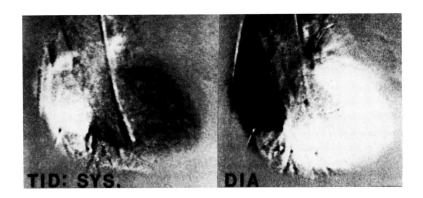

Figure 9-2 Time-interval difference intravenous contrast medium injection left ventriculograms are shown at end-systole (left) and end-diastole (right).

ography still requires contrast medium administration. Although the information gained from angiographic studies generally far outweighs their risk, some patients suffer from the systemic, vascular, renal, and cardiac effects of contrast media.[9] In addition, the hemodynamic effects of contrast material cannot be ignored in the analysis of cardiac function. The typical 20- to 60-mL dose of contrast material required for intravenous ventriculography significantly increases both right and left heart diastolic pressures, cardiac output, and heart rate and reduces mean arterial pressure. Although right-sided diastolic pressures are raised to a greater degree with intravenous compared with intraventricular contrast medium injection, the difference is not great and both routes produce similar left-sided hemodynamic changes. Multiple studies of ventricular function are especially hampered by the large contrast medium load used in intravenous studies. Combined with the reduction in patient motion artifact and the ability to demonstrate mitral insufficiency, the effects of a large bolus of contrast medium have led many to employ intraventricular contrast medium administration using a relatively small, 5- to 10-mL bolus of contrast material (Figure 9-1).[10] This approach substantially reduces contrast medium–induced hemodynamic changes and allows for the performance of multiple ventriculographic studies within a relatively short time period. Another approach to this problem has been the use of nonionic contrast media. Clinical studies have shown decreased patient toxicity; lessened alterations of myocardial contractility, relaxation, hemodynamics, and coronary blood flow; and absence of induced hypocalcemia and hypokalemia.[11] The use of nonionic agents may also reduce patient discomfort during injection and therefore, in turn, reduce patient motion.[12]

The quality of digitally subtracted images of central cardiac structures depends significantly on the site and nature of the bolus of contrast material. In general, the quality of any image is directly related to the peak concentration of iodine within the structure. This will, of course, vary for the different cardiac structures, as well as with different imaging hardware and subtraction techniques. Because the contrast medium bolus tends to become less compact as it transits the central circulation, centrally injected boluses theoretically have an advantage. For the most part, owing to occasional contrast medium extravasation and difficulties with injecting adequate amounts of contrast medium, central injections using catheters with multiple side-holes have replaced direct peripheral intravenous administration.[13–14] In an attempt to further increase the peak levophase iodine concentration, pulmonary artery and direct left ventricular, aortic root, and selective coronary artery injections have been performed. For left ventriculography, direct injection has the additional advantages of requiring considerably less contrast material and being able to demonstrate mitral valve insufficiency. Most angiographers have used between 5 and 10 mL of contrast agent, often diluted to an approximately 50% solution to achieve better distribution.[15]

VENTRICULOGRAPHY

Experimental Ventriculographic Studies

Several studies have demonstrated the validity of assessing left ventricular dynamics in experimental animals using digital approaches. Slutsky and associates demonstrated an excellent correlation between ejection fraction measured before and after mask mode subtraction ($r = 0.97$).[16] They found, however, that image subtraction produced consistent underestimations of both end-systolic and end-diastolic volumes. Radtke and colleagues reported that left ventricular wall mass and wall thickening could be assessed digitally through intracavitary contrast visualization of the endocardial boundary combined with identification of the epicardial boundary during myocardial opacification.[17–18] The end-systolic estimation of left ventricular mass correlated closely ($r = 0.94$) with post-mortem determinations. Wall thickening changes induced by both ischemia and inotropic agents were found to accurately reflect regional thickening measured by transmyocardial sonomicrometer crystals. In addition, myocardial perfusion defects caused by coronary occlusion were found to be demonstrable using digital enhancement.

Other groups have taken advantage of the space-density information present on digital subtraction studies. Gerber and co-workers analyzed regional wall motion through the use of parametric (functional) ejection shell and paradox images, demonstrating the effects of ischemia in a canine model.[19] Carey and associates performed digital ventriculographic determinations of stroke volume and cardiac output using area-length methods and demonstrated that these correlated well with indicator-dilution techniques.[20] Thus digital angiographic methodology can be used to assess ventricular function accurately in experimental animals, without the need for direct arterial access.

Clinical Ventriculographic Studies

Several groups have compared the results of standard clinical and intravenous digital estimations of ventricular volume and ejection fraction to determine potential clinical utility. Correlation coefficients between area-length measurements of ejection fraction using both standard and digital angiography have ranged from 0.75 to 0.98.[21–23] Some degree of underestimation of ejection fractions below 35% has been reported, and, as noted previously, image quality tends to be poorer in these instances. Those investigators demonstrating the best ejection fraction correlations generally used rapid central contrast medium injection.[21] Possibly because of the difficulty of detecting diluted contrast medium densities near the endocardial boundary, some investigators found systematic underestimations of end-diastolic and/or end-systolic volumes.[24] Clearly, the degree of image enhancement is a major factor in volumetric determinations, as is the method of edge detection. Volumetric estimations using the digital methodology have, in general, good correlations with standard ventriculography, with correlation coefficients ranging from 0.88 to 0.98. In contrast to the aforementioned studies, all of which used mask mode subtraction, Engels and colleagues[25] found a somewhat lower ejection fraction correlation ($r = 0.81$) using time-interval differencing in patients with at least 50% ejection fractions. It is likely that this is due to the limitation of this technique for visualizing nonmoving structures, that is, at end-diastole and end-systole. These studies suggest that using mask mode subtraction and rapid central venous contrast medium injections, left ventricular volume and ejection fraction determinations can be performed clinically with acceptable accuracy (Figure 9-3). Although hemodynamic changes are similar by both standard and digital methodologies, the latter has the advantages of visualization of the right ventricle, reduction in extrasystoles, ease of performing both simple and complex quantitative analyses, and avoidance of arterial entry.

Low–Contrast Medium Dose Direct Ventriculography

An alternative to the preceding approach is digital ventriculography using a small dose of contrast material administered directly into the left ventricle (see Figure 9-1). Although this requires arterial entry, the contrast medium burden can be substantially reduced, with attendant reductions in hemodynamic alterations and patient complications. In addition, owing to the reduced time span between the mask and successive images, patient motion artifact tends to be less. Clinical correlative studies have employed doses of contrast material ranging from 5 to 10 mL[26] and have found excellent ejection fraction correlations ($r = 0.91$ to 0.97) with those obtained using standard ventriculography. Some lack of correlation with standard volumetric determinations has been reported, and most investigators have found better visualization with 10 mL rather than 5 mL of contrast medium. In the author's experience, some degree of contrast medium dilution is preferable. Nichols and co-workers used 7 mL of contrast material diluted in 43 mL of saline solution and found a 0.97 correlation coefficient correspondence with end-diastolic and end-systolic volumes and ejection fraction assessed by standard ventriculography.[15] Similar results were reported by Seldin and associates, who additionally used semiautomated edge detection methods.[27] All of the aforementioned groups showed the expected lack of significant hemodynamic perturbations. Clearly, the major problem with this approach is inhomogeneity of contrast medium mixing due to the relatively small volume of the contrast medium bolus. This tends to be a greater problem in regions of reduced wall motion and for large hypokinetic ventricles. Despite this, the advantages, coupled with the ability to perform multiple ventriculographic studies, make low–contrast medium dose direct ventriculograms the digital approach of choice.

Ventriculographic Videodensitometry

In addition to the geometric ventriculographic determinations reported in the previous studies, other investigators have taken advantage of the density information present in digitally subtracted studies. Theoretically, the pixel density in subtracted images represents the line integral of the amount of contrast material traversed by the x-ray beam. In practice, even with logarithmic analog-to-digital conversion, a number of technical problems make absolute videodensitometric measurements extremely difficult. These

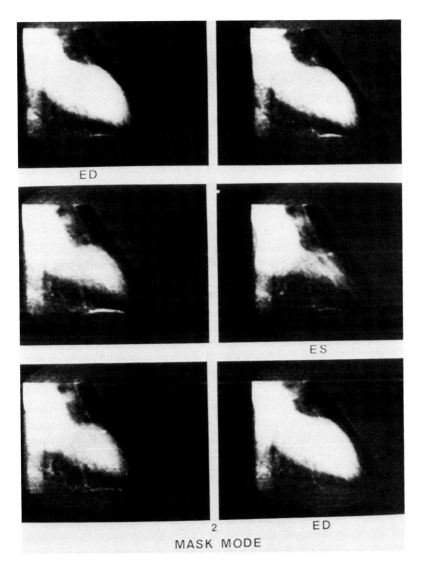

Figure 9-3 Sequential right anterior oblique frames of a left ventricle, obtained using mask mode subtraction and intravenous injection of 50 mL of contrast medium. End-diastole (ED) and end-systole (ES) are noted.

include nonlinearity of the image intensifier and plumbicon transfer functions, beam scatter and energy modification, geometric distortions of the cone-beam geometry, and veiling glare.[28] Moreover, these factors are not uniform over the field of view and therefore cannot be corrected by measurements performed at a single location. Variations in attenuation of the overlying noncardiac structures further complicate this problem. As mentioned earlier, inhomogeneous distribution of contrast medium within the cardiac chambers can also be a significant problem. Despite these difficulties, accurate estimations of ventricular volumes and ejection fractions have been reported.[29–30] These approaches do not require the precise edge detection necessary for geometric approaches.

Regional Ventricular Function

The interobserver variability of assessing regional wall motion by subjective visual means is quite substantial for both standard and digital ventriculograms. Recently, the innately quantitative nature of digital ventriculographic studies has been used by several investigators in attempts to reduce this subjectivity. At present, there is no standard approach to quantitating wall motion. Both long-axis and ventricular centroid frames of reference, mask mode and time-interval differencing subtraction techniques, and intravenous and low-dose direct ventricular contrast medium injections have been employed. The actual wall motion has been described in both radial and segmental area parameters. This approach allows for the calculation of statistical ranges of normal, developed on regional bases. Using these methods, good correlations with wall motion assessed by standard ventriculography have been reported, although normal apical and basal segment variability tends to be large, and diaphragmatic artifact reduces the accuracy of analysis of the inferior segments.[22,25] Recently, computer analysis of left ventricular curvature has been reported, which has the advantage of being independent of spatial coordinates.[31]

Stress Ventriculography

One of the greatest advantages of radionuclide ventriculography over the standard contrast method is the ability to compare the results of studies at rest and during exercise. Although many patients with significant coronary artery disease have normal resting ventricular function, most of these studies become abnormal during exercise. Intravenous contrast medium injection digital ventriculography has been employed in a manner similar to exercise radionuclide testing. This approach offers considerably greater spatial resolution in comparison with radionuclide ventriculography, but patient motion and breathing during stress can be problematic. Goldberg and associates reported their experience with 31 patients who were evaluated with exercise intravenous digital ventriculography.[32] In only two instances were the studies of such quality that they could not be interpreted. Ejection fraction changes were similar to those demonstrated by radionuclide ventriculography, and all patients developed new or increased wall motion abnormalities.

In other centers, patient motion has been found to be excessive during exercise, and investigators have turned to atrial pacing. Tobis and colleagues combined atrial pacing with low–contrast dose direct digital ventriculography and found a specificity of 83% and a sensitivity of 93% for the detection of significant coronary artery disease, using a 5% increase in ejection fraction as the criterion.[33] Wall motion abnormalities developed or worsened in 80% of the patients. Using intravenous ventriculography, atrial pacing, and quantitative wall motion analysis, Mancini and co-workers found significant increases in end-systolic volumes and decreases in ejection fraction in the immediate postpacing phase.[34] This allowed comparison of postpacing and baseline studies at equivalent heart rates. A similar study performed by Johnson and associates demonstrated 100% specificity and 82% sensitivity for the detection of coronary disease using subjective wall motion analysis in patients with normal resting wall motion.[35] These investigations suggest that digital approaches provide information on exercise ventricular function that is both clinically applicable and accurate.

Parametric Flow Reserve Imaging

Digital angiography offers several advantages for coronary arteriography in addition to those provided for left ventriculography. Over the past 30 years, standard selective coronary arteriography has become the gold standard by which coronary anatomy has been evaluated. Coronary arteriography is being increasingly recognized, however, as a poor index of clinical atherosclerotic pathophysiology. Considerable observer variability in the visual interpretation of coronary stenoses,[36] lack of correlation between the angiographic and pathologic appearance of lesions,[37] and inability of the arteriogram to predict hemodynamic consequences of individual stenoses[38] have been reported. Digital angiography offers coronary arteriography the advantages of coronary stenosis quantification and coronary flow and flow reserve assessment. Whereas digital angiography facilitates the performance of quantitative coronary angiography (Figure 9-4; see color insert), the latter does not appear to be able to differentiate significant from insignificant stenoses of moderate severity. This problem has led to interest in using the arteriogram to measure coronary flow or flow reserve directly.

Recent digital radiography advances in this area have been based on the analog videodensitometric approaches or Rutshauser et al. and Smith et al., who used selective and aortic root injections to determine absolute coronary blood flow.[39–40] This method uses determinations of proximal arterial segment length and diameter to estimate vascular volume. Together with measurement of the time required for an arterial bolus to transit the segment, arterial flow can be calculated. This method has been recently shown to be highly accurate and capable of assessing phasic flow.[41] This approach is generally limited to proximal predivisional coronary segments whose courses need to be relatively perpendicular to the x-ray beam (e.g., coronary bypass grafts). Additionally, precise determinations of vessel diameter are required. In an attempt to reduce some of these problems, Foerster et al. developed a method for measuring relative coronary flow ratios by comparing the videodensitometric areas under the baseline and hyperemic condition transit curves.[42] This approach is more suited to distal and circuitous arteries and requires that equal contrast medium boluses be administered subselectively under the two conditions. The resultant parameter, coronary flow reserve, is an important hemodynamic index of the severity of coronary stenosis.[43] The concept of using ECG-gated digital subtraction coronary arteriography to measure regional blood flow was first suggested by Robb et al.[44] Electrocardiogram gating allows more precise measurement of contrast medium density through subtraction of pre–contrast medium injection densities. Most approaches to measuring coronary flow utilize inflow (wash-in) rather than washout parameters since contrast media are poor indicator-dilution substances due to their alteration of intrinsic coronary flow and vascular volume, as well as their diffusibility, density, and adherence to vessel surfaces.[45] Hodgson et al. have reported that selectively injected contrast medium, however, altered coronary flow in a predictable manner, under both baseline and hyperemic conditions.[46]

A videodensitometric approach has been developed to assess relative regional coronary flow using ECG-gated mask mode subtracted digital arteriography.[47-48] In this technique, parametric images are generated that display the timing and density of the contrast medium bolus as it transits the coronary circulation. Color coding is used to depict the cycle following injection, in which contrast appears in each pixel and intensity is used to depict the maximal contrast density within each pixel. Pairs of parametric images, termed contrast medium appearance pictures, are acquired under baseline and papaverine-induced hyperemic conditions (Figure 9-5; see color insert). Regional flow reserve values are calculated using inflow videodensitometric principles, with appearance time and cumulative density being used as the transit time and volume parameters, respectively. This has proved to be more accurate than using transit time information alone, as was done initially.[48] Validation experiments have demonstrated this technique to be reasonably accurate and reproducible.[48-49]

This technique is technically demanding, requiring atrial pacing, ECG-gated power contrast medium administration, and complete cessation of patient motion. An alternate approach has been suggested by Elion et al.,[50] who combined parametric digital imaging with the indicator-dilution method first suggested by Foerster et al.[42] This technique requires subselective contrast medium administration and necessitates precise alignment and cessation of motion for both the baseline and hyperemic studies. Clinical interpretation of regional flow reserve information is also complicated by the many factors that can alter this relative parameter, independent of the effects of focal coronary stenosis.[51-52] The commonly encountered variables of myocardial hypertrophy, hypertension, prior myocardial infarction, collateral vessels, vasoactive drugs, endothelial damage, prolonged ischemia, and altered preload and afterload significantly affect (usually reduce) flow reserve values measured by any technique.[45] Whereas normal flow reserve assessment suggests absence of hemodynamically significant lesions, these multiple factors must be excluded when abnormally low values are encountered. Differentiation of normal and abnormal regional flow reserves is also aided by use of maximal hyperemic stimuli, such as intracoronary papaverine.

Several cardiac catheterization laboratories are gaining clinical experience with the parametric imaging technique. A close correlation between flow reserve values determined using the parametric and the Doppler velocity catheter methodologies has been reported. Specific concern for the use of regional coronary flow reserve determinations as a tool to evaluate the efficacy of angioplasty has arisen due to findings of transiently elevated resting coronary blood flow in the postdilation period. Despite these limitations, considerable pathophysiological information has been acquired from flow reserve measurements obtained during routine cardiac catheterization.

SUMMARY

The assessment of coronary atherosclerosis is a complex process involving appreciation of symptomatology, exercise capacity, coronary anatomy, and blood flow and ventricular function. Traditional visual assessment of selective coronary arteriography is fraught with inaccuracy and observer variability. Digital angiography has simplified the performance and assessment of left ventriculography and allows for the assessment of regional coronary flow reserve. The latter advances our ability to assess the significance of individual coronary lesions. Coronary flow reserve information must be interpreted with regard to the multiple factors that can affect it.

REFERENCES

1. Robb GP, Steinberg I. Visualization of the chambers of the heart, pulmonary circulation, and great blood vessels in man. *Am J Radiol.* 1938;1:1-17.

2. Frost MM, Fisher HD, Nudelman S, et al. A digital video acquisition system for extraction of subvisual information in diagnostic medical imaging. *Proc SPIE.* 1977;127:208-215.

3. Heintzen PH, Brennecke R, Bursch JH. Automated videoangiographic image analysis. *IEEE Comput Cardiol.* 1975;8:56-64.

4. Kruger RA, Mistretta CA, Houk TL, et al. Computerized fluoroscopy in real time for noninvasive visualization of the cardiovascular system. *Radiology.* 1979;130:49-57.

5. Wood EH, Sturm RE, Sanders JJ. Data processing in cardiovascular physiology with particular reference to roentgen videodensitometry. *Mayo Clin Proc.* 1984;39:849-865.

6. Reiderer SJ, Kruger RA. Intravenous digital subtraction: A summary of recent developments. *Radiology.* 1983;147:633-638.

7. Vogel R, LeFree M, Foster R, et al. Digital left ventriculography: Determination of the benefit from post-acquisition processing. *IEEE Comput Cardiol.* 1982:219-222.

8. Crummy AB. Computerized fluoroscopy: Digital subtraction for intravenous angiocardiography. *Am J Radiol.* 1980;135:1131-1140.

9. Mancini GBJ, Ostrander DR, Slutsky RA, et al: Intravenous vs. left ventricular injection of ionic contrast material: Hemodynamic implications for digital subtraction angiography. *Am J Cardiol.* 1983;140:425-430.

10. Kronenberg MW, Price RR, Smith CW, et al. Evaluation of left ventricular performance using digital subtraction angiography. *Am J Cardiol.* 1983;51:837-842.

11. Higgins CB, Gerber KH, Mattrey RA. Evaluation of the hemodynamic effects of intravenous administration of ionic and nonionic contrast materials. *Radiology.* 1982;142:681-686.

12. Sackett JR, Mann FA. Contrast media for digital subtraction arteriography of cerebral arteries. In: Mistretta CA, Crummy AB, Strother CM, et al. *Digital Subtraction Arteriography: An Application of Computerized Fluoroscopy.* Chicago: Year Book Medical Publishers, Inc; 1982:23-25.

13. Modie MT, Weinstein MA, Pavlicek W, et al. Intravenous digital subtraction angiography: Peripheral versus central injection of contrast material. *Radiology*. 1983;147:711–715.

14. Saddekni S, Sos TA, Sniderman KW, et al. Optimal injection technique for intravenous digital subtraction angiography. *Radiology*. 1984;150:655–659.

15. Nichols AB, Martin EC, Fles TP, et al. Validation of the angiographic accuracy of digital left ventriculography. *Am J Cardiol*. 1983;51:224–230.

16. Slutsky RA, Mancini GBJ, Norris S, et al. Digital intravenous ventriculography: Comparison of volumes from mask-mode and nonsubtracted images with thermodilution and sonocardiometric measurements. *Invest Radiol*. 1983;18:327–334.

17. Radtke W, Bursch JH, Brennecke R, et al. Visualization of the left ventricular wall by digital angiocardiography. *Eur Heart J*. 1981;2:135–142.

18. Radtke W, Bursch JH, Brennecke R, et al. Assessment of myocardial mass and infarction size by digital angiocardiography. In: Heintzen PH, Brennecke R, eds. *Digital Imaging in Cardiovascular Radiology*. Stuttgart: Georg Thieme Verlag; 1983:233–240.

19. Gerber KH, Slutsky RA, Ashburn WL, et al. Detection and assessment of severity of regional ischemic left ventricular dysfunction by digital fluoroscopy. *Am Heart J*. 1982;104:27–35.

20. Carey PH, Slutsky RA, Ashburn WL, et al. The validation of cardiac output by digital intravenous ventriculography in dogs: Correlation with thermodilution estimates. *Radiology*. 1982;143:623–626.

21. Goldberg HL, Borer JS, Moses JW, et al. Digital subtraction intravenous left ventricular angiography: Comparison with conventional intraventricular angiography. *J Am Coll Cardiol*. 1983;1:858–862.

22. Nissen SE, Booth D, Waters J, et al. Evaluation of left ventricular contractile pattern by intravenous digital subtraction ventriculography: Comparison with cineangiography and assessment of interobserver variability. *Am J Cardiol*. 1983;52:1293–1298.

23. Tobis JM, Nalcioglu O, Johnston WD, et al. Left ventricular imaging with digital subtraction angiography using intravenous contrast injection and fluoroscopic exposure levels. *Am Heart J*. 1982;104:20–27.

24. Vas R, Diamond GA, Forrester JS, et al. Computer-enhanced digital angiography: Correlation of clinical assessment of left ventricular ejection fraction and regional wall motion. *Am Heart J*. 1982;104:732–739.

25. Engels PHC, Ludwig JW, Verhoeven LAJ. Left ventricular evaluation by digital video subtraction angiography. *Radiology*. 1982;144:471–474.

26. Tobis JM, Nalcioglu O, Johnston WD, et al. Correlation of 10-milliliter digital subtraction ventriculograms compared with standard cineangiograms. *Am Heart J*. 1983;105:946–952.

27. Seldin DW, Esser PD, Nichols AB, et al. Left ventricular volume determined from scintigraphy and digital angiography by a semi-automated geometric method. *Radiology*. 1983;149:809–813.

28. Shaw CG, Ergun DL, Myerowitz PD, et al. A technique of scatter and glare correction for visodensitometric studies in digital subtraction videoangiography. *Radiology*. 1982;142:209–213.

29. Bursch J, Heintzen PH, Simon R. Videodensitometric studies by a new method of quantitating the amount of contrast medium. *Eur J Cardiol*. 1974;11:437–446.

30. Tobis J, Nalcioglu O, Seibert JA, et al. Measurement of left ventricular ejection fraction by videodensitometric analysis of digital subtraction angiograms. *Am J Cardiol*. 1983;52:871–875.

31. Mancini GBJ, LeFree MT, Vogel RA. Curvature analysis of normal ventriculograms: Fundamental framework for the assessment of shape changes in man. *Comput Cardiol*. 1985:141–144.

32. Goldberg HL, Moses JW, Borer JS, et al. Exercise left ventriculography utilizing intravenous digital angiography. *J Am Coll Cardiol*. 1982;2:1092–1098.

33. Tobis J, Nalcioglu O, Johnston WD, et al. Digital angiography in assessment of ventricular function and wall motion during pacing in patients with coronary artery disease. *Am J Cardiol*. 1983;51:668–675.

34. Mancini GBJ, Peterson KL, Gregoratos G, et al. Effects of atrial pacing on global and regional left ventricular function in coronary heart disease assessed by digital intravenous ventriculography. *Am J Cardiol*. 1984;53:456–461.

35. Johnson RA, Wasserman AG, Leibhoff RH, et al. Intravenous digital left ventriculography at rest and with atrial pacing as a screening procedure for coronary artery disease. *J Am Coll Cardiol*. 1983;2:905–910.

36. Fisher LD, Judkins MP, Lesperance J, et al. Reproducibility of coronary arteriographic reading in the coronary artery. Surgery study (CASS). *Cathet Cardiovasc Diagn*. 1982;8:565–575.

37. Grondin CM, Dyrda I, Paternac A, et al. Discrepancies between cineangiographic and post-mortem findings in patients with coronary artery disease and recent myocardial revascularization. *Circulation*. 1974;49:703–708.

38. White CW, Wright CB, Doty DB, et al. Does visual interpretation of the coronary arteriogram predict the physiological importance of a coronary stenosis? *N Engl J Med*. 1984;310:819–824.

39. Rutshauser W, Bussmann W, DiNoseda G, et al. Blood flow measurement through single coronary arteries by roentgen densitometry, I: A comparison of flow measured by a radiographic technique applicable in the intact organism and by electromagnetic flow meter. *Am J Roentgenol*. 1970;109:12–20.

40. Smith HC, Frye RL, Donald De, et al. Roentgen videodensitometric measure of coronary blood flow. Determination from simultaneous indicator-dilution curves at selected sites in the coronary circulation and in coronary artery-saphenous vein grafts. *Mayo Clin Proc*. 1971;46:800–806.

41. Spiller P, Schmiel FK, Politz B, et al. Measurement of systolic or diastolic flow rates in the coronary artery system by x-ray densitometry. *Circulation*. 1983;68:337–347.

42. Foerster JM, Link DP, Lantz BMT, et al. Measurement of coronary reactive hyperemia during clinical angiography by video dilution technique. *Acta Radiol*. 1981;22:209–216.

43. Gould KL, Lipscomb K, Hamilton GW. Physiologic basis for assessing critical coronary stenosis. Instantaneous flow response and regional distribution during coronary hyperemia as measures of coronary flow reserve. *Am J Cardiol*. 1974;33:87–94.

44. Robb RA, Wood EH, Ritman EL, et al. Three-dimensional reconstruction and display of the working canine heart and lungs by multi-planar x-ray scanning videodensitometry. *Comput Cardiol*. 1974:151–163.

45. Vogel RA. Radiographic assessment of coronary blood flow parameters. *Circulation*. 1985;72:460–465.

46. Hodgson JMcB, Mancini GBJ, LeGrand V. Characterization of changes in coronary blood flow during the first six seconds after intracoronary contrast injection. *Invest Radiol*. 1985;20:246–252.

47. Vogel RA, LeFree M, Bates E, et al. Application of digital techniques to selective coronary arteriography: Use of myocardial appearance time to measure coronary flow reserve. *Am Heart J*. 1983;107:153–164.

48. Hodgson JMcB, LeGrand V, Bates ER, et al. Validation in dogs of a rapid digital angiographic technique to measure relative coronary blood flow during routine cardiac catheterization. *Am J Cardiol*. 1985;55:188–193.

49. Cusma JT, Toggart EJ, Folts JD, et al. Digital subtraction angiographic imaging of coronary flow reserve. *Circulation*. 1987;75:461–472.

50. Elion JL, Nissen SE, DeMaria AN. Functional imaging of coronary flow reserve. *Am J Card Imaging*. 1987;1:103–110.

51. Klocke FJ. Measurements of coronary blood flow and degree of coronary stenosis. Current clinical implications and continuing uncertainties. *J Am Coll Cardiol*. 1983;1:31–41.

52. Hoffman JI. Maximal coronary flow and the concept of coronary vascular reserve. *Circulation*. 1984;70:153.

Chapter 10

Echocardiographic Evaluation of the Left Ventricle

Nelson B. Schiller, MD

PHYSICAL PRINCIPLES AND INSTRUMENTATION IN ECHOCARDIOGRAPHY

Echocardiography or cardiac ultrasound uses inaudible high-frequency sound to form images of cardiac structures and measure intravascular blood flow. High-frequency sound penetrates non–air-containing structures and reflects from interfaces formed by structures that differ in acoustic impedance (tissue density × speed of sound propagation). These reflections can be detected at the skin surface and used to construct a real-time image (25 to 30 frames per second) of the insonicated object, much as a navigator uses a depth finder to locate the sea bed.

As muscle and valves have different acoustic impedances, it is possible to image these contiguous structures by their different reflectances. Since ultrasound studies are usually conducted without using parenteral agents and are entirely noninvasive, ultrasound provides a risk-free method of organ imaging.

There are three modalities of cardiac ultrasound in use today. The more elaborate instruments now manufactured offer all three combined into one package. The three modalities are M-mode, two-dimensional, and Doppler echocardiography. (See Chapter 1.)

M-Mode Echocardiography

M-mode echocardiography was the first of these techniques to find wide application to cardiac diagnosis (Figures 10-1 and 10-2).[1–3] Here, an ultrasound generating and receiving transducer is placed on the anterior chest wall imaging window to provide an airless pathway to the heart, and a single beam of ultrasound is directed into the heart. Structures lying along that single beam send back signals to the stationary transducer, and returning signals are recorded onto a moving strip of paper in a location representing the distance of the reflecting object from the transducer (i.e., by the transit time of ultrasound to and from the target). Besides their characteristic locations, cardiac structures have unique motion patterns that identify them; alterations in these patterns identify pathology and its extent.

An advantage of M-mode echocardiography is that it allows sampling of moving cardiac structures at high rates (approaching 1000 per second), providing images of striking time and depth resolution. Its major disadvantage is its one-dimensionality and potential to produce unrepresentative samples of structures. An example of insufficient sampling is found in M-mode echocardiograms of the left ventricle. The normal left ventricle is a truncated ellipse, and the dilated left ventricle, a sphere. Since an M-mode study only provides minor axis images and no information about the left ventricle's long axis, misinformation about left ventricular cavity volume and myocardial noise are unavoidable.[4] Isolated or stand-alone M-mode instruments are becoming rare, and the standard echocardiographic examination is performed with an instrument capable of generating both two-dimensional and M-mode echocardiographic images. M-

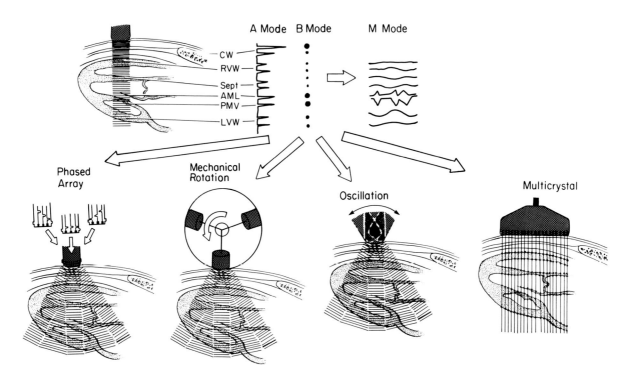

Figure 10-1 Instrumentation of echocardiography: M-mode and two-dimensional imaging methods. Top, The long axis of the left ventricle is diagrammed, showing the M-mode transducer applied to the chest wall with the ultrasound beam directed posteriorly. The stages in the production of an M-mode image are to the right of this diagram. Each intracardiac surface or interface formed by a blood-muscle, a blood-valve, or a muscle-pericardium boundary gives rise to reflections that are detected by the transducer during its listening mode. The excitation of the transducer gives rise to small electrical currents that are displayed in the order in which they are received. If the representation of these currents is a series of spikes whose height is proportional to the strength of the signal, the display is termed A-mode. If these spikes are modulated into dots whose size and brightness are proportional to signal strength, the display is called B-mode. Both A-mode and B-mode displays depict motion of structures in real time. However, in order to create a permanent record of the location, size, and motion of structures lying along the ultrasound beam, M-mode display was developed (see Figure 10-2). In M-mode, photosensitive paper is rolled or drawn across the B-mode display, creating a continuous record. RVW, right ventricular wall; sept, septum; AML, anterior mitral leaflet; PMV, posterior mitral valve; LVW, left ventricular wall. Bottom, Four methods of producing two-dimensional images are shown. In all, a single beam of ultrasound is played back and forth through a sector, enabling display of a cross-section of tissue in real time. In the phased array, an array of crystals with a relatively small diameter or "footprint" is excited. The excitation pattern allows the beam to be steered rapidly through the sector. Mechanically produced sector scans are produced by rotating three or four M-mode transducers past a window. This window is in contact with the skin surface. The motion required is supplied by a small electrical motor. In the oscillating type of mechanical two-dimensional instrument, a single M-mode–type crystal is rocked back and forth to create a two-dimensional display. In the multicrystal linear array, a long line of single transducer elements is directed at a target. The result is an image of the same size as the array. The intercostal spaces are usually too large to permit the use of this type of technology for cardiac imaging. *Source:* Courtesy of NH Silverman, San Francisco, Calif.

mode echocardiographic information, which remains valuable, is more reliable when obtained directly from these two-dimensional images.

Two-Dimensional Echocardiography

Two-dimensional echocardiography is the dominant ultrasound imaging method (Figure 10-1). A two-dimensional image is a real-time tomograph created by rapidly sweeping an ultrasound beam back and forth across cardiac structures. Depending on depth of targeted structures, it is capable of scanning a 60° to 90° scan plane at 25 to 60 frames per second. The resolution of the image depends on how deeply or laterally into the image information is situated (the deeper and more off-axis, the poorer the resolution) and upon the transducer carrier frequency. Higher frequencies have higher resolution but poorer penetration. In children, where depths are shallow and anatomic windows large, highly resolved images are usually obtained because conditions are ideal for employing high frequencies. As engineering improves, manufacturers are equipping instruments with transducers with frequencies of 5 to 10 MHz (1 MHz = 1,000,000 cycles per second). Recently, the standard imaging frequency was 2.25 MHz, which was satisfactory for adults but poor for children. Now most adults can be imaged with 3.5 MHz instruments, with higher frequencies being used for pediatric applications. The recent introduction of annular array technology and large-element phased arrays (96 and 128) may enable the use of frequencies as high as 5 or 7 MHz in adults.[5]

Doppler Echocardiography

Doppler echocardiography has undergone rapid propagation in the last 6 years. Essentially, this technique allows the measurement of intracardiac flow velocities by detecting changes in the carrier frequency of reflected ultrasound returning to the transducer. When ultrasound strikes moving red blood cells in the heart, it is not only reflected back to the transducer but its frequency is proportionately altered depending on the velocity of the red blood cells (Figure 10-3).[8] (See also Chapter 2.)

This velocity difference is displayed in real time as a function of time and simultaneously converted into an audible signal. The direction of the flow relative to the detecting transducer is also shown by displaying the spectral shift as lying above or below an arbitrary baseline (Figure 10-4). Progress in microprocessors has made the real-time generation of spectral information a major factor in the rapid proliferation of the Doppler technique. In addition to the velocity and duration of flow, the real-time spectral display provides information about the intensity of flow. The Doppler equation governing these relationships is

$$V = \frac{C}{2 F_O \times \cos \emptyset} F_d,$$

where C is velocity of sound, F_O is carrier frequency, COS is the cosine of the angle of interrogation, V is velocity, and F_d is frequency shift (Figure 10-5).

Two Doppler modalities are required for diagnostic examinations, pulsed wave and continuous wave; these are used interactively as each has unique qualities. Pulsed wave Doppler is able to examine a very limited region of flow and localize flow disturbances. For example, placement of the Doppler sample volume on the atrial surface of the mitral valve can detect even minimal mitral insufficiency. Pulsed wave Doppler has the disadvantage of accurately measuring only flow velocities in the normal range (<2 m/sec in adult patients). Rapidly moving flow, characteristic of valve lesions, cannot be measured by pulsed Doppler because rapid velocity induces aliasing. Aliasing produces an effect analogous to that seen in a motion picture of an airplane propeller; as the propeller increases speed, it appears to spin in the opposite direction because the slow frame rate of the film produces an aliased image. The relationships governing the limitations of pulsed wave Doppler are called the Nyquist limit and are defined by flow theorems.[8]

Continuous wave Doppler overcomes the aliasing problem and can be used to accurately measure high-velocity flow within the heart. For example, in critical aortic stenosis, velocities as high as 6 m/sec can be

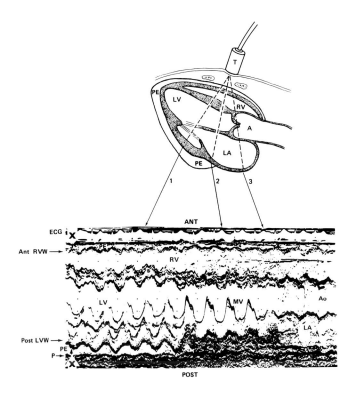

Figure 10-2 M-mode echocardiogram from a patient with a moderate pericardial effusion. Top, Diagrammatic representation of the long axis of the heart. The transducer (T) sits on the anterior chest wall in the precordial parasternal window, and the beam is swept from apex to base. The appearance of the echocardiogram at three beam positions on the diagram is designated by the arrows connecting the diagram to the echocardiogram. Bottom, The pericardial effusion (PE) is seen as an echo-free space posterior to the left ventricle. This space diminishes and finally disappears as the base of the heart is approached. A and Ao, aorta; Ant, anterior; ECG, electrocardiogram; LA, left atrium; LV, left ventricle; MV, mitral valve; P, pericardium; PE, pericardial effusion; Post, posterior; RV, right ventricle; W, wall. Source: Reproduced, with permission, from M Sokolow and MB McIlroy: *Clinical Cardiology*, 4th edition, copyright Lange Medical Publications, 1986 (Courtesy of NB Schiller).

There are two major types of two-dimensional echocardiographic imaging systems, mechanically driven large crystals (Figure 10-1) and electronically driven phased crystal arrays (32 to 128) (Figure 10-1).[6] The mechanical systems have enjoyed slightly better resolution but require bulky, motor-containing transducers that are prone to failure with extended use. The electronic phased array is now the predominant form of instrument, and its ascendancy is accelerating with the spread of Doppler ultrasound techniques. It appears to be easier for manufacturers to incorporate Doppler technology into two-dimensional instruments if the instrument is a microprocessor-driven phased array than if it is a mechanical system. In a parallel development, prototypes of annular arrays are beginning to appear, which combine features of both mechanical and phased array technology.[7]

REFLECTED SOUND FROM TARGET

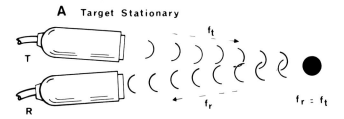

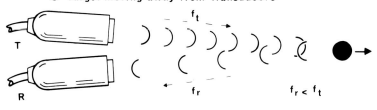

DOPPLER SHIFT OR FREQUENCY $(f_d) = f_r - f_t$

Figure 10-3 Pairs of transmitting and receiving transducers illustrate the Doppler effect as it occurs in relationship to targets that are stationary, moving toward, and moving away from the transducers. A, The carrier frequency (f_t) from the transmitting transducer (T) strikes the target (e.g., a red blood cell) and is reflected back to the receiving transducer (R) at the reflected frequency (f_r), which is unaltered from f_t. B, f_r is a higher frequency than the original transmitted frequency. An increase in frequency is typically seen when a target is moving toward the transducer. C, The carrier frequency, after being reflected from a target moving away from the transducers, is reduced. In all cases, the extent to which the carrier frequency is increased or reduced is proportional to the velocity of the target. *Source:* Reprinted from *Echocardiography*, ed 4 (p 30) by H Feigenbaum with permission of Lea & Febiger, © 1986.

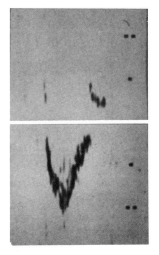

Figure 10-4 A typical Doppler signal from intracardiac blood flow. The moving red blood cells were traveling away from the transducer and resulted in a drop in the carrier frequency when reflected sound returned to the transducer. By convention, flow away from the transducer is displayed below the baseline, and that toward the transducer, above. Due to high-speed microprocessors, this spectral information can be displayed in real time. Note that the Doppler signal portrays the entire period of flow, showing acceleration, peak flow, and deceleration. If the same event is viewed from a window where the flow runs toward the transducer, the same waveform will be displayed above the baseline. *Source:* Courtesy of Johnson & Johnson Ultrasound Inc, © 1985.

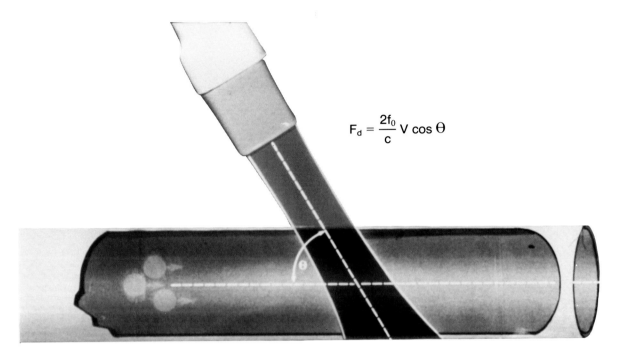

Figure 10-5 A diagrammatic illustration of a situation when the interrogating or insonicating beam lies at an angle to the flow of blood. Ideally the Doppler beam is placed parallel to blood flow. When the beam does not lie parallel, it is possible to introduce a correction into the calculation of flow velocity by measuring the cosine of the angle of interrogation and introducing this value into the Doppler equation. The version of the equation in this illustration merely represents a rearrangement of the equation given in the text. *Source:* Courtesy of Johnson & Johnson Ultrasound Inc, © 1985.

encountered. The disadvantage of continuous wave Doppler is that it samples everything lying along the path of the beam. Thus the ability of pulsed Doppler to precisely sample localized flow is lost. The advantage gained is that a great deal of quantitative information is contained in high-velocity jets, and continuous wave Doppler can extract that information. Normal flow velocities are given in Table 10-1.[8]

Color flow mapping is a major new development and is proliferating rapidly. A refinement of pulsed Doppler, this technique allows simultaneous visualization of blood flow throughout the field of interrogation. In the left ventricle, diastolic flow from the atrium is visualized as a color change of the pixels along the left ventricular inflow tract. If the left ventricle is viewed from the apex, mitral inflow is toward the transducer and will be coded in warm colors (red, orange, etc.). Left ventricular outflow, away from the transducer, will appear in cool colors (blue, green). The intensity of the color indicates the quantity of flow and the shade the velocity. Aliasing appears as reversals in direction (i.e., blue to red).

Because flow mapping allows rapid assessment of several regions and valves, it has advantages over more laborious mapping with a single pulsed Doppler sample volume, and its wide acceptance is assured.

Limits of Cardiac Echocardiography

It is important to be aware of the technical limitations of a cardiac ultrasound examination. Although some information can be obtained in nearly all patients, complete examinations are not always possible. For example, patients with obstructive airway disease and hyperinflated lungs are difficult to image; often only the subcostal imaging window will allow the echocardiographic technologist to obtain useful data. A similar situation is encountered in the very obese; fat conducts ultrasound poorly, and the resulting images can be of very poor quality. Patients in critical care situations on ventilators or postoperative patients also present a challenge to the skill of the technologist. Whenever echocardiographic data become limited to a few views or projections, they become less reliable. In a technically

Table 10-1 Peak Doppler Velocity Ranges for Children and Adults

	Children	Adults
Mitral flow	1.00 (0.8–1.2)	0.9 (0.4–1.3)
Tricuspid flow	0.60 (0.5–0.8)	0.5 (0.3–0.7)
Pulmonary artery	0.90 (0.7–1.1)	0.75 (0.6–0.9)
Left ventricle	1.00 (0.7–1.2)	0.90 (0.7–1.1)
Aorta	1.5 (1.2–1.8)	1.35 (1.0–1.7)

Source: Reprinted from *Doppler Ultrasound in Cardiology: Physical Principles and Clinical Application*, ed 2 (p 93) by L Hatle and B Angelson with permission of Lea & Febiger, © 1985.

optimal study, the interpretation can be facilitated by the plethora of redundant information that an echocardiographic examination characteristically provides. For example, in mitral stenosis the examiner has at least three views of the valve on two-dimensional imaging, the characteristic pattern of the disease on M-mode echocardiographic imaging and the high-velocity inflow pattern of the Doppler flow study. If the patient is technically difficult to study, only a small amount of this information might be available, and the characteristic diagnostic certainty is compromised.

EVALUATION OF THE LEFT VENTRICLE

Evaluation of the left ventricle is an important clinical application of echocardiography; a combined M-mode echocardiographic, two-dimensional Doppler evaluation can provide reliable information about left ventricular systolic function or global contractility, most commonly expressed as ejection fraction. Left ventricular size or volume, wall thickness, mass, and the motion of individual segments of the myocardium can also be assessed. Furthermore, an impression of the diastolic behavior of the left ventricle is obtained from a competently performed echocardiogram.

Assessment of Function (Global Contractility) of the Left Ventricle

Assessment by M-Mode Echocardiography

Assessment of left ventricular function requires images from several planes. Usually, the left ventricle is initially viewed in the precordial long-axis (Figure 10-6) and short-axis planes (Figure 10-7). M-mode tracings are generated from these two-dimensional images. Figures 10-8 and 10-9 show two-dimensional images on the left and M-mode images on the right. The line from which the M-mode image is generated is superimposed on the two-dimensional image.[9-10] Visual integration from the real-time nature of the images contributes to the information in the image and cannot be represented in these static examples. Technically, it is essential to obtain as much of the epicardium and endocardium as possible; failure to capture the actual boundaries of the convoluted endocardial surface is a common source of error in evaluation of left ventricular function.

All tomographic planes are recorded so that the cavity size is maximized. In addition to precordial views, the left ventricle is viewed from the apex in the two- and four-chamber long-axis planes and subcostally. These images are shown in Figures 10-10 to 10-14. Of these, M-mode images of the left ventricle can only be derived from the subcostal plane.

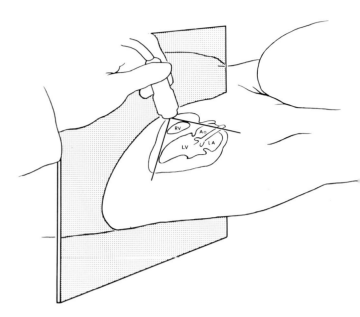

Figure 10-6 By directing the plane of the ultrasound beam from left hip to the right shoulder, the heart is sectioned along its long axis. This is usually the first view obtained in a standard ultrasound cardiac examination. The left ventricular base is shown, with the septum and posterior/inferior walls forming the perimeter of the chamber. Ao, aorta; LA, left atrium; LV, left ventricle; RV, right ventricle. *Source:* Reprinted from *Cardiology* by W Parmley and K Chatterjee (Eds) with permission of JB Lippincott Company, © 1987.

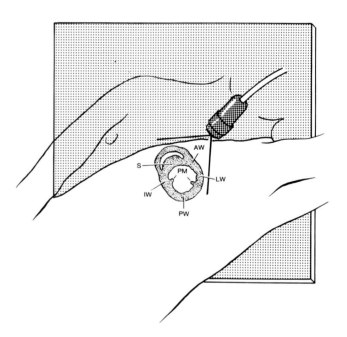

Figure 10-7 Short-axis view of the heart. In this view, the plane of the cut runs through or just below the minor axis of the left ventricle. AW, anteroseptal wall of the left ventricle; IW, inferior or diaphragmatic wall of the left ventricle; LW, lateral wall of the left ventricle; PW, posterior wall of the left ventricle; S, interventricular septum; PM, papillary muscles (posteromedial, right, and anterolateral, left). *Source:* Reprinted from *Cardiology* by W Parmley and K Chatterjee (Eds) with permission of JB Lippincott Company, © 1987.

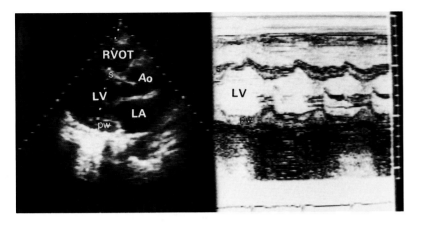

Figure 10-8 M-mode tracing taken through the minor axis of the left ventricle (right), generated directly from the long-axis two-dimensional view of the heart (left). The location of the line from which the M-mode tracing was generated is indicated by the dotted white line running through the base of the left ventricle (LV) and into the posterior wall (pw). Ao, aorta; LA, left atrium; RVOT, right ventricular outflow tract. *Source:* Reprinted from *Cardiology* by W Parmley and K Chatterjee (Eds) with permission of JB Lippincott Company, © 1987.

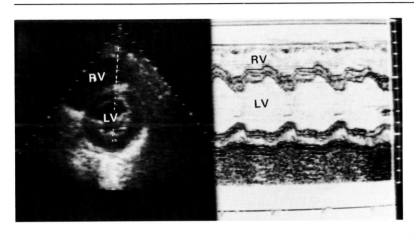

Figure 10-9 The same M-mode image of the left ventricle is obtained from a short-axis two-dimensional image of the left ventricle (LV). This is the preferred method of directing the M-mode beam (dotted line) through the two-dimensional image to ensure reproducible M-mode measurements of the left ventricle. RV, right ventricle. *Source:* Reprinted from *Cardiology* by W Parmley and K Chatterjee (Eds) with permission of JB Lippincott Company, © 1987.

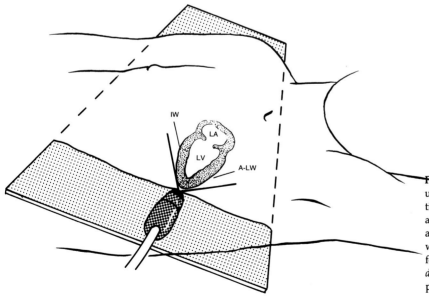

Figure 10-10 Transducer position and beam plane used in imaging the heart from the apex impulse location. This view is the two-dimensional, two-chamber apical view. It is similar to a right anterior oblique angiogram of the left ventricle (LV) in that the inferior wall (IW) and anterolateral walls (A-LW) are border forming. LA, left atrium. *Source:* Reprinted from *Cardiology* by W Parmley and K Chatterjee (Eds) with permission of JB Lippincott Company, © 1987.

The most useful M-mode indices of left ventricular function are fractional shortening of the left ventricle's minor axis,[11] mitral septal separation,[12–13] and aortic root motion.[10,14–22]

Fractional shortening is the distance the left ventricle moves from its maximum end-diastolic dimension to its minimum end-systolic dimension. Normally this cavity reduction is 30% or more. The interpretation of this index and of others depends not only on the health of the myocardium but also on its loading conditions and inotropic environment. Figure 10-9 illustrates an M-mode tracing with normal fractional shortening, and

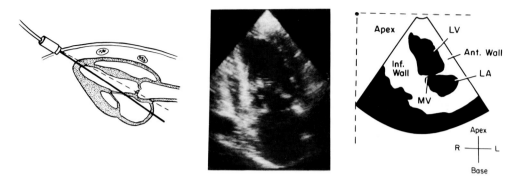

Figure 10-11 Diagrammatic illustration of the transducer location and beam direction used in forming the apical two-chamber view (left). In the center of the figure is an actual diastolic image from a patient with mild left ventricular hypertrophy and, to the right, a diagram of this image. The structure directly beneath the inferior wall is the diaphragm; beneath that is the liver. Ant. Wall, anterior wall; Inf. Wall, inferior wall; LA, left atrium; LV, left ventricle; MV, mitral valve. *Source:* Reprinted from *Cardiology* by W Parmley and K Chatterjee (Eds) with permission of JB Lippincott Company, © 1987.

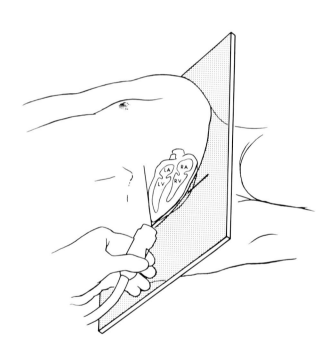

Figure 10-12 Transducer position and beam plane used to image the heart in the apical four-chamber view. The transducer is on the apex impulse location and is directed toward the right shoulder. LA, left atrium; LV, left ventricle; RA, right atrium; RV, right ventricle. *Source:* Reprinted from *Cardiology* by W Parmley and K Chatterjee (Eds) with permission of JB Lippincott Company, © 1987.

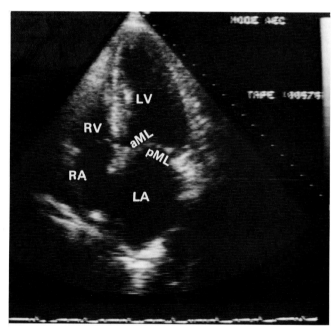

Figure 10-13 Stop-frame image of an apical four-chamber view. Note that the display shows the apex at the top of the screen, with the left ventricle (LV) to the viewer's right. Most laboratories choose to display the four-chamber view in this way, although some reverse the ventricles and others have the apex pointing downward. aML, anterior mitral leaflet; LA, left atrium; pML, posterior mitral valve leaflet; RA, right atrium; RV, right ventricle. *Source:* Reprinted from *Cardiology* by W Parmley and K Chatterjee (Eds) with permission of JB Lippincott Company, © 1987.

Figure 10-15 an M-mode cardiomyopathic tracing with a reduced index.

A potential source of error in the use of M-mode minor-axis fractional shortening is that the sample from which this assessment of global left ventricular function is made represents only a small sample of the structure.[23] In coronary artery disease, segments remote from this site may be abnormal, but this situation may not be reflected in the motion of the minor axis.

E-point septal separation (EPSS) and aortic root motion are M-mode indices of global left ventricular function. In EPSS, the proximity of the most anterior opening point of the mitral valve to the most posterior systolic location of the septum is measured. Normally the mitral valve opens to within 5 mm of the septum (Figure 10-16). When the ejection fraction falls, the amount of residual blood at end-systole increases, and the ventricle dilates. As a result, systolic stroke volume

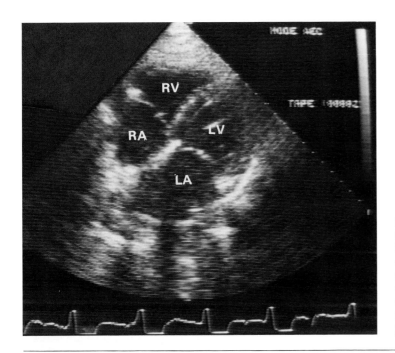

Figure 10-14 With the transducer in the subcostal position, a four-chamber view can be obtained. In this view, the lateral border of the left ventricle (LV) is delineated by the midlateral wall as opposed to the inferolateral wall on the apical four-chamber view. The apex of the left ventricle is rarely seen in this view, and the right ventricle (RV) is usually smaller than the left. Other abbreviations are explained in the legend for Figure 10-13. *Source:* Reprinted from *Cardiology* by W Parmley and K Chatterjee (Eds) with permission of JB Lippincott Company, © 1987.

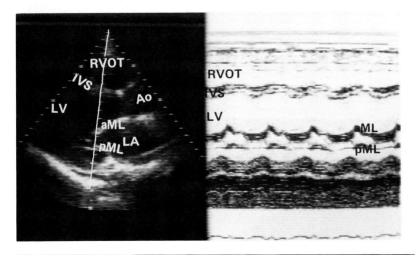

Figure 10-15 Spherical dilation and diffusely poor inward wall motion are reflected in both the two-dimensional (left) and the M-mode (right) images. Compare this figure with the normal ventricle in Figure 10-9. IVS, interventricular septum; other abbreviations are explained in legends for Figures 10-8 and 10-13. *Source:* Reprinted from *Cardiology* by W Parmley and K Chatterjee (Eds) with permission of JB Lippincott Company, © 1987.

is reduced, and the amount of blood coming across the mitral valve in diastole to replace it is also reduced. The mitral valve responds to decreased inflow by opening less. This combination of events results in the anterior excursion of the valve being decreased while the inward motion of the septum is also decreased; the two structures separate in proportion to the degree of dysfunction (Figure 10-17). In my laboratory, rather than make exact measurements of EPSS, I use it as a visual clue to the presence of left ventricular dysfunction; for exact quantitation I use quantitative two-dimensional or Doppler methods.

A third qualitative M-mode clue to left ventricular function is the motion of the aortic root at the base of the heart, because it is proportional to stroke volume. This phenomenon is created by the filling of the atrium behind the aortic root and the kinetic energy associated with each systole (Figures 10-18 and 10-19).[17] Normally

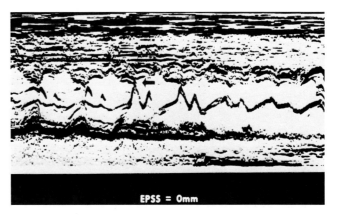

Figure 10-16 EPSS is used as an M-mode index of global left ventricular function. In a normal subject, the plane of the maximum inward septal incursion and that of the maximum anterior excursion of the mitral valve are roughly the same. *Source:* Reprinted from *Cardiology* by W Parmley and K Chatterjee (Eds) with permission of JB Lippincott Company, © 1987.

the aortic root moves forward more than 7 mm in systole (Figure 10-20). Caution in the interpretation of this index is advised because a low stroke volume may reflect decreased preload and not imply decreased contractility. If the aortic valve leaflets can be visualized simultaneously with the aortic root, it is a simple matter to derive systolic time intervals (pre–ejection period and left ventricular ejection time) from their movement.[24–25] The degree of aortic valve opening and the shape of its motion pattern are also clues to the stroke volume.

In recent years there has been interest in computer digitization of M-mode records.[26] In view of the unavailability of computers for performing these measurements, this area will not be covered.

Assessment by Two-Dimensional Echocardiography

Two-dimensional assessment of left ventricular function is performed both qualitatively and quantitatively. The quantitative approach is the most accurate echocardiographic means to evaluate left ventricular contractile function. It remains to be seen whether Doppler

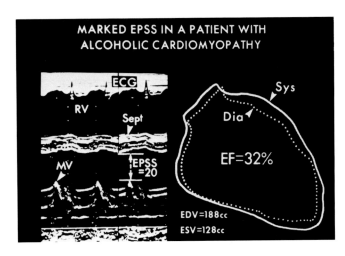

Figure 10-17 The M-mode echocardiogram on the left and the angiogram on the right are from a patient with advanced cardiomyopathy. The EPSS is 20 mm, and the ejection fraction (EF) is 32%. EDV, end-diastolic volume; ESV, end-systolic volume; MV, mitral valve; RV, right ventricle; Sept, septum; ECG, electrocardiogram; Dia, diastolic; Sys, systolic. *Source:* Reprinted from *Cardiology* by W Parmley and K Chatterjee (Eds) with permission of JB Lippincott Company, © 1987.

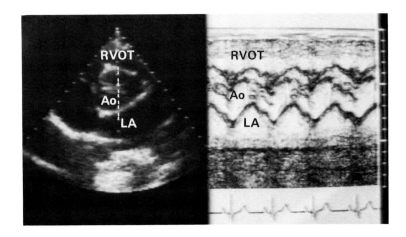

Figure 10-18 Normal systolic and diastolic aortic root motion in a patient with a bicuspid aortic valve. The aorta is imaged in the short-axis view, and on the two-dimensional image (left) appears circular. The M-mode image to the right is generated from the dotted line running through the aortic root (Ao) from the right ventricular outflow tract (RVOT) anteriorly to the left atrium (LA) posteriorly. The M-mode image shows that the anterior systolic motion of the root is 20 mm (scale to right of M-mode image = 1 cm per division). *Source:* Reprinted from *Cardiology* by W Parmley and K Chatterjee (Eds) with permission of JB Lippincott Company, © 1987.

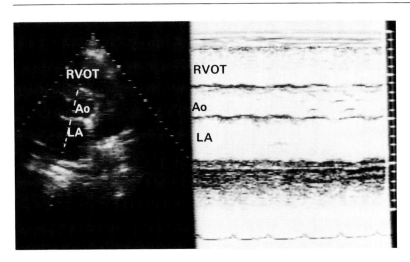

Figure 10-19 Depressed aortic root motion in a patient with cardiomyopathy. In comparison to Figure 10-18, note that the aortic root appears nearly motionless throughout the cardiac cycle. The decrease in root motion is a direct expression of the decrease in the stroke volume. If the stroke volume is depressed because of very small ventricular size, the aortic root motion will be depressed. In other words, the aortic root motion is independent of ejection fraction. Usually, of course, depressed stroke volume is associated with depressed ejection fraction. Abbreviations are explained in the legend for Figure 10-18. *Source:* Reprinted from *Cardiology* by W Parmley and K Chatterjee (Eds) with permission of JB Lippincott Company, © 1987.

Echocardiographic Evaluation of the Left Ventricle   139

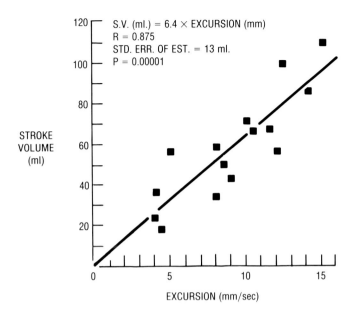

Figure 10-20 Quantitative relationship between aortic root excursion and angiographic stroke volume in 15 patients. *Source:* Reprinted from *Cardiology* by W Parmley and K Chatterjee (Eds) with permission of JB Lippincott Company, © 1987.

techniques will surpass this technique in accuracy and usefulness; in all likelihood they will be complementary.

In common practice, echocardiograms are evaluated like left ventricular cineangiograms. That is, the degree of emptying is judged by visual study of the contracting heart. While it has been shown that visual estimates of left ventricular ejection fraction are reliable,[27–28] we have been impressed by an occasional significant error arising when comparing visual estimation to direct quantitation.

The principle guiding quantitation of two-dimensional images is that a geometric model is chosen to represent the left ventricle. Then, using values obtained by planimetry of the endocardial surfaces and linear measurements of distances between them, left ventricular volume is calculated from an algorithm based on that model. Numerous algorithms have been validated, but these will not be discussed (Figure 10-21).[29] In our laboratory we use a form of Simpson's rule to calculate ventricular volumes.[4,30] This method, more properly termed the method of discs, is recommended by the American Society of Echocardiography Commit-

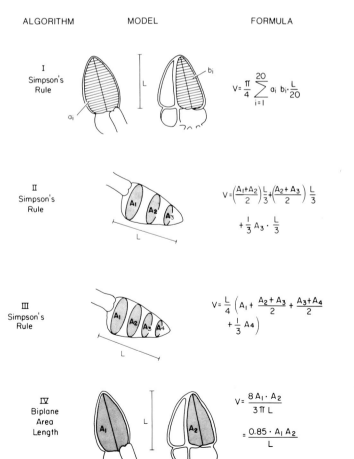

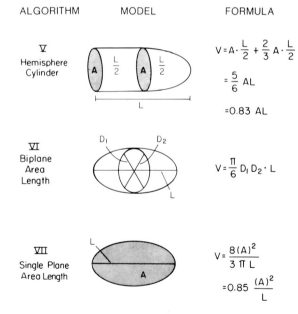

Figure 10-21 Seven algorithms used in the planimetric assessment of tomographic images of the left ventricle. The shaded or outlined areas represent the portions of the images of ventricle that must be measured or planimetered in order to satisfy the requirements of the various algorithms. The figures to the left are 2-chambered. The figures on the right are 4-chambered. In our laboratory, we use I (modified Simpson's rule, or method of disks); IV, biplane area length; or VII, single plane area length. Wherever possible, we prefer to use I. *Source:* Reprinted from *Two Dimensional Echocardiography in Congenital Heart Disease* by N Silverman and A Snider with permission of Appleton & Lange, © 1982.

tee on Quantitation of Two-Dimensional Echocardiograms.* This method is relatively independent of geometry in that it reconstructs the ventricle from 20 disk-shaped slices derived from biplane two-dimensional images assumed to be orthogonal to one another. This method has been validated by angiography and scintigraphy in a number of centers (Figure 10-22).[4,30–33] Its major shortcomings are that it underestimates angiographic volumes by approximately 25%, it can be time consuming, and it requires a dedicated on-line or off-line computer system. Nonetheless, as the price of these systems falls and the quality of two-dimensional images improves, the use of quantitative methods will grow.

Figure 10-23 illustrates paired two-dimensional images from which volumes have been computed. Note that the computer has superimposed an outline on the endocardial surface. This area is used by the computer to derive volumes. Normal sendentary population values for this technique are given in Table 10-2.[34]

An advantage of quantitative left ventricular two-dimensional echocardiography is that it provides volumes as well as the derived parameters of ejection fraction and stroke volume. The interpretation of quantitative data from a given study depends on knowledge of the clinical setting. With Doppler techniques it is possible to use this information in conjunction with knowledge of valvular function. Doppler also is an effective method of determining stroke volume.[35] Other indices, such as acceleration of systolic aortic flow and its peak velocity, appear to have utility as indices of left ventricular function.[36–37]

Assessment by Doppler Echocardiography

As in the quantitation of two-dimensional left ventricular echocardiograms, there are numerous proposed Doppler methods for determining the stroke volume of the left ventricle.[38–55] In our laboratory, we have used continuous wave Doppler combined with an M-mode image of the aortic valve to generate stroke volume information.[35] All of these methods are based on the use of the flow velocity integral and the cross-sectional area through which it is measured (Figure 10-24). The product of the duration of flow and its mean velocity is the distance blood travels during systole (stroke distance or flow velocity integral). The product of this distance and the cross-sectional area through which it flows is stroke volume. The product of stroke volume and heart rate is cardiac output.

---

*Schiller NB, Shah PM, Crawford M, et al. Recommendations for quantitation of the left ventricle by two-dimensional echocardiography. *J Am Soc Echocardiog*. 1989;2:358–367.

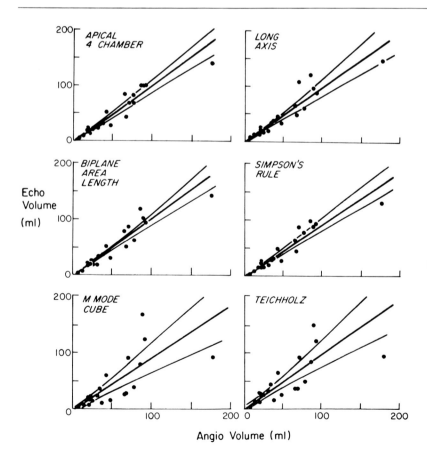

Figure 10-22 Relationship of angiographic left ventricular volumes to two-dimensional and M-mode methods. Top row, single plane area-length measurements of apical four-chamber view (left) and two-chamber long-axis view (right). Middle row, the two biplane methods. Note that Simpson's rule is the best of all methods, by a small margin. Bottom row, single-dimension extrapolations from M-mode images; these are far inferior as methods of predicting angiographic volume. This shortcoming is present despite the fact that this study was done with children with geometrically uniform ventricles. *Source:* Reprinted with permission from *Circulation* (1980;62:548), Copyright © 1980, American Heart Assocation.

Echocardiographic Evaluation of the Left Ventricle 141

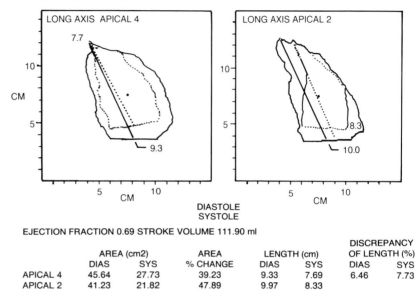

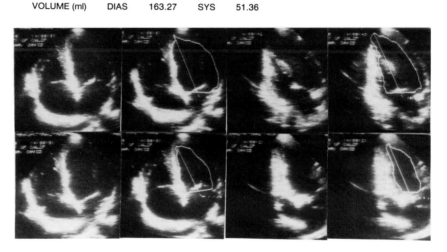

Figure 10-23 Quantitative two-dimensional echocardiography. A computer has been used to analyze digitized two- and four-chamber apical views by a Simpson's rule algorithm. In the bottom panel, two-chamber views in systole and diastole are shown on the right, and four-chamber views on the left, both with and without their endocardial outlines superimposed. Dias, diastole; Sys, systole. *Source:* Reprinted from *Cardiology* by W Parmley and K Chatterjee (Eds) with permission of JB Lippincott Company, © 1987.

Table 10-2 Normal Values for Left Ventricular End-Diastolic Volumes

Algorithm	Mean ± SD (Range) mL	Mean ± SD (Range) mL/m²
Four chamber area length		
Males	112 ± 27 (65–193)	57 ± 13 (37–94)
Females	89 ± 20 (59–136)	
Two chamber area length		
Males	130 ± 27 (73–201)	63 ± 13 (37–101)
Females	92 ± 19 (53–146)	
Simpson's biplane rule		
Males	111 ± 22 (62–170)	55 ± 10 (36–82)
Females	80 ± 12 (55–101)	

Source: Reprinted from *Cardiology* by W Parmley and K Chatterjee (Eds) with permission of JB Lippincott Company © 1987.

The Difficulty of Evaluating Left Ventricular Function

Differentiating among abnormalities of left ventricular contractile function is difficult. If left ventricular contraction is seen to be globally reduced, either by qualitative or quantitative assessment, a cardiomyopathy is likely. However, quantification of the precise type of cardiomyopathy is usually dependent upon knowledge of the clinical history and of features of left ventricular anatomy such as wall thickness. Unfortunately, our knowledge of the etiology of most cardiomyopathies remains rudimentary, and our ability to differentiate among them by echocardiographic means, limited. Our studies have shown that finding segmental abnormalities suggests an ischemic etiology. However, heterogeneity of involvement also occurs in nonischemic cardiomyopathy.[56] The finding of depressed contractility, particularly without dilation, can suggest a

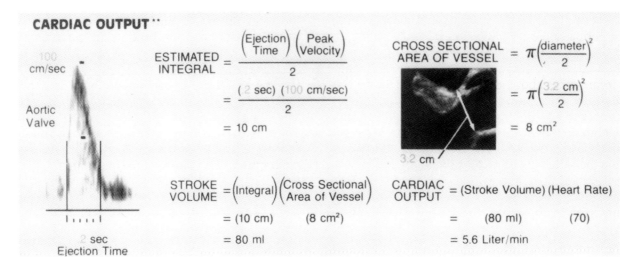

Figure 10-24 One of many methods for determining stroke volume by Doppler. In this method, the diameter of the left ventricular outflow tract is obtained from the two-dimensional long-axis view on the left. From this value, the area through which blood flows is calculated. A Doppler signal from the site from which cross-sectional area was determined provides the velocity profile of systolic flow at that location. Planimetry of that signal allows determination of the mean velocity and the duration of flow. The product of the mean velocity, duration of flow, and cross-sectional area is the volume of blood flow per beat, or the stroke volume. The product of the stroke volume and the heart rate provides the cardiac output. *Source:* Courtesy of Johnson & Johnson Ultrasound Inc.

restrictive cardiomyopathy or a secondary, nonmyocardial influence on global function. For example, inappropriately rapid heart rates can be associated with decreased ejection fraction without necessarily implying myocardial disease. Certain metabolic states can also be associated with myocardial depression. Acidosis is said to cause myocardial depression. Pharmacologic agents can also temporarily depress myocardial function. Anesthetic vapors, for example, are potent myocardial depressants.[57]

Assessment of Left Ventricular Size, Shape, Wall Thickness, and Mass

Left ventricular size can be measured with the same methods used to determine ejection fraction.

M-mode echocardiographic measurement of the left ventricle's minor-axis dimension is the oldest ultrasound technique for measuring the left ventricle.[58–59] While more informative than plane chest roentgenography and still in widespread use today, these methods can provide misleading information about left ventricular size because a single dimension is obtained from the minor axis and gives no information about the length of the chamber. Nonetheless, for years it was common practice to take the cube of this minor-axis dimension as an analog of left ventricular volume.[23] Currently the minor-axis dimension (uncubed) and its fractional shortening are the only acceptable ways of using this information. Because of the simplicity of these linear measurements, the majority of echocardiography laboratories rely on a single minor-axis left ventricular dimension to assess left ventricular size. Some laboratories prefer to take linear dimensions directly from the two-dimensional image.[60]

Two-dimensional methods of planimetry (particularly if they include all regions of the left ventricle) provide a reasonable estimate of left ventricular size. These measurements require the use of computer digitizing devices and knowledge of normal values (Table 10-3).[34] Additionally, as the population becomes

Table 10-3 Left Atrial Volumes Obtained from a Normal Population

	Males (mL)		Females (mL)		Volume Index (mL/m²)	
	Mean	0.9UCB*	Mean	0.9UCB	Mean	0.9UCB
Two-chamber view, single plane area length	50	82	36	57	24	41
Four-chamber view, single plane area length (two-chamber view)	41	64	34	60	21	36
Simpson's rule, biplane two-plus-four chamber views	41	65	32	52	21	32

*0.9UCB, 90% upper confidence bounds of the 95th percentile.

Source: Reprinted from *Cardiology* by W Parmley and K Chatterjee (Eds) with permission of JB Lippincott Company, © 1987.

more athletically active, further modification of these normal population data may be necessary.[61-71]

Our experience with clinical left ventricular volume determinations impressed us with how much additional information about left ventricular functional status these measurements add. This situation is not surprising, considering that volume bears a third-order relationship to dimension; very small increments in left ventricular diameter can represent large increments in volume when the left ventricle is dilated.

The shape of the left ventricle is seldom considered. We have been impressed by the spherical shape of cardiomyopathic chambers. In this regard, a "sphericity" index aids in differentiating among types of cardiomyopathy.

Although generalized shape changes of the left ventricle are rarely considered, local deformities in ischemic disease are commonly encountered and better appreciated.[72-77] Segmental remodeling is helpful in identifying ischemic myocardial damage. In evaluating left ventricular function, a segmental diastolic deformity or aneurysm makes assessment of global function difficult. If the aneurysm is fibrous, for example, the blood within it is noncompressible (the aneurysm does not expand), and the deformity exerts little negative influence on left ventricular performance. On the other hand, if the aneurysm is expansile, its influence on left ventricular performance may be important. Unfortunately, we have not yet developed methods to deal effectively quantitatively with these shape changes.

The wall thickness of the left ventricle has been an informative M-mode echocardiographic measurement. By itself, the linear thickness of the septum or the posterior wall has been used as an index of left ventricular hypertrophy (Figure 10-25).[78-80] The ratio of posterior wall thickness to septal thickness has been used[81] as an index of asymmetric hypertrophy. As in the case of left ventricular cavitary dimension, most laboratories use the simple linear measurement of wall thickness to assess indirectly left ventricular mass. With the replacement of stand-alone M-mode echocardiographs by two-dimensional instruments, the use of wall thickness as an index of hypertrophy has been questioned. The central point of this issue is that wall thickness is being used as an indirect expression of left ventricular mass or weight. If, for example, the weight of the left ventricle is normal but the preload or filling volume is greatly reduced, the wall will appear to be thickened in diastole. Similarly, if the cavity is dilated, the wall will appear to be thin, in spite of normal or even increased mass. For these reasons left ventricular mass is the preferred expression of left ventricular hypertrophy. All the methods proposed to measure left ventricular mass from M-mode echocardiography suffer from the same theoretical limitation as the cube method of estimating left ventricular volume from the minor-axis dimension. This limitation is imposed by the necessity of extrapolating the volume of the myocardium from a linear dimension and is most keenly realized in asymmetric hearts.[82-83] Working with more uniform hearts and in large populations where individual variations become less important, M-mode echocardiographic methods have given us valuable insight into the implications of ventricular hypertrophy, and

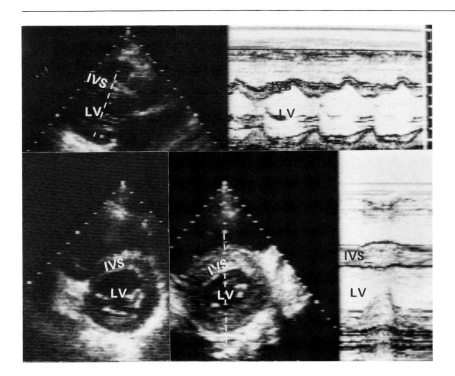

Figure 10-25 Left ventricular hypertrophy compared to normal. The top half of figure shows a long-axis two-dimensional image of the left ventricle (LV), with a dashed line indicating the plane from which the M-mode tracing at the right was generated. Note that the walls of the left ventricle (interventricular septum [IVS] and posterior) are thin (less than 1 cm). Bottom left, A short-axis view from the same normal patient as shown above. Bottom middle, A short-axis view from a patient with severe, slightly asymmetric hypertrophy. The line through this short-axis view indicates the plane of the M-mode tracing seen at the bottom right. Note the obvious difference between the M-mode tracings and between the short-axis views. Source: Reprinted from Cardiology by W Parmley and K Chatterjee (Eds) with permission of JB Lippincott Company, © 1987.

the sensitivity and specificity of electrocardiogram (ECG) criteria for hypertrophy in the hypertensive population.[84-89]

Based on these considerations, our own research, and other recent work, we feel that left ventricular mass should be measured directly from two-dimensional images; one method for determining left ventricular mass is illustrated in Figures 10-26 to 10-29. This method, termed the truncated ellipsoid method or the area-length method, has been recommended by the American Society of Echocardiography Committee on Quantitation of Two-Dimensional Echocardiograms.* Normal two-dimensional echocardiographic values for left ventricular mass are given in Table 10-4.[90] Our

---

*Schiller NB, Shah PM, Crawford M, et al. Recommendations for quantitation of the left ventricle by two-dimensional echocardiography. *J Am Soc Echocardiog.* 1989;2:358–367.

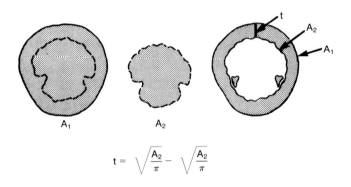

Figure 10-26 By planimetry of the epicardium and endocardium in the short-axis plane, mean wall thickness can be obtained, as shown. This approach circumvents the obvious problems associated with estimating wall thickness at only one or two spots. Knowledge of the inner area also allows simple back-calculation of the minor-axis radius. *Source:* Reprinted with permission from *Circulation* (1983;68:210), Copyright © 1983, American Heart Association.

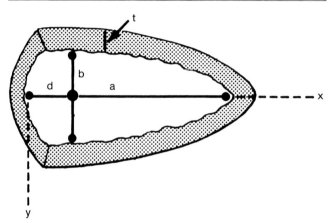

Figure 10-27 Using the two- or four-chamber view (whichever is longer), the semi-major axis (a) and the truncated semi-minor axis (d) are measured. Thickness (t) and minor-axis radius (b) can be calculated from the minor axis, as discussed in the legend for Figure 10-26. *Source:* Reprinted with permission from *Circulation* (1983;68:210), Copyright © 1983, American Heart Association.

$$V = \pi \left\{ (b+t)^2 \int_0^{d+a+t} \left[1 - \frac{(x-d)^2}{(a+t)^2}\right] dx - b^2 \int_0^{d+a} \left[1 - \frac{(x-d)^2}{a^2}\right] dx \right\} =$$

$$\pi \left\{ (b+t)^2 \left[\frac{2}{3}(a+t) + d - \frac{d^3}{3(a+t)^2}\right] - b^2 \left[\frac{2}{3}a + d - \frac{d^3}{3a^2}\right] \right\}$$

$$\text{MASS} = 1.05 \, V$$

Figure 10-28 The formula for left ventricular mass derived from the method of disks and based on a model of the left ventricle as a truncated ellipsoid. Basically, the formula calculates volume from the inner and outer shells of the left ventricle (epi- and endocardium). The difference of these two values is the volume of the myocardium, or, when multiplied by muscle density (1.05), its mass. *Source:* Reprinted with permission from *Circulation* (1983;68:210), Copyright © 1983, American Heart Association.

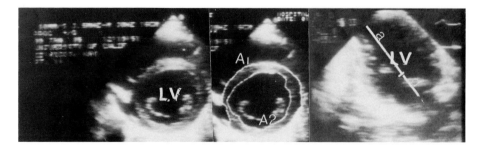

Figure 10-29 Although the algorithm shown in Figure 10-28 appears complicated, only three simple measurements are required to allow the computer to calculate left ventricular (LV) mass. The first two measurements are of the areas subtended by the epicardium and endocardium, A1 and A2; the third, of the semi-major axis (a); and the fourth, the truncated semi-major axis (d). *Source:* Reprinted from *Cardiology* by W Parmley and K Chatterjee (Eds) with permission of JB Lippincott Company, © 1987.

Table 10-4 Values for Left Ventricular Mass and Mass Index Derived from a Population of Normal, Sedentary Adults

	Males		Females	
	Mean	0.9UCB*	Mean	0.9UCB
Mass (g)	135	183	99	141
Mass index (mg/m^2)	71	94	62	89

*0.9UCB, 90% upper confidence bounds of the 95th percentile.

Source: Reprinted from Cardiology by W Parmley and K Chatterjee (Eds) with permission of JB Lippincott Company, © 1987.

observations in patients in whom elimination of hypertension follows renal transplantation have shown regression of hypertrophy within the 1st year following the acute therapeutic afterload reduction; in some, regression has exceeded 150 g.[91] Others have reported similar changes using M-mode echocardiography to follow patients receiving antihypertensive therapy.[84–89]

Assessment of Diastolic Function of the Left Ventricle

Many of the processes that affect systolic function or increase myocardial mass also influence diastolic function. In diastole, the healthy left ventricle should be able to fill within a wide range of blood volumes without elevating filling pressure. In health, most ventricular filling occurs in early diastole, a particularly important property at rapid heart rates when diastole is abbreviated.

Echocardiographically, diastolic function is mirrored in aortic root and mitral valve motion, left atrial size, and Doppler mitral inflow velocity patterns.

The aortic root, an intracardiac structure, sits astride the roof of the left atrium. Atrial volume changes during the cardiac cycle are reflected in the motion of the aortic root. In systole, the aortic root moves forward by the sudden expansion of the left atrium. The degree of atrial expansion is directly related to the degree of filling on the preceding beat (ventricular preload) and to the subsequent stroke volume.[92–93] In diastole, the aortic root moves posteriorly, returning from its anterior end-systolic position. Normally this return occurs immediately and is more rapid than anterior systolic motion.[94] Figure 10-30 shows a normal aortic root on the left and one from a patient with aortic stenosis and left ventricular hypertrophy on the right. When filling is slowed and a sinus rhythm is present, the posterior motion during active atrial emptying usually becomes exaggerated, and the positive filling phase, blunted.[94–97] Angio-

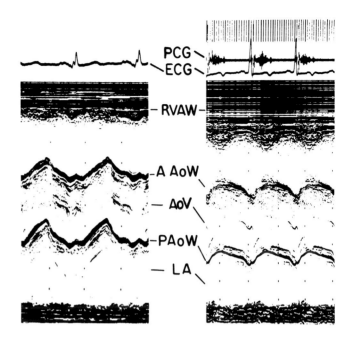

Figure 10-30 Decreased compliance and diastolic aortic root motion. A posterior aortic root echocardiogram from normal subject, left, is compared with one from a patient with left ventricular hypertrophy (right). PCG, phonocardiogram; ECG, electrocardiogram; RVAW, right ventricular anterior wall; A AoW, anterior aortic wall; AoV, aortic valve; PAoW, posterior aortic wall; LA, left atrium. Source: Reprinted with permission from Chest (1981;79:442), Copyright © 1981, American College of Chest Physicians.

graphic studies have shown that normally around 70% of atrial inflow occurs during the rapid filling phase of early diastole, and 30% during atrial contraction.[18] In states of decreased compliance or altered relaxation, this relationship is reversed so that the majority of filling occurs during atrial contraction. The early diastolic filling slope of the mitral valve is also reduced in a manner similar to that seen in mitral stenosis.

If a state of decreased compliance persists for an appreciable period, the left atrium will enlarge. If sinus rhythm is present, the mitral valve is normal, and no other chamber enlargement is present, the finding of isolated atrial enlargement should suggest an abnormality of left ventricular compliance.

Doppler demonstration of the velocity profile of left ventricular transmitral inflow appears to be an informative method of assessing left ventricular filling. Numerous studies have demonstrated that the relationship between peak velocity and deceleration in early diastole and late diastolic velocities during atrial contraction are altered. These alterations appear to be more sensitive than the M-mode mitral filling slope but are unreliable in mitral regurgitation because they are masked by exaggerated early filling velocity (Figures 10-31 and 10-32).[98–99]

146 COMPARATIVE CARDIAC IMAGING

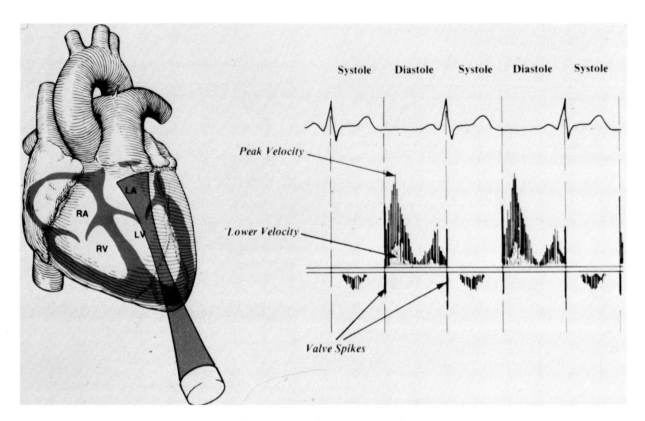

Figure 10-31 Normal mitral inflow Doppler tracing. Left, The Doppler beam is directed from the apex toward the base. Mitral inflow during diastole travels toward the transducer. Right, The Doppler frequency shift resulting from this diastolic transmitral flow. Note that the initial inflow signal of early diastole has the highest velocity, and the late diastolic atrial inflow signal the lowest. LA, left atrium; LV, left ventricle; RA, right atrium; RV, right ventricle. *Source:* Reprinted with permission of Hewlett Packard, Andover, MA.

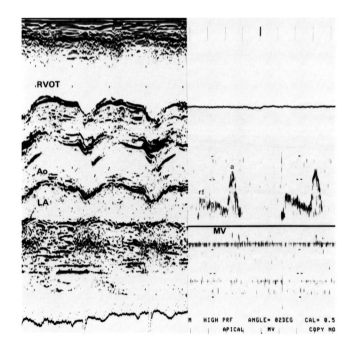

Figure 10-32 Decreased compliance by both aortic root motion and mitral flow. Left, An M-mode tracing of the aortic root, showing marked slowing of the early diastolic slope and exaggeration of posterior root motion during atrial contraction. Right, The Doppler mitral flow signal from the same patient. Note that the early diastolic inflow velocity is slow, and the a wave velocity is increased. Compare this tracing to the diagram of normal mitral flow in Figure 10-31. Ao, aorta; LA, left atrium; MV, mitral valve; RVOT, right ventricular outflow tract. *Source:* Reprinted from *Cardiology* by W Parmley and K Chatterjee (Eds) with permission of JB Lippincott Company, © 1987.

EVALUATION OF CORONARY ARTERY DISEASE AND RELATED ABNORMALITIES OF THE LEFT VENTRICLE

Myocardial infarction resulting from coronary artery disease produces segmental changes in the echocardiographic appearance of the left ventricle, ranging from hypokinesis to aneurysm.[100–115] Some segments are more commonly involved than others, and half of all abnormal segments are found in the distribution of the right coronary artery, and half in that of the left. It is important to appreciate the thickness of the given segment, the degree of scarring, changes in texture, and the degree to which each segment thickens with systole. These features can help to distinguish an old infarction from a recent event. The abrupt demarcation of an abnormally contracting segment typically seen in coronary disease can be ascribed to remodeling. The interface between contracting and akinetic tissue forms a visually distinctive image pattern. Most wall motion abnormalities arising from right coronary artery occlusion occur at the base of the inferior or diaphragmatic wall and are best seen with the apical two-chamber view.

Involvement of the septum, apex, and anteroseptal regions of the left ventricle is typical of occlusion of portions of the left coronary artery circulation, and lateral or free wall involvement typifies circumflex disease. The presence of a diastolic deformity, sharply demarcated, indicates aneurysm formation. Hypokinesis, particularly without wall thinning or remodeling, suggests ischemia rather than infarction. Figures 10-33 to 10-35 give the typical appearance of apical infarctions with aneurysm formation. Figure 10-36 represents an anteroseptal infarction. Figures 10-37 and 10-38 show typical locations and appearances of inferior infarction.

Echocardiography can detect most areas of myocardial ischemia or infarction. In view of its reliability, some laboratories have begun using it in conjunction with dynamic stress testing as a supplement to the ECG.[116–123] There are reports that it is superior in sensitivity and specificity to scintigraphic myocardial imaging with thallium 201 (^{201}Tl), but other studies, including our own, suggest that it is roughly equivalent to ^{201}Tl in its ability to detect areas of induced ischemia and their extent but that scintigraphy enjoys the advantage of being able to provide information about the timing of reperfusion and about prognosis. Future research will likely justify the use of echocardiography in this arena.

Echocardiography has a number of uses in the arena of acute myocardial infarction.[109–113] For example, a normal echocardiogram in the setting of chest pain of

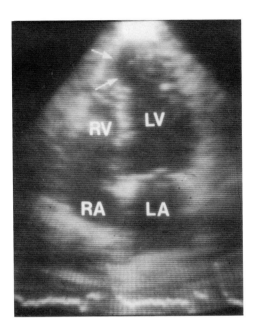

Figure 10-33 Four-chamber view of left ventricular apical aneurysm. The distribution of this lesion is typical of occlusion of the left anterior descending coronary artery. Abbreviations are explained in legend for Figure 10-13. *Source:* Reprinted from *Cardiology* by W Parmley and K Chatterjee (Eds) with permission of JB Lippincott Company, © 1987.

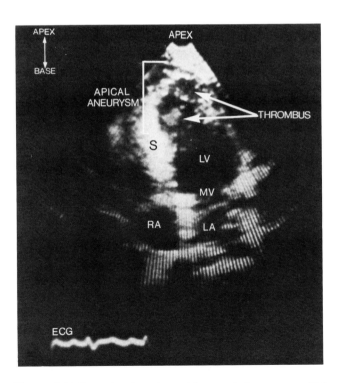

Figure 10-34 Four-chamber view with an apical aneurysm. This aneurysm shows intracavitary echoes consistent with a thrombus. MV, mitral valve; S, septum; other abbreviations are explained in the legend for Figure 10-13. *Source:* Courtesy of TA Ports, San Francisco, CA.

148 COMPARATIVE CARDIAC IMAGING

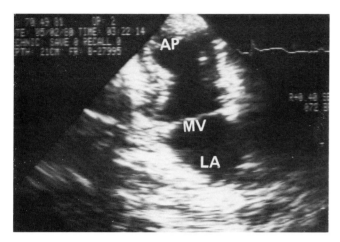

Figure 10-35 This apical aneurysm (AP) is seen in the two-chamber view, suggesting that it is fairly extensive. Abbreviations are explained in the legend for Figure 10-11. *Source:* Reprinted from *Cardiology* by W Parmley and K Chatterjee (Eds) with permission of JB Lippincott Company, © 1987.

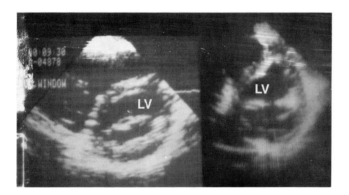

Figure 10-36 Normal short-axis view (left) contrasted with a short-axis view in a person with an anteroseptal infarction (right). Note the rather striking remodeling at the anteroseptal junction. LV, left ventricle. *Source:* Reprinted from *Cardiology* by W Parmley and K Chatterjee (Eds) with permission of JB Lippincott Company, © 1987.

uncertain etiology can provide helpful information for excluding myocardial infarction.[124–125] Some investigators have developed systems to estimate the degree of wall motion abnormality by systems of scoring.[126–130] These systems appear to offer independent information of patient prognosis in the setting of acute myocardial infarction.

Recognition of Complications of Myocardial Infarction

The common complications of acute myocardial infarction can usually be recognized by a bedside echocardiographic examination. In many of these situations, it is advantageous to employ both Doppler and contrast medium methods as well as standard imaging.

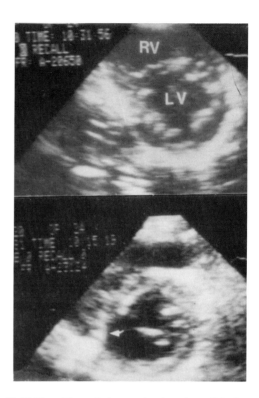

Figure 10-37 Top, Normal short-axis view (systolic). Bottom, Inferoposterior infarction. The arrow points to the area of segmental thinning and deformity. *Source:* Reprinted from *Cardiology* by W Parmley and K Chatterjee (Eds) with permission of JB Lippincott Company, © 1987.

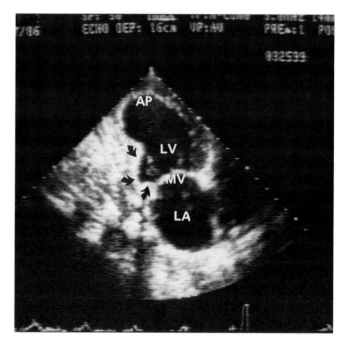

Figure 10-38 Two-chamber view of the left ventricle in a patient with a moderate to large inferior myocardial infarction. Note the basal location (arrow) of the area of infarction and the preserved apical (AP) geometry. Abbreviations are explained in the legend for Figure 10-11. *Source:* Reprinted from *Cardiology* by W Parmley and K Chatterjee (Eds) with permission of JB Lippincott Company, © 1987.

Left ventricular thrombi occur frequently in extensive anteroapical infarction (Figure 10-34) and rarely in inferior infarction.[131–148] They may appear in the early postinfarction period. Their echocardiographic appearance is protean, depending on their age and size. In general, older thrombi tend to have smooth cavitary surfaces and a texture resembling liver. Portions lying closer to the center of the chamber tend to be younger and highly reflective or luminescent. Very fresh or red thrombi are also found toward the center of the cavity. They are highly mobile and difficult to differentiate from the "pseudocontrast" appearance of slowly moving cavitary blood within left ventricular aneurysms. Often it is possible to appreciate several layers of differing texture in larger thrombi. The more mobile and irregular the surface of the thrombus, the more likely it is to be associated with emboli. Small thrombi can be difficult to differentiate from apical trabeculations, which are pronounced in cardiomyopathies.

Aneurysms are another complication of myocardial infarction.[77,149] Since remodeling is a feature of infarction, it is hard to define the point at which a segmental abnormality becomes an aneurysm. A simple, useful definition of aneurysm is a wall motion abnormality with diastolic deformity. Most form at the left ventricular apex, although an inferior basal aneurysm (Figure 10-39) is rarely encountered.

A pseudoaneurysm, as the name implies, resembles a true aneurysm but results from frank wall rupture, or cardiorrhexis. The sac of the aneurysm consists of pericardium, which contains or walls off the rupture. Since only small ruptures in the wall are compatible with survival, these aneurysms have a narrow neck. True aneurysms, on the other hand, are usually as wide at their necks as they are at their apices (Figure 10-40).[150–152] Often echocardiography is the first modality by which pseudoaneurysms are recognized; unrepaired, their prognosis is poor, making their recognition much more than a matter of taxonomy.

Abrupt myocardial rupture into the pericardium is usually fatal, and its echocardiographic recognition has little value. Slower accumulations can result in pericardial effusion and tamponade.[153] Most pericardial effusions complicating myocardial infarction are not due to wall rupture but to either inflammation at the site of epicardial necrosis or postmyocardial infarction pericardial inflammation (Dressler's syndrome).[154–155] Use of anticoagulants in myocardial infarction can result in progressive accumulation of pericardial blood, which may be recognized on serial echocardiograms.

Septal rupture leads to left-to-right shunting at the level of the ventricles (Figure 10-41),[156] and papillary muscle rupture to mitral insufficiency (Figure 10-42).[157–158] In the case of acute ventricular septal defect, Doppler and contrast medium studies can be diagnostic by demonstrating the appearance of the contrast agent on the left side after right-sided injection and by the Doppler detection of disturbed flow in and around the ventricular septal defect.

Right ventricular infarction can be identified by careful echocardiographic examination (Figure 10-41). With

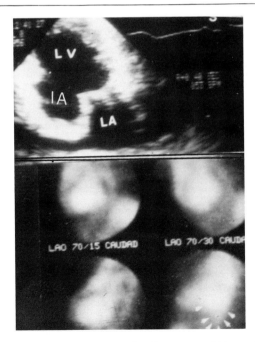

Figure 10-39 Top, A large inferior basal aneurysm (IA), nearly equal in size to the left atrium (LA), is seen in its usual location in the two-chamber apical view. Bottom, A blood pool scintigram of the left ventricle (LV), showing the aneurysm in the 70° left anterior oblique view (arrows). The contractility of uninvolved segments was excellent. *Source:* Courtesy of EH Botvinick, San Francisco, CA.

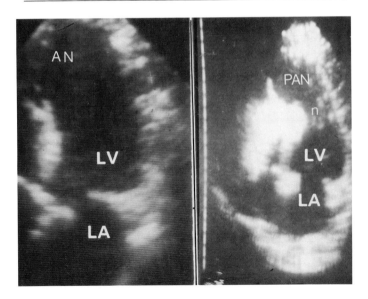

Figure 10-40 Left, An aneurysm (AN) of the left ventricular apex is shown in the apical four-chamber view. Note that the ventricle does not narrow at the opening to the aneurysm. Right, A pseudoaneurysm (PAN) identified by its connection to the left ventricle (LV) across a narrow neck (n). LA, left atrium. *Source:* Courtesy of R Clark, San Francisco, CA.

150 COMPARATIVE CARDIAC IMAGING

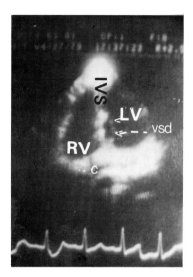

Figure 10-41 In the setting of a recent inferior myocardial infarction, this patient has developed two major complications, a ventricular septal defect (vsd) and a right ventricular infarction. The right ventricular infarction can be appreciated by the dilation of the right ventricle. In order to visualize the site of these abnormalities, posterior angulation of the transducer is necessary. In this example, the appearance of the coronary sinus (c) in the image is proof that this posterior angulation was performed by the sonographer. IVS, interventricular septum; RV, right ventricle; LV, left ventricle. *Source:* Reprinted from *Cardiology* by W Parmley and K Chatterjee (Eds) with permission of JB Lippincott Company, © 1987.

the accumulation of increasing numbers of recent studies of right ventricular function in inferior infarction,[159-167] the frequency of occurrence of this complication is now appreciated as being higher than previously thought. In fact, some studies suggest that as many as 40% of inferior infarctions are complicated by some degree of right ventricular dysfunction. However, it should be clear that there is a vast difference between depression of contractile function, which is often subclinical and echocardiographically subtle, and severe impairment of right ventricular performance. When extensive right ventricular infarction occurs, a low-output state with high mortality results. Since patency of the foramen ovale is present in 20% to 30% of normal individuals, the elevation of right-sided filling pressures accompanying severe right ventricular dysfunction can result in acute right-to-left shunting.[168] This can be easily appreciated with a simple saline echocardiographic contrast medium study that demonstrates the passage of microbubbles from right to left across the atrial septum.[169]

EVALUATION OF CARDIOMYOPATHIES

Three major types of cardiomyopathy are appreciated by echocardiography. The first type is dilated cardiomyopathy, the second is hypertrophic car-

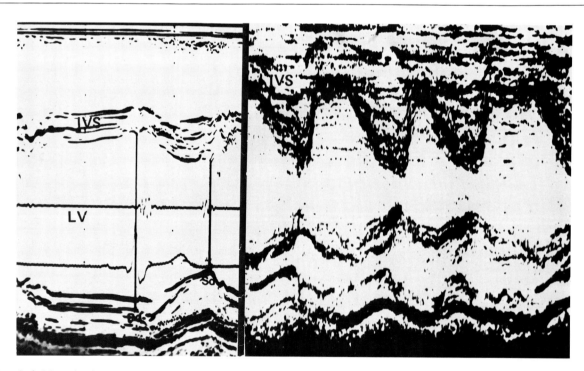

Figure 10-42 Left, M-mode of a normal left ventricle. Note that the motion of the septum (IVS) and its opposing posterior wall is fairly symmetrical. Right, Severe mitral insufficiency arising from ischemic damage to the inferior base of the left ventricle (LV). This damage has undermined the support of the posteromedial papillary muscle. The resulting mitral insufficiency has exaggerated the motion of the septum (IVS). The damaged inferior wall has only poor inward motion and stands in marked contrast to the septum. *Source:* Reprinted from *Cardiology* by W Parmley and K Chatterjee (Eds) with permission of JB Lippincott Company, © 1987.

diomyopathy, and the third, less frequently encountered and more difficult to identify by echocardiography, are the restrictive or infiltrative cardiomyopathies (see also Chapter 28). Dilated cardiomyopathies are readily identified by echocardiography in their advanced form but are more difficult to detect in their early stages. The most distinctive echocardiographic findings in these hearts are spherical cavitary dilation, normal or decreased wall thickness, and reduced inward endocardial systolic motion.[170–177] On M-mode echocardiography, additional features are mitral-septal E-point separation, poor mitral valve opening, poor aortic valve opening, and poor systolic aortic root motion. In addition to the sequelae of poor ejection fraction and low stroke volume, both M-mode and two-dimensional echocardiographic images usually demonstrate left atrial enlargement, and two-dimensional imaging also reveals four-chamber dilation. The involvement of the right heart is important because it implies pulmonary hypertension and/or right ventricular failure either secondary to pulmonary pressure elevation or to involvement of the right ventricular myocardium in the pathologic process. Quantitatively, the left ventricle often exceeds 250 mL in volume, while the left atrium can exceed 125 mL. The ejection fraction derived from the systolic and diastolic volume determinations at times falls below 20% but is usually between 20% and 30%. In spite of the low ejection fraction, cardiac output (stroke volume × heart rate) calculations frequently reveal normal values because patients with cardiomyopathy frequently have elevated heart rates. Figures 10-43 to 10-46 demonstrate some of the features of cardiomyopathy.

Doppler has been used in cardiomyopathy to measure decreased contractility. Both peak velocity and acceleration of velocity in the ascending aorta are decreased in primary myocardial dysfunction. The flow velocity integral (or stroke distance) is also frequently depressed in cardiomyopathy. This useful parameter is easily obtained by planimetry of the flow velocity curve of the pulmonary artery, left ventricular outflow tract, or aortic root.

When a cardiomyopathy is fully developed, echocardiography is a rapid and reliable method of diagnosis. However, this method offers less in establishing the etiology of this often mysterious condition. Once the diagnosis is known, performing echocardiograms at frequent intervals is also relatively unrewarding unless the laboratory is equipped to make reproducible Doppler determinations of pulmonary pressure and quan-

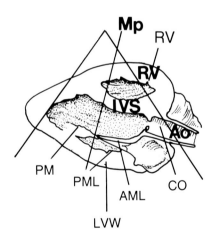

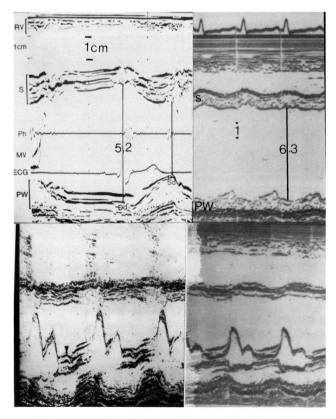

Figure 10-43 M-mode echocardiograms illustrating some of the cardinal features of cardiomyopathy. Normal M-mode tracings in the middle panels show the minor axis of the left ventricle (top middle) taken along the line (PML) superimposed on the anatomic diagram (top left). The panels on the right were taken from a patient with cardiomyopathy. The minor-axis dimension of the normal ventricle is 5.2 cm compared to 6.3 cm in the patient with cardiomyopathy. The fractional shortening of the normal heart [(LVEDd) 5.2-(LVESd) 3.4/5.2 = 35%] compared to that of the heart with cardiomyopathy (6.3-5.3/6.3 = 16%.) In the bottom middle panel the M-mode tracing was obtained from a more basal level, and the beam passed through the mitral valve. In the normal heart the mitral valve (MV), denoted by a small arrow, opens very near to the septum (S), while in the cardiomyopathic heart it is separated by nearly 2 cm. This separation is called E-point mitral-septal separation, or EPSS. IVS, interventricular septum; ECG, electrocardiogram; other abbreviations are explained in the legends for Figures 10-7 and 10-13. *Source:* Reprinted from *Cardiology* by W Parmley and K Chatterjee (Eds) with permission of JB Lippincott Company, © 1987.

152 Comparative Cardiac Imaging

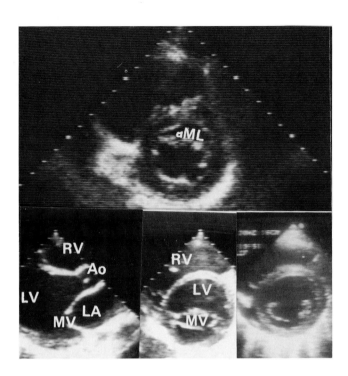

Figure 10-44 Top, a short-axis view of the left ventricle (LV) from a normal heart. The scale of this image is larger than for the images of the cardiomyopathic heart appearing below (the scale for each pair of dots is 1 cm). Note the relationship between the thickness of the myocardium and the muscle. Exaggeration of cavitary volume and thinning of myocardial segments are typical of cardiomyopathy: left bottom, long-axis view; middle bottom, short-axis view taken through the mitral valve (MV); right bottom, short-axis view through the papillary muscles. In comparison to normal, note the spherical nature of the left ventricle in the long-axis view and the relatively thin walls in both views. Ao, aorta; other abbreviations are explained in the legend for Figure 10-13. *Source:* Reprinted from *Cardiology* by W Parmley and K Chatterjee (Eds) with permission of JB Lippincott Company, © 1987.

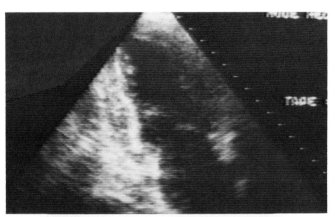

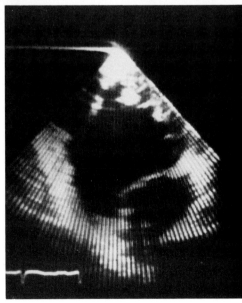

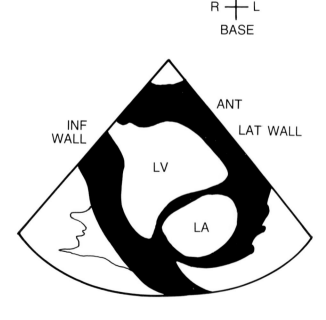

Figure 10-45 Top, A two-chamber apical view obtained from a normal heart. Bottom left, A two-chamber view from echocardiograph of a spherically dilated cardiomyopathic heart. Note the difference in shape between the normal and cardiomyopathic hearts. Ant Lat Wall, anterior lateral wall; other definitions are explained in the legend for Figure 10-11. *Source:* Reprinted from *Cardiology* by W Parmley and K Chatterjee (Eds) with permission of JB Lippincott Company, © 1987.

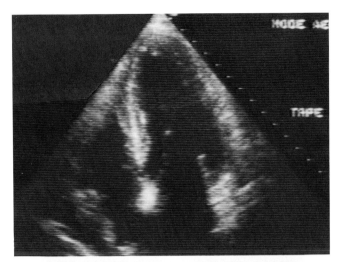

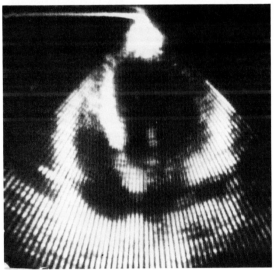

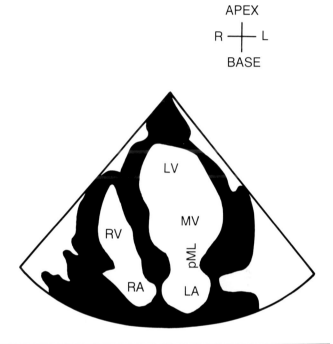

Figure 10-46 Top, A four-chamber view from a normal heart is shown in the upper panel. Bottom right and left, An image of a heart from a patient with cardiomyopathy is shown and diagrammed. Note that the cardiomyopathic heart is more spherical than its normal counterpart. This spherical configuration is fairly typical of cardiomyopathy. MV, mitral valve; other abbreviations are explained in the legend for Figure 10-13. *Source:* Reprinted from *Cardiology* by W Parmley and K Chatterjee (Eds) with permission of JB Lippincott Company, © 1987.

titative evaluations of chamber volumes and masses. In our laboratory, we attempted to distinguish among a group of patients with ischemic and primary cardiomyopathies using echocardiography and nuclear magnetic resonance imaging (NMR).[56] We evaluated the presence of segmental disease, remodeling of segmental abnormalities, thinning of segments, and sphericity as criteria. In patients with ischemic etiologies, segmental remodeling, thinning, and areas of preserved function tended to predominate. However, in some patients with no historic or angiographic evidence of ischemic disease, the echocardiographic and NMR images most closely suggested a segmental or ischemic process. We feel, therefore, that the features distinguishing one process from another blend into one another and that echocardiography is not a precise method of classifying cardiomyopathies. Recent work has suggested a widening role for echocardiography in the diagnosis of cardiomyopathy by showing that it is possible to perform endocardial biopsies under echocardiographic control.[178]

In South America, the finding of a congestive or a segmental cardiomyopathy suggests the possibility of Chagas' disease, a condition caused by human transmission of *Trypanosoma cruzi* from an infected insect that eventually leads to damaging infestation of the myocardium. In its congestive cardiomyopathic state, it is indistinguishable from any other cause of primary myocardial disease. In its segmental presentation, Chagas' disease presents as an apical aneurysm that, unlike coronary disease, spares the interventricular septum. In a rarer segmental presentation, the inferior base can be the site of segmentally isolated thinning and scar formation (Figures 10-47 and 10-48).[179]

Hypertrophic cardiomyopathies, asymmetric and symmetric, are characterized by increased left ventricular mass (wall thickness) without apparent etiology—in other words, by the presence of left ventricular hypertrophy (LVH) without a history of hypertension or aortic stenosis (see Chapter 28).[180–200] The echocardiogram in asymmetric septal hypertrophy (ASH) characteristically shows increased wall thickness

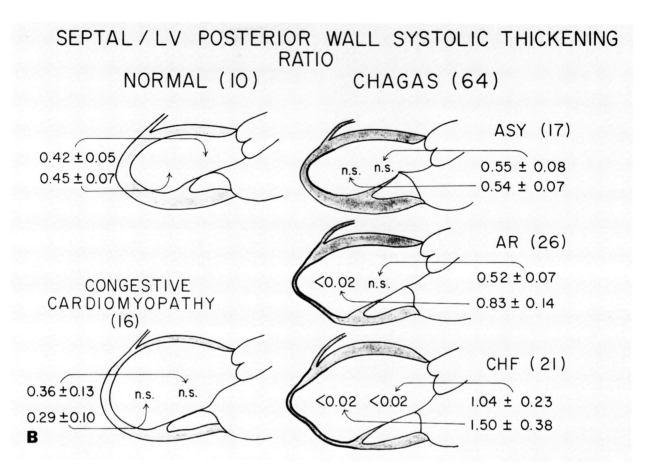

Figure 10-47 Diagrammatic comparison of Chagas' cardiomyopathy to idiopathic cardiomyopathy and normal hearts. The values next to each figure refer to the ratio of septal to posterior wall thickening at the ventricular level, indicated by the arrows. In normal hearts and idiopathic myopathies, the posterior wall thickening is greater than the septal at all levels. In Chagas' patients, those with arrhythmias (AR) and those with congestive cardiomyopathy (CHF) show a reversal of thickening ratios in the more apical segments. In other words, relative to the septum the posteroapical segments are more scarred and less contractile. This pattern of septal scarring appears to be unique to Chagas' heart disease. LV, left ventricle; n.s., not significant. *Source:* Reprinted with permission from *Circulation* (1980;62:787), Copyright © 1980, American Heart Association.

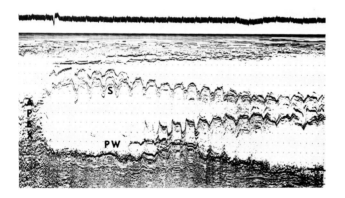

Figure 10-48 The pattern shown diagrammatically in Figure 10-47 is present on this M-mode sweep of a patient with advanced Chagas' cardiomyopathy. Note that the posterior wall (PW) shows very little sign of thickening, while the septum (S) is exaggerated. This picture is somewhat similar to that encountered in patients with inferior infarction and mitral regurgitation. *Source:* Reprinted with permission from *Circulation* (1980;62:787), Copyright © 1980, American Heart Association.

localized or most intense in the basal septum. The patterns of distribution of this hypertrophic state are not well understood because they follow an unpredictable pattern. Although it is clearly a heritable condition, the patterns of involvement differ among affected members of the same family.[183–184]

The echocardiogram is the most reliable means of making the diagnosis of hypertrophic cardiomyopathy. The usual features are thickening of a portion of the basal septum with sparing of the base and variable involvement of other portions of the myocardium (Figure 10-49). In a less common variation of ASH, the apex can become the site of the most intensive hypertrophy.[187–198] This variety is more difficult to identify by echocardiography because the apical myocardium is more difficult to image. When dynamic outflow tract obstruction accompanies ASH, the fully developed picture of hypertrophic obstructive cardiomyopathy (HOCM) or idiopathic hypertrophic subaortic stenosis

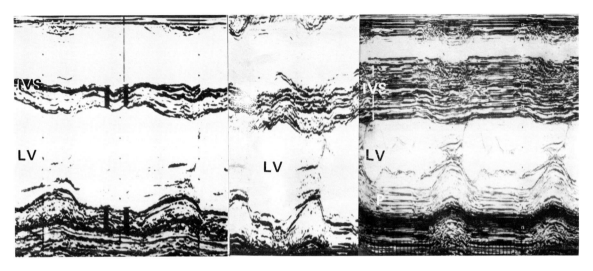

Figure 10-49 A normal left ventricle (left) is contrasted to mild symmetrical LVH (center) and marked asymmetric hypertrophy (right). Note that the a wave or presystolic ventricular filling is more exaggerated in the example of mild LVH. Exaggerated filling during the atrial phase of diastole can be a clue to decreased compliance. *Source:* Reprinted from *Cardiology* by W Parmley and K Chatterjee (Eds) with permission of JB Lippincott Company, © 1987.

(IHSS) is present. Echocardiographic identification is rather elementary when the condition is fully developed (Figure 10-50). However, when the obstruction is provocable (Figures 10-51 and 10-52) or mild, the task is much more difficult. The features of fully developed HOCM are ASH, systolic anterior motion of the mitral valve (SAM), crowding of the left ventricular outflow tract by the mitral apparatus and septum, and partial midsystolic closure or notching of the aortic valve. Calcification of the mitral annulus (MAC) is frequently found in HOCM, and in some patients the presence of MAC may be the only clue to the presence of dynamic outflow tract obstruction.

When obstructive or nonobstructive hypertrophic cardiomyopathy is suspected, it is usually desirable to perform an intervention or provocation during the echocardiographic examination. If Doppler is available (particularly continuous wave), simultaneous measurements of outflow tract systolic velocity with maneuvers may not only prove diagnostic of dynamic outflow tract obstruction but also indicate its severity (Figures 10-53 and 10-54).[8] At a minimum, laboratories without Doppler should have the patient first perform a Valsalva maneuver during M-mode echocardiographic imaging of the mitral valve base and left ventricular outflow tract area. If systolic anterior movement of the mitral valve results, it suggests the potential for dynamic obstruction. Inspection of the aortic valve at the peak of this provocation can provide secondary evidence of obstruction if there is midsystolic notching or closure (Figure 10-50). If a Valsalva maneuver fails to provoke obstruction, amyl nitrite inhalation should be used (Figures 10-53 and 10-54). During this more vigorous provocation, the same recordings are made. It is ideal during any intervention to perform auscultation; if technical problems prevent high-quality tracings from being obtained, the observations made during auscultation may be the only clinical data obtained. In some laboratories, a negative amyl nitrite provocation is followed by isoproterenol infusion.[200] When Doppler is available, provocations are accompanied by sampling of outflow velocity.

Restrictive cardiomyopathies are more difficult to diagnose with echocardiography than hypertrophic or congestive states. However, echocardiography is the most effective noninvasive means for the detection of this group of conditions. The most common restrictive state is the small, stiff heart of diabetes.[201] In the majority of diabetics this abnormality is clinically silent but can be detected by quantitative two-dimensional echocardiography. Amyloid heart disease is a condition that, although rare, is of clinical importance because of its poor prognosis. Echocardiographically, infiltrative amyloid myocardial disease is characterized by increased left ventricular wall thickness and a peculiar scintillating appearance of the myocardium.[202–206] Superficially, amyloid heart disease resembles simple LVH because in addition to the thickened, glittering myocardium, contractile function appears nearly normal and the left atrium is enlarged. Integration of these findings with the remainder of the clinical picture (e.g., history of chronic disease, low ECG voltage, neuropathy) allows a rational decision about the final step in the diagnostic process, a gingival, rectal, or myocardial biopsy. It should be noted that not all amyloid heart disease has typical echocardiographic features. We

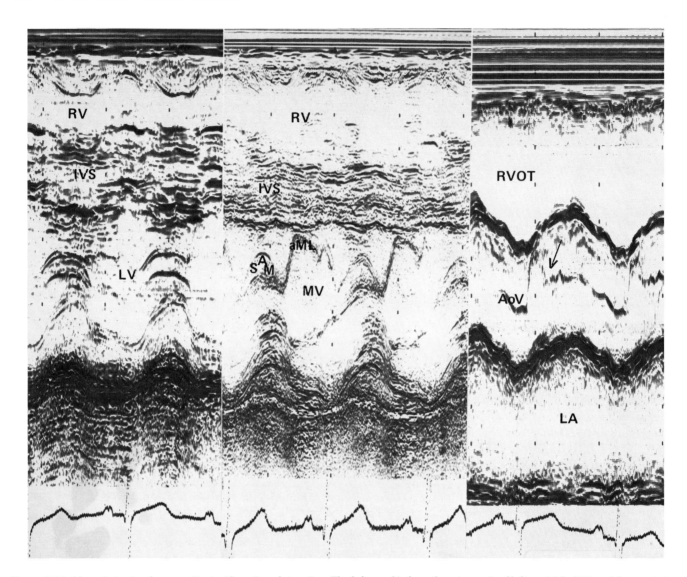

Figure 10-50 M-mode tracing from a patient with resting obstruction. The left panel is from the minor axis of left ventricle (LV) and demonstrates typical asymmetric ventricular hypertrophy of the interventricular septum (IVS). Systolic anterior motion of the mitral valve (MV) and crowded appearance of the ventricle are shown in the center panel. The degree of contact between the septum and the mitral valve during systole appears minimal, suggesting that obstruction is mild. However, the aortic valve (AoV) demonstrates striking midsystolic notching or closure (right panel, arrow), strongly suggesting that dynamic obstruction is significant. RV, right ventricle; other abbreviations are described in the legend for Figure 10-15. *Source:* Reprinted from *Cardiology* by W Parmley and K Chatterjee (Eds) with permission of JB Lippincott Company, © 1987.

have seen cases in which the echocardiogram failed to demonstrate increased left ventricular wall thickness or in which the appearance of the myocardium made differentiation between simple LVH and infiltrative cardiomyopathy very difficult. Endomyocardial fibrosis is a disease associated with restriction of left ventricular and right ventricular filling by obliteration of one or both cardiac apices by a thrombotic fibrocalcific process.[207] Recognition depends on a high level of clinical suspicion and a characteristic echocardiographic appearance.

FUNCTIONAL EVALUATION WITH STRESS TESTING

Technical factors have inhibited the use of dynamic exercise during two-dimensional echocardiography, but realization of its feasibility has led to growing interest in the technique. The major thrust for development has come from the assertion that the information from stress echocardiography is analogous to that from nuclear scintigraphic techniques. The major arena for the development of stress echocardiography has been

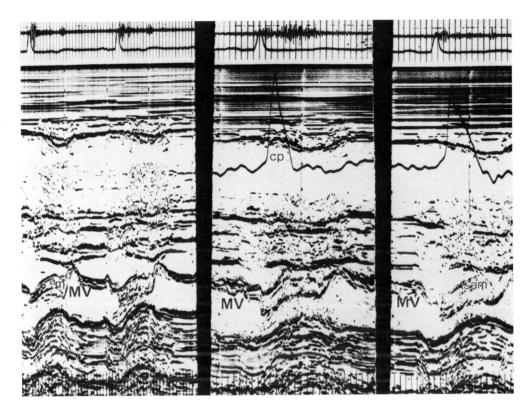

Figure 10-51 Provocable outflow tract obstruction. Simultaneous M-mode, phonocardiogram, and carotid pulse (cp) tracings during rest (left), peak amyl nitrite effect (center), and recovery (right). Note that the systolic anterior motion (sam) of the mitral valve (MV) increases markedly and lies in close apposition to the interventricular septum during the provocation. Note also that the murmur becomes more intense and the carotid pulse loses its dicrotic wave. During recovery the systolic anterior motion is much less marked and only barely makes contact with the septum. It can be inferred from this study that the patient has little if any resting subaortic gradient but has easily provoked dynamic obstruction. *Source:* Reprinted from *Cardiology* by W Parmley and K Chatterjee (Eds) with permission of JB Lippincott Company, © 1987.

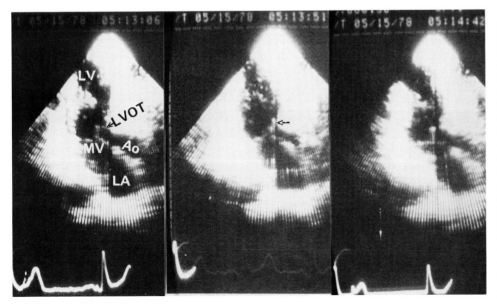

Figure 10-52 Two-chamber views of the mitral valve (MV) and left ventricular outflow tract (LVOT) demonstrate the site and appearance of systolic anterior motion of the mitral valve. Left, The patient is at rest, and the left ventricular outflow tract is open. Middle, The patient has been given amyl nitrite, and the recording was made at the drug's full effect. Note that the coapted mitral valve/chordae tip has moved into the left ventricular outflow tract and now abuts the septum (arrow). Right, The effect has passed, and the valve has returned to its resting position. Other abbreviations are explained in legend for Figure 10-15. *Source:* Reprinted from *Cardiology* by W Parmley and K Chatterjee (Eds) with permission of JB Lippincott Company, © 1987.

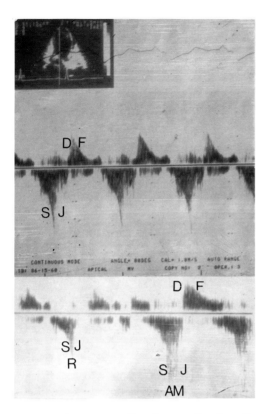

Figure 10-53 A continuous wave Doppler recording from the apical four-chamber view (top) records inflow across the mitral valve during diastole (DF) and the outflow tract systolic jet (SJ) typical of dynamic subaortic outflow obstruction. Features of the systolic jet include transient duration, giving the velocity profile the appearance of a horse's tail. At the very bottom of the figure, the systolic jet during rest (R) has a velocity of 2.5 m/sec, suggesting a 25 mm Hg subaortic gradient. During amyl nitrite (AM) provocation, the jet increases in velocity to over 4 m/sec, consistent with a gradient of at least 65 mm Hg. *Source:* Reprinted from *Cardiology* by W Parmley and K Chatterjee (Eds) with permission of JB Lippincott Company, © 1987.

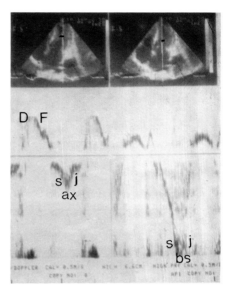

Figure 10-54 Dynamic subaortic outflow tract obstruction. Pulsed Doppler is used to localize the subaortic jet (sj). The location of the sample volume is shown by the black crosshatch on the line running through the four-chamber views in the top panel. In the left top panel, the sample volume is at the papillary muscle level, apical to the chordae, and the intraventricular velocity of the subaortic jet is low (ax). However, when the sample volume is moved basally (bs), below the contact point of the mitral valve with the septum, the subaortic jet suddenly accelerates. The high-velocity subaortic jet has the characteristic late systolic or horse's tail configuration, although this is partially obscured by aliasing at peak velocity. DF, diastole. *Source:* Reprinted from *Cardiology* by W Parmley and K Chatterjee (Eds) with permission of JB Lippincott Company, © 1987.

the evaluation of patients for coronary artery disease. There are numerous approaches to exercise evaluation of these patients.[116–123] Some perform upright bicycle exercise with the subject leaning forward, continuously imaging throughout the study.[208] Other groups[209] perform supine exercise with continuous subcostal imaging. Still others[116–123] have performed imaging immediately before and after upright treadmill stress testing. Regardless of which technique is used, some form of computer assistance in image acquisition and analysis is required. Indiana University has introduced the use of a continuous loop method of capturing a single, technically good frame from the control (rest) tracing and placing it side by side with one from the immediate poststress period. This method provides archiving of data in floppy disk format and allows rapid measurement of the parameters of ventricular function. Regardless of which method is used, the stress echocardiogram seeks to evaluate global left ventricular performance and segmental wall motion in response to stress. As in exercise angiography and scintigraphy, segments with compromised coronary circulation exhibit a loss of myocardial thickening. Normally, global and segmental function are augmented with stress, so the abnormal response to exercise is distinctly opposite to the physiological norm. From work done in this area, echocardiography has proved to be capable of detecting the deterioration in segmental and global function resulting from impaired perfusion. In our experience,[120] we found that echocardiographic imaging performed in conjunction with standard treadmill stress testing provides information equivalent to the data provided by perfusion scintigraphy. However, since echocardiography provides limited information about reperfusion rates, the future role of scintigraphy seems assured.

TRANSESOPHAGEAL ECHOCARDIOGRAPHY FOR INTRAOPERATIVE ASSESSMENT OF LEFT VENTRICULAR FUNCTION

The esophagus provides an airless pathway to the posterior aspect of the heart. In order to exploit this ideal ultrasound window, a flexible fiberoptic gastroscope was modified so that it housed a miniature phased array transducer at its tip and wiring in place of optical fibers.[210–224] The external controls permitted altering the angle at which the transducer abuts the mucosa, thereby making the ultrasound beam steerable. Many groups have used this method to produce images of improved resolution in patients presenting for routine ultrasound imaging. They have also used this approach to provide stable transducer positioning during dynamic exercise. Our institution and others have used this device as an intraoperative cardiovascular monitor. Employing the instrument's ability to provide a stable high-resolution short-axis left ventricular image, we have examined 2000 patients undergoing a variety of surgical procedures. In particular, we have directed our attention to global and segmental left ventricular function, left ventricular preload or filling, valvular function (with Doppler), and detection of intracardiac air. Perhaps its most important use is the intraoperative detection of ischemia. The highly resolved, full-thickness image of the endocardium allows study of segmental wall motion by observation of wall thickening rather than endocardial inward motion (Figure 10-55). A change in wall thickening is more sensitive to ischemia than intraoperative ECG, and if the change persists until the termination of surgery, there is a much greater chance of an intraoperative myocardial infarction. Other major changes occurring in the operating room include hypovolemia, easily recognized by a marked decrease in both systolic and diastolic cavity area. Interestingly, we have compared pulmonary capillary wedge pressure measurement changes with actual ventricular size changes and found wedge pressure changes unrepresentative of these volume changes. In particular, hemodynamically important hypovolemia may provoke little if any change in the wedge until the decrease in systolic and diastolic cavity area becomes rather advanced. The transesophageal echocardiographic image proved to be quite sensitive to these directional changes in blood volume. We and others use transesophageal echocardiography with simultaneous Doppler echocardiography and contrast medium studies to monitor mitral valve repair and mitral and aortic valve replacement. Along with the usual methods available to the surgeon, transesophageal echocardiography/Doppler allows an independent assessment of the degree of residual mitral regurgitation after valvuloplasty, the presence of paravalvular leak after valve implantation, or a change in mitral regurgitation after myocardial revascularization. It is also well suited for detecting intravascular air.

Outside of the operating room, we use transesophageal echocardiography to assess left ventricular function in patients in intensive care settings with technically inadequate surface studies. We also study selective outpatients suspected of prosthetic dysfunction or endocarditis. In these awake subjects, the procedure is performed in a manner similar to that of upper gastrointestinal endoscopy.

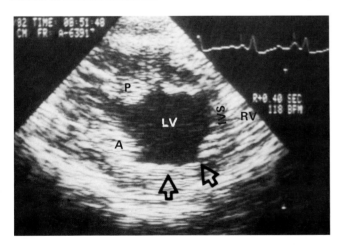

Figure 10-55 Short-axis view of the left ventricle (LV) obtained during surgery through a transesophageal transducer. Since imaging is from behind the heart, the posterior wall and posterior medial papillary muscle (P) are at the top of the image, and the anterior wall and anterolateral papillary muscle (A) are at the bottom. The arrows point to the anterior wall and its junction with the interventricular septum. During real-time examination this area was observed to have lost its normal systolic thickening, raising the possibility of intraoperative ischemia or infarction. *Source:* Reprinted from *Cardiology* by W Parmley and K Chatterjee (Eds) with permission of JB Lippincott Company, © 1987.

REFERENCES

1. Hertz CH. Ultrasonic engineering in heart diagnosis. *Am J Cardiol*. 1967;19:6–17.

2. Edler I. Ultrasound cardiogram in mitral valve disease. *Acta Chir Scand*. 1956;111:230–231.

3. Edler I, Gustafson A. Ultrasonic cardiogram in mitral stenosis. *Acta Med Scand*. 1957;159:85–90.

4. Schiller NB, Acquatella H, Ports TA, et al. Left ventricular volume from paired biplane two-dimensional echocardiographs. *Circulation*. 1979;60:547–555.

5. Melton HE Jr, Thurstone FS. Annular array design and logarithmic processing for ultrasonic imaging. *Ultrasound Med Biol*. 1978;4:1–12.

6. Silverman NH, Snider AR. *Two-Dimensional Echocardiography in Congenital Heart Disease*. Norwalk, Conn: Appleton-Century-Crofts; 1982:58.

7. Collins SM, Skorton DJ. *Cardiac Imaging and Image Processing*. New York: McGraw-Hill Book Co; 1986.

8. Hatle L, Angelsen B. *Doppler Ultrasound in Cardiology: Physical Principles and Clinical Applications*. 2nd ed. Philadelphia: Lea & Febiger; 1985:78.

9. Crawford MH, Petru MA, Amon KW, Sorensen SG, Vance WS. Comparative value of 2-dimensional echocardiography and radionuclide angiography for qualitating changes in left ventricular performance during exercise limited by angina pectoris. *Am J Cardiol*. 1984;53:42–46.

10. Weisse A, Jordan T. A comparison of M-mode left ventricular dimensions derived from parasternal long and short axis 2-D echocardiograms in normal and abnormal adults. *J Clin Ultrasound*. 1984; 3:51–55.

11. Feigenbaum H. *Echocardiography*. 4th ed. Philadelphia: Lea & Febiger; 1986:159.

12. Massie BM, Schiller NB, Ratshin RA, Parmley WW. Mitral septal separation: New echocardiographic index of left ventricular function. *Am J Cardiol*. 1977;39:1008–1016.

13. Child JS, Krivokapick J, Perloff JK. Effect of left ventricular size on mitral E point to ventricular septal separation in assessment of cardiac performance. *Am Heart J*. 1981;101:797–805.

14. Koenig W, Gehring J, Mathes P. M-mode echocardiography in the diagnosis of global and regional myocardial function in coronary heart disease: A study validated by quantitative left ventricular cineangiography. *J Clin Ultrasound*. 1984;3:165–173.

15. Engle SJ, DiSessa TG, Perloff JK, et al. Mitral valve E point to ventricular septal separation in infants and children. *Am J Cardiol*. 1983;52:1084–1087.

16. Djalaly A, Schiller NB, Poehlmann HW, Arnold S, Gertz EW. Diastolic aortic root motion in left ventricular hypertrophy. *Chest*. 1981;79:442–445.

17. Klausner S, Botvinick E, Schiller NB. Determination of LV stroke volume by aortic root motion: An echographic index of ventricular function. *Clin Res*. 1976;24:84. Abstract.

18. Bryhn M: Abnormal left ventricular filling in patients with sustained myocardial relaxation: Assessment of diastolic parameters using radionuclide angiography and echocardiography. *Clin Cardiol*. 1984;7:639–646.

19. O'Rourke RA, Hanrath P, Henry WN, et al. Report of the Joint International Society and Federation of Cardiology/World Health Organization task force on recommendations for standardization of measurements from M-mode echocardiograms—special report. *Circulation*. 1984;69(suppl 4): 854–857. Abstract.

20. Panidis IP, Ross J, Ren JF, Wiler M, Mintz G. Comparison of independent and derived M-mode echocardiographic measurements. *Am J Cardiol*. 1984;54:694–696.

21. Gardin JM, Tommaso CL, Talano JV. Echographic early systolic partial closure (notching) of the aortic valve in congestive cardiomyopathy. *Am Heart J*. 1984;107:135–142.

22. Kiulzi M, Gillam L, Gentile F, Newell J, Weyman A. Normal adult cross-sectional echocardiographic values: Linear dimensions and chamber areas. *Echocardiol*. 1984;1:403–426.

23. Teichholz LE, Kreulen T, Herman MV, Gorlin R. Problems in echocardiographic volume determinations: Echocardiographic-angiographic correlations in the presence or absence of asynergy. *Am J Cardiol*. 1976;37:7–11.

24. Vredevoe LA, Creekmore SP, Schiller NB. The measurement of systolic time intervals by echocardiography. *JCU*. 1974;2:99–104.

25. Spodick DM, Doi YL, Bishop RL, Hashimoto T. Systolic time intervals reconsidered. Reevaluation of the preejection period: Absence of relation to heart rate. *Am J Cardiol*. 1984;53:1667–1670.

26. Traill TA, Gibson DG, Brown DJ. Study of left ventricular wall thickness and dimension changes using echocardiography. *Br Heart J*. 1978;40:162–169.

27. Stamm RB, Carabello BA, Mayers DL, Martin RP. Two dimensional echocardiographic measurement of left ventricular ejection fraction: Prospective analysis of what constitutes an adequate determination. *Am Heart J*. 1982; 104:136–144.

28. Rich S, Sheikh A, Gallastegui J, Kondos GT, Mason T, Lam W: Determination of left ventricular fraction by visual estimation during real-time two-dimensional echocardiography. *Am Heart J*. 1982;104:603–606.

29. Wyatt HL, Meerbaum S, Heng MK, Gueret P, Corday E. Cross-sectional echocardiography, III: Analysis of mathematic models for quantifying volume of symmetric and asymmetric left ventricles. *Am Heart J*. 1980;100:821–828.

30. Silverman NH, Ports TA, Snider AR, Schiller NB, Carlsson E, Heilbron DC. Determination of left ventricular volume in children—echocardiographic and angiographic comparisons. *Circulation*. 1980;62:548–557.

31. Starling MR, Crawford MH, Sorensen SG, Levi B, Richards KL, O'Rourke RA. Comparative accuracy of apical biplane cross-sectional echocardiography and gated equilibrium radionuclide angiography for estimating left ventricular size and performance. *Circulation*. 1981;63:1075–1084.

32. Wyatt HL, Haendchen RV, Meerbaum S, Corday E. Assessment of quantitative methods for 2-dimensional echocardiography. *Am J Cardiol*. 1983;52:396–401.

33. Weiss JL, Eaton LW, Kallman CH, Maughan WL. Accuracy of volume determination by two-dimensional echocardiography: Defining requirements under controlled conditions in the ejecting canine left ventricle. *Circulation*. 1983;67:889–895.

34. Wahr DW, Wang YS, Schiller NB. Left ventricular volumes determined by two-dimensional echocardiography in a normal adult population. *J Am Coll Cardiol*. 1983;1:3, 863–868.

35. Bouchard A, Blumlein S, Schiller NB, et al. New method for the measurement of stroke volume and cardiac output by M-mode/continuous wave Doppler. *Circulation*. 1984;70(suppl) (II):684.

36. Gardin JM, Iseri LT, Elkayam U, et al. Evaluation of dilated cardiomyopathy by pulsed Doppler echocardiography. *Am Heart J*. 1983;106:1057–1065.

37. Elkayam U, Gardin JM, Berkley R, Hughes CA, Henry WL. The use of Doppler flow velocity measurement to assess the hemodynamic response to vasodilators in patients with heart failure. *Circulation*. 1983;67:377–383.

38. Goldberg SJ, Sahn DJ, Allen HD, Valdes-Cruz LM, Hoenecke H, Carnahan Y. Evaluation of pulmonary and systemic blood flow by two-dimensional Doppler echocardiography using fast Fourier transform spectral analysis. *Am J Cardiol*. 1982;50:1394–1400.

39. Nishimura RA, Callahan MJ, Schaff HV, Illstrup DM, Miller FA, Tajik AJ. Noninvasive measurement of cardiac output by continuous-wave Doppler echocardiography: Initial experience and review of the literature. *Mayo Clin Proc*. 1984;59:484–489.

40. Fisher DC, Sahn DJ, Friedman MJ, et al. The effect of variations on pulsed Doppler sampling site on calculation of cardiac output: An experimental study in open-chest dogs. *Circulation*. 1983;67:370–376.

41. Loeppky JA, Hoekenga DE, Greene ER, Luft UC. Comparison of noninvasive pulsed Doppler and Fick measurements of stroke volume in cardiac patients. *Am Heart J*. 1984;107:339–346.

42. Chandraratna PA, Nanna M, McKay C, et al. Determination of cardiac output by transcutaneous continuous-wave ultrasonic Doppler computer. *Am J Cardiol*. 1984; 53:234–237.

43. Magnin PA, Stewart JA, Myers S, Von Ramm O, Kisslo JA. Combined Doppler and phased-array echocardiographic estimation of cardiac output. *Circulation*. 1981;63:388–392.

44. Ihlen H, Amlie JP, Dale J, et al. Determination of cardiac output by Doppler echocardiography. *Br Heart J*. 1984;51:54–60.

45. Schuster AH, Nanda NC. Doppler echocardiographic measurement of cardiac output: comparison with a non-golden standard. *Am J Cardiol*. 1984;53:257–259.

46. Lewis JF, Kuo LC, Nelson JG, Limacher M, Quinones M: Pulsed Doppler echocardiographic determination of stroke volume and cardiac output: Clinical validation of two new methods using the apical window. *Circulation*. 1984; 70:425–431.

47. Schuster AH, Nanda NC. Doppler echocardiography, I: Doppler cardiac output measurements perspective and comparison with other methods of cardiac output determination. *Echocard.* 1984;1:45–54.

48. Gardin JM, Dabestani A, Matin K, Allfie A, Russell D, Henry W: Reproducibility of Doppler aortic valve blood flow measurements: Studies on intraobserver, interobserver and day to day variability in normal subjects. *Am J Cardiol.* 1984;54:1092–1098.

49. Rose JS, Nanna M, Rahimtoola SH, Elkayam U, McKay C, Chandraratna A. Accuracy of determination of changes in cardiac output by transcutaneous continuous-wave Doppler computer. *Am J Cardiol.* 1984;54:1099–1101.

50. Vargas-Barron J, Sahn DJ, Valdes-Cruz LM, et al. Clinical utility of two-dimensional Doppler echocardiographic techniques for estimating pulmonary to systemic blood flow ratios in children with L to R shunting ASD, VSD and PDA. *J Am Coll Cardiol.* 1984;3:169–178.

51. Labovitz A, Buckingham TA, Habermehl K, Nelson J, Kennedy H, Williams G. The effects of sampling site on the two-dimensional echo Doppler determination of cardiac output. *Am Heart J.* 1985;109:327–332.

52. Zhang Y, Nitter-Hauge S, Ihlen H, Myhre E. Doppler echocardiographic measurement of cardiac output using the mitral orifice method. *Br Heart J.* 1985;53:130–136.

53. Sahn DJ. Determination of cardiac output by echocardiographic Doppler methods: Relative accuracy of various sites for measurement. *J Am Coll Cardiol.* 1985;6:663–664.

54. Goldberg SJ, Dickinson DF, Wilson N. Evaluation of an elliptical area technique for calculating mitral blood flow by Doppler echocardiography. *Br Heart J.* 1985;54:68–75.

55. Trompler AT, Sold G, Vogt A, Kreuzer H. Noninvasive determination of stroke volume with spectral Doppler echocardiography. *Z Kardiol.* 1985;74:322–326.

56. Byrd BF III, Schiller NB, Botvinick EH, Bouchard A, Higgins CB. Magnetic resonance imaging and two-dimensional echocardiography in dilated cardiomyopathy. *Circulation.* 1985;72(suppl III):22.

57. Beaupre P, Cahalan M, Kremer P, Lurz F, Schiller NB. Contractility depression during anesthesia: Comparison of halothane, enflurane and isoflurance by transesophageal echocardiography. *Circulation.* 1983;68(suppl)(III):332.

58. Sahn DJ, DeMaria A, Kisslo J, Weyman A. Recommendations regarding quantitation in M-mode echocardiography: Results of a survey of echocardiographic measurements. *Circulation.* 1978;58:1072–1083.

59. Gibson DG. Measurement of left ventricular volumes in man by echocardiography—comparison with biplane angiographs. *Br Heart J.* 1971;33:614.

60. Schnittger I, Gordon EP, Fitzgerald PJ, Popp RL. Standardized intracardiac measurements of two-dimensional echocardiography. *J Am Coll Cardiol.* 1983;2:934–938.

61. Byrd BF III, Mickelson J, Bouchard A, Schiller NB, Botvinick EH. Left ventricular mechanics in distance runners. *Circulation.* 1984;70(suppl)(II):421.

62. Oakley D. Cardiac hypertrophy in athletes. *Br Heart J.* 1984;52:121–123.

63. Fagard R, Aubert A, Staessen J, Vanden Eynde E, Vanhees L, Amery A. Cardiac structure and function in cyclists and runners—comparative echocardiographic study. *Br Heart J.* 1984;52:124–129.

64. Shapiro LM. Physiological left ventricular hypertrophy. *Br Heart J.* 1984;52:130–135.

65. Csanady M, Gruber N. Comparative echocardiographic studies in leading canoe-kayak and handball sportmen. *Cor Vasa.* 1984;26:32–37.

66. Graettinger W. The cardiovascular response to chronic physical exertion and exercise training: An echocardiographic review. *Am Heart J.* 1984;108:1014–1019.

67. Shapiro LM, Smith RG. Effect of training on left ventricular structure and function. *Br Heart J.* 1983;50:534–539.

68. Hauser AM, Dressendorfer RH, Vos M, Hoshimoto T, Gordon S, Timmis G. Symmetric cardiac enlargement in highly trained endurance athletes: A two-dimensional echocardiographic study. *Am Heart J.* 1985;109:1038–1044.

69. Huston TP, Puffer JC, Rodney WM, Oberman A. The athletic heart syndrome. *N Engl J Med.* 1985;313:24–32.

70. Wolfe LA, Martin RP, Watson DD, Lasley R, Bruns D. Chronic exercise and left ventricular structure and function in healthy human subjects. *J Appl Physiol.* 1985;58:409–415.

71. Colan SD, Sanders SP, MacPherson D, Borow K. Left ventricular diastolic function in elite athletes with physiologic cardiac hypertrophy. *J Am Coll Cardiol.* 1985;6:545–549.

72. Arvan S, Varat MA. Persistent ST-segment elevation and left ventricular wall abnormalities: A two-dimensional echocardiographic study. *Am J Cardiol.* 1984;53:1542–1546.

73. Stamm RB, Gibson RS, Bishop HL, Carabello BA, Beller GA, Martin RP. Echocardiographic detection of infarct-localized asynergy and remote asynergy during acute myocardial infarction: Correlation with the extent of angiographic coronary disease. *Circulation.* 1983;67:233–244.

74. Weiss JL, Bulkley BH, Hutchins GM, Mason SJ. Two dimensional echocardiographic recognition of myocardial injury in man: Comparison with postmortem studies. *Circulation.* 1981;63:401–408.

75. Wong M, Shah PM. Accuracy of two-dimensional echocardiography in detecting left ventricular aneurysm. *Clin Cardiol.* 1983;6:250–254.

76. Roberts CS, MacLean D, Maroko P, Kloner R. Early and late remodeling of the left ventricle after acute myocardial infarction. *Am J Cardiol.* 1984;54:407–410.

77. Matsumoto M, Watanabe F, Goto A, et al. Left ventricular aneurysm and the prediction of left ventricular enlargement studies by two-dimensional echocardiography: Quantitative assessment of aneurysm size in relation to clinical course. *Circulation.* 1985;72:280–286.

78. Devereux RB, Reichek N. Echocardiographic determination of left ventricular mass in man. Anatomic validation of the method. *Circulation.* 1977;55:613–618.

79. Abbasi AS, MacAlpin RN, Eber LM, Pearce ML. Echocardiographic diagnosis of idiopathic hypertrophic cardiomyopathy without outflow obstruction. *Circulation.* 1972;46:897–904.

80. Casale PN, Devereux RB, Kligfield P, et al. Electrocardiographic detection of left ventricular hypertrophy: Development and prospective validation of improved criteria. *J Am Coll Cardiol.* 1985;6:572–580.

81. Feldman T, Borow KM, Neumann A, Lang R, Childers R: Relation of electrocardiographic R-wave amplitude to changes in left ventricular chamber size and position in normal subjects. *Am J Cardiol.* 1985;55:1168–1174.

82. Reichek N, Helak J, Plappert TA, St. John Sutton MG, Weber KT. Anatomic validation of left ventricular mass estimates from clinical two-dimensional echocardiography: Initial results. *Circulation.* 1983;67:348–352.

83. Helak JW, Reichek N. Quantitation of human left ventricular mass and volume by two-dimensional echocardiography: In vitro anatomic validation. *Circulation.* 1981;63:1398–1407.

84. Devereux RB, Casale PN, Eiserberg RR, Miller DH, Kligfield P. Electrocardiographic detection of left ventricular hypertrophy using echocardiographic determination of left ventricular mass as the

reference standard. Comparison of STD Crit, Comptrdx and MD interpret. *J Am Coll Cardiol.* 1984;3:82–87.

85. Godwin JD, Axel J, Adams JR, Schiller NB, Simpson PC, Gertz EW. Computed tomography: A new method for diagnosing tumor of the heart. *Circulation.* 1981;63:448–451.

86. Fernandez PG, Kim BK, Reichek N, et al. The correlation of changes in systolic blood pressure with regional anatomical regression of hypertensive left ventricular hypertrophy in patients on chronic antihypertensive therapy (more than 1 year). *Curr Med Res Opin.* 1984;8:720–733.

87. Kaul U, Mohan JC, Bhatia ML. Effects of labetalol on left ventricular mass and function in hypertension—an assessment by serial echocardiography. *Int J Cardiol.* 1984; 5:461–473.

88. Dunn FG, Oigman W, Ventura HO, Messerli FH, Kobrin I, Frohlich E. Enalapril improves systemic and renal hemodynamics and allows regression of left ventricular mass in essential hypertension. *Am J Cardiol.* 1984;53:105–108.

89. Nakashima Y, Fouad FM, Tarazi RC. Regression of left ventricular hypertrophy from systemic hypertension by enalapril. *Am J Cardiol.* 1984;53:1044–1049.

90. Byrd BF III, Wahr D, Wang YS, Bouchard A, Schiller NB. Left ventricular mass and volume/mass ratio in a normal population determined by two-dimensional echocardiography. *J Am Coll Cardiol.* 1985;6:1021–1025.

91. Himelman RB, Landzberg JS, Simonson JS, et al. Cardiac consequences for renal transplantation: Changes in left ventricular morphology and function. *J Am Coll Cardiol.* 1988;12:915–923.

92. Pratt RC, Parisi AF, Harrington JJ, Sasahara AA. The influence of left ventricular stroke volume on aortic root motion: An echocardiographic study. *Circulation.* 1976;53: 947–953.

93. Lalani AV, Lee SJK. Echocardiographic measurement of cardiac output using the mitral valve and aortic root echo. *Circulation.* 1976;54:738–743.

94. Kramer PH, Djalaly A, Poehlman H, Schiller NB. Abnormal diastolic left ventricular posterior wall motion in left ventricular hypertrophy. *Am Heart J.* 1983;106:1066–1069.

95. Ambrose JA, King BD, Teicholz LE, LeBlanc DT, Schwinger M, Stein JH. Early diastolic motion of the posterior aortic root as an index of left ventricular filling. *J Clin Ultrasound.* 1983;11:357–364.

96. Strunk BL, Fitzgerald JW, Lipton M, Popp RL, Barry WH. The posterior aortic wall echocardiogram: Its relationship to left atrial volume change. *Circulation.* 1976; 54:744–750.

97. Strunk BL, London EJ, Fitzgerald J, Popp RL, Barry WH. The assessment of mitral stenosis and prosthetic mitral valve obstruction, using the posterior aortic wall echocardiogram. *Circulation.* 1977;55:885–891.

98. Miyatake K, Okamoto M, Kinoshita N, et al. Augmentation of atrial contribution to left ventricular inflow with aging as assessed by intracardiac Doppler flowmetry. *Am J Cardiol.* 1984;53:586–589.

99. Rokey R, Kuo LC, Zoghbi WA, Limacher M, Quinones M. Determination of parameters of left ventricular diastolic filling with pulsed Doppler echocardiography: Comparison with cineangiography. *Circulation.* 1985;71:543–550.

100. Kerber RE, Abboud FM. Echocardiographic detection of regional myocardial infarction. *Circulation.* 1973;47: 997–1005.

101. Reeder GS, Seward JB, Tajik AJ. The role of two dimensional echocardiography in coronary artery disease. *Mayo Clin Proc.* 1982;57:247–258.

102. Visser CA, Durrer D. Echocardiographic determination of infarct size in acute myocardial infarction. *Prac Cardiol.* 1983;9: 225–231.

103. Pandian NG, Skorton DJ, Collins SM, et al. Myocardial infarct size threshold for two-dimesional echocardiographic detection: Sensitivity of systolic wall thickening and endocardial motion abnormalities in small versus large infarcts. *Am J Cardiol.* 1985;55:551–555.

104. Chen YZ, Sherrid MV, Dwyer EM. Value of two-dimensional echocardiography in evaluating coronary artery disease: A randomized blind analysis. *J Am Coll Cardiol.* 1985;5:911–917.

105. Ren JF, Kotler MN, Hakki AH, Panidis I, Mintz G, Ross J. Quantitation of regional left ventricular function by two-dimensional echocardiography in normals and patients with coronary artery disease. *Am Heart J.* 1985;110:552–560.

106. Freeman AP, Giles RW, Walsh WF, Fisher R, Murray I, Wilcken D. Regional left ventricular wall motion assessment: comparison of two-dimensional echocardiography and radionuclide angiography with contrast angiography in healed myocardial infarction. *Am J Cardiol.* 1985;56:8–12.

107. Quinones M, Roberts R. Role of two-dimensional echocardiography in acute myocardial infarction. *Echocardiography.* 1985;2:213–216.

108. Rasmussen S, Lovelace DE, Knoebel SB, et al. Echocardiographic detection of ischemic and infarcted myocardium. *J Am Coll Cardiol.* 1984;3:733–743.

109. Fraker TD Jr, Nelson AD, Arthur JA, Wilderson R. Altered acoustic reflectance on two-dimensional echocardiography as an early predictor of myocardial infarct size. *Am J Cardiol.* 1984;53: 1699–1702.

110. Chandraratna PA, Ulene R, Nimalasuriya A, Reid C, Kawanishi D, Rahimtoola S. Differentiation between acute and healed myocardial infarction by signal averaging and color encoding two-dimensional echocardiography. *Am J Cardiol.* 1985;56:381–384.

111. Weisman HF, Bush DE, Mannisi JA, Bulkley B. Global cardiac remodeling after acute myocardial infarction: A study in the rate mode. *J Am Coll Cardiol.* 1985;5:1355–1362.

112. Kinoshita Y, Shukuya M, Inagaki Y. Significance of a wall motion hinge point after myocardial infarction. *J Card Ultrasound.* 1983;2:235–241.

113. Nishimura R, Tajik A, Seward J. Cases from the Mayo clinic—distinctive two-dimensional echocardiographic appearance of septal infarct secondary to isolated occlusion of first septal perforator artery. *Echocardiography.* 1984;1:97–98.

114. Lapeyre AC, Steele PM, Kazmier FJ, Chesebro J, Vliestra R, Fuster V. Systemic embolism in chronic left ventricular aneurysm: Incidence and the role of anticoagulation. *J Am Coll Cardiol.* 1985; 6:534–538.

115. Friart A, Vandenbossche JL, Hamdan BA, Deuvaert F, Englert M. Association of false tendons with left ventricular aneurysm. *Am J Cardiol.* 1985;55:1425–1426.

116. Maurer G, Nanda NC. Two dimensional echocardiographic evaluation of exercise induced left and right ventricular asynergy: Correlation with thallium scanning. *Am J Cardiol.* 1981;48:720–727.

117. Crawford MH, Petru MA, Amon KW, Sorensen SG, Vanee WS. Comparative value of two dimensional echocardiography and radionuclide angiography for quantitating changes in left ventricular performance during exercise limited by angina pectoris. *Am J Cardiol.* 1984;53:42–46.

118. Robertson WS, Feigenbaum H, Armstrong WF, Dillon JC, O'Donnell J, McHenery PW. Exercise echocardiography: A clinically practical addition in the evaluation of coronary artery disease. *J Am Coll Cardiol.* 1983;2:1085–1091.

119. Ginzton LE, Conant R, Brizendine M, Lee F, Mena I, Laks MM. Exercise subcostal two dimensional echocardiography: A new method of segmental wall motion analysis. *Am J Cardiol.* 1984;53: 805–811.

120. Bersin R, Tubau JF, Merz R, Wolff A, Schiller NB. Diagnostic yield of echocardiography with routine treadmill testing. *Circulation.* 1985;72(suppl III):III49.

121. Child J. Stress echocardiography: A technique whose time has come. *Echocard*. 1984;1:107–110.

122. Heng MK, Simard M, Lake R, Udhoji V. Exercise two-dimensional echocardiography for diagnosis of coronary artery disease. *Am J Cardiol*. 1984;54:502–507.

123. Berberich SN, Zager J, Plotnick GD, Fisher M. A practical approach to exercise echocardiography: Immediate post exercise echocardiography. *J Am Coll Cardiol*. 1984;3:284–290.

124. Iskandrian AS, Hakki AH, Kotler MN, Segal B, Herling I. Evaluation of patients with acute myocardial infarction: Which test, for whom and why? *Am Heart J*. 1985;109:391–394.

125. Loh I, Hubert G. Use of 2-dimensional echocardiography in the differential diagnosis of AMI in the patient presenting with chest pain. *Pract Cardiol*. 1984;10:185–194.

126. Fujii J, Sawada H, Aizawa T, Kato K, Onoe M, Kuno Y. Computer analysis of cross sectional echocardiogram for quantitative evaluation of left ventricular asynergy in myocardial infarction. *Br Heart J*. 1984;51:139–148.

127. Kan G, Visser CA, Lie KI, Ferrer D. Measurement of left ventricular ejection fraction after acute myocardial infarction. *Br Heart J*. 1984;51:631–636.

128. Weyman AE, Franklin TD Jr, Hogan RD, et al. Importance of temporal heterogeneity in assessing the contraction abnormalities associated with acute myocardial ischemia. *Circulation*. 1984;70:102–112.

129. Gillam LD, Hogan RD, Foale RA, et al. A comparison of quantitative echocardiographic methods for delineating infarct-induced abnormal wall motion. *Circulation*. 1984;70:113–122.

130. Nishimura RA, Reeder GS, Miller FA Jr, et al. Prognostic value of predischarge two-dimensional echocardiogram after acute myocardial infarction. *Am J Cardiol*. 1984;53:429–432.

131. DeMaria AN, Bommer W, Neumann A, et al. Left ventricular thrombi identified by cross sectional echocardiography. *Ann Intern Med*. 1979;90:14–18.

132. Asinger RW, Mikell FL, Elsperger J, Hodges M. Incidence of left ventricular thrombosis after acute transmural myocardial infarction. Serial evaluation by two dimensional echocardiography. *New Engl J Med*. 1981;305:297–302.

133. Johannessen KA, Nordrehaug JE, von der Lippe G. Left ventricular thrombosis and cerebrovascular accident in acute myocardial infarction. *Br Heart J*. 1984;51:553–556.

134. Keren A, Billingham ME, Popp RL. Echocardiographic recognition and implications of ventricular hypertrophic trabeculations and aberrant bands. *Circulation*. 1984;70:836–842.

135. Meltzer RS, Visser CA, Kan G, Roelandt J. Two-dimensional echocardiographic appearance of left ventricular thrombi with systemic emboli after myocardial infarction. *Am J Cardiol*. 1984;53:1511–1513.

136. Visser CA, Kan G, Meltzer RS, et al. Embolic potential of left ventricular thrombus after myocardial infarction: A two-dimensional echocardiographic study of 119 patients. *J Am Coll Cardiol*. 1985;5:1276–1280.

137. Ezekowitz MD. Acute infarction, left ventricular thrombus and systemic embolization: An approach to management. *J Am Coll Cardiol*. 1985;5:1281–1282.

138. Arvan S. Persistent intracardiac thrombi and systemic embolization despite anticoagulant therapy. *Am Heart J*. 1985;109:178–181.

139. Kinney EL. The significance of left ventricular thrombi in patients with coronary heart disease: A retrospective analysis of pooled data. *Am Heart J*. 1985;109:191–194.

140. Singer E, Park Y. Splenic infarction following mural thrombus of the left ventricle in a patient with acute anteroseptal myocardial infarction. *Cardiol Rev Rep*. 1985;6: 835–850.

141. Rao A, Agatston A, Samet P. Multiple mural biventricular thrombi: Echocardiographic detection in congestive cardiomyopathy. *J Clin Ultrasound*. 1985;4:65–71.

142. Takamoto T, Kim D, Urie PM, et al. Comparative recognition of left ventricular thrombi by echocardiography and cineangiography. *Br Heart J*. 1985;53:36–42.

143. Lloret RL, Cortada X, Bradford J, Metz M, Kinney. Classification of left ventricular thrombi by their history of systemic embolization using pattern recognition of two-dimensional echocardiograms. *Am Heart J*. 1985;110:761–765.

144. Asinger RW, Mikell FL. Left ventricular thrombosis after myocardial infarction. *Primary Cardiol*. 1983;9:19–31.

145. Stratton JR. Mural thrombi of the left ventricle. *Chest*. 1983;83:166–168.

146. Visser CA, Kan G, David GK, Lie KI, Durrer D. Two-dimensional echocardiography in the diagnosis of left ventricular thrombus: A prospective study of 67 patients with anatomic validation. *Chest*. 1983;83:228–232.

147. Mongiardo R, Digaetano A, Pennestri F, Marion G, Mazzari B, Loperfido F. Left ventricular thrombus evolution: Assessment by two dimensional echocardiography. *G Ital Cardiol*. 1982;12:308–310.

148. Spirito P, Bellotti P, Gharela F, Domenicucci S, Sementa A, Vecchio C. Prognostic significance and natural history of left ventricular thrombi in patients with acute anterior myocardial infarction: a two-dimensional echocardiographic study. *Circulation*. 1985;72:774–780.

149. Wong M, Shah PM. Accuracy of two-dimensional echocardiography in detecting left ventricular aneurysm. *Clin Cardiol*. 1983;6:250–254.

150. Knowlton A, Grauer J, Plehn J, Liebson P. Ventricular pseudoaneurysm: A rare but ominous condition. *Cardiol Review*. 1985;6:508–513.

151. Loperfido F, Pennestri F, Mazzari M, et al. Diagnosis of left ventricular pseudoaneurysm by pulsed Doppler echocardiography. *Am Heart J*. 1985;110:1291–1293.

152. Kaul S, Josephson MA, Tei C, Wittig JH, Millman J, Shah PM. Atypical echocardiography and angiographic presentation of a postoperative pseudoaneurysm of the LV after repair of a true aneurysm. *J Am Coll Cardiol*. 1983;2:780–784.

153. Rath S, Eldar M, Shemesh Y, et al. Acute cardiac rupture and tamponade: Angiographic appearance. *Am J Cardiol*. 1985;55:588–589.

154. Wunderink RG: Incidence of pericardial effusions in acute myocardial infarctions. *Chest*. 1984;85:494–496.

155. Kaplan K, Davison R, Parker M, et al. Frequency of pericardial effusion as determined by M-mode echocardiography in acute myocardial infarction. *Am J Cardiol*. 1985;55: 335–337.

156. Missri JC, Spath EA, Stark S, Thomas DR, Jaskolka D, Martino A. Ventricular septal rupture detected by two dimensional echocardiography. *J Card Ultrasound*. 1983;2:259–264.

157. Nishimura RA, Schaff HV, Shub C, Gersh BJ, Edwards WD, Tajik AJ. Papillary muscle rupture complicating acute myocardial infarction: Analysis of 17 patients. *Am J Cardiol*. 1983;51:373–377.

158. Clements SP Jr, Story WE, Hurst JW, Graver J, Jones E. Ruptured papillary muscle, a complication of myocardial infarction: Clinical presentation, diagnosis, and treatment. *Clin Cardiol*. 1985;8:93–103.

159. Arditti A, Lewin RF, Hellman C, Sclarovsky S, Strasberg B, Agmon J. Right ventricular dysfunction in acute inferoposterior myocardial infarction. An echocardiographic and isotopic study. *Chest*. 1985;87:307–314.

160. Jugdutt BI, Sussex BA, Siraram CA, Rossall RE. Right ventricular infarction: Two-dimensional echocardiographic evaluation. *Am Heart J*. 1984;107:505–518.

161. Panidis IP, Kotler MN, Mintz GS, Ross J, Ren JF, Kutalek S. Right ventricular function in coronary artery disease as assessed by two-dimensional echocardiography. *Am Heart J*. 1984;107:1187–1194.

162. Panidis IP, Ren JF, Kotler MN, Mintz G, Iskandrian A, Ross J. Two-dimensional echocardiographic estimation of right ventricular ejection fraction in patients with coronary artery disease. *J Am Coll Cardiol*. 1983;2:911–918.

163. Lopez-Sendon J, Garcia-Fernandez MA, Coma-Canella I, Yanguela MM, Banuelos F. Segmental right ventricular function after acute myocardial infarction: Two-dimensional echocardiographic study in 63 patients. *Am J Cardiol*. 1983;51:390–396.

164. Kaul S, Hopkins JM, Shah PM. Chronic effects of myocardial infarction on right ventricular function: a noninvasive assessment. *J Am Coll Cardiol*. 1983;2:607–615.

165. Haupt HM, Hutchins GM, Moore GW. Right ventricular infarction: Role of the moderator band artery in determining infarct size. *Circulation*. 1983;67:1268–1272.

166. Baigrie RS, Hag A, Morgan CD, Rakowski H, Drobac M, Caughlin. The spectrum of right ventricular involvement in inferior wall myocardial infarction: A clinical, hemodynamic and noninvasive study. *J Am Coll Cardiol*. 1983;1:1396–1404.

167. Roberts N, Harrison DG, Reimer KA, Crain B, Wagner G. Right ventricular infarction with shock but without significant left ventricular infarction: A new clinical syndrome. *Am Heart J*. 1985;110:1047–1053.

168. Manno BV, Bemis CE, Carver J, Mintz GS. Right ventricular infarction complicated by right to left shunt. *J Am Coll Cardiol*. 1983;1:554–557.

169. Higgins JR, Sundstrom J, Gutman J, Schiller NB. Contrast echocardiography with quantitative Valsalva maneuver to detect patent foramen ovale. *Clin Res*. 1982;30:12A. Abstract.

170. Corya B, Feigenbaum H, Rasmussen S, Black MJ. Echocardiographic features of congestive cardiomyopathy compared with normal subjects and patients with coronary artery disease. *Circulation*. 1974;49:1153–1159.

171. Goldberg SJ, Valdes-Cruz LM, Sahn DJ, Allen HD. Two dimensional echocardiographic evaluation of dilated cardiomyopathy in children. *Am J Cardiol*. 1983;52:1244–1248.

172. Fortuin NJ, Pawsey CG. The evaluation of left ventricular function by echocardiography. *Am J Med*. 1977;63:1–9.

173. Hayakawa M, Yokota Y, Kumaki T, et al. Intracardiac flow pattern in dilated cardiomyopathy studied with pulsed Doppler echocardiography. *J Cardiog Phy*. 1983;13:317–326.

174. Unverferth DV, Magorien RD, Moeschberger ML, Baker P, Leier C. Factors influencing the one-year mortality of dilated cardiomyopathy. *Am J Cardiol*. 1984;54:147–152.

175. Hirota Y, Shimizu G, Kaku K, Saito T, Kino M, Kawamura K. Mechanisms of compensation and decompensation in dilated cardiomyopathy. *Am J Cardiol*. 1984;54:1033–1038.

176. Lambertz H. Functional tricuspid insufficiency in patients with severe heart failure: Follow-up study using echocardiography. *Z Kardiol*. 1984;73:159–163.

177. Gardin JM, Iseri LT, Elkayam U, et al. Evaluation of dilated cardiomyopathy by pulsed Doppler echocardiography. *Am Heart J*. 1983;106:1057–1065.

178. French JW, Popp RL, Pitlick PT. Cardiac localization of transvascular bioptome using 2-dimensional echocardiography. *Am J Cardiol*. 1983;51:219–223.

179. Acquatella H, Schiller NB, Puigbo JJ, et al. M-mode and two-dimensional echocardiography in chronic Chagas' heart disease. *Circulation*. 1980;62:787–799.

180. Henry WL, Clark CE, Roberts WC, Morrow AG, Epstein SE. Difference in distribution of myocardial abnormalities in patients with obstructive and non-obstructive asymmetric septal hypertrophy (ASH): echocardiographic and gross anatomic findings. *Circulation*. 1974;50:447–455.

181. Henry WL, Clark CE, Epstein SE. Asymmetric septal hypertrophy: The unifying link in the IHSS disease spectrum: Observation regarding its pathogenesis, pathophysiology, and course. *Circulation*. 1973;47:827–832.

182. Henry WL, Clark CE, Epstein SE. Asymmetric septal hypertrophy (ASH): Echocardiographic identification of the pathognomonic anatomic abnormality of IHSS. *Circulation*. 1973;47:225–233.

183. Ciro E, Nichols PF, Maron BJ. Heterogeneous morphologic expression of genetically transmitted hypertrophic cardiomyopathy. Two dimensional echocardiographic analysis. *Circulation*. 1983;67:1227–1233.

184. Maron BJ, Nichols PF III, Pickle LW, Wesley Y, Mulvihill J. Patterns of inheritance in hypertrophic cardiomyopathy: Assessment by M-mode and two-dimensional echocardiography. *Am J Cardiol*. 1984;53:1087–1094.

185. Ballester M, Rees S, Rickards AF, McDonald L. An evaluation of two-dimensional echocardiography in the diagnosis of hypertrophic cardiomyopathy. *Clin Cardiol*. 1984;7:631–638.

186. Maron BJ. Asymmetry in hypertrophic cardiomyopathy: the septal to free wall thickness ratio revisited. *Am J Cardiol*. 1985;55:835–838.

187. Maron BJ, Bonow RO, Seshagiri TN, Roberts WC, Epstein SE. Hypertrophic cardiomyopathy with ventricular septal hypertrophy localized to the apical region of the left ventricle (apical hypertrophic cardiomyopathy). *Am J Cardiol*. 1982;49:1838–1848.

188. Steingo L, Dansky R, Pocock WA, Barlow JB. Apical hypertrophic nonobstructive cardiomyopathy. *Am Heart J*. 1982;104:635–637.

189. Kereiakes DJ, Anderson DJ, Crouse L, Chatterjee K. Apical hypertrophic cardiomyopathy. *Am Heart J*. 1983;105:855–856.

190. Maron BJ, Spirito P, Chiarella F, Vecchio C. Unusual distribution of left ventricular hypertrophy in obstructive hypertrophic cardiomyopathy: Localized posterobasal free wall thickening in two patients. *J Am Coll Cardiol*. 1985;5:1474–1477.

191. Mori H, Ogawa S, Nakazawa H, et al. Apical hypertrophy as a part of the morphologic spectrum of hypertrophic cardiomyopathy. *J Cardiogr*. 1984;14:289–300.

192. Koga Y, Takashi H, Ifuku M, Itaya M, Adachi K, Toshima H. Hypertrophic cardiomyopathy with ventricular septal hypertrophy localized to the apical region of the left ventricle (apical ASH). *J Cardiogr*. 1984;14:301–310.

193. Vacek JL, Davis WR, Bellinger RL, McKiernan T. Apical hypertrophic cardiomyopathy in American patients. *Am Heart J*. 1984;108:1501–1506.

194. Koga Y, Itaya M, Takahahi H, Koga M, Ikeda H, Itaya K, Toshima H. Apical hypertrophy and its genetic and acquired factors. *J Cardiogr*. 1985;15:65–74.

197. Martin RP, Rakowski H, French J, Popp RL. Idiopathic hypertrophic subaortic stenosis viewed by wide-angle, phased-array echocardiography. *Circulation*. 1979;59:1206–1217.

198. Sugishita Y, Iida K, Matsuda M, et al. Apical hypertrophy and catecholamine. *J Cardiogr*. 1985;15:75–83.

199. Spirito P, Maron BJ. Patterns of systolic anterior motion of the mitral valve in hypertrophic cardiomyopathy: Assessment by two-dimensional echocardiography. *Am J Cardiol*. 1984;54:1039–1046.

200. Baragan J, Fernandez F, Thiron JM. Dynamic auscultation and phonocardiography. In: Morton E. Tavel, ed. *Regurgitation from the Great Vessels of the Base*. Bowie, Md: The Charles Press Publishers; 1979;143–154.

201. Kereiakes DJ, Naughton JL, Brundage B, Schiller NB. The heart in diabetes. *West J Med*. 1984;140:583–593.

202. Child JS, Levisman JA, Abbasi AS, MacAlpin RN. Echocardiographic manifestations of infiltrative cardiomyopathy: A report of seven cases due to amyloid. *Chest*. 1976;70:726–731.

203. Siqueira-Filho AG, Cunha CL, Tajik AJ, Seward JB, Schattenberg TT, Giuliani ER. M-mode and two dimensional echocardiographic features in cardiac amyloidosis. *Circulation*. 1981;63:188–196.

204. Roberts WC, Waller BF: Cardiac amyloidosis causing cardiac dysfunction: Analysis of 54 necropsy patients. *Am J Cardiol*. 1983;52:137–146.

205. Nicolosi GL, Pavan D, Lestuzzi C, Burelli C, Zardo F, Zanuttini D. Prospective identification of patients with amyloid heart disease by two dimensional echocardiography. *Circulation*. 1984;70:432–437.

206. Sedlis SP, Saffitz JE, Schwob VS, Jaffe A. Cardiac amyloidosis simulating hypertrophic cardiomyopathy. *Am J Cardiol*. 1984;53:969–970.

207. Acquatella H, Schiller NB, Puigbo JJ, Gomez-Mancebo JR, Suarez C, Acquatella G. Value of two dimensional echocardiography in endomyocardial disease with and without eosinophilia. A clinical and pathologic study. *Circulation*. 1983;67:1219–1226.

208. Crawford MH, Amon KW, Vance WS. Exercise 2-dimensional echocardiography. *Am J Cardiol*. 1983;51:1–6.

209. Ginzton LE, Laks MM, Brizendine M, Conant R, Mena I. Noninvasive measurement of the rest and exercise peak systolic pressure/end-systolic volume ratio: A sensitive two-dimensional echocardiographic indicator of left ventricular function. *J Am Coll Cardiol*. 1984;4:509–516.

210. Schluter M, Hanrath P. Transesophageal echocardiography: Potential advantages and initial clinical results. *Pract Cardiol*. 1983;9:149–182.

211. Hisanaga K, Hisanaga A, Nagata K, Ichi Y. Transesophageal cross-sectional echocardiography. *Am Heart J*. 1980;100:605–609.

212. Schluter M, Langenstein BA, Polster J, et al. Transesophageal cross-sectional echocardiography with phased array transducer system. Technique and initial clinical results. *Br Heart J*. 1982;48:67–72.

213. Schluter M, Langenstein BA, Hanrath P, et al. Assessment of transesophageal pulsed Doppler echocardiography in the detection of mitral regurgitation. *Circulation*. 1982; 66:784–789.

214. Matsumoto M, Oka Y, Strom J, et al. Application of transesophageal echocardiography to continuous intraoperative monitoring of left ventricular performance. *Am J Cardiol*. 1980;46:95–105.

215. Cahalan MK, Kremer PF, Beaupre PN, et al. Intraoperative myocardial ischemia detected by transesophageal 2-dimensional echocardiography. *Anesthesiology*. 1983;59(suppl 3): A164. Abstract.

216. Beaupre PN, Cahalan MK, Kremer PF, et al. Does pulmonary artery occlusion pressure adequately reflect left ventricular filling during anesthesia and surgery? *Anesthesiology*. 1983;59(suppl 3):A3. Abstract.

217. Beaupre PN, Roizen MF, Cahalan MK, Alpert RA, Cassorla L, Schiller NB. Hemodynamic and two-dimensional transesophageal echocardiographic analysis of an anaphylactic reaction in a human. *Anesthesiology*. 1984;60:482–484.

218. Shively B, Cahalan M, Benefiel D, Schiller NB. Interoperative assessment of mitral valve regurgitation by transesophageal Doppler echocardiography. *J Am Coll Cardiol*. 1986;7:228A. Abstract.

219. Beaupre PN, Kremer PF, Cahalan MK, Lurz FW, Schiller NB, Hamilton WK. Intraoperative detection of changes in left ventricular segmental wall motion by transesophageal echocardiography. *Am Heart J*. 1984;107(I): 1021–1023.

220. Smith JS, Cahalan MK, Benefiel DJ, et al. Intraoperative detection of myocardial ischemia in high risk patients: Electrocardiography vs two-dimensional transesophageal echocardiography. *Circulation*. 1985;72:1015–1021.

221. Shively B, Watters T, Benefiel D, Cahalan M, Botvinick EH, Schiller NB. The intraoperative detection of myocardial infarction by transesophageal echocardiography. *J Am Coll Cardiol*. 1986;7:2. Abstract.

222. Topol EJ, Humphrey LS, Blanck TJJ, et al. Characterization of post-cardiopulmonary bypass hypotension with intraoperative transesophageal echocardiography. *Anesthesiology*. 1983;59(suppl 13):A2. Abstract.

223. Martin RW, Colley PS. Evaluation of transesophageal Doppler detection of air embolism in dogs. *Anesthesiology*. 1983;58:117–123.

224. Furuya H, Suzuki T, Okumura F, Kishi Y, Uefuji T. Detection of air embolism by transesophageal echocardiography. *Anesthesiology*. 1983;58:124–129.

Chapter 11

Acute Myocardial Infarction Infarct Avid Scintigraphy

Elias H. Botvinick, MD, and Michael W. Dae, MD

INTRODUCTION

The diagnosis of acute myocardial infarction frequently can be made from the characteristic clinical presentation. However, additional criteria have improved the accuracy of diagnosis in both the classic and the more subtle or complicated cases. The development of reliable electrocardiography with diagnostic criteria for its clinical use[1-3] and the later recognition of elevated myocardial enzyme levels in the serum of patients with infarction[4,5] allowed improved diagnostic sensitivity and specificity. The subsequent introduction of the myocardium-specific (MB) fraction of creatine phosphokinase (CK)[6,7] provided an additional increment of specificity to the diagnosis of infarct. However, the extreme sensitivity of the enzymatic method,[8] the temporal pattern of enzyme release,[9] and the biologic factors governing serum enzyme levels[10] presented their own limitations. Paralleling these developments was the recognition of the prognostic implications of infarct localization and sizing.[11] More recently, the growth of acute therapeutic interventions has made even more critical the complete evaluation of infarcted, ischemic, and threatened zones[12,13]; yet neither electrocardiographic nor enzymatic methods can provide these important dimensions in the patient with acute myocardial infarction. Infarct avid agents provide a potentially important component of this picture.

DEVELOPMENT OF THE METHOD

Imaging the Inflammatory Response

Gallium citrate (^{67}Ga) has long been established as an imaging agent that localizes in regions of abscess and inflammation. Recently, indium 111–labeled leukocytes have been introduced as a more specific marker of the inflammatory response.[14] Although both ^{67}Ga- and ^{111}In-labeled leukocytes have been applied to the visualization of vegetations in endocarditis,[15,16] this appears not to have widespread application. Over the last several years gallium localization in the myocardium has been reported in myocarditis, and its resolution has been said to provide an indicator of responsiveness to steroid therapy.[17,18]

Technetium-Based Imaging Agents

The use of technetium 99m as a tracer in medicine was originally suggested because of the isotope's optimal physical properties,[19,20] and it is now widely used for clinical scintigraphy. Its popularity stems from the abundance of its 140-KeV gamma photon emission, highly advantageous for imaging by commercially available scintillation cameras; its relatively short

---

This text represents an update of the field and a consideration of the topic since originally assessed in a review article published previously (*West J Med.* 1977;127:464–478).

6-hour half-life, facilitating decreased radiation exposure; and its favorable decay scheme with an excellent parent–daughter relationship, allowing production from in-house molybdenum 99 generators and thus permitting a ready supply of the radiopharmaceutical.[21]

Specific Agents

While ^{67}Ga localizes in acute infarction, its soft tissue clearance[22,23] requires a 1- to 2-day interval between injection and imaging. Technetium 99m, when bound to tetracycline, labels acute infarction and has been shown to provide a qualitative measure of infarct size. However, clearance of the agent from the blood again permits image evaluation only after a 24-hour delay, reducing the radioactivity available for imaging. Hepatic labeling brings a poor target to the background ratio, making image interpretation difficult and decreasing diagnostic accuracy.[24] Technetium (Tc) 99m glucoheptinate localizes in acute infarction within hours of the event, but again with distracting hepatic uptake[25] and poor image resolution. Inability to image this radionuclide late after the event is another negative feature. While several chelates of technetium have been developed to image infarction, pyrophosphate (PYP) and imidodiphosphonate (IDP) appear to be the most useful of these infarct-localizing radionuclides.

Radionuclide Development

Much of the interest in clinical myocardial infarction scintigraphy was stimulated by Bonte and co-workers in 1973.[26] These investigators showed in dogs that ^{99m}Tc [stannous]-PYP, already in common use as a bone-imaging agent,[27] localizes in acutely infarcted myocardium. The use of bone-imaging agents for infarct labeling was based on earlier work by D'Agostino[28] and Shen and Jennings,[29] who described calcium accumulation in myocardial cells that had recently undergone necrosis.

Relationship of Localization to Cellular Events

Similar to its localization in bone, the localization of myocardial ^{99m}Tc-PYP appears to occur in regions of hydroxyapatite deposition in mitochondria and other subcellular fractions.[30–32] Localization is not dependent on calcium deposition or leukocytic infiltration and has been documented to be confined to regions of irreversible damage.[33–35] Recently, other phosphate compounds, particularly ^{99m}Tc-IDP, have been shown to have superior infarct affinity and accelerated blood clearance, providing advantages for infarct imaging.[36,37]

Relationship to Blood Flow

At higher levels of blood flow, ^{99m}Tc-PYP uptake appears to parallel the density of myocardial necrosis. However, ^{99m}Tc-PYP uptake falls precipitously below flow levels that are 30% to 40% of normal, regardless of the extent of cellular damage[30–32] (Figure 11-1). On the basis of such evidence it appears that ^{99m}Tc-PYP gains access into acutely infarcted myocardium via residual and collateral flow to the region. Some support for this hypothesis may be found in the "doughnut" pattern of ^{99m}Tc-PYP uptake seen in association with proximal occlusion of the left anterior descending coronary artery in animals and patients, and which has been related to a poor prognosis[38] (Figure 11-2). Some ascribe the central clear area to reduced radionuclide localization in a region of ischemic infarction, the area with greatest necrosis and least flow. Conversely, the peripheral zone with greater residual or collateral flow, as identified by less-intense necrosis, shows maximal ^{99m}Tc-PYP uptake on the scintigram. However, the origin of the pattern may relate to projectional factors as well.

Radionuclide Behavior

Although infarct tissue reveals increased radioactivity above normal regions as early as 4 to 6 hours after infarction, time is required for maturation of cellular processes, and a positive ^{99m}Tc-PYP image is not generally seen until at least 12 hours after the onset of necrosis. Localization relates as well in part to the development of collateral supply to the infarct zone. In most patients with infarction, cardiac radioactivity is maximum at 48 to 72 hours, becomes less intense by 6 to 7 days, and is usually absent by 10 to 14 days after the event.[26,33–35,39] An exception to this pattern of uptake lies in the case of infarct reperfusion, now commonly produced with angioplasty or thrombolytic therapy. Here uptake seems to be both immediate and intense and may appear to overestimate the extent of infarction.[40] Early localization of the radiotracer in infarcted tissue in association with reperfusion therapy or angioplasty appears to relate to early restoration of regional perfusion via the native or collateral vessels.[41] Reduced radionuclide avidity likely relates to progressive replacement of necrotic myocardium by granulation tissue and scar tissue with progressive reduction in calcium deposits in the area of damage.[32] Patterns of uptake, however, are somewhat variable: Some patients maintain low levels of increased radioactivity for several weeks or, in some cases, months after

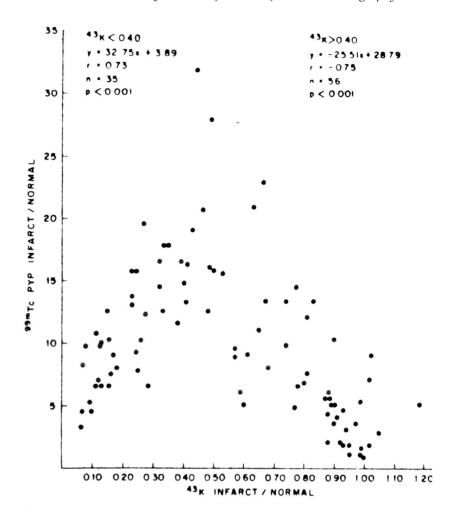

Figure 11-1 Relationship of technetium 99m pyrophosphate (^{99m}Tc-PYP) to perfusion. Shown is the relationship between ^{99m}Tc-PYP and potassium 43 (^{43}K) in infarcted and normal tissue. Although perfusion relates inversely to the extent of ^{99m}Tc-PYP localization and apparent infarct density at high flow levels, at low flow levels ^{99m}Tc-PYP density also falls. *Source:* Reprinted with permission from *Circulation* (1976;53:422), Copyright © 1976, American Heart Association.

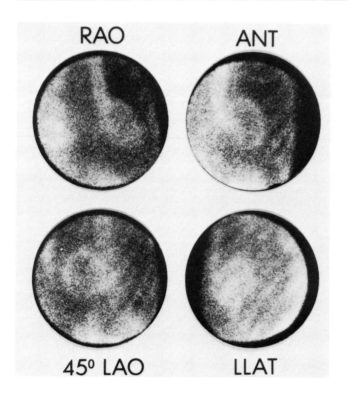

Figure 11-2 Doughnut configuration. Technetium 99m pyrophosphate images in multiple projections in a patient with a recent anterolateral infarction. The doughnut configuration has been associated with a large infarct area and a poor prognosis, and may relate to poor central infarct perfusion.

the event. Persistently positive scintigrams have been reported.[42]

Scintigraphic Method

Both ^{99m}Tc-PYP and ^{99m}Tc-IDP are made available by a variety of methods from inexpensive, prepackaged kits. These agents are generally used for skeletal imaging. Technetium 99m-PYP is prepared according to a modification of the method of Huberty and co-workers.[43] The material has a 100:1 ratio of PYP ions to stannous ions. Stannous ions permit the formation of stable PYP with the addition of ^{99m}Tc as pertechnetate, the ubiquitous, generator-produced anion. The radionuclide is quite stable, and in previous studies it has shown less than 3% ^{99m}Tc as free ionic pertechnetate. We currently use ^{99m}Tc-IDP because of its more rapid blood clearance and the related improved ratio of target to background.[37] Both ^{99m}Tc-PYP and ^{99m}Tc-IDP scintigrams are best acquired using 37-phototube, high-resolution scintillation cameras that are available commercially as both stationary and portable units. A 15- to 20-mCi dose of ^{99m}Tc tagged to 5 mg of stannous PYP is most commonly used. After an intravenous injection, approximately 50% of the dose is excreted by the kidneys and 50% is taken up in hydroxyapatite-bearing structures, mainly bone. Imaging is best carried out at least 2 hours after the intravenous administration of radionuclide, allowing adequate time for blood clearance. A 1-hour delay is likely sufficient for ^{99m}Tc-IDP imaging. A prolonged delay between injection and imaging could lead to loss of radionuclide integrity and increased levels of free pertechnetate. All images are obtained in the anterior, 45° left anterior oblique and left lateral projections at the 140-KeV ^{99m}Tc photopeak, employing a 20% window and taken to 300,000 counts. Because of the temporal factors involved in ^{99m}Tc-PYP uptake, the optimal time for imaging ^{99m}Tc-PYP or ^{99m}Tc-IDP is 2 to 3 days after the clinical event.

Image Interpretation

Criteria have been established for image interpretation.[44] Images are graded according to the intensity of radioactivity in the cardiac region on a scale from 0 to 4+, where 0 represents no increase in radioactivity beyond background; 1+ represents a slight, faint, or indefinite increase in cardiac radioactivity beyond background; 2+ represents a definite increase in cardiac radioactivity beyond background but less intense than that of bone; 3+ represents increased cardiac radioactivity beyond background and equal in intensity to that of bone; and 4+ represents increased radioactivity in the cardiac region more intense than that of bone. Images graded 0 or 1+ are generally called normal, while 2+ to 4+ images are called positive.

Image radioactivity is called discrete if it is confined to a localized cardiac region and diffuse if it is generally apparent in all regions[44] (Figure 11-3). The former is specific for myocardial damage; the latter, originally thought to be related to subendocardial infarction, is nonspecific for acute infarction.[45]

Studies strongly suggest a relationship of the diffuse pattern of uptake with blood pool labeling. The frequent visualization of femoral vasculature in bone scans, or the heart in noncoronary patients with dimin-

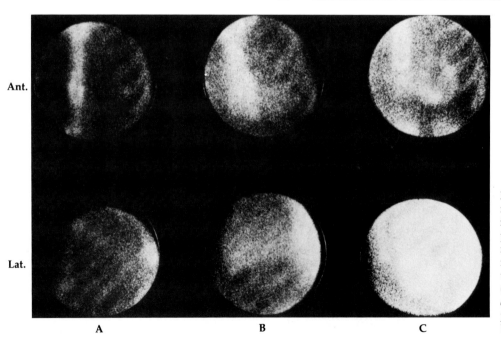

Figure 11-3 Infarct imaging. Technetium 99m pyrophosphate images in anterior (top) and lateral (bottom) projections. The images demonstrate normal bony uptake (A); diffuse, nonspecific uptake (B); and focal uptake, specific for acute infarction (C). *Source:* Reprinted from *Cardiology* by W Parmley and K Chatterjee (Eds) with permission of JB Lippincott Company, © 1987.

ished tissue attenuation on the left side of the chest after left mastectomy, supports this possibility.[45] Previous attempts to relate blood levels to the diffuse pattern of ^{99m}Tc-PYP uptake showed no difference when compared with normals. More sensitive methods of radionuclide analysis will be required to detect accurately the quantitatively small but clinically significant differences in blood radioactivity. Factors that could play a role in the production of the diffuse pattern of uptake include camera-field nonuniformity, radiopharmaceutical instability, tagging inefficiency, variation in blood clearance, image timing after radionuclide administration, and delay in imaging after infarction. However, some patients with diffuse uptake will have suffered an acute infarction, and totally disregarding these cases will reduce test sensitivity. Accepting the 2+ intensity classification as only equivocal, Berman and co-workers[46] demonstrated a dramatic improvement in test specificity with only 3% false-positive test results and preserved sensitivity. The development and application of more specific imaging agents could help resolve this problem. Furthermore, diagnostic security increases with the intensity of uptake. Diffuse image abnormalities generally are faint and often relate to radiolabeling of the blood pool with homogeneous uptake in the distribution of the ventricular cavities with radiolabeling from the cardiac apex to the sternum.

"Diffuse" should not be confused with "generalized" myocardial uptake. The latter occasionally is localized to widespread subendocardial, "shell," infarction but is projected in these ungated images throughout the left ventricular myocardium (Figure 11-4A and B). Blood pool imaging combined with computer comparison or computer subtraction of background and bony structures has been suggested to distinguish myocardial radioactivity from cavitary or skeletal radioactivity[46-48] and may be helpful in 10% to 15% of cases. However, choice of the optimal radiopharmaceutical, care in image acquisition, and the interpretation of multiple projections are more practical approaches. Rotating

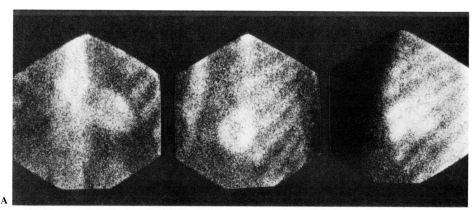

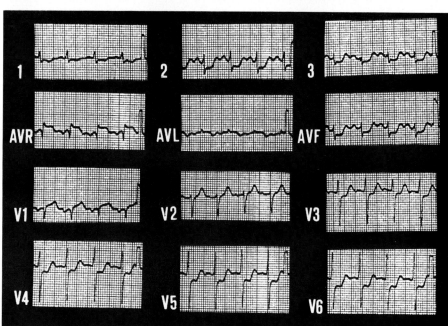

Figure 11-4 Extensive technetium 99m pyrophosphate (^{99m}Tc-PYP) uptake. The extensive pattern of ^{99m}Tc-PYP uptake illustrated in multiple projections (A) is related to a poor prognosis. In surviving patients it indicates a widespread subendocardial or shell infarction. The related electrocardiogram (B) shows widespread ST segment depression without Q waves. *Source:* Reprinted with permission from *Western Journal of Medicine* (1977;127:464–479), Copyright © 1977, California Medical Association.

slant-hole tomography at the bedside and single-photon emission-computed tomography with a stable and specialized camera have recently been shown to aid diagnostic accuracy[49] (Figure 11-5).

Diagnostic Accuracy

The most important factor in the potential clinical use of infarct avid imaging is the diagnostic accuracy of the test. Initial experimental work in animals documented the sensitivity of infarct scintigraphy to small amounts of tissue damage. Early animal studies revealed the ability of the method to detect regional transmural infarction as small as 3 g.[47,50] Parkey and co-workers[44] demonstrated positive discrete uptake in all of 23 consecutive patients admitted with transmural infarction. Massie and co-workers[51] found a direct relationship between the frequency of infarct visualization and infarct size as revealed by enzymatic testing. These workers also demonstrated reduced sensitivity in subendocardial infarction, while others suggested a "diffuse" pattern of radionuclide distribution in subendocardial infarction. Soon a variety of investigators reported false-positive ^{99m}Tc-PYP scintigrams in association with valvular disease,[52] unstable angina pectoris,[53,54] stable angina pectoris,[55,56] and heart failure after cardiopulmonary bypass and at a time remote from past infarction.[57,58] Although occasionally patients in these series revealed discrete uptake associated with a punctate area of valvular calcification, associated with unstable angina and enzyme release, or associated with postinfarction pericarditis, most of these reported "false-positive" scintigrams were of the relatively low-intensity, nonspecific "diffuse" pattern. Many of the others, with discrete uptake, could largely be explained by actual myocardial necrosis.

Malin and co-workers[56] and Lyons and co-workers[57] reported decreased specificity of ^{99m}Tc-PYP for acute infarction due to a significant number of positive images weeks and months after the event. Studies by Ahmed and co-workers[54,58] corroborated the accuracy of the discrete pattern of uptake for acute infarction, but they also found a large percentage of patients with left ventricular aneurysm with discrete radionuclide accumulation and suggested that this was the cause of the false-positive images. Others have reported the occurrence of abnormal scintigrams in patients with left ventricular wall motion abnormalities and unstable angina pectoris.[42,59,60] In many cases these nonspecific abnormalities were diffuse and could be managed as described above. The exact significance of discrete radionuclide accumulation in these cases is not clear, but it may represent regional radionuclide deposition in areas of focal calcification or in spotty areas of myocardial necrosis. A study of ^{99m}Tc-PYP scintigrams performed in more than 50 patients 9 days to 10 years after a documented transmural infarction revealed only 2 patients with discrete uptake, both of whom had extensive prior infarction and aneurysm.[61] Another pathologic study revealed evidence of ongoing necrosis in each of 52 patients with remote infarction, and scintigraphic myocardial uptake prior to demise in the absence of a new coronary event.[62]

Discrete uptake appears to be specific for acute infarction and correlates with electrocardiographic and pathologic localization of the infarct. The scintigraphic area of ^{99m}Tc-PYP uptake has been shown to relate directly to infarct area and weight in animals[33,34,50] and agrees closely with enzymatic and functional measures of infarct size in patients.[63,64] One report suggests a diagnostic sensitivity for transmural infarction of 85% to 90%.[65] However, smaller transmural and subendocardial infarctions are less easily resolved, and the

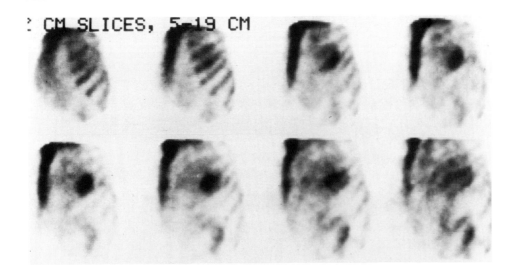

Figure 11-5 Rotating slant-hole tomography of technetium 99m pyrophosphate. Six planar images acquired at 60° angles in the 45° left anterior oblique projection were used to reconstruct the tomograms shown. The images are displayed here at 2-cm separations spanning a depth from 5 to 19 cm. The patient has an obvious acute apicolateral infarction. *Source:* Reprinted from *Cardiology* by W Parmley and K Chatterjee (Eds) with permission of JB Lippincott Company, © 1987.

method has a sensitivity for subendocardial infarction of 60% to 70%.[66] The presence of discrete uptake cannot itself differentiate transmural infarction from subendocardial infarction. The discrete pattern is not specifically related to transmural necrosis and may be seen as well in association with subendocardial infarction. However, there is a direct relationship between image sensitivity and infarct size and density, therefore providing an increased sensitivity of the method for transmural infarction.

Clinical Settings for Reduced Accuracy

"False-positive" scintigrams related to discrete ^{99m}Tc-PYP uptake are not common. Localized ^{99m}Tc-PYP uptake in the absence of acute infarction or associated left ventricular aneurysm has been reported with valvular calcification,[67] pericarditis,[68] tumor infiltration,[69] and penetrating and nonpenetrating trauma. Many of these conditions are actually related to myocardial necrosis, although not to an acute coronary event.

Occasionally, scintigrams will be reported negative simply because of the relationship of the acute event to the time of imaging. Scintigrams reported negative when performed within 24 hours of the event may be found positive when repeated 2 to 3 days after the event. Similarly, scintigrams reported negative when performed longer than 1 week after the event cannot exclude the diagnosis of acute infarction. The specificity of image abnormalities in the presence of prior infarction and aneurysm can be clarified by relating the regional uptake to electrocardiographic abnormalities and noting variation on serial imaging. Intense image abnormalities are most frequently true-positive findings related to an acute event, whereas faint abnormalities are less specific.

"Primary" cardiac amyloidosis provides an exception to this analysis[70] (Figure 11-6). Here, the infiltrative process is frequently associated with dense accumulation of the radiotracer. While this may relate in part to an associated element of necrosis, there appears also to be a relationship with the amyloid deposit itself. Images in such cases may be quite impressive and, as in all circumstances, care must be taken to relate image findings to the clinical presentation.

Relationship to Other Diagnostic Methods

Although it is quite sensitive to the presence of acute infarction, infarct avid scintigraphy is less sensitive than other methods such as serial electrocardiographic changes, induced contraction abnormalities, serum enzyme release, and the presence of perfusion scintigraphic defects. However, neither wall motion nor perfusion scintigraphic abnormalities are specific for acute infarction; and electrocardiographic changes also may be nonspecific, concealed, or mimicked by a multitude of conduction abnormalities, drug or electrolyte effects, pericarditis, or other concomitant conditions. Infarct avid imaging maintains its specificity in these settings.

The availability of portable scintillation cameras, the simplicity of the technique, and the marginal advantage and related optional nature of computer processing make infarct avid scintigraphy practical in community hospitals as well as large medical centers. Although by current standards this is not an expensive technology, clinicians should demand that the improvement in infarct diagnosis and evaluation be sufficient to justify the additional expense.

The presence of cellular damage and pathophysiologic evidence of necrosis does not itself determine clinical management. Infarction with extensive necrosis would likely generate a conservative approach. However, prolonged or recurrent episodes of chest pain with evidence of only minimal necrosis in the setting of unstable angina, even with evidence of enzyme release, should encourage an aggressive approach. Acute myocardial infarction scintigraphy adds to the specificity of infarction diagnosis and provides information relating to infarct size, localization, and prognosis. The method often provides better understanding of the clinical presentation and a more rational approach to patient management. Infarct scintigraphy appears to be of greatest

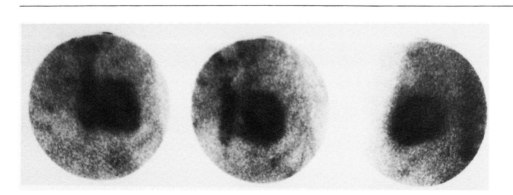

Figure 11-6 Technetium 99m pyrophosphate (^{99m}Tc-PYP) in amyloidosis. Shown are markedly abnormal ^{99m}Tc-PYP images in (left to right) anterior, left anterior oblique, and left lateral projections in a patient with primary amyloidosis. *Source:* Courtesy of R Lull, San Francisco, CA.

value for infarct diagnosis in patients whose acute electrocardiographic and enzymatic test results are nonspecific, unavailable, or confusing.

SPECIFIC CLINICAL APPLICATIONS

Diagnosis of Perioperative Infarction

The diagnosis of perioperative infarction can be difficult, especially after coronary artery bypass graft surgery. Although the appearance of new Q waves after bypass surgery seems significant, the diagnosis may remain in doubt.[71] The presence of prior infarction, conduction abnormalities, nonspecific ST-T abnormalities, and pericarditis-related repolarization changes frequently makes electrocardiographic findings nonspecific.[72] In this setting, MB-CK is often elevated postoperatively, unassociated with a clinical event. Even the MB-CK fraction may not be a useful clinical indicator of significant myocardial necrosis following coronary artery bypass surgery.[52,71-74] Several studies have demonstrated infarct scintigraphy to be a useful adjunct for the diagnosis of perioperative infarction following revascularization.[52,71-74] Preoperative images may be of added utility in gauging the extent of perioperative necrosis, especially in the presence of a known or potential and relatively recent preoperative event. In this case the great specificity and limited sensitivity of the method works to clinical advantage. Patients with negative postoperative images generally have an excellent prognosis and a benign postoperative course regardless of associated electrocardiographic or enzymatic findings. Positive images generally relate to new contraction abnormalities and a reduced ejection fraction; occasionally they direct attention to the cause of new symptoms after surgery.

Evaluation after Cardioversion

Another important diagnostic subgroup is composed of patients in whom cardioversion or defibrillation has been carried out. For this group, frequently comprising patients suffering unexpected hemodynamic collapse or survivors of out-of-hospital sudden death, confirmation or exclusion of acute infarction is of considerable importance for both acute and chronic management. Yet classification in this population may be difficult when persistent enzyme and electrocardiographic abnormalities could relate to the trauma of resuscitation or prior, unrelated, infarction or conduction abnormality. Although false-positive studies have been reported with cardioversion,[75,76] the method appears to maintain its specificity after coronary artery bypass graft surgery, even with direct electrical defibrillation,[77] and

in the setting of catheter ablation of ectopic electrical foci or pathways. In fact, the literature generally fails to document any significant evidence of myocardial necrosis in patients related to transthoracic defibrillation regardless of the energy or the number of shocks delivered. However, care must be taken to avoid confusion with chest wall uptake, and, in turn, paddles must be placed judiciously to permit visualization of myocar-

Infarct Localization

The imaging method permits accurate infarct localization. While of some importance for prognostic value, this also facilitates the differentiation of current infarction from prior infarction. With serial study, infarct imaging may permit the identification of infarct extension and its differentiation from other postinfarction pain syndromes.[65,78]

A number of studies have demonstrated the relationship between right ventricular ^{99m}Tc-PYP uptake, right ventricular wall motion abnormalities and inappropriate elevation of right-sided pressures[79,80] (Figure 11-7). Such scintigraphic findings parallel pathologic findings and are seen in roughly one third of all acute inferior infarctions. Technetium 99m-PYP imaging is the only specific, direct, noninvasive method of

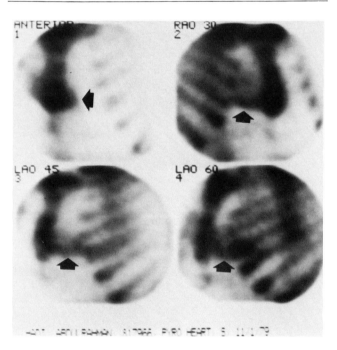

Figure 11-7 Technetium 99m pyrophosphate images in a patient with right ventricular infarction. The presence of right ventricular infarction is shown by the horizontal extension of radioactivity from the inferior left ventricular wall to the sternum in the left anterior oblique (LAO) projection (arrows) and the extension of uptake to the right of the sternum in the right anterior oblique (RAO) projection.

diagnosing right ventricular infarction; this is of considerable clinical importance for both diagnostic and therapeutic reasons.[81]

Prognostic Value

Several reports have documented the relationship between infarct size and the development of heart failure and the inverse relationship between estimates of infarct size and subsequent survival.[82-84] Imaging methods have been assessed for their ability to evaluate infarct size and to point to ways to limit its extent and improve the related prognosis. The ^{99m}Tc-PYP image infarct area correlates well with the weight and projected area of infarction in living dogs.[33,50] In patients, the infarct area correlates inversely with the stroke volume index and with morbidity and mortality post-infarction[63] (Figure 11-8). Although the infarct image area is able to differentiate infarct survivors from nonsurvivors, it cannot further subdivide the survivors, as can the size of the perfusion scintigraphic abnormality and the left ventricular ejection fraction[84-89] (Figure 11-9). Although some workers should have shown a correlation between CK enzyme infarct size and the projected image infarct area,[86-88] it is not surprising, owing to differences in their release dynamics, to note

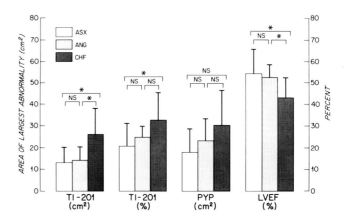

Figure 11-9 Scintigraphic prognosis. Perfusion defect size, measured with myocardial perfusion scintigraphy (Tl-201), and left ventricular ejection fraction (LVEF) on blood pool scintigraphy were the best discriminators of asymptomatic patients (ASX) from those who suffered recurrent angina (ANG) or heart failure (CHF) after infarction. PYP is technetium 99m pyrophosphate image infarct size; NS, not significant. *Source:* Reprinted with permission from *Circulation* (1982;66;960), Copyright © 1982, American Heart Association.

some disagreement in the magnitude of the amount of enzyme released and the magnitude of the image infarct size.[64]

New Agent

Recently interest has focused on the ^{111}In-labeled Fab fragment of monoclonal antimyosin antibody to tag acute infarction[90-93] (Figure 11-10). The agent appears to be specific for acute infarction. However, it requires 24 to 48 hours for clearance of background and it deposits significant radioactivity in the liver, which might interfere with visualization of inferior infarction. Animal and patient studies thus far conducted indicate excellent diagnostic accuracy and acceptable imaging characteristics. Most enticing is the independence of agent localization from regional flow. This could make localization a purer function of infarct density and presents an excellent prospect for accurate infarct sizing. Additionally, this imaging method may be applicable to the diagnosis of cardiac transplant rejection.[94] The diagnostic accuracy, sensitivity, and specificity of this new agent have not yet been established. The high energy and long half-life of the ^{111}In label make adaptation of the radiopharmaceutical to ^{99m}Tc an important consideration.

Noninvasive Evaluation of Myocardium at Ischemic Risk

Other studies have demonstrated the prognostic value of infarct imaging. The relationship appears to be qualitative, as negative images relate to a benign prog-

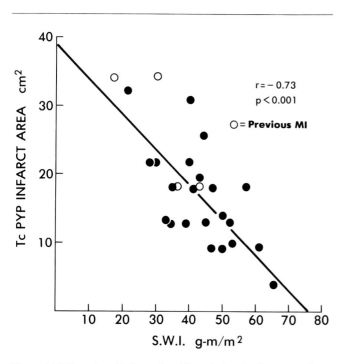

Figure 11-8 Functional infarct size. The relationship between "hot" spot technetium 99m pyrophosphate (^{99m}Tc-PYP) image infarct size and stroke work index (S.W.I), an index of left ventricular function that correlates closely with prognosis. MI, Myocardial infarction. *Source:* Reprinted with permission from *Circulation* (1978;57:307), Copyright © 1978, American Heart Association.

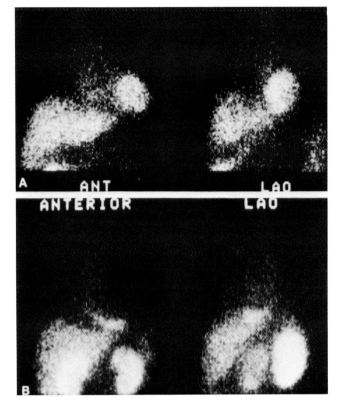

Figure 11-10 New infarct imaging agent. Anterior (ANT) and left anterior oblique (LAO) projections of the precordium made 24 hours after the administration of indium 111 antimyosin antibody in patients with acute anterior (A) and inferior (B) myocardial infarctions. In both cases, angiography confirmed the totally occluded state of the involved coronary artery. *Source:* Reprinted with permission from "Acute Myocardial Infarct Imaging with Indium-III Labeled Monoclonal Antimyosin Fab," by BA Khaw, T Yasuda, HK Gold, et al. in *Journal of Nuclear Medicine* (1987;28:1671), Copyright © 1987, Society of Nuclear Medicine.

nosis and large abnormalities relate to a poor prognosis. However, the size of the related perfusion abnormality and the left ventricular ejection fraction appear to relate better to prognosis in patients both with and without prior infarction. This is likely due to the fact that prognosis relates to the full extent of myocardium at ischemic risk, including regions of acute infarction as well as prior infarction and ongoing ischemia.[12,13,85–87,89] Infarct imaging can quantitate only one of these parameters and lacks the full prognostic impact of the scintigraphic indicators of perfusion and function. Nonetheless, the method provides one aspect of evaluation of the profile of a myocardium at ischemic risk, upon which much of clinical decision making in ischemic heart disease is based. A small infarction related to extensive ischemic regions may bear a prognosis not dissimilar to that for an extensive infarction. Of course, the therapeutic options available in the former case may be extensive, and our ability to influence the clinical course favorably may be significant when reversible ischemic abnormalities dominate. Differences in the sizes of abnormalities on perfusion or blood pool and infarct images similarly may give clues to the full extent of myocardium at ischemic risk and the related prognosis.

Of current interest and concern is the potential application of these methods to the evaluation of candidates for and patients after invasive angioplasty or thrombolytic therapy of acute infarction. Previous studies have demonstrated enhanced uptake of infarct avid agents after reperfusion of acutely obstructed infarct vessels in both animals and patients.[40] The specific role of infarct avid agents and the role of scintigraphic perfusion imaging agents in the assessment and management of patients before and after reperfusion therapy remain to be developed.[41]

In addition to infarct avid imaging, the quantitative scintigraphic assessment of the resting left ventricular ejection fraction, the extent of regional akinesis, and the depth and distribution of perfusion abnormalities have been well correlated with infarct prognosis.[95–97] Serial changes in the ejection fraction have also demonstrated prognostic value, as might have been expected.[98] However, Gibson and co-workers[99] have shown that multiple scintigraphic perfusion defects represent a poor prognosis in patients with otherwise uncomplicated inferior infarction, and Becker and co-workers[100] have demonstrated the complementary prognostic value of scintigraphic perfusion defect size and left ventricular ejection fraction. Furthermore, reversible perfusion defects seen after an injection of the perfusion agent in a patient at rest indicate ongoing ischemia consistent with a tight coronary lesion, and suggest aggressive management.[101] "Stress" perfusion scintigraphy performed in association with dynamic exercise has demonstrated great prognostic value in assessment of the postinfarct patient prior to discharge or during the home recovery period.[102] Recently perfusion scintigraphy performed in association with pharmacologic stress has been suggested as a safe alternative to dynamic exercise stress and may be of value in the evaluation of the postinfarct patient, specifically when dynamic exercise is not possible or practical.[103] The wealth of imaging methods is apparent. The choice depends on the specific clinical question and setting. Often a combination of complementary modalities or methods may be applied appropriately.

REFERENCES

1. Parkinson J, Bedford DE. Successive changes in the electrocardiogram after cardiac infarction (coronary thrombosis). *Heart.* 1928;14:195–239.

2. Pardee HEB. The significance of an electrocardiogram with a large Q in lead 3. *Arch Intern Med.* 1930;46:470–481.

3. Pardee HEB. Nomenclature and description of the electrocardiogram. *Am Heart J.* 1940;20:655–666.

4. Vincent WR, Rapport E. Serum creatine phosphokinase in the diagnosis of acute myocardial infarction. *Am J Cardiol*. 1961;15:17–26.

5. Jennings RB, Kaltenback JP, Smetters GW. Enzymatic changes in acute myocardial ischemic injury. *Arch Pathol*. 1957;64:10–16.

6. Van Der Veen KJ, Willebrands AF. Isoenzymes of CPK in normal and pathologic sera. *Clin Chim Acta*. 1966;13:312–316.

7. Roberts P, Sobel BE. Isoenzymes of CPK and the diagnosis of acute myocardial infarction. *Ann Intern Med*. 1973;79:741–743.

8. Klein MS, Shell WE, Sobel BE. Serum creatine phosphokinase (CPK) isoenzymes following intramuscular injections, surgery and myocardial infarction: Experimental and clinical studies. *Cardiovasc Res*. 1973;7:412–418.

9. Sobel BE, Shell WE. Serum enzyme determinations in the diagnosis and assessment of myocardial infarction. *Circulation*. 1972;45:471–482.

10. Roberts R, Karlsberg RP, Sobel B. Factors retarding the decline of plasma CK activity after myocardial infarction. *Am J Cardiol*. 1977;39:316. Abstract.

11. Sobel BE, Bresnahan GF, Shell WE, et al. Estimation of infarct size in man and its relation to prognosis. *Circulation*. 1982;46:640–648.

12. Schustor EH, Bulkley BH. Early post-infarction angina: Ischemia at a distance and ischemia in the infarct zone. *N Engl J Med*. 1981;19;1101.

13. Bulkley BH, Silverman KJ, Wisfeldt ML, et al. Pathologic basis of thallium 201 scintigraphic defects in patients with fatal myocardial injury. *Circulation*. 1979;60:785–793.

14. Sfakianakis GN, Al-Sheikh W, Heal A, et al. Comparisons of scintigraphy with In-111 Leukocytes and GA-67 in the diagnosis of occult sepsis. *J Nucl Med*. 1983;23:618–625.

15. Wiseman J, Rouleau J, Rigo P, et al. Gallium-67 myocardial imaging for the detection of bacterial endocarditis. *Radiology*. 1976;120:135–140.

16. Riba AL, Thakur MI, Gottschalk A, et al. Imaging experimental infective endocarditis with indium-111 labeled blood cellular components. *Circulation*. 1979;59:336–342.

17. O'Connell JB, Fowles RE, Robinson JA, et al. Clinical and pathological findings of myocarditis in 2 families with dilated cardiomyopathy. *Am Heart J*. 1983;107:127–135.

18. O'Connell JB, Robinson JA, Henkin RE, et al. Immunosuppressive therapy in patients with congestive cardiomyopathy and myocardial uptake of gallium-67. *Circulation*. 1981;64:780–788.

19. Richards P. A survey of the production at Brookhaven National Laboratory of radioisotopes for medical research. In: *Transactions of the Fifth Nuclear Congress*. New York: Institute of Electrical and Electronic Engineering; 1960:225–244.

20. Harper PV, Lathrop KA, McArdle RJ. The use of technetium 99m as a clinical scanning agent for thyroid, liver and brain. In: *Medical Radioisotope Scanning*. Vienna, Austria: International Atomic Energy Agency; 1964:33–45.

21. Tucker WD, Green MW, Weiss AJ. *Transactions of the Fourth Annual Meeting of the American Nuclear Society*. New York: Academic Press; 1958;1(no 1);11:160–161.

22. Schor RA, Massie BM, Botvinick EH, et al. Gallium-67 uptake in silent myocardial infarction. *Radiology*. 1978;129:117–121.

23. Kramer RJ, Goldstein RE, Hirschfeld JW, et al. Accumulation of gallium-67 in regions of acute myocardial infarction. *Am J Cardiol*. 1974;33:851–866.

24. Holman BL, Lesch M, Zweiman FG, et al. Detection and sizing of acute myocardial infarcts with 99mTc (SN) tetracycline. *N Engl J Med*. 1974;291:159–163.

25. Rossman DJ, Rouleau J, Strauss HW, et al. Detection and size estimation of acute myocardial infarction using 99mTc glucoheptinate. *J Nucl Med*. 1975;16:980–987.

26. Bonte FJ, Parkey RW, Graham KD, et al. A new method for radionuclide imaging of myocardial infarcts. *Radiology*. 1973;110:473–474.

27. Subramanian G, McAfee JG. A new complex of 99mTc for skeletal imaging. *Radiology*. 1971;99:192–196.

28. D'Agostino AN. An electron microscopic study of cardiac necrosis produced by 9α-fluorocortisol and sodium phosphate. *Am J Pathol*. 1964;45:633–644.

29. Shen AC, Jennings RB. Myocardial calcium and magnesium in acute ischemic injury. *Am J Pathol*. 1972;67:441–456.

30. Coleman RE, Klein MS, Ahmed SA, et al. Mechanisms contributing to myocardial accumulation of technetium-99m stannous pyrophosphate after coronary occlusion. *Am J Cardiol*. 1977;39:55–59.

31. Zaret BL, DiCola UC, Donbedial RK, et al. Dual radionuclide study of myocardial infarction. *Circulation*. 1976;53:422–428.

32. Rivas F, Cobb FR, Bache RJ, et al. Relationship between blood flow to ischemic regions and extent of myocardial blood flow to ischemic regions and extent of myocardial infarction. *Circ Res*. 1976;38:439–447.

33. Buja LM, Parkey RW, Dees JH, et al. Morphologic correlates of technetium-99m stannous pyrophosphate imaging of acute myocardial infarcts in dogs. *Circulation*. 1975;52:596–607.

34. Buja LM, Tofe AJ, Kulkarni PV, et al. Sites and mechanisms of localization of technetium-99m phosphorous radiopharmaceuticals in acute myocardial infarcts and other tissues. *J Clin Invest*. 1977;60:724–732.

35. Schelbert HR, Ingwall JS, Sybers HD, et al. Uptake of infarct imaging agents in reversibly and irreversibly injured myocardium in cultured fetal mouse heart. *Circ Res*. 1976;39:860–868.

36. Subramanian G, McAfee JG, Blair RJ. Technetium 99m labeled stannous imidodiphosphate, a new radiodiagnostic agent for bone scanning; comparison with other 99mTc complexes. *J Nucl Med*. 1975;16:1137–1142.

37. Joseph SP, Ell PJ, Ross P. 99mTc imidodiphosphonate: A superior radiopharmaceutical for in vivo positive myocardial infarct imaging. *Br Heart J*. 1978;40:325–241.

38. Rude RE, Parkey RW, Bonte FJ, et al. Clinical implications of the technetium-99m stannous pyrophosphate myocardial scintigraphic "doughnut" pattern in patients with acute myocardial infarcts. *Circulation*. 1979;59:721–725.

39. Bonte FJ, Parkey RW, Graham KD, et al. Distributions of several agents useful in imaging myocardial infarcts. *J Nucl Med*. 1975;16:132–135.

40. Wheelan K, Wolfe C, Corbett J, et al. Early positive technetium-99m stannous pyrophosphate images as a marker of reperfusion in patients receiving thrombolytic therapy for acute infarction. *Am J Cardiol*. 1985;56:252–260.

41. Beller GA. Role of myocardial perfusion imaging in evaluating thrombolytic therapy for acute myocardial infarction. *J Am Coll Cardiol*. 1987;9:661–674.

42. Gertz EW, Rollo FD, Wisneski JA. Lack of specificity of Tc-99 pyrophosphate scans for acute myocardial infarction. *Clin Res*. 1976;24:83-A. Abstract.

43. Huberty JP, Hattner RS, Pavell MR. A 99mTc pyrophosphate kit: a convenient, economical and high quality skeletal-imaging agent. *J Nucl Med*. 1974;15:124–126.

44. Parkey RW, Bonte FJ, Meyer SL, et al. A new method for radionuclide imaging of acute MI in humans. *Circulation*. 1974;50:540–546.

45. Prasquier R, Taradash MR, Botvinick EH, et al. The specificity of the diffuse pattern of cardiac uptake in myocardial infarction imaging with technetium-99m stannous pyrophosphate. *Circulation.* 1977;55:61–66.

46. Berman DS, Amsterdam DS, Hines H, et al. New approach to interpretation of technetium-99m pyrophosphate scintigraphy in the detection of acute myocardial infarction. *Am J Cardiol.* 1977;39:341–346.

47. Stokely EM, Buja IM, Lewis SE, et al. Measurement of acute myocardial infarcts in dogs with 99mTc-stannous pyrophosphate scintigrams. *J Nucl Med.* 1976;17:1–5.

48. Parkey RW, Bonte FJ, Buja LM, et al. Myocardial infarct imaging with technetium-99m phosphates. *Semin Nucl Med.* 1977;7:15–28.

49. Holman BL, Goldhaber SZ, Kirsch CM, et al. Measurement of infarct size using single photon emission computed tomography and 99mTc-pyrophosphate: A description of the method and a comparison with patient prognosis. *Am J Cardiol.* 1982;50:503–508.

50. Botvinick EH, Shames D, Lappin H, et al. Noninvasive quantitation of myocardial infarction with technetium-99 pyrophosphate. *Circulation.* 1975;52:909–915.

51. Massie BM, Botvinick EH, Werner JA, et al. Myocardial infarction scintigraphy with technetium 99m stannous pyrophosphate: An insensitive test for nontransmural myocardial infarction. *Am J Cardiol.* 1979;43:186–193.

52. Righetti A, O'Rourke RA, Schelbert HR, et al. Usefulness of preoperative and postoperative Tc-99m (SN) pyrophosphate scans in patients with ischemic and valvular heart disease. *Am J Cardiol.* 1977;39:43–49.

53. Dosky MS, Meyer SL, Platt MR, et al. Unstable angina; clinical, angiographic and myocardial scintigraphic observations. *Circulation.* 1975;51,52(suppl 2):II–89.

54. Ahmed M, Dubiel JP, Logan KW, et al. Limited clinical diagnostic specificity of technetium-99m stannous pyrophosphate myocardial imaging in acute myocardial infarction. *Am J Cardiol.* 1977;39:50–54.

55. Willerson JT, Parkey RW, Bonte FJ, et al. Technetium stannous pyrophosphate myocardial scintigrams in patients with chest pain of varying etiology. *Circulation.* 1975;51:1046–1052.

56. Malin FR, Rollo FD, Gertz EW. Sequential pyrophosphate myocardial scintigraphy after myocardial infarction. *Circulation.* 1976;53,54(suppl 2):II-207.

57. Lyons KP, Olson HG, Brown WT, et al. Persistence of an abnormal pattern on 99mTc pyrophosphate myocardial scintigraphy following acute myocardial infarction. *Clin Nucl Med.* 1976;1:253–257.

58. Ahmed M, Dubiel JP, Verdon TA, et al. Technetium 99m stannous pyrophosphate myocardial imaging in patients with and without left aneurysm. *Circulation.* 1976;53:833–838.

59. Wisneski JA, Rollo FD, Gertz EF. Tc-99m pyrophosphate myocardial accumulation in the absence of acute infarction: Correlation with left ventricular wall motion abnormalities. *Circulation.* 1976;53,54(suppl 2):II–81. Abstract.

60. Soin JS, Burdine JA, Beal W. Myocardial localization of 99mm Tc-pyrophosphate without evidence of acute myocardial infarction. *J Nucl Med.* 1975;16:944–946.

61. Botvinick EA, Shames DM, Sharpe DN, et al. The specificity of technetium stannous pyrophosphate myocardial scintigrams in patients with prior myocardial infarction. *J Nucl Med.* 1978;129:117–125.

62. Poliner LR, Buja LM, Parkey RW, et al. Clinicopathologic findings in 52 patients studied by technetium-99m stannous pyrophosphate myocardial scintigraphy. *Circulation.* 1979;59:257.

63. Henning H, Schelbert H, Righetti A, et al. Tc-99m-pyrophosphate imaging for sizing acute myocardial infarction. *Clin Res.* 1977;25:91A. Abstract.

64. Sharpe DW, Botvinick EH, Shames DM, et al. The clinical estimation of acute myocardial infarct size with 99mTc pyrophosphate. *Circulation.* 1978;57:307–314.

65. Werner JA, Botvinick EH, Shames DM, et al. Acute myocardial infarction: Clinical application of technetium 99m stannous pyrophosphate infarct scintigraphy. *West J Med.* 1977;127:464–478.

66. Willerson JT, Parkey RW, Bonte FJ, et al. Acute subendocardial infarction in patients. *Circulation.* 1975;51:436–441.

67. Klein MS, Weiss AN, Roberts R, et al. Technetium-99m stannous pyrophosphate scintigrams in normal subjects, patients with exercise-induced ischemia and patients with a calcified valve. *Am J Cardiol.* 1977;39:360–363.

68. Fleg JL, Siegel BA, Roberts R. Detection of pericarditis with 99m technetium pyrophosphate images. *Am J Cardiol.* 1977;39:273. Abstract.

69. Harford W, Weinberg MN, Buja LM, et al. Positive 99m Tc-stannous pyrophosphate myocardial image in a patient with carcinoma of the lung. *Radiology.* 1977;122:747–748.

70. Braun SD, Lisbona R, Novales-Diaz JA, et al. Myocardial uptake of 99mTc-phosphate tracers in amyloidosis. *Clin Nucl Med.* 1979;6:244–248.

71. Sternberg L, Wisneski JA, Ullyot DJ, et al. Significance of new Q waves after aortocoronary bypass surgery. *Circulation.* 1975; 52:1037–1044.

72. Righetti A, Crawford MH, O'Rourke RA, et al. Detection of perioperative myocardial damage after coronary artery bypass graft surgery. *Circulation.* 1977;5:173–177.

73. Coleman RE, Klein MS, Roberts R, et al. Improved detection of myocardial infarction with technetium-99m stannous pyrophosphate and serum MB creatine phosphokinase. *Am J Cardiol.* 1976;37:732–735.

74. Klausner SC, Botvinick EH, Shames DM, et al. The application of radionuclide infarct scintigraphy to diagnose perioperative myocardial infarction following revascularization. *Circulation.* 1977; 56:173–180.

75. Pugh BR, Buja LM, Parkey RW, et al. Cardioversion and "false positive" technetium-99m stannous pyrophosphate myocardial scintigrams. *Circulation.* 1976;54:399–403.

76. DiCola UC, Freedman GS, Downing SE, et al. Myocardial uptake of technetium-99m stannous pyrophosphate following direct current countershock. *Circulation.* 1976;54:980–986.

77. Werner JA, Botvinick EH, Shames DM, et al. Diagnosis of acute myocardial infarction, following cardioversion: Accurate detection with Tc-99m pyrophosphate scintigraphy. *Circulation.* 1977; 56:III–63. Abstract.

78. Dressler W. The post myocardial infarction syndrome: A report on 44 cases. *Arch Intern Med.* 1959;103:28–42.

79. Sharpe N, Botvinick EH, Shames DM, et al. The non-invasive diagnosis of right ventricular infarction. *Circulation.* 1978;57:483–491.

80. Wackers FJT, Lie KI, Busemann-Sokole E, et al. Prevalence of right ventricular involvement in inferior wall infarction assessed with myocardial imaging with thallium-201 and technetium-99m pyrophosphate. *Am J Cardiol.* 1978;42:358–362.

81. Cohn JN, Guiha NA, Broder MI, et al. Right ventricular infarction: Clinical and hemodynamic features. *Am J Cardiol.* 1974; 33:209–217.

82. Page DL, Caulfield JB, Kastor JA, et al. Myocardial changes associated with cardiogenic shock. *N Engl J Med.* 1971;285:133–137.

83. Swan HJC, Forrester JS, Diamond G, et al. Hemodynamic spectrum of myocardial infarction and cardiogenic shock: A conceptual model. *Circulation*. 1972;45:1097–1110.

84. Sobel BE, Bresnahan GF, Shell WE, et al. Estimation of infarct size in man and its relation to prognosis. *Circulation*. 1972;46:640–648.

85. Holman BL, Chisholm RJ, Braunwald E. The prognostic implications of acute myocardial infarct scintigraphy with 99mTc-pyrophosphate. *Circulation*. 1978;57:320–327.

86. Botvinick E, Perez-Gonzales J, Dunn R, et al. Scintigraphic parameters of infarct size and long term clinical course in complicated myocardial infarction without heart failure. *Am J Cardiol*. 1983;51:1045–1054.

87. Perez-Gonzales J, Botvinick E, Dunn R, et al. The late prognostic value of acute scintigraphic measurements of myocardial infarction size. *Circulation*. 1982;66:960–970.

88. Sobel B, Shell WE, Roberts R. An improved basis for enzymatic estimation of infarct size. *Circulation*. 1975;52:743–754.

89. Warnowicz MA, Parker H, Cheitlin M. Prognosis of patients with acute pulmonary edema and normal ejection fraction after myocardial infarction. *Circulation*. 1983;67:330–338.

90. Khaw BA, Beller GA, Haber E. Experimental myocardial infarction imaging following intravenous administration of iodine 131 labeled antibody (FAB') 2 fragments specific for cardiac myosin. *Circulation*. 1978;57:743–751.

91. Berger H, Alderson L, Becker L, et al. Multicenter trial of In-111 antimyosin for infarct avid imaging. *J Nucl Med*. 1986;27:967–972.

92. Khaw AB, Yasuda T, Gold HK, et al. Acute myocardial infarct imaging with indium-111 labeled monoclonal antimyosin Fab. *J Nucl Med*. 1987;28:1671–1680.

93. Berger H, Lahire A, Leppo J, et al. Antimyosin imaging in patients with ischemic chest pain: Initial results of phase III multicenter trial. *J Nucl Med*. 1988;29:1671–1680.

94. LaFrance ND, Frazier C, Ravert H, et al. In-111 antimyosin monoclonal antibody (AMAB) in detecting rejection of cyclosporin treated heart transplants. *J Nucl Med*. 1987;28:663–670.

95. Misbach GA, Botvinick EH, Tyberg J, et al. The functional implications of scintigraphic measures of ischemia and infarction. *Am Heart J*. 1982;106:996–1004.

96. Nicod P, Corbett JR, Firth BG, et al. Prognostic value of resting and exercise radionuclide ventriculography after acute myocardial infarction in high risk patients with single and multivessel disease. *Am J Cardiol*. 1983;52:32–40.

97. Brown KA, Boucher CA, Okada RD, et al. Prognostic value of exercise thallium-201 in patients presenting for evaluation of chest pain. *J Am Coll Cardiol*. 1983;1:994–1002.

98. Schelbert HR, Henning H, Ashburn WL, et al. Serial measurements of left ventricular ejection fraction by radionuclide angiography early and late after myocardial infarction. *Am J Cardiol*. 1976;38:707–714.

99. Gibson RS, Crampton RS, Watson DD, et al. Precordial ST segment depression during acute inferior myocardial infarction: Clinical scintigraphic and angiographic correlations. *Circulation*. 1982;66:732–741.

100. Becker LC, Silverman KJ, Bulkley BH, et al. Comparison of early thallium-201 scintigraphy and gated blood pool imaging for predicting mortality in patients with acute myocardial infarction. *Circulation*. 1983;67:1272–1281.

101. Brown KA, Okada RD, Boucher CA, et al. Serial thallium-201 imaging at rest in patients with stable and unstable angina pectoris: Relationship of myocardial perfusion at rest to presenting clinical syndrome. *Am Heart J*. 1983;106:70–77.

102. Gibson RS, Watson DD, Craddock GB, et al. Prediction of cardiac events after uncomplicated myocardial infarction: A prospective study comparing predischarge exercise thallium-201 scintigraphy and coronary angiography. *Circulation*. 1983;68:321–330.

103. Leppo JA, O'Brien J, Rothender JA, et al. Dipyridamole thallium-201 scintigraphy in the prediction of future events after acute myocardial infarction. *N Engl J Med*. 1984;310:1014–1020.

Chapter 12

Evaluation of Myocardial Perfusion Using Planar Thallium 201 Imaging at Rest and During Exercise

Sanjiv Kaul, MD

BACKGROUND

Since 1974, when it was first used in patients (F.J.T. Wackers, MD), thallium 201 has become the most widely used radionuclide for the assessment of myocardial perfusion in the clinical setting. Thallium is a metallic element in group III-A of the periodic table.[1] The in vitro half-life of ^{201}Tl is 73 hours. It emits mercury x-rays of 69 to 83 KeV, which is the lower end of the energy spectrum of a gamma camera.[2] Imaging of ^{201}Tl can be achieved by using a high-resolution camera with a low-energy collimator.[3] Optimal planar images can be obtained with an intravenous injection of 2.0 mCi of ^{201}Tl followed by data collection 5 minutes later. Data should be collected in each view for 10 minutes. Delayed images should be obtained from 2 to 4 hours after obtaining the initial images.

Kinetics of Thallium 201

At normal rates of perfusion, approximately 85% of ^{201}Tl is extracted by myocytes in the first pass.[4] The uptake of ^{201}Tl by myocytes involves both a passive diffusion and an active process utilizing the Na$^+$, K$^+$ pump. The latter step requires adenosine triphosphate.[3] If there is loss of cell membrane integrity, ^{201}Tl uptake will be reduced despite adequate blood flow. Irreversibly damaged myocardial tissue cannot concentrate ^{201}Tl intracellularly.[5] Approximately 4% of the ^{201}Tl injected intravenously is concentrated in the myocardium under normal conditions.[3] Within physiologic ranges of coronary blood flow there is a linear relationship between myocardial uptake of ^{201}Tl and regional myocardial blood flow measured using radiolabeled microspheres (Figure 12-1).[6–8]

Redistribution

Following initial uptake, there is a continuous exchange of ^{201}Tl between the intracellular compartment and the blood pool that recirculates through the myocardium.[4,9–17] Therefore, a region of the myocardium that has received less ^{201}Tl initially because of reduced blood flow will continue to accumulate ^{201}Tl over time as long as the myocardial cells are viable. In contrast, a region that has received adequate ^{201}Tl initially will show a net loss of ^{201}Tl over time in proportion to the decrease in blood levels of ^{201}Tl. Therefore, the disparity noted in the ^{201}Tl activity on the initial image between regions of normal and decreased flow will tend to become less over time. This phenomenon is known as "redistribution" and can be seen when a delayed image (taken 2 to 4 hours after the injection of ^{201}Tl) is compared with the initial image (taken 5 minutes after the injection of ^{201}Tl) (Figure 12-2). This reversal of the initial disparity can be either "complete" (indicating that the entire region of the myocardium showing the initial defect is viable) or "partial" (implying that some viable tissue is present within the region of the initial defect).

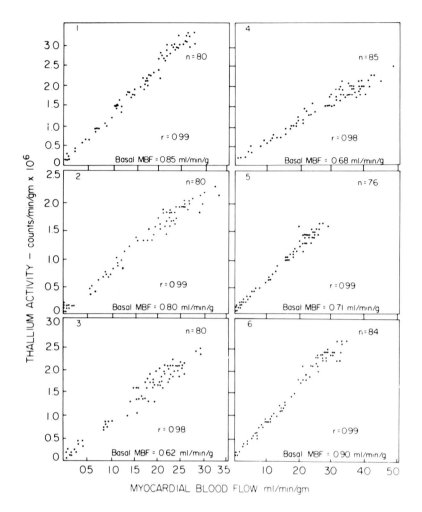

Figure 12-1 Relationship between thallium 201 activity (y axis) and regional myocardial blood flow (MBF) (x axis) in six dogs during exercise. A linear relationship is noted between MBF and ^{201}Tl activity over the entire physiologic range of MBF. *Source:* Reprinted with permission from *Circulation* (1980;61:797–801), Copyright © 1981, American Heart Association.

Persistent Defect

When ^{201}Tl uptake is reduced in the initial image with no redistribution noted in the delayed image, a "persistent" defect is said to be present.[9] Persistent defects can be of several grades depending upon the initial uptake of ^{201}Tl in that region. Severely reduced uptake, such that there is virtual absence of ^{201}Tl activity in that region in both the initial and the delayed images (Figure 12-3), has been correlated with previous myocardial infarction, Q waves on the electrocardiogram, and akinetic or dyskinetic left ventricular regions on cineangiography.[18,19]

Because redistribution implies the presence of hypoperfused viable myocardium, the absence of redistribution unfortunately has been interpreted as the lack of viability.[20,21] The lack of redistribution in an image showing a severe initial defect, as is illustrated in Figure 12-3, probably does imply the absence of viable tissue in that region. However, when initial defects are not severe (reduced but definite uptake of ^{201}Tl) and ^{201}Tl activity is present in the delayed image without any evidence of redistribution (Figure 12-4), there is probably a mixture of viable and nonviable tissue. If there is only nonviable tissue in these regions, there should be no uptake of ^{201}Tl in the first place, and certainly there should be no reason for the myocardium to retain ^{201}Tl several hours later. This supposition is supported by normal uptake of ^{201}Tl in such regions following revascularization.[22,23] In addition, there is strong experimental evidence that myocardial ^{201}Tl activity in a canine model of coronary occlusion followed by reperfusion reflects the presence or absence of viable non-necrotic myocardium.[24–26]

Reverse Redistribution

Other than "persistent defect" and "redistribution," another abnormality that can be noted on ^{201}Tl images is "reverse redistribution." This phenomenon is said to occur when there is normal uptake of ^{201}Tl in the initial image, but a defect occurs in the delayed image[27] (Figure 12-5). Reverse redistribution has different implications under different conditions. In patients with coronary artery disease without a recent infarction, this phenomenon implies the presence of a severely stenotic

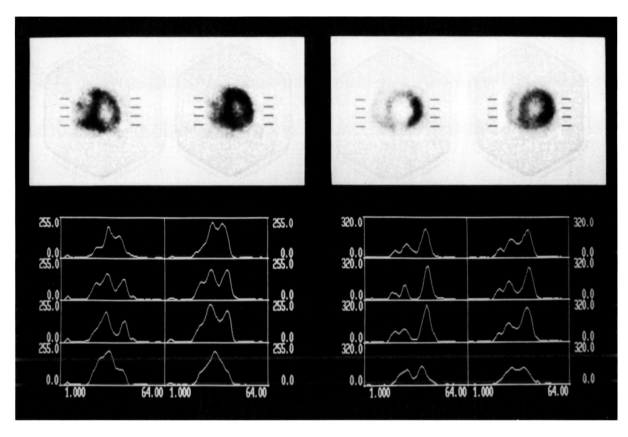

Figure 12-2 "Complete" redistribution in the anterolateral wall (anterior projection) and "partial" redistribution in the interventricular septum (45° left anterior oblique projection). The initial defect noted in the interventricular septum shows less fill-in in the 2-hour image compared with that noted in the anterior wall.

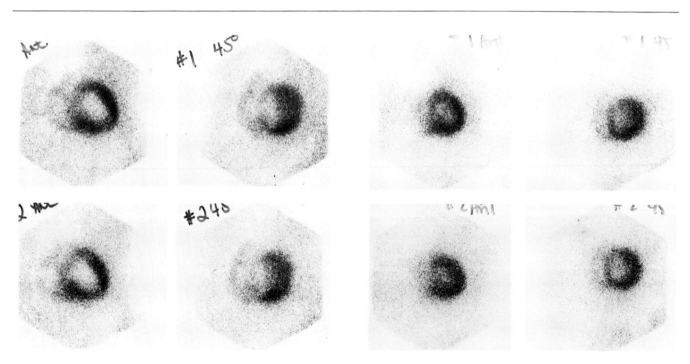

Figure 12-3 A severe "persistent" defect in the interventricular septum in the 45° left anterior oblique projection. In the initial defect almost no counts are seen and there is no fill-in in this region even at 2 hours.

Figure 12-4 A mild "persistent" defect in the interventricular septum in the 45° left anterior oblique projection; no fill-in is noted in this region in the delayed image. The thallium counts within the defect are not as low as those noted in Figure 12-3.

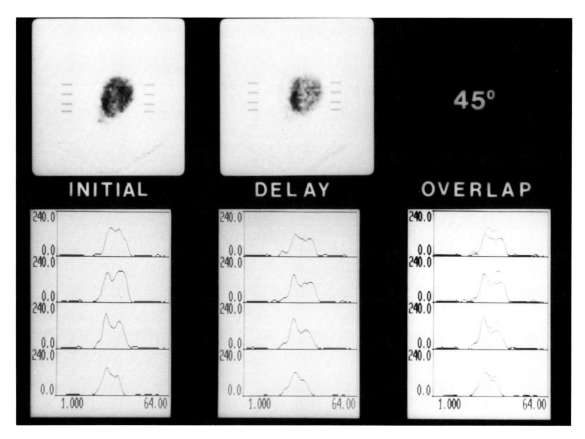

Figure 12-5 The phenomenon of "reverse redistribution." In the initial image, the counts in both the posterolateral wall and the interventricular septum appear to be equal, although a defect is noted in the inferoapical region. However, in the 2-hour image, the counts in the posterolateral wall are significantly less than those in the interventricular septum, resulting in a defect in the posterolateral wall in the delayed images.

vessel supplying the contralateral wall. In patients in whom an image is obtained after a recent myocardial infarction, reverse redistribution implies successful reperfusion (either spontaneous or related to thrombolytic therapy).[28,29] In both cases reverse distribution denotes areas that are at jeopardy for a future event.

Lung Uptake of Thallium 201

Lung activity of ^{201}Tl has conventionally been assessed visually.[30–32] Recently, however, quantitation of the lung-heart ^{201}Tl ratio has been shown to be of major prognostic importance.[33–35] To assess the lung-heart ratio accurately, because the anterior projection provides the best view of lung activity and because lung activity has a propensity to decrease much faster than the myocardial activity, one should obtain the anterior projection first. Figure 12-6 shows lung activity in anterior and in 45° and 70° left anterior oblique projections in a patient immediately after exercise; the interval between each view is approximately 10 minutes. Note the decrease in ^{201}Tl activity in the lungs over time.

Increased lung uptake of ^{201}Tl during exercise implies exercise-induced pulmonary edema and has been cor-

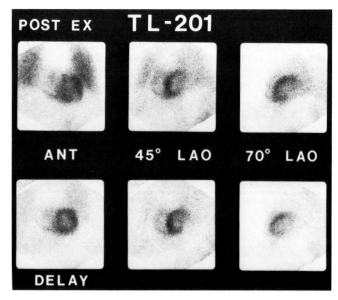

Figure 12-6 Initial and delayed images in anterior (ANT) and in 45° and 70° left anterior oblique (LAO) projections. The intervals between the three initial images are 10 minutes each. Increased lung uptake of thallium 201 (TL-201) is noted best in the anterior projection and diminishes rapidly by the time the 70° left anterior oblique projection is obtained. POST EX, after exercise. *Source:* Reprinted from *Clinical Cardiac Imaging* (pp 275–283) by DD Miller et al. (Eds) with permission of McGraw-Hill Book Company, © 1987.

related with the presence of multivessel disease, number of myocardial segments demonstrating redistribution, number of myocardial segments showing abnormal motion, presence of large infarcts, and heart failure.[30,36–38] Lung uptake of ^{201}Tl is influenced by the maximal heart rate achieved during exercise.[39] Abnormally high lung uptake has also been correlated with larger infarct size and lower ejection fractions in patients with acute myocardial infarction who have undergone a submaximal exercise test.[35]

THALLIUM 201 IMAGING AT REST

Thallium 201 imaging at rest is not performed as frequently as it should be. There are several situations in which ^{201}Tl imaging can be performed at rest to provide important clinical information (Table 12-1).

Detection and Sizing of Acute Myocardial Infarction

Thallium 201 imaging has been used in the coronary care unit for the detection, sizing, and localization of acute myocardial infarction[40–42] (Figure 12-7). In this regard, there is good correlation between the ^{201}Tl score and post-mortem findings[41] (Figure 12-8). As would be expected, the ^{201}Tl score provides important prognostic information in patients with acute myocardial infarction[43] (Figure 12-9). More important, the ^{201}Tl score is useful in patients admitted to the hospital with nondiagnostic clinical and electrocardiographic features.[44,45] In such patients, ^{201}Tl imaging can identify those likely to have suffered an infarction.

Figure 12-10 depicts ^{201}Tl findings in patients referred to the coronary care unit with chest pain. Only those with acute ischemic syndromes (specifically, acute myocardial infarction) demonstrated definite abnormalities on ^{201}Tl imaging.[44] On the basis of these findings it has been suggested that patients with nondiagnostic electrocardiograms and abnormal ^{201}Tl findings may be transferred to the coronary care unit for further care.[44,45] From these data one can extrapolate

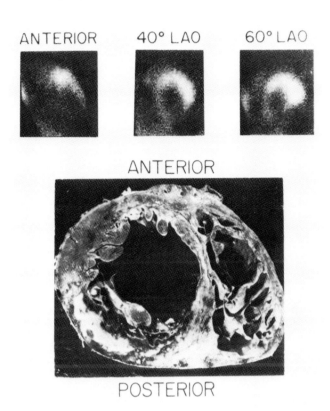

Figure 12-7 An example of thallium 201 and post-mortem correlation in a patient with both anterior and inferior infarction. The anterior infarct is healed, but the posterior infarct is acute. Severe defects are noted in the ^{201}Tl images in both regions. LAO, Left anterior oblique projection. *Source:* Reprinted with permission from *Circulation* (1979;60:785), Copyright © 1979, American Heart Association.

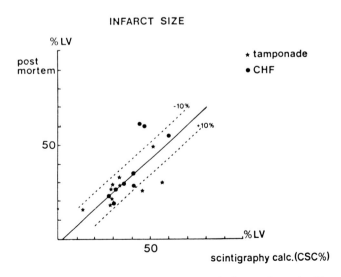

Figure 12-8 Relationship between the size of infarct as determined by post-mortem examination (*x* axis) and that determined antemortem by planar thallium 201 imaging. Patients with cardiac tamponade and congestive heart failure (CHF) are also shown. LV, left ventricle. *Source:* Reprinted with permission from *Circulation* (1977;56:72), Copyright © 1977, American Heart Association.

Table 12-1 Indications for Rest Thallium 201 Imaging

- Detection and sizing of acute myocardial infarction
- When left ventricular dysfunction exceeds infarct size
- Chest pain after reperfusion therapy
- Unstable angina pectoris
- Prinzmetal angina
- Assessment of myocardial viability

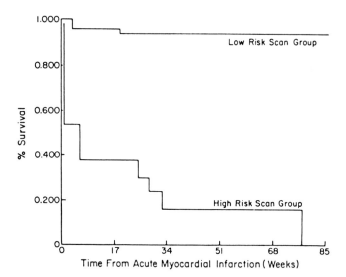

Figure 12-9 Actuarial survival rates in patients after acute myocardial infarction who have a high-risk versus a low-risk thallium 201 scan. *Source:* Reprinted with permission from *Circulation* (1980;61:996), Copyright © 1980, American Heart Association.

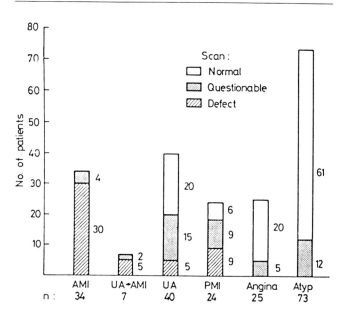

Figure 12-10 The results of thallium 201 scans versus final diagnosis in patients with chest pain syndromes and equivocal electrocardiograms admitted to the hospital. Definite defects were noted in most of the patients with acute myocardial infarction (AMI) and those with unstable angina (UA) progressing to acute infarction (UA-AMI). Patients with angina and those with atypical chest pain (Atyp) had either questionable or normal scans. PMI, postmyocardial infarction. *Source:* Reprinted with permission from *British Heart Journal* (1979;41:111), Copyright © 1979, British Cardiac Society.

that ^{201}Tl imaging at rest would have a high sensitivity for the detection of acute myocardial infarction in patients coming to the emergency department with chest pain, and could therefore help determine patient disposition.

When Left Ventricular Dysfunction Exceeds Infarct Size

There are several instances when patients with no evidence of prior infarction present with acute infarction and have left ventricular dysfunction far in excess of the ongoing infarction judged from creatine kinase levels or electrocardiography. Such patients might have ischemia in myocardial beds remote from the infarct zone.[46–48] If untreated, such patients have a high mortality rate.[49] Thallium 201 images during rest have been shown to be useful in such patients.[50] The presence of redistribution in segments remote from the infarct zone confirms ongoing ischemia, and immediate revascularization results in improvement of left ventricular function. Figure 12-11 illustrates initial and delayed ^{201}Tl images during rest in such a patient at the time of hospital admission; Figure 12-12 illustrates ^{201}Tl images during exercise after revascularization. Only a small infarct is noted in the anteroapical and distal interventricular septal regions in the exercise image; all other areas that showed redistribution on the rest images prior to surgery now show normal perfusion.

Chest Pain after Reperfusion Therapy

There is no role for routine cardiac catheterization in all patients undergoing intravenous reperfusion ther-

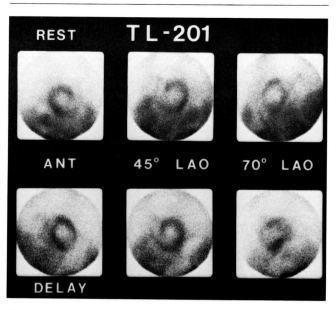

Figure 12-11 Rest thallium 201 (TL-201) images in a patient with Q waves in V_1 to V_4 and a peak creatine kinase level of 1196 IU/dL whose left ventricular dysfunction was far in excess of infarct size (left ventricular ejection fraction of 0.30). This patient showed a mild persistent defect in the lower interventricular septum and anterior wall with partial redistribution to these regions as well as redistribution to other beds. ANT, anterior projection; LAO, left anterior oblique projection. *Source:* Reprinted with permission from *American Heart Journal* (1988;115:749–753), Copyright © 1988, CV Mosby Company.

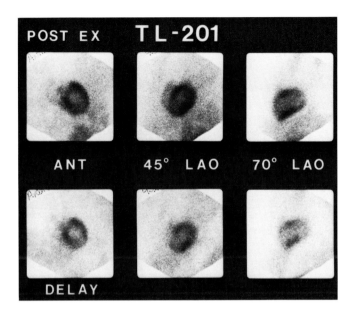

Figure 12-12 Exercise thallium 201 (TL-201) imaging in the patient whose rest images are shown in Figure 12-11 after the patient underwent bypass surgery to all three major coronary arteries. The rest left ventricular ejection fraction improved to 0.45. There is only a mild persistent apical and lower septal defect. Perfusion to all other regions is normal. POST EX, after exercise; ANT, anterior projection; LAO, left anterior oblique projection. *Source:* Reprinted with permission from *American Heart Journal* (1988;115:749–753), Copyright © 1988, CV Mosby Company.

apy.[51] Only about half of the patients with open infarct-related vessels show improvement in regional function.[49] However, a patient who develops chest pain after reperfusion therapy might have ischemia within the infarct zone or in a remote myocardial bed. Because regional wall motion abnormalities persist for 5 to 10 days after successful reperfusion,[49] their presence does not indicate the absence of viable myocardium. Uptake of ^{201}Tl in the infarct zone that persists in the delayed image can be said to indicate viable myocardium. Similarly, rest redistribution in remote beds indicates ischemia in those beds. In either case, ^{201}Tl imaging will help in determining the therapeutic strategy.

There is controversy regarding the optimal timing of ^{201}Tl imaging after reperfusion therapy. It has been suggested that, because of the occurrence of hyperemia or tissue edema after reperfusion, ^{201}Tl might accumulate in areas that are not viable if it is injected immediately after reperfusion therapy.[52,53] This phenomenon is not observed if ^{201}Tl is injected approximately 48 hours after reperfusion.[52] Other studies have demonstrated that patients who have achieved successful reperfusion show a decrease in defect size in images obtained serially over time compared with those who do not show a change.[54]

Unstable Angina Pectoris or Prinzmetal Angina

Patients who have unstable or Prinzmetal angina show redistribution on delayed images compared with those who suffer an acute transmural infarction; the latter demonstrate persistent defects.[55–59] In patients with unstable angina, revascularization has been shown to reverse the ^{201}Tl uptake abnormalities. Patients with atypical chest pain who develop ^{201}Tl uptake abnormalities during intravenous infusion of ergonovine maleate respond to medical therapy, in contrast to those who do not show such abnormalities.[60] The latter patients might not have Prinzmetal angina, and ^{201}Tl images during rest could help in decisions about patient management.

THALLIUM 201 IMAGING DURING EXERCISE FOR DETECTION OF CORONARY ARTERY DISEASE

Qualitative versus Quantitative Assessment

Thallium 201 imaging during exercise is the most widely used form of myocardial perfusion imaging. Although the initial qualitative assessment of ^{201}Tl images provided an acceptable sensitivity for the detection of coronary artery disease[19,61–67] (Table 12-2), the inter- and intraobserver errors were large.[68,69] Computer-assisted quantitative imaging now has been available for almost a decade. Several algorithms have been used for analysis of planar ^{201}Tl images, and all have been shown to offer a higher sensitivity for the detection of coronary artery disease than visual analysis (Table 12-3) with a smaller observer variability.[70–74] All programs utilize background subtraction, alignment of initial and delayed images, and determination of regional myocardial ^{201}Tl activity. In Appendix 12-A is listed the program developed at the Massachusetts General Hospital. Because there is both temporal and spatial heterogeneity in background, all programs utilize interpolative background subtraction.[75]

Table 12-2 Sensitivity and Specificity of Thallium 201 Imaging Using Visual Analysis

Year	Authors	Reference	Sensitivity (%)	Specificity (%)
1977	Bailey et al	61	73	100
1977	Ritchie et al	62	76	92
1978	Verani et al	63	79	97
1978	Bodenheimer et al	19	66	100
1979	McCarthy et al	64	85	79
1979	Pohost et al	65	87	75
1979	Vogel et al	66	74	96
1979	Borer et al	67	81	90
	Average		78	91

Table 12-3 Sensitivity and Specificity of Thallium 201 Imaging Using Quantitative Analysis

Year	Authors	Reference	Sensitivity (%)	Specificity (%)
1979	Burow et al	70	88	92
1981	Berger et al	71	91	90
1981	Maddahi et al	72	93	91
1984	Massie et al	73	95	85
1986	Kaul et al	74	90	80
	Average		91	88

Whereas the program developed at the University of Virginia depicts activity along horizontal planes transecting the left ventricle at seven myocardial segments in two views[76] (Figure 12-2), that developed at the Cedars-Sinai Medical Center depicts myocardial ^{201}Tl activity as a profile around the circumference of the heart.[72] The program developed at the Massachusetts General Hospital depicts average ^{201}Tl activity in 15 myocardial segments in three views[74] (Figure 12-13). In each program myocardial activity is defined as a percentage of the activity in the area showing maximal counts in that view. All programs use values for abnormality based upon data from normal subjects. In addition to the improvement in the overall sensitivity of ^{201}Tl images for the detection of coronary artery disease, quantitative imaging provides other benefits: it is more specific for detecting initial defects; it is more sensitive for the detection of redistribution; and it is more accurate in determining the lung-heart ratio of ^{201}Tl.[77] Therefore, computer-assisted programs should be used routinely to interpret exercise ^{201}Tl images.[78]

The Normal Thallium 201 Image

Figure 12-14 illustrates initial and delayed images in a normal subject. There is significant heterogeneity in ^{201}Tl uptake. Table 12-4 illustrates the mean activity in each of five segments in each view in both the initial and

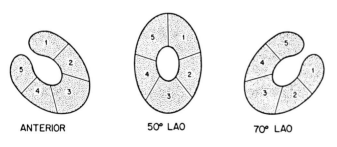

Figure 12-13 Program developed at the Massachusetts General Hospital, which depicts myocardial thallium 201 activity as an average activity within each of 15 myocardial segments in three views. LAO, Left anterior oblique projection. *Source:* Reprinted with permission from *Journal of the American College of Cardiology* (1986;7:527–537), Copyright © 1986, American College of Cardiology.

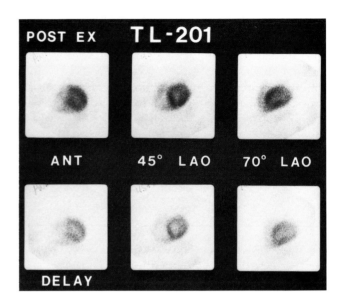

Figure 12-14 Normal exercise thallium 201 (TL-201) images. Reduced ^{201}Tl activity can be noted in the base of the heart. ANT, anterior projection; LAO, Left anterior oblique projection; POST EX, after exercise. *Source:* Reprinted from *Clinical Cardiac Imaging* (pp 275–283) by DD Miller et al. (Eds) with permission of McGraw-Hill Book Company, © 1987.

Table 12-4 Regional Thallium Myocardial Activity in 55 Normal Subjects Where the Myocardium Is Divided Into Five Segments in Each View

Segment	Initial Image (%)	Delayed Image (%)
Anterior view		
1	70 ± 8*	32 ± 5*
2	91 ± 4	40 ± 7
3	89 ± 5	36 ± 7
4	87 ± 6	36 ± 6
5	62 ± 8*	26 ± 6*
50° LAO† view		
1	73 ± 8*	33 ± 7*
2	92 ± 4	38 ± 7
3	88 ± 5	36 ± 7
4	84 ± 5	36 ± 7
5	65 ± 9*	30 ± 8*
70° LAO view		
1	71 ± 6*	30 ± 5*
2	90 ± 4	38 ± 5
3	89 ± 6	36 ± 6
4	90 ± 5	38 ± 6
5	72 ± 9*	32 ± 6*

*$P<0.01$ compared with adjacent segments in the same view. Values are mean ± 1 SD and are expressed as a percentage of the highest-count three-pixel region of the myocardium in that view.
†LAO, Left anterior oblique.

Source: Adapted with permission from *Journal of the American College of Cardiology* (1986;7:527–537), Copyright © 1986, American College of Cardiology.

delayed images in 55 normal subjects.[74] This heterogeneity underscores the importance of defining normal values for each segment in any laboratory. The basal segments show lower activity than that in the other segments, which may be related to several factors. These segments are near the left ventricular outflow tract and there may be less myocardial tissue in these segments; the great vessels overlying these segments may attenuate the actual counts; and the basal segments are farther from the detector and subject to more attenuation. The normal lung-heart ratio of ^{201}Tl is 0.37 ± 0.07.[36]

Factors Affecting the Sensitivity of the Test

Presence of Prior Infarction

The presence of prior infarction has a significant effect on the sensitivity of the exercise ^{201}Tl test. When ^{201}Tl images are analyzed in patients with single-vessel disease in whom normal myocardium is clearly delineated from abnormal myocardium (in contrast to patients with multivessel disease), the sensitivity is higher in those who have had a previous infarction.[78] Similar results have been obtained in patients with multivessel disease.[79]

Extent of Coronary Artery Disease

When all patients with coronary artery disease are assessed for sensitivity, those with single-vessel disease are the usual source of false-negative results.[74] Even in patients without a prior infarction, the sensitivity of ^{201}Tl imaging is related to the number of diseased vessels, being higher in those with multivessel and left main disease[78,80] (Table 12-5). Thallium 201 imaging has a significantly higher sensitivity than ST segment depression for the detection of coronary artery disease for all subsets of patients (Table 12-5).

Location of Disease

In patients with single-vessel disease the sensitivities for the detection of coronary artery disease are similar for left anterior descending, left circumflex, and right coronary arteries.[78] In other reports where poor sensitivity has been reported for right coronary and left circumflex disease, the patients had multivessel disease. Such studies of patients with multivessel disease have two major limitations. First, in these studies regions of the myocardium are usually assigned to different vessels a priori, which may be an incorrect assumption because of the large variability in the location of ^{201}Tl uptake abnormalities.[72,81,82] For example, while most of the defects in the interventricular septum in patients with single-vessel disease can be attributed to the left anterior descending artery, the inferior, inferoposterior, and posterolateral beds can be supplied by either the right or left circumflex coronary arteries.[77] Second, if a patient stops exercising as a result of development of ischemia in one vascular territory, another vascular territory might not show a defect because of relatively higher flow to that region compared with the region where ischemia developed.

Severity of Disease

As would be expected, initial defects are more likely to be present when the degree of stenosis is severe.[76] The more severe the stenosis, the more apparent is the blood flow mismatch. Similarly, the more severe the mismatch in the initial image, the more likely it is for redistribution to be present on the delayed image as long as the myocardium being supplied by the stenotic artery is viable.[76]

Table 12-5 Sensitivity of Qualitative and Quantitative Thallium 201 Imaging and Exercise Electrocardiography in Relation to the Number of Vessels with Coronary Artery Disease

Parameter	No (%)				
	One Vessel (n = 56)	Two Vessels (n = 40)	Three Vessels (n = 40)	Left Main (n = 13)	Total (n = 152)
ST segment depression on exercise test	33 (59)*	26 (65)*	33 (77)*	11 (85)	103 (68)*
Qualitative thallium image	40 (71)	31 (77)	38 (88)	11 (85)	120 (79)
Quantitative thallium image	46 (82)	37 (92)	42 (98)	13 (100)	138 (91)

*Significantly ($P<0.05$) lower than quantitative thallium imaging.

Source: Reprinted with permission from Journal of the American Medical Association (1986;255:508–511), Copyright © 1986, American Medical Association.

Degree of Effort During Exercise

Unlike the prevalence of ST segment depression, which has been shown to vary with the degree of effort, the relationship between the prevalence of ^{201}Tl imaging parameters and effort has not been well established. To test whether different degrees of effort would affect the prevalence of initial defects in any one patient would require exercising the same patient to various degrees of effort and acquiring ^{201}Tl images at these stages. This latter approach has been tried in a small number of patients, and the results suggest that the sensitivity of initial defects increases at higher degrees of effort.[83] Another study has also found that the sensitivity of ^{201}Tl imaging is superior to that of exercise electrocardiography only in patients who do not achieve 85% or higher of predicted heart rate.[71]

However, in a recent study of a large number of patients undergoing symptom-limited exercise testing, the level of effort did not appear to influence the overall ^{201}Tl test results except in those with single-vessel disease who exercised to a high workload. Thallium 201 imaging was found to be superior to exercise electrocardiography at all levels of effort[84] (Figure 12-15). The decrease in the prevalence of initial defects at high workloads in patients with single-vessel disease may occur because in these patients either the stenosis is of only borderline hemodynamic significance, which does not result in ischemia at these workloads, or the region of ischemia is small and therefore not detected on planar imaging. The decrease in the prevalence of redistribution at high workloads in patients with single-vessel coronary artery disease probably occurs because patients with multivessel coronary artery disease become ischemic at lower levels of effort and therefore stop exercising. Those who continue to exercise may not have stenoses sufficient enough to cause ischemia or may have only small regions of ischemia that are not detected on ^{201}Tl imaging. The decrease in prevalence of redistribution in patients with prior infarction at higher workloads also noted in this study may occur because such patients may have no or little inducible ischemia and can exercise to a higher degree of effort because they are not limited by angina or shortness of breath. Such patients would manifest initial defects since they have regional myocardial hypoperfusion secondary to scarring.

Number of Observers Analyzing the Images

Because of large inter- and intraobserver variability, the opinions of individual observers either alone or in

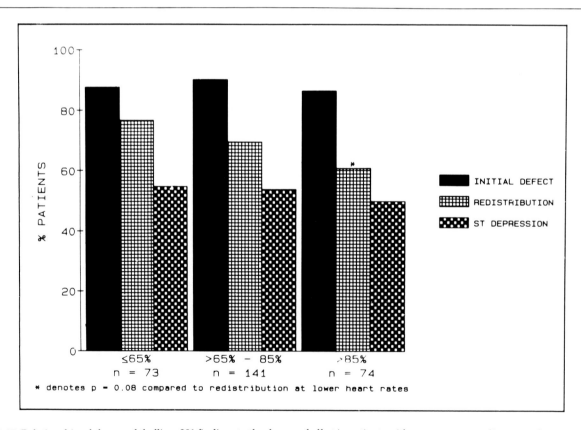

Figure 12-15 Relationship of abnormal thallium 201 findings to the degree of effort in patients with coronary artery disease undergoing symptom-limited exercise testing. The prevalence of ^{201}Tl abnormalities is not significantly influenced by the level of effort. *Source:* Reprinted with permission from *American Journal of Cardiology* (1989;63:160–165), Copyright © 1989, Cahners Publishing Company.

consensus are more likely to produce false-negative results than is an average score of multiple observers.[69] An average score of multiple observers is almost as good as quantitative ^{201}Tl imaging in terms of sensitivity of initial defects[76]; however, it is not feasible for multiple observers to interpret routinely ^{201}Tl images in the clinical setting.

Factors Affecting the Specificity of the Test

Artifacts

Because of frequent artifacts (Table 12-6 and Figures 12-16 to 12-20) present on planar ^{201}Tl images, a fully automated program can provide an unacceptably low specificity. Therefore, it is essential that an experienced interpreter assess both analog and computer-generated data at the same time to assess the presence of artifacts. Because of the high false-positive rate, exclusion of isolated basal segments in the 45° left anterior projection has been shown to increase specificity without affecting sensitivity.[74,85] Such isolated basal defects have no adverse prognostic consequence and should be interpreted as a variation of normal.[85]

True False-Positives

Other than artifacts that can decrease the specificity of exercise ^{201}Tl imaging for coronary artery disease, causes for true false-positive findings include left ventricular hypertrophy,[86] dilated cardiomyopathy,[87] sarcoidosis,[88] myocardial tumors,[89] and sickle cell anemia.[90] In the case of dilated cardiomyopathy, the perfusion defects follow a patchy, rather than a segmental, pattern.

Referral Bias

The specificity of ^{201}Tl imaging has decreased in recent years because of referral bias. A number of patients with chest pain syndromes who are sent for exercise ^{201}Tl imaging may not have significant coronary artery disease on coronary angiography. However, these patients are not entirely normal. On follow-up patients with "normal" coronary arteries and normal thallium 201 images do much better than those with "normal" coronary arteries and abnormal thallium 201 images.[91] The latter group has a higher cardiac event rate and is more likely to have "subcritical" stenosis on angiography.[92]

Problems with Assessing Myocardial Thallium 201 Clearance

The use of myocardial clearance of ^{201}Tl in the detection of coronary artery disease has several theoretical advantages. Both the amount of ^{201}Tl entering the myocardium and the rate of clearance from the myocardium are related to myocardial blood flow. Zones with the best perfusion have the most ^{201}Tl delivered initially and have the largest myocardial–blood pool gradient resulting in the fastest clearance. Zones with lower perfusion have less ^{201}Tl delivered initially and clear more slowly. However, although the sensitivity of

Table 12-6 Possible Artifacts Present on Planar Thallium 201 Images

- Breast attenuation (Figure 12-16)
- Upper septal thinning (Figure 12-17A)
- Isolated septal abnormality in the presence of left bundle-branch block (Figure 12-17B)
- Isolated washout abnormalities without any numerically significant defects (Figure 12-18)
- Presence of artificially low counts in the interventricular septum in the 45° left anterior oblique projection because of abnormally high uptake in the posteromedial papillary muscle (Figure 12-19)
- Exaggerated apical thinning (Figure 12-20)
- Artificially low inferior wall counts in the anterior image due to an enlarged or hypertrophied right ventricle
- Artificially low counts in the inferoposterior wall in the 70° left anterior oblique image due to diaphragmatic attenuation

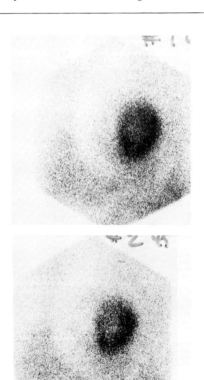

Figure 12-16 Decreased thallium 201 activity due to breast attenuation. The breast shadow can be seen as a halo around the heart and is causing decreased ^{201}Tl activity in the upper interventricular septum in both the initial and delayed images.

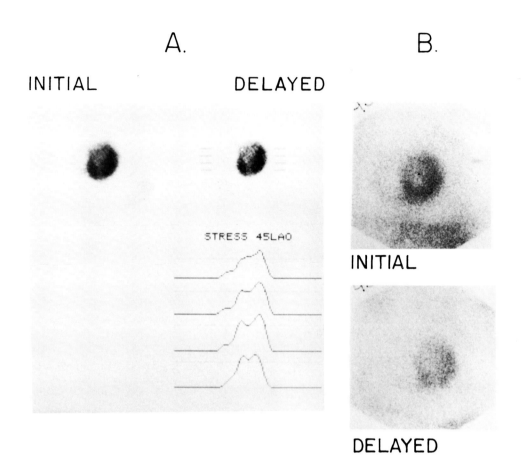

Figure 12-17 A, Isolated upper septal defect in the initial image in the 45° left anterior oblique (LAO) projection. No redistribution was noted in the delayed images (not shown here). This isolated finding is not associated with coronary artery disease and has no adverse prognostic implications on long-term follow-up.[86] B, isolated septal defect in the presence of left bundle-branch block without coronary artery disease, the mechanism of which is unknown.

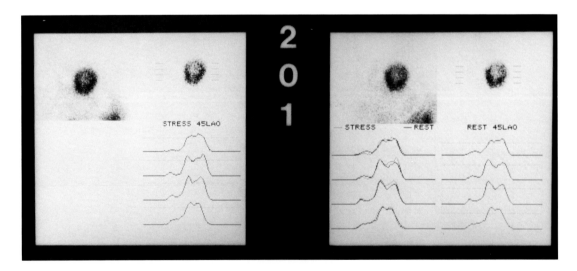

Figure 12-18 Isolated washout abnormality in the interventricular septum without any numerically significant defect. This finding is a frequent source of false-positive thallium 201 scans. It has been shown to have no adverse prognostic implications on long-term follow-up.[86] 45LAO, Left anterior oblique projection at 45°.

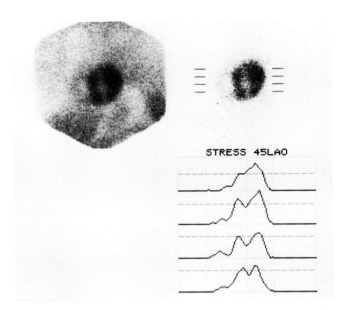

Figure 12-19 Artificially low counts in the interventricular septum in the 45° left anterior oblique projection (45LAO) due to a "hot" posteromedial papillary muscle.

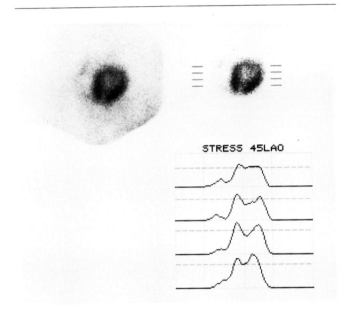

Figure 12-20 Exaggerated apical thinning. 45LAO, 45° left anterior oblique projection.

absolute clearance values for detecting coronary artery disease is good (92%), the specificity is very poor (16%).[74]

Because the myocardial clearance of ^{201}Tl is influenced by the peak heart rate achieved during exercise[93,94] (Figure 12-21), an attempt has been made to improve the specificity of clearance by correcting it for the peak exercise heart rate. Although this correction improves the specificity of clearance for the detection of coronary artery disease significantly, it does not add to the value of initial defects and redistribution in the

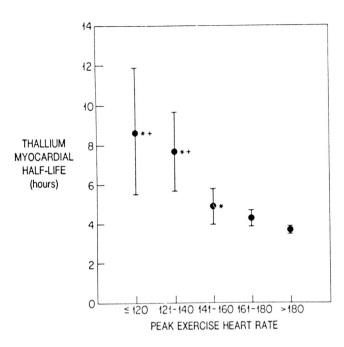

Figure 12-21 Relationship between the half-life of thallium 201 in the myocardium (y axis) and the peak exercise heart rate (x axis) in subjects without coronary artery disease. The higher the heart rate, the faster the clearance. *Source:* Reprinted with permission from "Influence of Peak Heart Rate on Normal Thallium-201 Myocardial Clearance" by S Kaul et al in *Journal of Nuclear Medicine* (1986; 27[1]:26–30), Copyright © 1986, Society of Nuclear Medicine.

Table 12-7 Factors That May Affect Myocardial Clearance of Thallium

- Muscle mass of the patient
  Patients with larger muscle mass may have lower blood levels of thallium and therefore faster clearance.
- Prandial state of the patient
  Patients who have eaten may have slower thallium clearance. This is thought to be related to the glucose, insulin, and intracellular potassium changes induced by eating.
- Peak exercise heart rate
  The higher the heart rate, the higher the cardiac output; therefore, more thallium enters the heart and washes out more rapidly.
- Delay of thallium at site of injection
  Delay of thallium (not extravasation) is seen in a significant number of patients in whom it is injected in a vein distal to the antecubital vein. From the site of injection, thallium is then released slowly into the blood stream, causing perturbations in the blood level of thallium and therefore in its clearance from the myocardium.

Source: Reprinted with permission from *Seminars in Nuclear Medicine* (1987;17:131–144), Copyright © 1987, W.B. Saunders Co.

detection of coronary artery disease.[94] Because variables other than heart rate influence the clearance of ^{201}Tl from the myocardium (Table 12-7), and because it would be impossible to correct for all of these variables, a relative value rather than an absolute value for clearance might be more useful in the clinical setting. As each segment in the myocardium "sees" the same blood level of ^{201}Tl, any difference in the clearance of ^{201}Tl from a particular segment, compared with other segments, should be related to myocardium-specific factors such as myocardial ischemia or loss of myocardial cellular integrity.

There is significant heterogeneity in the mean segmental ^{201}Tl clearance, even in normal subjects. The mean intersegmental variability is 36% ± 13% with a maximal variability of 98%.[95] The basal segments demonstrate the slowest clearance; apical segments, the fastest.[95] When clearance is determined to be abnormal in any segment if it is slower than 98% or more as compared with the fastest of the 15 segments in the heart, the specificity of abnormal clearance becomes similar to that of redistribution.[95] Thus, although the original impetus to use clearance was based on the hope that absolute clearance values would enhance the detection of coronary artery disease, it is the relative abnormality of clearance that might be useful in the clinical setting.

Combination of Different Variables To Optimize Detection of Coronary Artery Disease

Because a number of imaging variables are obtained during exercise ^{201}Tl testing, the use of logistic regression analysis has been recommended to increase the specificity of the test.[74] In this manner receiver-operator curves can be generated and optimal sensitivity and specificity can be defined for each laboratory. Table 12-8A lists the sensitivity and specificity of each imaging variable and that provided by the logistic regression model for the detection of coronary artery disease. Table 12-8B shows the relationship between the probability threshold derived by using this model and the sensitivity, specificity, and diagnostic accuracy of the test for detection of coronary artery disease. On the basis of this relationship, a cutpoint can be selected for the probability threshold to optimize the detection of coronary artery disease in any given setting.

Other electrocardiographic and hemodynamic variables complement ^{201}Tl imaging variables for the detection of coronary artery disease. Table 12-9 illustrates the sensitivity and specificity of variables found to be important when using logistic regression analysis in patients with coronary artery disease and no previous myocardial infarction.[80] None of the variables alone has an optimal diagnostic accuracy for the detection of coronary artery disease. However, a logistic model that incorporates these same variables affords optimal diagnostic accuracy. In comparison, as shown in Table 12-10, if the sensitivity and specificity of each variable are added in a stepwise manner, the sensitivity increases to almost 100% while the specificity declines to almost 0%.

Bayesian analysis has also been used in conjunction with exercise ^{201}Tl imaging to optimize the detection of coronary artery disease.[96-100] The conventional application of Bayes' theorem uses the sensitivity and specificity of a test to provide the post-test probability of a disease. The pretest probability of disease is usually assessed by the patient's age, gender, and type of chest pain (typical/atypical). The presence of typical chest pain in an older man makes the pretest likelihood of coronary artery disease high, and a positive ^{201}Tl stress test makes the post-test likelihood virtually certain. On the other hand, if the pretest likelihood is in the moderate range (40% to 50%), a negative exercise ^{201}Tl test reduces the post-test probability of coronary artery disease to 10%. Another simplified way to use the Bayes theorem is by assigning weights to tests.[101] These weights are derived from both the sensitivity and specificity data, and a single value can determine the post-test probability.

When more than one test is used, bayesian analysis utilizes the post-test probability of disease derived from one test as the pretest probability for the next test. This serial use of information from different tests assumes that the tests being used are conditionally independent, which may not be true. For example, during exercise, both ST segment depression on the electrocardiogram and redistribution on the delayed ^{201}Tl images reflect

Table 12-8

A. Sensitivity and Specificity of Thallium Imaging Variables and the Logistic Regression Model in Detection of Coronary Artery Disease

	Sensitivity (%) (n = 281)	Specificity (%) (n = 44)
Initial defect	95	66
Redistribution	60	91
Logistic regression model	90	80

B. Relationship between Five Values of Probability and Sensitivity, Specificity, and Diagnostic Accuracy

Probability Threshold	Sensitivity (%)	Specificity (%)	Diagnostic Accuracy (%)
0.51	95	50	87
0.69	90	80	88
0.81	85	80	85
0.88	80	90	80
0.95	70	95	75

Source: Adapted with permission from *Journal of the American College of Cardiology* (1986;7:527–537), Copyright © 1986, American College of Cardiology.

Table 12-9 Sensitivity and Specificity of Exercise Thallium Stress Test Variables and the Logistic Regression Model in Detection of Coronary Artery Disease in Patients without a Prior Myocardial Infarction

Variable	Sensitivity (%) (n = 152)	Specificity (%) (n = 44)	Diagnostic Accuracy (%) (n = 196)
Angina during exercise test	53 (35)*	37 (84)*†	90 (46)*
ST segment depression during exercise test	103 (68)*	35 (80)	138 (70)*
Abnormal qualitative thallium image	120 (79)*†	21 (48)*	141 (72)
Abnormal quantitative thallium image	138 (91)*†	29 (67)*	167 (85)*†
Logistic regression model	138 (91)	36 (82)*	175 (89)

*$P<0.01$, significant difference between indicated test and test above.
†$P<0.01$, compared with all tests above.

Source: Reprinted with permission from *Journal of the American Medical Association* (1986;255:508–511), Copyright © 1986, American Medical Association.

Table 12-10 Additive Ability of Clinical, Electrocardiographic, and Thallium Imaging Data To Detect Coronary Artery Disease When One Abnormality Is Required for the Diagnosis

Parameter	Sensitivity (%) (n = 152)	Specificity (%) (n = 44)	Diagnostic Accuracy (%) (n = 196)
1. History of typical angina	97 (64)	29 (66)	126 (64)
2. 1 or angina during exercise test	110 (72)	24 (55)	134 (68)
3. 1 or 2 or ST segment depression during exercise test	132 (87)*	17 (39)*	149 (76)*
4. 1 or 2 or 3 or abnormal qualitative thallium image	146 (96)*	6 (14)*	152 (78)
5. 1 or 2 or 3 or 4 or abnormal quantitative thallium image	150 (99)	4 (9)	154 (79)

*$P<0.01$, compared with previous combinations of parameters.

Source: Reprinted with permission from *Journal of the American Medical Association* (1986;255:508–511), Copyright © 1986, American Medical Association.

myocardial ischemia. These two tests, therefore, may not provide independent information regarding the presence of exercise-induced ischemia, in which case the serial application of Bayes' theorem could miscalculate the probability of coronary artery disease. An approach has been developed that overcomes this limitation by allowing simultaneous calculation of multiple disease states and multiple test outcomes.[102] Other approaches, such as discriminant function analysis, have also been used to determine the presence and extent of disease by combining imaging and other exercise variables.[103]

EXERCISE THALLIUM IMAGING FOR DETERMINATION OF HIGH-RISK PATIENTS AND PROGNOSIS

The value of a test lies not only in its ability to detect the presence of disease but also in its ability to determine prognosis in a patient. This latter ability is of paramount importance in terms of management decisions. There are three clinical scenarios in which exercise ^{201}Tl imaging has been shown to provide important prognostic information.

Ambulatory Patients with Stable Angina

Thallium 201 imaging is comparable to coronary angiography in determining prognosis in ambulatory symptomatic patients with coronary artery disease.[33,34,104,105] Table 12-11 lists the ^{201}Tl imaging variables that have been shown to predict an adverse prognosis in this patient population. Figure 12-6 illustrates a high-risk ^{201}Tl scan. The combination of ^{201}Tl imaging variables with other electrocardiographic and hemodynamic variables (Table 12-11) is superior to any test alone in predicting long-term survival rates.[33,105]

The number of myocardial segments showing redistribution on delayed images has been shown to be an excellent predictor of prognosis in several studies.[33,104-106] The cardiac event rate has been related exponentially to the number of segments showing redistribution[106] (Figure 12-22). Patients without evidence of redistribution on delayed images have much better event-free survival rates than do those who demonstrate redistribution.[33,104-106] In several studies comparing both angiographic and catheterization variables, only the presence of redistribution predicted the occurrence of future nonfatal myocardial infarction.[33,105,106] Coronary anatomy and rest and exercise left ventricular

Table 12-11 Exercise Thallium 201 Variables Found To Yield Important Prognostic Information

A. In Ambulatory Patients with Chest Pain

Imaging Variables	Nonimaging Variables
• Increased lung-heart thallium ratio	• Poor heart rate response to exercise
• Multiple areas showing redistribution	• Ventricular premature beats during exercise
• Multiple severe perfusion defects	
• Left ventricular dilatation	

B. In Post-Myocardial Infarct Patients Who Have Not Received Thrombolytic Therapy

Imaging Variables	Nonimaging Variables
• Multiple initial defects	• Failure to achieve >85% predicted heart rate
• Redistribution	• ST segment depression at <70% predicted heart rate
• Lung-heart thallium ratio	• Decrease in blood pressure during exercise

C. In Post-Myocardial Infarct Patients Who Have Received Thrombolytic Therapy

Imaging Variables
• Mild hypoperfusion
• Redistribution
• Reverse redistribution

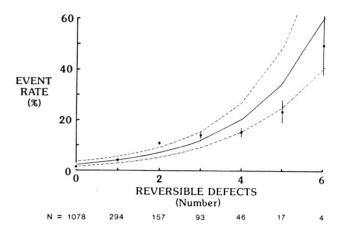

Figure 12-22 Relationship of cardiac event rate (y axis) to number of segments with redistribution on exercise thallium 201 images. An exponential relationship is noted. *Source:* Reprinted with permission from *Journal of the American College of Cardiology* (1986;7:464–471), Copyright © 1986, American College of Cardiology.

ejection fraction were not as useful in this regard. These latter variables are important as predictors of mortality.[104]

The lung-heart ratio of ^{201}Tl has also been shown to be a strong predictor of future events.[33,34,107] In these studies it was a better predictor of events than was the presence of redistribution. In two studies in which angiographic and ^{201}Tl were compared, the heart-lung ratio was a better predictor than the number of diseased vessels.[33,34] Figure 12-23 illustrates the event-free rate in 204 patients with stable symptoms who were followed for a period of 4 to 9 years. The presence of multivessel disease (Figure 12-23A) had a slightly lower predictive power for cardiac events than did the presence of an increased lung-heart ratio of ^{201}Tl (Figure 12-23B). In comparison, the presence of redistribution on delayed ^{201}Tl images (Figure 12-23C) was not as powerful a predictor of future events as either lung-heart ^{201}Tl ratio or coronary anatomy, and the presence of ST segment depression on the electrocardiogram was the least powerful of these four predictors[33] (Figure 12-23D). The superiority of the lung-heart ratio of ^{201}Tl for predicting future events (especially mortality) should not be surprising. Increased lung uptake of ^{201}Tl during exercise indicates the presence of pulmonary edema and thus, in the setting of coronary artery disease, the presence of left ventricular dysfunction. The presence of increased lung uptake of ^{201}Tl, therefore, should provide the same information as a decrease in the left ventricular ejection fraction during exercise. Other ^{201}Tl predictors found to correlate with the occurrence of future events include the number of defects on the initial images, the degree of initial defect (mild or severe),[108,109] and left ventricular dilatation.[32]

Several studies have indicated that patients with chest pain and normal exercise ^{201}Tl images have an excellent prognosis even if underlying coronary artery disease has been demonstrated angiographically. One study reported a 0.1% event rate in 500 patients with normal ^{201}Tl images,[109] while another reported a yearly cardiac mortality rate of 0.5% in 345 patients with chest pain and normal ^{201}Tl images.[110] Similarly, no deaths and two nonfatal infarctions were reported during a 2-year follow-up of 95 patients with normal exercise ^{201}Tl images.[111] Patients with chest pain and insignificant coronary artery disease on angiography who had normal exercise ^{201}Tl images had no events in a 4- to 8-year follow-up in one study as compared with 4 of 24 patients with insignificant coronary artery disease but abnormal ^{201}Tl images who had nonfatal cardiac events.[91]

High-risk exercise ^{201}Tl findings are more prevalent than high-risk exercise electrocardiographic findings in patients with left main and three-vessel disease.[32,112] In one study, high-risk ^{201}Tl images were seen in 38 of 40 patients with three-vessel coronary artery disease. In these same patients, only 15 (38%) had ST segment depression equal to or greater than 2 mm on the electrocardiogram within 6 minutes of exercise.[32] In another study, 70% of patients with 50% or more left main stenosis and 50% of patients with proximal three-

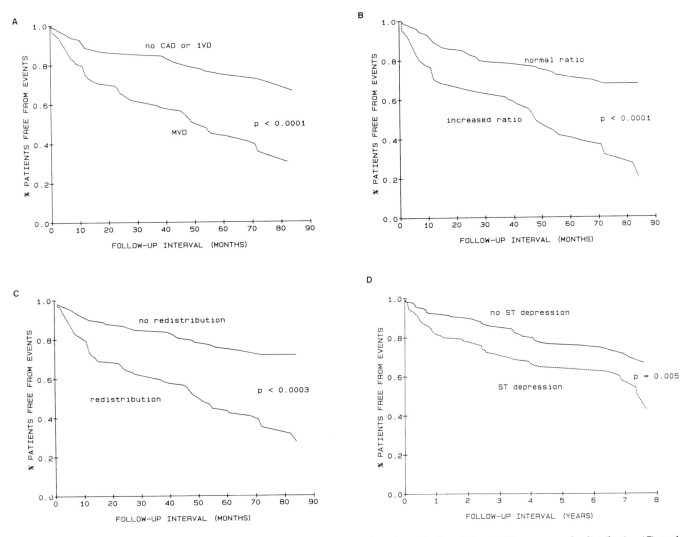

Figure 12-23 The event-free survival rates based upon coronary anatomy (A), lung-heart thallium 201 ratio (B), presence of redistribution (C), and presence of ST segment depression (D). The lung-heart ratio of ^{201}Tl is the best predictor of future events, followed closely by the number of diseased vessels. CAD, coronary artery disease; 1VD, one vessel disease; MVD, multiple vessel disease. *Source:* Reprinted with permission from *Journal of the American College of Cardiology* (1988;12:25–34), Copyright © 1988, American College of Cardiology.

vessel disease had high-risk ^{201}Tl images that were more prevalent than high-risk electrocardiographic findings during exercise stress testing.[112] Thallium 201 variables may be more predictive of future cardiac events than is exercise electrocardiography alone owing to better detection of multivessel ischemia by the former.

Patients with Uncomplicated Infarction Who Do Not Receive Thrombolytic Therapy

Patients at high risk for subsequent cardiac events after an uncomplicated acute myocardial infarction can be identified by using ^{201}Tl imaging in conjunction with low-level treadmill exercise testing.[35,113–115] A prospective study in which such patients were followed for a mean interval of 15 months found that, while 49% of patients with future events had ST segment depression and 60% had angina during exercise, 86% had high-risk ^{201}Tl images[113] (Figure 12-24). Table 12-11 lists the high-risk ^{201}Tl findings. Coronary angiography did not add to the prognostic power of the test. Future nonfatal ischemic events were better predicted by ^{201}Tl imaging than by coronary angiography. In another study, a combination of ^{201}Tl imaging and hemodynamic and electrocardiographic findings (Table 12-11) complemented each other in identifying three-vessel and left main disease in patients who had recently suffered a myocardial infarction.[114] In a more recent study, the lung-heart ratio of ^{201}Tl during predischarge exercise testing was found to relate closely with the extent and severity of coronary artery disease, residual ischemia, and resting left ventricular systolic function.[35]

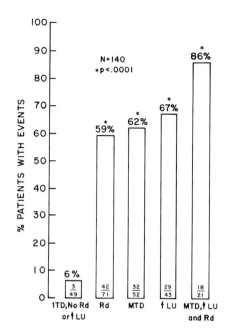

Figure 12-24 The relationship between cardiac event rate and thallium 201 abnormalities during submaximal exercise stress test in patients who were discharged from the hospital after an uncomplicated myocardial infarction. 1TD, one thallium defect; Rd, redistribution; MTD, multiple thallium defects; LU, lung uptake. *Source:* Reprinted with permission from *Circulation* (1983;68:321–336) Copyright © 1983, American Heart Association.

Patients with Uncomplicated Infarction Who Receive Thrombolytic Therapy

In addition to the findings listed above, patients who receive thrombolytic therapy during an infarction can show another ^{201}Tl image finding postinfarction that reflects the presence of viable myocardium.[28,30] This finding is "reverse redistribution" and has been alluded to earlier (Figure 12-5). The exact mechanism of reverse redistribution in such patients is not known. Some have postulated that hyperemic blood flow in the noninfarcted epicardial tissue of the reperfused zone results in increased initial ^{201}Tl uptake in that region. During the interval between the initial and delayed images there is faster washout of ^{201}Tl from this region as compared with that in normal areas, resulting in the appearance of a defect in the delayed images (reverse redistribution).[28] In the canine model, regional hyperemia following coronary reperfusion has been reported to result in accelerated ^{201}Tl clearance[116] and may be related to an initial accumulation of thallium in the interstitial space. Although the exact mechanism of reverse redistribution remains unclear, reverse redistribution is associated with improvement in regional ventricular function.[29]

The incidence of reverse redistribution in patients receiving thrombolytic therapy depends upon the definition of the term. If reverse redistribution is said to occur in a segment showing decreased counts in the delayed image, even if there is a partial defect in the initial image, the incidence can be as high as 75%.[28] However, if reverse redistribution is said to be present when there is no numerically significant defect in the initial image but a significant defect in the delayed images, the incidence is about 25%.[29] Reverse redistribution is usually seen only in the inferoapical and low posterolateral regions. One possible explanation for the reverse redistribution observed predominantly in these regions is that these regions are thin and there is also less tissue overlay. The interventricular septum in the ^{201}Tl images, on the other hand, represents a summation image of the right ventricle and of both the anterior and posterior interventricular septa. Therefore, heterogeneity of blood flow between the epicardium and endocardium is less likely to be detected in this region. It has also been suggested that reverse redistribution could be an artifact of background subtraction.[117] However, it is readily observed in the analog nonbackground-subtracted images.

SPECIFIC SITUATIONS IN WHICH TO USE EXERCISE THALLIUM IMAGING

Ambulatory Patients with Chest Pain

Because ^{201}Tl imaging provides a better sensitivity for the detection of coronary artery disease than does exercise electrocardiography at all levels of workload in symptomatic patients,[84] and because ^{201}Tl imaging is superior to exercise in terms of risk stratification,[33,34,104,106] it can be argued that all patients undergoing exercise stress testing should undergo exercise ^{201}Tl imaging if (1) ^{201}Tl imaging is readily available and (2) the quality of the ^{201}Tl imaging is superior. If the latter criterion is not met, adding ^{201}Tl imaging routinely to exercise may not be useful. In addition, ^{201}Tl imaging is expensive.

There are selected patients in whom ^{201}Tl imaging definitely offers more diagnostic information (Table 12-12). This group includes patients with electrocardiographic abnormalities such as ST segment depression or left bundle-branch block at rest, left ventricular hypertrophy, or Wolff-Parkinson-White syndrome, even in the absence of ST segment abnormalities; those on digitalis therapy, even if their ST segments appear isoelectric at rest; and those who demonstrate ST segment motion on hyperventilation. The incidence of a false-positive exercise test is unacceptably high in women,[118,119] and it may be best to use ^{201}Tl imaging in conjunction with exercise testing in all women. In addition, patients who do not exercise to 85% or more of the maximal predicted heart rate and who have no ST segment changes on exercise are

Table 12-12 Situations in Which Thallium 201 Imaging Is Definitely Superior to Exercise Electrocardiographic Testing

- Nondiagnostic stress test because of inability to reach >85% maximal predicted heart rate
- ST depression on hyperventilation
- Testing of women, who have an unacceptably high incidence of false-positive exercise tests
- Presence of left ventricular hypertrophy, Wolff–Parkinson–White syndrome, or digitalis therapy even if there are no baseline ST segment abnormalities
- Baseline ST segment abnormalities
- Left bundle-branch block
- Differentiation of true-negative from false-negative exercise electrocardiograms in patients with probability of coronary artery disease
- Differentiation of ischemia in anterior wall versus regional dys-synergy in posterior wall in patients with posterior infarction who have exercise-induced ST segment depression in anterior leads

deemed to have a nondiagnostic test. In such patients ^{201}Tl imaging adds important additional information.[71,84]

Patients with a low likelihood of coronary artery disease who have ST depression on exercise may have a false-positive test. In such patients, ^{201}Tl imaging can be used to determine whether the test is truly positive or false positive.[120] This information may be especially useful in patients for whom it is considered important to exclude the presence of any coronary artery disease, such as airline pilots, personnel in the armed forces, or athletes. ST segment depression on the anterior leads during exercise in patients with prior posterior myocardial infarction may be due to either ischemia in the anterior wall or regional dys-synergy of the infarcted posterior myocardium. Exercise ^{201}Tl imaging has been shown to be able to differentiate between these two mechanisms and may be useful in this setting.[121]

After Uncomplicated Myocardial Infarction

On the basis of data presented earlier, important prognostic information can be obtained by using ^{201}Tl imaging in patients who have had an uncomplicated myocardial infarction. This is particularly true in patients with Q-wave infarction who did not receive thrombolytic therapy and in all patients who did receive thrombolytic therapy. Because the 15- to 18-month cardiac event rate in patients with non–Q-wave infarction is twice as high as in patients with Q-wave infarction (67% versus 33%), these patients probably should undergo cardiac catheterization directly. Although ^{201}Tl imaging has been shown to be effective in identifying the high-risk patients within this group,[122] it is perhaps less cost-effective to have them undergo exercise ^{201}Tl imaging when most of them will require angiography on the basis of the results of the ^{201}Tl stress test.

In contrast, because the event rate is lower in patients with Q-wave infarction and in those who have received thrombolytic therapy, it would be more cost-effective to have such patients undergo exercise ^{201}Tl imaging in order to determine those who need revascularization and thus coronary angiography. The presence of multiple ^{201}Tl-imaged defects, redistribution, or increased lung uptake of ^{201}Tl should identify patients with uncomplicated Q-wave myocardial infarction who would potentially benefit from a revascularization procedure. Similarly, those showing redistribution or reverse redistribution on ^{201}Tl imaging after thrombolytic therapy or those who demonstrate normal perfusion or only mild hypoperfusion would likely benefit from a revascularization procedure.

Assessing Myocardial Viability Prior to Revascularization

Patients who show normal or nearly normal wall motion in regions of infarction have viable myocardium and will therefore benefit from a revascularization procedure. However, it is difficult to assess myocardial viability in the presence of reduced wall motion. For example, patients with chronic ischemia ("hibernating myocardium") show regional dysfunction despite having viable myocardium.[123] Similarly, patients who have had successful reperfusion therapy show reduced regional function despite the presence of salvaged myocardium ("stunned myocardium").[49] Exercise ^{201}Tl imaging can be useful in determining the presence of viability in these situations.

It is useful to combine a modality that examines regional function (such as two-dimensional echocardiography) with ^{201}Tl imaging to determine the presence of viability. Abnormal ^{201}Tl patterns in the presence of normal or nearly normal regional function should be ignored, and the myocardium within that region should be considered to be viable.[20] In contrast, when there is regional dysfunction, ^{201}Tl imaging patterns can be useful in differentiating viable tissue from scarred tissue. In the case of prior infarctions (without reperfusion therapy) and chronically ischemic myocardium, the presence of a mild persistent defect or redistribution is a sign of viable myocardium. These patients will show improved regional and global function and their ^{201}Tl abnormalities will reverse after successful revascularization.[22,23] In patients with postischemic dysfunction, the presence of normal perfusion or mild hypoperfusion and the presence of reversed or delayed redistribution indicates viable myocardium.[29]

Assessing Success of Revascularization

If a revascularization procedure has been successful, ^{201}Tl imaging abnormalities should reverse[124-133] (Figure 12-25). The persistence of abnormalities several months after revascularization indicates incomplete or unsuccessful revascularization and has a greater diagnostic value than do defects noted shortly after revascularization; the latter might reverse over time.[134] Thallium 201 imaging has been used after angioplasty to identify patients who are likely to suffer adverse ischemic events.[135] In most of these patients, the abnormal ^{201}Tl findings represent inadequate revascularization. In the case of bypass surgery, the occurrence of a new defect indicates a sizable perioperative infarct. If chest pain recurs after revascularization, a normal perfusion pattern indicates that the chest pain probably is not ischemic in nature. Reversal of ^{201}Tl abnormalities after angioplasty, followed by their recurrence, usually indicates restenosis of the vessel.[136] Thallium 201 abnormalities might precede symptoms by several months.

It is frequently difficult to identify the "culprit" vessel in a patient with multivessel disease during angiography. In this setting, ^{201}Tl imaging can be used to define the culprit vessel and to determine the success of the dilation after the procedure.[137] In patients who show no evidence of redistribution after dilation of the culprit vessel, the need for angioplasty of another lesion is very rare.[137] These data suggest that angioplasty of the culprit vessel detected on exercise ^{201}Tl imaging might be an acceptable alternate approach to multivessel dilation in patients with multivessel disease.

DIPYRIDAMOLE THALLIUM 201 IMAGING

Mechanism of Action

Dipyridamole is a phosphodiesterase inhibitor that causes an increase in cyclic adenosine 3', 5'-monophosphate levels leading to vascular smooth-muscle dilation.[138] The peak vasodilatory effect of dipyridamole is noted 4 to 5 minutes after an intravenous infusion and lasts for as long as 15 to 30 minutes.[138] Dipyridamole-^{201}Tl imaging has been suggested as an alternate approach to exercise imaging,[18,139] especially in patients who are unable to exercise. Being a potent coronary vasodilator, dipyridamole causes a several-fold increase in blood flow through normal coronary arteries without changing the flow to a similar extent in arteries with critical stenosis. This disparity in flow results in a mismatch of ^{201}Tl uptake in different regions of the myocardium. In this regard dipyridamole achieves the same objective as exercise. Reduced reserve becomes manifest during exercise before actual ischemia develops and might be the reason why ^{201}Tl imaging is superior to exercise electrocardiography for the detection of coronary artery disease. If the myocardium supplied by the stenotic vessel is viable, redistribution is noted in the delayed images after dipyridamole infusion.

Method of Administration

Although oral dipyridamole (300 mg) has been shown to be as effective as intravenous dipyridamole

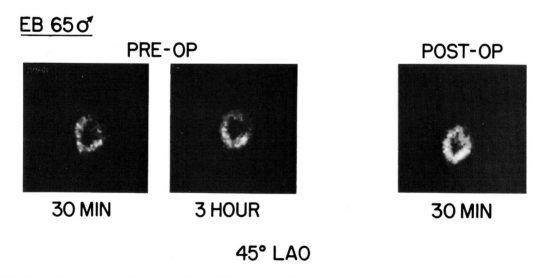

Figure 12-25 Initial and delayed preoperative images in the 45° left anterior oblique (LAO) projection of a patient who had multivessel disease (left and middle panels). After bypass surgery (right panel), there is better perfusion of both the posterolateral wall and the interventricular septum.

(0.56 mg/kg)[140] because of the ease of reversing side effects, the intravenous approach is probably better. The intravenous form of dipyridamole has not yet been approved for this use by the Food and Drug Administration. The patient is placed in a supine position prior to infusion of the drug.[18] The drug is infused over a 4-minute period, after which the table is tilted semi-upright or the patient is asked to sit up.[18] Three minutes later ^{201}Tl is injected intravenously, the patient is placed back in the supine position, and imaging is initiated.

Side Effects

Dipyridamole has caused noncardiac side effects such as nausea, vomiting, headache, and dizziness in about half the patients receiving the drug.[18,141] In addition, chest pain develops in about one fourth of the patients, of whom about one third also have electrocardiographic changes. Half of the patients with electrocardiographic changes (about one fifth of all patients) experience chest pain. Ventricular dysrhythmias develop in about 10% of the patients and are seen almost exclusively in those with multivessel disease.[141]

These symptoms can be reversed with an intravenous infusion of 50 to 200 mg of aminophylline. Fifty milligrams of aminophylline reverses the side effects in about three fourths of all patients.[141] Thallium 201 injection should precede aminophylline administration within about 2 to 5 minutes, so that myocardial uptake of ^{201}Tl is achieved before reversal of the dipyridamole effect.[141] Aminophylline is also a phosphodiesterase inhibitor and a competetive antagonist of dipyridamole. For this reason dipyridamole-^{201}Tl imaging should not be performed in patients who are taking theophylline-related drugs for respiratory or other symptoms. If dipyridamole imaging is to be performed in such patients, these drugs should be discontinued for at least 48 hours prior to the administration of dipyridamole.

Specific Situations in Which Dipyridamole Thallium 201 Imaging Provides Useful Information

Despite sensitivity similar to that of exercise testing for the detection of coronary artery disease,[139] currently dipyridamole-^{201}Tl imaging is indicated primarily for patients who are unable to exercise. This modality has been attempted with success in three particular clinical situations.

In Patients Who Cannot Exercise but Are Suspected of Having Coronary Artery Disease

Patients with physical limitations such as severe peripheral vascular disease, amputation, arthritis, or limiting respiratory disease are candidates for dipyridamole-^{201}Tl imaging. In such patients, the presence of initial defects and redistribution offer a high sensitivity for the detection of coronary artery disease.[142] Because the myocardial clearance of ^{201}Tl is slower after dipyridamole administration than after exercise, it has been found not to be of much clinical importance.[142] In addition, because the background activity (especially hepatic, splenic, and splanchnic) is much higher after dipyridamole administration than after exercise (Figure 12-26), quantitation using background subtraction tends to "oversubtract" and create false-positive results. Therefore, quantitation has been found not to be particularly more useful than expert visual analysis in this situation.

In Patients Referred for Major Vascular Surgery

Because most patients with peripheral vascular disease have coronary artery disease, the risk for intraoperative cardiac events is high if these patients undergo major vascular surgery, especially that requiring aortic cross-clamping.[143,144] The latter procedure increases afterload to the heart, resulting in greater myocardial oxygen demand. In the presence of critical coronary stenoses, this increase in demand could result in myocardial ischemia and infarction.[144] As a result, several institutions routinely perform coronary angiography in these patients; if significant coronary artery disease is present, bypass surgery is performed either prior to or at the time of the vascular procedure.[145] Because these patients cannot exercise on a treadmill,

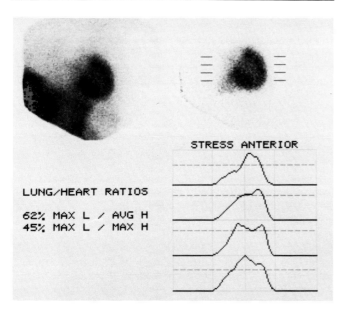

Figure 12-26 Rest dipyridamole scan in a patient with multivessel coronary artery disease. Note the increased splanchnic activity.

dipyridamole-^{201}Tl imaging has been attempted and found to be successful in identifying patients who are at increased risk for a cardiac event either in the operating room or shortly thereafter.[146] The presence of redistribution is particularly useful in this regard, and when combined with other clinical variables can offer a high degree of certainty in identifying high-risk patients.[147] On the basis of these results, dipyridamole-^{201}Tl imaging is now being performed routinely in such patients in several institutions.

In Patients with Acute Myocardial Infarction

Dipyridamole–thallium 201 imaging has been safely performed in patients who have suffered a recent myocardial infarction in order to identify those who are at risk for a subsequent cardiac event after discharge. The presence of redistribution is a strong predictor of events (mortality, unstable angina, and death) in such patients.[148] This method of risk stratification after acute infarction may be particularly suited to patients who are unable to exercise.

REFERENCES

1. Weast RC, ed. *Handbook of Chemistry and Physics.* Boca Raton, Fla: CRC Press; 1986–1987:B38.

2. Lebowitz E, Greene MV, Fairchild R, et al. Thallium-201 for medical use. *J Nucl Med.* 1975;16:151–155.

3. Beller GA. Nuclear cardiology: Current indications and clinical usefulness. *Curr Probl Cardiol.* 1985;10:1–76.

4. Grunwald AM, Watson DD, Holzgrefe HH, et al. Myocardial thallium-201 kinetics in normal and ischemic myocardium. *Circulation.* 1981;64:610–618.

5. Goldhaber SZ, Newell JB, Ingwall JS, et al. Effects of reduced coronary flow on thallium-201 accumulation and release in an in vitro rat heart preparation. *Am J Cardiol.* 1983;51:891–896.

6. Nielsen AT, Morris KG, Murdock R, et al. Linear relationship between the distribution of thallium-201 and blood flow in ischemic and non-ischemic myocardium during exercise. *Circulation.* 1980; 61:797–801.

7. Beller GA, Holzgrefe HH, Watson DD. Effects of dipyridamole-induced vasodilation on myocardial uptake and clearance of thallium-201. *Circulation.* 1983;68:1328–1338.

8. Mays AE, Cobb FR. Relationship between regional myocardial blood flow and thallium-201 distribution in the presence of coronary arterial stenosis and dipyridamole-induced vasodilation. *J Clin Invest.* 1984;73:1359–1366.

9. Pohost GM, Zir LM, Moore RH, et al. Differentiation of transiently ischemic from infarcted myocardium by serial imaging after a single dose of thallium-201. *Circulation.* 1977;55:294–302.

10. Beller GA, Watson DD, Ackell P, et al. Time course of thallium-201 redistribution after transient myocardial ischemia. *Circulation.* 1980;61:791–797.

11. Gerry JL, Becker LC, Flaherty JT, et al. Evidence for a flow-independent contribution to the phenomenon of thallium redistribution. *Am J Cardiol.* 1980;45:58–62.

12. Bergman S, Hack SN, Sobel BE. "Redistribution" of myocardial thallium-201 without reperfusion: Implications regarding absolute quantification of perfusion. *Am J Cardiol.* 1982;49:1691–1698.

13. Okada RD, Jacobs ML, Daggett WM, et al. Thallium-201 kinetics in nonischemic myocardium. *Circulation.* 1982;65:70–77.

14. Nishiyama H, Adolph R, Gabel M, et al. Effect of coronary blood flow on thallium-201 uptake and washout. *Circulation.* 1982; 65:534–542.

15. Pohost GM, Okada RD, O'Keefe DB, et al. Thallium redistribution in dogs with severe coronary artery stenosis of fixed caliber. *Circ Res.* 1981;48:439–446.

16. Leppo J, Rosenkrantz J, Rosenthal R, et al. Quantitative thallium-201 redistribution with a fixed coronary stenosis in dogs. *Circulation.* 1981;63:632–639.

17. Okada RD, Leppo JA, Strauss HW, et al. Mechanisms and time course for the disappearance of thallium-201 defects at rest in dogs: Relation of time to peak activity to myocardial blood flow. *Am J Cardiol.* 1982;49:699–706.

18. Leppo J, Boucher CA, Okada RD, et al. Serial thallium-201 myocardial imaging after dipyridamole infusion: Diagnostic utility in detecting coronary stenoses and relationship to regional wall motion. *Circulation.* 1982;66:649–657.

19. Bodenheimer MM, Banka VS, Fooshee C, et al. Relationship between regional myocardial perfusion and the presence, severity and reversibility of asynergy in patients with coronary heart disease. *Circulation.* 1978;58:789–795.

20. Kiess MC, Fung AY, Thompson CR, et al. Can thallium scan alone determine viability post thrombolytic therapy? *Circulation.* 1988; 78(suppl 2):91. Abstract.

21. Brunken RC, Mody FV, Hawkins RA, et al. Positron tomography detects glucose metabolism in segments with 24 hour tomographic thallium defects. *Circulation.* 1988;78(suppl 2):91. Abstract.

22. Gibson RS, Watson DD, Taylor GJ, et al. Prospective assessment of regional myocardial perfusion after coronary revascularization surgery by quantitative thallium-201 scintigraphy. *J Am Coll Cardiol.* 1983;1:804–815.

23. Liu P, Kiess MC, Okada RD, et al. The persistent defect on exercise thallium imaging and its fate after myocardial revascularization: Does it represent scar or ischemia? *Am Heart J.* 1985;110: 996–1001.

24. Granato JE, Watson DD, Flanagan TL, et al. Myocardial thallium-201 kinetics during coronary occlusion and reperfusion: Influence of method of reflow and timing of thallium-201 administration. *Circulation.* 1986;73:150–160.

25. Granato JE, Watson DD, Flanagan TL, et al. Myocardial thallium-201 kinetics and regional flow alterations with 3 hours of coronary occlusion and either rapid reperfusion through a totally patent vessel or slow reperfusion through a critical stenosis. *J Am Coll Cardiol.* 1987;9:109–118.

26. Melin JA, Wijns W, Keyux A, et al. Assessment of thallium-201 redistribution versus glucose uptake as predictors of viability after coronary occlusion and reperfusion. *Circulation.* 1988; 77:927–934.

27. Hecht HS, Hopkins JM, Rose JG, et al. Reverse redistribution: worsening of thallium-201 myocardial images from exercise to redistribution. *Radiology.* 1981;140:177–181.

28. Weiss AT, Maddahi J, Lew AS, et al. Reverse redistribution of thallium-201: A sign of nontransmural myocardial infarction with patency of the infarct related coronary artery. *J Am Coll Cardiol.* 1986; 7:61–67.

29. Touchstone DA, Beller GA, Nygaard TW, et al. Functional significance of predischarge exercise thallium-201 findings following intravenous streptokinase therapy during acute myocardial infarction. *Am Heart J.* 1988;116:1500–1507.

30. Boucher CA, Zir LM, Beller GA, et al. Increased lung uptake of thallium-201 during exercise myocardial imaging: Clinical, hemo-

dynamic and angiographic implications in patients with coronary artery disease. *Am J Cardiol*. 1980;46:189–196.

31. Gibson RS, Watson DD, Carabello BA, et al. Clinical implications of increased lung uptake of thallium-201 during exercise scintigraphy 2 weeks after myocardial infarction. *Am J Cardiol*. 1982; 49:1586–1593.

32. Canhasi B, Dae M, Botvinick E, et al. Interaction of "supplementary" scintigraphic indicators of ischemia and stress electrocardiography in the diagnosis of multivessel coronary disease. *J Am Coll Cardiol*. 1985;6:581–588.

33. Kaul S, Finkelstein DM, Homma S, et al. Superiority of quantitative exercise thallium-201 variables in determining long-term prognosis in ambulatory patients with chest pain: A comparison with cardiac catheterization. *J Am Coll Cardiol*. 1988;12:25–34.

34. Miller DD, Kaul S, Strauss HW, et al. Increased exercise thallium-201 lung uptake: A noninvasive prognostic index in two-vessel coronary artery disease. *Can J Cardiol*. 1988;4:270–276.

35. Al-Khwaja I, Lahiri A, Rodrigues EA, et al. Clinical significance of exercise-induced pulmonary uptake of thallium 201 in uncomplicated myocardial infarction. *Am J Cardiogr Imag*. 1988; 2:135–141.

36. Homma S, Kaul S, Boucher CA. Correlates of lung/heart ratio of thallium-201 in coronary artery disease. *J Nucl Med*. 1987; 28:1531–1535.

37. Kushner FG, Okada RD, Kirshenbaum HD, et al. Lung thallium-201 uptake after stress testing in patients with coronary artery disease. *Circulation*. 1981;63:341–347.

38. Bingham JB, McKusick KA, Strauss HW, et al. Influence of coronary artery disease on pulmonary uptake of thallium-201. *Am J Cardiol*. 1980;46:821–826.

39. Brown KA, Boucher CA, Okada RD, et al. Quantification of pulmonary thallium-201 activity after upright exercise in normal persons: Importance of peak heart rate and propranolol usage in defining normal values. *Am J Cardiol*. 1984;53:1678–1682.

40. Wackers FJT, Busemann-Sokole E, Samson G, et al. Value and limitations of thallium-201 scintigraphy in the acute phase of myocardial infarction. *N Engl J Med*. 1976;295:1–5.

41. Wackers FJT, Becker AE, Samson G, et al. Location and size of acute transmural myocardial infarction estimated from thallium-201 scintiscans: A clinical pathological study. *Circulation*. 1977;56:72–78.

42. Bulkley BH, Silverman KJ, Weisfeldt ML, et al. Pathologic basis of thallium-201 scintigraphic defects in patients with fatal myocardial injury. *Circulation*. 1979;60:785–792.

43. Silverman KJ, Becker LC, Bulkley BH, et al. Value of early thallium-201 scintigraphy for predicting mortality in patients with acute myocardial infarction. *Circulation*. 1980;61:996–1003.

44. Wackers FJT, Lie KI, Liene KL, et al. Potential value of thallium-201 scintigraphy as a means of selecting patients for the coronary care unit. *Br Heart J*. 1979;41:111–117.

45. Van der Wieken LR, Kan G, Belfer AJ, et al. Thallium-201 scanning to decide CCU admissions in patients with nondiagnostic electrocardiograms. *Int J Cardiol*. 1983;4:285–295.

46. Gascho JA, Lesnefsky AJ, Mahanes MS, et al. Effects of acute left anterior descending artery occlusion on regional myocardial blood flow and wall thickening in the presence of a circumflex stenosis in dogs. *Am J Cardiol*. 1984;54:399–406.

47. Schwartz JS, Cohn JN, Bache RJ. Effects of coronary occlusion on flow in the distribution of a neighboring stenotic coronary artery in the dog. *Am J Cardiol*. 1983;52:89–95.

48. Homans DC, Sublett E, Elsperger KJ, et al. Mechanisms of remote myocardial dysfunction during coronary artery occlusion in the presence of multivessel disease. *Circulation*. 1986;784:588–596.

49. Touchstone DA, Beller GA, Nygaard TW, et al. Effects of successful intravenous reperfusion therapy on regional myocardial function and geometry in man: A tomographic assessment using two-dimensional echocardiography. *J Am Coll Cardiol*.

50. Smucker ML, Beller GA, Watson DD, et al. Left ventricular dysfunction in excess of the size of infarction: A possible management strategy. *Am Heart J*. 1988;115:749–753.

51. Topol EJ, Califf RM, George BS, et al. A randomized trial of immediate versus delayed elective angioplasty after intravenous tissue plasminogen activator in acute myocardial infarction. *N Engl J Med*. 1987;317:581–588.

52. Okada RD, Pohost GM. The use of pre intervention and post intervention thallium imaging for assessing the early and late effects of experimental coronary arterial reperfusion in dogs. *Circulation*. 1986;69:1153–1160.

53. Forman R, Kirk ES. Thallium-201 accumulation during reperfusion of ischemic myocardium: Dependence on regional blood flow rather than viability. *Am J Cardiol*. 1984;54:659–663.

54. DeCoster PM, Melin JA, Detry JR, et al. Coronary artery reperfusion in acute myocardial infarction: Assessment by pre- and post-infarction thallium-201 perfusion imaging. *Am J Cardiol*. 1985; 55:889–895.

55. Maseri A, Parodi O, Severi S, et al. Transient transmural reduction of myocardial blood flow demonstrated by thallium-201 scintigraphy as a cause of variant angina. *Circulation*. 1976;54:280–288.

56. Parodi O, Marzullo P, Neglia D, et al. Transient predominant right ventricular ischemia caused by coronary spasm. *Circulation*. 1984;70:170–177.

57. Berger BC, Watson DD, Burwell LR, et al. Redistribution of thallium at rest in patients with stable and unstable angina and the effect of coronary artery bypass graft surgery. *Circulation*. 1979; 60:1114–1125.

58. Brown KA, Okada RD, Boucher CA, et al. Serial thallium-201 imaging at rest in patients with stable and unstable angina pectoris: Relationship of myocardial perfusion at rest to presenting clinical syndrome. *Am Heart J*. 1983;106:70–77.

59. Smitherman TC, Osborn RC, Narahara KA. Serial myocardial scintigraphy after a single dose of thallium-201 in men after acute myocardial infarction. *Am J Cardiol*. 1978;42:177–182.

60. DiCarlo LA, Botvinick EH, Canhasi BS, et al. Value of noninvasive assessment of patients with atypical chest pain in suspected coronary spasm using ergonovine infusion with thallium-201 scintigraphy. *Am J Cardiol*. 1984;54:744–748.

61. Bailey IK, Griffith LSC, Rouleau J, et al. Thallium-201 myocardial perfusion imaging at rest and during exercise: Comparative sensitivity to electrocardiography in coronary artery disease. *Circulation*. 1977;55:79–87.

62. Ritchie JL, Trobaugh GB, Hamilton GW, et al. Myocardial imaging with thallium-201 at rest and during exercise: Comparison with coronary arteriography and resting and stress electrocardiography. *Circulation*. 1977;56:66–71.

63. Verani MS, Marcus M, Razzak MA, et al. Sensitivity and specificity of thallium-201 perfusion scintigrams during exercise for the diagnosis of coronary artery disease. *J Nucl Med*. 1978;19:773–782.

64. McCarthy DM, Blood DK, Sciacca RR. Single dose myocardial perfusion imaging with thallium-201: Application in patients with nondiagnostic electrocardiographic stress test. *Am J Cardiol*. 1979; 43:899–906.

65. Pohost CM, Boucher CA, Zir LM, et al. The thallium stress test: The qualitative approach revisited. *Circulation*. 1979;59(suppl 2): 49. Abstract.

66. Vogel RA, Kirch DL, LeFree MT, et al. Thallium-201 myocardial perfusion scintigraphy: Results of standard and multipinhole tomographic technique. *Am J Cardiol*. 1979;43:787–793.

67. Borer JS, Bacharach SL, Green MV. Sensitivity of stress radionuclide angiography and stress thallium perfusion scannings in detecting coronary artery disease. *Am J Cardiol*. 1979;43:431. Abstract.

68. Trobaugh GB, Wackers FJT, Sokole ED, et al. Thallium-201 myocardial imaging: An inter-institutional study of observer variability. *J Nucl Med*. 1978;19:359–363.

69. Okada RD, Boucher CA, Kirschenbaum HK, et al. Improved diagnostic accuracy of thallium-201 stress test using multiple observers and criteria derived from interobserver analysis of variance. *Am J Cardiol*. 1980;46:619–624.

70. Burow RD, Pond M, Schefer AW, et al. "Circumferential profiles": A new method for computer analysis of thallium-201 myocardial perfusion images. *J Nucl Med*. 1979;20:771–777.

71. Berger BC, Watson DD, Taylor GJ, et al. Quantitative thallium-201 exercise scintigraphy for detection of coronary disease. *J Nucl Med*. 1981;22:585–593.

72. Maddahi J, Garcia EV, Berman DS, et al. Improved noninvasive assessment of coronary artery disease by quantitative analysis of regional stress myocardial distribution and washout of thallium-201. *Circulation*. 1981;64:924–935.

73. Massie BM, Wisneski JA, Hollenberg M, et al. Quantitative analysis of seven pinhole tomographic thallium-201 scintigrams: Improved sensitivity for estimation of the extent of coronary involvement by evaluation of radiotracer uptake and clearance. *J Am Coll Cardiol*. 1984;3:1178–1186.

74. Kaul S, Boucher CA, Newell JB, et al. Determination of quantitative thallium parameters that optimize detection of coronary artery disease. *J Am Coll Cardiol*. 1986;7:527–537.

75. Goris ML, Daspit SG, McLaughlin P, et al. Interpolative background subtraction. *J Nucl Med*. 1976;17:744–747.

76. Kaul S, Chesler DA, Okada RD, et al. Computer versus visual analysis of exercise thallium-201 images: A critical appraisal in 325 patients with chest pain. *Am Heart J*. 1987;114:1129–1137.

77. Kaul S, Chesler DA, Boucher CA, et al. Quantitative aspects of myocardial perfusion imaging. *Semin Nucl Med*. 1987;17:131–144.

78. Kaul S, Keiss MC, Liu P, et al. Comparison of exercise electrocardiography and quantitative thallium imaging for one-vessel coronary artery disease. *Am J Cardiol*. 1985;56:257–261.

79. Rigo P, Bailey IK, Griffith LSC, et al. Stress thallium-201 myocardial scintigraphy for the detection of individual coronary artery lesions in patients with and without previous myocardial infarction. *Am J Cardiol*. 1981;48:209–216.

80. Kaul S, Newell JB, Chesler DA, et al. Value of computer analysis of exercise thallium images in the noninvasive detection of coronary artery disease. *JAMA*. 1986;255:508–511.

81. Massie BM, Botvinick EH, Brundage BH. Correlation of thallium-201 scintigrams with coronary anatomy: Factors affecting region by region sensitivity. *Am J Cardiol*. 1979;49:616–622.

82. Rigo P, Bailey IK, Griffith LSC. Value and limitations of segmental analysis of stress thallium myocardial imaging for localization of coronary artery disease. *Circulation*. 1980;61:973–981.

83. McLaughlin PR, Martin RP, Doherty P, et al. Reproducibility of thallium-201 myocardial imaging. *Circulation*. 1977;55:497–503.

84. Esquivel L, Pollock SG, Beller GA, et al. Effect of the degree of effort on the sensitivity of the exercise thallium-201 stress test in symptomatic coronary artery disease. *Am J Cardiol*. 1989;63:160–165.

85. Esquivel L, Kaul S, Watson DD, et al. The effect of excluding isolated basal segmental defects and washout abnormalities on thallium imaging on long-term prognosis. *Circulation*. 1988;78(suppl 2):90. Abstract.

86. Vellinga T, Akhtar R, Krubsack AJ, et al. Does hypertension influence thallium-201 imaging in detecting coronary artery disease? *Circulation*. 1988;78(suppl 2):190. Abstract.

87. Dunn RF, Uren RF, Sadick N, et al. Comparison of thallium-201 scanning in idiopathic dilated cardiomyopathy and severe coronary artery disease. *Circulation*. 1982;66:804–810.

88. Makler PT, Lavine SJ, Denenberg BS, et al. Redistribution on the thallium scan in myocardial sarcoidosis. *J Nucl Med*. 1981;22:428–432. Concise Communication.

89. McDonnel TJ, Becker LC, Bulkley BH. Thallium imaging in cardiac lymphoma. *Am Heart J*. 1981;101:809–814.

90. Manno BV, Burka ER, Hakki AH, et al. Biventricular function in sickle cell anemia: Radionuclide angiographic and thallium-201 scintigraphic evaluation. *Am J Cardiol*. 1983;52:585–587.

91. Kaul S, Okada RD, Pohost GM, et al. How valid is it to classify coronary artery disease based on the number of diseased vessels? *Circulation*. 1986;74(suppl 2):473. Abstract.

92. Kaul S, Newell JB, Chesler DA, et al. Quantitative thallium imaging findings in patients with normal coronary angiographic findings and clinically normal subjects. *Am J Cardiol*. 1986;57:509–572.

93. Massie BM, Wisneski J, Kramer B, et al. Comparison of myocardial thallium-201 clearance after maximal and submaximal exercise: Implications for diagnosis of coronary artery disease. *J Nucl Med*. 1982;23:381–385. Concise Communication.

94. Kaul S, Chesler DA, Pohost GM, et al. Influence of peak heart rate on normal thallium-201 myocardial clearance. *J Nucl Med*. 1986;27:26–30.

95. Kaul S, Chesler DA, Newell JB, et al. Regional variability in myocardial clearance of thallium-201 and its importance in determining the presence or absence of coronary artery disease. *J Am Coll Cardiol*. 1986;8:95–99.

96. Weintraub WS, Madeira SW, Bodenheimer MM, et al. Critical analysis of the application of Bayes' theorem to sequential testing in the noninvasive diagnosis of coronary artery disease. *Am J Cardiol*. 1984;54:43–49.

97. Patterson RE, Eng C, Horowitz SF. Practical diagnosis of coronary artery disease: A Bayes' theorem normogram to correlate clinical data with noninvasive exercise tests. *Am J Cardiol*. 1984;53:252–256.

98. Diamond GA, Staniloff HM, Forrester JS, et al. Computer-assisted diagnosis in the noninvasive evaluation of patients with suspected coronary artery disease. *J Am Coll Cardiol*. 1983;1:444–455.

99. Epstein SE. Implications of probability analysis on the strategy used for the noninvasive detection of coronary artery disease: Role of single or combined use of exercise electrocardiographic testing, radionuclide angiography, and myocardial perfusion imaging. *Am J Cardiol*. 1980;46:491–499.

100. Melin JA, Piret LJ, Vanbutsele RJM, et al. Diagnostic value of exercise electrocardiography and thallium myocardial scintigraphy in patients without previous myocardial infarction: A bayesian approach. *Circulation*. 1981;63:1019–1024.

101. Rembold CM, Watson DD. Post test probability calculated by weights: A simple form of Bayes theorem. *Ann Intern Med*. 1988;108:115–120.

102. Pollock S, Watson DD, Finkelstein DM, et al. A simple approach to evaluating multiple test outcomes and multiple disease states in relation to the exercise thallium-201 stress test. *Am J Cardiol*. 1989;64:466–470.

103. McCarthy DM, Sciacca RR, Blood DK, et al. Discriminant function analysis using thallium-201 scintiscans and exercise stress test variables to predict the presence and extent of coronary artery disease. *Am J Cardiol*. 1982;49:1917–1926.

104. Brown KA, Boucher CA, Okada RD, et al. Prognostic value of exercise thallium-201 imaging in patients presenting for evaluation of chest pain. *J Am Coll Cardiol*. 1984;1:994–1001.

105. Kaul S, Lilly DR, Gascho GA, et al. Prognostic utility of the exercise thallium-201 test in ambulatory patients with chest pain: comparison with cardiac catheterization. *Circulation*. 1988;77:745–758.

106. Ladenheim MC, Pollock BH, Rozanski A, et al. Extent and severity of myocardial hypoperfusion as predictors of prognosis in patients with suspected coronary artery disease. *J Am Coll Cardiol.* 1986;7:464–471.

107. Gill JB, Ruddy TD, Newell JB, et al. Prognostic importance of thallium uptake by the lungs during exercise in coronary artery disease. *N Engl J Med.* 1987;317:1485–1489.

108. Iskandrian AS, Hakki AH, Kane-Marsch S. Prognostic implications of exercise thallium-201 scintigraphy in patients with suspected or known coronary artery disease. *Am Heart J.* 1985;110:135–143.

109. Staniloff HM, Forrester JS, Berman DS, et al. Prediction of death, myocardial infarction, and worsening chest pain using thallium scintigraphy and exercise electrocardiography. *J Nucl Med.* 1986;27:1842–1848.

110. Pamelia FX, Gibson RS, Watson DD, et al. Prognosis with chest pain and normal thallium-201 exercise scintigrams. *Am J Cardiol.* 1985;55:920–926.

111. Wackers FJ, Russo DS, Russo D, et al. Prognostic significance of normal qualitative planar thallium-201 stress scintigraphy in patients with chest pain. *J Am Coll Cardiol.* 1985;6:27–32.

112. Nygaard TW, Gibson RS, Ryan JM, et al. Prevalence of high-risk thallium-201 scintigraphic findings in left main coronary artery stenosis: Comparison with patients with multi- and single-vessel coronary artery disease. *Am J Cardiol.* 1984;53:462–469.

113. Gibson RS, Watson DD, Craddock GB, et al. Prediction of cardiac events after uncomplicated myocardial infarction: A prospective study comparing predischarge exercise thallium-201 scintigraphy and coronary arteriography. *Circulation.* 1983;68:321–336.

114. Patterson RE, Horowitz SF, Eng C, et al. Can noninvasive exercise test criteria identify patients with left main or three-vessel coronary disease after a first myocardial infarction? *Am J Cardiol.* 1983;51:361–371.

115. Hung J, Goris ML, Nash E, et al. Comparative value of maximal treadmill testing, exercise thallium myocardial perfusion scintigraphy, and exercise radionuclide ventriculography for distinguishing high- or low-risk patients soon after acute myocardial infarction. *Am J Cardiol.* 1984;53:1221–1227.

116. Okada RD. Kinetics of thallium-201 reperfused canine myocardium after coronary artery occlusion. *J Am Coll Cardiol.* 1984;3:1245–1251.

117. Brown KA, Benoit L, Clements JP, et al. Fast washout of thallium-201 from area of myocardial infarction: Possible artifact of background subtraction. *J Nucl Med.* 1987;28:945–949.

118. Linhart JW, Laws JC, Satinsky JD. Maximum treadmill exercise electrocardiography in female patients. *Circulation.* 1974;50:1173–1178.

119. Sketch MR, Mohiuddin MM, Lynch TD, et al. Significant sex differences in the correlation of electrocardiographic exercise testing and coronary angiograms. *Am J Cardiol.* 1975;36:139–173.

120. Guiney TE, Pohost GM, McKusick KA, et al. Differentiation of false- from true-positive ECG responses to exercise stress by thallium-201 perfusion imaging. *Chest.* 1981;80:4–10.

121. Gibson RS, Beller GA, Kaiser DL. Prevalence and clinical significance of painless ST-segment depression during early postinfarction exercise testing. *Circulation.* 1987;75(suppl 2):II-36–39.

122. Gibson RS, Beller GA, Gheorghiade M, et al. The prevalence and clinical significance of residual myocardial ischemia two weeks after uncomplicated non-Q infarction: A prospective natural history study. *Circulation.* 1986;73:1187–1197.

123. Rahimtoola SH. A perspective on the three large multicenter randomized clinical trials of coronary bypass surgery for chronic stable angina. *Circulation.* 1985;72(suppl 5):123–135.

124. Hirzel HO, Nuesch K, Siale RG, et al. Thallium-201 exercise myocardial imaging to evaluate myocardial perfusion after coronary bypass surgery. *Br Heart J.* 1980;43:426–435.

125. Ritchie JL, Narahara KA, Trobaugh JB, et al. Thallium-201 myocardial imaging before and after coronary revascularization: Assessment of regional myocardial blood flow and graft patency. *Circulation.* 1977;56:830–836.

126. Verani MS, Marcus ML, Spoto G, et al. Thallium-201 myocardial perfusion scintigrams in the evaluation of aorto-coronary saphenous bypass surgery. *J Nucl Med.* 1978;19:765–772.

127. Greenberg BH, Hart R, Botvinick EH, et al. Thallium-201 myocardial perfusion scintigraphy to evaluate patients after coronary bypass surgery. *Am J Cardiol.* 1978;42:167–176.

128. Sbarbaro JA, Karunaratne H, Cantez S, et al. Thallium-201 imaging and assessment of aorto-coronary artery bypass graft patency. *Br Heart J.* 1979;42:553–561.

129. Robinson TS, Williams BT, Webb-Peploe MM, et al. Thallium-201 myocardial imaging and assessment of results of aorto-coronary bypass surgery. *Br Heart J.* 1979;42:455–462.

130. Hirzel HO, Nuesch K, Gruentzig AR, et al. Short- and long-term changes in myocardial perfusion after percutaneous transluminal coronary angioplasty assessed by thallium-201 exercise scintigraphy. *Circulation.* 1981;63:1001–1007.

131. Scholl JM, Chaitman BR, David PR, et al. Exercise electrocardiography and myocardial scintigraphy in the serial evaluation of the results of percutaneous transluminal coronary angioplasty. *Circulation.* 1982;66:380–390.

132. Wijns W, Serruys PW, Simoons ML, et al. Predictive value of early maximal exercise test and thallium scintigraphy after successful percutaneous transluminal coronary angioplasty. *Br Heart J.* 1985;53:194–200.

133. Verani MS, Tadros S, Raizner AE. Quantitative analysis of thallium-201 uptake and washout before and after transluminal coronary angioplasty. *Int J Cardiol.* 1986;13:109–124.

134. Mayanari DE, Knudtson M, Kloiber R, et al. Segmental thallium-201 myocardial perfusion studies after successful percutaneous transluminal coronary artery angioplasty: Delayed resolution of exercise-induced scintigraphic abnormalities. *Circulation.* 1988;77:86–95.

135. Miller DD, Liu P, Strauss HW, et al. Prognostic value of computer-quantitated exercise thallium imaging early after percutaneous transluminal coronary angioplasty. *J Am Coll Cardiol.* 1987;10:275–283.

136. Stuckey TD, Burwell LR, Nygaard TW, et al. Value of quantitative exercise thallium-201 scintigraphy for predicting angina recurrence after percutaneous transluminal coronary angioplasty. *Am J Cardiol.* March 1989;63(suppl 9):517–521.

137. Breisblatt WM, Barnes JV, Weiland F, et al. Incomplete revascularization in multivessel percutaneous transluminal coronary angioplasty: The role of stress thallium-201 imaging. *J Am Coll Cardiol.* 1988;11:1183–1190.

138. West JW, Bellet S, Manzoli VC, et al. Effects of Persantine (RA8), a new coronary vasodilator, on coronary blood flow and cardiac dynamics in the dog. *Circ Res.* 1962;10:35–44.

139. Josephson MA, Brown BG, Hecht HS, et al. Noninvasive detection and localization of coronary stenoses in patients: Comparison of resting dipyridamole and exercise thallium-201 myocardial perfusion imaging. *Am Heart J.* 1982;103:1008–1018.

140. Homma S, Callahan RJ, Amaer B, et al. Usefulness of oral dipyridamole suspension for stress thallium imaging without exercise in the detection of coronary artery disease. *Am J Cardiol.* 1986;57:503–508.

141. Homma S, Gilliland Y, Guiney T, et al. Safety of intravenous dipyridamole for stress testing with thallium imaging. *Am J Cardiol.* 1987;59:152–154.

142. Ruddy TD, Dighero HR, Newell JB, et al. Quantitative analysis of dipyridamole-thallium images for the detection of coronary artery disease. *J Am Coll Cardiol.* 1987;10:142–149.

143. Thompson JE, Garrett WV. Peripheral-arterial surgery. *N Engl J Med.* 1980;302:491–503.

144. Attia RR, Murphy JD, Snider M, et al. Myocardial ischemia due to infrarenal aortic cross-clamping during aortic surgery in patients with coronary artery disease. *Circulation.* 1976;53:961–965.

145. DeBakey ME, Lawrie GM. Combined coronary artery and peripheral vascular disease: recognition and treatment. *J Vasc Surg.* 1984;1:605–607.

146. Boucher CA, Brewster DG, Darling RC, et al. Determination of cardiac risk of dipyridamole-thallium imaging before peripheral vascular surgery. *N Engl J Med.* 1985;312:389–394.

147. Eagle K, Coley CM, Newell JB, et al. Combining clinical and thallium data optimizes preoperative assessment of cardiac risk before major vascular surgery. *Ann Intern Med.* 1989;110:859–866.

148. Leppo JA, O'Brien J, Rothendler JA, et al. Dipyridamole-thallium-201 scintigraphy in the prediction of future cardiac event after acute myocardial infarction. *N Engl J Med.* 1984;310:1014–1018.

Appendix 12-A

Computer Algorithm Used for Analysis of Planar Thallium Images

The program described here was developed at the Massachusetts General Hospital, using a minicomputer (VAX 11/780; Digital Equipment Corporation, Maynard, MA) interfaced to a frame buffer[1] (DeAnza Systems, Sunnyvale, CA) and employs the following steps.

PLACEMENT OF REGION OF INTEREST

Placement of the region of interest is the only step requiring observer interaction (Figure 12-A1). An elliptical region of interest is placed around the left ventricle such that it lies one pixel outside the myocardial activity. The myocardial thickness is then defined, and a second (smaller) ellipse, concentric with the first ellipse and outlining the endocardial surface of the left ventricle, is formed automatically. The operator then defines the valve plane and outflow tract of the left ventricle. An elliptical region of interest closely approximates the left ventricular shape and defines the long axis of the left ventricle such that segmental activity of thallium 201 in one heart corresponds to activity in the same part of another heart, irrespective of its position within the thorax.[2]

REGISTRATION OF SERIAL IMAGES

The images are first smoothed to attenuate Poisson noise. A search for the shift and rotation of the delayed image is then performed to maximize the correlation of the activities between the points within the initial image and the corresponding points in the delayed image. The correlation (C) is defined by the following equation:

$$C = \frac{\Sigma (A_n)(B_n)}{\sqrt{(\Sigma A_n^2)(\Sigma B_n^2)}}$$

where A_n and B_n are the counts in corresponding positions within the ellipse in the initial and delayed images, respectively. The delayed image is shifted and rotated to allow maximal correlation with the initial image.

BACKGROUND SUBTRACTION

A background ellipse (Figure 12-A2) is formed automatically around the left ventricle, lying two pixels

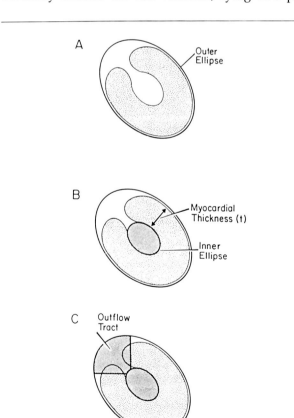

Figure 12-A1 Method of placement of region of interest on a thallium 201 image (see text for details). *Source:* Reprinted with permission from *Journal of the American College of Cardiology* (1986;7:527–537), Copyright © 1986, American College of Cardiology.

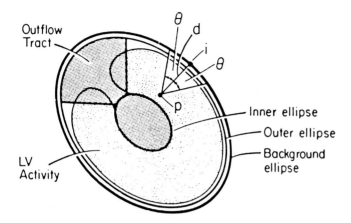

Figure 12-A2 Method of background subtraction (see text for details). LV, left ventricle. *Source:* Reprinted with permission from *Journal of the American College of Cardiology* (1986;7:527–537), Copyright © 1986, American College of Cardiology.

outside the outer ellipse placed by the observer. All points on the perimeter of this ellipse are considered as background. For each pixel (p) within the myocardial region of interest, the background (B) is estimated from the weighted sum of the activity values of all points (A_i) on the perimeter of the background ellipse, using the following equation:

$$B_p = \frac{\Sigma (A_i)(W_{ip})}{\Sigma (W_{ip})}$$

where W_{ip} (weighting constant) is related to the distance (d_{ip}) between the myocardial pixel (p) and background pixel (i), and the angle (θ_{ip}) between p and two perimeter points on either side of i. The weighting constant is derived using the following equation:

$$W_{ip} = (\theta_{ip})(1 - d_{ip}/L_{max})$$

where L_{max} is the length of the long axis of the background ellipse. In this manner, the weighting factor falls off linearly with distance (thereby giving less weight to distant background activity) and with the angle subtended by adjacent background pixels (to compensate for the fact that there are more distant pixels than nearby pixels in the background). Estimated background is then subtracted from the myocardial activity.

SEGMENTATION OF THE LEFT VENTRICULAR MYOCARDIUM AND DETERMINATION OF REGIONAL MYOCARDIAL ACTIVITY

One hundred twenty-eight equally spaced points are defined along the outer ellipse. A tangent is defined for each of these points. The computer then searches inward from each point at right angles to its tangent until it reaches the inner ellipse. The average of three consecutive highest-count pixels along this linear search is defined as myocardial activity within that region. The average highest-count pixel region of the myocardium in each view is set to 100, and all other activity is expressed as a percentage of this activity. The myocardium (excluding the outflow tract) is divided into five equal segments, and activity is averaged within each segment to reflect segmental ^{201}Tl activity (see Figure 12-13 in text).

DETERMINATION OF REDISTRIBUTION

Redistribution is considered to be present when using either of two approaches (Figure 12-A3). In the first approach (normalization algorithm), the segment with maximal activity in the initial image is identified (Figure 12-A3A). The activity in the same segment in the delayed image is then increased to match the activity in the initial image (Figure 12-A3B). Activity in all other segments is also increased by the same factor. With this approach there should be no significant difference in the normalized delayed and initial activity in any segment that is normal (Figure 12-A3B). Similarly, no significant difference is noted between the normalized delayed and initial activity in any segment with decreased initial uptake that does not demonstrate redistribution (Figure 12-A3C). In contrast, in any segment demonstrating redistribution, the normalized delayed activity is higher than the initial activity (Figure 12-A3D). Although this method measures even the smallest degree of redistribution, it assumes that the segment with the highest count rate in the initial image is normal. Because this may not necessarily be true, a second approach (comparison algorithm), which does not make such an assumption, is also used. In this approach, redistribution is said to be present in the delayed image if a segment that shows decreased uptake in the initial image has a "normal" ratio to the segment with the maximal counts in the delayed image. This "normal" ratio is defined for each segment by using data from normal asymptomatic subjects. Redistribution is considered present only if uptake in the initial image is reduced.

DETERMINATION OF REGIONAL MYOCARDIAL THALLIUM 201 CLEARANCE

A monoexponential curve is fitted between the average segmental activity in the initial and delayed images. In this manner, as long as the time is known between

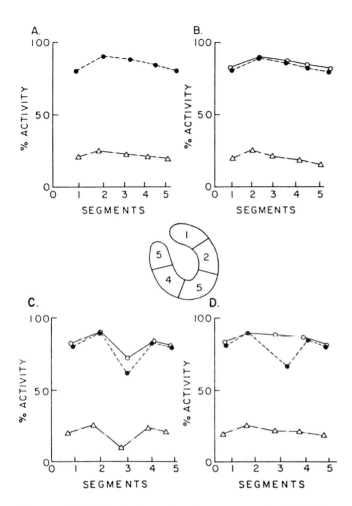

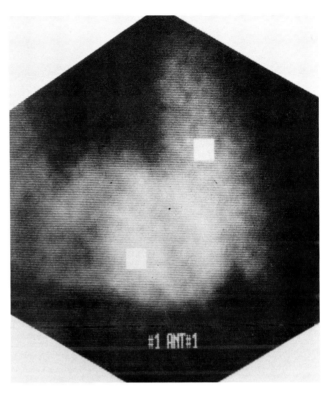

Figure 12-A3 Method of assessing the presence of redistribution using the program developed at the Massachusetts General Hospital. A depicts activity in five segments in the initial and delayed images; B, the effect of normalization of delayed activity in a patient with normal thallium 201 activity; C, the effect of normalization of delayed activity in a patient with an infarct; and D, the effect of normalization in a patient with redistribution (see text for details). *Source:* Reprinted with permission from *Seminars in Nuclear Medicine* (1987;17:131), Copyright © 1987, Grune & Stratton Inc.

the first and second images, the half-life ($t_{1/2}$) of thallium can be determined using the following equation:

$$t_{1/2} = \frac{(t_2 - t_1)\ln 2}{\ln (A t_1)/(A t_2)}$$

where A_{t1} and A_{t2} are measured segmental ^{201}Tl activity at times t_1 and t_2. The advantage of this approach over that of determining percentage washout is that it is not necessary to have the same interval between initial and delayed images in every patient.

DETERMINATION OF THE LUNG-HEART RATIO OF THALLIUM

The operator identifies an area over the lung in the initial image in the anterior view and measures activity per pixel in that area (Figure 12-A4). An area is then chosen over the myocardial segment showing the maximal counts in the same image. Lung activity is then expressed as a percentage of the myocardial activity.

Figure 12-A4 Method of determining the lung-heart ratio of thallium 201 (see text for details). *Source:* Reprinted with permission from *Seminars in Nuclear Medicine* (1987;17:131), Copyright © 1987, Grune & Stratton Inc.

REFERENCES

1. Kaul S, Boucher CA, Newell JB, et al. Determination of quantitative thallium parameters that optimize detection of coronary artery disease. *J Am Coll Cardiol.* 1986;7:527–537.

2. Horowitz SF, Machac J, Levin H, et al. Effect of variable left ventricular vertical orientation on planar myocardial perfusion images. *J Nucl Med.* 1986;27:694–700.

Chapter 13

Evaluation of Left Ventricular Function in Coronary Artery Disease: Radioisotope Imaging at Rest and During Stress

Richard J. Peterson, MD, and Robert H. Jones, MD

INTRODUCTION

As early as 1926 Blumgart and colleagues[1-3] utilized a radioactive tracer injected into an arm vein to measure the velocity of venous return to the heart (Figure 13-1). Today, first-pass radionuclide angiocardiography and multigated equilibrium acquisition play a prominent role in the management of patients with coronary artery disease. Unlike thallium scanning techniques, which assess myocardial perfusion (Chapter 12), first-pass and multigated techniques assess parameters of cardiac function. The techniques have been used in three general areas: (1) *diagnosis* of coronary artery disease; (2) *assessment of function* relative to the impact of a given therapy; and, most recently, (3) *determination of prognosis*.

The accepted "gold standard" in the evaluation of coronary artery disease has been selective arteriography, which defines anatomic irregularities within coronary arteries. It has become increasingly apparent that determination of the anatomic severity of a coronary artery lesion does not fully define its clinical relevance. Important additional prognostic information can be derived from radioisotope studies designed to evaluate cardiac function, particularly when studies are performed during stress conditions such as exercise.

PERSPECTIVE

The concept that abnormalities in myocardial perfusion may alter cardiac function has evolved over the past century. During basal conditions, myocardial energy stores are limited and oxygen extraction from blood is nearly maximal. Roy,[4] in 1879, observed that an abrupt change in the coronary perfusate of an excised heart preparation from blood to saline caused rapid diminution in contractile force. In 1935, Tennant and Wiggers[5] further demonstrated that left ventricular contraction was altered within a few beats of ligation of a coronary artery. Subsequent animal experiments have supported the concept of a dynamic progression of functional abnormalities that begin within seconds after coronary artery occlusion.[6] In 1967 Herman et al.[7] reported on a series of patients studied during and after periods of spontaneous anginal pain and first demonstrated reversible left ventricular dysfunction as an indicator of myocardial ischemia in humans.

Measurements of myocardial function at rest, particularly the left ventricular ejection fraction, provide an index of the extent of myocardial fibrosis due to previous myocardial infarction. Although the left ventricular ejection fraction during rest carries prognostic information, recognition of myocardial ischemia prior to permanent damage provides an optimal approach to management. In patients with stable coronary artery disease, myocardial ischemia rarely occurs at rest, as metabolic demands are low. However, blood flow to regions distal to narrowed coronary vessels may not increase adequately to meet metabolic demands during exercise. Patients with myocardial ischemia have decreased ejection fractions during the stress of exercise, primarily as a result of an increase in end-diastolic

212 COMPARATIVE CARDIAC IMAGING

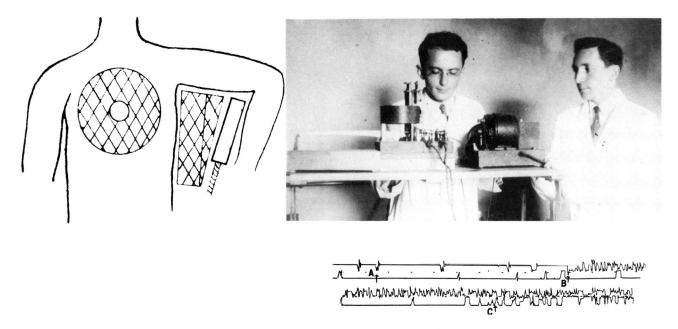

Figure 13-1 Herrmann L. Blumgart and Soma Weiss, pictured in their laboratory. Diagrams of their early work depict the relationship of a patient to a detector head and transit of isotope in the pulmonary circulation. *Source:* Reproduced from *The Journal of Clinical Investigation*, 1927, 4, pp. 399–425, by copyright permission of the American Society for Clinical Investigation. Laboratory photo courtesy of Ruth Freiman, Archivist, Beth Israel Hospital.

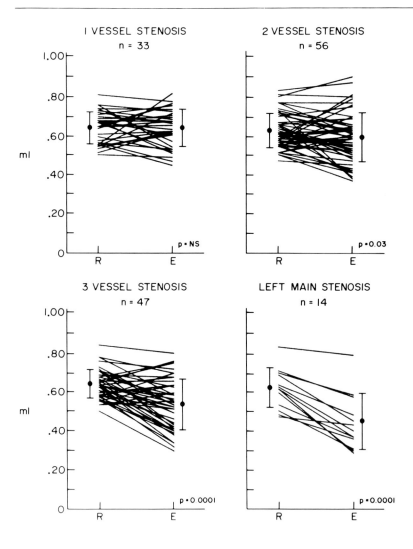

Figure 13-2 Left ventricular ejection fractions at rest (R) and exercise (E) from 150 men with chest pain and coronary artery disease as assessed by arteriography. *Source:* Reprinted with permission from *Circulation* (1981;64:592), Copyright © 1981, American Heart Association.

volume with a constant or slightly increased stroke volume.

A relationship can be defined between the anatomic severity of disease and the magnitude of functional alterations. In patient groups with one-, two-, or three-vessel disease, a progressive decrease in the left ventricular ejection fraction during exercise occurs as the anatomic severity of disease increases[8] (Figure 13-2). However, a patient with a single-vessel left anterior descending coronary artery stenosis may show greater left ventricular dysfunction than other patients with three-vessel stenosis. Peterson et al.[9] evaluated the relationship of the severity of coronary artery stenosis and the associated pressure gradient to the magnitude of exercise-induced left ventricular dysfunction in patients with isolated proximal left anterior descending coronary artery lesions. For lesions whose stenosis measurements were extreme (either minimal or markedly severe), the measurement of stenotic severity generally predicted the patient's functional response during exercise. The greatest individual variation was in the patient group with stenosis of intermediate severity (Figure 13-3). Dynamic factors not accounted for by stenosis measurements alone may play a role in determining why patients with lesions of similar stenotic severity and location have different functional responses to exercise.

TECHNIQUE

First-Pass Radionuclide Angiocardiography

Radionuclide angiocardiograms are performed by utilizing a collimated high-count-rate multicrystal gamma camera placed over the precordium in either an anterior or right anterior oblique position. Images most frequently are obtained while the subject is in the upright position at rest and again during peak exercise on a bicycle ergometer. Technetium 99m pertechnetate (10 to 15 mCi) is injected rapidly via a standard Teflon intravenous catheter inserted into either an external jugular vein or an antecubital vein and flushed with a bolus of 20 mL of normal saline. Transit through the cardiac and pulmonary circulation is then imaged at 25-millisecond intervals for 1 minute. Background correction is made just prior to the exercise acquisition, which is performed when the patient reaches 85% of maximal predicted heart rate. These techniques have been described in detail elsewhere.[10–12] The total of 30 mCi introduced for both studies produces a whole-body radiation dose of 0.36 rad, which is a small amount compared with the exposure dose of cardiac catheterization.

As isotope transit from each chamber is imaged; not only count intensity, but spatial and temporal data are

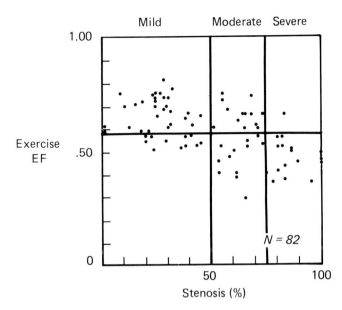

Figure 13-3 Eighty-two individual comparisons of stenotic severity and exercise left ventricular ejection fraction (EF) made before and after coronary angioplasty in 41 patients with isolated left anterior descending coronary artery disease. *Source:* Reprinted with permission from *Journal of the American College of Cardiology* (1987;10:255), Copyright © 1987, American College of Cardiology.

available for beat-to-beat analysis. The left ventricular ejection fraction is determined from count changes between end-systole and end-diastole, averaged from four to eight contractions during the time of maximal tracer concentration in the ventricle. Images can be visualized in cine-format for analysis of regional wall motion. Definition of the area of the perimeter and length of the major axis of the left ventricular end-diastolic image permits calculation of the end-diastolic volume by standard geometric formulas. From the heart rate, the left ventricular ejection fraction, and the end-diastolic volume data, other parameters such as stroke volume, end-systolic volume, and cardiac output can be determined. These parameters have been validated for accuracy and reproducibility.[10,13]

Multigated Equilibrium Acquisition

Multigated equilibrium radionuclide studies image autologous red blood cells that have been labeled in vivo or in vitro with ^{99m}Tc pertechnetate.[14] The reinfused cells then mix with the entire circulating blood pool. Count data acquired with a single-crystal or multicrystal scintillation camera centered over the cardiac region are synchronized with the electrocardiogram. Several hundred beats are imaged serially. This process requires 1 to 3 minutes of data acquisition for each study. Data processed from each beat are divided into 16 to 24 segments based on the RR interval of the electrocardiogram. Framing intervals vary from 20 to 50 milliseconds depending on the heart rate. Composite images are thus "gated" with the electrocardiogram and depict an average cardiac cycle over the period studied. Cardiac function must be stable, motion must be minimized, and tracer activity must be relatively constant in order to achieve reliable data.[15] Proper subtraction of background counts is also important. The oblique projection is most commonly used to maximize spatial separation of the right and left ventricular chambers. The ventricular ejection fraction is calculated from composite count curves of end-diastole and end-systole. Cine-format images of composite cycles are generated to evaluate left ventricular wall motion.

Both first-pass and multigated radionuclide techniques are very effective in determining ventricular function at rest and during exercise.[16,17] Calculated left ventricular ejection fractions are comparable with both techniques, and both techniques have been shown to correlate well with data derived from cardiac catheterization.[18]

NORMAL RESPONSE TO EXERCISE

The normal response to exercise is an augmentation of cardiac output, primarily a result of an increase in heart rate. Left ventricular end-diastolic volume remains relatively unchanged with upright exercise, whereas end-systolic volume decreases. This results in an increase in ejection fraction and stroke volume. The direction of change is quite consistent in normal subjects. Variation is observed in the magnitude of change, which is influenced by the level of exercise achieved and the state of physical conditioning as well as by age, gender, and systemic blood pressure.

Interpretations vary with regard to what constitutes a normal left ventricular response to exercise. A commonly applied criterion for a "normal study" has been an increase in the left ventricular ejection fraction of at least 0.05 from rest to exercise. Rozanski et al.[19] have emphasized the limitations of such criteria. From a large multicenter analysis of patients undergoing either a first-pass or multigated equilibrium radionuclide study at rest and during exercise, an enhanced definition of a normal response for an individual study was defined (Figure 13-4).

BAYESIAN CONCEPTS

Radionuclide studies are often interpreted as yielding either a positive or a negative result. However, additional information is derived in consideration of the amount of abnormality in important variables such as

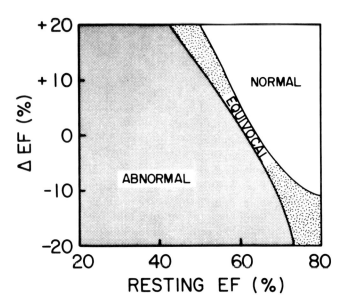

Figure 13-4 Rest left ventricular ejection fraction (EF) versus change in ejection fraction from rest to exercise based on 2400 rest–exercise combinations at a 50% confidence interval. "Normal," "abnormal," and "equivocal" responses are defined. *Source:* Reprinted with permission from *Journal of the American College of Cardiology* (1985;5:242), Copyright © 1985, American College of Cardiology.

the exercise ejection fraction. Results may be analyzed as continuous variables based on relative probability rather than as a categorical result. Application of Bayes' theorem of conditional probability to continuous variables derived from radionuclide evaluation of coronary artery disease was first proposed by Diamond and Forrester.[20] Christopher et al.[21] found that the bayesian model enhanced both post-test probability of disease and prognostic information for individual patients with symptoms suggestive of coronary artery disease. An important finding was that none of the 29 patients who had a post-test probability of disease less than 0.95 by the continuous bayesian model died during the 35-month interval following the radionuclide study.

PREVALENCE

Variations in the prevalence of coronary artery disease in a population of patients studied influence the apparent sensitivity of a given test. Rozanski et al.[22] reported a declining sensitivity of multigated exercise radionuclide studies in their studies. Most patients with angiographically normal coronary arteries examined from 1978 to 1979 had normal ejection fraction responses (94%). Angiographically normal patients studied in a subsequent period, 1980 to 1982, demonstrated a dramatically lower rate of normal ejection fraction responses (49%).

Two factors were considered in the analysis of this phenomenon. In initial evaluations of radionuclide studies, researchers tended to compare patients with known normal coronary anatomy with those with the most severe disease. Clinicians tended to refer patients in these two extreme subsets after cardiac catheterization to evaluate the usefulness of the new test modality. With this "pretest referral bias" the subset of normal patients had a very low pretest probability of coronary artery disease; therefore, the early sensitivities were high. Subsequently, increasing numbers of patients presenting diagnostic dilemmas were referred for study. As the pretest probability of disease increased, a higher proportion of abnormal ventricular responses occurred and the apparent sensitivity of the test fell accordingly.

A "post-test referral bias" began to appear as clinicians gained confidence in the reliability of radionuclide studies. Patients with an abnormal exercise radionuclide study tended to undergo cardiac catheterization. Rozanski and colleagues[22] argued that if carried to the extreme such that only patients with positive radionuclide studies were subjected to cardiac catheterization, the test would appear 100% sensitive and 0% specific. All patients with disease would have had a positive radionuclide study, but all of those with normal catheterization studies would also have had a positive radionuclide study. Therefore, as radionuclide angiocardiography was employed in a broader population as a decision criterion for catheterization, its diagnostic accuracy appeared to diminish in comparison to early use of the test in selected populations.

SELECTION FOR CATHETERIZATION

Recent studies employing radionuclide techniques during exercise have demonstrated that abnormalities in left ventricular function often occur before electrocardiographic ST segment changes or the onset of angina.[23] Campos et al.[24] evaluated the accuracy of rest and exercise radionuclide angiocardiography (RNA) and exercise treadmill testing (ETT) for the diagnosis of left main or three-vessel coronary artery disease in 544 patients. The ETT and RNA sensitivities were similar (88% versus 92%), but ETT was more specific than RNA. Campos et al.[24] proposed that patients with suspected coronary artery disease first be screened with ETT. Patients with a negative or indeterminate ETT would then undergo a rest and exercise RNA study; those with a positive ETT would be referred directly for cardiac catheterization. This approach effectively separated patients into groups of high or low probability with respect to extensive coronary artery disease. Jones et al.[8] found that RNA variables most diagnostic for coronary artery disease were a left ventricular exercise ejection fraction at least 6% less than predicted, an increase of greater than 20 mL in end-systolic volume with exercise, and the appearance of an exercise-induced wall motion abnormality.

The selection of patients who might benefit from coronary arteriography to assess the patency of saphenous vein grafts after coronary artery bypass procedures is often difficult. Loop and colleagues[25–27] from the Cleveland Clinic reported a cumulative vein graft patency of only 45% in a large group of patients studied with arteriography 5 or more years (mean 88 months) after operation. Although data from the Coronary Artery Surgery Study demonstrated a vein graft patency rate of 82% at 5 years, Bourassa et al.[28] found that the attrition rate of vein grafts due to atherosclerosis increased significantly between 6 and 11 years after operation. Noninvasive tests that reflect recurrent ischemia would aid selection of patients for catheterization. Austin et al.[29] found that resting left ventricular function was not altered after coronary artery bypass grafting but that exercise function was significantly improved. Radionuclide techniques for assessment of exercise cardiac function on a serial basis may have an impact on the long-term management of patients after coronary artery bypass grafting as more is known about its relative prognostic importance.

PROGNOSIS

The prognosis of patients with coronary artery disease managed medically can be related to the anatomic severity of the disease. Groups of patients with left main or three-vessel disease generally have a poorer long-term survival outcome than do patients with single-vessel disease. Randomized studies have demonstrated that surgery has a significant positive impact on survival in patient groups with left main or three-vessel coronary artery disease.[30–34] Therefore, decisions to manage an individual patient medically or surgically often have been based on the number of coronary vessels involved. However, it is a common clinical occurrence to observe a catastrophic event in an individual patient with a single stenotic coronary artery (Figure 13-5). For these patients management decisions might be improved by the use of additional functional criteria.

It has been hypothesized that in patients with angiographically demonstrated coronary artery disease, left ventricular dysfunction during exercise reflects ischemia. These patients should benefit from coronary artery bypass grafting. One might further hypothesize that patients with angiographically defined coronary artery disease but little or no ischemic response during exercise would profit little from revascularization. In order to test this hypothesis, 875 consecutive patients were followed who fit the criteria of a technically adequate rest and exercise RNA study

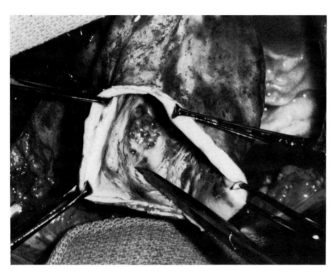

Figure 13-5 Intraoperative photograph of a left ventricular aneurysm. This patient had single-vessel coronary artery disease, an isolated lesion of the left anterior descending coronary artery.

within 90 days of catheterization, stenosis of at least 75% of one or more coronary arteries, no left ventricular aneurysm or surgically significant valvular heart disease, and a resting left ventricular ejection fraction of greater than 0.20.[35]

The 875 patients were separated into four groups. Patients in group 1 had normal rest and normal exercise ventricular function. While there were no surgical deaths in this group, they did not benefit from operation in terms of either survival or complete pain relief (Figure 13-6). Group 2, the largest subset of patients studied, had normal rest and abnormal exercise left ventricular function. Roughly equal numbers of patients in this group had been managed either medically or surgically. Also, roughly equal numbers had single-, double-, and triple-vessel disease; 27 had left main anatomic disease. The survival at 4 years was superior for the surgically treated group, as was the achievement of complete pain relief (Figure 13-7). Patients in group 3 demonstrated abnormal rest yet normal exercise ventricular function. Therefore, these patients did not reflect the physiology of ischemia. Again, among patients in this group roughly equal numbers had single-, double-, and triple-vessel disease. These patients experienced improved pain relief early after surgery. However, this effect was diminished 4 years later and may represent an early placebo response. At 4 years, survival favored medically treated patients in this group (Figure 13-8). Perhaps most likely to benefit from coronary artery bypass grafting are patients with marginal ventricular function at rest, indicative of a significant area of nonfunctional myocardium from previous infarction; during exercise these patients have a further decrease in the ejection fraction, demonstrating an ischemic response of the remaining myocardial tissue. Group 4 consisted of 236 such patients. In this group approximately half of the

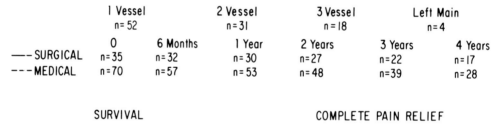

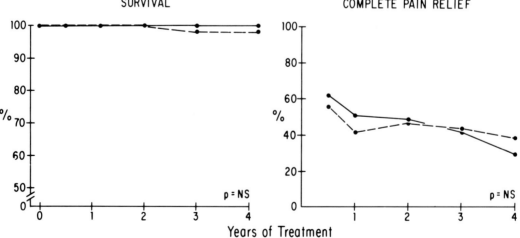

Figure 13-6 Group 1: Incidence of survival and complete pain relief in 105 patients with normal left ventricular function at rest and during exercise. *Source:* Reprinted with permission from *Seminars in Nuclear Medicine* (1987;17:99), Copyright © 1987, Grune & Stratton Inc.

Left Ventricular Function in Coronary Artery Disease 217

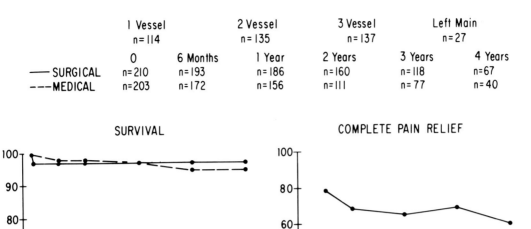

Figure 13-7 Group 2: Incidence of survival and complete pain relief in 413 patients with normal rest and abnormal exercise left ventricular function.
Source: Reprinted with permission from *Seminars in Nuclear Medicine* (1987;17:100), Copyright © 1987, Grune & Stratton Inc.

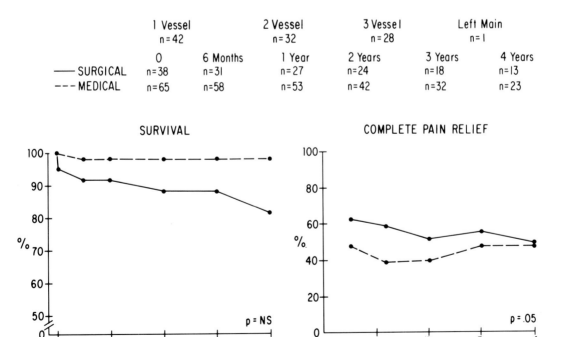

Figure 13-8 Group 3: Incidence of survival and complete pain relief in 103 patients with abnormal rest and normal exercise left ventricular function.
Source: Reprinted with permission from *Seminars in Nuclear Medicine* (1987;17:100), Copyright © 1987, Grune & Stratton Inc.

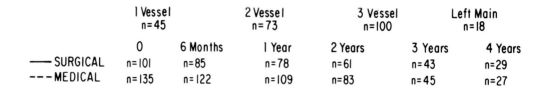

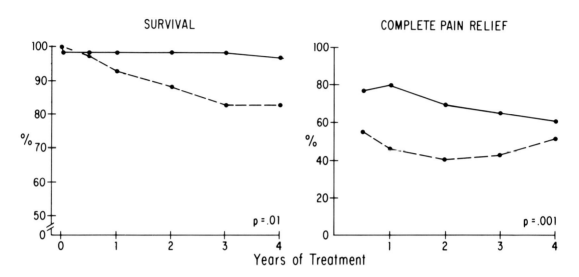

Figure 13-9 Group 4: Incidence of survival and complete pain relief in 236 patients with abnormal rest and abnormal exercise left ventricular function. *Source:* Reprinted with permission from *Seminars in Nuclear Medicine* (1987;17:100), Copyright © 1987, Grune & Stratton Inc.

patients had only single- or double-vessel disease. Medically treated patients demonstrated significantly poorer survival as well as less pain relief than did patients treated surgically (Figure 13-9).

These data suggest that patients who do not demonstrate ischemia by exercise RNA will not benefit from coronary artery bypass grafting in terms of either survival or complete pain relief. The Coronary Artery Surgery Study confirmed that the presence of exercise-induced angina identified patients who had a survival advantage if assigned to surgical therapy.[31] Several recent studies employing multivariable analysis have demonstrated that the exercise ejection fraction is the single most important prognostic indicator in patients with coronary artery disease who are managed medically[36] (Figure 13-10). Jones et al.[37] employed χ^2 analysis to assess the relative significance of variables known to provide prognostic information. The exercise ejection fraction was a better predictor of prognosis than the resting ventricular function, exercise electrocardiographic changes, or the duration of exercise. A positive exercise RNA study was shown to be the single most powerful prognostic indicator for patients with coronary artery disease, even more powerful than three-vessel or left main anatomic disease[37] (Figure 13-11).

The severity of an abnormal exercise ejection fraction correlates with survival. Very low ejection fractions

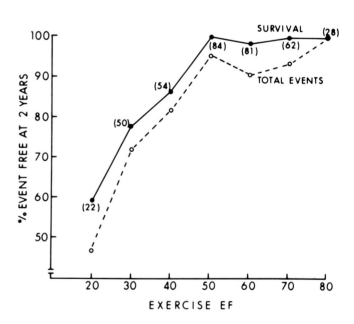

Figure 13-10 Two-year survival and total cardiac event-free rates as a function of exercise left ventricular ejection fraction (EF) rounded to the nearest 10. Numbers in parentheses are the numbers of patients within each exercise EF subgroup. *Source:* Reprinted with permission from *American Journal of Cardiology* (1984;53:21), Copyright © 1984, Cahners Publishing Company.

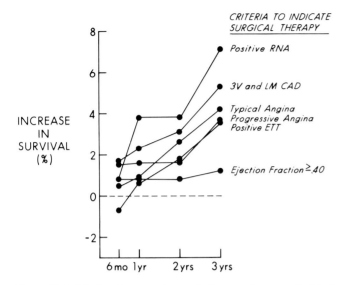

Figure 13-11 Maximal increase in survival over a 3-year period that could be achieved using specific criteria to indicate surgical therapy and absence of those criteria to indicate medical therapy. CAD, coronary artery disease; V, vessel; LM, left main coronary artery. *Source:* Reprinted with permission from *Annals of Surgery* (1983;197:749), Copyright © 1983, JB Lippincott Company.

correlate with a progressively poorer prognosis. Pryor et al.[36] demonstrated that patients with exercise ejection fractions of less than 0.35 have a greater incidence of major cardiac events than do patients with less severely depressed or normal ejection fractions, regardless of their angiographic anatomy. This was independent of whether the exercise ejection fraction was greater than or less than the resting ejection fraction.

PROGNOSIS AFTER ACUTE MYOCARDIAL INFARCTION

The use of infarct avid agents to evaluate alterations of myocardial perfusion after an acute infarction is discussed in Chapter 11. Radionuclide angiographic determination of the left ventricular ejection fraction at rest also relates to prognosis when applied to patients who have had an acute myocardial infarction.[38] A left ventricular ejection fraction at rest of less than 0.30 portends a dramatically decreased short-term and long-term survival after an acute myocardial infarction.[39,40]

Postinfarction determination of the left ventricular ejection fraction during exercise yields even stronger prognostic information. Morris et al.[41] studied 106 consecutive patients who had had an acute myocardial infarction. Both the rest and exercise ejection fraction variables correlated inversely with subsequent mortality and added prognostic information to the clinical assessment. When both variables were included in a bivariable model, only the exercise value remained significant. The severity of exercise-induced left ventricular ejection fraction abnormalities also related to mortality in a continuous manner (Figure 13-12).

EMERGING TECHNOLOGIES

Several recent technologic advances have expanded the use of radionuclide studies in the evaluation of ventricular function in coronary artery disease. With portable, high-count, multicrystal imaging cameras, routine studies of ventricular function can be performed intraoperatively and in the intensive care unit (Figure 13-13). Currently, inotropic support and volume requirements of patients undergoing coronary artery bypass grafting are managed primarily by measurements of blood pressure and cardiac output. However, cardiac output may remain unchanged while significant changes occur in the left ventricular ejection fraction, end-diastolic volume, or end-systolic volume that reflect important underlying changes in cardiac physiology. Recent work at Duke University by Purut, Sell, and colleagues (unpublished data) has coupled radionuclide data with simultaneous intraventricular catheter measurements of left ventricular pressures to generate pressure–volume work loops in patients immediately after coronary artery bypass grafting and subsequently in the intensive care unit (Figure 13-14). Subtle changes in ventricular function may thus be detected and influence management of volume or inotropic support. This approach may prove to be of

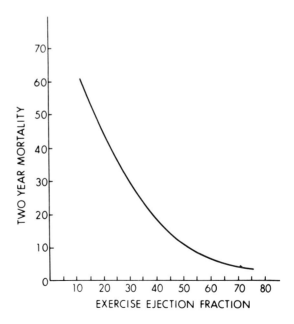

Figure 13-12 Unadjusted Cox model of 2-year mortality as a function of left ventricular exercise ejection fraction. *Source:* Reprinted with permission from *American Journal of Cardiology* (1985;55:322), Copyright © 1983, Cahners Publishing Company.

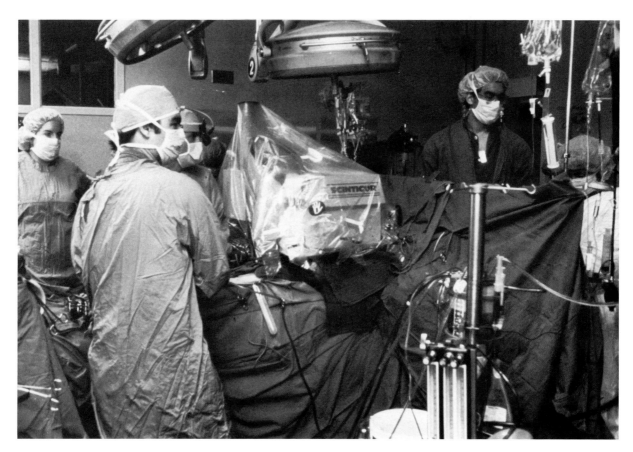

Figure 13-13 Assessment of left ventricular ejection fraction, end-systolic volume, end-diastolic volume, and wall motion in the operating room immediately after separation from cardiopulmonary bypass.

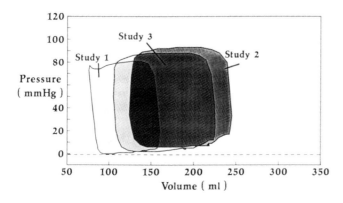

Figure 13-14 Pressure–volume loops from a single patient just prior to initiation of cardiopulmonary bypass (Study 1) and at 17 minutes (Study 2) and 31 minutes (Study 3) after reperfusion from ischemic arrest. *Source:* Courtesy of TL Sell, CM Purut, R Silva, and RH Jones, unpublished observations, Durham, NC.

particular importance in the increasing numbers of patients with marginal ventricular function undergoing surgical interventions.

New ultra-short half-life isotopes currently coming into clinical use will allow many serial studies to be performed in the same patient, yet provide dramatic reduction in the total radiation dose. Three-dimensional imaging will likely develop as advanced computer technology continues to impact this field dramatically.

SUMMARY

Further investigations are indicated to evaluate prognosis in patients with coronary artery disease. However, previous animal experiments (and now many clinical studies) confirm that exercise-induced left ventricular dysfunction is one of the most sensitive and specific indicators of myocardial ischemia. A positive exercise radionuclide study has been shown to be the single most significant prognostic indicator for patients with coronary artery disease, even more reliable than any single anatomic descriptor. These principles may be applied to the large population of patients selected for medical or surgical management of coronary artery disease.

REFERENCES

1. Blumgart HL, Yens OC. Studies on the velocity of blood flow, I: The method utilized. *J Clin Invest.* 1927;4:1–13.

2. Blumgart HL, Weiss S. The velocity of venous blood to the right heart in man. *Proc Soc Exp Biol Med.* 1926;23:694–696.

3. Blumgart HL, Weiss S. Studies on the velocity of blood flow, VII: The pulmonary circulation time in normal resting individuals. *J Clin Invest.* 1927;4:399–425.

4. Roy CS. On the influences which modify the work of the heart. *J Physiol (Lond).* 1879;1:452–496.

5. Tennant R, Wiggers CJ. The effect of coronary occlusion on myocardial contraction. *Am J Physiol.* 1935;112:351–361.

6. Forrester JS, Wyatt HL, DaLuz PL, et al. Functional significance of regional ischemic contraction abnormalities. *Circulation.* 1976;54:64–70.

7. Herman MV, Heinle RA, Klein MD, et al. Localized disorders in myocardial contraction: asynergy and its role in congestive heart failure. *N Engl J Med.* 1967;277:222–232.

8. Jones RH, McEwan P, Newman GE, et al. Accuracy of diagnosis of coronary artery disease by radionuclide measurement of left ventricular function during rest and exercise. *Circulation.* 1981;64:586–601.

9. Peterson RJ, King SB III, Fajman WA, et al. Relation of coronary artery stenosis and pressure gradient to exercise-induced ischemia before and after coronary angioplasty. *J Am Coll Cardiol.* 1987;10:253–260.

10. Upton MT, Rerych SK, Newman GE, et al. The reproducibility of radionuclide angiographic measurements to left ventricular function in normal subjects at rest and during exercise. *Circulation.* 1980;62:126–132.

11. Rerych SK, Scholz PM, Newman GE, et al. Cardiac function at rest and during exercise in normals and in patients with coronary artery disease. *Ann Surg.* 1978;187:449–464.

12. Port S, McEwan P, Cobb FR, et al. Influence of resting left ventricular function on the left ventricular response to exercise in patients with coronary artery disease. *Circulation.* 1981;63:856–863.

13. Anderson PAW, Rerych SK, Moore TE, et al. Accuracy of left ventricular end-diastolic dimension determinations obtained by radionuclide angiocardiography. *J Nucl Med.* 1981;22:500–505.

14. Zaret BL, Strauss HW, Hurley PJ, et al. A noninvasive scintiphotographic method for detecting regional ventricular dysfunction in man. *N Engl J Med.* 1971;284:1165–1170.

15. Berger HJ, Zaret BL. Nuclear cardiology. *N Engl J Med.* 1981;305:855–865.

16. Marshall RC, Berger HJ, Reduto LA, et al. Variability in sequential measures of left ventricular performance assessed with radionuclide angiocardiography. *Am J Cardiol.* 1978;41:531–536.

17. Wackers FJ III, Berger HJ, Johnstone DE, et al. Multiple gated cardiac blood pool imaging for left ventricular ejection fraction: Validation of the technique and assessment of variability. *Am J Cardiol.* 1979;43:1159–1166.

18. Beller GA, Gibson RS. Sensitivity, specificity, and prognostic significance of noninvasive testing for occult or known coronary artery disease. *Prog Cardiovasc Dis.* 1987;4:241–270.

19. Rozanski A, Diamond GA, Jones R, et al. A format for integrating the interpretation of exercise ejection fraction and wall motion and its application in identifying equivocal responses. *J Am Coll Cardiol.* 1985;5:238–247.

20. Diamond GA, Forrester JS. Improved interpretation of a continuous variable in diagnostic testing: Probabilistic analysis of scintigraphic rest and exercise left ventricular ejection fractions for coronary disease detection. *Am Heart J.* 1981;102:189–195.

21. Christopher TD, Konstantinow G, Jones RH. Bayesian analysis of data from radionuclide angiocardiograms for diagnosis of coronary artery disease. *Circulation.* 1984;69:65–72.

22. Rozanski A, Diamond GA, Berman D, et al. The declining specificity of exercise radionuclide ventriculography. *N Engl J Med.* 1983;309:518–522.

23. Upton MT, Rerych SK, Newman GE, et al. Detecting abnormalities in left ventricular function during exercise before angina and ST-segment depression. *Circulation.* 1980;62:341–349.

24. Campos CT, Chu HW, D'Agostino HJ Jr, et al. Comparison of rest and exercise radionuclide angiocardiography and exercise treadmill testing for diagnosis of anatomically extensive coronary artery disease. *Circulation.* 1983;67:1204–1210.

25. Loop FD. CASS continued. *Circulation.* 1985;72(suppl 2):1–6.

26. Lytle BW, Loop FD, Cosgrove DM, et al. Long-term (5 to 12 years) serial studies of internal mammary artery and saphenous vein coronary bypass grafts. *J Thorac Cardiovasc Surg.* 1985;89:248–258.

27. Loop FD, Lytle BW, Cosgrove DM, et al. Influence of the internal-mammary-artery graft on 10-year survival and other cardiac events. *N Engl J Med.* 1986;314:1–6.

28. Bourassa MG, Fisher LD, Campeau L, et al. Long-term fate of bypass grafts: The Coronary Artery Surgery Study (CASS) and Montreal Heart Institute experiences. *Circulation.* 1985;72(suppl 5):71–78.

29. Austin EH, Oldham HN Jr, Sabiston DC Jr, et al. Early assessment of rest and exercise left ventricular function following coronary artery surgery. *Ann Thorac Surg.* 1983;35:159–169.

30. CASS principal investigators and their associates: Coronary Artery Surgery Study (CASS). A randomized trial of coronary artery bypass surgery: survival data. *Circulation.* 1983;68:939–950.

31. Ryan TJ, Weiner DA, McCabe CH, et al. Exercise testing in the Coronary Artery Surgery Study randomized population. *Circulation.* 1985;72(suppl 5):31–38.

32. Detre KM, Takaro T, Hultgren H, et al. Long-term mortality and morbidity results of the Veterans Administration randomized trial of coronary artery bypass surgery. *Circulation.* 1985;72(suppl 5):84–89.

33. Varnauskas E, and the European coronary surgery study group. Survival, myocardial infarction, and employment status in a prospective randomized study of coronary bypass surgery. *Circulation.* 1985;72(suppl 5):90–101.

34. Passamani E, Davis KB, Gillespie MJ, Killip T, and the CASS principal investigators and their associates. A randomized trial of coronary artery bypass surgery. *N Engl J Med.* 1985;312:1665–1671.

35. Jones RH. Use of radionuclide measurements of left ventricular function for prognosis in patients with coronary artery disease. *Semin Nucl Med.* 1987;17:95–103.

36. Pryor DB, Harrell FE Jr, Lee KL, et al. Prognostic indicators from radionuclide angiography in medically treated patients with coronary artery disease. *Am J Cardiol.* 1984;53:18–22.

37. Jones RH, Floyd RD, Austin EH, et al. The role of radionuclide angiocardiography in the preoperative prediction of pain relief and prolonged survival following coronary artery bypass grafting. *Ann Surg.* 1983;197:743–754.

38. Borer JS, Miller D, Schreiber T, et al. Radionuclide cineangiography in acute myocardial infarction: Role in prognostication. *Semin Nucl Med.* 1987;18:89–94.

39. Shaw PK, Pichler M, Berman DS, et al. Left ventricular ejection fraction determined by radionuclide ventriculography in early stages of first transmural myocardial infarction. *Am J Cardiol.* 1980;45:542–546.

40. The Multicenter Postinfarction Research Group. Risk stratification and survival after myocardial infarction. *N Engl J Med.* 1986;309:331–336.

41. Morris KG, Palmeri ST, Califf RM, et al. Value of radionuclide angiography for predicting specific cardiac events after acute myocardial infarction. *Am J Cardiol.* 1985;55:318–324.

Chapter 14

Evaluation of Acute Myocardial Infarction by Computed Tomography

Bruce H. Brundage, MD, and Eva V. Chomka, MD

For 15 years major research efforts have been expended to develop a method that could accurately measure acute myocardial infarct size. These efforts have been further stimulated by the increasing use of therapeutic strategies to limit infarct size. In order to judge the efficacy of these therapies, an accurate method of measuring infarct size is essential. This need has become more urgent in the past 10 years with the advent of thrombolytic agents that re-establish flow in acutely thrombosed vessels.[1] Coronary angioplasty and emergent bypass surgery have also been employed to reperfuse the myocardium supplied by the thrombosed vessel.[2,3] The value of a method that can measure infarct size before and after therapy would be increased significantly if it could also assess regional myocardial blood flow and function. The lack of reflow obviously results in necrotic myocardium, but the value of reflow in preserving myocardial function after infarction is still not fully understood. Ultrafast computed tomography is a new technology that shows promise in meeting these needs.

CONVENTIONAL (SLOW) COMPUTED TOMOGRAPHY

Animal Studies

In 1976 the first reports of the use of conventional x-ray transmission computed tomography to image myocardial infarcts appeared. Imaging time was slow because a head scanner was used and infarcts could be visualized only in the nonbeating heart. These early studies usually imaged infarcts without contrast medium enhancement.[4-7] Periods of coronary occlusion as brief as 60 minutes followed by 45 minutes of reperfusion produced an area of infarction that could be recognized as decreased x-ray attenuation when compared with normal myocardium.[6] The same observations were made after 2 to 6 hours of coronary ligation without reperfusion. The decreased attenuation of myocardium on the computed tomography scan has been correlated with increased water content (edema) of the infarct region.[6]

Recognition of the decreased attenuation of infarcted myocardium without contrast medium enhancement in a beating heart by conventional computed tomography scanners has not been possible.[8,9] Therefore, investigators have used contrast medium enhancement to delineate infarcts in beating dog hearts.[8-10] Higgins and co-workers[11,12] have carefully studied experimental infarction in dogs with contrast medium. In nonbeating dog hearts they found an increased iodine concentration (delayed enhancement) in infarcted myocardium compared with viable myocardium from 10 minutes to 3 hours after the administration of contrast medium. Others have identified enhancement of portions of the infarct as soon as 1.5 minutes after the injection of contrast medium.[9] Higgins et al.[12] did not find an increased concentration of iodine until the infarct was at least 2 hours old. The earliest that Doherty et al.[8] observed delayed enhancement in beating dog hearts

---

This work was supported in part by a grant from the National Heart, Lung and Blood Institute (NIH RO1-32232).

was 6 hours after infarction, and this phenomenon was observed in only three of eight dogs with infarcts less than 12 hours old. Higgins et al.[12] described three types of infarct images discriminated by computed tomography in nonbeating dog hearts after contrast medium enhancement. Global enhancement, where the entire infarct had a higher Hounsfield number than did normal myocardium, was seen in 40% to 50% of animals. Partial enhancement, where the epicardial perimeter of the infarct had a higher Hounsfield number, was observed in 25% to 30%. No enhancement was detected in 25% to 30%. When contrast medium enhancement was not present, the area of infarction had a lower computed tomography number than did normal myocardium.

After an intravenous injection of contrast medium, Abraham and Higgins[13] demonstrated an increased iodine concentration in necrotic myocardial cells by scanning electron microscopy using energy-dispersive x-ray microanalysis. This work suggests that the loss of cell membrane integrity after ischemic injury permits intracellular accumulation of iodine. Normally, myocardial cells have a high potassium and a low sodium and chloride concentration compared with blood. After infarction, sodium and chloride ions enter the cell and potassium leaves the cell. Of particular note is the fact that these electrolyte shifts are first detected about 6 hours after infarction, similar to the time when delayed contrast medium enhancement is first observed by computed tomography. This shift in electrolytes reaches its maximum more than 12 hours after infarction. Jennings et al.[14] observed that "An irreversibly injured cell in contact with the freely flowing blood supply of the transient ischemic system loses potassium and gains sodium quickly." While in the center of the infarct where blood flow is reduced, the diffusion of electrolytes is slow in both directions. This observation is consistent with the reported findings that delayed contrast medium enhancement is often greatest at the epicardial margins of the infarct,[8] where collateral blood flow in dogs is often greatest.[15]

Several investigators have correlated infarct size as determined by conventional computed tomography with post-mortem measurements. The earliest studies in nonbeating hearts found a highly significant correlation but a tendency to underestimate the infarct size.[7,12] More recent studies in beating dog hearts found an excellent correlation ($r = 0.98$) between infarct volume as measured by conventional computed tomography and post-mortem weight[9] or volume.[8]

Human Studies

Images from a number of patients with clinically diagnosed acute myocardial infarction have been obtained by using conventional computed tomography and contrast medium enhancement.[16] Infarcts were identified in 61%, typically as a low-density region (perfusion deficit) produced by contrast medium enhancement (increased density) of the normal myocardium (Figure 14-1). A reasonably good correlation ($r = 0.75$) was demonstrated between the volume of the infarct as determined by computed tomography and the area under the total creatine kinase enzyme curve obtained in patients with acute myocardial infarctions.[17] However, as in the early animal studies, computed tomography underestimated infarct size when compared with the enzyme release determined volume.

In patients with infarctions and in whom no perfusion defect could be identified by contrast medium–enhanced computed tomography scanning, the area under the creatine kinase curve was 50% less than in patients whose infarct was visualized. This finding suggests that small infarcts often cannot be detected by conventional computed tomographic scanners, presumably because of poor temporal resolution.

In summary, human studies using conventional computed tomography indicate that myocardial infarction, particularly involving the septum and anterior and lateral walls, can be imaged in some patients. Inferior infarcts are more difficult to visualize because of the parallel orientation of the inferior ventricular wall to the scanning x-ray beam, their smaller size, and cardiac motion.

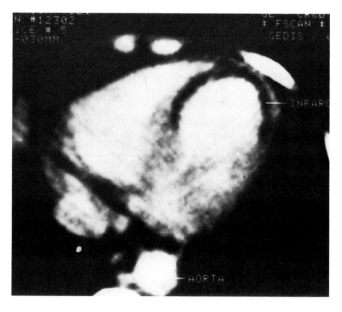

Figure 14-1 The acute anterior wall myocardial infarction is well delineated by conventional computed tomography as a crescent of low density during contrast medium enhancement. *Source:* Reprinted with permission from *American Heart Journal* (1984;108:1516), Copyright © 1984, CV Mosby Company.

IMAGING ACUTE MYOCARDIAL INFARCTION IN HUMANS BY ULTRAFAST COMPUTED TOMOGRAPHY

Considerable experience in imaging acute myocardial infarction in humans by ultrafast computed tomography has been gained at the University of Illinois at Chicago over the past 5 years. More than 150 patients have been scanned between 1 hour and several days after onset of symptoms. Infarction typically is recognized during iodine contrast medium enhancement as a region of decreased density compared with normal myocardium (Figure 14-2). There is a linear correlation between regional myocardial iodine concentration and myocardial blood flow, so areas of little or no flow appear as lucent defects.[18] A preliminary study comparing the detection of acute myocardial infarction by ultrafast computed tomography with technetium (Tc) 99m pyrophosphate scanning and thallium 201 myocardial scintigraphy indicates that ultrafast computed tomography is both more specific and more sensitive than either radioisotope method.[19] Further studies involving larger numbers of patients are needed to confirm these initial findings.

There are no autopsy studies comparing the ultrafast computed tomographic representation of acute myocardial infarction with the actual heart, as have been done in animals. However, a patient who had an ultrafast computed tomography scan several days after an acute myocardial infarction underwent cardiac transplantation 3 weeks later; this provided a unique opportunity to compare the scan results with the actual heart (Figures 14-3A and B). The depiction of the infarct by ultrafast computed tomography correlated well with the explanted heart. Even regional differences in the transmural extent of the infarction were well visualized.

Correlation of the estimation of infarct size by creatine phosphokinase (CK)–myocardium-specific (MB) (CK-MB) time-release curve analysis with ultrafast computed tomography measurements of infarct volume has been made.[20] A good ($r = 0.78$; $P < 0.01$), but far from perfect, correlation was found. Furthermore, ultrafast computed tomography measurements of infarct size were consistently less than the CK-MB estimates. The reasons for the difference are unclear, but since the ultrafast computed tomography detection of acute myocardial infarction is presumed to be dependent on the no-reflow phenomenon, regions of irreversibly damaged myocardium with preserved flow may not be detected.

With the potential for accurate determination of the transmural extent of acute infarction by ultrafast computed tomography, interesting results have been found when compared with the presence or absence of Q

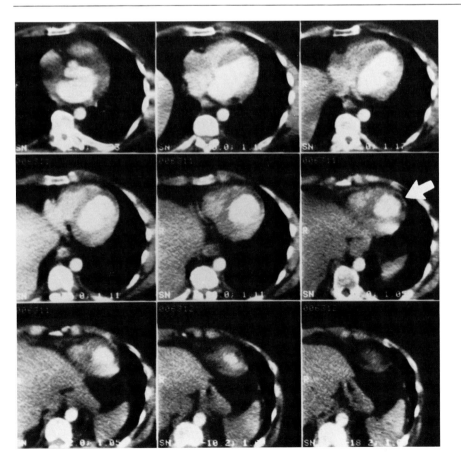

Figure 14-2 Adjacent short-axis ultrafast computed tomography scans define the extent of an anterior wall infarct during contrast medium enhancement. The area of infarct is recognized by its typical low density (arrow) compared with normal myocardium. Note the infarction on adjacent scans.

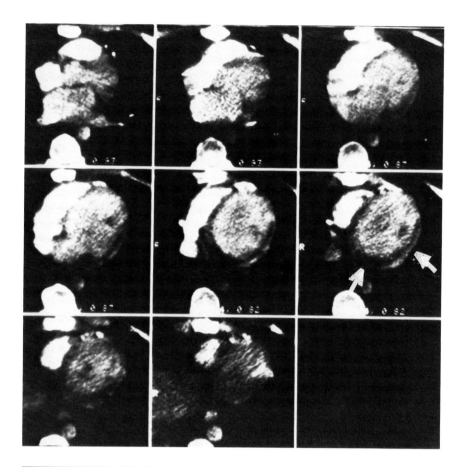

Figure 14-3A A large posterolateral wall infarction (arrows) is detected by contrast medium-enhanced ultrafast computed tomography scanning.

waves on the electrocardiogram. As pathologists have indicated previously, the ultrafast computed tomography scan confirms that the electrocardiogram cannot predict the transmural extent of infarction.[21] The Q waves are often associated with nontransmural infarction (54%), and transmural infarction occurs quite frequently in the absence of Q waves on the electrocardiogram (33%).

In earlier animal studies, several investigators have commented on the delayed "paradoxical" contrast medium enhancement of acute infarction using conventional computed tomography scanning. However, experience with ultrafast computed tomography has not found this phenomenon to be common in infarcts in humans.[16,22] The reason for the difference is not clear. Also, animal studies of nonbeating hearts detected areas of acute infarction as low-density regions, presumably because of myocardial edema. This finding has not been observed in ultrafast computed tomographic images of the beating heart in animals or humans, probably because the density differences between normal and edematous myocardium are too small.

These are still some technical limitations to imaging acute myocardial infarction with ultrafast computed tomography. These are primarily related to the problem of partial voluming of the inferior wall. Although the scanner couch permits axial tilting of the patient, in large or obese individuals with increased thoracic diameters, the degree of tilt (15° to 20°) may be insufficient to provide true short-axis views of the left ventricle. It is hoped that future scanner modifications will correct this limitation. Probably 30° of cranial axial tilt will be sufficient for the vast majority of patients. Another potential limitation of ultrafast computed tomography for imaging acute infarcts is the effect of spontaneous early reperfusion. Currently, it is unknown whether this process will diminish the sensitivity of the technique for detecting particularly the small inferior infarction.

Ultrafast computed tomography has great potential for evaluating patients with myocardial infarction. The excellent spatial and temporal resolution of the technique can provide clear images of acute infarcts. Analysis of these images may improve our ability to predict morbidity and early mortality after infarction. The value of ultrafast computed tomographic images in differentiating jeopardized, stunned, or hibernating myocardium from irreversibly damaged myocardium is at present unknown.

LIMITATIONS OF OTHER INFARCT-SIZING TECHNIQUES

In the early 1970s Sobel, Shell, and associates developed a method of judging infarct size by measuring the amount of CK released from infarcted myocardium into plasma.[23,24] Numerous studies have validated a significant correlation between the total amount of enzyme released and the size of experimentally produced infarcts in animals as measured at autopsy.[25,26] Recently, a human autopsy study also reported a good correlation between the amount of enzyme released and infarct size as determined at autopsy.[27] The measurement of enzyme release, usually the CK isoenzyme MB, has been used to estimate infarct size in a number of clonical trials to judge the value of a variety of interventions, including nitroglycerin, nifedipine, β-blockers, clonidine, and streptokinase.[28–33] The enzyme method can be used to evaluate such therapies only by dividing patients into control and treatment groups, as there is no way to estimate how large the infarct would be had the treatment not been given. Attempts to estimate infarct size from the upstroke of the CK-MB enzyme time release curve have proved to be inaccurate.[34] Furthermore, recent studies have shown that enzyme release overestimates infarct size after reperfusion by thrombolytic agents because more enzyme is washed out of the infarcted tissue.[35,36] A recent review suggests that this overestimation can be compensated for by developing new correction factors, but the ability of CK-MB release to judge the myocardial salvage potential of thrombolytic therapy remains unproved.[37]

A comparison of several reports that used CK or CK-MB release curves to estimate infarct size in humans shows an extremely wide range for the average infarct size.[38–43] These data suggest that the quantitative accuracy of the enzyme method is suspect (Figure 14-4).

Palmeri and others have recently popularized a QRS scoring system for assessing acute myocardial infarct size.[44–46] They demonstrated a significant correlation between the electrocardiographic method and infarct size, but the correlation was relatively weak compared with other methods. As with enzyme measurements, the QRS system can be employed to evaluate infarct-limiting therapy only by dividing patient populations into control and treatment groups.

Radioisotope imaging techniques have been used extensively to image acute myocardial infarction. During the 1970s, research evaluating ^{99m}Tc-pyrophosphate imaging and ^{201}Tl myocardial perfusion scintigraphy suggested that these methods could be used for measuring infarct size.[47] However, after more than a decade of development, the accuracy of these isotopic techniques for evaluating infarct size has been disappointing.[48,49] The poor spatial resolution of radioisotope imaging probably always will limit the use of these methods when precise measurement of infarct size is required. The development of single-photon emission computed tomography has improved the accuracy of both the ^{201}Tl and ^{99m}Tc-pyrophosphate methods, but the image resolution is still less than required.[50,51]

Positron emission tomography (PET) is another radioisotope technique that has been extremely useful for the study of myocardial infarction.[52] This imaging modality requires an on-site cyclotron for manufacture of the short half-life isotopes necessary to study cardiac metabolism during acute ischemia. However, PET images, like other radioisotope images, have limited spatial resolution and this limits the study of the distribution of infarcted myocardium within the ventricular wall.[53] Therefore, PET scanning may not be sensitive enough to quantitate the effects of interventional therapies on infarct size. Furthermore, the expense of PET scanning will probably prevent its use in many hospitals, thereby limiting its clinical applicability.

Echocardiography is an extremely valuable imaging tool for the assessment of patients with ischemic heart disease, including those with acute myocardial infarction. However, the ability of echocardiography to judge infarct size has been limited largely to indirect forms of evaluation, such as assessment of regional wall motion and thickening, which are nonspecific and of limited

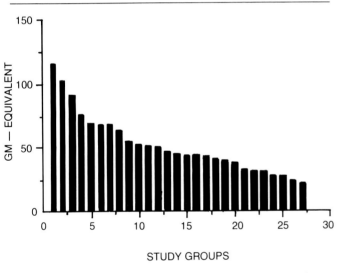

Figure 14-4 The range of myocardial infarct sizes in humans as determined by CK-MB enzyme testing shows a six-fold difference between largest and smallest reported average infarct size. The large difference in reported infarct sizes raises doubts about the accuracy of the enzyme technique.

Note: Key for clinical features of study groups: 1, Anterior myocardial infarction, ejection fraction 33%; 2, dead; 3, Killip class III to IV; 4, dead, small percentage with autopsy correlation; 5, hemodynamically stable; 6, streptokinase failure; 7, nitroglycerin controls; 8, Killip class II to III; 9, streptokinase, late reperfusion; 10, inferior myocardial infarction, ejection fraction 60%; 11, nitroglycerin controls; 12, clonidine controls; 13, infarct survivors; 14, nitroglycerin treatment; 15, Killip I (controls); 16, nifedipine treatment; 17, dead—autopsy correlation; 18, Killip II to III, nitroglycerin treatment; 19, nifedipine study (placebo); 20, clonidine treatment; 21, NYHA class I to II; 22, nitroglycerin treatment; 23, nifedipine study (placebo and treated); 24, Killip I, nitroglycerin treatment; 25, nitroglycerin controls; 26, streptokinase, early reperfusion; 27, nitroglycerin treatment.

sensitivity.[54] Also, the requirement for an adequate acoustic window limits its use in patients with large chests and hyperinflated lungs. Experimental work with contrast (echoreflectant) agents to evaluate myocardial perfusion is promising, as are studies of the acoustic differences of normal and infarcted myocardium.[55,56] However, at the present time there is no clear indication that either of these lines of research will soon provide a clinically useful and accurate method for measuring acute myocardial infarct size.

Nuclear magnetic resonance (NMR) is a new and rapidly developing tomographic imaging technique that already has been employed to evaluate acute myocardial infarction.[57] Most data have been obtained in nonbeating animal hearts, and experience in humans is very limited.[58,59] The NMR device is very expensive and usually requires extensive site preparation. The strong magnetic field required for imaging prohibits the presence of any metal objects in the scanner area, thereby limiting its use in acutely ill patients. Experimental studies indicate that the detection of acute myocardial infarction in the beating heart may be difficult even with use of contrast agents.[60] Also, NMR must be gated to the electrocardiogram in order to acquire enough signal from many cardiac cycles to create a clear image. Gating is time consuming and may limit the use of NMR in critically ill patients with acute myocardial infarction. Significant dysrhythmias may also interfere with image acquisition. However, NMR may prove to be a valuable tool for the study of myocardial metabolism during ischemia.[61]

The clinician is fortunate to have many imaging modalities available with which to evaluate the patient with acute myocardial infarction. Each has strengths and limitations and should only be employed with these in mind. Ultrafast computed tomography may join the other imaging modalities and find its own clinical niche in the evaluation of acute myocardial infarction. Its strength is the ability to assess both myocardial blood flow and regional ventricular function. However, more clinical experience is necessary before its role can be defined fully.

REFERENCES

1. Mathey DG, Kuck KH, Tilsner V, et al. Non-surgical coronary artery recanalization in acute transmural myocardial infarction. *Circulation.* 1981;63:489–497.

2. DeWood MA, Spores J, Berg R Jr, et al. Acute myocardial infarction: A decade of experience with surgical reperfusion in 701 patients. *Circulation.* 1983;68:8–16.

3. Meyer J, Merx W, Schmitz H, et al. Percutaneous transluminal coronary angioplasty immediately after intracoronary streptolysis of transmural myocardial infarction. *Circulation.* 1982;66:905–913.

4. Ter-Pogossian MM, Weiss ES, Coleman RE, et al. Computed tomography of the heart. *Am J Roentgenol.* 1976;127:79–90.

5. Adams DF, Hessel SJ, Judy PF, et al. Computed tomography of the normal and infarcted myocardium. *Am J Roentgenol.* 1976;126:786–791.

6. Powell WJ, Wittenberg J, Maturi RA, et al. Detection of edema associated with myocardial ischemia by computerized tomography in isolated, arrested canine hearts. *Circulation.* 1977;55:46–52.

7. Siemers PT, Higgins CB, Schmidt W, et al. Detection, quantitation and contrast enhancement of myocardial infarction utilizing computerized axial tomography. Comparison with histochemical staining and 99m pyrophosphate imaging. *Invest Radiol.* 1978;13(2):103–109.

8. Hessel SJ, Adams DF, Judy PF, et al. Detection of myocardial ischemia in vitro by computed tomography. *Radiology.* 1978;127(suppl 2):413–418.

9. Gray WR, Buja LM, Hagler HK, et al. Computed tomography for localization and sizing of experimental acute myocardial infarcts. *Circulation.* 1978;58(3):497–504.

10. Powell WJ, Wittenberg J, Miller SW, et al. Assessment of drug intervention on the ischemic myocardium: Serial imaging and measurement with computerized tomography. *Am J Cardiol.* 1979;44:46–52.

11. Higgins CB, Sovak M, Schmidt W, et al. Differential accumulation of radiopaque contrast material in acute myocardial infarction. *Am J Cardiol.* 1979;43(1):47–51.

12. Higgins CB, Siemers PT, Schmidt W, et al. Evaluation of myocardial ischemic damage of various ages by computerized transmission tomography. *Circulation.* 1979;60(suppl 2):284–291.

13. Newell JD, Higgins CB, Abraham JL, et al. Computerized tomographic appearance of evolving myocardial infarctions. *Invest Radiol.* 1980;15(3):207–214.

14. Jennings RB, Sommers HM, Kaltenbach JP, et al. Electrolyte alterations in acute myocardial ischemic injury. *Circ Res.* 1964;14(suppl I):260.

15. Cohen MV. The functional value of coronary collaterals in myocardial ischemia and therapeutic approach to enhance collateral flow. *Am Heart J.* 1978;95(suppl 3):396–404.

16. Kramer PH, Goldstein JA, Herfkens RJ, et al. Imaging of acute myocardial infarction in man with contrast-enhanced computed transmission tomography. *Am Heart J.* 1984;108:1514–1522.

17. Brundage BH. The measurement of acute myocardial infarct size by CT. In: Califf RM, Wagner GS, eds. *Acute Coronary Care 1986.* Boston: Martinus Nijhoff; 1985:133–144.

18. Wolfkiel CJ, Ferguson JL, Chomka EV, et al. Measurement of myocardial blood flow by ultrafast computed tomography. *Circulation.* 1987;76:1262–1273.

19. Brundage BH, Hart K, Chomka E, et al. Detection, location and sizing of acute myocardial infarction by ultrafast computed tomography. *Clin Res.* 1985;33:172A. Abstract.

20. Brundage BH, Wolfkiel C, Chomka E, et al. Comparison of infarct size estimation by CK-MB analysis with infarct volume determined by ultrafast CT. *Circulation.* 1985;72(suppl 3):413.

21. Brundage BH, Chomka E, Hart K, et al. Evaluation of Q and non-Q wave myocardial infarcts by ultrafast CT. *Circulation.* 1985;72(suppl 3):413.

22. Masuda Y, Yoshida H, Morooka N, et al. The usefulness of x-ray computed tomography for the diagnosis of myocardial infarction. *Circulation.* 1984;70:217–225.

23. Sobel BE, Bresmahan GF, Shell WE, Yoder RD. Estimation of infarct size in man and its relation to prognosis. *Circulation.* 1972;46:640–648.

24. Roberts R, Henry PD, Sobel BE. An improved basis for enzymatic estimation of infarct size. *Circulation.* 1975;52:743–754.

25. Shell WE, Kjekshus JK, Sobel BE. Quantitative assessment of the extent of myocardial infarction in the conscious dog by means of

analysis of serial changes in serum creatine phosphokinase (CPK) activity. *J Clin Invest*. 1971;50:2614–2625.

26. Shell WE, Lavelli JF, Covell JW, Sobel BE. Early estimation of myocardial damage in conscious dogs and patients with evolving acute myocardial infarction. *J Clin Invest*. 1973;52:2579–2590.

27. Hackel DB, Reimer KA, Ideker RE, et al. Comparison of enzymatic and anatomic estimates of myocardial infarct size in man. *Circulation*. 1984;70:824–835.

28. Bussmann W-D, Passek D, Seidel W, Kaltembach M. Reduction of CK and CK-MB indexes of infarct size by intravenous nitroglycerin. *Circulation*. 1981;63:615–622.

29. Sirnes PA, Overskeid K, Pederson TR, et al. Evaluation of infarct size during the early use of nifedipine in patients with acute myocardial infarction: The Norwegian Nifedipine Multicenter Trial. *Circulation*. 1984;70:638–644.

30. The International Collaborative Study Group. Reduction of infarct size with the early use of timolol in acute myocardial infarction. *N Engl J Med*. 1984;310:9–15.

31. Yusuf S, Sleight P, Rossi P, et al. Reduction in infarct size, arrhythmias and chest pain by early intravenous beta blockade in suspected acute myocardial infarction. *Circulation*. 1983;67(Suppl I):32–41.

32. Zochowski RJ, Lada W. Intravenous clonidine in acute myocardial infarction in men. *Int J Cardiol*. 1984;6:189–201.

33. Schwarz F, Faure A, Katus H, et al. Intracoronary thrombolysis in acute myocardial infarction: An attempt to quantitate its effect by comparison of enzymatic estimate of myocardial necrosis with left ventricular ejection fraction. *Am J Cardiol*. 1983;51:1573–1578.

34. Thygesen, Horder M, Petersen PH, Nielsen BL. Limitation of enzymatic models for predicting myocardial infarct size. *Br Heart J*. 1983;50:70–74.

35. Tamaki S, Murekami T, Kadota K, et al. Effects of coronary artery reperfusion on relation between creatine kinase-MB release and infarct size estimated by myocardial emission tomography with thallium-201 in man. *J Am Coll Cardiol*. 1983;2:1031–1038.

36. Blanke H, Hardenberg D, Cohen M, et al. Patterns of creatine kinase release during acute myocardial infarction after nonsurgical reperfusion: Comparison with conventional treatment and correlation with infarct size. *J Am Coll Cardiol*. 1984;3:675–680.

37. Roberts R, Ishikawa Y. Enzymatic estimation of infarct size during reperfusion. *Circulation*. 1983;68:83–89.

38. Bleifeld W, Mathey D, Hamrath P, Buss H, Effert S. Infarct size estimated from serial serum creatine phosphokinase in relation to left ventricular hemodynamics. *Circulation*. 1977;55:303–311.

39. Korewicki J, Kraska T, Opolski G, et al. Beneficial effects of intravenous nitroglycerin on hemodynamics and enzymatically estimated infarct size. *Eur Heart J*. 1984;5:697–704.

40. Mueller JE, Morrison J, Stone PH, et al. Nifedipine therapy for patients with threatened and acute myocardial infarction: A randomized, double-blind, placebo-controlled comparison. *Circulation*. 1984;69:740–747.

41. Jugdutt BI, Sussex BA, Warnica JW, Rossall RE. Persistent reduction in left ventricular asynergy in patients with acute myocardial infarction by intravenous infusion of nitroglycerin. *Circulation*. 1983;68:1264–1273.

42. Jaffey AS, Geltman EM, Tiefenbrunn AJ, et al. Reduction of infarct size in patients with inferior infarction with intravenous glyceryl trinitrate: A randomised study. *Br Heart J*. 1983;49:452–460.

43. Hirsowitz GS, Lakier JB, Marks DS, et al. Comparison of radionuclide and enzymatic estimate of infarct size in patients with acute myocardial infarction. *J Am Coll Cardiol*. 1983;1:1405–1412.

44. Palmeri ST, Harrison DG, Cobb FR, et al. A QRS scoring system for assessing left ventricular function after myocardial infarction. *N Engl J Med*. 1982;306:4–9.

45. Wagner GS, Freye CJ, Palmeri ST, et al. Evaluation of a QRS scoring system for estimating myocardial infarct size, I: Specificity and observer agreement. *Circulation*. 1982;65:342–347.

46. Seino Y, Staniloff HM, Shell WE, et al. Evaluation of a QRS scoring system in acute myocardial infarction: Relation to infarct size, early stage left ventricular ejection fraction and exercise performance. *Am J Cardiol*. 1983;52:37–42.

47. Henning SH, Schelbert HR, Righetti A, Ashburn WL, O'Rourke RA. Dual myocardial imaging with technetium-99m pyrophosphate and thallium-201 for detecting, localizing, and sizing acute myocardial infarction. *Am J Cardiol*. 1977;40:147–155.

48. Sharpe DN, Botvinick EH, Shanes DM, et al. The clinical estimation of acute myocardial infarct size with 99m-technetium pyrophosphate scintigraphy. *Circulation*. 1978;57:307–313.

49. Wackers FJT, Becker AE, Samson G, et al. Location and size of acute transmural infarction estimated from thallium-201 scintiscans: A clinicopathological study. *Circulation*. 1977;56:72–78.

50. Corbett JR, Lewis M, Willerson JT, et al. 99m Tc-Pyrophosphate imaging in patients with acute myocardial infarction: Comparison of planar imaging with single-photon tomography with and without blood pool overlay. *Circulation*. 1984;69:1120–1128.

51. Tamaki S, Nakajima H, Murakami T, et al. Estimation of infarct size by myocardial emission computed tomography with thallium-201 and its relation to creatine kinase-MB release after myocardial infarction in man. *Circulation*. 1982;66:994–1001.

52. Ter-Pogossian MM, Hoffman EJ, Weiss ES, et al. Positron emission reconstruction tomography for the assessment of regional myocardial metabolism by the administration of substrates labeled with cyclotron-produced radionuclides. In: Harrison DC, Sandler H, Miller HA, eds. *Proceedings from the Conference on Cardiovascular Imaging and Image Processing Theory and Practice*. Palos Verdes Estates, Calif: Society of Photo-Optical Instrumentation Engineers; 1975;72:277–282.

53. Ter-Pogossian MM, Klein MS, Markham J, et al. Regional assessment of myocardial metabolic integrity in vivo by positron-emission tomography with "C-labeled palmitate." *Circulation*. 1980;61:242–255.

54. Pandion NG, Skorton DJ, Collins SM, et al. Myocardial infarct size threshold for two-dimensional echocardiographic detection: Sensitivity of systolic wall thickening and endocardial motion abnormalities in small versus large infarcts. *Am J Cardiol*. 1985;55:551–555.

55. Skorton DJ, Melton HE Jr, Pandian NG, et al. Detection of acute myocardial infarction in closed-chested dogs by analysis of regional two-dimensional echocardiographic gray-level distributions. *Circ Res*. 1983;52:36–44.

56. Kemper A, Force T, Kloner R, et al. Contrast echocardiographic estimation of regional blood flow following coronary occlusion. *Clin Res*. 1985;33:199A. Abstract.

57. Higgins CB, Herfkens R, Lipton MJ, et al. Nuclear magnetic resonance imaging of acute myocardial infarction in dogs: Alterations in magnetic relaxation times. *Am J Cardiol*. 1983;52:184–188.

58. Higgins CB, Lanzer P, Stask D, et al. Assessment of cardiac anatomy using nuclear magnetic resonance imaging. *J Am Coll Cardiol*. 1985;5:775–815.

59. Fisher MR, McNamara MT, Higgins CB. Acute myocardial infarction: MR evaluation in 29 patients. *Am J Radiol*. 1987;148:247–251.

60. Johnston DL, Lui P, Brady TJ, et al. Gadolinium-DPTA as a myocardial perfusion agent during magnetic resonance imaging: Potential applications and limitations. *J Am Coll Cardiol*. 1985;5:476.

61. Nunally RL, Bottomley PA. ^{31}P NMR studies of myocardial ischemia and its response to drug therapies. *J Comput Assist Tomogr*. 1981;5:296–297.

Chapter 15

The Evaluation of Myocardial Blood Flow by Ultrafast Computed Tomography

Christopher J. Wolfkiel, PhD, Eva V. Chomka, MD, and Bruce H. Brundage, MD

Measurement of myocardial blood flow by ultrafast computed tomography can be achieved with an imaging system that has high spatial and temporal resolution[1] by employing indicator dilution principles. Until the mid-1980s no commercially available tomographic imaging system could reduce the motion effects of the heart sufficiently to make an accurate measurement of indicator concentration within the myocardium. The Imatron C-100 ultrafast computed tomography (CT) scanner solves this problem by magnetically steering an electron beam on a series of concurrent semicircular tungsten target rings surrounding the imaging field. By using a high-power electron gun, images can be acquired in 50 milliseconds, thus minimizing the artifacts of cardiac motion. When the scanning sequence is programmed to scan multiple levels at the same point in the cardiac cycle, the apparent motion of the heart is stopped in the same manner as a stroboscope. As a consequence, any region defined in one image will correspond (by pixel coordinates) to the same region in other images of the same level. Thus the changes in the measured CT density of the myocardium due to the entry of a radio-opaque contrast material after intravenous injection can be quantified easily.

Measurement of the concentration of contrast material (diatrizoate sodium and diatrizoate meglumine) by ultrafast CT is similar to measurement of other indicators in the circulation, such as heat and dyes. This area of study has been referred to as indicator dilution theory.[2] Initially developed in the late 1800s by Stewart[3] to measure the transit times of an indicator in the circulation, indicator dilution theory was extended in the 1920s and the 1930s to include measurement of blood volume and cardiac output.[4,5] The Stewart-Hamilton equation for calculation of cardiac output is also used in ultrafast CT cardiac output measurements.[6,7] The Stewart-Hamilton equation is

$$CO = \frac{kQ}{A}$$

where CO = cardiac output, Q = amount of injected indicator, A = area under a blood-pool time-density curve, and k = constant of proportionality.

Computed tomography of cardiac output using the Stewart-Hamilton equation can be made by knowing the values of the amount of indicator (Q), the area under the time-density curve (A), and the constant of proportionality (k). The amount of indicator injected can be derived from the volume of contrast material injected and the known concentration of iodine. The relationship of the concentration of the indicator to CT number is linear over the complete range of dilutions.[7] These relationships are included in the constant of proportionality used in the Stewart-Hamilton equation. The area under a time-density curve from a region of interest (ROI) within the central blood pool and the cardiac output is shown in Figure 15-1.

Computed tomography measurements of cardiac output have been compared with other indicator dilution methods in several laboratories with very good

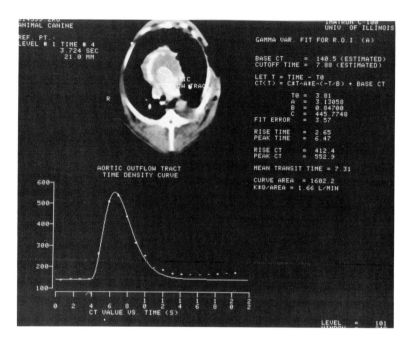

Figure 15-1 Left ventricular region of interest time-density curve. The data points on the graph correspond to the Hounsfield unit (HU) of the outlined region of interest (ROI) for each scan. The curve is produced by a bolus of intravenously injected contrast material. A measure of the area under the curve is made by use of a gamma-function fit of the data points.

results; correlation coefficients of 0.89 to 0.95 were obtained when compared with thermodilution measurements.[6,7] Additionally, the same cardiac output measurement can be obtained from ROIs anywhere in the circulation imaged.[7] Right-side and left-side cardiac output measurements generally can be made in the same scanning sequence (Figure 15-2). This ability indicates a great potential for the application of ultrafast CT in cases that require a precise measurement of intracardiac shunts.

There are several important assumptions of measuring cardiac output by indicator dilution theory that apply to measurement of myocardial blood flow as well. As described by Meier and Zeirler[2] in 1954, they are as follows:

1. The bolus injection mixes completely and instantaneously.
2. The indicator has the same transit characteristics as blood.
3. The indicator does not affect the blood flow.
4. The distribution of transit times through the circulatory path is constant. This is known as stationarity.
5. The concentration of the indicator can be measured accurately.

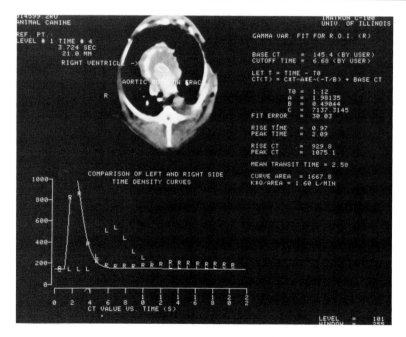

Figure 15-2 Right- and left-sided time-density curves. Two time-density curves representing ROIs from the right (R) and left (L) ventricle, respectively. The areas under each curve are within 5% of each other.

6. The measurement is based on the first pass of the indicator (that is, recirculation is accounted for).
7. The indicator does not wash out until after the peak concentration in the myocardium.

The equation used to measure blood flow by CT is derived from a single compartment model with a single inflow.[8,9] The outflow is assumed to be zero until after the peak indicator concentration (assumption 7). The equation is

$$F/V = \frac{MYC_{peak} - MYC_{base}}{\text{Blood-pool curve area}} \times 60$$

where F/V = the flow per unit volume of the outlined ROI, MYC_{peak} = the peak CT number of the myocardium, and MYC_{base} = the CT number of the myocardium before contrast enters. The units of CT blood flow measurement are milliliters per minute per volume of tissue. The volume of tissue is defined by the user-outlined region of interest of myocardium. A time-density graph (Figure 15-3) from animal data shows typical values of the peak and base myocardial CT numbers for an intravenous bolus administration (0.35 mL/kg). The last three assumptions of the single-compartment model must be analyzed with respect to a CT myocardial blood flow measurement.

The assumption of accurate measurement of indicator concentration within the myocardium would seem to be inherent (assumption 5); however, certain radiologic phenomena limit this assumption. As in any radiologic imaging technique, the external detection of radiation is a complex function of radiation type, dose, and material properties. The two phenomena that have been presumed to affect CT images are beam hardening and photon scatter.[10] Beam hardening occurs when the polychromatic x-rays are not attenuated evenly. Photon scatter is caused by interaction of the x-rays with the electron densities being measured. The net result of these phenomena is that the CT image is not "flat." Flatness of the image refers to the relationship between CT number and the actual attenuation factor of the material. For example, in a flat image the CT number for a homogeneous material would be the same anywhere in the region corresponding to the material. For heterogeneous materials (including animals and patients), neighboring regions of high and low density can be artifactually different from other high and low regions; hence, the flatness of the image cannot be assumed.

However, the assumption of accurate measurement of myocardial indicator concentration is not an assumption that the image is flat (that is, a CT value of 35 in the anterior myocardium connotes the same indicator concentration as a CT value of 35 in the septum). Rather, it is assumed that the linear relationship between the change in CT number due to the indicator concentration is constant throughout the image and for every image in the time series scan. The imaging process is not assumed to produce flat images; it is assumed to produce consistent images for changes only due to the indicator concentration in the region of interest. There are two instances where this assumption has been observed to be in doubt for the myocardium. First, when the ventricular chamber is maximally enhanced, some areas of myocardium appear to be decreasing in CT number when they could only be increasing. Second, the image composition is different in the baseline and peak myocardial scans because other organs move into the scanning plane (for example, diaphragm and liver). Thus the indicator concentration measurement is considered to be accurate if the subsequent images have the same image composition as the unenhanced image and there are no highly concentrated blood-pool regions present. For these reasons it is especially important to minimize the bolus amount of contrast medium and to prevent respiratory motion.

There are the consequences of flow through a closed system that need to be taken into account for accurate myocardial blood flow measurements: recirculation (assumption 6) and early washout of indicator from the myocardium (assumption 7). Because the circulatory system is closed, organ systems with short pathways that receive a significant amount of cardiac output will feed back a portion of the indicator before it has been completely washed out of the central circulation. This physiologic phenomenon will produce an increase in

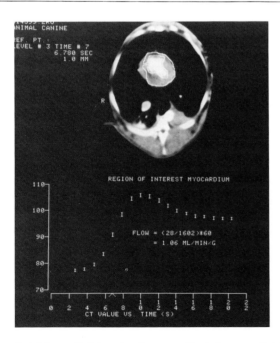

Figure 15-3 Myocardial time-density curve. Typical myocardial time-density curve for the entire myocardium imaged in one slice. The change from baseline is 28.3 HU produced by an input LV curve area of 1602, which is equal to a flow of 1.06 mL/min/g.

the indicator dilution curve area. However, monoexponential decay of washout curves can be used to model the first-pass washout. The presence of recirculation generally can be identified by using a gamma-variate fit to the early portion of the data.[11]

The measure of accumulation of indicator in the myocardium assumes that only the first pass of the indicator is measured. Regional myocardial flow is based on the principle that the first pass of the indicator is fractioned in accordance to regional flow. This is known as the Sapirstein principle,[12] which has been used to measure myocardial flow in other cardiac imaging systems.[13,14] In a series of animal experiments we did not observe significant recirculation of indicator before the myocardium reached peak concentration in the normal range of flow. When pharmacologic increases in cardiac flow were produced, the peak CT value of the myocardial curve used for flow calculation was defined before recirculation was detected in the left ventricle.

The Sapirstein principle assumes that there is no washout of the indicator from the tissue (assumption 7). When the indicator is composed of radiolabeled tracer microspheres, this assumption is valid because the small plastic spheres are trapped in the small arterioles. In the case of radio-opaque contrast material, the indicator actually is composed of dissolved iodine ions, which are assumed to have the same volume of distribution as plasma and as such will flow into the venous drainage of the heart. The single-compartment model assumes that the peak myocardial iodine concentration is reached before this occurs. The current spatial resolution of the C-100 ultrafast CT scanner is such that the only venous structure that can be visualized reliably is the coronary sinus; hence, the washout assumption can be tested only for a global flow measurement.

A comprehensive comparison of measurements of myocardial blood flow by CT and by microspheres has shown many interesting results.[9] In an anesthetized canine model, two separate experimental procedures were performed. The first procedure measured myocardial blood flow by CT and microspheres during a control period and during pharmacologic coronary artery dilation. The second procedure measured blood flow during a control period and during temporary left anterior descending coronary artery (LAD) occlusion. The results of these experiments showed that ultrafast CT could measure myocardial blood flow accurately ($r = 0.95$) for large sections of myocardium over a limited range of flow (0.4 to 1.6 mL/min per gram). Flow measurements during extreme vasodilation (>2.5 mL/min per gram) did not correlate statistically. Comparison of regional flows were limited by imprecise registration of CT and microsphere regions and thus showed reduced correlation ($r = 0.63$).

The second experimental procedure produced ischemia detectable by both CT and microspheres. The correlation of the measured ischemic flow was linear. The ischemic regions identified by CT were defined by repeated ROI tracings of slightly different size and location until the largest homogeneous region of reduced flow had been found. A homogeneous region of flow was defined as the last ROI tracing made before a significant change (that is, positive deflection) was produced in the time-density graph. This detection algorithm was used in order to ensure that a homogeneous region with the minimal flow was found. The microsphere data showed a distinct region of minimal flow in the same region of the heart as the CT-identified ischemic region. However, absolute geometric comparisons were limited by visualization of CT myocardial flow and the arbitrarily limited size of the microsphere regions.

The potential clinical significance of these results is that, in general, accurate measurements of myocardial flow can be made for large sections of myocardium for flow ranges from ischemic to approximately twice normal flow. The current flow model has not been shown to measure flow at higher rates, probably because of early washout of indicator, which has been observed from the coronary sinus when it has been imaged properly. An alternative model, used in conjunction with aortic root injections, accounts for the change in blood volume within the myocardium as the explanation for inaccurate high flow measurements has been proposed by Wang et al.[15] Using the dynamic spatial reconstructor, they have shown that the proportion of blood in the myocardium increases approximately logarithmically with blood flow as measured by microspheres (7% to 20%). Wang et al.[15] calculated myocardial blood flow as the ratio of myocardial blood volume to the indicator transit time to the volume of myocardium. Coupled with the blood volume information, a simple model of the mechanisms of increased myocardial blood flow is revealed: At normal flow rates the means of flow increases include vasodilation and capillary recruitment; at higher flow rates, when the coronary arterial bed is maximally dilated, further flow increases are a consequence of the blood volume passing more quickly through the organ. This model supports the observation by Wolfkiel et al.[9] that washout of indicator from the myocardium is the primary source of error in the single-compartment model of myocardial blood flow.

The actual flow resolution of the system has yet to be defined adequately. Whereas the image resolution can be defined in terms of the spatial frequency of an input phantom,[16] the smallest region possible for an accurate measurement of change in indicator concentration is not known. An estimation of the flow resolution can be made from an analysis of the CT image statistics.

Assuming that pixels from an ROI can be considered as samples of a random variable and that the next time-sequential data point represents another set of samples, a statistical test of the distribution of CT numbers within the ROIs can be made in order to test for significant difference.

The statistical level of significance of the test quotient depends on the number of pixels in the ROI; for regions of 30 pixels or more, a normal distribution can be assumed and a large sample test of means can be used; for regions of fewer than 30 pixels, a T distribution can be used.[17] For example:

$$Z \text{ or } T = \frac{\text{Mean}_1 - \text{Mean}_2}{\sqrt{\frac{\text{Var}_1 + \text{Var}_2}{N}}}$$

for two regions: (1) mean 35 ± 10 HU(Hounsfield unit), 30 pixels; (2) mean 41 ± 10 HU, 30 pixels.

$$Z = \frac{41 - 35}{\sqrt{\frac{100 + 100}{30}}} = 2.3237 \quad P<0.005$$

Thus, the mean CT values of the two regions can be considered significantly different. Reversing the test quotient asks the question: What is the smallest significant difference at $P<0.005$?

$$1.96 = \frac{\delta}{\sqrt{\frac{100 - 100}{30}}}$$
$$\delta = 1.96 \times \sqrt{200/30}$$
$$= 5.06 \text{ HU}$$

If the area under the input curve were 1500 CT numbers per second, then the estimated minimum flow difference that could be detected would be 0.20 mL/min per gram for regions of 30-pixel size (approximately 0.25 mL using a 260-mm reconstruction circle) and with a standard deviation of 10 HU.

The relationship between statistically different means and the number of pixels in a sample for different standard deviations is shown in Figure 15-4. This analysis assumes that each ROI in the time sequence has the same number of pixels in each sample and the same standard deviation. Also shown in Figure 15-4 are examples of the standard deviations of ROIs from the myocardium from a flow measurement, from a canine flow measurement, and a human myocardial flow measurement.

Preliminary analysis of these data suggests that there are three controllable factors that relate to the flow resolution of a myocardial flow measurement by ultrafast CT. First, the size of the ROI can be considered a variable similar to the magnification factor produced by

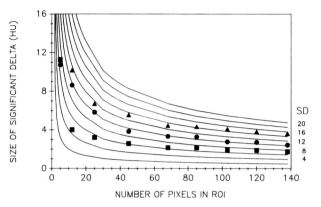

Figure 15-4 The relationship between the size of an ROI, noise, and flow resolution. The predicted smallest statistically significant change in myocardial density (delta) is shown as a function of the size of the region (number of pixels) and the amount of noise as measured by the standard deviation of the pixels in the region. The squares represent data from a canine experiment; the circles, from a female patient; and the triangles, from a male patient who was tilted such that the imaging plane included the diaphragm. This comparison shows the decrease in image quality for larger subjects and predicts a corresponding loss in flow resolution.

choice of the reconstruction circle or spatial smoothing. Second, the noise in the image as represented by the standard deviation of a homogeneous ROI is a complex function of the region size, location, radiation dose, and image structure. As summarized by Pentlow,[18] an approximation of the relationship between the noise of an image and these factors is

$$\text{Noise} \sim \frac{1}{\sqrt{B \cdot D \cdot h \cdot w^3}}$$

where B = fractional transmission of radiation by a patient, D = maximal surface radiation dose, h = slice thickness, and w = pixel width. Maximization of any of these parameters theoretically will improve the flow resolution of the measurement by reduction of the noise in the image. In order to compensate for the greater body mass of patients as compared with the dogs used in the experiments described, while preserving the spatial resolution, a comparable increase in radiation dose would be necessary to achieve similar flow resolution.

Third, the *delta*, or change in myocardial density, can be controlled by the input of contrast material from the left ventricle. The area under the time-density curve from the left ventricle scales the flow resolution. For a given myocardial flow state, the maximal increase in myocardial HU is limited by the amount of contrast material used. The greater the amount of contrast material used, the proportionately greater the increase in the myocardial time density. Assuming that heterogenous flow distribution within the myocardium is reflected by

ROIs of different density change, the degree to which these differences can be detected is a function of the ratio of the minimal myocardial HU difference to the maximal myocardial HU difference.

Thus the flow resolution is a combination of the fixed significant difference and the variable full-scale change produced by the input contrast bolus. The full-scale myocardial change is proportional to the amount of contrast in the input bolus as measured by the input time-density curve area. For example, for a minimally statistically significant delta of 1 HU detected for a given ROI, if the input curve area could produce a maximal myocardial delta of only 10 HU, the flow resolution could be defined as the smallest flow difference detectable (1/10 of the flow calculation). Table 15-1 shows sample calculations for the same flow measurement but different flow resolutions. In canine measurements the flow resolution was estimated at 0.05 mL/min per gram for typical input areas of 2000 HU-second, deltas of 25 HU, and standard deviations of 5 HU for regions of 100 pixels (approximately 1 mL of tissue).

The visualization of flow differences within the myocardium is difficult with a monochrome representation of single image. However, pseudocolor images can be produced that represent the delta of each pixel versus time. Arbitrarily assigning a color to values of delta for myocardial pixels results in a display of the distribution of the contrast material as it enters the myocardium. An example of this process is shown in Figure 15-5. The LAD-occluded images clearly show a difference from the unoccluded images. Although interpretation of these images for regional flow information is constrained by flow resolution issues, this display algorithm clearly provides advantages over a monochromatic representation of the data. These advantages include visualization of regional differences and characterization of transit time variability within the myocardium.

In summary, the potential for clinical regional myocardial blood flow measurements has been demonstrated in animal experiments. However, several issues remain to be resolved before the technique can become clinically useful. First, measurement of high flows produced by exercise or pharmacologic coronary artery dilation cannot be measured using the current flow algorithm. A method accounting for the regional washout of indicator must be developed before accurate clinical estimations of coronary reserve are possible. Visualization of flow patterns with monochromatic images is not possible, and would restrict widespread use of ultrafast CT as a tool for measuring myocardial blood flow. Pseudocolor representation of the flow data appears to resolve this issue. The flow resolution issue remains to be more accurately defined before widespread clinical use is feasible. As with any new measurement technique, accuracy is an integral factor in its clinical acceptance. Further progress on these issues should be expected to lead to an increase in the use of ultrafast CT for clinical myocardial blood flow measurements.

REFERENCES

1. Boyd DB. Computerized transmission tomography of the heart using scanning electron beams. In: Higgins CH, ed. *Computed Tomography of the Heart and Great Vessels*. New York: Futura; 1983: 45–52.

2. Meier P, Zeirler KL: On the theory of the indicator-dilution method for the measurement of blood flow and volume. *J Appl Physiol*. 1954;12:731–744.

3. Stewart GN. Researches on the circulation time in organs and on the influences which affect it. *J Physiol (Lond)*. 1894;15:177–221.

4. Stewart GN. Researches on the circulation time and on the influences which affect it, V: The circulation time of the spleen, kidney, intestine, heart (coronary circulation) and retina with some further observations on the time of the lesser circulation. *Am J Physiol*. 1921;58:278–295.

5. Hamilton WF, Moore JW, Kinsman et al. Studies on the circulation, IV: Further analysis of the injection method, and of changes in hemodynamics under physiologic and pathologic conditions. *Am J Physiol*. 1931;99:534–551.

6. Herfkens RJ, Axel L, Lipton MJ, et al. Measurement of cardiac output by computed transmission tomography. *Invest Radiol*. 1982;17:550–553.

7. Wolfkiel CJ, Ferguson JL, Chomka EV, et al. Determination of cardiac output by ultrafast computed tomography. *Am J Physiol Imaging*. 1986;1:117–121.

8. Rumberger JA, Feiring AJ, Lipton MJ, et al. Use of ultrafast computed tomography to quantitate regional myocardial perfusion: a preliminary report. *J Am Coll Cardiol*. 1987;9:59–69.

9. Wolfkiel CJ, Ferguson JL, Chomka EV, et al. Measurement of myocardial blood flow by ultrafast computed tomography. *Circulation*. 1987;76:1262–1273.

10. Joseph PM: Artifacts in computed tomography. In: Newton TM, Potts DG, eds. *Radiology of the Skull and Brain: Technical Aspects of Computed Tomography*. St Louis, Mo: CV Mosby; 1980:3980–3982.

11. Thompson HH, Starmer CF, Whalen RE, et al. Indicator transit time considered as gamma variate. *Circ Res*. 1964;14:502–512.

12. Sapirstein LA. Regional blood flow by fractional distribution of indicators. *Am J Physiol*. 1958;193:161–169.

13. Shah A, Schelbert HR, Schwaiger M, et al. Measurement of regional myocardial blood flow with N-13 ammonia and positron-

Table 15-1 Sample Flow Resolutions

Input Area	Maximal Delta Observed	Flow Measurement	Minimal Delta Detectable	Flow Resolution*
800	10	0.75	2.0	0.15
1600	20	0.75	2.0	0.08
800	10	0.75	4.0	0.30
1600	20	0.75	4.0	0.15

*Theoretical differences in flow resolution for the same size region produced by different combinations of input-curve area and minimally significant deltas.

emission tomography in intact dogs. *J Am Coll Cardiol.* 1986;5: 192–199.

14. Strauss HW, Harrison K, Langer JK, et al. Thallium-201 for myocardial imaging: Relation of thallium-201 to regional myocardial perfusion. *Circulation.* 1975;51:641–653.

15. Wang T, Xuesi W, Chung N, et al. Myocardial blood flow estimated by synchronous, multislice, high-speed computed tomography. *IEEE Trans Biomed Eng.* 1989;8:70–77.

16. Assimakopoulus PA, Boyd DP, Jaschke W, et al. Spatial resolution analysis of computed tomographic images. *Invest Radiol.* 1986;21:260–271.

17. Miller I, Freund JE. *Probability and Statistics for Engineers.* Englewood Cliffs, NJ: Prentice-Hall; 1985:218–222.

18. Pentlow KS: Dosimetry in computed tomography. In: Newton TM, Potts DG, eds. *Radiology of the Skull and Brain: Technical Aspects of Computed Tomography.* St Louis, Mo: CV Mosby; 1980:4237.

Chapter 16

Evaluation of Left Ventricular Function by Exercise Bicycle Ergometry Ultrafast Computed Tomography

Eva V. Chomka, MD, and Bruce H. Brundage, MD

Coronary artery disease remains a major cause of mortality and morbidity in the United States despite ongoing advances. Approximately 6 million persons in the United States have symptomatic coronary artery disease.[1] An estimated 280,000 coronary bypass operations and 176,000 coronary angioplasties were performed in 1987, and the volume of angioplasties is increasing annually.[1] More elderly patients and occasionally sicker patients, some even in cardiogenic shock, are undergoing percutaneous transluminal coronary angioplasty. Coronary angioplasty has a 30% restenosis rate at 6 to 8 months; as the number of cases studied increases, the number of potential redo evaluations escalates.[1] The acceptance of internal mammary artery bypass has also augmented the value and use of coronary bypass in addition to standard saphenous vein bypass grafting.[2]

As clinicians, we face the dilemma of whether the patient has physiologically significant coronary artery disease. First of all, there are significant discrepancies in interpreting coronary artery stenoses subjectively by coronary angiography.[3] Second, anatomic evaluation is not always adequate for making clinical decisions. Physiologic assessment of the effects of coronary stenoses is especially important in deciding whether anatomy truly represents physiologic significance, such as in patients with single-vessel disease, complex multivessel disease, or collateral disease.

CARDIOVASCULAR RESPONSE TO EXERCISE

Previous exercise studies have revealed that the cardiovascular response to exercise is dependent on several factors, including the type of exercise performed. Ideally, an isotonic type of exercise is performed that leads to the expenditure of aerobic energy. Stress electrocardiography,[4] radionuclide angiography,[5,6] echocardiography,[7-9] thallium scintigraphy,[5] and ultrafast computed tomography (CT)[10,11] rely on isotonic exercise. Isometric exercise such as handgrip, alone or in conjunction with cold pressor, has also been used.[12,13] Atrial pacing and the administration of dipyridamole thallium are alternate methods of inducing cardiac stress, using increased heart rate or vasodilation.[14-16] Typically, sufficient aerobic exercise will augment cardiac output by increasing both heart rate and stroke volume until a desired effect is attained. Isometric exercise can be performed easily; however, it does not necessarily achieve the same physiologic effects as isotonic exercise.

Stress electrocardiography is the most commonly used form of exercise for detection of ischemia. Patients who have multiple risk factors for coronary artery disease and a classic history for angina pectoris have a high bayesian probability for ischemia and pose the least problem for detection of ischemia. Women notoriously have a large number of false-positive exercise electrocardiograms, as evidenced by ST depression, and may undergo subsequent coronary arteriography that documents normal coronary arteries. Patients who are taking digoxin may have ST depression secondary to therapy without ischemia. Persons with left ventricular hypertrophy and bundle-branch blocks may have secondary repolarization changes that make the diagnosis of ischemia impossible on the basis of electrocardiography alone.[4]

ALTERNATIVE METHODS OF CARDIOVASCULAR EVALUATION

Alternative modalities have been developed to overcome the limitations of electrocardiographic evaluations. Physiologic evaluation of the impact of coronary artery stenosis on left ventricular function by noninvasive techniques is an important clinical area. Radionuclide angiographic equilibrium studies during rest and exercise have been performed for many years to evaluate the global ejection fraction and to address qualitatively some aspects of regional function. There have been problems with cardiac chamber overlap and depth attenuation of images.[5] Single-photon emission CT promises to improve resolution, ease problems with overlap, and provide more localization. An ejection fraction increase of 5 percentage units or greater is considered a normal response. Localized regional analysis is available; however, since resolution is limited, complete base-to-apex evaluation is impossible. The right ventricle also poses problems for radionuclide angiography because of difficulty in discriminating the right ventricle from the left ventricle.

Thallium scintigraphy was developed for use in conjunction with exercise treadmill electrocardiography. Thallium 201 is useful in evaluating relative myocardial perfusion. Limitations include tissue overlap, such as overlap of breast and diaphragm; having to give an injection to the patient at peak exercise; the usual requirement of having to move the patient to a gamma nuclear scanner; and the need to rescan 3 to 4 hours after exercise.[5] Patients may be deprived of food for 7 to 8 hours. Single-photon emission CT also promises improved results; however, exercise studies have not yet been done.

Exercise echocardiography is another modality that can evaluate function response.[7,8] At present, however, the operator must be able to scan the area of interest during ischemia; otherwise the abnormality will be missed. This method does not evaluate the entire left ventricle at one time, so regional abnormalities may be missed. Dynamic exercise studies are more difficult than rest evaluations since the method is affected by chest wall motion in addition to cardiac motion. Supine, upright bicycle, and post-treadmill exercise echocardiography have been used. Exercise Doppler echocardiography has potential in assessing change in aortic velocity as a marker of left ventricular dysfunction.[9]

Alternate stress induction methods such as handgrip,[12] cold pressor,[13] and atrial pacing[14] have been used in conjunction with echocardiography. All have major limitations because of the less-than-maximal exercise physiologic responses to these stimuli. Different limits of responsiveness include handgrip strength, sole chronotropic response of atrial pacing without major preload or afterload alterations, and pain secondary to cold exposure.

ULTRAFAST CT BICYCLE ERGOMETRY

In the search for a technique with enhanced image resolution, therefore resulting in more regional left ventricular data, we developed ultrafast CT bicycle ergometry at the University of Illinois (C-100 ultrafast CT scanner, Imatron-Picker Inc., San Francisco, Calif; and model 8450 cardiac stress test ergometer, Engineering Dynamics Corporation, Lowell, Mass) (Figure 16-1). This modality has evolved considerably over the last several years. The initial setup consisted of a bicycle ergometer without a bicycle seat or shoulder stabilization. Exercise scans were performed within 10 to 15 seconds of cessation of exercise to stabilize the patient to a position identical with that of the rest images.

Techniques

A small intravenous 18-gauge catheter is inserted into an antecubital vein. If there are no medial antecubital veins, the external jugular vein is used as an alternative. Patient monitoring involves a single modified lead V_5 to

Figure 16-1 C-100 ultrafast CT scanner and model 8450 cardiac stress test bicycle ergometer with a patient positioned in the short-axis view for acquisition of exercise images.

obtain the tallest possible QRS signal for the ultrafast CT scan triggering. Patient positioning is accomplished by initial laser localization of the horizontal and vertical axes of the body. The patient's point of maximal left ventricular impulse is palpated, or the left ventricular position is estimated. The scanning table is tilted a maximum of +17° to +20° to allow for bicycle ergometry. The table is slanted to the right to approximate the left ventricular short axis, +13° to +15°. An eight-level localization scan is performed to identify precisely the left ventricular base. Scanning starts at the base and proceeds to the left ventricular apex.

Standard contrast medium in amounts of 25 to 30 mL is injected for rest studies and 30 to 40 mL for exercise studies. A preparation of diatrizoate meglumine and diatrizoate sodium (Renografin-76, Squibb Diagnostics) is used for most patients. If a patient has left ventricular failure, borderline renal insufficiency, or a history of contrast medium reactions, a nonionic agent such as iopamidol (Isovue-350, Squibb Diagnostics) is used.

Since we prefer to use small-bolus injections for ventricular function assessment, a method of predetermining circulation time is beneficial. In the initial trials of ultrafast CT scanning, we scanned in the flow mode. Images were acquired every several heart beats to measure the interval from injection to left ventricular cavity opacification. However, this required at least 6 minutes and was impractical during exercise assessment. An alternate method was developed by performing ear densitometry, using a Waters D402A densitometer (Waters Instruments, Rochester, Minn) and 5 mg of indocyanine green, simultaneously with CT flow determination of time to left ventricular contrast medium appearance.[17] A formula (contrast peak time = 0.65 ± 11 indocyanine green peak time) is used for the rest evaluation.[15]

For exercise studies we add 2 seconds, since there appears to be greater variability due to patients' increased respiratory effort that may delay venous return. An actual densitometer sample recording is depicted in Figure 16-2. The relationship between the left ventricular time and ear appearance time is depicted schematically in Figure 16-3.

Initial scanning is performed to evaluate rest left ventricular function from the base of the left ventricle in the 50-ms/image cine-mode to the apex. The patient is then exercised with 25-W/s increments of ergometer resistance every 2 minutes until peak exertion. Debilitated or postmyocardial infarction patients are exercised with 10-W/s increments every 2 minutes.

Data Analysis

Parameters analyzed in current ultrafast CT exercise studies are shown in Table 16-1.

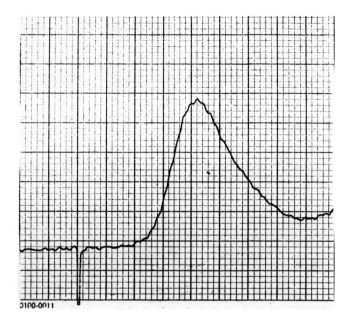

Figure 16-2 Ear indocyanine green densitometry. Each major division represents 2 seconds. The indicator dilution curve peaks at 9 seconds; therefore, the peak left ventricular enhancement would be predicted to occur at 6 seconds.

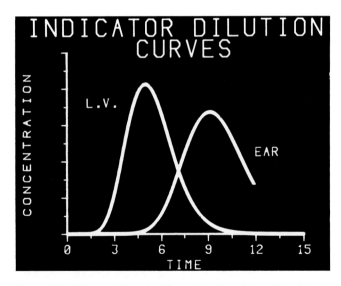

Figure 16-3 Schematic drawing illustrating the relationship of simultaneous left ventricular (L.V.) ultrafast CT and ear densitometry indicator dilution curves.

Table 16-1 Ultrafast CT Exercise Evaluation

- Visual assessment of regional wall motion for abnormality
- Global left ventricular ejection fraction at rest
- Rest and exercise ejection fraction at comparable levels
- Quantitative regional wall motion analysis of each 30° segment within each level

Visual Assessment of Regional Wall Motion for Abnormality

Initial evaluation involves visual identification of regional wall motion abnormality as the ultrafast CT scan is played in a closed-loop movie of one cardiac cycle. A preliminary evaluation of normal versus abnormal wall motion is made (Figures 16-4 and 16-5).

Global Left Ventricular Ejection Fraction

Rest and exercise studies are analyzed with a computerized regional wall motion analysis program. Each cross-sectional image of the left ventricle from base to apex is analyzed by tracing the inner boundaries of the left ventricular wall at end-systole and end-diastole, excluding the papillary muscles. Tomographic ejection

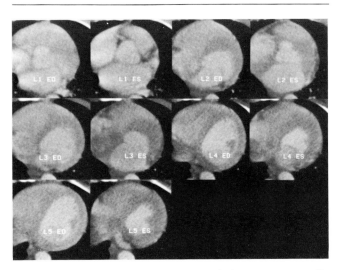

Figure 16-4 Cross-sectional images of the contrast medium–enhanced upper left ventricle at end-diastole (ED) and end-systole (ES) in patient A with normal wall motion.

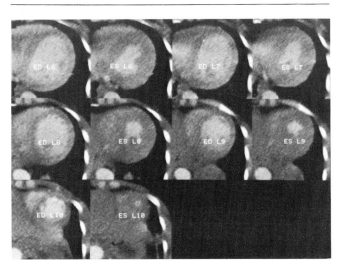

Figure 16-5 Cross-sectional images of the contrast medium–enhanced lower left ventricle at end-diastole (ED) and end-systole (ES) in patient A with normal wall motion.

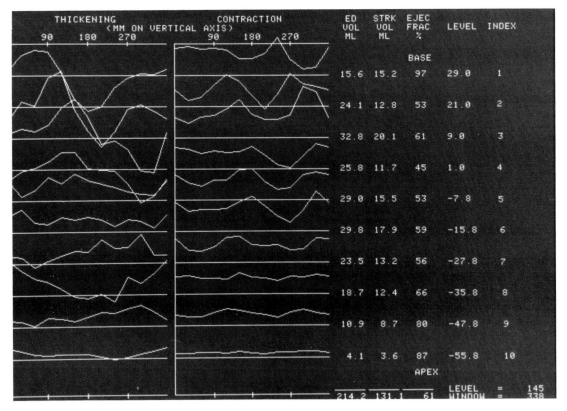

Figure 16-6 Rest global and tomographic-level left ventricular ejection fractions (EJEC FRAC) in patient A. ED VOL, end-diastolic volume; STRK VOL, stroke volume.

fractions and end-diastolic and end-systolic volumes are calculated at each level. After summation of the volumes at each tomographic (8 mm) level, a global ejection fraction is calculated at rest and exercise. The ejection fraction is considered abnormal during exercise if it fails to increase 5 percentage units above the rest value. Usually 7 to 10 scan levels are necessary to include the entire left ventricle for evaluation of global function (Figure 16-6). Dilated left ventricles may require up to 16 levels.

Rest and Exercise Ejection Fractions at Comparable Levels

Comparable rest and exercise levels are analyzed for segmental ejection fractions and compared directly (Figures 16-7 and 16-8).

Quantitative Wall Motion Analysis

Quantitative analysis can be performed by using a computer program. The subsegmental ejection fraction can be calculated by placing the 0° starting point on the posterior wall; the right ventricular and left ventricular junction is 180°. The computer assists in generating counterclockwise 30° radii, using the endocardial end-diastolic centroid as the origin of the radii for the end-systolic and end-diastolic images. The mitral and aortic valve segments, which consist of regions not bordered by myocardium, are excluded from the analysis. The divisions used in our evaluations are as follows: the lateral wall was 60° to 150°, the anteroseptal wall was 150° to 300°, and the posterior wall was 300° to 60° (Figures 16-9 and 16-10).

Corresponding cross-sectional images of the left ventricle are analyzed at rest and during exercise. Internal left ventricular landmarks (aortic valve, mitral valve, anterolateral papillary muscle, posteromedial papillary muscle, and apex) are used to ensure that similar levels are compared during rest and exercise. To compare different segments within the same patient and results among different patients, the ejection fraction of each segment is normalized with the use of the following formula:

$$\text{Normalized ejection fraction (EF)} = \frac{\text{Segmental EF} - \text{Global EF}}{\text{Global EF}}$$

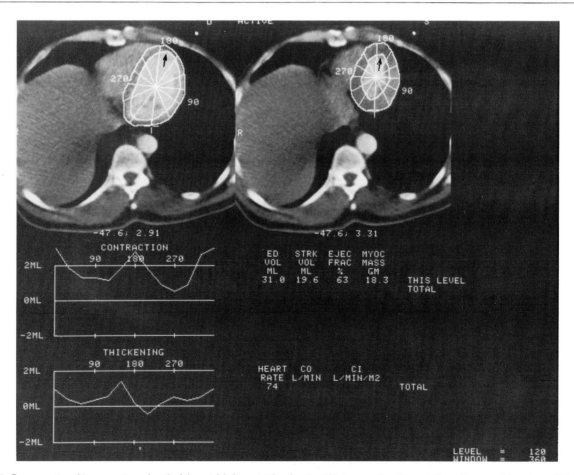

Figure 16-7 Cross-sectional images at one level of the mid-left ventricle of patient B at rest using the end-diastolic centroid as the radial origin. ED VOL, end-diastolic volume; EJEC FRAC, ejection fraction; MYOC MASS, myocardial mass; STRK VOL, stroke volume; CO, cardiac output; CI, cardiac index.

244 COMPARATIVE CARDIAC IMAGING

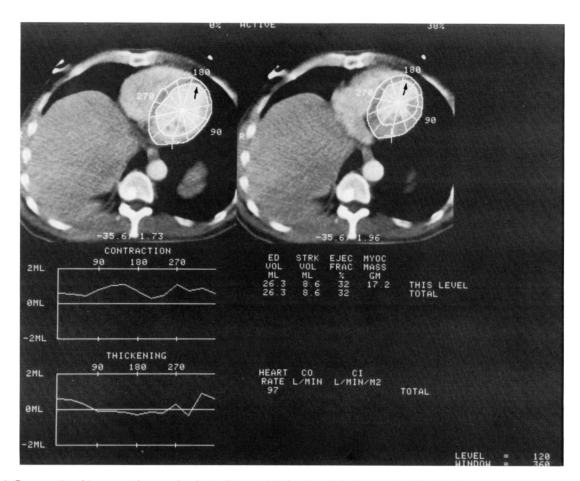

Figure 16-8 Cross-sectional images at the same level as in Figure 16-7 of patient B during exercise. The development of anterior wall ischemia is shown (arrows), at a heart rate of 97 beats per minute with 32% tomographic ejection fraction. Abbreviations are defined in legend of Figure 16-7.

ANGLE	STROKE VOL ML	EJEC FRAC %	MYOC MASS GM
0 TO 30	1.59	60.0	2.69
30 TO 60	1.07	48.9	2.18
60 TO 90	0.90	45.0	1.70
90 TO 120	1.25	51.0	1.20
120 TO 150	1.26	46.3	1.60
150 TO 180	1.92	57.2	2.03
180 TO 210	1.96	65.4	2.62
210 TO 240	1.14	58.1	2.51
240 TO 270	0.68	46.6	2.87
270 TO 300	0.78	39.9	2.79
300 TO 330	1.59	49.3	2.65
330 TO 360	1.75	55.0	3.44

LEVEL = 145
WINDOW = 338

Figure 16-9 Cross-sectional 30° radius ejection fractions for the images shown in Figure 16-7. Abbreviations are defined in legend of Figure 16-7.

Regional ejection fractions greater than 2 SD from the normalized value are considered abnormal.

Clinical Experience

In our initial exercise evaluation of 20 patients who also had coronary angiography, global left ventricular ejection fractions ranged from 20% to 75%. Thirteen men and seven women were evaluated, ranging in age from 38 to 72 years (mean age 54 years). Eighteen patients were taking cardiac medications at the time of the initial evaluation. The peak workload was 75 W/s, which is equivalent to functional class I. The exercise scan was measured only at the mid-left ventricle, since the patients moved considerably during the bicycling. Patients were stratified according to no disease or one-, two-, or three-vessel disease by angiography. In patients with no vessel disease and one-vessel disease the ejection fraction either stayed the same or increased. In patients with two-vessel disease, the ejec-

LEVEL −35 6 MM (6)

DIASTOLE			SYSTOLE	
VOLUME TO INNER WALL ML	VOLUME TO OUTER WALL ML	ANGLE	VOLUME TO INNER WALL ML	VOLUME TO OUTER WALL ML
1.88	3.47	0 TO 30	1.28	3.47
1.78	2.73	30 TO 60	1.24	2.73
1.90	2.64	60 TO 90	1.48	2.53
2.29	3.19	90 TO 120	1.47	2.25
2.87	3.71	120 TO 150	1.81	2.54
2.95	4.14	150 TO 180	1.84	2.89
2.17	3.46	180 TO 210	1.48	2.49
1.65	2.94	210 TO 240	1.36	2.51
1.66	3.09	240 TO 270	1.14	2.36
2.55	4.10	270 TO 300	1.46	3.31
2.87	5.55	300 TO 330	2.14	4.49
2.96	5.17	330 TO 360	2.05	5.15

ANGLE	STROKE VOL ML	EJEC FRAC %	MYOC MASS GM
0 TO 30	0.60	31.8	1.66
30 TO 60	0.54	30.3	0.99
60 TO 90	0.43	22.5	0.77
90 TO 120	0.82	35.9	0.95
120 TO 150	1.07	37.2	0.88
150 TO 180	1.11	37.6	1.25
180 TO 210	0.70	32.1	1.36
210 TO 240	0.29	17.5	1.35
240 TO 270	0.52	31.4	1.50
270 TO 300	1.09	42.7	1.62
300 TO 330	0.73	25.5	2.81
330 TO 360	0.91	30.8	2.32

LEVEL = 120
WINDOW = 360

Figure 16-10 Cross-sectional 30° radius ejection fractions during exercise for the images shown in Figure 16-8. Abbreviations are defined in legend of Figure 16-7.

tion fraction increased in half the patients and decreased in half. All of the patients with three-vessel disease had a decrease in the ejection fraction. Of the twenty patients, thirteen patients had an ejection fraction response that increased and seven patients had a response that stayed the same or decreased. Changes in the end-systolic dimension of the mid-left ventricular level were also evaluated. Ten patients had a normal response and 10 other patients had an increase in the end-systolic dimension, which is considered an abnormal response.[10]

In a more recent study of 31 patients who underwent cardiac catheterization with ventriculography and coronary arteriography for evaluation of chest pain,[10] exercise ultrafast CT was performed within 1 month of the cardiac catheterization and interpreted without knowledge of the angiographic results. Exercise treadmill electrocardiography and thallium scintigraphy were also performed in some patients. Coronary angiograms were analyzed by Brown's method of quantitative angiography. In this study there were 13 men and 18 women, with a mean age of 52 ± 8 years. Of these, 15 patients had normal coronary arteries and 16 patients had coronary artery disease. Exclusion criteria included previous myocardial infarction, coronary artery bypass grafting, and coronary angiography. The results of this study are summarized in Tables 16-2 to 16-6.[11]

Table 16-4 indicates heart rate, workload, and exercise duration for patients with coronary artery disease and normal coronary arteries. Fifteen patients had significant (50%) reduction in the diameter of one or more coronary arteries. Single-vessel disease was found in twelve patients and multivessel disease in three. Normal coronary arteries were found in sixteen patients (Tables 16-3 and 16-5). Ten of the fifteen patients with coronary artery disease and five of the twelve with normal coronary arteries developed chest pain during exercise ultrafast CT ($P<0.05$). All studies were performed while the patients were taking their usual medications, including β-blockers, calcium blockers, or calcium blockers and a nitrate.

The rest ejection fraction ranged from 50% to 79% in all patients from whom complete data were obtained. Four patients were less than optimally positioned during exercise and were excluded from analysis. The mean rest ejection fraction of the 12 patients without coronary artery disease was 68% ± 6% (SD) and the mean at peak exercise was 75% ± 6% ($P<0.001$). In 10

Table 16-2 Results of Exercise Ultrafast CT in 15 Patients with Coronary Artery Disease

Patient No.	Levels Analyzable	Ejection Fraction (%)	No. of Abnormal Segments During Exercise	Size of Abnormal Segmental Changes (°)
1	4	−9*	3	0 to 60
				330 to 360
2	3	−8	7	60 to 120
3	7	−5	1	300 to 330
4	5	−19	4	120 to 240
5	6	+11	0	0
6	7	−9	9	0 to 30
				240 to 360
7	4	−9	7	150 to 300
8	6	−10	8	150 to 300
9	7	−4	10	0 to 60
10	5	−10	3	180 to 270
11	5	−7	2	180 to 240
12	6	−1	3	120 to 180
13	5	−6	5	180 to 270
14	7	−7	5	150 to 210
15	5	−10	11	150 to 300

*−, Negative; +, positive.

Source: Reprinted with permission from *Journal of the American College of Cardiology* (1989;13:1077), Copyright © 1989, American College of Cardiology.

Table 16-3 Results of Quantitative Coronary Angiography and Other Stress Tests in 15 Patients with Coronary Artery Disease

		Quantitative Coronary Artery Disease				Other Stress Tests	
		% Stenosis		Minimal Diameter	Minimal Area	Treadmill Electrocardiographic	Thallium 201
Patient No.	Vessel*	By Diameter	By Area	(mm)	(mm^2)	Results†	Results†
1	LCx	100				+	+
2	LAD	100				+	+
3	RCA	76	94	0.81	0.51	0	0
4	LAD	88	98	0.18	0.02	−	+
5	LCx	56	80	0.70	0.38	−	0
	RCA	57	82	0.72	0.41		
	Diag	73	93	0.30	0.07		
6	RCA	93	99	0.15	0.02	−	0
7	LAD	100				−	+
8	LAD	81	96	0.41	0.13	−	−
9	LCx	100				0	0
	RCA	74	93	0.62	0.30		
10	LAD	88	98	0.22	0.04	+	+
11	LAD	74	93	0.46	0.16	−	−
12	LAD	64	87	1.09	0.93	0	0
	RCA	67	89	0.63	0.31		
13	RCA	74	93	0.56	0.24	+	0
14	LAD	100				−	0
15	LAD	97	99	0.06	0.01	0	0

*LAD, Left anterior descending coronary artery; LCx, left circumflex coronary artery; RCA, right coronary artery; Diag, diagonal branch.
†+, Positive; −, negative; 0, not done.

Source: Reprinted with permission from *Journal of the American College of Cardiology* (1989;13:1077). Copyright © 1989, American College of Cardiology.

Table 16-4 Exercise Ultrafast CT

Measurement	Coronary Artery Disease	Normal Coronary Arteries
Heart rate	103 ± 9	113 ± 20
Workload (W/s)	45 ± 18	50 ± 22
Duration (min)	8 ± 2	9 ± 3

Table 16-5 Coronary Anatomy (70%)

Involvement	No. of Patients	Coronary Vessel*		
		LAD	LCx	RCA
One vessel	10	7	1	2
Two vessels	3	3	1	2
Three vessels	2	2	2	2

Normal coronary arteries 16

*LAD, left anterior descending coronary artery; LCx, left circumflex coronary artery; RCA, right coronary artery.

Table 16-6 Combined Results

	No. of Patients	Sensitivity	Specificity	Predictive Value
Function evaluation				
Ejection fraction	27	83%	86%	87%
Ultrafast CT wall motion	27	86%	92%	92%
Abnormality evaluation				
Exercise treadmill	24	75%	58%	64%
Exercise thallium	15	88%	67%	78%

(83%) of the patients, the ejection fraction increased 5 percentage units over rest; the mean increase was 7% ± 4% in this group. In the two other patients in this group, who were not taking β-blockers, a change of less than 5 percentage units was observed in one; there was no change in another (see Figure 16-11).

The mean ejection fraction at rest of the 15 patients with coronary artery disease was 65% ± 7%. During exercise, the ejection fraction decreased in 14 patients (93%) with a mean ejection fraction during exercise of 60% ± 7% ($P<0.001$). The mean decrease was 7% ± 3%. Only one patient with coronary artery disease had an increase in the ejection fraction during exercise. Patients with left anterior descending stenosis had a greater decline in the exercise ejection fraction than did those without ($P<0.05$). β-Blockers did not alter the ejection fraction response (see Figure 16-11).

The intraobserver variability for measuring the global ejection fraction on 2 different days was 1.46% ± 1.05%.

The quantitative method for the analysis of left ventricular wall motion was performed in 26 of 27 patients. Regional left ventricular contraction was found to be somewhat heterogeneous at rest. The distribution of the normalized segmental ejection fractions of patients with normal coronary arteries is shown in Figure 16-12. This heterogeneity decreased during exercise in patients with normal coronary arteries.

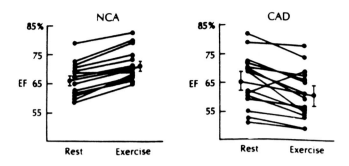

Figure 16-11 Individual values of the global ejection fractions (EF) and the mean values at rest and during exercise of the 12 patients with normal coronary arteries (NCA) (left) and the 15 patients with coronary artery disease (CAD) (right). *Source:* Reprinted with permission from *Journal of the American College of Cardiology* (1989;13:1077), Copyright © 1989, American College of Cardiology.

Twelve (86%) of the fourteen analyzable patients with coronary artery disease and one patient without coronary artery disease developed regional wall motion abnormalities or worsening of rest abnormalities during exercise (Table 16-2). Regional wall motion became abnormal in the anteroseptal region during exercise in eight (89%) of nine patients with significant left anterior descending lesions. Figure 16-13 shows the distribution of the regional ejection fractions at rest in a patient with coronary artery disease. The abnormality of regional contraction is demonstrated by the fact that the rest distribution of subsegmental ejection fractions broadens during exercise.

Of the 27 patients undergoing exercise ultrafast CT, 19 (70%) also underwent a treadmill stress test (Table 16-3). The electrocardiographic response was abnormal in 5 (45%) of 11 patients with angiographically demonstrated coronary artery disease and normal in 4 (50%) of 8 with normal coronary arteries. Ten patients (37%) also underwent thallium 201 myocardial scin-

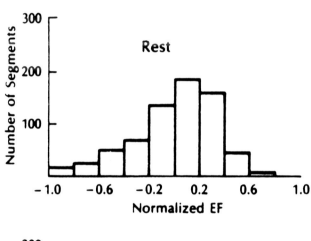
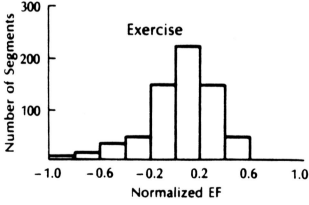

$$\text{Normalized EF} = \frac{\text{Segment EF} - \text{Global EF}}{\text{Global EF}}$$

Figure 16-12 The distribution of the normalized segmental ejection fractions (EF) of the 12 patients with normal coronary arteries at rest (top) and after exercise (bottom). The heterogeneity of the left ventricular contraction is represented by the spread of the normalized ejection fractions. *Source:* Reprinted with permission from *Journal of the American College of Cardiology* (1989;13:1078), Copyright © 1989, American College of Cardiology.

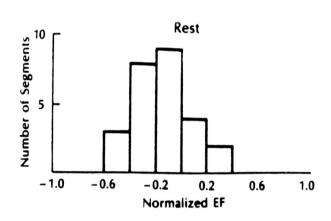
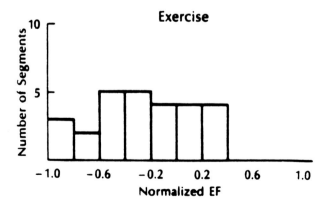

Figure 16-13 Distribution of the normalized segmental ejection fraction (EF) of the anterior wall segments of a patient with an 88% left anterior descending coronary artery stenosis. A normal contraction pattern is observed at rest (top), which becomes abnormal during exercise (bottom) as represented by the increased spread of the normalized segments to the left. *Source:* Reprinted with permission from *Journal of the American College of Cardiology* (1989;13:1078), Copyright © 1989, American College of Cardiology.

tigraphy. The scan was interpreted as abnormal in 5 (71%) of 7 patients with abnormal coronary angiography and normal in 2 (68%) of 3 patients with normal coronary arteries.

The combined results are shown in Table 16-6.

We have since extended our experience to scanning patients before and after coronary angioplasty. Regional wall motion evaluations were performed both qualitatively and quantitatively, at rest and during peak exercise.[18] Clinical decisions regarding culprit vessels and lesions are now being made routinely on the basis of these results.

Myocardial blood flow may also be evaluated with exercise ultrafast CT.[19] Ten to thirteen images are acquired over 20 to 30 cardiac cycles at six to eight levels of the heart. One of the current limitations of exercise blood flow evaluation is that the patient has a difficult time sustaining a breath hold for the entire imaging period (10 to 15 seconds) after exercise. Measurement of myocardial blood flow is arduous and requires several hours for a complete evaluation. A color-coded evaluation of blood flow is under development that will help to facilitate this evaluation and provide for immediate visual assessment of blood flow patterns.

COMMENTS ON EXERCISE METHODS

The exact volumetric responses of the left ventricle during rest and exercise are modified by the patient's position and the exercise protocol. Ultrafast exercise evaluation is performed in a semisupine position. There have been many previous studies using radionuclide angiography to measure relative ventricular volume and define the exercise response to bicycle ergometry. Protocol differences using radionuclide angiography make direct interstudy comparisons impossible.[20–23] Rodrigues et al[23] recently reported higher heart rates when patients were upright rather than supine during radionuclide ventriculography. No difference in left ventricular ejection fractions was observed in the same normal persons undergoing evaluations in both positions. There is consensus that left ventricular end-diastolic volumes increase during ischemia in both supine and upright positions.[21,22]

The ejection fraction may decrease during exercise owing to conditions other than ischemia, such as increased afterload in hypertensive patients. There is also considerable overlap of the ejection fraction response in normal persons and patients with coronary artery disease, as demonstrated by Osbakken et al.[24]

Evaluation of regional wall motion is an indirect assessment of what has transpired on a molecular level. First, a certain amount or critical mass of myocardial tissue must be involved before abnormalities can be detected by physiologic imaging evaluations of wall motion. Second, reperfusion that occurs after thrombolysis and/or angioplasty intervention for myocardial infarction may precede improvement in wall motion abnormalities. Stunned myocardium, as defined by Braunwald and Kloner,[25] may be present after any acute event, and a return to normal left ventricular function may occur much later. Assessment of perfusion and metabolism holds the key to more direct physiologic evaluation.

Three modalities can demonstrate wall motion abnormalities during exercise: radionuclide angiography, echocardiography, and ultrafast CT. The largest experience is with radionuclide angiography. Radionuclide angiography acquires images over several minutes. Exercise echocardiography can scan only one specific region of the heart at a time, and total evaluation may take several minutes. Ultrafast CT scans the entire left ventricle within several seconds; therefore, it offers a greater probability of detecting localized ischemia by wall motion.

Exercise thallium myocardial scintigraphy is preferred over exercise radionuclide angiography in many institutions. Thallium may be better for localizing disease than radionuclide angiography because images are obtained in multiple projections. Exercise echocardiography is limited in obese individuals and those with pulmonary disease. Identifying regional wall motion abnormalities is dependent on adequately scanning the involved area at the time of ischemia. It may help to know which vessel is involved. Ultrafast CT is able to evaluate most to all of the left ventricle during exercise.

Every technique has some limitations. Premature ventricular beats and atrial fibrillation cause particular problems for acquisition of radionuclide angiographic images. With both echocardiography and ultrafast CT fewer problems are encountered with dysrhythmias since individual heartbeats are assessed. Imaging or excluding the right ventricle poses problems for many modalities, including radionuclide angiography, but not for ultrafast CT. Ultrafast CT provides clear apical delineation, and breast shadows, obesity, and the diaphragm do not cause attenuation problems resulting in loss of border definition. Technology is under development to increase the immediate scanning memory and allow acquisitions at more than eight levels.

We have demonstrated that myocardial blood flow can be measured by ultrafast CT.[19] Easier methods of measurement will make the modality more clinically applicable.

The strength of ultrafast CT in relation to other modalities lies in its excellent resolution and tomographic acquisition, which allow multilevel cardiac imaging with clear anatomic landmarks within several seconds. There is no other currently available modality that can approach this resolution and regional capability during exercise.

Exercise ultrafast CT is evolving and demonstrates promise because of its high resolution, which allows refined regional wall motion evaluation. There is also a future for ultrafast CT evaluation of myocardial blood flow during exercise.

REFERENCES

1. ACC/AHA Task Force Report. Guidelines for percutaneous transluminal coronary angioplasty. *J Am Coll Cardiol.* 1988; 12:529–545.

2. Loop FD, Lytle BW, Cosgrove DM, et al. Influence of the internal mammary graft on 10 year survival and other cardiac events. *N Engl J Med.* 1986;314:1–6.

3. White CW, Wright CB, Doty OB, et al. Does visual interpretation of the coronary arteriogram predict the physiologic importance of a coronary stenosis? *N Engl J Med.* 1984;310:819–829.

4. Ellestad M. *Stress Testing: Principles and Practice.* Philadelphia, Pa: FA Davis Co; 1986:263, 291–294.

5. Iskandrian A. Thallium-201 myocardial imaging and radionuclide ventriculography: Theory, technical considerations and interpretations. In: *Nuclear Cardiac Imaging: Principles and Applications.* Philadelphia, Pa: FA Davis Co; 1987:81–102.

6. Iskandrian AS, Hakki AH. Left ventricular function in patients with coronary heart disease in the presence or absence of angina pectoris during exercise radionuclide ventriculography. *Am J Cardiol.* 1984;53:1239.

7. Robertson WS, Feigenbaum H, Armstrong WF, et al. Exercise echocardiography: A clinically practical addition to the evaluation of coronary artery disease. *J Am Coll Cardiol.* 1983;6:1083–1091.

8. Feigenbaum H. *Echocardiography.* Philadelphia, Pa: Lea & Febiger; 1986:471–475.

9. Teague SM, Mark DB, Albert D, et al. Exercise Doppler aortovenography in coronary insufficiency. *Clin Res.* 1984;32:211A. Abstract.

10. Chomka EV, Fletcher McK, Stein M, et al. Ultrafast computed tomography during exercise bicycle ergometry. *J Am Coll Cardiol.* 1986;7:154A. Abstract.

11. Roig E, Chomka EV, Castaner A, et al. Exercise ultrafast computed tomography for the detection of coronary artery disease. *J Am Coll Cardiol.* 1989;13:1073–1081.

12. Mitamura H, Ogawa S, Hori S, et al. Two dimensional echocardiographic analysis of wall motion abnormalities during handgrip exercise in patients with coronary artery disease. *Am J Cardiol.* 1981; 48:711–719.

13. Gondi B, Nanda NC. Cold pressor test during two dimensional echocardiography: Usefulness in detection of patients with coronary artery disease. *Am Heart J.* 1984;107:278–285.

14. Chapman PD, Doyle TP, Troup PJ, et al. Stress echocardiography with transesophageal atrial pacing: Preliminary report of a new method for detection of ischemic wall motion abnormalities. *Circulation.* 1984;70:445–450.

15. Gould KL, Westcott RJ, Albro PC, et al. Noninvasive assessment of coronary stenoses by myocardial imaging during pharmacologic coronary vasodilation, II: Clinical methodology and feasibility. *Am J Cardiol.* 1978;41:279–287.

16. Leppo J, Boucher CA, Okada RD, et al. Serial thallium-201 myocardial imaging after dipyridamole infusion: Diagnostic utility in detecting coronary stenoses and relationship to regional wall motion. *Circulation.* 1982;66:649–657.

17. Chomka EV, Wolfkiel CJ, Brundage BH: Indocyanine green ear densitometry to predict left ventricular contrast enhancement during ultrafast computed tomography. *Clin Res.* 1986;34:289A. Abstract.

18. Kondos GT, Chomka EV, Roig E, et al. Evaluation of percutaneous coronary angioplasty using rest and exercise ultrafast CT. *Circulation.* 1988;78(suppl):II–398.

19. Wolfkiel CJ, Ferguson JL, Chomka EV, et al. Measurement of myocardial blood flow by ultrafast computed tomography. *Circulation.* 1987;76:1262–1273.

20. Poliner LR, Dehmer GH, Lewis SE, et al. Left ventricular performance in normal subjects: A comparison of the response to exercise in the upright and supine positions. *Circulation.* 1980;62:528–534.

21. Manyari DE, Kostuk WJ. Left and right ventricular function at rest and during bicycle exercise in the supine and sitting positions in normal subjects and patients with coronary artery disease. *Am J Cardiol.* 1983;51:36–42.

22. Thadani U, West R, Mathew T, et al. Hemodynamics at rest and during supine and sitting bicycle exercise in patients with coronary artery disease. *Am J Cardiol.* 1977;39:776–783.

23. Rodrigues EA, Maddahi J, Brown H, et al. Responses of left and right ventricular ejection fractions to aerobic and anaerobic phases of upright and supine exercise in normal subjects. *Am Heart J.* 1989; 118:319–324.

24. Osbakken MD, Boucher CA, Okada RD, et al. Spectrum of global left ventricular responses to supine exercise: Limitations in the use of ejection fraction in identifying patients with coronary artery disease. *Am J Cardiol.* 1988;51:28–35.

25. Braunwald E, Kloner RA. The stunned myocardium: prolonged postischemic ventricular dysfunction. *Circulation.* 1982; 66:1146–1149.

Chapter 17

Evaluation of Left Ventricular Aneurysm and Thrombus by Ultrafast Computed Tomography

Eulalia Roig, MD, Eva V. Chomka, MD, and Bruce H. Brundage, MD

Left ventricular aneurysm is a frequent sequela of transmural myocardial infarction; its formation is related to the magnitude of ventricular necrosis during the acute event.[1] Ventricular aneurysms frequently involve the apex and anterior wall of the left ventricle, occurring in inferior or posterior locations in only 10% of cases.[2,3] They were first reported at necropsy and are found in 5% to 20% of patients dying after a myocardial infarction.[4,5] Mural thrombi are present in the aneurysmal sac in 20% to 60% of these patients.[6–8]

Noninvasive assessment of ventricular aneurysms and intraventricular thrombi can be performed well by ultrafast computed tomography (CT). This technique provides for accurate evaluation of left ventricular function[9] by scanning the left ventricle using the cine-mode protocol, which permits the acquisition of 17 scans per second during one cardiac cycle. The cross-sectional images obtained can be analyzed in real time, allowing an accurate evaluation of regional wall motion.[10] The assessment of consecutive tomographic levels of the left ventricle provides information about the extent and severity of regional dyskinesis. Ventricular aneurysm is defined by ultrafast CT as a well-circumscribed area of the left ventricle that is thinner than normal myocardium and that is either dyskinetic or akinetic during real-time wall motion assessment (Figure 17-1). Ventricular thrombus is identified as a filling defect that is visualized within the ventricular cavity adjacent to a dyskinetic or akinetic segment (Figure 17-2) and is present in adjacent tomographic levels (Figure 17-3).

The association of a ventricular aneurysm and a mural thrombus is a common finding.[6,7] Aneurysm formation can be observed as soon as 2 days after a myocardial infarction. The combination of thrombogenic factors associated with acute myocardial infarction (damaged endocardium and the localized stasis of blood due to segmental dysfunction) sets the stage for the formation of a mural thrombus.[11] Ultrafast CT can be performed safely soon after a myocardial infarction,[12,13] allowing the detection of early aneurysm formation and the diagnosis of an associated ventricular thrombus (Figure 17-4).

The development of mural thrombosis in chronic aneurysms has been reported to be associated with specific histologic patterns of the aneurysm's wall. Hochman et al[14] described the absence of fibroelastosis in the scar tissue associated with thrombus formation. Ultrafast CT is also useful in the evaluation of chronic ventricular aneurysms to ascertain the presence or absence of thrombi.

Cardiac catheterization is widely used for the assessment of left ventricular wall motion in patients surviving myocardial infarction.[15] Intraventricular thrombi usually are diagnosed by angiography when an intracavitary filling defect is seen associated with an area of dyskinesis or akinesis. However, the low sensitivity and low specificity of left ventriculography for the diagnosis of ventricular thrombi are well recognized.[16] Reeder et al,[8] in a series of patients undergoing cardiac surgery, found the sensitivity and specificity to be 31%

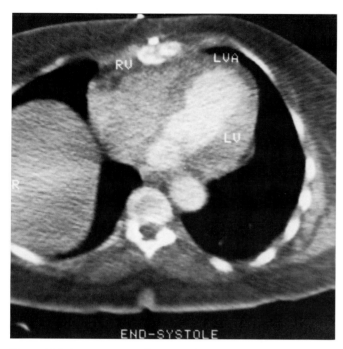

Figure 17-1 A cross-sectional image of the mid-left ventricle (LV) with contrast medium enhancement demonstrates the dyskinetic anterior wall at end-systole. This finding is characteristic of ventricular aneurysms. RV, right ventricle; LVA, left ventricular aneurysm.

and 75%, respectively. The failure of contrast medium to opacify dyskinetic areas, the masking of filling defects by the overlying dense contrast material (planar imaging), and the difficulty in distinguishing small clots from prominent trabecular patterns are among the causes for the low sensitivity.

Noninvasive techniques have also been used for the evaluation of left ventricular thrombi. Mode M echocardiography was the first technique used for this purpose, but it has limited application because the apex of the left ventricle is not well visualized and this is a likely site for thrombus formation.[17]

Two-dimensional echocardiography has emerged as the most commonly used noninvasive technique for evaluation of left ventricular thrombi. This tool permits evaluation of the left ventricle in real time; therefore, wall motion abnormalities, which are nearly always at the foundation of ventricular thrombi, usually can be recognized easily. Two-dimensional echocardiography often is capable of visualizing most of the left ventricle, although imaging the apex is sometimes difficult. It has a good sensitivity for detecting left ventricular thrombi[18-20]; however, false-positive findings have been reported to be relatively common.[19-21] Intraventricular masses can be simulated by artifacts caused by reverberations between the transducer and the chest wall or dense apical myocardial scar. Sometimes it is difficult to identify thrombus margins with echocardiography because of the beam width, which may include image information from adjacent functioning myocardium.

Asinger et al[21] have noted that images of muscular trabeculae, aneurysmal shelves, chordal structures, and prominent papillary muscles may simulate thrombi. In their study, technically satisfactory echocardiograms could not be obtained in 11% of patients because of poor visualization of the apex. Ports et al[22] easily identified ventricular aneurysms by two-dimensional echocardiography, but they noted that mural

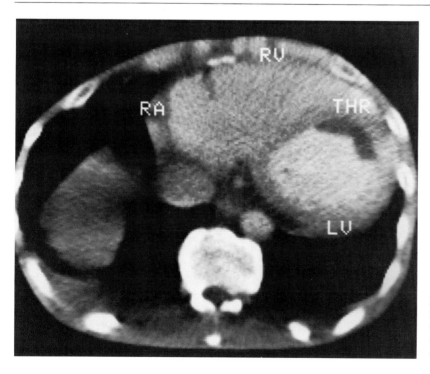

Figure 17-2 Contrast medium enhancement of the mid-left ventricle (LV) defines a filling defect adherent to the anterior wall that is characteristic of a ventricular thrombus. RA, right atrium; RV, right ventricle; THR, thrombus.

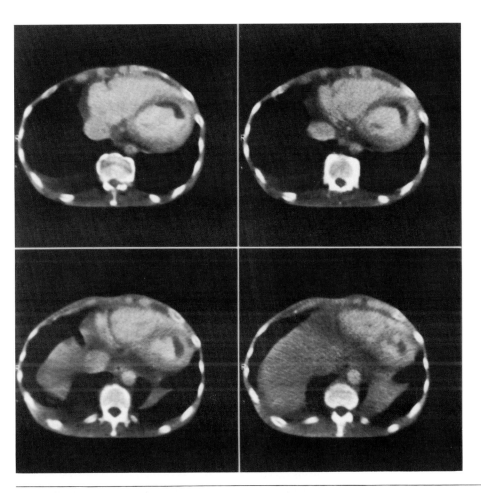

Figure 17-3 This series of contrast medium–enhanced ultrafast CT scans depicts the longitudinal extent and spatial orientation of a left ventricular thrombus from the mid–left ventricle to the apex.

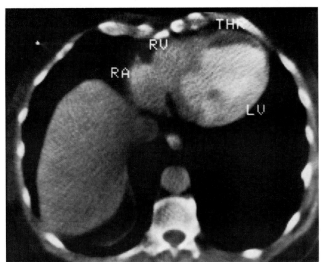

Figure 17-4 Contrast medium enhancement of the mid-left ventricle (LV) during real-time ultrafast CT scanning permits visualization of a dyskinetic anterior wall at end-systole with a mural thrombus (THR) adherent to the aneurysm wall. RA, right atrium; RV, right ventricle.

thrombi within them were not always detected. Finally, Stratton et al.[19] observed that small apical thrombi are not always evident by two-dimensional echocardiography.[19]

Indium 111 platelet imaging, which identifies sites of active intravascular platelet deposition, has been used to detect left ventricular thrombi.[23,24] Although ^{111}In is more specific than two-dimensional echocardiography (100% versus 80%), it is less sensitive (67% versus 97%).[25] The fact that platelet imaging detects only recent thrombosis may account for this low sensitivity.

Filling defects created by left ventricular thrombi have been visualized with radionuclide angiography in patients with anterior or apical aneurysms.[26] However, the sensitivity of this technique has not been assessed prospectively, and the poor spatial resolution of radioisotope imaging makes it unlikely that radionuclide angiography will receive wide clinical use.

Recent reports[27–30] have suggested that CT can diagnose intracardiac thrombi more accurately than can two-dimensional echocardiography. Transverse consecutive slices of all cardiac chambers can be obtained by conventional tomography; but its long scan time results in a significant motion artifact, and images of the heart in real time cannot be obtained.

Ultrafast CT, with its capability of obtaining 50-millisecond scans virtually free of motion artifacts, provides cross-sectional images of the whole heart that can be analyzed in real time. Left ventricular wall motion can be assessed qualitatively by identifying akinesis or dyskinesis of the ventricular wall after a myocardial infarction from a continuous-loop movie. Recently the ability to measure regional wall motion abnormalities has also been shown.[10] Ultrafast CT allows complete visualization of the left ventricular wall and cavity through consecutive tomographic levels. The papillary muscles are easily identified by their characteristic shape and location within the ventricular cavity, and these features are helpful in differentiating them from intraventricular thrombi. Ultrafast CT has better spatial resolution than does two-dimensional echocardiography and is reported to be especially useful in cases where questionable masses have been suggested by two-dimensional echocardiography.[31]

Apical dyskinesis of the left ventricle is better visualized when the left ventricle is scanned in the long-axis plane; however, thrombi at the apex are often well visualized in the short-axis plane as well (Figure 17-5).

A ventricular aneurysm may have calcium in the wall adjacent to a mural thrombus (Figure 17-6). Cardiac calcification is easily detected by ultrafast CT because of its high density.

Ultrafast CT requires only a small amount of contrast medium, usually 25 to 30 mL administered via a peripheral vein, to provide good enhancement of both ventricles. The diagnosis of intraventricular thrombi can be made by a single scanning sequence using the flow-mode protocol (see Chapter 4). Briefly stated, this protocol gates scanning to the electrocardiogram, and each scan is obtained at the same point of the cardiac cycle for every heartbeat or multiple of that interval for up to 20 sequences. The other commonly used scanning protocol is known as the cine-mode. It scans one target repetitively for one cardiac cycle at a rate of 17 scans per second. Scanning is triggered by the electrocardiogram. This mode provides real-time cross-sectional images of the beating heart that may provide additional information about thrombus motion.

Although intraventricular thrombi frequently are associated with acute anterior myocardial infarction, systemic embolization is uncommon, occurring in fewer than 5% to 10% of patients.[32–34] However, recent data suggest that mobile thrombi that protrude into the left ventricular cavity are more prone to embolization than are sessile thrombi adjacent to the aneurysm wall.[35–39] Confident identification of patients with ventricular thrombi by ultrafast CT after an acute myocardial infarction is important, because they may benefit from anticoagulant therapy.

As previously mentioned, real-time scanning provides useful information about thrombus size and motion. A mural thrombus with a mobile margin that protrudes into the left ventricular cavity is shown in Figure 17-7. The high spatial resolution of ultrafast CT provides excellent definition of most small thrombi. An example of a small mobile thrombus located at the left ventricular apex is shown in Figure 17-8.

The digital format of ultrafast CT makes measurement of thrombus size simple. Recent thrombi are frequently larger and more mobile than are old thrombi, which are laminated and more organized within the aneurysmal sac[30]; therefore, the measurement of thrombus size may help in determining its relative age.

Although systemic embolization is considered to be less frequent in patients with an organized mural thrombus located in an old ventricular aneurysm[33] than it is in patients with a recent thrombus formed after an acute myocardial infarction, it is possible that increased mobility at the border of the thrombus could be associated with disruption and embolization. In 8 of the 11 patients with ventricular thrombi reported by Stratton and Resnick,[39] the embolic event occurred long

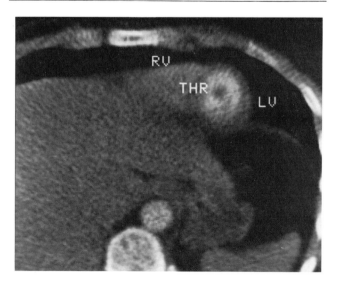

Figure 17-5 A short-axis image of the contrast medium–enhanced ventricular apex (LV) clearly detects a small thrombus (THR) as a filling defect. RV, right ventricle.

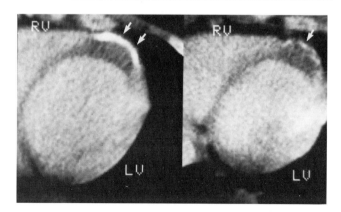

Figure 17-6 Two consecutive tomographic levels of the contrast medium–enhanced mid-left ventricle (LV) permit the visualization of a mural thrombus adherent to the anterior wall and associated calcification of the aneurysm wall (arrows). RV, right ventricle.

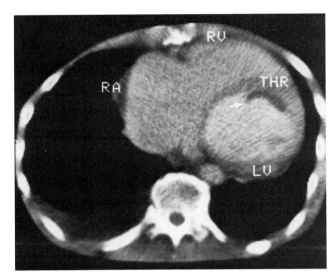

Figure 17-7 A single image from a real-time cross-section at a movie of the mid-left ventricle (LV) shows a mural filling defect (THR) with a freely mobile edge (arrow). RA, right atrium; RV, right ventricle.

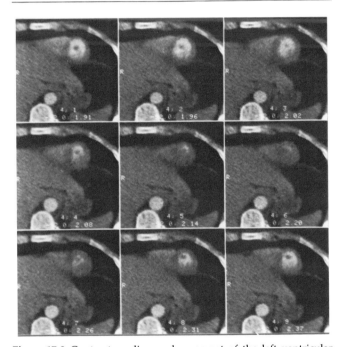

Figure 17-8 Contrast medium enhancement of the left ventricular apex during real time (left to right and top to bottom) tracks a mobile thrombus throughout the cardiac cycle. Note that the thrombus moves from the center to the edge of the cavity. Frame 1 is end-diastole and frame 6 is end-systole.

after the myocardial infarction. Anticoagulation therapy in these patients is still controversial[34,40–43]; thus the identification, characterization, and measurement of intraventricular thrombi by ultrafast CT may help direct a rational therapeutic approach.

In patients with dilated cardiomyopathy, the stasis of blood in the cardiac apex as a consequence of severe impairment of the systolic function has been suspected as the major cause of thrombus formation.[44] Ultrafast CT is useful in the evaluation of left ventricular function and the assessment of intraventricular thrombi in these patients as well. Furthermore, ultrafast CT is able to diagnose atrial thrombi, especially those localized at the left atrial appendage[31] in cardiomyopathic patients with atrial fibrillation.

Ventricular thrombi may resolve spontaneously or after anticoagulation therapy[45,46]; therefore, ultrafast CT is a minimally invasive technique that may be useful for the serial evaluation of intracardiac thrombi to help clarify the clinical course of these patients.

In summary, ultrafast CT is an excellent tool for the evaluation of wall motion and ventricular function after myocardial infarction. The dyskinetic or akinetic left ventricular wall can be identified accurately and thrombi in the aneurysmal sac easily detected. Ultrafast CT has better spatial resolution than does two-dimensional echocardiography, allowing the detection of small thrombi. Its tomographic format permits good visualization of the entire left ventricle, including the apex, and the real-time images obtained using the cine-mode protocol provide useful information about thrombus motion.

REFERENCES

1. Heras M, Sanz G, Azqueta M, et al. Natural history and determinants of prognosis in patients with left ventricular aneurysm. Presented at the Tenth World Congress of Cardiology; 1986; Abstract 407.

2. Swan HJ, Magnusson PT, Buchbinder NA, et al. Aneurysm of the cardiac ventricle: Its management by medical and surgical intervention. *West J Med*. 1978;129:26–40.

3. Dubnow MH, Burchell HB, Titus JL. Postinfarction ventricular aneurysm: A clinicomorphologic and electrocardiographic study of 80 cases. *Am Heart J*. 1965;70:753–760.

4. Schlicter J, Hellerstein HK, Katz LN. Aneurysm of the heart: A correlative study of one hundred and two proved cases. *Medicine (Baltimore)*. 1954;33:43–86.

5. Abrams DL, Edelist A, Luraia MH, et al. Ventricular aneurysm: A reappraisal based on a study of sixty-five consecutive autopsied cases. *Circulation*. 1963;27:164–169.

6. Phares WS, Edwards JE, Burchell BH. Cardiac aneurysms: Clinicopathologic studies. *Proc of Staff Meet Mayo Clin*. 1953;28:264–271.

7. Garvin CF. Mural thrombi in the heart. *Am Heart J*. 1941;21:713–720.

8. Reeder GS, Lengyel M, Tajik AJ, et al. Mural thrombus in left ventricular aneurysm: Incidence, role of angiography, and relation between anticoagulation and embolization. *Mayo Clin Proc*. 1981;56:77–81.

9. Rich S, Chomka EV, Stagl R, et al. Ultrafast computed tomography to determine left ventricular ejection fraction. *Am Heart J*. 1986;112:392–396.

10. Roig E, Castaner A, Chomka EV, et al. Exercise ultrafast CT: A sensitive and specific test for coronary artery disease. *Clin Res*. 1987;35:320A. Abstract.

11. Yater WM, Welsh PP, Stapleton JF, et al. Comparison of clinical and pathologic aspects of coronary artery disease in men of various age groups: A study of 950 autopsied cases from the Armed Forces Institute of Pathology. *Ann Intern Med*. 1951;34:352.

12. Brundage BH, Chomka EV, Hart K, et al. Evaluation of Q wave and non Q wave myocardial infarcts by ultrafast CT. *Circulation.* 1985;72(suppl 3):1651.

13. Brundage BH, Wolfkiel C, Chomka EV, et al. Comparison of infarct size estimation by CK-MB analysis with infarct volume determined by ultrafast CT. *Circulation.* 1985;72(suppl 3):1652.

14. Hochman JS, Platia EB, Bulkley BH. Endocardial abnormalities in left ventricular aneurysms: A clinicopathologic study. *Ann Intern Med.* 1984;100:29–35.

15. Sanz G, Castaner A, Betriu A, et al. Determinants of prognosis in survivors of myocardial infarction. *N Engl J Med.* 1982;306:1065–1070.

16. Hamby RI, Wisoff BG, Davison ET, et al. Coronary artery disease and left ventricular mural thrombi: Clinical, hemodynamic and angiocardiographic aspects. *Chest.* 1974;66:488–494.

17. Niehues B, Heuser L, Jansen W, et al. Noninvasive detection of intracardiac tumors by ultrasound and computed tomography. *Cardiovasc Intervent Radiol.* 1983;6:30–36.

18. Visser CA, Kan G, David GK, et al. Two-dimensional echocardiography in the diagnosis of left ventricular thrombus: A prospective study of 67 patients with anatomic validation. *Chest.* 1983;83:228–232.

19. Stratton JR, Lighty GW, Pearlman AS, et al. Detection of left ventricular thrombus by two-dimensional echocardiography: Sensitivity, specificity and causes of uncertainty. *Circulation.* 1982;66:156–166.

20. DeMaria AN, Bommer W, Neumann A, et al. Left ventricular thrombi identified by cross-sectional echocardiography. *Ann Intern Med.* 1979;90:14–18.

21. Asinger RW, Mikell FL, Sharma B, et al. Observations on detecting left ventricular thrombus with two dimensional echocardiography: Emphasis on avoidance of false positive diagnoses. *Am J Cardiol.* 1981;47:145–156.

22. Ports TA, Cogan J, Schiller NB, et al. Echocardiography of left ventricular masses. *Circulation.* 1979;58:528–536.

23. Ezekowitz MD, Kellerman DJ, Smith EO, et al. Detection of active left ventricular thrombosis during acute myocardial infarction using indium-111 platelet scintigraphy. *Chest.* 1984;86:35–39.

24. Ezekowitz MD, Wilson DA, Smith EO, et al. Comparison of indium-111 platelet scintigraphy and two-dimensional echocardiography in the diagnosis of left ventricular thrombi. *N Engl J Med.* 1982;306:1509–1513.

25. Seabold JE, Schroder E, Conrad GR, et al. Indium-111 platelet scintigraphy and two-dimensional echocardiography for detection of left ventricular thrombus: Influence of clot size and age. *J Am Coll Cardiol.* 1987;9:1057–1066.

26. Stratton JR, Ritchie JL, Hammermeister KE, et al. Detection of left ventribular thrombi with radionuclide angiography. *Am J Cardiol.* 1981;48:565–572.

27. Goldstein JA, Schiller NB, Lipton MJ, et al. Evaluation of left ventricular thrombi by contrast-enhanced computed tomography and two-dimensional echocardiography. *Am J Cardiol.* 1986;57:757–760.

28. Tomoda H, Hoshiai M, Furuya H, et al. Evaluation of intracardiac thrombus with computed tomography. *Am J Cardiol.* 1983;51:843–852.

29. Tomoda H, Hoshiai M, Furuya H, et al. Evaluation of left ventricular thrombus with computed tomography. *Am J Cardiol.* 1981;48:573–577.

30. Nair CK, Sketch MH, Mahoney PD, et al. Detection of left ventricular thrombi by computerized tomography: A preliminary report. *Br Heart J.* 1981;45:535–541.

31. Roig E, Chomka EV, Helgason CM, et al. Ultrafast computed tomography in the diagnosis of cardiac thrombus in patients with systemic embolization. *Clin Res.* 1987;35:320A. Abstract.

32. Asinger RW, Miekell FL, Elsperger J, et al. Incidence of left ventricular thrombosis after acute transmural myocardial infarction. *N Engl J Med.* 1981;305:297–302.

33. Fruedman MJ, Carlson K, Marcus FI, et al. Clinical correlations in patients with acute myocardial infarction and left ventricular thrombus detected by two-dimensional echocardiography. *Am J Med.* 1982;72:894–898.

34. Meltzer RS, Visser CA, Fuster V. Intracardiac thrombi and systemic embolization. *Ann Intern Med.* 1986;104:689–698.

35. Meltzer RS, Visser CA, Kan G, et al. Two-dimensional echocardiographic appearance of left ventricular thrombi with systemic emboli after myocardial infarction. *Am J Cardiol.* 1984;53:1511–1513.

36. Haugland JM, Asinger RW, Miekell FL, et al. Embolic potential of left ventricular thrombi detected by two-dimensional echocardiography. *Circulation.* 1984;70:588–598.

37. Kinney EL. The significance of left ventricular thrombi in patients with coronary heart disease: A retrospective analysis of pooled data. *Am Heart J.* 1985;109:191–194.

38. Johannessen K, Nordrehaug JE, Lippe G. Left ventricular thrombosis and cerebrovascular accident in acute myocardial infarction. *Br Heart J.* 1984;51:553–556.

39. Stratton JR, Resnick AD. Increased embolic risk in patients with left ventricular thrombi. *Circulation.* 1987;75:1004–1011.

40. Gueret P, Dubourg O, Ferrier A, et al. Effects of full-dose heparin anticoagulation on the development of left ventricular thrombosis in acute transmural myocardial infarction. *J Am Coll Cardiol.* 1986;8:419–426.

41. Tramarin R, Pozzoli M, Febo O, et al. Two-dimensional echocardiographic assessment of anticoagulant therapy in left ventricular thrombosis early after acute myocardial infarction. *Eur Heart J.* 1986;7:482–492.

42. Frandsen EH, Egeblad H, Mortensen SA. Transience of left ventricular thrombosis. *Br Heart J.* 1983;49:193–194.

43. Arvan S. Persistent intracardiac thrombi and systemic embolization despite anticoagulant therapy. *Am Heart J.* 1985;109:178–181.

44. Gottdiener JS, Gay JA, VanVoorhees L, et al. Frequency and embolic potential of left ventricular thrombus in dilated cardiomyopathy: Assessment by two-dimensional echocardiography. *Am J Cardiol.* 1983;52:1281–1285.

45. Visser CA, Kan G, Meltzer RS, et al. Long-term follow-up of left ventricular thrombus after acute myocardial infarction: A two-dimensional echocardiographic study in 96 patients. *Chest.* 1984;86:532–536.

46. Keating EC, Gross SA, Schlamowitz RA, et al. Mural thrombi in myocardial infarctions: Prospective evaluation by two-dimensional echocardiography. *Am J Med.* 1983; 74:989–995.

Chapter 18

Evaluation of Coronary Artery Disease by Nuclear Magnetic Resonance Imaging

George E. Wesbey, MD

In comparison to the presently widespread applications in the ideal stationary milieu of the brain and spine, greater technical problems and perhaps even greater clinical promise face nuclear magnetic resonance (NMR) imaging applications in coronary artery disease. This chapter reviews both the present and the future clinical applications of gated proton NMR imaging in the assessment of the anatomic, functional, and biophysical manifestations of atherosclerotic heart disease. Regional myocardial metabolic analysis of high-energy phosphates by spatially localized (1 to 3 mL) phosphorus 31 NMR spectroscopy in ischemic heart disease eventually may reach clinical utility as well, in an effort to assess directly the cellular energy reserve in regional myocardial ischemia and to evaluate the metabolic response to interventions.[1]

However, global imaging of individual millimolar cardiac metabolites by NMR with adequate (1 cm) spatial resolution with less than 1 hour of imaging will never be clinically feasible. Phosphocreatine myocardial imaging by ^{31}P NMR will never rival proton NMR imaging. In this regard, positron emission tomography is the unrivaled leader in myocardial metabolic imaging and stands a far greater chance of clinical application, especially with the development of lower-cost cyclotrons and generator-produced positron-emitting radiopharmaceuticals such as rubidium 82. Other review articles encompassing all cardiovascular NMR applications, including NMR spectroscopy, are recommended for the interested reader.[2-4]

A basic physical principle in NMR underlies the tremendous potential and present clinical promise exhibited in the cardiovascular system. As flowing blood moves through an image slice at sufficient velocity, spin-echo signal acquisition is disrupted.[2] Because blood flowing in blood vessels and chambers of the heart appears black on most spin-echo images, an optimal situation results in high natural contrast between the blood pool and the walls of vessels and cardiac chambers. This means that no exogenous contrast media are required to depict the cardiovascular system and its anatomy with NMR. Nonetheless, paramagnetic contrast media may have a role in myocardial proton NMR imaging, such as in the demonstration of regional myocardial perfusion.

PHYSIOLOGIC GATING AND ACQUISITION TECHNIQUES

Gating is required to provide adequate NMR images of the heart, and the magnetic fields require a nonferromagnetic, physiologic, signal-sensing circuit. Electronically isolated electrocardiographic (ECG) circuits containing little ferromagnetic material and low-resistance electrodes are preferred. The ECG signal is not affected by the radiofrequency (RF) fields, but the static magnetic field does lead to peaking of the T wave of the ECG, the so-called magnetohydrodynamic effect, which increases linearly with the static magnetic field strength. This effect is simply due to a moving conductor (flowing blood) in a magnetic field and is *not* an electrophysiologic effect on the myocardial conduction system. With the use of ECG gating, the pulse repetition rate is defined by the RR interval. If the heart rate is 60 beats per minute and signal acquisition is performed with every other beat, the TR time is 2 seconds. The choice of gating to every beat versus every other beat can affect image quality dramatically. A technically important flow-artifact reduction technique used in high-field (1.5 T) NMR cardiac imaging uses RF spatial presaturation and works best with gating to every heartbeat.[5] With this technique, 14 single-echo (TE = 20 milliseconds) slices can be acquired in the axial, sagittal, coronal, or long-axis/short-axis[6] planes through the heart (assumes a heart rate of 60 beats per minute; with a heart rate of 100 beats per minute, 10 sections can be obtained). These sections, usually 5 or 10 mm thick, can encompass the entire left ventricle with an imaging time of approximately 6 to 10 minutes. In patients who exhibit severe atrial fibrillation with a very irregular ventricular response or who exhibit frequent premature ventricular contractions, adequate cardiac image quality cannot be obtained with ECG-gated NMR imaging.

As with all NMR studies, patients with pacemakers (temporary or permanent) and intracranial aneurysm clips should be excluded from cardiac NMR imaging.

The ability of patients with diminished cardiac ejection fractions to dissipate effectively any RF power deposition in excess of the present Food and Drug Administration guideline of 0.4 W/kg (whole-body average) is presently unknown and is under active investigation.

NORMAL ANATOMY, PHYSIOLOGY, AND BIOPHYSICS

With optimized cardiac image sections orthogonal to the long and short axis, accurate wall thickness and chamber size measurements are feasible.[6] With ECG-gated spin-echo cardiac NMR, the presence of high-signal intensity in the blood pool of the left atrium, the left ventricle, and (less commonly) the right ventricle can be a normal finding in healthy young adults during the slow-flow diastolic phase for that particular chamber[2] (Figure 18-1). Many different imaging techniques can be applied by the radiologist to determine whether the blood pool signal in the cardiovascular system is physiologic (slowly flowing blood) or pathoanatomic (for example, a clot). These techniques include analysis of even-numbered versus odd-numbered echoes,[7-9] phase-sensitive reconstruction,[7-9] temporal evaluation of cardiosynchronous intensity changes with ECG gating,[10,11] ECG-gated digital subtraction projection angiography of pulsatile flow,[12] velocity-sensitive

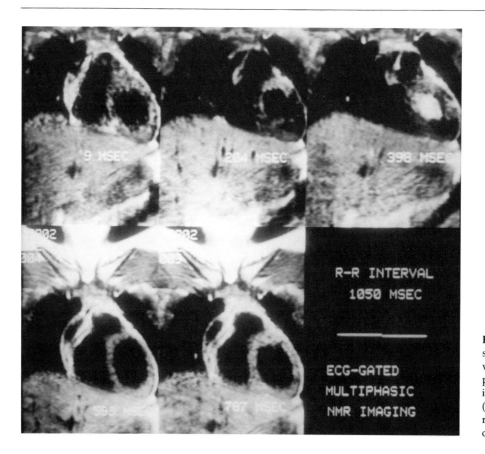

Figure 18-1 Five-slice coronal ECG-gated spin-echo NMR imaging study in a normal volunteer. Each slice varies in the temporal phase of the cardiac cycle. Notice the high intraluminal signal within the left ventricle (LV) cavity during the phase of isovolumic relaxation, a normal finding due to slow, coherent blood flow.

phase display techniques,[13] and many more methods currently in evolution. By analyzing the same myocardial slice at different phases of the cardiac cycle, the dynamics of myocardial wall thickening can be characterized.[14] Several studies have validated the accuracy of ECG-gated spin-echo cardiac NMR in the measurement of in vivo cardiac chamber dimensions, wall thickening, chamber volumes, myocardial mass, and ejection fractions.[14–22]

Gated NMR images of the normal heart are able to display the internal architecture of the heart, demonstrating such normal anatomic structures as the papillary muscles of the left ventricle, the moderator band of the right ventricle, and the proximal coronary arteries (Figure 18-2). This internal topography of the cardiac chambers is displayed with high resolution because of the natural contrast between flowing blood in the cardiac chambers and the relative medium intensity of the myocardium.

NMR ANGIOGRAPHY IN CORONARY ARTERY DISEASE

Current NMR imaging techniques are unsatisfactory for the diagnosis of atherosclerotic lesions in native coronary arteries, but they have proved useful in the detection of coronary artery fistulas and aneurysms.[23] NMR imaging has been used to evaluate the patency of coronary artery bypass grafts.[24–27] The image quality of the proximal coronary arteries is improved with surface-coil imaging.[24] In a study of 37 grafts, NMR correctly identified all 29 patent proximal anastomoses confirmed as being open by conventional coronary arteriography.[24] Of the 29 patent grafts, 8 were sequential anastomoses; the status of the distal anastomosis was correctly identified by NMR in 5 of these. In the 8 occluded grafts, 5 were seen by NMR; 2 of these 5 were incorrectly interpreted to be patent. With blind interpretation (knowledge of surgery but not of graft angiography) of 20 patients with 47 grafts, 26 of 29 grafts were correctly identified as patent (90% sensitivity); 13 of 18 occluded grafts were correctly classified (specificity 72%).[25] When results from 3 patients were excluded because of technically poor gating and degraded image quality, sensitivity and specificity were 92% and 85%, respectively.[25] In another investigation, NMR identified the presence of 89% of left coronary grafts, 83% of right coronary grafts, and 45% of internal mammary grafts.[26] White et al,[27] in a blind study of 72 grafts, found that 10% of grafts could not be assessed for patency because they were visualized at only one transaxial anatomic section. In the remaining 90% of grafts identified adequately, the accuracy values for definitive NMR evaluation were 91% for patency determination and 72% for occlusion determination.

Although selective NMR coronary angiograms have not as yet been developed, Wedeen et al[12] demonstrated an exciting new application of magnetic resonance by employing projective imaging of pulsatile arterial flow in the lower extremities using a digital subtraction technique analogous to radiographic digital subtraction angiography. Techniques analogous to this[28] may someday be used for NMR imaging of the proximal coronary arteries, perhaps in screening the asymptomatic population. With the technique of Wedeen et al,[12] signals arising in all structures except vessels that carry pulsatile flow are eliminated by means of a velocity-dependent phase contrast, ECG gating, and image subtraction. As in conventional invasive arteriography, projection imaging enables the vascular tree to be visualized in a well-recognized format. Not only does this give physicians a familiar, vertically oriented, view of the arterial circulation, but the problems of three-dimensionally integrating the information from a conventional NMR multislice sectional imaging experiment are obviated. The only difference between radiographic digital subtraction angiography and the digital subtraction magnetic resonance angiography of Wedeen et al[12] is in the replacement of exogenous pharmaceutical contrast material delivered by invasive intra-arterial catheterization with endogenous, physiologic, velocity-dependent phase contrast by the NMR signal. Surface-coil projective imaging of the trifurcation of the popliteal artery by this technique revealed excellent anatomic detail of the sural branches of the popliteal artery, which are about 1 mm in diameter.[12] A problem with this technique is that the diminished pulsatility of atherosclerotic vessels gives

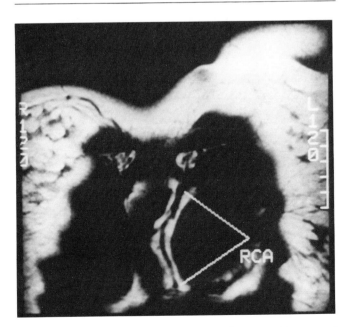

Figure 18-2 Coronal ECG-gated NMR image demonstrating a normal right coronary artery (RCA) in the right atrioventricular groove.

diminished phase contrast between systolic and diastolic acquisitions.[29]

Atherosclerotic deposits in the aorta are seen as eccentric thickening of the aorta, characterized by focal discrete lesions protruding into the lumen of the aorta.[8] Wesbey et al[7] studied the ability of NMR imaging to detect aortic, iliac, and femoral stenoses and occlusions on a 1983 model 0.35T scanner. Multislice imaging was obtained from the infrarenal aorta to the femoral bifurcation in 24 patients, all of whom had undergone intraarterial angiography within 14 days of NMR imaging. Arterial stenoses and occlusions in these vessels as detected by NMR imaging correlated with angiographic findings in 91% of instances. Protrusional atherosclerotic plaques, occlusions, and stenoses in the aortoiliac region were demonstrated accurately on NMR images; complications of previous vascular surgery, such as the development of aneurysms at sites of previous anastomoses or endarterectomy, were also identified. Femoral stenoses were not well detected in this study performed in the pre–surface-coil imaging era.

FUNCTIONAL EVALUATION OF CORONARY ARTERY DISEASE BY NMR

New, clinically feasible, techniques using ECG gating of gradient-echo NMR signals excited by limited flip-angle RF pulses promise to extend cardiovascular NMR beyond anatomic imaging into the realm of functional imaging.[30–32] Cine-NMR imaging of the heart has been shown to calculate left ventricular ejection fractions accurately, as validated by radiographic contrast left ventriculography.[30] This technique boasts both high temporal resolution (32 time frames per cardiac cycle) and high spatial resolution (1.5 to 3.0 mm), with a reasonable imaging time of 9 minutes. Validation of both right and left ventricular stroke volumes as measured by cine-NMR has been achieved.[31] This new technique can depict absent or decreased wall thickening in myocardial infarction.[32] Recent breakthroughs in cine-NMR imaging have been reported by Rzedzian and Pykett,[33] with a 40-millisecond scan time, 64 × 128 voxels over a 40-cm field of view, and an excellent signal-to-noise ratio of 30:1.

CANINE STUDIES OF CORONARY OCCLUSION BY NMR

In the area of ischemic heart disease, cardiovascular NMR has shown promising experimental results. Williams et al[34] were the first to show an elevation of proton T1 relaxation time with canine myocardium infarcted in vitro after 1 to 2 hours of coronary occlusion. Early in vitro NMR studies of excised canine hearts 24 hours after left anterior descending (LAD) coronary artery ligation revealed high signal intensity in the area of myocardial infarction.[35] This was due to a long T2 relaxation time that was associated with an elevated water content in the infarcted myocardium.[35] After 6 hours of LAD occlusion, in vitro T2-weighted NMR imaging accurately estimated infarct size over a wide range (3% to 29% of the left ventricular [LV] mass), in comparison to triphenyl-tetrazolium-chloride staining estimates.[36] With 4 hours of LAD occlusion, Canby et al[37] found no statistically significant increase in severely ischemic (less than 5% of preocclusion flow) myocardial T1 as compared with nonischemic myocardium in vitro. Moderately ischemic (5% to 50% of preocclusion flow) myocardium demonstrated statistically significant increases in T1 and T2 as compared with nonischemic myocardium.[37] A statistically significant increase was found in severely ischemic (less than 5% of preocclusion flow) myocardial T2 as compared with nonischemic myocardium in vitro.[37]

These in vitro studies were confirmed in vivo with ECG-gated NMR imaging 1 to 7 days after LAD ligation in dogs[38] (Figure 18-3). The pathologically confirmed infarcted myocardium had a statistically significantly (66% ± 27% [SD]) higher NMR signal intensity in images of TE = 28 milliseconds as compared with nor-

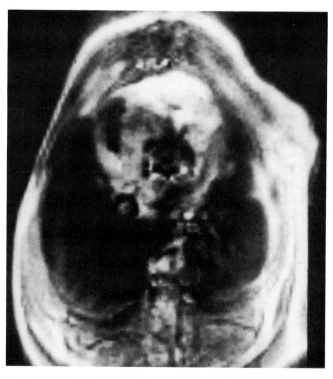

Figure 18-3 Transaxial ECG-gated NMR image of canine myocardium in vivo at 7 days after ligation of the left anterior descending coronary artery (echo delay time 56 milliseconds). Notice the high signal intensity in the anterior wall of the LV, which conformed to the region of myocardial infarction at the post-mortem examination.[38]

mal myocardium, owing to a prolonged T2 relaxation time.

Pflugfelder et al[39] demonstrated the ability of ECG-gated cardiac in vivo NMR to detect focal increases in NMR myocardial signal intensity with less than 60 minutes of LAD or circumflex occlusion in dogs. However, Tscholakoff et al[40] continuously obtained images from dogs in vivo from 5 minutes to 5 hours after LAD occlusion and did not find statistically significant increases in infarcted myocardial signal intensity or in T2 until 3 hours after occlusion (29% ± 8% [SD] with TE = 60 milliseconds and gating to every other heartbeat; 36.8 ± 4.2 milliseconds to 49.2 ± 9.9 milliseconds, respectively). An immediate (5 minutes postocclusion) 55% decrease in anterior wall systolic wall thickness was observed.[40] Pflugfelder et al[41] found peak infarct image intensity at 4 days to 13 days after LAD ligation (62% ± 24% with TE = 30 milliseconds) in a serial study over the first 20 days postocclusion. Infarct intensity increase relative to normal myocardium was less on day 20 (12% ± 19%) than on day 1 (30% ± 24%).[41] NMR measurements of LV mass and infarct mass 3 days and 21 days after LAD ligation correlated well with postmortem measurements.[42]

CANINE STUDIES OF CORONARY OCCLUSION AND REPERFUSION BY NMR

With 3 hours of coronary artery occlusion followed by 1 hour of reperfusion, Johnston et al[43] documented that regional in vitro NMR relaxation times were proportional to myocardial blood flow as measured by radiolabeled microspheres. In an in vitro reperfusion study by Askenase et al[44] with 15 minutes of occlusion, ischemic zone T1 was 532 milliseconds; nonischemic zone T1 was 395 milliseconds. After 30 minutes of reperfusion, ischemic zone T1 increased to 565 milliseconds; but after 1 hour of reperfusion, the T1 had declined to 448 milliseconds, decreasing infarct-to-normal NMR image contrast.[44] With 15 minutes of coronary occlusion followed by 15 minutes of reperfusion, Canby and Pohost[45] found small but statistically significant increases in in vitro T1, T2, total tissue water, and extracellular water in the ischemic reperfused myocardium as compared with the nonischemic myocardium. No significant changes were observed with 15 minutes of occlusion only[45] in this nonimaging study.

Miller et al[46] concluded that viable ischemic myocardium following 3 hours of occlusion and 1 hour of reperfusion (with mannitol pretreatment to enhance myocardial salvage) does not create sufficient intrinsic NMR image contrast at 1.4 T to permit imaging in vitro, despite statistically significant differences in blood flow, T1, and T2. This same ischemia-reperfusion protocol without mannitol pretreatment,[47] imaged in vivo at 0.15 T and in vitro at 1.4 T in the same laboratory, resulted in significant (31% ± 17% in vivo; 25% ± 10% in vitro) increases in NMR signal intensity (TE = 60 milliseconds) from the reperfused myocardium. The different results, compared with those of the study by Miller et al,[46] are explainable by the mannitol-induced reduction in reperfused infarct-to-normal NMR contrast.[48] Small, insignificant signal increases were seen 3 hours after occlusion in the ischemic myocardium.

With 1 hour of occlusion, followed by continuous imaging during 5 hours of reperfusion, Tscholakoff et al[49] found the following significant increases in reperfused infarct signal intensity in vivo relative to normal myocardial signal intensity (TE = 60 milliseconds, gating to every other beat): 56% ± 17% at 30 minutes after reperfusion, 89% ± 32% at 60 minutes after reperfusion, and 65% ± 29% at 300 minutes after reperfusion. T2 increases paralleled the signal intensity increases in the reperfused infarcts. No studies of the correlation between the size or severity of histologic changes in the infarct at post-mortem examination and the extent of signal intensity increase seen in vivo were performed. Thus the increased NMR signal intensity in reperfused infarcts is not specific for infarction.[50]

With 2 hours of occlusion followed by reperfusion and in vivo NMR imaging immediately and 5 days and 21 days after reperfusion, Wisenberg et al[50] concluded that reperfusion produces an immediate, abrupt, marked rise in myocardial T1 and T2 values as compared with nonreperfused infarcted myocardium. At days 5 and 21, with reperfusion, normalization of T2 occurs (compared with nonischemic myocardium) while T1 remains elevated; without reperfusion, both T1 and T2 remain elevated.

CANINE STUDIES OF MYOCARDIAL ISCHEMIA WITH GADOLINIUM

Because of the nonspecificity of increased NMR signal intensity on T2-weighted images and because of the lack of a clear relationship between regional myocardial blood flow and in vivo NMR relaxation times or image signal intensity, efforts have focused on the development of intravenously administered paramagnetic contrast agents that increase the diagnostic specificity of the NMR study. Since the aqueous ion of the rare earth element gadolinium provides the strongest proton relaxation enhancement (shortening of relaxation times), most pharmaceutical development of NMR imaging has focused on this compound. The dimeglumine salt of gadolinium chelated to diethylene-triaminepentaacetic acid (DTPA)[51,52] has undergone the most extensive testing since its conception in 1981, and US Food and Drug Administration approval for clinical use came in 1988. The agent distributes to the extracellular space in a fashion similar to iodinated radiographic contrast agents and the radiopharmaceutical technetium-DTPA. In more than 3000 human studies performed worldwide, no adverse reactions have been reported.

Initial studies of Gd-DTPA employed ex vivo imaging of excised hearts.[53-56] In a study of the use of Gd-DTPA in acute canine myocardial infarction (24-hour-old LAD occlusion), significant enhancement of relaxation rate was noted in both normal and infarcted myocardium. Greater enhancement of relaxation rate in normal myocardium was noted at 90 seconds after the injection of Gd-DTPA (0.35 mmol/kg) in comparison to 5 minutes after injection, compatible with an early perfusion phase. Greater regional infarct relaxation enhancement was seen at 5 minutes after injection than at 90 seconds after injection.

In a subsequent study of canine myocardial ischemia with only 60 seconds of LAD occlusion (no reflow), McNamara et al[54] found significant relaxation enhancement in normal myocardium (compared with ischemic myocardium) 1 minute after the injection of Gd-DTPA (0.5 mmol/kg). McNamara et al[55] later studied the effects of Gd-DTPA (0.5 mmol/kg) in postischemic reperfusion. Reversible myocardial injury was produced in nine dogs by LAD occlusion for 15 minutes followed by 24 hours of reperfusion; irreversible injury was produced in nine dogs by 1 hour of occlusion followed by 24 hours of reperfusion. In the group with reversible injury, there were no significant regional differences in NMR image intensity or relaxation times 5 minutes after injection of Gd-DTPA ($n = 6$) or without Gd-DTPA administration ($n = 3$). In the group with irreversible injury but no Gd-DTPA ($n = 3$), infarct intensity on the T2-weighted image (TR = 2000 milliseconds; TE = 56 milliseconds) was 20.1% ± 3.2% greater than adjacent normal myocardium. With Gd-DTPA administration, the infarct intensity on the T1-weighted image (TR = 500 milliseconds; TE = 28 milliseconds) was 31.3% ± 9.4% greater than the image on adjacent normal myocardium at 5 minutes postinjection.

Peshock et al[56] studied the effect of Gd-DTPA on occlusive and reperfused canine infarcts. The LAD was occluded for 3 hours (occlusive infarcts, $n = 10$, group I) or for 2 hours, with a 1-hour reperfusion (reperfused infarcts, $n = 10$, group II). Each group was further divided into hearts that received no Gd-DTPA (groups IA and IIA) and hearts that received an injection of Gd-DTPA (0.34 mmol/kg) at 2 hours 5 minutes postocclusion (group IB, ex vivo imaging at 55 minutes postinjection) and the same dose after 5 minutes of reperfusion (group IIB, ex vivo imaging at 55 minutes postinjection). Reperfused hearts given Gd-DTPA (group IIB) demonstrated a significant increase in contrast enhancement of the ischemic myocardium (39% ± 9% intensity increase postinjection) compared with occlusive infarcts (8% ± 3%) on T1-weighted images (TR = 500 milliseconds; TE = 28 milliseconds). Microsphere measurements of regional myocardial blood flow in the injured myocardium validated effective occlusion (less than 5% of preocclusion flow) as well as the expected hyperemic response (150% of preocclusion flow) after reperfusion. Peshock et al[56] concluded that Gd-DTPA–enhanced NMR imaging allows the detection of reperfusion early in the course of acute infarction.

These ex vivo studies provided two potentially very important clinical applications for Gd-DTPA–enhanced in vivo myocardial NMR imaging: (1) distinction of irreversibly damaged infarcted myocardium from reversibly injured ischemic myocardium and (2) distinction of reperfused infarcted myocardium from occlusive myocardial infarcts. The first application eventually could rival thallium 201 stress perfusion scintigraphy[57]; the second would fill a void in clinical cardiology by providing a noninvasive means of assessing the adequacy of the myocardial response to percutaneous transluminal coronary angioplasty or thrombolytic therapy.

In vivo NMR myocardial imaging studies employing Gd-DTPA have been less numerous than excised heart studies. Peshock et al[56] studied two dogs with in vivo NMR imaging before LAD occlusion, 1 hour after occlusion, immediately after reperfusion (2 hours postocclusion), and 1 hour after reperfusion (55 minutes after the administration of Gd-DTPA [0.34 mmol/kg]). Easily visible Gd-DTPA enhancement was seen in the reperfused region in vivo as compared with preinjection images, but quantitative analysis was not reported.

Rehr et al[58] obtained images from 10 dogs at 1 to 5 days after LAD occlusion (no reperfusion). In vivo imaging for 2 hours continuously following the intravenous administration of Gd-DTPA (0.34 mmol/kg) demonstrated that peak contrast enhancement of infarcted myocardium took place at 1 hour postinjection, consistent with delayed clearance of Gd-DTPA from the infarct. The 24- to 48-hour-old infarcts increased in intensity 21% after contrast; the 4- to 5-day-old infarcts were enhanced 20%.

Tscholakoff et al[59] conducted the most extensive clinically relevant study reported to date of in vivo myocardial NMR imaging using Gd-DTPA. This was the first study to use a clinically proved safe dose of Gd-DTPA (0.1 mmol/kg); all other myocardial Gd-DTPA studies had used doses of 0.34 to 0.5 mmol/kg. Worldwide clinical trials have all been conducted with doses no greater than 0.2 mmol/kg; Food and Drug Administration approval for doses greater than 0.2 mmol/kg seems highly unlikely. The study by Tscholakoff et al[59] compared Gd-DTPA–enhanced (0.1 mmol/kg) 5-hour-old LAD-occluded infarcts with infarcts receiving 1 hour of occlusion and 4 hours of reperfusion. In vivo

NMR imaging was performed before and 5 minutes after Gd-DTPA administration (images were obtained from both groups at 5 hours after the start of occlusion). On the most T1-weighted imaging sequence (gated to every heartbeat, TR = 600 to 900 milliseconds; TE = 30 milliseconds), statistically significant ($P<0.05$) contrast enhancement occurred in the acutely reperfused infarcts (49% intensity increase postinjection) but not in the occluded infarcts (30% increase, not statistically significant). As opposed to those in the studies by Rehr et al[58] and Peshock et al,[56] the postcontrast images of Tscholakoff et al[59] were obtained 5 minutes after injection rather than the more optimal time (for peak contrast enhancement) of 55 minutes after injection.

In summary, Gd-DTPA–enhanced myocardial NMR imaging demonstrates greater contrast enhancement in acutely reperfused myocardial infarcts than in nonreperfused acute infarcts. Gating to every heartbeat (effective TR = 600 to 1000 milliseconds) with an echo time (TE) of 28 to 30 milliseconds is far from the optimal T1 weighting necessary to demonstrate best Gd-DTPA–induced T1 shortening and NMR signal intensity increases. Better pulse sequences providing greater myocardial T1 weighting in vivo are now available and should be provided in future clinical and canine myocardial Gd-DTPA studies comparing postischemic reperfusion with occlusive infarction. These studies also need to correlate the relationship between in vivo Gd-DTPA–induced intensity increases and microsphere measurements of regional myocardial blood flow to validate the concept of Peshock et al[56] that Gd-DTPA serves as a marker of reperfusion.

ACUTE MYOCARDIAL ISCHEMIA IN HUMANS

The canine studies were soon followed by confirmation in humans of the ability of ECG-gated NMR to detect acute myocardial infarction. In a study of nine patients, McNamara et al[60] visualized acute myocardial infarcts 5 to 12 days after coronary occlusion (Figures 18-4 and 18-5). Johnston et al,[61] using short-axis NMR imaging, studied 18 patients for a mean of 10 days (± 5 days [SD]) after acute myocardial infarction (MI). Of the 18 studies, 14 were of sufficient diagnostic quality to permit interpretation. The authors did not state whether any of the patients were treated acutely with thrombolytic therapy or percutaneous transluminal coronary angioplasty. Signal intensity increased from the acutely infarcted ± 16% relative to noninfarcted myocardium on TE = 30-millisecond images ($P<0.01$) and 39% ± 12% (range 25% to 66%) on TE = 60-millisecond images ($P<0.001$). All 14 patients had increased signal intensity in the infarcted myocardium on the TE = 60-millisecond (odd echo) images. NMR segments containing increased signal intensity were shown to correspond to both the ECG region of the acute MI and regions of severe hypokinesia or akinesia on contrast left ventriculography.

In studying nine normal volunteers, Johnson et al[61] found one patient with NMR changes compatible with MI. Filipchuk et al[62] found increased myocardial signal in 83% of volunteers. This study lacked any quantitative analysis of signal intensity in patients with MI and normal subjects. Dilworth et al[63] found interobserver

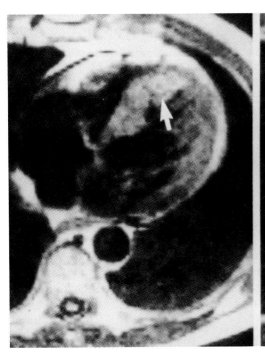
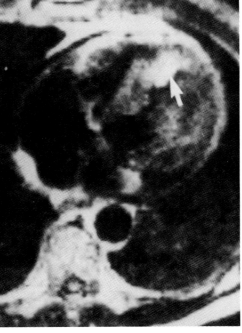

Figure 18-4 TE = 28-millisecond (left image) and TE = 56-millisecond (right image) ECG-gated NMR images of a patient at 48 hours after a transmural anteroseptal myocardial infarction. There is greater increased myocardial signal intensity in the infarcted myocardium on the TE = 56-millisecond image, due to the increased T2 relaxation time in the infarct.

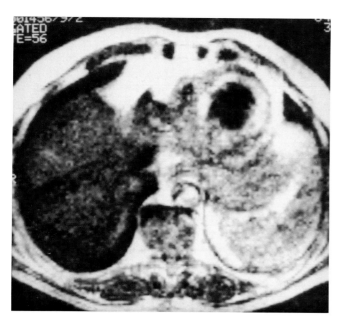

Figure 18-5 Transaxial ECG-gated NMR image of an acute subendocardial myocardial infarction (TE = 56 milliseconds) characterized by concentric, nontransmural high signal intensity.

and intraobserver variability in intensity measurements of normal myocardium of 25.8% and 15.0%, respectively; in infarcted myocardium, 12.4% and 9.1%. In a study of nine patients at 3 to 5 days post-MI, the mean TE = 28-millisecond infarct intensity had significantly increased 37% ± 13% relative to normal myocardium. Seven of the nine patients were treated with thrombolytic agents or percutaneous transluminal coronary angioplasty a mean of 4.5 hours from the onset of chest pain. At 10 to 14 days post-MI, the TE = 28-millisecond increase was 51% ± 17%.

Fisher et al[64] studied 29 patients with recent (3 to 17 days old) myocardial infarcts; four patients were excluded from analysis because of nondiagnostic studies. The mean T2 relaxation time of infarcted myocardium (79 ± 22 milliseconds) calculated from a double spin-echo sequence (TE 28/56 or 30/60) was significantly prolonged in comparison to normal myocardium (43.9 ± 9 milliseconds). Second-echo image intensity from infarcted myocardium was 65.6% ± 34.0% greater than that from normal myocardium; the first-echo increase was 27.5% ± 18.7%. In 20 normal subjects, no difference in T2 between the anterolateral (40.3 ± 5.7 milliseconds) and septal (39.5 ± 7.4 milliseconds) myocardium was noted. The first- and second-echo regional myocardial intensity differences in the normal subjects were 9.1% ± 7.4% and 15.0% ± 13.3 milliseconds, respectively. The increased signal intensity in the infarcted myocardium in humans resolves in approximately 3 months postinfarction.[65] Dinsmore et al[66] found no significant difference in ischemic myocardial NMR signal intensity between reperfused (tissue plasminogen activator or streptokinase) or nonreperfused ischemic myocardium in 19 patients. Infarct location and size were concordant with LV angiographic locations and estimates of infarct size.

The major pitfall in identifying acute MI by NMR imaging is the presence of high signals from the slowly moving LV blood pool adjacent to dysfunctional infarcted myocardium, rendering the infarct isointense with the blood pool, resulting in overestimation of infarct size. This phenomenon occurs much more commonly in symmetric double-echo studies (for example, TE 30/60) but can be reduced by use of a single late echo (TE 60)[61] and RF spatial presaturation.[5] An even worse pitfall can result in simulation of an acute infarct in normal myocardium. Physiologic increases in the LV blood pool signal, contractile motion of the heart, and respiratory motion can be mismapped in the phase-encoding direction, resulting in "ghost" accumulation in myocardium. A simple technical solution to this potential pitfall is to "swap" the phase-encoding and frequency-encoding gradient axes, moving the ghosts to a new location. Other solutions include routine use of RF spatial presaturation,[5] respiration-ordered phase encoding, and bipolar motion-compensation gradients, in addition to ECG gating.

CLINICAL SEQUELAE OF ACUTE MYOCARDIAL ISCHEMIA

Chronic myocardial infarctions in humans have been demonstrated as focal regions of decrease in wall thickness of the left ventricle[67,68] (Figure 18-6). Often the transition between wall thinning and normal myocar-

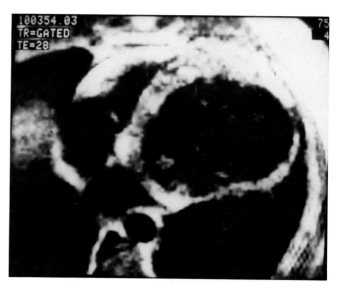

Figure 18-6 Transaxial ECG-gated NMR image of a remote posterolateral LV myocardial infarction, characterized by marked wall thinning and LV chamber enlargement.

dial wall thickness is rather abrupt. This provides an imaging estimate of the left ventricle involved by the previous myocardial infarction. Ventricular aneurysms can also be demonstrated by NMR. The detection of intraventricular thrombi is possible by careful analysis of the behavior of signal from within the ventricular chamber[69,70] (Nemanich J, Stratton J, Wesbey GE. In preparation, 1989) (Figure 18-7).

The optimal measurement of myocardial wall thickness for accurate characterization of LV mass is now achieved by cardiac NMR imaging oriented orthogonal to the cardiac axis to avoid tangential sections to the myocardial fibers resulting in overestimation or underestimation of wall thickness.[6]

SUMMARY OF NMR IMAGING IN CORONARY ARTERY DISEASE

NMR imaging has to date shown significant potential for evaluation of the anatomic, physiologic, and biophysical manifestations of coronary artery disease. In comparison to two-dimensional echocardiography, cardiac NMR imaging provides differential characterization of normal and pathologic myocardium, provides a larger field of view, and is not dependent on operator technique or the patient's body habitus. Imaging of patients with chronic obstructive pulmonary disease or severely obese patients can be difficult with two-dimensional echocardiography. Patients with pacemakers, intracranial aneurysm clips, life-support devices, or severe dysrhythmias are currently excluded from NMR imaging of *any* part of the body, including the heart. The major advantages of cardiac NMR imaging over ultrafast cardiac x-ray computed tomography are the absence of ionizing radiation, the ability to obtain direct long-axis images, and the lack of need to inject iodinated contrast medium (and its attendant risks) to distinguish cardiac anatomy from the blood pool and to identify blood vessels.

For the study of physiologic abnormalities in the cardiovascular system, NMR techniques have not yet reached widespread clinical application. However, cine-magnetic resonance imaging (MRI) software and hardware upgrade packages disseminated worldwide will soon provide an extensive evaluation of functional imaging by MRI in relation to other noninvasive cardiac imaging modalities. Although NMR is a totally noninvasive modality for obtaining excellent anatomic images of the heart and blood vessels, two-dimensional echocardiography is far less expensive, equally as noninvasive, and a much more readily available and portable cost-effective medical technology that can produce clinically adequate anatomic information about the heart with more functional information derived from echo-Doppler studies than from present NMR studies. Thus at present NMR is not cost-competitive with two-dimensional echocardiography in the evaluation of myocardial anatomy and physiology in humans. Patients with cardiac disease who are too ill to tolerate 30 to 60 minutes supine cannot be studied by NMR, such as patients with severe congestive heart failure.

At present, gated NMR can be offered as a supplemental anatomic examination of the myocardium when two-dimensional echocardiography leaves questions of anatomic or tissue characterization unresolved. Even if current physiologic studies advance such with the cine-NMR imaging techniques, it is doubtful that anatomic and physiologic cardiovascular data derived from NMR studies would be cost-competitive with two-dimensional echocardiography. Thus it is the author's opinion that the major realistic application for cardiovascular NMR in clinical cardiology is in evaluation of regional myocardial perfusion after thrombolytic therapy or angioplasty of acute myocardial ischemia. Since ventricular function may take days to weeks to recover after ischemia and reperfusion, modern cardiovascular medicine has no tools available with which to judge the immediate success of interventions aimed at preserving a myocardium at risk. Cine-NMR imaging with Gd-DTPA may fill this void and provide the high spatial resolution necessary to characterize transmural gradients in myocardial perfusion.

Recent progress in NMR coronary angiography[71] has nurtured the exciting concept of risk-free, completely noninvasive outpatient screening for coronary artery disease. This dream awaits further technical improvements to overcome the formidable challenges of imag-

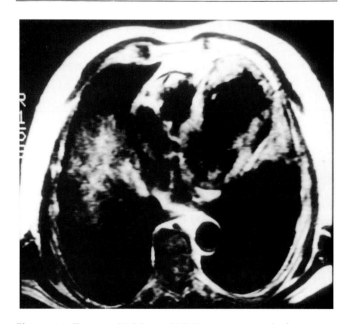

Figure 18-7 Transaxial ECG-gated NMR image at 1 week after anteroseptal myocardial infarction. A mural thrombus is adjacent to the infarcted myocardium.

ing the small lumina and the complex anatomy of the coronary arteries in the presence of cardiac and respiratory motion.

REFERENCES

1. Bottomley PA, Herfkens RJ, Smith LS, et al. Altered phosphate metabolism in myocardial infarction: P-31 MR spectroscopy. *Radiology*. 1987;165:703–707.
2. Higgins CB. Overview of MR of the heart-1986. *AJR*. 1986; 146:907–918.
3. Wesbey GE. Cardiovascular and pulmonary magnetic resonance imaging. In: Wehrli F, ed. *Biomedical Magenetic Resonance Imaging*. New York, NY: VCH Publishers, Inc; 1987:chap 7.
4. Osbakken M. Clinical uses of nuclear magnetic resonance spectroscopy in cardiovascular disease. *Am J Cardiac Imag*. 1987;1:242–253.
5. Felmlee JP, Ehman RL. Spatial presaturation: a method for suppressing flow artifacts and improving depiction of vascular anatomy in MR imaging. *Radiology*. 1987;164:559–564.
6. Dinsmore RE, Wismer GL, Levine RA, et al. Magnetic resonance imaging of the heart: Positioning and gradient angle selection for optimal imaging planes. *AJR*. 1984;143:1135–1142.
7. Wesbey GE, Higgins CB, Amparo EG, et al. Peripheral vascular disease: Correlation of MR imaging and angiography. *Radiology*. 1985;156:733–739.
8. Wesbey GE, Higgins CB, Valk PE, et al. Magnetic resonance applications in atherosclerotic vascular disease. *Cardiovasc Intervent Radiol*. 1986;8:342–350.
9. von Schulthess GK, Augustiny N. Calculation of T2 values versus phase imaging for the distinction between flow and thrombus in MR imaging. *Radiology*. 1987;164:549–554.
10. Lallemand DP, Gooding CA, Wesbey GE, et al. Magnetic resonance imaging of the aorta and pulmonary circulation: Initial experience with ECG-gating. *Ann Radiol (Paris)*. 1985;28:289–298.
11. Lallemand DP, Wesbey GE, Gooding CA, et al. Cardiosynchronous MRI intensity changes of the great vessels and pulmonary circulation. *Ann Radiol. (Paris)*. 1985;28:299–304.
12. Wedeen VJ, Meuli RA, Edelman RR, et al. Projective imaging of pulsatile flow by magnetic resonance. *Science*. 1985;230:946–949.
13. Dinsmore RE, Wedeen V, Rosen B, et al. Phase offset technique to distinguish slow blood flow and thrombus on MR images. *AJR*. 1987;148:634–636.
14. Fisher MR, von Schulthess GK, Higgins CB. Multiphasic cardiac magnetic resonance imaging: Normal regional left ventricular wall thickening. *AJR*. 1985;145:27–30.
15. Stratemeier EJ, Thompson R, Brady TJ, et al. Ejection fraction determination by MR imaging: Comparison with left ventricular angiography. *Radiology*. 1986;158:775–777.
16. Friedman BJ, Waters J, Kwan OL, et al. Comparison of magnetic resonance imaging and echocardiography in determination of cardiac dimensions in normal subjects. *J Am Coll Cardiol*. 1985;5: 1369–1376.
17. Markiewicz W, Sechtem U, Kirby R, et al. Measurement of ventricular volumes in the dog by nuclear magnetic resonance imaging. *J Am Coll Cardiol*. 1987;10:170–177.
18. Sechtem U, Sommerhoff BA, Markiewicz W, et al. Regional left ventricular wall thickening by magnetic resonance imaging: Evaluation in normal persons and patients with global and regional dysfunction. *Am J Cardiol*. 1987;59:145–151.
19. Dilworth LR, Aisen AM, Mancini J, et al. Determination of left ventricular volumes and ejection fraction by nuclear magnetic resonance imaging. *Am Heart J*. 1987;113:24–32.
20. Caputo GR, Tscholakoff D, Sechtem U, et al. Measurement of canine left ventricular mass by using MR imaging. *AJR*. 1987; 148:33–38.
21. Florentine MS, Grosskreutz CL, Chang W, et al. Measurement of left ventricular mass in vivo using gated magnetic resonance imaging. *J Am Coll Cardiol*. 1986;8:107–112.
22. Keller AM, Peshock RM, Malloy CR, et al. In vivo measurement of left ventricular mass using nuclear magnetic resonance imaging. *J Am Coll Cardiol*. 1986;8:113–117.
23. Paulin S, von Schulthess GK, Fossel E, et al. MR imaging of the aortic root and proximal coronary arteries. *AJR*. 1987;148:665–670.
24. Holmvang G, Edelman R, Dinsmore R, et al. Coronary graft patency by NMR. *Circulation*. 1986;167(suppl 2):42. Abstract.
25. Rubinstein RI, Askenase AD, Thickman D, et al. Magnetic resonance imaging to evaluate patency of aortocoronary bypass grafts. *Circulation*. 1987;76:786–791.
26. Gomes AS, Lois JF, Drinkwater DC, et al. Coronary artery bypass grafts: Visualization with MR imaging. *Radiology*. 1987; 162:175–179.
27. White RD, Caputo GR, Mark AS, et al. Coronary artery bypass graft patency: Noninvasive evaluation with MR imaging. *Radiology*. 1987;164:681–686.
28. Nayler GL, Firmin DN, Longmore DB. Blood flow imaging by magnetic resonance imaging. *J Comput Assist Tomogr*. 1986;10: 715–722.
29. Meuli RA, Wedeen VJ, Geller SC, et al. MR gated subtraction angiography: Evaluation of lower extremities. *Radiology*. 1986; 159:411–418.
30. Utz JA, Herfkens RJ, Heinsimer JA, et al. Cine MR determination of left ventricular ejection fraction. *AJR*. 1987;148:839–843.
31. Sechtem U, Pflugfelder PW, Gould RG, et al. Measurement of right and left ventricular volumes in healthy individuals with cine MR imaging. *Radiology*. 1987;163:697–702.
32. Sechtem U, Pflugfelder PW, White RD, et al. Cine MR imaging: Potential for the evaluation of cardiovascular function. *AJR*. 1987; 148:239–246.
33. Rzedzian RR, Pykett IL. Instant images of the human heart using a new, whole-body MR imaging system. *AJR*. 1987;149:245–250.
34. Williams ES, Kaplan JI, Thatcher F, et al. Prolongation of proton spin-lattice times in regionally ischemic tissue from dog hearts. *J Nucl Med*. 1980;21:449–453.
35. Higgins CB, Herfkens R, Lipton MJ, et al. Nuclear magnetic resonance imaging of acute myocardial infarction in dogs: Alterations in magnetic relaxation times. *Am J Cardiol*. 1983;52:184–188.
36. Rokey R, Verani NS, Bolli R, et al. Myocardial infarct size quantification by MR imaging early after coronary artery occlusion in dogs. *Radiology*. 1986;158:771–774.
37. Canby RC, Reeves RC, Evanochko WT, et al. Proton nuclear magnetic resonance relaxation times in severe myocardial ischemia. *J Am Coll Cardiol*. 1987;10:412–420.
38. Wesbey G, Higgins CB, Lanzer P, et al. Imaging and characterization of acute myocardial infarction by gated nuclear magnetic resonance. *Circulation*. 1984;69:125–130.
39. Pflugfelder PW, Wisenberg G, Prato FS, et al. Early detection of canine myocardial infarction by magnetic resonance imaging in vivo. *Circulation*. 1985;71:587–594.
40. Tscholakoff D, Higgins CB, McNamara MT, et al. Early phase myocardial infarction: Evaluation by MR imaging. *Radiology*. 1986;159:667–672.
41. Pflugfelder PW, Wisenberg G, Prato FS, et al. Serial imaging of canine myocardial infarction by in vivo nuclear magnetic resonance. *J Am Coll Cardiol*. 1986;7:843–849.

42. Caputo GR, Sechtem U, Tscholakoff D, et al. Measurement of myocardial infarct size at early and late time intervals using MR imaging: An experimental study in dogs. *AJR.* 1987;149:237–243.

43. Johnston DL, Brady TJ, Ratner AV, et al. Assessment of myocardial ischemia using proton magnetic resonance: Effect of a three hour coronary occlusion with and without reperfusion. *Circulation.* 1985;71:595–601.

44. Askenase AD, Thickman DI, Brown T, et al. Time sequence for reversal of ischemic changes seen with MRI in excised canine hearts. *Circulation.* 1986;74(suppl 2):II-318. Abstract.

45. Canby RC, Pohost GM. Changes in proton NMR relaxation times with reflow after a brief myocardial ischemic insult. *J Am Coll Cardiol.* 1987;9:73A. Abstract.

46. Miller DD, Kantor HL, Aretz T, et al. Proton magnetic resonance for the detection of viable myocardium within a reperfused ischemic risk area. *J Am Coll Cardiol.* 1986;7:197A. Abstract.

47. Johnston DL, Liu P, Rosen BR, et al. In vivo detection of reperfused myocardium by nuclear magnetic resonance imaging. *J Am Coll Cardiol.* 1987;9:127–135.

48. Miller DD, Kantor HL, Johnston DL, et al. Effect of mannitol on magnetic resonance relaxation parameters in canine myocardial infarction. *J Am Coll Cardiol.* 1986;7:174A. Abstract.

49. Tscholakoff D, Higgins CB, Sechtem U, et al. MRI of reperfused myocardial infarct in dogs. *AJR.* 1986;146:925–930.

50. Wisenberg G, Prato FS, Carroll ES, et al. Serial changes in myocardial T1 and T2 values with and without reperfusion. *J Am Coll Cardiol.* 1986;7:173A. Abstract.

51. Weinmann HJ, Brasch RC, Press WR, et al. Characteristics of gadolinium DTPA complex: A potential NMR contrast agent. *AJR.* 1984;142:619–624.

52. Brasch RC, Weinmann HJ, Wesbey GE. Contrast-enhanced NMR imaging: Animal studies using gadolinium-DTPA complex. *AJR.* 1984;142:625–630.

53. Wesbey GE, Higgins CB, McNamara MT, et al. Effect of gadolinium-DTPA on the magnetic relaxation times of normal and infarcted myocardium. *Radiology.* 1984;153:165–169.

54. McNamara MT, Higgins CB, Ehman RL, et al. Acute myocardial ischemia: Magnetic resonance contrast enhancement with gadolinium-DTPA. *Radiology.* 1984;153:157–164.

55. McNamara MT, Tscholakoff D, Revel D, et al. Differentiation of reversible and irreversible myocardial injury by MR imaging with and without gadolinium-DTPA. *Radiology.* 1986;158:765–769.

56. Peshock RM, Malloy CR, Buja LM, et al. Magnetic resonance imaging of acute myocardial infarction: Gadolinium diethylenetriamine pentaacetic acid as a marker of reperfusion. *Circulation.* 1986;74:1434–1440.

57. Miller DD, Holmvang G, Gill JB, et al. Detection of coronary stenoses by continuous paramagnetic contrast infusion during dipyridamole-induced hyperemia: The nuclear magnetic imaging stress test. *Circulation.* 1986;74(suppl 2):319. Abstract.

58. Rehr RB, Peshock RM, Malloy CR, et al. Improved in vivo magnetic resonance imaging of acute myocardial infarction after intravenous paramagnetic contrast agent administration. *Am J Cardiol.* 1986;57:864–868.

59. Tscholakoff D, Higgins CB, Sechtem U, et al. Occlusive and reperfused myocardial infarcts: Effect of gadolinium-DTPA on ECG-gated MR imaging. *Radiology.* 1986;160:515–520.

60. McNamara MT, Higgins CB, Schechtmann N, et al. Detection and characterization of acute myocardial infarction in man with the use of gated magnetic resonance. *Circulation.* 1985;71:717–724.

61. Johnston DL, Thompson RC, Liu P, et al. Magnetic resonance imaging during acute myocardial infarction. *Am J Cardiol.* 1986;57:1059–1065.

62. Filipchuk NG, Peshock RM, Malloy CR, et al. Detection and localization of recent myocardial infarction by magnetic resonance imaging. *Am J Cardiol.* 1986;58:214–219.

63. Dilworth LR, Aisen AM, Mancini J, et al. Serial nuclear magnetic resonance imaging in acute myocardial infarction. *Am J Cardiol.* 1987;59:1203–1205.

64. Fisher MR, McNamara MT, Higgins CB. Acute myocardial infarction: MR evaluation in 29 patients. *AJR.* 1987;148:247–251.

65. Ahmad M, Johnson RF, Fawcett D, et al. Magnetic resonance and thallium-201 myocardial imaging: Segment to segment comparison in patients with myocardial infarction and ischemia. *J Am Coll Cardiol.* 1987;9:75A. Abstract.

66. Dinsmore RE, Johns JA, Yasuda T, et al. Characterization of myocardial signal intensity by magnetic resonance imaging in proven normal and infarcted myocardium. *J Am Coll Cardiol.* 1986;7:174A. Abstract.

67. Higgins CB, Lanzer P, Stark D, et al. Imaging by nuclear magnetic resonance in patients with chronic ischemic heart disease. *Circulation.* 1984;69:523–530.

68. McNamara MT, Higgins CB. Magnetic resonance imaging of chronic myocardial infarctions in man. *AJR.* 1986;146:315–320.

69. Dooms GC, Higgins CB. MR imaging of cardiac thrombi. *J Comput Assist Tomogr.* 1986;10:415–420.

70. Gomes AS, Lois JF, Child JS, et al. Cardiac tumors and thrombus: Evaluation with MR imaging. *AJR.* 1987;149:895–900.

71. Alfidi RJ, Masaryk TJ, Haacke EM, et al. MR angiography of peripheral, carotid, and coronary arteries. *AJR.* 1987;149:1097–1110.

Chapter 19

Imaging Techniques for the Evaluation of Coronary Artery Bypass Grafts

Timothy M. Bateman, MD, and James S. Whiting, PhD

Many patients with ischemic heart disease will, during the course of their illness, be benefited by a revascularization procedure. Coronary artery bypass graft surgery (CABG) is one such technique that is both effective and well established for certain patient subsets. Its short- and long-term success is markedly influenced by maintained patency of the bypass graft, which in turn is related to such factors as noninjurious intraoperative handling of bypass conduits; the integrity of proximal and distal anastomoses; use of internal mammary and sequential saphenous vein bypass grafts; antiplatelet medications; and avoidance of postoperative smoking, hypertension, and hypercholesterolemia. Nevertheless, even with careful attention to all of the above, CABG is not a cure for what is generally a progressive pathologic vascular process that frequently invades the grafts themselves.

Closure of a bypass graft is sometimes silent, but there is a definite relationship between patency of bypass grafts and deleterious clinical events such as sudden death, myocardial infarction, progressive angina, ventricular dysrhythmias, deteriorating ventricular function, and a variety of quality-of-life indices including employment and function classification.[1–4] Furthermore, bypass graft closure occurs frequently: up to 20% of saphenous vein grafts are occluded by 1 year and as many as 50% by 10 years after bypass operation.[4–7] Both of the above considerations attest to the need for a highly accurate and widely available means to determine bypass graft patency. Symptom status is unreliable, as the first clinical sign of a diseased graft can be unstable angina, myocardial infarction, or sudden death. Electrocardiographic changes during treadmill exercise testing are only moderately sensitive and specific indications of graft disease. Imaging techniques are therefore most frequently relied upon to provide diagnostic information about graft status. This information can be obtained inferentially by evaluation of exercise thallium scintigrams or exercise radionuclide ventriculograms, or directly by invasive graft angiography. With newer and improved technologies, direct assessment with minimally invasive digital subtraction angiography or x-ray computed tomography and noninvasive nuclear magnetic resonance imaging has been gaining increasing clinical attention. This chapter reviews the available data and discusses the clinical applicability of each of the above approaches to imaging coronary bypass grafts.

INDIRECT EVALUATION OF BYPASS GRAFTS

Extensive literature addresses the capabilities of exercise thallium 201 scintigraphy and exercise radionuclide ventriculography for studying bypass grafts. Both provide indirect and hence inferential information about graft status, the presumption being that exercise-induced ischemia or new regions of infarction identify

---

This work was supported in part by Ischemic Heart Disease SCOR Grant No. 17651 from the National Institutes of Health, Bethesda, Md.

diseased bypass grafts. Radionuclide tests appear to be particularly useful when employed in sequential fashion.

Thallium 201 Scintigraphy

Ritchie et al[8] assessed regional myocardial perfusion with planar and visually interpreted ^{201}Tl scintigraphy in 20 patients before (11 with exercise, 9 at rest) and after (all underwent exercise scintigraphy) CABG; all had coronary and graft angiography at 3 to 15 months after CABG and within 4 weeks of scintigraphic evaluation. In the 7 patients who had a new persistent or new exercise-induced reversible ^{201}Tl perfusion defect, only 54% of the grafts were patent. By comparison, in the 13 patients who showed no new defects with postoperative exercise imaging, 26 of 30 (87%) of the bypass grafts were patent. Regional defects did not always correlate with bypass graft status, but when the preoperative scintigraphic images were available for comparison, it was possible to differentiate ischemia in the nongrafted areas from that arising from premature graft closure. From this study, one can conclude that when patients undergo both preoperative and postoperative evaluation, and when the postoperative assessment is performed early after CABG, visually assessed planar exercise ^{201}Tl scintigraphy can separate patients with occluded grafts from patients with maintained patent bypass grafts. Late after CABG, one would anticipate an increasing likelihood of a random distribution between new disease and graft closure as the etiology for newly appearing exercise-induced perfusion defects.

Greenberg et al[9] evaluated 25 patients early after coronary bypass surgery, all of whom had also had invasive angiography, to determine the accuracy of the scintigraphic findings in patients who had not had pre-CABG ^{201}Tl reference studies. Each of the 10 patients with "positive" thallium studies had flow-limiting (75% or more) obstructions in at least one vessel, but in only 80% was the culprit lesion in a bypass graft; in the others the narrowing was either distal to the graft insertion or in the distribution of an ungrafted vessel. Furthermore, 5 of the 15 patients who had normal studies had one or more closed grafts. Greenberg et al[9] concluded that post-CABG exercise thallium scintigraphy had good sensitivity and was highly specific for the diagnosis of coronary stenosis but was equivocal about the etiology of the ischemia (graft closure versus other causes). Compared with chest pain (sensitivity 60%, specificity 20%) and stress electrocardiography (sensitivity 60%, specificity 86%, not counting 25% of the patients who had equivocal or suboptimal tests), thallium was the most useful technique.

Pfisterer et al[10] determined the accuracy of serial myocardial perfusion scintigraphy with ^{201}Tl for predicting graft patency early and later after coronary bypass surgery by comparing the results of thallium testing with temporally related graft angiography. All of their patients had exercise ^{201}Tl scintigraphy and angiography performed preoperatively and at 2 weeks and 1 year after operation. The sensitivity and specificity for identifying closed grafts were 80% and 88%, respectively. The scintigraphic finding indicative of a closed graft was either a new reversible perfusion defect or a persistent "new scar." Occluded grafts were correctly localized by scintigraphy in only 61%. Of interest was the fact that two thirds of patients developed new, nonreversible, apical perfusion defects corresponding to the site of apical venting during cardiopulmonary bypass surgery.

The addition of quantitative analysis to rest and exercise thallium scintigrams and the performance of tomographic rather than planar imaging have not been studied specifically for their ability to determine bypass graft patency. Both studies could offer significant improvements, as previous investigations have indicated the need for more accurate detection of myocardial hypoperfusion and better localization of disease to specific vascular territories.

Radionuclide Ventriculography

More recently, Higginbotham et al[11] evaluated biplane rest and exercise radionuclide ventriculography for assessing individual bypass grafts. Although they attempted to evaluate the status of 59 grafts, radionuclide prediction of graft status was precluded for 27 grafts because there was normal wall motion both preoperatively and postoperatively in the region subtended by a graft or because a subtended region was not represented on the biplane radionuclide angiogram. An assumption here was that, if a region was bypassed but had no abnormalities on the preoperative study, persisting normalcy in a postoperative study would prohibit accurate evaluation of perfusion in the distribution of that vessel. Of the 32 predictions that were made, 78% were correct, including 93% accuracy in identifying open grafts and 67% accuracy in predicting grafts that were either closed or had significant (75% or more) narrowings. Most incorrect predictions of graft inadequacy were due to new septal wall motion abnormalities postoperatively. Thus exercise radionuclide ventriculography does not appear to be as effective as exercise thallium scintigraphy in determining the patency of bypass grafts, probably for two reasons. First, septal motion is often abnormal in most patients in the early postoperative period and may persist for up to 2 years.[12] Second, radionuclide ventriculography, even when performed in two views, may be somewhat less sensitive than exercise thallium scintigraphy for detecting flow-limiting stenoses.

Strengths and Weaknesses of Scintigraphic Methods for Evaluating Bypass Grafts

The studies reviewed indicate that there are some laudable strengths and some formidable weaknesses in the use of exercise radionuclide studies for determining the patency of bypass grafts (Table 19-1). The favorable aspects are that the tests are noninvasive, relatively inexpensive, and widely available, and they can be performed serially. Because they evaluate myocardial hypoperfusion, they provide unique and clinically relevant information beyond the extrapolated assumption about whether a bypass graft is patent or occluded. For example, it can be reasonably argued that the patency of a bypass graft is largely irrelevant if a patient can exercise to a high level without evidence for exercise-induced ischemia in the distribution of that graft. Furthermore, by measuring the severity and extent of the hypoperfusion created by closure of a bypass graft, the radionuclide tests provide a clinically important assessment of the prognostic importance of the event. Independent of symptomatology, the finding of a substantial perfusion defect in the distribution of a vessel that has been revascularized would suggest the need for invasive evaluation and intervention if possible.

The weaknesses of these approaches to determining bypass graft patency or occlusion are numerous. The first weakness is that the extrapolation about graft patency from the scintigraphic findings will be anticipated to be inaccurate in a large number of situations. The existence of distal coronary disease or incomplete revascularization after bypass surgery renders the findings of exercise-induced hypoperfusion a nonspecific indicator of graft closure. The corollary is that many patients without exercise-induced ischemia in the distribution of a bypassed vessel may well have an occluded graft if the graft had been placed into a vessel with a non-flow-limiting stenosis. This is commonly done when important vessels such as the left anterior descending (LAD) coronary artery contain angiographically recognized stenoses even though they are not yet flow-limiting. Bypass grafts are also frequently placed into regions of a myocardial scar on the reasoning that the bypass graft may provide collateral support to other

scintigraphy will not be able to determine the status of a bypass graft that has been inserted into a vessel subtending an infarcted region, because a fixed perfusion defect will pertain whether the bypass graft is patent or occluded.

A second problem is the inability of exercise thallium scintigraphy or radionuclide ventriculography to separate the various grafts that lead to adjacent myocardial regions. For example, a reversible perfusion defect of the lateral wall of the left ventricle could be consistent with an occluded graft to the LAD, diagonal, or circumflex marginal coronary arteries. A reversible septal defect in a left anterior oblique-45° planar scintigram could be consistent with exercise-induced ischemia of the inferior septum (often supplied by the right or left circumflex coronary arteries) or the anterior septum (almost always supplied by the LAD coronary artery).

A third weakness of exercise radionuclide evaluation of bypass grafts is that the accuracy of the findings is critically dependent upon the level of exercise achieved. Submaximal exercise can lead to false-negative findings, as there is a progressive decrement in the elicitation of exercise-induced ischemia in patients with known flow-limiting stenoses with lesser degrees of performed exercise. Furthermore, when exercise is limited by a significant end point (2 mm or greater ST segment depression, chest pain, or shortness of breath), a single region of myocardial hypoperfusion may represent discovery of the worst flow-limiting narrowing while missing one or more other areas that may be potentially ischemic at higher levels of exercise and may be fed by occluded bypass grafts. Finally, independent of the various reasons, at least the four studies reviewed have proved that the sensitivity and specificity of presently available radionuclide techniques are not optimal for determining whether bypass grafts are patent. Furthermore, best results require sequential studies before and after CABG.

DIRECT EVALUATION OF BYPASS GRAFTS

Invasive Techniques

Selective Graft Angiography

Direct graft angiography is the most commonly applied method to determine whether coronary artery bypass grafts are patent or occluded and is the accepted "gold standard." Accuracy is high (not necessarily 100%) for determining graft patency or occlusion, but avoidance of several pitfalls is critical or erroneous conclusions about graft status will be drawn.

Cannulation of bypass graft orifices can be difficult, so that inability to engage the graft orifice does not necessarily equate with bypass graft occlusion. Three important considerations are knowledge of which

Table 19-1 Strengths and Weaknesses of Scintigraphic Methods for Evaluating Patency of Aortocoronary Bypass Grafts

Strengths	Weaknesses
Noninvasive	Inferential
Widely available	Not highly accurate
Permit serial evaluation	Dependent on level of exercise attained
Evaluate extent and severity of ischemia associated with graft closure	Best results from serial studies

grafts were placed during surgery, the usual sites of proximal anastomoses, and the value of aortic root angiograms. Surgeons do not always indicate the origin of the bypass grafts with metallic (radiodense) markers, so that graft angiography is always best performed when operative information about the numbers of grafts is known. Knowledge of the site of graft origin is also critical, as there is no positioning of the proximal anastomoses of the bypass grafts that is universally subscribed to. By convention, it is common to place the circumflex origin highest up and on the left lateral or posterolateral side of the ascending aorta and to place diagonal and LAD grafts sequentially lower and slightly more anteriorly on the ascending aorta. Grafts to the right coronary artery and its branches typically come off the right side of the ascending aorta and are closest to the aortic valve. Unless an occluded stump is visualized, it is essential that the angiographer perform an aortic root injection while carefully inspecting the ascending aorta for faint filling of patent but difficult-to-locate bypass grafts. Biplane aortic root angiography is best because a single-plane view can miss grafts that are coming off in the image projection plane.

Filling of a bypass graft with contrast material usually indicates that the bypass graft is patent. Slow contrast filling can occur through permeation of a fresh or soft thrombus in a functionally occluded graft. Such cases are recognized by extending imaging until the contrast medium has cleared from the graft. Figure 19-1 shows a bypass graft that is well opacified with contrast medium, but it is functionally occluded; opacification of this freshly thrombosed graft persisted for several minutes.

Misidentification of LAD and first or second diagonal grafts, or first and second obtuse marginal bypass grafts, is a potential problem in patients with multiple grafts to similar myocardial regions and in those who have highly diseased native vessels. Careful attention to the appearance of the recipient vessel can help to avoid this confusion.

A clue to detecting a patent graft without a cannulable orifice is recognition of dye regurgitating in retrograde fashion up the bypass graft after native vessel injection. Alternatively, if the bypass graft flow is better than that in the native vessel, one might visualize diluted opacification of the distal native vessel.

Aortic Root Digital Subtraction Angiography

Bypass graft patency can be determined accurately by nonselective aortic root injection of contrast medium. Optimally, a mask-subtracted digital technique is employed in which images are obtained just prior to and again at the time of aortic root opacification with contrast medium. The first study is subtracted from the later acquisition. Assuming no patient movement, the regions filling with contrast medium should be seen starkly outlined against a dark background. Guthaner et al[13] emphasized that injection volumes of 40 to 45 mL at rates of at least 20 mL/s were required to obtain a diagnostic image, and that several projections were necessary, as each provided complementary information. Whiting et al[14] compared aortic root digital subtraction angiography (DSA) with selective graft angiography and reported a diagnostic accuracy of 100% in a study that included 11 patent and 6 occluded grafts.

Aortic root DSA can be an important adjunct to selective graft angiography when direct cannulation proves difficult, as it permits identification of patent grafts and confirmation of occlusions. It has not emerged as a clinically important "stand-alone" test for determining graft status primarily because it requires arterial catheterization and injection of contrast medium. As such, the risk and expense are nearly the same as with selective graft angiography, while not providing the same amount of information about the native arteries, the detailed appearance of the bypass grafts, and their proximal and distal anastomoses. Development of smaller catheters, newer imaging agents, and better image enhancement capabilities may result in expanded use in the future.

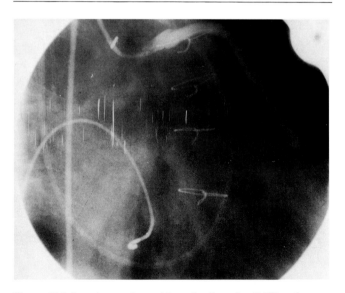

Figure 19-1 Invasive angiographic evaluation of an LAD saphenous vein bypass graft. The graft completely fills with contrast medium. However, two findings indicate that it is technically occluded: there is no distal runoff, and continued imaging showed that the contrast medium did not clear from the graft for several minutes, indicating contrast medium staining of a fresh thrombus.

Minimally Invasive Techniques

Three minimally invasive techniques have been evaluated for their ability to assess bypass graft status. They do not require invasive intra-arterial access, but they do not require the intravenous injection of potentially toxic contrast agents and significant exposure to ionizing radiation. Two techniques are mainly of research or historic interest but the third, ultrafast computed tomography (CT), holds some promise of evolving into a clinically relevant and relatively commonplace diagnostic test for determining graft patency.

Intravenous DSA

The patency of bypass grafts can be evaluated by injecting 30 to 50 mL of contrast medium into the superior vena cava or femoral vein and performing DSA. Several acquisitions in differing image planes are necessary, so that exposure to contrast material and to x-rays is significant. The accuracy of the technique has been reported by Whiting et al[14] to be suboptimal for clinical use because of poor sensitivity. Although all 13 occluded grafts were assessed correctly by intravenous DSA, 4 of 14 patent grafts were not seen and were incorrectly assumed to be occluded. Therefore, graft visualization was invariably associated with patency, but lack of visualization was insensitive for graft occlusion. The authors hypothesized that the limited sensitivity reflected some or all of the following: the size of the grafts, their anatomic positioning relative to the plane of the projection image, and suboptimal image resolution. Furthermore, registration artifact caused by patient motion or breathing during the acquisition resulted in linear streaks that could be mistaken for patent bypass grafts; the high specificity in this study reflected careful attention to such details as electrocardiographic triggering of the exposures and detailed knowledge of the surgery. The authors concluded that intravenous DSA was not suitable for definitive determination of bypass graft patency. In a confirmative study, Guthaner et al[13] found that intravenous DSA identified only 41% of known patent saphenous vein grafts. They also emphasized that the technique visualized only portions of the graft.

X-Ray CT

The patency of saphenous vein bypass grafts can be determined by contrast medium–enhanced nondynamic x-ray CT with a clinically acceptable sensitivity and specificity as compared with angiography.[15-19] Most studies have used total-body scanners without additional equipment for dynamic scanning or electrocardiographic gating, and with exposure times per image of 1 to 5 seconds. One or more scout images are acquired to determine the appropriate scanning levels. Radiographic contrast material (30 to 50 mL) is then injected over 3 to 4 seconds into a peripheral vein, using a hand or infusion pump; scanning is begun 5 to 8 seconds later. Several scans are acquired at different levels to permit visualization of grafts at more than one site and to obviate confusion over grafts that are patent for 2 to 3 cm proximal to an occlusion. Radiation exposure is considerable, ranging from 1 to 5 rad depending upon the number of acquisitions at each level. Appropriately located and appropriately appearing structures that opacify with contrast medium are judged to be patent bypass grafts, whereas nonvisualization of a graft known to have been placed at surgery is equated with occlusion. With these criteria, accuracies range from 48% to 100%.

Knowledge of which grafts were placed at surgery is essential for accurate results. Because distal anastomoses are seldom visualized, the interpreter must either know from the operative report the order in which left-sided grafts exit the aortic root or make assumptions about this.

Daniel et al[19] have summarized the limitations of standard x-ray CT for assessing bypass grafts. Several limitations arise from the lengthy (1 to 5 seconds) acquisition times per image of nondynamic CT scanners. Short grafts or the distal end of sequential grafts can be missed because of a greater problem with graft motion in relationship to their attachment to the heart. Grafts that are adherent to the pericardial surface of the heart are frequently missed, again because of blurring caused by cardiac motion; the inability to observe movement of the contrast medium through the graft means that all opacified structures potentially can be mistaken for bypass grafts. Some of the most frequent offenders are the atrial appendage, pulmonary veins, and native coronary arteries. Another significant problem with the technique, limiting its acceptance, is that nondynamic x-ray CT does not provide information about the proximal or distal anastomoses, about the presence or absence of stenoses within the bypass graft, or about bypass graft flows. Third, considerable experience is necessary to obtain quality images with optimal graft opacification and to interpret the results of the study correctly. Finally, metallic surgical clips result in scatter artifact and can preclude determination of patency if each image cut includes a clip. Clips are often used with internal mammary artery bypasses. No studies have addressed the accuracy of nondynamic CT to determine the patency of internal mammary artery grafts.

Ultrafast CT

The recent introduction of the ultrafast CT scanner (Imatron C-100) has reawakened interest in x-ray CT as a practical method for evaluating bypass grafts. The ultrafast CT scanner permits high-speed (50 ms/image), rapid serial acquisitions (up to 17 images per second) at multiple levels,[20] achieved through elimination of all mechanically moving parts and substitution of an electromagnetically steered electron beam that is passed along four 210° tungsten target rings. Each target generates fan beams of photons that pass through the patient to a double array of 864 stationary luminescent crystal silicon photodiode detectors arranged in a semicircle above the patient. From these data, a computer reconstructs four pairs of 0.8-cm thick images with 0.4-cm gaps between image pairs.[20] The resolution is not as good as it is with standard CT (1.5 to 2 mm spatial discrimination versus 0.5 to 1 mm), but for determining whether bypass grafts are patent or occluded, this relative loss in spatial resolution does not appear to be clinically important.

Bateman et al[21] used an ultrafast CT scanner for determining the patency of bypass grafts in comparison with invasive graft angiography. Several different acquisition protocols were tried before a seemingly optimized approach was arrived at. In the protocol studied, patients were positioned supine in the scanner, perpendicular to its gantry. A localization scan was employed so that the highest scan would be at the highest necessary level. For patients with saphenous vein bypass grafts, this was just cephalad to the uppermost metallic clip on the ascending aorta, or in the absence of clips at the inferior margin of the aortic arch. When internal mammary grafts were also present, scanning was started 1 cm above the cephalad border of the aortic arch.

Images were acquired after peripheral arm vein injection of a bolus of contrast medium, using an 18-gauge angiocath with a Med-rad injector that delivered 7 to 9 mL/s for 5 seconds. For patients with presumed normal circulation times, the scanning sequence began approximately 7 seconds after injection. In patients with congestive heart failure or pulmonary disease, it was necessary to measure circulation time before injection of contrast medium. At each level, 10 temporally sequential images were then acquired at multiples of electrocardiographic R waves, depending upon heart rate. Figure 19-2 shows the relationship of imaging planes from an ultrafast CT graft study to the standard positions of left circumflex, LAD, and right coronary artery saphenous vein bypass grafts. A second, more caudally placed, acquisition series was acquired if the lowest scan was not at least 4 cm along the length of a graft.

Figure 19-3 shows a typical eight-level ultrafast CT graft study in a patient with two saphenous vein aortocoronary grafts. Both grafts are visualized at a number of tomographic levels spanning their extent from aortic through distal anastomoses. The relationship of the image planes to graft locations is such that the grafts are seen in longitudinal projection as they exit from the aortic root and in cross-section in sequentially more caudal planes. Figure 19-4 shows a stump of an LAD graft, and Figure 19-5 demonstrates an occluded and a patent saphenous vein graft. Chron-

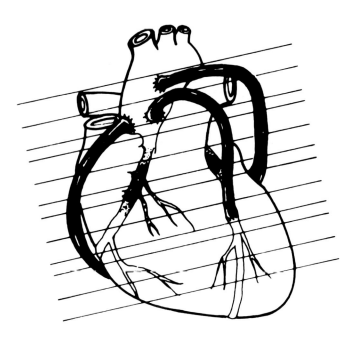

Figure 19-2 Customary location of aortocoronary saphenous vein grafts in relation to the eight imaging planes obtained by an ultrafast CT bypass graft study. *Source:* Reprinted with permission from *Journal of the American College of Cardiology* (1986;8:696), Copyright © 1986, American College of Cardiology.

ically occluded grafts are frequently visible as structures of sufficient density to contrast with low-density mediastinal fat; they may contain dense, organized fibrous thrombi or may be calcified, but they do not opacify with contrast agents.

In our study, ultrafast CT correctly identified 46 of 48 patent and 31 of 32 occluded grafts, for an overall diagnostic accuracy of 96%. Several shortcomings were clarified in this study. First, for optimal accuracy it was essential to know which grafts were placed and to obtain excellent images of the entire ascending aorta, because it was often impossible to identify the recipient vessel. Further experience has shown particular confusion with differentiating diagonal from LAD insertion sites, the various obtuse branches of the circumflex, and posterolateral from posterior descending insertion sites into the distal right coronary artery. Second, the distal limbs of sequential grafts are occasionally identifiable, but motion artifact resulting in image blur often precluded their visualization. As such, visualization of distal limbs of sequential grafts can be equated with patency, while nonvisualization is an indeterminate finding. Third, as with nondynamic CT, metallic clips produce scatter artifact; the amount of resulting "noise" is less and becomes a problem only if a large number of clips are used along the entire length of a graft. Finally, the accuracy of ultrafast CT determination of patency of internal mammary artery bypass grafts is not yet certain; because of the generally smaller diameter of these grafts (2 to 4 mm versus 3 to 6 mm for saphenous vein

Imaging Techniques for Evaluation of Coronary Artery Bypass Grafts   275

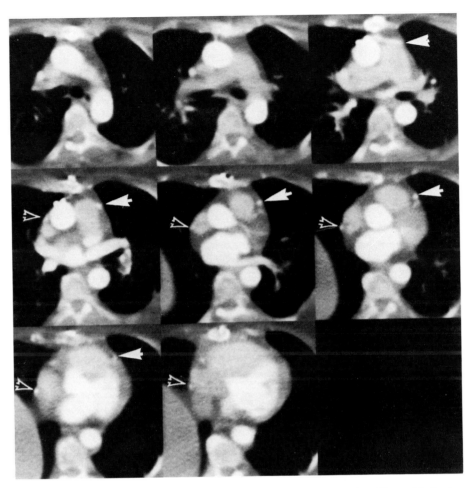

Figure 19-3 Eight-level ultrafast CT study spanning the distance from the aortic arch (top left) to the midventricle (bottom right) and showing two patent saphenous vein bypass grafts. The LAD graft (closed arrow) is seen in longitudinal projection in frame 3 and in cross-section in frames 4 through 7, where it is anastomosing with the native LAD vessel. The right coronary artery graft (open arrow) is visualized in frames 4 through 8, where it is several centimeters proximal to the distal anastomosis (note native right coronary artery in the atrioventricular groove).

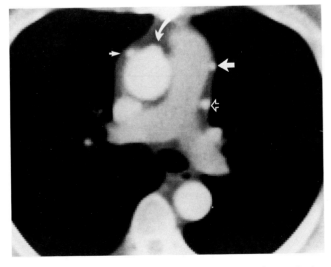

Figure 19-4 Stump of an occluded LAD graft (curved arrow) visualized with ultrafast CT. Three patent grafts (circumflex, open arrow; diagonal, closed arrow; right coronary artery, small arrow) are also seen. *Source:* Reprinted with permission from *Circulation* (1987; 75:1020). Copyright © 1987, American Heart Association.

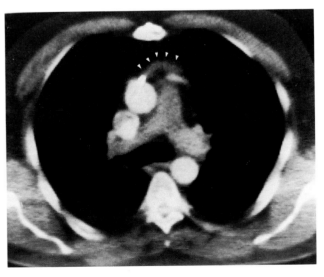

Figure 19-5 Ultrafast CT image at peak aortic opacification showing one patent and one occluded (closed arrows) saphenous vein graft.

conduits), and the more frequent use of metallic clips, their assessment will likely be more difficult.

A multicenter study[22] has further established the high accuracy of ultrafast CT for assessing saphenous vein bypass grafts. Five centers participated in this project, each performing the studies according to their own protocols, but interpretation was highly standardized. In a pilot group, almost 29% of the studies were technically inadequate, but a second component of the study had 94% adequate images, reflecting improved imaging protocols. The overall accuracy was 92% (sensitivity 93.4% and specificity 88.9%) for 91 patent and 36 occluded grafts.

The clinical relevance of knowing whether bypass grafts are patent or occluded in the absence of further information about the presence or absence of ischemia or details of the native circulation deserves comment. Altered clinical decision making based solely on this knowledge has been demonstrated,[23] most frequently for assessing graft status in patients whose presentation would not ordinarily mandate invasive catheterization and in patients who are unwilling to undergo such procedures.

A limitation is that partially obstructed grafts cannot be differentiated from normal grafts by anatomic criteria. A promising solution employs physiologic information about graft flows. The speed of ultrafast CT does permit analysis of the mean rate at which the peak concentration of the contrast bolus moves through a bypass graft. A time-density curve can be generated by placing a region of interest over portions of the graft and measuring contrast density in this region during each of the CT acquisitions. A typical time-density curve for a widely patent graft contrasts with the flat curve of an occluded graft (Figure 19-6). Whiting et al[24] have proposed that there may be sufficient information in such curves to quantitate graft flows; such an approach would be important for physiologically differentiating healthy grafts from diseased but patent grafts. Limited resolution of ultrafast CT prevents this distinction from anatomic criteria alone. The methodology of Whiting et al.[24] for measuring absolute graft flows uses graft cross-sectional area, measured distances along the length of a graft, and the transit time of the peripherally injected contrast bolus density peak between these known distances. In a phantom model they have demonstrated accuracy to within 20% of absolute flows.[24] Rumberger et al[25] have shown that flow reserve can be measured in grafts placed in animals and have suggested that this approach may help to differentiate stenotic grafts from healthy grafts in humans.

Several researchers have been exploring the capabilities of ultrafast CT to quantitate myocardial perfusion,[26,27] and, in addition, it appears possible to evaluate the presence or absence of myocardial ischemia in the distribution of a graft based on subtended regional wall motion at rest and at peak exercise.[28] While these investigations require further research, the information so provided could make ultrafast CT an extremely powerful tool for simultaneously evaluating the anatomic and physiologic integrity of bypass grafts and the functional implications of any graft pathologic processes.

In conclusion, ultrafast CT is a promising technique for the routine assessment of saphenous vein bypass grafts. It has the highest accuracy of any approach, with the exception of invasive angiography, for determining whether grafts are patent or occluded. As performed today, it has limited resolution, so that anatomic details about anastomoses and luminal obstructions are not discernible. Promising research indicates the possibility of circumventing these limitations through perhaps more important physiologic information about graft flow and myocardial perfusion and function in regions subtended by grafts.

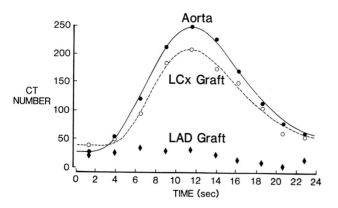

Figure 19-6 Time-density coordinates, fitted to a gamma-variate function, for separate regions of interest within the aorta, a patent left circumflex (LCx) graft, and an occluded LAD graft. The fitted curve of the patent graft resembles that of the aorta; the flat curve indicates lack of contrast medium enhancement. *Source:* Reprinted with permission from *Circulation* (1987;75:1022), Copyright © 1987, American Heart Association.

Noninvasive Techniques

Nuclear Magnetic Resonance Imaging

Several authors[29–34] have demonstrated that aortocoronary bypass grafts can be visualized by nuclear magnetic resonance (NMR) imaging. This technique is inherently attractive because it does not require injected contrast agents or ionizing radiation. Images are generated from hydrogen (proton) resonance in response to static and changing magnetic fields and radiofrequency pulses. Without cardiac and respiratory gating, images are variably degraded because there can be substantial loss of signal from moving structures and changing positions of target structures within the pixels of interest when data are acquired at different times in the cardiac or respiratory cycles. Gating is thus essential for imaging structures adherent to the heart (such as bypass grafts). Furthermore, NMR presents the necessity of using nonferromagnetic electronic circuitry. For optimal interrogation of vascular regions of interest, gradient angles can be altered without changing patient position within the scanner so that sagittal, coronal, and oblique planes can be obtained. A number of studies have demonstrated that NMR imaging of patients who have had coronary bypass surgery is safe and, although the metallic wires used to stabilize the sternum create substantial artifacts, these artifacts for the most part do not interfere with the interpretation of bypass graft patency. Nuclear magnetic resonance imaging is contraindicated in patients with intracranial aneurysm clips, pacemakers, or automatic implantable cardioverters/defibrillators.

With the spin-echo technique, blood flow results in a signal void. Gomes et al[31] acquired spin-echo electrocardiographically gated NMR images in 20 patients with previous bypass surgery. They used an echo time of 28 milliseconds and a repetition time determined by heart rate. The image section thickness was 9 mm at 12-mm intervals. They reduced the intersection gap by obtaining interlaced images and adjusting the imaging level by 2 to 4 mm once the grafts had been detected. Patent grafts were seen as tubular structures with a low internal signal arising from the aorta approximately at the level of the bifurcation of the main pulmonary artery. Grafts were assumed to be occluded if they were not identified. There was no angiographic correlation.

Rubinstein et al[32] subsequently applied the spin-echo technique in 20 patients who underwent graft angiography. They found that the sensitivity for detecting patent grafts was 90% and the specificity (for detecting occluded grafts) was only 72%. Total acquisition time was approximately 1 hour. In considering the low specificity, the authors emphasized that a nonvisualized bypass graft could occur with a technically poor acquisition or with a graft that for some reason had diminished flow so that a signal void was created.

White et al[33] also used spin-echo NMR to evaluate graft patency in 28 patients who underwent graft angiography. They used transaxial imaging, with 10 sequential images of 10 or 11 mm thick, from the level of the upper portion of the ascending aorta to the level of the left ventricular apex. A single spin-echo multiphasic technique (echo time 30 milliseconds) was used to obtain images in five or six phases of the cardiac cycle at each anatomic level. The total imaging time was approximately 45 minutes. Three studies were technically inadequate because of gating problems, motion artifact, and claustrophobia. Angiography in the 25 patients with technically adequate NMR studies indicated that 50 bypass grafts were patent and 22 were occluded. Ability to assess bypass patency or occlusion was possible with NMR for 90% of the grafts. The sensitivity for detecting patent grafts was 86%, whereas the specificity for determining an occluded graft was only 59%. The authors indicated several potential problems in the assessment of bypass graft status by spin-echo NMR. Metal hemostatic clips, bands of mediastinal fibrosis or thickened pericardium, small pericardial collections of air or fluid, native coronary arteries, and coronary veins all showed highly localized signal voids that closely resembled the signal voids of the small-diameter patent bypass grafts. They also emphasized the relatively high percentage of degraded studies resulting from distortion artifact caused by the metallic rings placed surgically to indicate the origin of bypass grafts, and blurring artifact caused by suboptimal electrocardiographic gating or by respiratory or body motion. Another problem is that the signal (or degree of signal void with spin-echo NMR) observed from flowing blood is nonlinearly related to its velocity. Therefore, it is anticipated that this technique cannot differentiate an occluded graft from one with restricted flow due to disease, anastomotic problems, recipient distal vessel disease, or low regional flow demand.

In a subsequent study, White et al[34] used cine-NMR imaging to evaluate the status of bypass grafts. The cine-format differs from spin-echo NMR both in the method of acquiring images and in the appearance of patent vascular structures. Cine-images are obtained by gradient-recalled acquisition in the steady state (the so-called GRASS technique), which uses a low flip-angle, a fast repetition rate, gradient-refocused echoes, and cardiac gating. With this technique, bypass grafts are seen as structures of high signal intensity at one or more levels. The study population consisted of 10 patients with a total of 28 bypass grafts (14 patent); 13 of 14 patent and 12 of 14 occluded grafts were correctly assessed, for an overall accuracy of 89%. Figure 19-7 shows two patent saphenous vein bypass grafts as visualized by cine-NMR imaging.

These data suggest that cine-NMR imaging achieves greater accuracy than does the spin-echo technique for evaluating coronary bypass grafts. Presumably the increased signal intensity from flowing blood rather than the spin-echo signal void can minimize many of the possible false-positive results and the greater problem with motion artifacts associated with the spin-echo technique. Cine-NMR could emerge as an attractive method for routine evaluation of the patency of saphenous vein bypass grafts. It is noninvasive and does not require ionizing radiation or contrast materials, as does ultrafast CT. While imaging time is protracted at present, it is likely that a complete bypass graft study could be done within an acceptable time frame of approximately 30 minutes. The cine-NMR

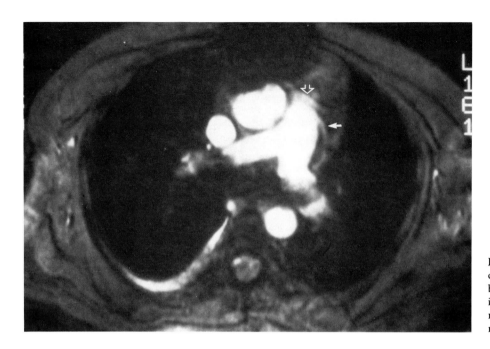

Figure 19-7 Patent LAD (open arrow) and circumflex (closed arrow) saphenous vein bypass grafts in a transaxial cine-NMR image acquired with the GRASS (gradient-recalled acquisition in steady state) technique.

technique cannot yet be used to differentiate partially occluded bypass grafts from widely patent bypass grafts, and the image quality with the cine-NMR technique is somewhat degraded relative to the better signal-to-noise ratio in standard static spin-echo or ultrafast CT images. Only portions of the grafts are visualized, so that direct characterization of the graft lumen or anastomosis is not possible at present.

SUMMARY

Table 19-2 compares and contrasts the various approaches to determining whether coronary artery bypass grafts are patent and the clinical importance and limitations of the observations made by each of the tests. Radionuclide methods are most widely available and convenient but are not highly accurate. Ultrafast CT and cine-NMR are highly accurate but are not yet widely available; furthermore, both techniques are expensive and neither can detect diseased but patent grafts or assess the proximal or distal anastomoses. As such, optimal noninvasive or minimally invasive methods have not yet been discovered. The value of graft angiography is also limited, as it does not evaluate myocardial perfusion, is invasive, and carries the greatest risk to the patient. For the present, most clinicians will continue to use more than one form of imaging technology to determine the status of bypass grafts. While this approach is somewhat inefficient, the clinical significance of bypass graft status and the relatively high rates of premature bypass graft closure underscore the importance of careful and serial evaluation of all patients who have undergone coronary artery bypass surgery.

Table 19-2 Comparative Features of Methods To Evaluate Patency of Bypass Grafts

	Method						
	Noninvasive			Minimally Invasive		Invasive	
Feature	Thallium 201	Radionuclide Angiography	Cine-NMR	Intravenous DSA	Ultrafast CT	Selective Graft Angiography	Aortic Root DSA
Contrast required	No	No	No	Yes	Yes	Yes	Yes
X-ray exposure	+ +	+	0	+ + +	+ + +	+ + + +	+ + +
Expense	+ +	+	+ + +	+ +	+ + +	+ + + +	+ + +
Accuracy	+ +	+	+ +	Inadequate	+ + +	+ + +	+ + +
Convenient for serial application	+ +	+ +	+ +	+	+ +	0	0

CT, computed tomography; DSA, digital subtraction angiography; NMR, nuclear magnetic resonance.

REFERENCES

1. Wolf NM, Kreulen TH, Bove AA, et al. Left ventricular function following coronary bypass surgery. *Circulation*. 1978;58:63–70.

2. Campeau L, Lesperance J, Hermann JR, et al. Loss of the improvement of angina between 1 and 7 years after aortocoronary bypass surgery: Correlations with changes in vein grafts and in coronary arteries. *Circulation*. 1979;60(suppl 1):1–5.

3. Peduzzi P, Hultgren HN. Effect of medical vs surgical treatment of symptoms in stable angina pectoris. The Veterans Administration Cooperative Study of Surgery for Coronary Arterial Occlusive Disease. *Circulation*. 1979;60:888–892.

4. Rahimtoola SH. Coronary bypass surgery for chronic angina—1981. *Circulation*. 1982;65:225–241.

5. Campeau L, Enjalbert M, Lesperance J, et al. Atherosclerosis and late closure of aortocoronary saphenous vein grafts: Sequential angiographic studies at 2 weeks, 1 year, 5 to 7 years, and 10 to 12 years after surgery. *Circulation*. 1983;68(suppl 2):1–7.

6. Chesebro JH, Clements IP, Fuster V, et al. A platelet-inhibitor drug trial in coronary artery bypass operations: Benefit of perioperative dipyridamole and aspirin therapy on early postoperative vein graft patency. *N Engl J Med*. 1982;307:73–78.

7. Chesebro JH, Fuster V, Elveback LR, et al. Effect of dipyridamole and aspirin on late vein-graft patency after coronary bypass operation. *N Engl J Med*. 1984;310:209–214.

8. Ritchie JL, Narahara KA, Trobaugh BG, et al. Thallium-201 myocardial imaging before and after coronary revascularization: Assessment of regional myocardial blood flow and graft patency. *Circulation*. 1977;56:830–836.

9. Greenberg BH, Hart R, Botvinick EH, et al. Thallium-201 myocardial perfusion scintigraphy to evaluate patients after coronary artery surgery. *Am J Cardiol*. 1978;42:167–176.

10. Pfisterer M, Emmenegger H, Schmitt HE, et al. Accuracy of serial myocardial perfusion scintigraphy with thallium-201 for prediction of graft patency early and late after coronary artery bypass surgery. *Circulation*. 1982;66:1017–1024.

11. Higginbotham MB, Belkin RN, Morris KG, et al. Value and limitation of biplane rest and exercise radionuclide angiography for assessing individual bypass grafts: A prospective study. *J Am Coll Cardiol*. 1986;7:1003–1014.

12. Vignola PA, Boucher CA, Curfman GD, et al. Abnormal interventricular septal motion following surgery: Clinical, surgical, echocardiographic and radionuclide correlates. *Am Heart J*. 1979;97:27–34.

13. Guthaner DF, Wexler L, Bradley B. Digital subtraction angiography of coronary grafts. *AJR*. 1985;145:1185–1190.

14. Whiting JS, Nivatpumin T, Pfaff M, et al. Assessing the coronary circulation by digital angiography: Bypass graft and myocardial perfusion imaging. In: Heintzen PH, Brennecke R, eds. *Digital Imaging in Cardiovascular Radiology*. New York, NY: Thieme-Stratton; 1983:205–211.

15. Brundage BH, Lipton JR, Herfkens RJ, et al. Detection of coronary bypass grafts by computed tomography. *Circulation*. 1980;61:826–831.

16. Moncada R, Salinas M, Churchill R, et al. Patency of saphenous aortocoronary bypass grafts demonstrated by computed tomography. *N Engl J Med*. 1980;303:503–505.

17. Guthaner DR, Brody WR, Ricci M, et al. The use of computed tomography in the diagnosis of coronary artery bypass graft patency. *Cardiovasc Intervent Radiol*. 1980;3:3–8.

18. Daniel WG, Dohring W, Lichtlen PR, et al. Noninvasive assessment of aortocoronary bypass graft patency by computed tomography. *Lancet*. 1980;1:1023–1024.

19. Daniel WG, Dohring W, Stender HS, et al. Value and limitations of computed tomography in assessing aortocoronary bypass graft patency. *Circulation*. 1983;67:983–987.

20. Sethna DH, Bateman TM, Whiting JS, et al. Comprehensive and quantitative cardiac assessment using cine-CT: Description of a new clinical diagnostic modality. *Am J Card Imag*. 1987;1:18–28.

21. Bateman TM, Gray RJ, Whiting JS, et al. Prospective evaluation of ultrafast computed tomography for determination of coronary bypass graft patency. *Circulation*. 1987;75:1018–1024.

22. Stanford W, Brundage B, MacMillan R, et al. Sensitivity and specificity of assessing bypass graft patency with ultrafast computed tomography: Results of a multicenter study. *J Am Coll Cardiol*. 1988;12:1–7.

23. Bateman TM, Gray RJ, Whiting JS, et al. Ultrafast computed tomographic evaluation of aorto-coronary bypass graft patency. *J Am Coll Cardiol*. 1986;8:693–698.

24. Whiting JS, Bateman TM, Sethna DH, et al. Quantitation of saphenous vein bypass graft flow using intravenous contrast ultrafast-CT. *Circulation*. 1986;74:11–41. Abstract.

25. Rumberger JA, Feiring AJ, Hiratzka LE, et al. Determination of changes in coronary bypass graft flow rate using cine-CT. *J Am Coll Cardiol*. 1986;7:155A. Abstract.

26. Rumberger JA, Feiring AJ, Lipton MJ, et al. Use of ultrafast computed tomography to quantitate regional myocardial perfusion: A preliminary report. *J Am Coll Cardiol*. 1987;9:59–62.

27. Chomka EV, Wolfkiel CJ, Caludio J, et al. Combined perfusion and functional imaging of the left ventricle in patients with recent myocardial infarction by ultrafast computed tomography. *J Am Coll Cardiol*. 1987;9(suppl):159A. Abstract.

28. Chomka EV, Fletcher M, Stein M, et al. Ultrafast computed tomography during bicycle ergometry. *J Am Coll Cardiol*. 1986;7(suppl):154A. Abstract.

29. Herfkens RJ, Higgins CB, Hricak H, et al. Nuclear magnetic resonance imaging of the cardiovascular system: Normal and pathologic findings. *Radiology*. 1983;147:749–759.

30. Holmvang G, Edeman R, Dinsmore R, et al. Coronary bypass graft patency by NMR. *Circulation* 1986;74:11–42. Abstract.

31. Gomes AS, Lois JF, Drinkwater DC, et al. Coronary artery bypass grafts: Visualization with MR imaging. *Radiology*. 1987;162:175–179.

32. Rubinstein RI, Askenase AD, Thickman D, et al. Magnetic resonance imaging to evaluate patency of aortocoronary bypass grafts. *Circulation*. 1987;76:786–791.

33. White RD, Caputo GR, Mark AS, et al. Coronary artery bypass patency: Noninvasive evaluation with MR imaging. *Radiology*. 1987;164:681–686.

34. White RD, Pflugfelder PW, Lipton MJ, et al. Coronary artery bypass grafts: Evaluation of patency with cine MR imaging. *Am J Radiol*. 1988;150:1271–1274.

Part III
Valvular Heart Disease

Chapter 20

Evaluation of Valvular Heart Disease by Contrast Angiography

Charles R. McKay, MD, and Maleah Grover-McKay, MD

Contrast angiography, along with hemodynamic measurements, provides the complementary information necessary for the diagnosis and interpretation of cardiac valve dysfunction. Only the angiographic evaluation is discussed in this chapter. General hemodynamic evaluation of valve disease is discussed elsewhere.[1-3] The techniques of angiography are elaborated in Chapter 6. Therefore, this chapter delineates (1) how to obtain angiograms in optimal projections for making a specific diagnosis and (2) how to interpret these studies to determine the presence and severity of valvular heart diseases.

FLUOROSCOPY AND CONTRAST CINEANGIOGRAPHY OF THE LEFT VENTRICLE

Fluoroscopy to detect native valve calcification and to define prosthetic valve function usually is performed just prior to contrast angiographic evaluation. Contrast left ventriculography is performed using intracardiac injection of radiopaque contrast material via catheters placed directly into the left ventricular chamber. Cinefilm images obtained just prior to and during contrast ventriculography permit analysis of the left ventricular wall motion, chamber volume, and aortic and mitral valve motion throughout the cardiac cycle. Evaluation of left ventricular function and volume is indicated in all patients with valvular heart disease to determine the effects of valvular disease on the ventricle and also to diagnose concomitant coronary, pericardial, or myopathic disease.

The "standard" views used in performing left ventriculography have changed as the radiographic equipment has become more sophisticated. Movable image intensifiers on "C" or "U" arms developed for coronary arteriography[4] also can be used to obtain right and left anterior oblique views with cephalad or caudal angulation during ventriculography. These views allow orientation of the long axis of the interventricular septum parallel to the plane of the image intensifier, which permits unobstructed imaging of the basilar and apical septa,[5] identification of septal wall motion abnormalities, location of membranous and muscular ventricular septal defects, location of left ventricular outflow tract abnormalities, identification of abnormal systolic and diastolic motion of the anterior leaflet of the mitral valve, and identification of systolic mitral valve prolapse.

The technical quality of left ventricular angiograms can be improved by (1) placing catheters in the cavity to avoid excessive premature ventricular beats and entanglement in mitral valve chordae; (2) injecting sufficient contrast medium volume in patients with valve regurgitation and septal defects to fill the entire left ventricle rapidly; (3) positioning the image tube off the standard 30° right anterior oblique (RAO) or 60° left anterior oblique (LAO) views to eliminate overlap of the spine and mediastinal background structures; and

(4) placing the shutters around the heart and excessive lung in the image. Attention to these details helps to eliminate artifacts and improve image quality.

Qualitative Interpretation of Left Ventriculograms

The qualitative or subjective assessment of left ventriculograms may include evaluation of (1) overall left ventricular size and function; (2) segmental wall motion; (3) cardiac valve morphology and function; and (4) the presence of concomitant abnormalities, such as left ventricular aneurysm, ventricular septal defects, or intracavitary masses.

Experienced angiographers may be able to differentiate small, normally contracting left ventricles and very large, poorly contracting left ventricles. However, studies have shown that routine qualitative assessments of left ventriculograms do not accurately identify ventricular size or dysfunction in all patients.[6] It is difficult to assess subjectively the relative diastolic volume, to view the two-dimensional planar images, and to determine intuitively the three-dimensional changes in ventricular volume.[7]

Segmental wall motion, assessed subjectively, usually consists of a simple comparison of the relative timing and extent of contraction of different wall segments during systole. However, cardiac contraction involves complex rotation of the heart in addition to shortening of both longitudinal and radially oriented muscle fibers. These complex contractions result in regional differences in wall motion even in normal ventricles.[8,9] For all the above reasons, final interpretations of left ventricular function in adult patients optimally should be based on formal quantitative analyses of the angiographic images.

Although qualitative interpretation of left ventricular function has the limitations discussed above, the qualitative assessment of valve morphology and valve function by fluoroscopy, left ventriculography, and aortography can be very helpful. With respect to the aortic valve, the location and extent of calcium in the valve annulus and in the aortic valve cusps can be seen readily on fluoroscopy.[10] The location of obstruction in the left ventricular outflow tract during systole can be identified by using contrast ventriculography. Doming and thickening of the noncalcified bicuspid aortic valve and a narrow stream of contrast entering the aorta are present during systole. Especially in the LAO views with cranial angulation, supravalvular and subvalvular membranes can be identified. In patients with mitral valve stenosis, restricted motion of the anterior leaflet is seen (Figure 20-1). In patients with classic hypertrophic cardiomyopathy, septal bulging into the outflow tract

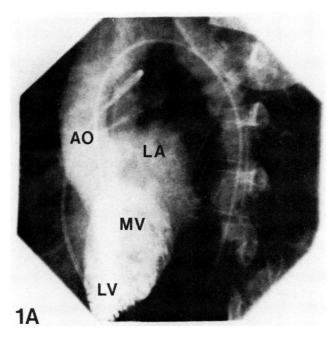

Figure 20-1A Cineangiogram of the left ventricle in the 60° LAO/20° cranial view in a patient with rheumatic mitral disease. The left ventricular apex (LV) is displaced downward and is not superimposed onto the mitral orifice (MV), which is restricted but open during diastole. Contrast medium partially fills the left atrium (LA) owing to moderate mitral regurgitation. The sinuses of Valsalva, aortic valve cusps, and ascending aorta (AO) are visible without being superimposed on the left ventricle or left atrium.

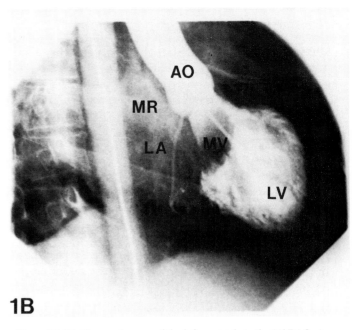

Figure 20-1B Cineangiogram of the left ventricle in the 30° RAO view in a patient with rheumatic mitral disease. The mitral valve is displaced anteriorly into the left ventricle (LV) during diastole, and the posterior leaflet of the mitral valve (MV) creates a scalloped filling defect against the posterior LV endocardium. An eccentric stream of contrast medium regurgitates across the mitral valve (MR) and is seen in the superior wall of the left atrium (LA). AO, ascending aorta.

can be seen along with systolic anterior motion of the mitral valve.

With respect to the mitral valve, annular calcification and nodular calcification of the anterior and posterior leaflets can be assessed easily in the angulated LAO view during fluoroscopic inspection. During contrast medium injection in the RAO view, the leaflets of stenotic rheumatic mitral valves show restricted anterior motion and a diastolic filling defect near the atrioventricular groove[11] (Figure 20-1). In these patients, the shortening of the chordae tendineae also can be identified. In patients with mitral valve prolapse, individual portions of the posterior leaflet of the mitral valve can be seen readily to prolapse into the left atrium and the presence of mitral regurgitation identified.[5,12] Congenital abnormalities of the mitral valve, including parachute mitral valve, can be identified readily in the LAO view with cranial angulation. In this view, the interventricular septum, mitral valve leaflets, chordae tendineae cordis, and papillary muscle structures are separated and well seen.

Other qualitative findings confirmed by left ventriculography are helpful. Ventricular septal defects are identified and located readily in the basal or apical septum by using the LAO cranial angulated view. The extent of the calcified pericardium can be seen in constrictive pericarditis in standard RAO and LAO views. Left ventriculography also may identify left ventricular mural wall thrombi or left ventricular wall aneurysms in the apex or inferior basal walls using RAO views.

Quantitative Interpretation of Left Ventriculograms

Quantitation preferably is performed on at least two views of the left ventricle.[13] Two simultaneous views can be obtained with a biplane system. The standard projections are 30° RAO and 60° LAO. Quantitative volumes can also be obtained in the LAO cranial views.[14] The technical aspects of measuring and calculating quantitative volumes are discussed in detail elsewhere.[2,15] Briefly stated, the area of the left ventricular volume image is compared with a reference image, usually a calibration centimeter grid placed at the same distance from the image tube and x-ray tube as the left ventricle. Chamber areas and axes are measured on the images and chamber volume is determined by assuming a spheroid of revolution or by a Simpson's rule calculation.[16,17] From the end-diastolic and end-systolic volumes, the ejection fraction and stroke volumes can be calculated. If one determines net forward stroke volume (FSV) by the Fick or thermodilution methods and total stroke volume (TSV) by ventriculography, the regurgitant volume (RV) per beat (calculated as TSV minus FSV) and regurgitant fraction (RF) can be determined as $RF = (TSV - FSV)/TSV$.[18,19] Quantitative analyses of left ventriculograms in patients with valvular heart disease can yield accurate determinations of (1) the volume of the left ventricular chamber during systole and diastole and (2) the overall function, the regional left ventricular systolic function,[20,21] and the myocardial mass,[22] as well as the severity of mitral valve regurgitation.

FLUOROSCOPY AND CINEAORTOGRAPHY

Fluoroscopy of the aortic valve may reveal a heavily calcified aortic valve in middle-aged and elderly patients with significant aortic valve gradients. However, elderly patients with sclerotic nonobstructive valves and patients with renal failure also may have some valve calcification. Younger adults with bicuspid or rheumatic disease may have significant obstruction without calcification. Although the aortic valve cusps usually are identified in the standard LAO and RAO views, the cusps are best seen tangentially using a 45° LAO projection with 10° to 15° cranial angulation and an RAO projection with steep angulation to image directly the orifice and cuspal opening in patients with congenital bicuspid aortic valve[23] or in patients with rheumatic disease (Figure 20-2).

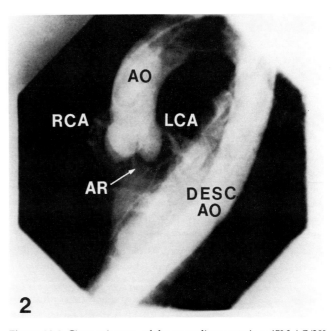

Figure 20-2 Cineangiogram of the ascending aorta in a 45° LAO/20° cranial angulated view in a patient with rheumatic aortic valve disease. As the aortic root (AO) fills with contrast medium, the proximal portions of the right coronary artery (RCA) and left coronary artery (LCA) are seen. A small, concentric stream of contrast medium is seen to regurgitate through the aortic valve (AR, arrow) during diastole. After several cardiac cycles, the aortic arch and descending aorta (DESC AO) also are seen.

Evaluation of Aortic Regurgitation by Aortography

Aortic regurgitation is best evaluated by injection of contrast medium into the aortic root with images obtained in both the RAO and LAO projections. Artifacts of catheter placement can be avoided by placing the catheter down onto the valve and pulling it back slightly to the level of the sinotubular junction. Catheters placed too close to the valve leaflets cause artifactual regurgitation, and aortography performed with the catheter too far from the valve may underestimate the degree of regurgitation. If severe aortic regurgitation is present, adequate opacification of the left ventricle may be obtained just from the aortic root injection. In the RAO projection, these images can also be used with a calibration good for quantitative ventriculography.

Qualitative Interpretation of Aortograms

Cine-films of the aortic valve taken in RAO and LAO views just prior to injection of contrast medium permit identification of calcium in the aortic valve cusps, the sinuses of Valsalva, the ascending aortic wall, and the coronary arteries. During aortic contrast medium injection, the size of the aortic root and the presence of aneurysms, dissection, or patent bypass grafts can be identified. Contrast flowing into the sinuses of Valsalva demonstrates the presence of aneurysms, right ventricular fistulas, aortic annulus abscesses, aortic valve incompetence, and the position of the proximal coronary arteries. Contrast medium flowing forward demonstrates dissection or pseudoaneurysms of the aortic arch and coarctation or patent ductus arteriosus.

Quantitative Interpretation of Aortograms

Quantitative analyses of the aortic root structures are possible when calibration grids or catheters not in metal markers are placed in the image field at the same level as the aortic valve. These analyses are not performed routinely but can be helpful. Knowledge of the aortic annulus size before surgery may be helpful in anticipating the size or type of aortic valve prosthesis likely to be needed. It is also helpful in choosing balloon sizes for balloon aortic valvuloplasty in order to avoid overdilating the aortic annulus. Fusiform aneurysms in patients with Marfan syndrome or syphilitic aneurysms of the aortic root can be measured and followed serially by using the above quantitative techniques.

Quantitative interpretation of aortic regurgitation is an important indication for aortography. Quantitative assessment of aortic regurgitation remains very difficult when only clinical and echocardiographic Doppler examinations are performed. The grade of regurgitation can be compared with the quantitative regurgitant fraction, the left ventricular end-diastolic volume, and the systolic ventricular function to determine the severity of the insufficiency.

AORTIC STENOSIS

The fluoroscopic and cineangiographic findings associated with different etiologies of aortic valve disease are listed in Table 20-1. Aortic stenosis in younger patients is most often due to congenital abnormalities of the aortic valve, commonly a bicuspid valve. In older patients with bicuspid valves, deposition of calcium on the aortic valve cusps can lead to progressive stenosis. Rheumatic aortic valve disease in young adults is associated with commissural fusion, leaflet thickening, and calcific nodules. In patients with connective tissue disease such as scleroderma and rheumatoid arthritis, aortic stenosis may result from nodular thickening of the valve cusps and calcific involvement of the proximal aorta. Fluoroscopy and angiography therefore may be helpful in determining the etiology of valvular obstruction and in differentiating valvular disease from other causes of left ventricular outflow obstruction due to supravalvular or subvalvular obstruction.

Evaluation of Aortic Stenosis by Fluoroscopy

Although some diffuse and annular calcification of the aortic valve is seen on fluoroscopy in many adult patients without significant aortic valve obstruction, the presence of extensive cuspal calcification is often associated with significant aortic valve stenosis with or without regurgitation.[24] Cuspal calcification and cusp motion are best evaluated on biplanar fluoroscopy or cine-imaging just prior to injections for aortography or left ventriculography.

Evaluation of Aortic Stenosis by Left Ventriculography

Left ventriculography using 30° RAO and angulated LAO projections can image the left ventricular outflow tract and identify the level of obstruction; determine the presence and severity of associated mitral regurgitation; and evaluate left ventricular myocardial mass, chamber volume, and systolic function. The left ventricular mass increases in response to the increased left ventricular pressure with increasing stenosis. Left ventricular chamber dilation and abnormal systolic and diastolic function indicate a poor prognosis without surgical treatment.[25] The quantitative evaluation of left ventricular function is necessary to evaluate surgical

Table 20-1 Fluoroscopic and Cineangiographic Findings Often Associated with Various Etiologies of Aortic Valve Disease

Finding	Rheumatic	Bicuspid	Senile Calcific	Endocarditis	Connective Tissue
Ascending aorta					
Dilated	+	+	+ +	+	+ + +
Aneurysm	0	0	0	+	+ + +
Calcified	0	0	+	0	0
Aortic annulus					
Enlarged	0	0	0	+	+ + +
Calcified	0	+ +	+ + +	0	0
Aortic cusps					
Calcified	+	+ + +	+ + +	0	0
Thickened/immobile	+	+	+ +	0	0
Prolapsed	0	0	0	+ +	+ + +
Regurgitant	0 or +	0 or +	0 or +	+ +	+ + +
Left ventricles					
aWall thickness	+ +	+ + +	+ + +	+	+
aDiastolic volume	0 or +	0 or +	0 or +	+ + +	+ + +
bEjection fraction	0 or +	0	0 or +	0 or +	+
Other angiographic findings	Mitral disease, atrial thrombus	Left dominant coronary, aortic coarctation	Associated coronary artery disease	Annulus abscess, vegetations	Fusiform aneurysm, aortic and mitral cusp prolapse

*0, Finding not present; + to + + +, finding present to prominent.

risk and to determine the contribution of dysfunction due to coexisting aortic regurgitation or mitral valve disease. The use of low-osmolality contrast agents, which cause less hemodynamic perturbation, may improve the safety of performing these contrast medium injections, especially in patients with left ventricular failure.

Evaluation of Aortic Stenosis by Aortography

Aortography remains an essential part of the evaluation of patients with aortic stenosis not only to delineate further the morphology of the aortic valve and ascending aorta, but also to assess the presence and severity of aortic regurgitation. Prior to aortic valve surgery, aortography can identify dissections, aneurysms, and aortic valve vegetations that can cause perioperative complications.

AORTIC REGURGITATION

The causes of aortic regurgitation can be grouped into pathologic processes that damage either the aortic valve or the aortic root.[26] Processes that damage the valve cusps and result in aortic regurgitation include bacterial endocarditis (leaflet destruction or perforation), calcification and retraction of bicuspid valve leaflets, and rheumatic and degenerative calcification of tricuspid valve leaflets. Trauma or proximal aortic dissection can cause acute, severe aortic regurgitation due to diastolic prolapse of a previously normal aortic cusp into the left ventricle. Processes that can cause aortic annulus and aortic root dilation include connective tissue diseases (Marfan syndrome, Ehlers-Danlos syndrome, and pseudoxanthoma elasticum) and diseases associated with aortitis (syphilis, ankylosing spondylitis, and Reiter's syndrome) (Table 20-1). Fluoroscopy and aortography can identify specific abnormalities of the aortic valve cusps, aortic wall, and the left ventricle, as well as measure the degree of aortic regurgitation.[27]

Evaluation of Aortic Regurgitation by Fluoroscopy

Calcification of the proximal ascending aorta without cuspal calcification may be seen in syphilitic aortitis and also in atherosclerotic and rheumatoid diseases. Heavily calcified annuli[24] and cusps and angiographic regurgitation indicate mixed aortic stenosis and regurgitation in rheumatic, bicuspid, or senile calcific valves.

Evaluation of Aortic Regurgitation by Left Ventriculography

In acute aortic regurgitation, the left ventricular volume may be normal, whereas in chronic aortic regurgitation the left ventricular diastolic volume is enlarged even when the ejection fraction remains normal. Left ventriculography can determine the presence and severity of coexisting mitral valve disease. Such determination is necessary to interpret left ventricular chamber enlargement properly and to plan for definitive surgical treatment. Since the timing of surgery and the prognosis for patients with chronic aortic regurgitation are dependent on left ventricular function, quantitative assessment of the end-diastolic volume and ejection fraction is important in these patients.[25]

Left ventriculography is also helpful in confirming the quantitative assessment of aortic regurgitation by aortography (Table 20-2) and determining whether left ventricular enlargement is consistent with the amount of aortic regurgitation. Accuracy of the measurements of these regurgitant fractions can be improved in a given laboratory by determining the correspondence between stroke volume by the Fick, thermodilution, and angiographic methods in patients *without* regurgitation, correcting the angiographic stroke volume by a regression equation,[27] and using this corrected angiographic stroke volume to calculate the regurgitant fraction.

The regurgitant fraction is small (0 to 0.3) in patients with mild (1+) or moderate (2+) regurgitation, and the ventricles of these patients have normal or only slightly enlarged end-diastolic volumes. Very large end-diastolic volumes can be due to coexistent mitral regurgitation, severe aortic stenosis, hypertension, coronary artery disease, or ventricular septal defects.

Larger regurgitation fractions (0.4 to 0.8) are found in patients with severe (3+ to 4+) isolated aortic regurgitation or combined mitral and aortic regurgitation.

Evaluation of Aortic Regurgitation by Aortography

Criteria for quantitative evaluation of aortic insufficiency by aortography are given in Table 20-2.[27-29] For properly placed catheters, small traces of regurgitation just below the aortic valve cusps in the left ventricular outflow tract may be seen in many normal aortograms. The criteria are based on the amount of regurgitant contrast medium and the clearing or filling of the ventricle in one or more beats.[28] Although the formal reading of the study should adhere to these criteria, it is clear that the adequacy of injection and catheter placement is important. In addition, in patients with large left ventricles, the evaluation of aortic regurgitation by aortography is more variable when compared with quantitative regurgitant volumes.[29] Therefore, both qualitative and quantitative evaluations are useful in interpreting the severity of regurgitation.

MITRAL STENOSIS

Mitral stenosis is present when morphologic changes in the mitral valve apparatus impede blood flow from the left atrium to the left ventricle. Changes in physiologic importance occur when mean diastolic pressure in the left atrium is higher than that in the left ventricle. Although there are several causes of mitral valve obstruction, the most frequent cause in adults is mitral valve stenosis secondary to rheumatic fever. Rheumatic mitral stenosis can be diagnosed by contrast angiography and hemodynamic evaluation, and it can be differentiated from atrial myxoma, atrial thrombus, congenital mitral stenosis, mitral annular calcification, and other infrequently encountered causes of mitral valve obstruction.

Evaluation of Mitral Stenosis by Fluoroscopy

The standard radiographic projections (RAO 30°, LAO 60°, and LAO with cranial angulation) yield important information about calcification of the mitral valve annuli and valve leaflets and the presence of calcified masses or thrombi in the left atrium or appendage. Severe mitral annular calcification out of proportion to leaflet calcification is an important finding in patients without mitral valve obstruction who are elderly or who have other diseases such as chronic renal disease. Measurement of mitral valve leaflet calcification in rheumatic mitral stenosis is difficult because dystrophic calcification may be distributed diffusely across the leaflets and may not be visible by fluoroscopy. However, the presence of extensive, visible calcification on the mitral valve leaflets is a clinically relevant finding, as treatment by valve replacement rather than repair is usually necessary in these cases.

Table 20-2 Grading of Aortic Regurgitation by Contrast Aortography

Quantitative Grade	Qualitative Definition	Regurgitant Fraction
0	No regurgitation	0 to 0.2
1+	Regurgitant stream in LV* outflow tract only; clears on next beat	0.1 to 0.2
2+	Regurgitation into LV chamber; not progressive; persists slightly on next beat	0.1 to 0.3
3+	Progressively fills entire LV in 2 to 3 beats; LV density less than aortic root density	0.2 to 0.4
4+	Regurgitant stream fills entire LV on single beat; contrast medium density in LV = aorta	0.4 to 0.8

*LV, Left ventricle.

Evaluation of Mitral Stenosis by Left Ventriculography

Evaluation of Mitral Valvular, Subvalvular, and Annular Morphology

Detailed evaluation of mitral valve morphology during contrast angiography of the left ventricle can be very informative but is not reported routinely. Table 20-3 outlines the expected findings at different levels of the mitral valve apparatus in various abnormal conditions associated with obstruction of the left ventricular inflow tract. In rheumatic mitral stenosis, the posterior leaflet motion is best seen in the RAO projection, as it traps contrast medium against the left ventricular posterior wall at the annulus and forms a scalloped filling defect as the ventricle fills during diastole. In the LAO or cranially angulated LAO projection, the normal anterior leaflet demonstrates an M-shaped motion into the left ventricular inflow tract during diastole. This finding is obvious in nonrheumatic valves and in inflow tract obstruction due to a myxoma or thrombus. Conversely, rheumatic valves demonstrate "doming" throughout diastole; congenital "parachute" mitral valves have large, redundant leaflets with doming and prolapse and a single papillary muscle at the base of the ventricle. Heavy, crescentic calcification of the posterior annulus with some posterior leaflet calcium and immobility is often a hallmark of mitral annular calcification in the elderly. Anterior leaflet mobility may be normal in these patients, in contrast to rheumatic valves. Stenotic rheumatic mitral valves may show variable distribution of calcium throughout the annulus and leaflets. Nodular calcified areas in the leaflets can be seen on fluoroscopy and left ventriculography and can be used to track the leaflet motion during diastole. Rheumatic subvalvular disease causes shortening and thickening of the chordae, which draw the papillary muscle close to the leaflets. These findings are important in evaluating whether a patient is a candidate for mitral commissurotomy or valve replacement and are best assessed angiographically in the RAO projection.[30] These detailed angiographic findings help to determine the etiology of obstruction to the left ventricular inflow tract and can also help to determine which therapy (such as balloon valvuloplasty, mitral valve repair, or valve replacement) may be appropriate.

Table 20-3 Fluoroscopic and Cineangiographic Findings Often Associated with Mitral Valve Obstruction*

Level of Mitral Valve	Rheumatic Mitral Stenosis	Myxoma or Thrombus	Mitral Annular Calcification	Congenital Mitral Stenosis
Annulus				
Calcium	0 to +++	0	+++	0
Anterior leaflet				
Calcium	0 or +	0	+	0
Normal diastolic motion (M-shaped)	0	+	0 or +	0
Diastolic doming	+	0	0	0 or +
Systolic prolapse	0 or +	0	0	+++
Posterior leaflet				
Calcium	0 to ++	0	0	0
Contrast medium entrapment	+	0	0	0
Prolapse	0 or +	0	0	+++
Mitral regurgitation	0 or +	0	0 or +	++
Chordae				
Normal	0	+	+	0
Thick/short	++	0	0	0
Long, redundant	0	0	0	++
Papillary muscle				
Close to leaflet	++	0	0	Single
Other findings	Associated aortic valve disease	Filling defect in LV inflow tract during diastole	Classic crescentic annular calcium in LAO view	Large redundant anterior and posterior leaflets; parachute valve

*0, Finding absent; + to +++, finding present to prominent; LAO, left anterior oblique; LV, left ventricle.

Evaluation of Left Ventricular Function in Mitral Stenosis

Quantitative left ventriculography in patients with mitral valve stenosis can determine left ventricular systolic function (ejection fraction) and volume. Patients with decreased systolic function may have associated regurgitant valve lesions, coronary disease, or rheumatic cardiomyopathy. Therapy can be planned on the basis of these findings. For example, in patients with heavily calcified mitral valves without regurgitation and with small left ventricular end-diastolic volumes, insertion of a low-profile mechanical valve may prevent left ventricular outflow tract obstruction or posterior annulus tears associated with high-profile valves.

Evaluation of Left Atrial Size and Function in Mitral Stenosis

Evaluation of left atrial size and function is not performed routinely during contrast angiography and requires either trans-septal catheterization with contrast medium injection directly into the left atrium or contrast medium injection into the pulmonary artery with cine-images obtained as the contrast agent passes into the left atrium and left ventricle. The latter is particularly useful in evaluating patients with cor triatriatum, left atrial thrombus, and left atrial myxoma.

Evaluation of Coronary Arteries in Mitral Stenosis

Coronary arteriography in patients with mitral stenosis can be reserved for those older than 40 years of age or for younger patients with chest pain.[31] Approximately 20% of patients with mitral stenosis and atrial fibrillation will have thrombi in the left atrial appendage. These may be visible by neovascularization of the thrombi and filling of these vessels from the left atrial branch of the circumflex coronary artery.

MITRAL REGURGITATION

Alteration of the morphology of mitral valve leaflet or distortion of the subvalvular apparatus from many different etiologies can cause poor leaflet coaptation during systole and allow regurgitation of blood into the left atrium. Acute, severe regurgitation of blood is poorly tolerated and often requires urgent evaluation. Chronic or progressive regurgitation is more insidious and often is apparent only after secondary changes in left ventricular function or pulmonary vasculature produce symptoms. Fluoroscopy and cineangiography can be very helpful in determining the acuity and the etiology of mitral regurgitation. These examinations may also determine the primary morphologic changes in the mitral valve and the presence and severity of changes in cardiopulmonary function secondary to the regurgitation.

Evaluation of Mitral Regurgitation by Fluoroscopy

Fluoroscopy of the regurgitant mitral valve can demonstrate annular and leaflet calcification in rheumatic disease or mitral annular calcification in the elderly patient. Precise measurement of calcification is difficult, but the presence of dense dystrophic leaflet calcification is a helpful finding, as it excludes the feasibility of mitral valve repair.

Evaluation of Mitral Regurgitation by Left Ventriculography

Mitral regurgitation can be caused by morphologic changes at several locations in the mitral valve apparatus that can be differentiated by left ventriculography. The presence of thickened mitral valve leaflets with associated stenosis and diffuse or nodular calcification of the leaflets is consistent with rheumatic disease. Thin, focal, eccentric streams of contrast medium may be seen regurgitating into the left atrium during systole. In the RAO projection, superior or inferior eccentric contrast medium streams are seen; in the angulated LAO projections, anterior septal or posterior wall streams are seen easily. These eccentric streams can be very difficult to identify by color Doppler examinations. Mitral valve prolapse is diagnosed in the RAO projection by systolic bulging of the posterior mitral valve leaflet or of both leaflets across the atrioventricular groove into the left atrium. Mitral valve prolapse may be an isolated finding or may be associated with prolapse of the aortic tricuspid and pulmonic valves in patients with Marfan syndrome and other connective tissue diseases. The specific morphologic findings include redundancy of the mitral valve leaflet tissue or of chordal tissue, and enlargement of the mitral valve annulus.

Acute, severe mitral regurgitation usually is due to chordal rupture, leaflet tears, or ruptured papillary muscle; it may be seen in patients with a history of bacterial endocarditis, mitral valve prolapse, connective tissue disease, or recent acute myocardial infarction. Left ventriculography classically demonstrates severe mitral regurgitation with a normal-sized left atrium and left ventricle.

Chronic, moderate to severe mitral regurgitation is associated with multiple etiologies. Patients with mild to moderate mitral regurgitation may be relatively asymptomatic for long periods of time. Mitral regurgitation is self-perpetuating because it increases left ventricular volumes continuously. Because of the increased volumes in the left ventricular chamber, the mitral valve chordae and the base of the papillary muscles are held away from the mitral valve annulus during systole. This process further exacerbates poor leaflet coaptation and increases regurgitant volume. Therefore, patients may present late in the course of the disease, when the chronic volume overload has already produced left ventricular enlargement, impaired ventricular function, and pulmonary hypertension.

The severity of mitral regurgitation can be graded quantitatively by using both left ventricular angiography and aortography (Table 20-4).[12,18,19,22] Assessing the amount of regurgitation by the quantitative grade is clinically useful and has been shown to correlate roughly with quantitative determinations of regurgitant stroke volume and regurgitant fraction (Table 20-4). Angiographic grades from 1+ to 4+ are based on (1) the amount of contrast medium streaming into the left atrium, (2) whether the contrast medium clears entirely from the left atrium during the next diastole, and (3) whether the contrast medium fills the left atrium on successive beats or fills it completely in a single beat after the injection of contrast medium into the left ventricle.

PULMONIC VALVE DISEASE

Congenital pulmonic stenosis is rare in adult patients. Right ventriculography in the 30° RAO or posteroanterior and lateral projections delineates pulmonic valve stenosis and the presence of infundibular stenosis, which is common in these patients. The posteroanterior projection with cranial angulation is useful in observing pulmonary valve doming and in identifying peripheral pulmonary artery stenosis. These lesions are amenable to balloon valvuloplasty and angioplasty.

Table 20-4 Grading of Mitral Regurgitation by Contrast Left Ventriculography

Quantitative Grade	Quantitative Definition*	Quantitative Regurgitant Fraction
0	No regurgitation	0 to 0.2
1+	Regurgitant stream near mitral valve; LA clears on next diastole	0.1 to 0.2
2+	Regurgitation to mid-LA; not progressive; contrast medium persists in LA on next diastole	0.1 to 0.3
3+	Progressively fills entire LA in 2 to 3 beats; LA density less than LV density	0.2 to 0.4
4+	Regurgitant contrast medium stream fills LA back to pulmonary veins on a single beat; density in LA equals density in LV	0.4 to 0.8

*LA, Left atrium; LV, left ventricle.

Acquired pulmonic stenosis is rare in patients with a history of rheumatic disease. In patients with carcinoid disease, fibrous plaques have been reported to cause pulmonic stenosis. Rarely, obstruction of the right ventricular outflow tract may be caused by a large aneurysm in the sinus of Valsalva or by a cardiac tumor.

Congenital pulmonic regurgitation also is rare, is manifest early in childhood, and is tolerated for long periods of time. The causes of acquired pulmonic regurgitation include bacterial endocarditis, connective tissue disease, and diseases that cause pulmonary hypertension or dilation of the pulmonary artery. Measurement of pulmonic regurgitation is not performed by angiography because a catheter is placed across the pulmonic valve into the main pulmonary artery for injection of contrast medium, and calculation of regurgitant volumes by right ventriculography is very difficult. If pulmonary angiography is performed, care must be taken to position the catheter properly with all side holes above the valve to avoid creating artifactual pulmonic regurgitation.

TRICUSPID VALVE DISEASE

Tricuspid disease in adults is unusual. The causes of tricuspid stenosis or regurgitation (or both) include Ebstein's anomaly, endocarditis, rheumatic disease, right ventricular enlargement with pulmonary hypertension or right ventricular failure, carcinoid disease, vegetations or tumors, and pericardial constriction. Contrast medium can be injected into the enlarged right atrium as cine-images are acquired in the 30° RAO projection. Tricuspid valve stenosis, obstructing masses, and apical displacement of the valve into the right ventricle, seen in Ebstein's anomaly, can be identified. Right ventriculography can demonstrate tricuspid regurgitation, but measurement of the regurgitation is hampered by the necessity of placing the catheter across the tricuspid valve and by the frequent occurrence of premature ventricular contractions.[32]

MIXED VALVE DISEASE

To evaluate a patient with both stenosis and regurgitation of the same valve or with dysfunction of more than one valve, each valvular lesion first should be analyzed separately as outlined above; the relative effects of the valvular disease on ventricular and pulmonary function then can be compared and interpreted.

Quantitative left ventriculography is used in patients with mixed valve disease to determine not only left ventricular systolic function but also the effects of the valve lesions on left ventricular volumes. In patients with mixed aortic stenosis and regurgitation, the angiographic stroke volume can be used in the Gorlin equation to assess more correctly the aortic valve orifice. In both aortic stenosis and regurgitation, even though the left ventricular mass is increased and the ejection fraction may be normal or decreased, the end-diastolic volume will be normal or only slightly enlarged in hearts with predominantly stenotic aortic valves. It is important to identify this subgroup of patients, as ventricular function may improve markedly after valve replacement, as it does in patients with isolated aortic stenosis. In contrast, hearts with predominantly regurgitant valves will have greatly enlarged diastolic volumes, only some of which will improve and then rarely return to normal after valve replacement.

In patients with mixed mitral stenosis and regurgitation, the angiographic stroke volume may be used in the Gorlin equation to assess the area of the mitral valve orifice. The angiographic stroke volume may be difficult to determine accurately in patients with atrial fibrillation. Symptomatic patients with these mixed lesions should undergo early catheterization and angiographic evaluation. Because the low-impedance regurgitant flow across the mitral valve acts to reduce afterload on the left ventricle, the total afterload on these ventricles is reduced.[25] Therefore, the ejection fraction is not an accurate indicator of left ventricular systolic function. Ejection fractions characteristically decrease after successful valve replacement in these cases. However, the enlarged left ventricular volume and regurgitant volume are helpful in assessing the degree and importance of mitral regurgitation. Patients with mixed mitral stenosis and regurgitation should therefore be identified early and treated by valve repair or replacement before irreversible ventricular dysfunction occurs.

PROSTHETIC VALVE DISEASE

Prosthetic valve dysfunction can be ruled out definitively by cardiac catheterization using angiographic and hemodynamic evaluation. Prosthetic valve regurgitation may be due to torn sutures, late prosthetic degeneration, valve component malfunction, or thrombosis. Dehiscence of the sewing ring may be due to prosthetic degeneration or endocarditis. Supravalvular aortic angiography is useful in determining the degree of aortic regurgitation and whether the regurgitation is periprosthetic or due to leaflet abnormalities.

Prosthetic valve dehiscence can be diagnosed when the prosthetic sewing ring is noted to be displaced away from the valvular annulus or when an abnormal valvular rocking motion is noted during systole. Calcification of the bioprosthetic leaflets and poor motion of disk or ball valve occluders are important findings that raise the possibility of valve dysfunction. Soon after surgery valve malfunction due to thrombosis or interference by several chordae tendina is also possible. However, because subvalvular and supravalvular masses may also cause valve obstruction while the valves retain normal disk or ball valve motion on fluoroscopy, the normal fluoroscopic motion of prosthetic valve occluders does not rule out valve dysfunction.

With respect to the degree of regurgitation, the seating of the bioprosthetic valve cusps or the mechanical disk or ball occluders is normally associated with slight regurgitation of contrast medium in all prosthetic valves. This centrally located stream is graded no more than trace or 1+, and it is also seen on careful examination of cineangiograms of normally functioning prosthetic valves. Valve dysfunction is associated with obvious increases in the amount of regurgitation.

Valve dysfunction is also associated with eccentric location of the regurgitant stream. The eccentricity is caused by bioprosthetic or mechanical occluder thrombosis or malfunction, by bioprosthetic cusp degeneration or perforation, or streaming around the valve between the sewing ring and the annular tissue. It is important to identify these periprosthetic eccentric regurgitant streams, as often they generate strong shear forces and are associated with red blood cell fragmentation and anemia. Periprosthetic regurgitation also is associated with endocarditis and perivalvular myocardial abscesses. Biplanar angiograms in optimally angulated projections, in conjunction with hemodynamic evaluation, can identify definitively causes for prosthetic valve malfunction.

SUMMARY

Contrast angiography and hemodynamic evaluations provide definitive and complementary information that can be used for patients with valvular heart disease to determine the presence and severity of valve lesions and the effects of valve dysfunction on ventricular and circulatory function. This information is essential in planning medical, interventional, or surgical treatment of patients with valvular heart disease.

REFERENCES

1. Grossman W, ed. *Cardiac Catheterization and Angiography*. Philadelphia, Pa: Lea & Febiger; 1986:3–545.

2. Yang SS, Bentivoglio LG, Maranhao V, et al., eds. *From Cardiac Catheterization Data to Hemodynamic Parameters*. Philadelphia, Pa: FA Davis Co; 1978:1–519.

3. Braunwald E, ed. Valvular heart disease. In: *Heart Disease: A Textbook of Cardiovascular Medicine*. Philadelphia, Pa: WB Saunders Co; 1988:1023–1092.

4. Grover M, Slutsky R, Higgins C, et al. Terminology and anatomy of angulated coronary arteriography. *Clin Cardiol*. 1984;7:37–43.

5. Elliott LP, Green CE, Rogers WJ, et al. Advantages of the caudocranial left anterior oblique left ventriculogram in adult heart disease. *Am J Cardiol*. 1982;49:369–380.

6. Chaitman BR, DeMots H, Bristow JD, et al. Objective and subjective analysis of left ventricular angiograms. *Circulation*. 1975; 42:420–425.

7. Wisneski JA, Deil CM, Wyse DG, et al. Left ventricular ejection fraction calculated from volumes and areas: Underestimation by Arla method. *Circulation*. 1981;63:149–151.

8. Lew WYW. Time-dependent increase in left ventricular contractility following acute volume loading in the dog. *Circ Res*. 1988; 63:635–647.

9. Weiss RM, Shonka MD, Kinzey JE, et al. Effects of loading alterations on the pattern of heterogeneity of regional LV function. *FASEB J*. 1988;2:A1494. Abstract.

10. Szamosi A, Wassberg B. Radiologic detection of aortic stenosis. *Acta Radiol Diagn*. 1983;24:201.

11. Baron MG. The angiocardiographic diagnosis of valvular stenosis. *Circulation*. 1971;44:143–154.

12. Baron MG. Angiocardiographic evaluation of valvular insufficiency. *Circulation*. 1971;43:599–605.

13. Rackley CE, Hood WP Jr. Quantitative angiographic evaluation and pathophysiologic mechanisms in valvular heart disease. *Prog Cardiovasc Dis*. 1973;15:427–447.

14. Rogers WJ, Smith LR, Bream PR, et al. Quantitative axial oblique contrast left ventriculography: Validation of the method by demonstrating improved visualization of regional wall motion and mitral valve function with accurate volume determinations. *Am Heart J*. 1982;103:185–190.

15. Collins SM, Skorton DJ, eds. *Cardiac Imaging and Image Processing*. New York, NY; McGraw-Hill Book Co; 1986:1–451.

16. Dodge HT, Kennedy JW, Petersen JL. Quantitative angiographic methods in the evaluation of valvular heart disease. *Prog Cardiovasc Dis*. 1973;16:1–23.

17. Davila JC, San Marco ME. An analysis of the fit of mathematical models to the measurement of left ventricular volume. *Am J Cardiol*. 1966;18:31–42.

18. Sandler H, Dodge HT, Hay RE, et al. Quantitation of valvular insufficiency in man by angiocardiography. *Am Heart J*. 1963; 65:501–513.

19. Miller GAH, Kirklin JW, Swan HJC. Myocardial function and left ventricular volumes in acquired valvular insufficiency. *Circulation*. 1965;31:374–384.

20. Gelberg HJ, Brundage BH, Glantz S, et al. Quantitative left ventricular wall motion analysis: Comparison of area, chord, and radical methods. *Circulation*. 1979;59:991–1000.

21. Sheehan FH, Schofer J, Mathey DG, et al. Measurement of regional wall motion from biplane contrast ventriculograms: A comparison of the 30 degree right anterior oblique and 60 degree left anterior oblique projections in patients with acute myocardial infarction. *Circulation*. 1986;74:796–804.

22. Kennedy JW, Trenholmse SE, Kasser IS. Left ventricular volume and mass from single plane cine angiocardiograms: A comparison of anteroposterior and right anterior oblique methods. *Am Heart J*. 1970;80:343–352.

23. Folger GM Jr, Sabbath HN, Stein PD. Evaluation of the anatomy of congenitally malformed aortic valves by orifice-view aortography. *Am Heart J*. 1980;100:152–159.

24. Glancy DL, Freed TA, O'Brien KP, et al. Calcium in the aortic valve. *Ann Intern Med*. 1969;71:245–250.

25. Ross J: Afterload mismatch in aortic and mitral valve disease: Implications for surgical therapy. *J Am Coll Cardiol*. 1985;5:811–826.

26. Waller BF. Rheumatic and nonrheumatic conditions producing valvular heart disease. In: *Valvular Heart Disease: Comprehensive Evaluation and Management*. Frankl WS, Brest A, eds. Philadelphia, Pa: FA Davis Co; 1986:3–104.

27. Hunt D, Baxley WA, Kennedy JW, et al. Quantitative evaluation of cineaortography in the assessment of aortic regurgitation. *Am J Cardiol*. 1973;31:696–700.

28. Taubman JO, Goodman DJ, Steiner RE. The value of contrast studies in the investigation of aortic valve disease. *Clin Radiol*. 1966; 17:23–31.

29. Croft CH, Lipscomb K, Mathis K, et al. Limitations of qualitative angiographic grading in aortic or mitral regurgitation. *Am J Cardiol*. 1984;53:1593–1598.

30. Akins CW, Kirklin JK, Block PC, et al. Preoperative evaluation of subvalvular fibrosis in mitral stenosis: A predictive factor in conservative vs. replacement surgical therapy. *Circulation*. 1979;60 (suppl. 1):I-71–I-76.

31. Chun PK, Gertz E, Davia J, et al. Coronary atherosclerosis in mitral stenosis. *Chest*. 1982;81:36–40.

32. Pepino CJ, Nichols WW, Selby JH. Diagnostic tests for tricuspid insufficiency: How good? *Cathet Cardiovasc Diagn*. 1979;5:1.

Chapter 21

Echocardiographic Evaluation of Valvular Heart Disease

Eric K. Louie, MD

INTRODUCTION

Beginning in the mid-1950s with the initial efforts by Edler and Hertz to evaluate mitral stenosis with the "ultrasonic cardiogram," valvular heart disease has provided a major impetus for the development of M-mode and subsequently two-dimensional echocardiography. Combined with Doppler echocardiographic techniques of measurement of intracardiac blood flow, it is now possible to diagnose, characterize, and, in many cases, quantify noninvasively the major congenital and acquired forms of valvular heart disease. The analysis of Doppler frequency-shifted reflected ultrasound is treated separately in Chapter 22, although in clinical application it has become an integral and inseparable component of echocardiographic imaging of valvular heart disease. On the one hand Doppler techniques provide functional information that complements the structural data acquired by ultrasonic imaging, while on the other hand echocardiographic localization of the interrogating Doppler beam and sample volume has greatly facilitated the Doppler examination and provided the opportunity to examine new physiologic questions. In this chapter discussion is limited to echocardiographic imaging of valvular and myocardial structures and the physiologic significance that can be inferred from such measurements, but the reader should be aware that meaningful interpretation of these data requires integration with concurrently acquired Doppler echocardiographic measurements.

M-mode echocardiography provides high temporal and spatial resolution, but the spatial resolution is limited to one dimension and the spatial orientation is dependent on two-dimensional echocardiographic targeting that is concurrent but not simultaneous. We will explore applications of this technique in the measurement of cardiac chamber size and function in valvular diseases but will focus primarily on two-dimensional echocardiography as the imaging tool for the qualitative anatomic characterization of valvular heart disease. While the axial spatial resolution and temporal resolution of two-dimensional echocardiography cannot match those of M-mode echocardiography, the added information provided by the spatial orientation of targets within the tomographic imaging plane results in an extremely powerful noninvasive technique for the real-time assessment of cardiac structures. Several limitations of the technique must be kept in mind, however, particularly when attempting to quantify the data. First, the heart exhibits significant linear as well as rotational motion within the thorax such that for any given echocardiographic imaging plane the cardiac structures visualized during *different* parts of the cardiac cycle may not be identical. In addition, the apparent motion of any cardiac structure relative to the transducer represents the net effect of the intrinsic motion of that structure with respect to the cardiac center of mass *and* the translational and rotational movement of the heart as a whole. Second, although one of the strengths of the

---

Supported in part by College of Medicine, Committee on Research Grant No. 48612, University of Illinois.

two-dimensional echocardiographic technique is its ability to obtain images from multiple tomographic planes, in routine practice these planes are defined by internal anatomic landmarks and there is no external frame of reference to ascertain precisely the relative orientation of the various tomographic planes or to ensure their reproducibility. Finally, the acoustic characteristics of thoracic structures surrounding the heart may limit the windows available for imaging and in a minority of patients may prevent the acquisition of useful data.

In this chapter we examine the application of these techniques to study valvular abnormalities of the inflow and outflow tracts of both ventricles. In the interest of illustrating principles of the application of these techniques to cardiac diagnosis, left-sided valvular lesions are treated in greater detail, although most of the observations generalize to right-sided valvular lesions. In general the echocardiographic examination can provide information about three aspects of any given valvular heart condition: (1) pathologic valvular anatomy, (2) structural changes in the chamber upstream of the abnormal valve, and (3) structural changes in the chamber (or vessel) downstream of the abnormal valve. In addition to characterizing the congenital or acquired anatomic abnormalities of the disordered valve, echocardiographic imaging permits the assessment of associated structural cardiac abnormalities (for example, complex congenital malformations) and the exclusion of nonvalvular heart disease (for example, hypertrophic cardiomyopathy) that may mimic primary valvular heart disease. With the renewed interest in conservative and reparative valvular surgery and the recent development of percutaneous balloon valvuloplasty, the ability to characterize valvular disease and to determine the size of valvular annuli noninvasively has assumed increasing importance. Since valvular lesions can result in structural changes in the cardiac chambers resulting from abnormal pressure or volume loading, echocardiographic techniques have been used to measure chamber size and wall thickness. These data, in conjunction with invasively determined intracardiac pressures (or noninvasively determined estimates of these pressures), permit the calculation of wall stress, which has particular significance for the assessment of ventricular function in valvular heart disease.

VENTRICULAR OUTFLOW TRACT: AORTIC AND PULMONIC VALVES

Valvular Anatomy

As outlined in Table 21-1, a wide variety of congenital and acquired abnormalities of the aortic and pulmonic valves can be visualized and diagnosed by two-dimen-

Table 21-1 Common Outflow Tract Abnormalities Diagnosed by Echocardiography

Congenital Malformations of the Semilunar Valves

- Unicuspid aortic valve
- Biscuspid aortic valve
- Pulmonic stenosis: fusion of cusps with systolic doming
- Multileaflet semilunar valve of truncus arteriosus
- Aortic valve prolapse
- Malsuspension of the aortic valve due to aortic root dilation (annuloaortic ectasia, Marfan syndrome)

Acquired Disorders of the Semilunar Valves

- Calcific aortic valve disease
- Rheumatic aortic valve disease
- Aortic valve endocarditis and its complications (flail cusp, annular abscess)
- Aortic root dissection with involvement of the aortic valve; aortic root dilation with malsuspension of the aortic valve.

Nonvalvular Conditions Mimicking Valvular Disease

- Supravalvular aortic stenosis
- Discrete subvalvular aortic stenosis
- Hypertrophic cardiomyopathy with obstruction of the left ventricular outflow tract
- Muscular subpulmonic obstruction of the outflow tract

sional echocardiography, and significant nonvalvular causes of outflow tract obstruction can be excluded readily. Congenitally bicuspid aortic valves (Figure 21-1) serve as a good example of the use of two-dimensional echocardiography to distinguish abnormal anatomic patterns. A diastolic short-axis image of the normal trileaflet aortic valve reveals the Y-shaped pattern of the coaptation margins of the three cusps that subsequently open independently during systole. By contrast, the image of the bicuspid aortic valve shows only a single coaptation line demarcating two cusps, usually of unequal size; the absence of a third cusp is confirmed by the systolic motion of only two leaflets. In as many as 75% of patients with aortic stenosis the two-dimensional echocardiographic images are of sufficient technical quality to distinguish bicuspid from tricuspid aortic valves with a sensitivity of 78%, specificity of 96%, and diagnostic accuracy of 93%.[1] The superiority of two-dimensional echocardiography over M-mode echocardiography in identifying bicuspid aortic valves, a task where spatial orientation is critical, is illustrated by the lower sensitivity (55%) and specificity (71%) of the M-mode echocardiographic criterion of eccentric diastolic closure of the aortic valve.* Patients with bicuspid aortic valves also may be demonstrated to have

---

*Closure is considered eccentric when diastolic aortic radius (distance from closure line to nearest aortic wall) is greater than 1.3.

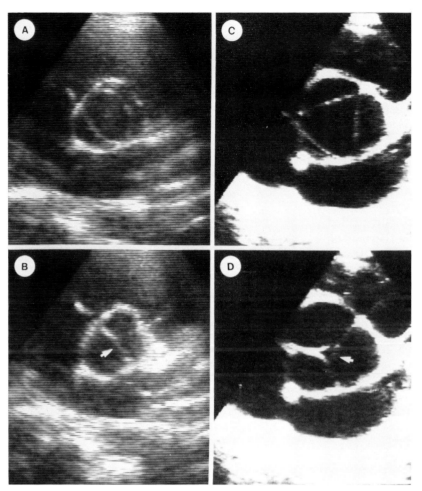

Figure 21-1 Stop-frame short-axis two-dimensional echocardiograms of the aortic valve. A and B show systolic (A) and diastolic (B) frames from a patient with a bicuspid aortic valve. Note the single coaptation line (arrow in B) between the two cusps and the two cusp margins forming the orifice (A). C and D show systolic (C) and diastolic (D) frames from a patient with a tricuspid aortic valve. Note the Y-shaped pattern (arrow in D) formed by the coaptation of the three cusps and the triangle-shaped orifice comprising the margins of the three leaflets (C).

aortic valve prolapse (defined as diastolic displacement of the leaflets into the left ventricular cavity inferior to the plane of insertion of the cusps).[2,3] Quantitating aortic valve prolapse by the maximal diastolic displacement of the aortic cusp or by the area subtended by the prolapsing portion of the leaflets relative to their insertion permits the discrimination of patients with normal aortic valves from those with bicuspid aortic valves with prolapse. Despite this potential for precise definition of anatomic detail, conclusions about the functional significance of these structural observations (for example, the presence of aortic regurgitation) must be made with caution. While the volume subtended by the prolapsing leaflet was significantly greater in patients with aortic regurgitation due to prolapse than in those without aortic regurgitation,[2] no single measurement could distinguish patients with aortic regurgitation from those without aortic regurgitation because of the wide overlap between these two groups. Clearly, Doppler echocardiographic detection of aortic regurgitation has an important complementary role in evaluating these patients.

In the adult population aged beyond the fifth decade calcification of stenotic aortic valves is common, and the absence of such calcification (as assessed by cinefluoroscopy) significantly decreases the likelihood of hemodynamically important obstruction to blood flow. Accordingly there has been much interest in the detection, description, and quantification of the intense echocardiographic signals emanating from calcification of the aortic valve by M-mode echocardiographic[4] and two-dimensional echocardiographic techniques.[5–7] The echocardiographic approach to detecting tissue calcification is limited in clinically available equipment by the fact that the intensity of the reflected ultrasound is gain-dependent, is subject to attenuation as the signal transits through the thorax, and can be standardized internally only relative to the intensity of signals reflected from other (presumably normal) structures. Accordingly, various investigators have used operator-dependent incremental rejection schemes[5,7] to judge the presence of calcium by the persistence of its echocardiographic signals at a level of signal suppression where only the aortic wall is visualized. The "blooming" of the margins of the presumed calcified structures in the video image and a degree of subjectivity in the interpretation of these images[7] prevent quantitative analysis. Notwithstanding these limitations, carefully performed studies comparing two-dimensional echocardiography against cinefluoroscopic detection of calcified deposits 1.5 mm in diameter have demonstrated a sensitivity of 76% and a specificity of 89% for detecting aortic valvular calcification by echocardiography.[6] In one study the area of the calcified aortic valve tissue measured from short-axis *tomographic* views obtained preoperatively by two-dimensional echocardiography correlated ($r = 0.79$) with the area of calcification measured from *planar* radiographs of the operatively excised tissue.[7]

The detection of vegetations of the aortic valve in patients with infective endocarditis is another useful application of two-dimensional echocardiography that exploits the technique's ability to display structural details with a high degree of spatial definition and acceptable temporal resolution.[8] For the technique to be sensitive, however, multiple tomographic planes through the valve must be examined so that an adequate sampling of the entire three-dimensional structure has been achieved. In addition, in the severely deformed or calcified valve, the multitude of bright and potentially confusing echoes emanating from the valve may conceal a relatively small vegetation. Of particular clinical significance in aortic valve endocarditis is the ability of two-dimensional echocardiography to image the anatomic consequences of contiguous spread of the infection to the aortic annulus, proximal interventricular septum, and anterior mitral leaflet.[9,10] Early noninvasive detection of paravalvular abscess formation (see Figure 21-2) in such patients has obvious implications for surgical management.

A logical outgrowth of the ability to visualize and characterize the structure of the aortic valve is to ask whether the two-dimensional echocardiogram can provide an estimate of the anatomic orifice size of the stenotic aortic valve, which could separate patients meaningfully into subsets according to the hemodynamic severity of their outflow tract obstruction. Because of the complex three-dimensional geometry of the trileaflet aortic valve, echocardiographic imaging of the complete orifice in a single systolic short-axis frame is not achieved consistently[11]; hence most investigators have resorted to measuring the maximal separation of the systolic aortic cusp as an index of anatomic orifice size.[11-13] The presence of valvular calcification in older patients renders this measurement less precise unless careful attention is directed toward minimizing signal gain so that blooming of the bright echoes from the calcified tissue does not obscure the actual leaflet margins. As the anatomic orifice in patients with acquired aortic stenosis is often irregular, the spatial orientation provided by two-dimensional echocardiography or two-dimensional echocardiographic targeting of M-mode recordings of the aortic orifice is essential to the measurement of the maximal cusp separation.[12] In

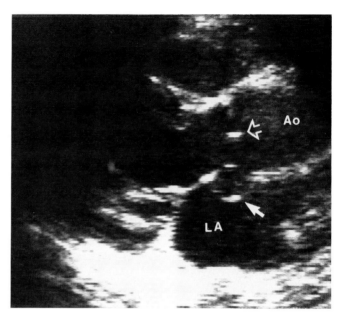

Figure 21-2 Stop-frame diastolic long-axis two-dimensional echocardiogram from a patient with an abscess of the aortic annulus. At the level of the aortic valve (open arrow) an echolucent outpouching (solid arrow) of the posterior aortic wall protrudes into the left atrium (LA). This cavity, which communicates with the aortic root (Ao) via a narrow neck, was demonstrated at surgery to represent an abscess tracking posteriorly into the fibrous continuity between the aortic and mitral valves.

younger patients with congenital aortic stenosis, multiple angulations in the short-axis plane and confirmation from long-axis views of the aortic valve are required to determine that the maximal cusp separation is measured at the apex of the domed stenotic valve.[12-13] With these techniques linear correlations between maximal cusp separation and peak systolic aortic valve gradient ($r = 0.88$) or aortic valve area derived by the Gorlin equation ($r = 0.80$) have been achieved in children (mean age 9 years) with congenital aortic stenosis[13]; however, the technique has not enjoyed a similar degree of success in a predominantly adult population (mean age 58 years).[11] Nonetheless, as seen in Figure 21-3, measurement of the maximal cusp separation enabled separation of patients with critical aortic stenosis* from normal persons in studies of children[13] and adults.[11] The measurement did not clearly distinguish mild to moderate aortic stenosis from critical aortic stenosis, and the overlap was greater for adult patients. Recent experience with transesophageal two-dimensional echocardiography[14] suggests that direct visualization of the entire valvular orifice may be feasible with this technique and that planimetry of the anatomic

---

*Critical aortic stenosis in children is defined as a peak systolic gradient of 75 mm Hg or greater[13]; in adults, as an aortic valve area less than 0.75 cm^2.[11]

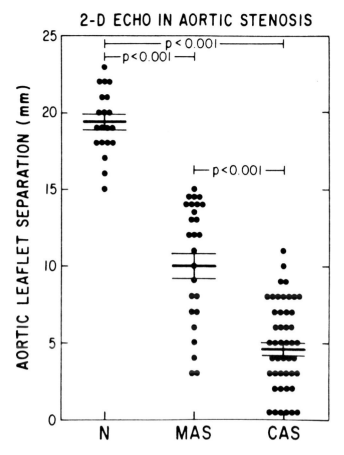

Figure 21-3 Values for individual, mean, and standard error of the mean for maximal aortic leaflet separation in normal subjects (N), patients with moderate aortic stenosis (MAS), and patients with critical aortic stenosis (CAS). By definition, critical aortic stenosis was distinguished from moderate aortic stenosis by a peak systolic pressure gradient of 75 mm Hg or greater in children and an aortic valve area of less than 0.75 cm² in adults. Normal subjects are clearly separated from patients with aortic stenosis, but there is considerable overlap between patients with moderate stenosis and those with critical aortic stenosis. *Source:* Reprinted with permission from *Circulation* (1980; 62:304–312). Copyright © 1982, American Heart Association.

orifice area correlates well with hemodynamic orifice size as calculated by the Gorlin equation ($r = 0.92$). Advances in the Doppler echocardiographic measurement of transvalvular pressure gradients and the derivation of hemodynamic orifice area have overshadowed these direct anatomic approaches to estimating valve area; however, the measurement of transvalvular flow necessary to the calculation of effective valve area from the Doppler data is still dependent on the two-dimensional echocardiographic measurement of the anatomic orifice size at some reference site (for example, left ventricular outflow tract or mitral annulus). In addition, the two-dimensional echocardiographic assessment of cusp separation provides important confirmatory evidence for assessment of the severity of aortic stenosis, particularly when the Doppler study is technically limited.

Chamber Enlargement and Hypertrophy Proximal to the Aortic and Pulmonic Valves

Particular interest has been directed toward the morphologic alterations of the left ventricle in aortic stenosis and regurgitation. Not only do these structural adaptations provide an indirect assessment of the abnormal pressure or volume load imposed and hence are reflective of the severity of valvular disease, but they also provide insights into the degree of functional compensation (for example, wall stress, stress-dimension relationships) to the abnormal loading states that may have important prognostic implications for the timing of valvular replacement or repair. M-Mode echocardiography has been the primary tool used to assess these changes in left ventricular geometry because it provides sufficient spatial resolution in the axial dimension to permit accurate measurement of changes in left ventricular minor axis dimension and wall thickness and sufficient temporal resolution to permit the timing of the measurements with simultaneously acquired intracardiac pressures. Two-dimensional echocardiographic imaging is also important in these studies because it provides spatial orientation for targeting of the M-mode cursor to ensure proper alignment with cardiac structures. In addition, measurements obtained directly from the two-dimensional echocardiogram permit measurement of axes other than the left ventricular minor axis. Knowledge of changes in the long axis of the left ventricle is critical to geometric models used to calculate left ventricular volume and circumferential (as opposed to meridional) wall stress. In addition, measurement of the short and long axes of the left ventricle is required for the assessment of wall stress-cavity dimension relationships, since meridional stress (derived from the short-axis dimension) should be related to long-axis dimension and shortening while circumferential stress (derived from the long-axis dimension) should be related to short-axis dimension and shortening.[15] Finally, two-dimensional echocardiography has been useful for demonstrating in several studies that left ventricular geometry may not change in a symmetric fashion. In aortic regurgitation, for instance, with progressive left ventricular dilation the chamber assumes a more spherical shape with disproportionate increases in the short-axis dimension,[15,16] which may result in an overestimation of volume changes by M-mode echocardiographic techniques.

In aortic stenosis, systolic pressure loading of the left ventricle results in predominant wall thickening without chamber enlargement (concentric hypertrophy with increased wall thickness-radius ratio), which results in normalization of peak systolic and end-diastolic wall stress despite elevation of peak systolic and end-diastolic pressures.[17] By assuming that systolic wall stress (σ) is normalized to some common value in patients with hemodynamically compensated aortic stenosis, several investigators have attempted to estimate left ventricular systolic pressure (P) from echocardiographic measurements of wall thickness (h) and short-axis cavity radius (r),[18,19] since in simplified form σ is proportional to (P)(r)/h. Having estimated left ventricular systolic pressure, these investigators then proposed that the transaortic systolic pressure gradient could be derived from the simultaneous measurement of the brachial artery pressure by sphygmomanometry. While this approach has an intuitive appeal, several problems have limited its practical application. First, the method does not apply to patients in whom systolic stress is not appropriately "normalized" by concentric hypertrophy, and the identification of aortic stenosis by noninvasive techniques in this subset of patients is problematic. Second, invasive studies suggest that the degree of compensatory hypertrophy and the resultant reduction in systolic wall stress may be greater in patients with congenital aortic stenosis than it is in patients with acquired aortic stenosis.[20] These observations suggest that there are quantitative differences in the hypertrophic response to pressure overload in congenital aortic stenosis as compared with that in acquired aortic stenosis and hence the relationship between left ventricular pressure and the left ventricular wall thickness-radius ratio also may differ. Finally, even among patients with normal left ventricular systolic function, systolic wall stress exhibits a considerable range of variation resulting in a poor correlation between left ventricular systolic pressure as measured at catheterization and left ventricular systolic pressure as predicted from wall thickness-radius ratios and an assumed constant systolic wall stress.[21]

Because of the success of continuous-wave Doppler measurement of instantaneous transaortic pressure gradients, interest in the analysis of left ventricular geometry for the sole purpose of predicting left ventricular systolic pressure has waned. Recent preliminary work, however, has suggested the possibility of successfully estimating instantaneous left ventricular systolic pressure from continuous-wave Doppler measurement of the transaortic pressure gradient and simultaneous estimation of central aortic pressure derived from time-corrected, calibrated carotid pulse tracings.[22] If the potential of this technique is realized, it may be possible to use this noninvasively measured estimate of left ventricular systolic pressure to calculate systolic wall stress from echocardiographically determined cavity dimensions and wall thickness. Examination of the meridional wall stress-myocardial fiber shortening relationships by this combination of Doppler and echocardiographic techniques could provide a relatively preload-independent, afterload-adjusted index of myocardial contractility[23,24] applicable to the study of systolic function in patients with aortic stenosis. This line of investigation is of particular interest in view of the demonstration by invasive techniques that examination of the wall stress-ejection fraction relationships can distinguish patients with aortic stenosis in congestive heart failure due to afterload mismatch from those with intrinsic depression of myocardial contractility.[25]

Pulmonic stenosis presumably results in right ventricular adaptations similar to those observed in the left ventricle in aortic stenosis; however, the unique geometry of the right ventricle and the lack of convenient models to approximate its shape have been obstacles to pursuing detailed analyses of wall stress-fiber shortening relationships in right ventricular pressure overload. One striking echocardiographic observation in right ventricular pressure overload is that the normal, nearly circular contour of the left ventricle in the short-axis view is distorted by leftward ventricular septal displacement, resulting in a flattened septum; in patients with severe right ventricular systolic hypertension, there is a reversal of the normal curvature of the ventricular septum.[26,27] These findings are most marked at end-systole[26] and persist into early diastole.[27] Normalized end-systolic ventricular septal curvature correlates well ($r = 0.86$) with the transventricular systolic pressure gradient and presumably is a reflection of the reduction or reversal of the normal trans-septal pressure gradient in right ventricular pressure overload.[26]

In aortic regurgitation, diastolic volume loading of the left ventricle results in chamber enlargement with mild to moderate wall thickening (eccentric hypertrophy with normal to decreased wall thickness-radius ratio) such that peak systolic wall stress is nearly normal but end-diastolic wall stress is increased.[17] By contrast to aortic stenosis, where deterioration in systolic function is a late finding usually preceded by symptoms attributable to obstruction of the left ventricular outflow tract, aortic regurgitation may take a more insidious and indolent course, presenting with left ventricular systolic dysfunction. Accordingly, much effort has been devoted toward examining the value of echocardiographic indices of chamber dilation and fiber shortening to define prognostic subsets and to detect the onset of left ventricular systolic dysfunction. Initial M-mode echocardiographic studies of symptomatic patients undergoing aortic valve replacement for aortic regurgitation suggested that a preoperative left ventricular end-systolic dimension of greater than 55 mm and a minor-axis fractional shortening of less than 25% identified a high-risk subset of patients with an increased incidence of perioperative deaths and congestive heart failure terminating in late postoperative death.[28] These observations raised the possibility that serial tracking of such indices of left ventricular systolic function might identify asymptomatic patients at risk for the development of systolic dysfunction and symptoms of congestive heart failure who might benefit from earlier intervention.[29] Follow-up studies suggested that asymptomatic patients with normal minor-axis fractional shortening and moderate to severe aortic regurgitation have a favorable prognosis and that a policy of delaying valvular replacement until the onset of symptoms or the development of left ventricular systolic dysfunction does not result in increased operative mortality or poor operative results as assessed by symptomatic response to surgery and postoperative reduction in left ventricular end-diastolic dimension.[30]

Other investigators studying symptomatic patients with aortic regurgitation have not been able to confirm the value of a preoperative end-systolic dimension of greater than 55 mm and a minor-axis fractional shortening of less than 25% as predictors of postoperative death[31,32] and have emphasized that this subset of patients may indeed have good symptomatic relief and normalization of echocardiographically determined left ventricular dimensions postoperatively.[31,33] The explanations for the apparent discrepancies between these studies are not entirely clear but may relate to (1) improved techniques of myocardial preservation during cardioplegic arrest in the more recent studies, (2) intertest variability of echocardiographic measurements,[34] and (3) limitations of assessing end-systolic volume and stroke volume with the one-dimensional M-mode echocardiographic technique.[15,16] While preoperative left ventricular fractional shortening and end-systolic dimension are significant predictors (by univariate life-table analysis) of survival after aortic valve replacement for aortic regurgitation,[35] it is probably unrealistic to expect such measurements to separate out dichotomously patients at high and low risk for operative survival, and it is not surprising that there is significant overlap in these subsets of patients.[35] Additionally, as indices of systolic function, neither the left ventricular end-systolic dimension nor the minor-axis fractional shortening is adjusted for afterload. By analogy to the evaluation of systolic function in aortic stenosis, it may be important to evaluate myocardial fiber shortening in the context of afterload (assessed by systolic wall stress) to define more closely the intrinsic myocardial contractile function in these patients.[15,36]

Since aortic regurgitation now can be detected easily by Doppler echocardiographic techniques, the recognition by M-mode echocardiography of diastolic high-frequency vibrations or fluttering of the ventricular septum and anterior mitral leaflet[37] due to direction of the aortic regurgitant jet upon these structures[38] has lost some of its diagnostic utility. Nonetheless, in acute, severe, aortic regurgitation where the rapid volume overloading of a noncompliant left ventricle results in elevation of left ventricular end-diastolic pressure to levels equivalent to left atrial pressure and aortic diastolic pressure, M-mode echocardiography provides a graphic demonstration of these physiologic events by demonstrating premature diastolic closure of the mitral valve[39,40] and presystolic opening of the aortic valve.[40]

Structural Changes in the Great Vessels Distal to the Aortic and Pulmonic Valves

In both aortic stenosis and pulmonic stenosis the poststenotic dilation of the proximal great vessel can be demonstrated echocardiographically. Similarly, enlargement of the ascending aorta due to increased left ventricular stroke output in aortic regurgitation also can be visualized. Of particular interest, however, is the use of two-dimensional echocardiography to define abnormalities of the proximal aortic root that result in valvular regurgitation[41-43] (see also Chapter 22). A dramatic example of such a situation occurs in Marfan syndrome,[44] where massive dilation of the aortic root is associated with aortic regurgitation resulting from malsuspension and central malcoaptation of the aortic cusps (Figure 21-4).

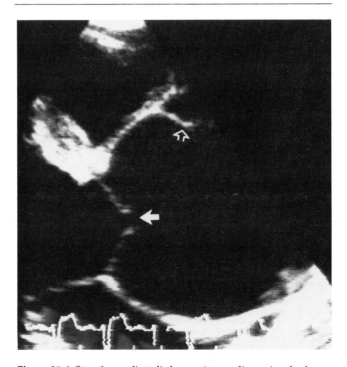

Figure 21-4 Stop-frame diastolic long-axis two-dimensional echocardiogram from a patient with aortic root dilation, aortic regurgitation, and Marfan syndrome. The aortic valve leaflets (solid arrow) fail to close completely during diastole because of malsuspension of the valve in the dilated aortic root (maximal transverse diameter 8.0 cm). Incidental note is made of a small intimal tear (open arrow) of the anterior wall of the dilated aortic root. Calibration marks are 1.0 cm apart.

VENTRICULAR INFLOW TRACT: MITRAL AND TRICUSPID VALVES

Valvular Anatomy

Two-dimensional echocardiography has been especially successful in delineating the wide spectrum of pathologic disorders afflicting the mitral and tricuspid valves (Table 21-2). The various forms of congenital mitral deformities are distinguished by the enumeration of the papillary muscles associated with the atrioventricular valve as well as by demonstration of structural alterations of the valve leaflets themselves.[45-48] In addition, the echocardiogram can demonstrate associated congenital malformations, such as the many variations of the endocardial cushion defect that accompany cleft mitral valves. Ebstein's anomaly of the tricuspid valve results from abnormal displacement of the attachment of the septal (88% of the time) and posterior (69% of the time) leaflets of the tricuspid valve toward the apex of the right ventricle, resulting in a portion of the anatomic right ventricle functioning as right atrium from a hemodynamic standpoint.[49] Early M-mode echocardiographic observations in the disease were dependent on indirect measurements such as increased right ventricular end-diastolic dimension and delayed closure of the tricuspid valve relative to the mitral valve. Two-dimensional echocardiography by contrast permits the direct demonstration of the apical displacement of the septal tricuspid leaflet insertion.[50]

Prolapse of the mitral and tricuspid valves has been studied extensively echocardiographically, and the availability of a noninvasive tool for the diagnosis of this condition has permitted epidemiologic surveillance for the disease and the putative establishment of associations with a wide range of systemic disorders. Earlier investigators using M-mode echocardiography demonstrated abnormal posterior systolic motion of anterior or posterior (or both) mitral leaflets (see Figure 21-5A) in patients with angiographic evidence of mitral valve prolapse[51] and noted that the timing of this midsystolic to late systolic posterior bowing of the leaflets coincided with the phonocardiographically timed midsystolic click.[52] Subsequent investigators demonstrated that holosystolic or late systolic posterior displacement of the mitral leaflets could be demonstrated in mitral valve prolapse, but the apparent separation of the anterior and posterior leaflets did not bear a strict relationship to the presence or absence of mitral regurgitation.[53]

A variety of other M-mode echocardiographic findings have been described in this syndrome, including (1) large anterior mitral leaflet diastolic excursion, (2) abnormal posterior mitral leaflet motion during diastole, (3) early systolic anterior mitral leaflet motion followed by late systolic posterior motion, and (4) increased thickness and density of the leaflets.[54] Given the presence of a mobile midsystolic click (with or without late systolic murmur) as a definition for mitral valve prolapse, the M-mode echocardiographic criteria of late systolic or holosystolic abnormal posterior leaflet motion appear to be the best diagnostic criteria for mitral valve prolapse (85% sensitivity, 99% specificity).[54] In a study of symptomatic patients with moderate to severe mitral regurgitation, the pathologic diagnosis of mitral valve prolapse (based on leaflet area and basal-to-distal length of leaflets) was predicted by similar M-mode echocardiographic criteria with a sensitivity of 88% and a specificity of 82%.[55] The relatively high prevalence in adults of mitral valve prolapse (diagnosed by M-mode echocardiography) and recent population studies demonstrating the lack of a true association with symptoms previously attributed to the syndrome (for example, atypical chest pain, dyspnea) challenge the assumption that abnormal motion or anatomic position of the leaflets can be equated with clinically relevant disease.[56,57] Furthermore, among 237 minimally symptomatic or asymptomatic patients with M-mode echocardiographic evidence for mitral valve prolapse observed at the Mayo Clinic, actuarial 8-

Table 21-2 Common Inflow Tract Abnormalities Diagnosed by Echocardiography

Congenital Malformations of the Atrioventricular Valves

- Congenitally malformed mitral valves
  - Parachute mitral valve with single papillary muscle
  - Arcade mitral valve with multiple papillary muscles
  - Cleft mitral valve
  - Double-orifice mitral valve
- Mitral valve prolapse
- Ebstein's anomaly of the tricuspid valve
- Tricuspid valve prolapse

Acquired Disorders of the Atrioventricular Valves

- Acute rheumatic fever
- Rheumatic mitral and tricuspid valve disease
- Mitral annular calcification
- Disorders primarily of the submitral apparatus
  - Flail mitral leaflet
  - Papillary muscle dysfunction
- Systemic disorders that may involve the atrioventricular valves
  - Systemic amyloidosis
  - Idiopathic hypereosinophilic syndrome
  - Carcinoid syndrome
- Mitral and tricuspid valve endocarditis

Nonvalvular Conditions Mimicking Valvular Disease

- Cor triatriatum, cor triatriatum dexter
- Supravalvular membranes
- Atrial myxomas
- Hypertrophic cardiomyopathy with systolic anterior motion of the mitral valve resulting in mitral regurgitation

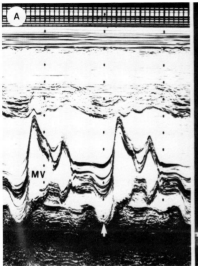
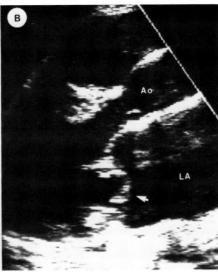

Figure 21-5 M-Mode and two-dimensional echocardiogram of the mitral valve (MV) from a patient with mitral valve prolapse. The M-mode echocardiogram (A) demonstrates late systolic posterior motion of the mitral valve (solid arrow). The stop-frame systolic long-axis two-dimensional echocardiogram (B) demonstrates prolapse of the posterior mitral leaflet (solid arrow) into the left atrium (LA) and relatively normal position of the anterior mitral leaflet relative to the mitral annulus. Ao, aortic root.

year survival (88%) was not significantly different from that of matched controls.[58]

A variety of two-dimensional echocardiographic phenomena have been found in patients with systolic clicks and M-mode echocardiographic evidence for mitral valve prolapse, including (1) whiplike excursion of the anterior mitral leaflet, (2) exaggerated mitral annular motion, (3) apparent redundancy of leaflet tissue, and (4) superior displacement of the mitral leaflets or coaptation point above the plane of the mitral annulus.[59] The displacement of the mitral leaflets or coaptation point into the left atrium (Figure 21-5B) generally is accepted as the best criterion for diagnosing prolapse, with a sensitivity of 87% and a specificity of 97%, versus auscultation and M-mode echocardiography.[59] Despite the advantages of improved spatial orientation and structural definition of the mitral valve offered by the two-dimensional echocardiographic technique, we are still faced with the dilemma of deciding whether a geometric definition of the leaflet-annular relationship uniquely defines a disease syndrome with characteristic pathologic and clinical associations. The demonstration of a 35% prevalence of mitral valve prolapse in children and adolescents (ranging from 10 to 18 years of age) using these two-dimensional echocardiographic criteria suggests that there is a range of *normal* systolic superior displacement of the mitral leaflets that may be difficult to discern from that seen in the mitral valve prolapse syndrome.[60] Recent studies of the geometry of the mitral annulus suggest that it cannot be modeled by a Euclidian plane but rather assumes the saddle-shaped conformation of a hyperbolic paraboloid.[61] These observations challenge the validity of equating superior systolic displacement of the mitral leaflets in the apical four-chamber and two-chamber views and offer an explanation for the apparent increased frequency of systolic displacement of the leaflets in the apical four-chamber view.[61]

It is apparent from these studies that, although echocardiographic techniques can define a wide variety of structural relationships between the mitral valve and its annulus, further epidemiologic and clinical correlations are required to define structural parameters that may distinguish individuals with myxomatous degeneration of the mitral valve and an unfavorable prognosis from persons with a normal life expectancy and a normal degree of systolic superior mitral leaflet motion. In this regard the longitudinal study[58] from the Mayo Clinic made the provocative observation that patients with mitral valve prolapse and apparent thickening of the anterior mitral leaflet (assessed by M-mode echocardiography) had a significantly increased incidence (10.3%) of untoward complications (sudden death, endocarditis, and cerebral embolic events) in comparison to the incidence (0.7%) among persons without leaflet thickening.[58] Further prospective evaluation of this and other echocardiographic signs of mitral valve prolapse is needed to identify structural variables predictive of poor outcome. Evidence is also accumulating suggesting that hemodynamic deterioration, risk of endocarditis, and the prevalence of important ventricular dysrhythmias[62] may be related primarily not to mitral valve prolapse per se but rather to the severity of the associated mitral regurgitation; thus Doppler echocardiography[63] may be especially helpful in defining a high-risk subset of patients with clinically important disease. Ultimately we may expect too much of the technique to think that a geometric definition of valvular architecture can be used in isolation to define a disease syndrome. A more reasonable approach might be to define patient subgroups by a *collection* of clinical variables (including echocardiographic signs) to avoid the *monolithic* reference standard of a single echocardiographic measurement.[64]

Of the acquired diseases of the mitral valve, rheumatic mitral stenosis must stand out as a landmark in echocardiographic diagnosis. Although the echocardiographic findings in acute rheumatic fever are nonspecific and indirect (for example, chamber enlargement and pericardial effusion),[65] echocardiography has played a major role in characterizing the abnormalities of the mitral leaflets and the subvalvular apparatus that evolve subsequently. The initial M-mode echocardiographic description of a reduced diastolic closing velocity (EF slope) of the anterior mitral leaflet reflective of prolonged early diastolic filling was the first recognized sign of mitral stenosis[66] (Figure 21-6). Recognition of the associated abnormal anterior diastolic motion of the posterior mitral leaflet provided a means of distinguishing mitral stenosis from other conditions associated with impaired early diastolic filling and reduced mitral valve EF slope.[67] Two-dimensional echocardiographic imaging of the stenotic mitral valve demonstrates marked thickening of the leaflets with reduced mobility and intense echodensities in the leaflets, probably representing calcification (Figure 21-6A). The leaflet tips demonstrate greater deformity and restriction of excursion than do the more proximal leaflet bodies, resulting in diastolic tethering of the tips and bowing of the bodies of the leaflets into the left ventricle[68] (Figure 21-6B). Diastolic anterior motion of the anterior leaflet toward the ventricular septum pulls the posterior leaflet along with it, resulting in the appearance of concordant motion of the two leaflets previously described by M-mode echocardiography[68] (Figure 21-6C). In addition to providing a qualitative description of rheumatic mitral stenosis, two-dimensional echocardiography has been used to measure anatomic mitral orifice size, which can be compared with measurements made at the time of mitral valve replacement[69] (Figure 21-6D), orifice area determined from excised valves,[70] and hemodynamic estimates of mitral valve area.[68,70–72] The accuracy of the technique is critically dependent on transducer alignment in the parasternal short-axis plane such that the tips of the leaflets, which usually form the limiting orifice, are clearly imaged and cross-sections through more basal portions of the funnel-shaped mitral valve are avoided. In addition, attention to adjustment of signal gain is important, as echo dropout will enlarge the apparent anatomic ori-

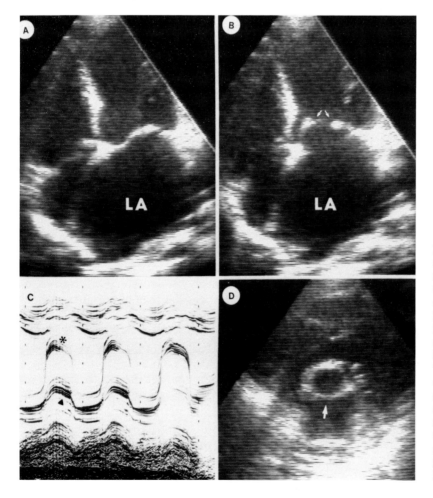

Figure 21-6 M-Mode and two-dimensional echocardiograms of the mitral valve from a patient with rheumatic mitral stenosis. Stop-frame apical four-chamber two-dimensional echocardiograms in systole (A) and diastole (B) show the thickened echo-dense mitral leaflets and the enlarged left atrium (LA). In the diastolic image (B) the tips of the mitral leaflets (pair of small arrows) do not have full excursion, and the more proximal bellies of the leaflets bow into the left ventricle. The M-mode echocardiogram (C) of the thickened echo-dense mitral valve demonstrates a retarded diastolic closing velocity (EF slope, asterisk) and abnormal anterior diastolic motion of the posterior mitral leaflet (▲). The stop-frame diastolic short-axis two-dimensional echocardiogram at the level of the mitral leaflet tips (D) reveals the narrowed mitral orifice (arrow). An estimate of anatomic orifice area can be obtained by planimetry of this orifice.

fice, while image saturation due to excessive gain will make the orifice appear smaller.[71] With attention to these details in patients with technically acceptable studies, good linear correlations ($r = 0.85$ to 0.95) can be obtained for the mitral valve area compared with those obtained by anatomic or hemodynamic techniques.[68–72]

Increasing interest in conservative mitral valve surgery (for example, reconstructive valve repair and mitral commissurotomy) and percutaneous balloon valvuloplasty has presented new diagnostic challenges to the echocardiographer in terms of (1) selection of patients suitable for these procedures, (2) sizing of the mitral annulus, (3) guidance of balloon catheters, and (4) anatomic and functional assessment of outcomes. For instance, preliminary results suggest that after balloon valvuloplasty for mitral stenosis an increase in the transverse diameter at the mitral commissures and increases in the anterior and posterior commissural opening angles as assessed by echocardiography may be predictive of relief of obstruction.[73] On the other hand, two-dimensional echocardiographic planimetry of the mitral valve area after surgical commissurotomy has resulted in less reliable prediction of hemodynamically measured orifice size than that provided by Doppler techniques.[72]

The recognition in some elderly patients with systolic murmurs of dense echocardiographic bands behind the posterior leaflet of the mitral valve (by M-mode echocardiography) and echocardiographically dense masses in and about the base of the mitral valve (by two-dimensional echocardiography) has elevated the diagnosis of "mitral annular calcification" from an infrequently diagnosed clinical pathologic correlation[74] to a commonly recognized degenerative disorder of the mitral valve in the elderly. Echocardiographic and pathologic correlations have shown that "annular" calcification is somewhat a misnomer, as the abnormal deposits of dystrophic calcification are most commonly located in the angle between the posterior mitral leaflet and the left ventricular wall rather than in the mitral ring itself.[75,76] Although it is commonly associated with mild mitral regurgitation,[77,78] obstruction to mitral inflow has also been documented in a small (8%) percentage of patients.[76,78] Contiguous involvement of the base of the anterior mitral leaflet, the aortic valve, and the proximal interventricular septum also occurs.

As discussed previously for the semilunar valves, infective endocarditis also may affect the atrioventricular valves, resulting in vegetations, ruptured chordae tendineae, and flail atrioventricular valves. Of particular clinical importance is the diagnosis of infective vegetations of the tricuspid valve[79,80] in febrile patients who abuse drugs intravenously. These patients often have multiple potential etiologies for fever, and the relatively low right ventricular systolic pressures may make recognition of tricuspid insufficiency difficult; hence the two-dimensional echocardiographic identification of the bulky frondlike masses that commonly appear during the course of tricuspid endocarditis may be of great assistance in confirming the diagnosis of right-sided endocarditis (Figure 21-7).

Several relatively uncommon systemic disorders may involve the heart and the atrioventricular valves secondarily, resulting in acquired regurgitation or stenosis that can be detected echocardiographically. In systemic amyloidosis with cardiac involvement, in addition to detecting (1) marked thickening of the ventricular walls, (2) thickening of the interatrial septum, (3) "granular sparkling" echoes from the myocardium,

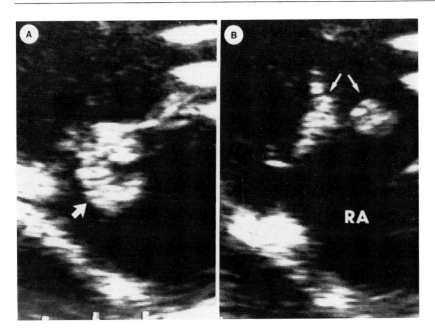

Figure 21-7 Stop-frame long-axis two-dimensional echocardiogram of the right ventricular inflow tract from a patient with tricuspid valve endocarditis. During systole (A) a large, bulky vegetation attached to the posterior tricuspid valve leaflet (arrow) prolapses into the right atrium. During diastole (B) both anterior and posterior leaflets are noted to be thickened (pair of small arrows) as a result of infective endocarditis. The apparent change in shape and size of the tricuspid valve vegetation during different portions of the cardiac cycle occurs as the mass moves in and out of the tomographic plane. RA, right atrium.

(4) left atrial enlargement, and (5) pericardial effusions,[81-83] two-dimensional echocardiography frequently detects thickening and distortion of the atrioventricular valves and subvalvular structures[81,82] that result from amyloid deposition into the leaflets and are associated with valvular regurgitation. In the idiopathic hypereosinophilic syndrome, two-dimensional echocardiography can demonstrate localized thickening of the posterior mitral valve leaflet with restriction of its motion.[84] Confirmatory pathologic studies have suggested that this echocardiographic appearance results from organizing fibrothrombotic material deposited over damaged endocardium in the submitral angle. Subsequent adherence of the posterior mitral leaflet to endocardium and distortion of the valvular architecture result in significant mitral regurgitation. In the malignant carcinoid syndrome the endocardial lesions are restricted to the right side of the heart, and two-dimensional echocardiography demonstrates right ventricular enlargement and thickening of the tricuspid (and occasionally pulmonic) valve.[85] Typically the tricuspid valve appears thickened, retracted, and fixed in a semiopen position that results in not only valvular regurgitation but also significant stenosis.

Mitral regurgitation also can occur in primary disease of the subvalvular apparatus (with normal leaflet architecture), resulting in abnormal tethering and positioning of the leaflets during systole. The most dramatic examples are seen with the sudden rupture of chordae tendineae in a patient with myxomatous degeneration of the mitral valve but may also occur when the chordae of a normal valve are disrupted by infective endocarditis (Figure 21-8) or when the head of a papillary muscle is ruptured during the course of an acute myocardial infarction. In the most severe cases, the flail mitral leaflet and attached chordae fling with a whiplike motion, prolapsing into the left atrium and failing to coapt with the normally tethered leaflet.[86,87]

Smaller degrees of mitral regurgitation may result from less marked disorders of leaflet coaptation resulting from distortion of the geometric relationships of the submitral apparatus in patients with papillary muscle dysfunction without actual discontinuity of subvalvular structures. Ischemic injury to a papillary muscle may result in its failure to shorten appropriately during systole; the ensuing disproportion between chamber size and degree of mitral tethering during systole results in the echocardiographic appearance of prolapse of the mitral valve into the left atrium.[88] On the other hand, marked left ventricular dilation or aneurysm formation in myocardium subjacent to a papillary muscle may have the reverse effect, pulling the tips of the mitral leaflets into the left ventricular cavity and preventing the normal motion of the leaflets toward the mitral annulus that occurs with normal mitral closure.[88] These conditions are recognized echocardiographically by the association of the left ventricular wall motion abnormalities and abnormal systolic mitral leaflet positioning.

The presence of "incomplete mitral leaflet closure" (defined as failure of the mitral leaflets to reach the mitral annulus at peak superior excursion) was examined in a group of patients with prior myocardial infarction and new murmurs of mitral regurgitation. This systolic mitral leaflet pattern identified 20 of these 22 patients with the papillary muscle dysfunction syndrome in conjunction with the presence of dyskinetic left ventricular wall motion subjacent to the papillary muscles.[89] Although normal persons were never shown to have these echocardiographic findings,[89] recent studies in patients with incomplete mitral leaflet closure and ischemic cardiomyopathy did not demonstrate an increased prevalence of mitral regurgitation (79% versus 65% [not significant]) relative to persons matched for left ventricular size and fractional shortening.[90] From these data it is apparent that (1) incomplete mitral leaf-

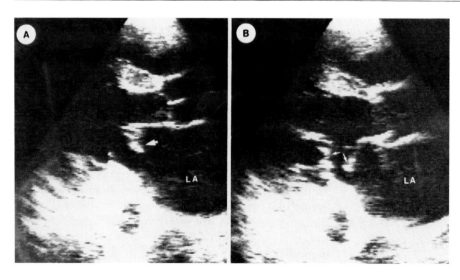

Figure 21-8 Stop-frame long-axis two-dimensional echocardiograms of the mitral valve from a patient with a flail valve due to endocarditis. During diastole (A) a large vegetation attached to the anterior mitral valve (arrow) swings into the inflow tract from the left atrium (LA). During systole (B) the flail anterior mitral leaflet and its associated vegetation prolapse into the left atrium, and the anterior and posterior mitral leaflets (pair of small arrows) fail to coapt properly, resulting in mitral regurgitation.

let closure alone does not ensure the presence of mitral regurgitation, (2) left ventricular dilation with decreased systolic function can be associated with mitral regurgitation in the absence of incomplete mitral leaflet closure, and (3) the *combination* of regional dyskinetic left ventricular wall motion and abnormal mitral leaflet closure is required to identify the papillary muscle dysfunction syndrome correctly.

Chamber Enlargement Upstream of the Mitral and Tricuspid Valves

Left atrial enlargement secondary to mitral stenosis or regurgitation can be assessed by measuring its minor-axis diameter by M-mode echocardiography[91] or by planimeterizing its cavity area imaged by two-dimensional echocardiography.[92,93] Although the left atrium can be massively enlarged in either severe mitral stenosis or regurgitation, exaggerated systolic expansion of the left atrium[93] with end-systolic displacement of the interatrial septum toward the right atrium is particularly characteristic of severe acute mitral regurgitation.[94] Left atrial volume can be estimated from single-plane[92] or biplane[93] planimetry of the left atrial cross-sectional area by two-dimensional echocardiography, and an emptying volume (left atrial end-systolic volume less end-diastolic volume) can be determined. In one study, 3+ and 4+ mitral regurgitation assessed angiographically could be distinguished from lesser degrees of regurgitation by a left atrial emptying volume of more than 40 mL, and the left atrial emptying volume correlated linearly ($r = 0.85$) with the regurgitant fraction determined by cardiac catheterization.[93] By contrast the enlarged left atrium of a patient with mitral stenosis does not show marked phasic changes in volume and exhibits nearly normal left atrial emptying volumes.[92] Assessment of left atrial size also may be helpful in predicting therapeutic responses to cardioversion in these patients, who are prone to develop atrial fibrillation.[95]

The dilated left atrium in a patient with mitral valve disease and atrial fibrillation is especially prone to thrombus formation and subsequent systemic embolization. Although atrial thrombi can be imaged by two-dimensional echocardiography, systematic and complete interrogation of this dilated chamber, including the atrial appendage, may not always be possible, and other tomographic imaging techniques may have superior sensitivity.[96] The recent finding by two-dimensional echocardiography of dynamic intracavitary echoes within the left atrial blood pool[97-99] has raised the speculation that hemorrheologic conditions conducive to thrombus formation might be identified by echocardiography[97]; such identification could point to patients at risk for systemic embolization.[99]

In analogous fashion, right atrial volume has been estimated from planimetry of the right atrial area in the apical four-chamber plane. The right atrial emptying volume (right atrial end-systolic volume less end-diastolic volume) was more than 40 mL in all patients with severe tricuspid regurgitation (assessed by contrast echocardiography) and allowed separation of these patients from normal persons and those with mild tricuspid regurgitation.[100]

Chamber Enlargement and Hypertrophy Downstream of the Mitral and Tricuspid Valves

In patients with isolated mitral stenosis the left ventricle is subject neither to pressure nor to volume overload; if anything, it receives inadequate preload. Consequently M-mode echocardiographic indices of chamber size (end-diastolic dimension and end-systolic dimension and mass [short-axis muscle cross-sectional area]) are normal preoperatively and do not change significantly after mitral valve replacement.[101] By contrast, patients with mitral regurgitation experience left ventricular volume overload and subsequent chamber enlargement with eccentric hypertrophy.[17] As a result, left ventricular end-diastolic dimension, end-systolic dimension, and muscle cross-sectional area derived by M-mode echocardiography are increased in these patients in comparison to normal persons.[101] The response to mitral valve replacement in these patients is heterogeneous: Some patients exhibit a favorable course characterized by reduction in left ventricular cavity size, regression in myocardial mass, and small reductions in ejection fraction (70% ± 5% to 59% ± 10%); others experience clinical deterioration as reflected by lack of reduction of cavity size, absence of regression of hypertrophy, and a severe reduction of ejection fraction (57% ± 5% to 26% ± 6%). In one study preoperative M-mode echocardiographic measurements correctly distinguished patients with a favorable surgical outcome (preoperative left ventricular end-diastolic dimension less than 6.5 cm, end-systolic dimension greater than 4.3 cm, and a muscle cross-sectional area of 24.2 ± 6.5 cm^2) from those with progressive postoperative deterioration (preoperative left ventricular end-diastolic dimension greater than 7.0 cm, end-systolic dimension greater than 5.0 cm, and a muscle cross-sectional area of 31.6 ± 4.4 cm^2.[101]

Other investigators using M-mode echocardiography and calibrated carotid pulse tracings have complemented these geometric and dimensional indices of left ventricular function by calculating simultaneous left ventricular meridional wall stress and assessing stress-dimension-shortening relationships in patients with mitral regurgitation.[102,103] These investigators have demonstrated that in patients with mitral regurgitation the presence of a left ventricular end-systolic dimension greater than 2.6 cm/m^2, a minor-axis fractional shortening less than 31%, or an end-systolic wall-stress index of more than 195 mm Hg preoperatively defines a subgroup of patients who will not experience a postoperative reduction in left ventricular end-diastolic dimension or mass and who will remain symptomatic despite continued medical therapy.[102]

CONTRAST AND INTRAOPERATIVE ECHOCARDIOGRAPHY

Small amounts (5 to 10 mL) of agitated saline contrast medium have been injected in a peripheral vein of the upper extremity in conjunction with echocardiographic imaging to detect tricuspid regurgitation.[104,105] Two-dimensional echocardiographically targeted M-mode tracings of the inferior vena cava near its junction with the hepatic veins were used to time (relative to the jugular venous pulse and electrocardiogram) and detect the abnormal appearance of saline contrast medium in patients with tricuspid regurgitation. The detection of saline contrast medium in the inferior vena cava coincident with the V wave of the jugular venous pulse and end-systole signified the presence of tricuspid regurgitation, and the effect was enhanced with inspiration.[104,105] The technique provided better separation of patients with tricuspid regurgitation from normal persons than did simple measurement of the diameter of the inferior vena cava. The usefulness of this technique as compared with Doppler echocardiography is discussed in Chapter 22.

Renewed interest in conservative valvular surgery (for example, mitral commissurotomy and annuloplasty) and percutaneous balloon valvuloplasty has rekindled efforts to find simple techniques for assessing residual regurgitation after valvular repair. The saline contrast angiogram has been applied successfully in the intraoperative evaluation of conservative mitral valve surgery to detect regurgitation requiring further attention.[106] Intraventricular injection of saline contrast medium with simultaneous imaging of the left side of the heart, using a two-dimensional echocardiographic probe placed directly upon the epicardium, permits the rapid identification and semiquantitation of valvular regurgitation prior to chest closure but after the cardiopulmonary bypass procedure. A similar technique could be employed with precordial two-dimensional echocardiography to assess results after balloon valvuloplasty without requiring the additional hemodynamic burden of angiographic contrast.

SUMMARY

Echocardiographic imaging permits the noninvasive assessment of detailed valvular structures with high spatial resolution in real time and offers avenues for quantitating the hemodynamic impact of valvular disease on the cardiac chambers. Measurements of chamber size and wall thickness combined with simultaneously acquired systemic arterial blood pressure allow inferences to be drawn with regard to the abnormal loading of the left ventricle in valvular heart disease and place ejection phase indices in the context of loading conditions so that intrinsic myocardial contractility can be evaluated. The noninvasive nature of the technique lends itself to serial evaluation of patients with valvular heart disease so that prognostication and longitudinal follow-up can be achieved. In conjunction with Doppler echocardiographic techniques for simultaneously measuring disturbances of blood flow, echocardiography provides the unique opportunity to provide noninvasively an accurate structural and functional assessment of valvular heart disease.

REFERENCES

1. Brandenburg RD Jr, Tajik AJ, Edwards WD, et al. Accuracy of two-dimensional echocardiographic diagnosis of congenitally bicuspid aortic valve: Echocardiographic anatomic correlation in 115 patients. *Am J Cardiol*. 1983;51:1469–1473.

2. Stewart WJ, King ME, Gillam LD, et al. Prevalence of aortic valve prolapse with bicuspid aortic valve and its relation to aortic regurgitation: A cross-sectional echocardiographic study. *Am J Cardiol*. 1984;54:1277–1282.

3. Shapiro LM, Thwaites B, Westgate C, et al. Prevalence and clinical significance of aortic valve prolapse. *Br Heart J*. 1985;54:179–183.

4. Dancy M, Leech G, Leatham A. Comparison of cine fluoroscopy and M-mode echocardiography for detecting aortic valve calcification: Correlation with severity of stenosis of non-rheumatic etiology. *Br Heart J*. 1984;51:416–420.

5. Rubler S, King ML, Tarkoff DM, et al. The role of aortic valve calcium in the detection of aortic stenosis: An echocardiographic study. *Am Heart J*. 1985;109:1049–1058.

6. Wong M, Tei C, Shah PM. Sensitivity and specificity of two-dimensional echocardiography in the detection of valvular calcification. *Chest*. 1983;84:423–427.

7. Wong M, Tei C, Sadler N, et al. Echocardiographic observations of calcium in operatively excised stenotic aortic valves. *Am J Cardiol*. 1987;59:324–329.

8. Martin RP, Meltzer RS, Chin BL, et al. Clinical utility of two-dimensional echocardiography in infective endocarditis. *Am J Cardiol*. 1980;46:379–385.

9. Scanlan JG, Seward JB, Tajik AJ. Valve ring abscess in infective endocarditis: Visualization with wide angle two-dimensional echocardiography. *Am J Cardiol*. 1982;49:1794–1800.

10. Zabalgoitia-Reyes M, Mehlman DJ, Talano JV. Persistent fever with aortic valve endocarditis. *Arch Intern Med*. 1985;145:327–328.

11. DeMaria AN, Bommer W, Joye J, et al. Value and limitations of cross-sectional echocardiography of the aortic valve in the diagnosis and quantification of valvular aortic stenosis. *Circulation*. 1980;62:304–312.

12. Weyman AE, Feigenbaum H, Dillon JC, et al. Cross-sectional echocardiography in assessing the severity of valvular aortic stenosis. *Circulation*. 1975;52:828–834.

13. Weyman AE, Feigenbaum H, Hurwitz RA, et al. Cross-sectional echocardiographic assessment of the severity of aortic stenosis in children. *Circulation*. 1977;55:773–778.

14. Hofman T, Kasper W, Meinertz T, et al. Determination of aortic valve orifice area in aortic valve stenosis by two-dimensional transesophageal echocardiography. *Am J Cardiol*. 1987;59:330–335.

15. Sutton MG, Plappert TA, Hirshfeld JW, et al. Assessment of left ventricular mechanics in patients with asymptomatic aortic regurgitation: A two-dimensional echocardiographic study. *Circulation*. 1984;69:259–268.

16. Vandenbossche JL, Kramer BL, Massie BM, et al. Two-dimensional echocardiographic evaluation of the size, function and shape of the left ventricle in chronic aortic regurgitation: Comparison with radionuclide angiography. *J Am Coll Cardiol*. 1984;4:1195–1206.

17. Grossman W, Jones D, McLaurin LP. Wall stress and patterns of hypertrophy in the human left ventricle. *J Clin Invest*. 1975;56:56–64.

18. Aziz KU, van Grondelle A, Paul MH, et al. Echocardiographic assessment of the relation between left ventricular wall and cavity dimension and peak systolic pressure in children with aortic stenosis. *Am J Cardiol*. 1977;40:775–780.

19. Schwartz A, Vignola PA, Walker HJ, et al. Echocardiographic estimation of aortic-valve gradient in aortic stenosis. *Ann Intern Med*. 1978;89:329–335.

20. Assey ME, Wisenbaugh T, Spann JF Jr, et al. Unexpected persistence into adulthood of low wall stress in patients with congenital aortic stenosis: Is there a fundamental difference in the hypertrophic response to a pressure overload present from birth? *Circulation*. 1987;75:973–979.

21. DePace NL, Ren J, Iskandrian AS, et al. Correlation of echocardiographic wall stress and left ventricular pressure and function in aortic stenosis. *Circulation*. 1983;67:854–859.

22. Borow KM, Neumann A, Briller R, et al. Can the modified Bernoulli equation be used to accurately determine intraventricular pressures throughout systole in patients with valvular aortic stenosis? *J Am Coll Cardiol*. 1987;9:236A. Abstract.

23. Borow KM, Green LH, Grossman W, et al. Left ventricular end-systolic stress-shortening and stress-length relations in humans: Normal values and sensitivity to inotropic state. *Am J Cardiol*. 1982;50:1301–1308.

24. Colan SD, Borow KM, Neumann A. The left ventricular end-systolic wall stress-velocity of fiber shortening relation: A load independent index of myocardial contractility. *J Am Coll Cardiol*. 1984;4:715–724.

25. Carabello BA, Green LH, Grossman W, et al. Hemodynamic determinants of prognosis of aortic valve replacement in critical aortic stenosis and advanced congestive heart failure. *Circulation*. 1980;62:42–48.

26. King ME, Braun H, Goldblatt A, et al. Interventricular septal configuration as a predictor of right ventricular systolic hypertension in children: A cross-sectional echocardiographic study. *Circulation*. 1983;68:68–75.

27. Louie EK, Rich S, Brundage BH. Doppler echocardiographic assessment of impaired left ventricular filling in patients with right ventricular pressure overload due to primary pulmonary hypertension. *J Am Coll Cardiol*. 1986;8:1298–1306.

28. Henry WL, Bonow RO, Borer JS, et al. Observations on the optimum time for operative intervention for aortic regurgitation, I: Evaluation of the results of aortic valve replacement in symptomatic patients. *Circulation*. 1980;61:471–483.

29. Henry WL, Bonow RO, Rosing DR, et al. Observations on the optimum time for operative intervention for aortic regurgitation, II: Serial echocardiographic evaluation of asymptomatic patients. *Circulation*. 1980;61:484–492.

30. Bonow RO, Rosing DR, McIntosh CL, et al. The natural history of asymptomatic patients with aortic regurgitation and normal left ventricular function. *Circulation*. 1983;68:509–517.

31. Fioretti P, Roelandt J, Bos RJ, et al. Echocardiography in chronic aortic insufficiency: Is valve replacement too late when left ventricular end-systolic dimension reaches 55 mm? *Circulation*. 1983;67:216–221.

32. Daniel WG, Hood WP Jr, Siari A, et al. Chronic aortic regurgitation: Reassessment of the prognostic value of preoperative left ventricular end-systolic dimension and fractional shortening. *Circulation*. 1985;71:669–680.

33. Fioretti P, Roelandt J, Sclavo M, et al. Postoperative regression of left ventricular dimensions in aortic insufficiency: A long-term echocardiographic study. *J Am Coll Cardiol*. 1985;5:856–861.

34. Szlachcic J, Massie BM, Greenberg B, et al. Inter-test variability of echocardiographic and chest x-ray measurements. Implications for decision making in patients with aortic regurgitation. *J Am Coll Cardiol*. 1986;7:1310–1317.

35. Bonow RO, Picone AL, McIntosh CL, et al: Survival and functional results after valve replacement for aortic regurgitation from 1976 to 1983: Impact of preoperative left ventricular function. *Circulation*. 1985;72:1244–1256.

36. Branzi A, Lolli C, Piovaccari G, et al. Echocardiographic evaluation of the response to afterload stress test in young asymptomatic patients with chronic severe aortic regurgitation: Sensitivity of the left ventricular end-systolic pressure volume relationship. *Circulation*. 1984;70:561–569.

37. Grayburn PA, Smith MD, Handshoe R, et al. Detection of aortic insufficiency by standard echocardiography, pulsed Doppler echocardiography, and auscultation: A comparison of accuracies. *Ann Intern Med*. 1986;104:599–605.

38. Louie EK, Mason TJ, Shah R, Bieniarz T, Moore AM. Determinants of anterior mitral leaflet fluttering in pure aortic regurgitation from pulsed Doppler study of the early diastolic interaction between the regurgitant jet and mitral inflow. *Am J Cardiol*. 1988;61:1085–1091.

39. Botninick EH, Schiller NB, Wickramasekaran R, et al. Echocardiographic demonstration of early mitral valve closure in severe aortic insufficiency: Its clinical implications. *Circulation*. 1975;51:836–847.

40. Meyer T, Sareli P, Pocock WA, et al. Echocardiographic and hemodynamic correlates of diastolic closure of mitral valve and diastolic opening of aortic valve in severe aortic regurgitation. *Am J Cardiol*. 1987;59:1144–1148.

41. Imaizumi T, Orita Y, Koiwaya Y, et al. Utility of two-dimensional echocardiography in the differential diagnosis of the etiology of aortic regurgitation. *Am Heart J*. 1988;115:1118–1119.

42. Tucker CR, Fowles RE, Calin A, et al. Aortitis in ankylosing spondylitis: Early detection of aortic root abnormalities with two-dimensional echocardiography. *Am J Cardiol*. 1982;49:680–686.

43. Matthew T, Nanda NC. Two-dimensional and Doppler echocardiographic evaluation of aortic aneurysm and dissection. *Am J Cardiol*. 1984;54:379–385.

44. Liu MW, Louie EK, Levitsky S. Color flow Doppler assessment of aortic regurgitation complicated by aneurysmal dilation and dissection of the ascending aorta in the Marfan syndrome. *Am J Med*. 1988;115:1118–1119.

45. Vitarelli A, Landolina G, Gentile R, et al. Echocardiographic assessment of congenital mitral stenosis. *Am Heart J*. 1984;108:523–531.

46. Grenadier E, Sahn DJ, Valdes-Cruz LM, et al. Two-dimensional echo Doppler study of congenital disorders of the mitral valve. *Am Heart J*. 1984;107:319–325.

47. DiSegni E, Bass JL, Lucas RV Jr, et al. Isolated cleft mitral valve: A variety of congenital mitral regurgitation identified by two-dimensional echocardiography. *Am J Cardiol.* 1983;51:927–931.

48. Trowitzsch E, Bano RA, Burger BM, et al. Two-dimensional echocardiographic findings in double orifice mitral valve. *J Am Coll Cardiol.* 1985;6:383–387.

49. Nihoyannopoulos P, McKenna WJ, Smith G, et al. Echocardiographic assessment of the right ventricle in Ebstein's anomaly: Relation to clinical outcome. *J Am Coll Cardiol.* 1986;8:627–635.

50. Gussenhoven EJ, Stewart PA, Becker AE, et al. "Off setting" of the septal tricuspid leaflet in normal hearts and in the hearts with Ebstein's anomaly. *Am J Cardiol.* 1984;54:172–176.

51. Dillon JC, Hains CL, Chang S, et al. Use of echocardiography in patients with prolapsed mitral valve. *Circulation.* 1971;43:503–507.

52. Kerber RE, Lsaeff DM, Hancock EW. Echocardiographic patterns in patients with the syndrome of systolic click and late systolic murmur. *N Engl J Med.* 1971;284:691–693.

53. Popp RL, Brown OR, Silverman JF, et al. Echocardiographic abnormalities in the mitral valve prolapse syndrome. *Circulation.* 1974;49:428–435.

54. Haikal M, Alpert MA, Whiting RB, et al. Sensitivity and specificity of M-mode echocardiographic signs of mitral valve prolapse. *Am J Cardiol.* 1982;50:185–190.

55. Waller BF, Maron BJ, Del Negro AA, et al. Frequency and significance of M-mode echocardiographic evidence of mitral valve prolapse in clinically isolated pure mitral regurgitation: Analysis of 65 patients having mitral valve replacement. *Am J Cardiol.* 1984;53:139–147.

56. Savage DD, Devereux RB, Garrison RJ, et al. Mitral valve prolapse in the general population, 2: Clinical features: The Framingham study. *Am Heart J.* 1983;106:577–581.

57. Devereux RB, Kramer-Fox R, Brown WT, et al. Relation between clinical features of the mitral prolapse syndrome and echocardiographically documented mitral valve prolapse. *J Am Coll Cardiol.* 1986;8:763–772.

58. Nishimura RA, McGoon MD, Shub C, et al. Echocardiographically documented mitral valve prolapse: Long-term follow-up of 237 patients. *N Engl J Med.* 1985;313:1305–1309.

59. Alpert MA, Carney RJ, Flakes GC, et al. Sensitivity and specificity of two-dimensional echocardiographic signs of mitral valve prolapse. *Am J Cardiol.* 1984;54:792–796.

60. Warth DC, King ME, Cohen JM, et al. Prevalence of mitral valve prolapse in normal children. *J Am Coll Cardiol.* 1985;5:1173–1177.

61. Levine RA, Triulzi MO, Harrigan P, et al. The relationship of mitral annular shape to the diagnosis of mitral valve prolapse. *Circulation.* 1987;75:756–767.

62. Kligfield P, Hochreiter C, Kramer H, et al. Complex arrhythmias in mitral regurgitation with and without mitral valve prolapse: contrast to arrhythmias in mitral valve prolapse without mitral regurgitation. *Am J Cardiol.* 1985;55:1545–1549.

63. Panidis IP, McAllister M, Ross J, et al. Prevalence and severity of mitral regurgitation in the mitral valve prolapse syndrome: a Doppler echocardiographic study of 80 patients. *J Am Coll Cardiol.* 1986;7:975–981.

64. Perloff JK, Child JS, Edwards JE. New guidelines for the clinical diagnosis of mitral valve prolapse. *Am J Cardiol.* 1986;57:1124–1129.

65. Vardi P, Markiewicz W, Weiss Y, et al. Clinical-echocardiographic correlations in acute rheumatic fever. *Pediatrics.* 1983;71:830–834.

66. Edler I. The diagnostic use of ultrasound in heart disease. *Acta Med Scand Suppl.* 1955;308:32–36.

67. Duchak JM Jr, Chang S, Feigenbaum H. The posterior mitral valve echo and the echocardiographic diagnosis of mitral stenosis. *Am J Cardiol.* 1972;29:628–632.

68. Nichol PM, Gilbert BW, Kisslo JA. Two-dimensional echocardiographic assessment of mitral stenosis. *Circulation.* 1977;55:120–128.

69. Henry WL, Griffith JM, Michaelis LL, et al. Measurement of mitral orifice area in patients with mitral valve disease by real-time two-dimensional echocardiography. *Circulation.* 1975;51:827–835.

70. Wann LS, Weyman AE, Feigenbaum H, et al. Determination of mitral valve area by cross-sectional echocardiography. *Ann Intern Med.* 1978;88:337–341.

71. Martin RP, Rakowski H, Kleiman JH, et al. Reliability and reproducibility of two-dimensional echocardiographic measurement of the stenotic mitral valve orifice area. *Am J Cardiol.* 1979;43:560–568.

72. Smith MD, Handshoe R, Handshoe S, et al. Comparative accuracy of two-dimensional echocardiography and Doppler pressure half-time methods in assessing severity of mitral stenosis in patients with and without prior commissurotomy. *Circulation.* 1986;73:100–107.

73. Reid C, McKay C, Chandraratna P, et al. Mechanism of increase in mitral valve area by double balloon catheter balloon valvuloplasty in adults with mitral stenosis: Echocardiographic-Doppler correlation. *J Am Coll Cardiol.* 1987;9:217A. Abstract.

74. Korn D, DeSanctis RW, Sell S. Massive calcification of the mitral valve ring. *N Engl J Med.* 1962;267:900–909.

75. D'Cruz I, Panetta F, Cohen H, et al. Submitral calcification or sclerosis in elderly patients: M-mode and two-dimensional echocardiography in "mitral annulus calcification." *Am J Cardiol.* 1979;44:31–38.

76. Osterberger LE, Goldstein S, Khaja F, et al. Functional mitral stenosis in patients with massive mitral annular calcification. *Circulation.* 1981;64:472–476.

77. Mellino M, Salcedo EE, Lever HM, et al. Echocardiographically quantified severity of mitral annulus calcification: Prognostic correlation to related hemodynamic, valvular, rhythm, and conduction abnormalities. *Am Heart J.* 1982;103:222–225.

78. Labowitz AJ, Nelson JG, Windhorst DM, et al. Frequency of mitral valve dysfunction from mitral annular calcium as detected by Doppler echocardiography. *Am J Cardiol.* 1985;55:133–137.

79. Kisslo J, von Ramm OT, Haney R, et al. Echocardiographic evaluation of tricuspid valve endocarditis: an M-mode and two-dimensional study. *Am J Cardiol.* 1976;38:502–507.

80. Come PC, Kurland GS, Vine HS. Two-dimensional echocardiography in differentiating right atrial and tricuspid valve mass lesions. *Am J Cardiol.* 1979;44:1207–1212.

81. Sigueira-Filho AG, Cunha CLP, Tajik AJ, et al. M-mode and two-dimensional echocardiographic features in cardiac amyloidosis. *Circulation.* 1981;63:188–196.

82. Hongo M, Ikeda S. Echocardiographic assessment of the evaluation of amyloid heart disease: A study with familial amyloid polyneuropathy. *Circulation.* 1986;73:249–256.

83. Falk RH, Plehn JF, Decring T, et al. Sensitivity and specificity of the echocardiographic features of cardiac amyloidosis. *Am J Cardiol.* 1987;59:418–422.

84. Gottdiener JS, Maron BJ, Schooley RT, et al: Two-dimensional echocardiographic assessment of the idiopathic hypereosinophilic syndrome: Anatomic basis of mitral regurgitation and peripheral embolization. *Circulation.* 1983;67:572–578.

85. Cullahan JA, Wroblewski EM, Reeder GS, et al. Echocardiographic features of carcinoid heart disease. *Am J Cardiol.* 1982;50:762–768.

86. Mintz GS, Kotler MN, Parry WR, et al. Statistical comparison of M-mode and two-dimensional echocardiographic diagnosis of flail mitral leaflets. *Am J Cardiol.* 1980;45:253–259.

87. DePace NL, Mintz GS, Ren JF, et al. Natural history of the flail mitral leaflet syndrome: A serial two-dimensional echocardiographic study. *Am J Cardiol.* 1983;52:789–795.

88. Ogawa S, Hubbard FE, Mardelli TJ, et al. Cross-sectional echocardiographic spectrum of papillary muscle dysfunction. *Am Heart J.* 1979;97:312–321.

89. Godley RW, Wann SL, Rogers EW, et al. Incomplete mitral leaflet closure in patients with papillary muscle dysfunction. *Circulation.* 1981;63:565–571.

90. Kinney EL, Frangi MJ. Value of two-dimensional echocardiographic detection of incomplete mitral leaflet closure. *Am Heart J.* 1985;109:87–90.

91. Loperfido F, Pennestri F, Digaetano A, et al. Assessment of left atrial dimensions by cross-sectional echocardiography in patients with mitral valve disease. *Br Heart J.* 1983;50:570–578.

92. Gehl LG, Mintz GS, Kotler MN, et al. Left atrial volume overload in mitral regurgitation: A two-dimensional echocardiographic study. *Am J Cardiol.* 1982;49:33–38.

93. Ren JF, Kotler MN, DePace NL, et al. Two-dimensional echocardiographic determination of left atrial emptying volume: A noninvasive index in quantifying the degree of non-rheumatic mitral regurgitation. *J Am Coll Cardiol.* 1983;2:729–736.

94. Tei C, Tanaka H, Nakoo S, et al. Motion of the interatrial septum in acute mitral regurgitation: Clinical and experimental echocardiographic studies. *Circulation.* 1980;62:1080–1088.

95. Henry WL, Morganroth J, Pearlman AS, et al. Relation between echocardiographically determined left atrial size and atrial fibrillation. *Circulation.* 1976;53:273–279.

96. Helgason CM, Chomka E, Louie E, et al. The potential role for ultrafast cardiac computed tomography in patients with stroke. *Stroke.* 1989;20:465–472.

97. Beppu S, Nimura Y, Sakakibara H, et al. Smoke-like echo in the left atrial cavity in mitral valve disease: Its features and significance. *J Am Coll Cardiol.* 1985;6:744–749.

98. Iliceto S, Antonelli G, Sorino M, et al. Dynamic intracavitary left atrial echoes in mitral stenosis. *Am J Cardiol.* 1985;55:603–606.

99. Daniel WG, Nellesson U, Nonnast-Daniel B, et al. Left atrial spontaneous echo contrast in mitral valve disease: An indicator of increased thromboembolic risk. *J Am Coll Cardiol.* 1986;7:31A. Abstract.

100. DePace NL, Ren JF, Kotler MN, et al. Two-dimensional echocardiographic determination of right atrial emptying volume: A noninvasive index in quantifying the degree of tricuspid regurgitation. *Am J Cardiol.* 1983;52:525–529.

101. Schuler G, Peterson KL, Johnson A, et al. Temporal response of left ventricular performance to mitral valve surgery. *Circulation.* 1979;59:1218–1231.

102. Zile WR, Gaasch WH, Carroll JD, et al. Chronic mitral regurgitation: Predictive value of preoperative echocardiographic indexes of left ventricular function and wall stress. *J Am Coll Cardiol.* 1984;3:235–242.

103. Zile MR, Gaasch WH, Levine HJ. Left ventricular stress-dimension-shortening relations before and after correction of chronic aortic and mitral regurgitation. *Am J Cardiol.* 1985;56:99–105.

104. Meltzer RS, Van Hoogenhuyze D, Serruys PW, et al. Diagnosis of tricuspid regurgitation by contrast echocardiography. *Circulation.* 1981;63:1093–1099.

105. Wise NK, Myers S, Fraker TD, et al. Contrast M-mode ultrasonography of the inferior vena cava. *Circulation.* 1981;63:1100–1109.

106. Goldman ME, Mindich BP, Teichholz LE, et al. Intraoperative contrast echocardiography to evaluate mitral valve operations. *J Am Coll Cardiol.* 1984;4:1035–1040.

Chapter 22

Doppler Echocardiographic Evaluation of Valvular Heart Disease

Eric K. Louie, MD

Prior to the development and widespread integration of Doppler techniques into the echocardiographic examination, assessment of valvular function was limited to inferences drawn from imaging of valvular structure (including measurement of anatomic orifice size) and quantitation of the secondary morphologic effects of abnormal pressure and volume loading on the cardiac chambers (see Chapter 21). Spectral analysis of the Doppler frequency shift resulting from the interaction between the interrogating Doppler ultrasound beam and moving blood has now provided a method of characterizing normal and abnormal blood flow velocities in patients with valvular heart disease. In addition to the qualitative recognition of abnormal Doppler spectral patterns resulting from disturbed blood flow and their distinction from normal patterns, several characteristics of intracardiac blood flow can now be quantitated: (1) the magnitude (speed) of blood flow, (2) the direction of blood flow, (3) the degree of "organization" or "disturbance" of the flow regimen, (4) the timing of flow velocity transients relative to the motion of cardiac structures or the electrocardiogram, and (5) the spatial distribution within the cardiac chambers of abnormal flow velocities. Although this chapter focuses on the use of these Doppler measurements to characterize valvular heart disease, in clinical practice the performance and interpretation of the Doppler study are inextricably tied to the concurrent two-dimensional and M-mode echocardiographic examinations (see Chapter 21).

Comprehensive assessment of a patient with valvular heart disease requires the integration of blood flow data from the Doppler examination with morphologic data from the echocardiographic examination. In addition, both performance and interpretation of the Doppler examination often are facilitated by two-dimensional echocardiographic guidance of the Doppler beam and spatial localization of the flow disturbance observed relative to the cardiac structures. In this chapter the use of those techniques to detect and quantitate obstruction and incompetence of the cardiac valves is explored with particular reference to the aortic and mitral valves (although many of the principles may be generalized to the evaluation of right-sided cardiac valves), and the application of these principles to the evaluation of bioprosthetic and mechanical prosthetic heart valve functions is examined briefly.

BLOOD FLOW VELOCITY MEASUREMENT BY THE DOPPLER PRINCIPLE

To a first approximation we can consider the moving red blood cells as scatterers or reflectors of the interrogating Doppler ultrasound signal that impart a small shift in the frequency of the signal, referred to as the Doppler frequency shift, f_D. For a given carrier frequency, f, of the interrogating ultrasound signal, f_D is

---

Supported in part by College of Medicine, Committee on Research Grant No. 48612, University of Illinois.

related to the magnitude of the velocity of blood flow by the equation

$$f_D = 2f\left(\frac{v \cos \theta}{c}\right) \quad (1)$$

where c is the speed of sound in biologic tissues (approximately 1560 m/s) and θ is the angle of incidence between the interrogating Doppler ultrasound beam and the direction of blood flow. From this relationship we can appreciate that the Doppler frequency shift is directly proportional to the component of blood flow velocity parallel to the interrogating Doppler beam, $v \cos \theta$. Accordingly, blood flow parallel to the ultrasound beam but moving away from the ultrasound source (cos [180°] = −1) will result in a negative Doppler shift, while blood flow moving toward the ultrasound source (cos [0°] = 1) will result in a positive Doppler shift. Blood flow perpendicular to the interrogating beam (cos [90°] = 0) will result in no Doppler frequency shift.

Even under conditions of undisturbed blood flow, these ultrasound reflectors do not all move at a uniform speed or in a common direction within any finite sample volume. Thus the composite Doppler shift signal from the sampled volume of blood flow represents only a statistical approximation of the distribution of instantaneous velocities of these ultrasound reflectors. Spectral analysis (for example, fast Fourier transform analysis) is required to sort out this composite signal so that the frequency distribution of Doppler shifts can be characterized. For instance, Figure 22-1 illustrates a pulsed Doppler spectrum (plotted as velocity [or as f_D] versus time) sampled from the left ventricular outflow tract with the Doppler transducer (ultrasound source) positioned at the left ventricular apex. During systole (simultaneous electrocardiographic rhythm strip) the spectrum assumes an inverted parabolic waveform below the zero baseline, representing negative Doppler shifts from systolic blood flow moving away from the transducer. During diastole no Doppler signals are recorded, reflecting the absence of flow across the aortic valve during this phase of the cardiac cycle. The envelope of this pulsed Doppler spectrum reflects the instantaneous distribution of Doppler frequency shifts, and the width of the envelope represents the range of Doppler frequency shifts detected. The intensity of the spectrum is proportional to the number of ultrasound reflectors (roughly speaking, the number of red blood cells) contributing to the particular Doppler shift. Hence the time-varying modal Doppler frequency shift is represented by the most intense (in this case the brightest) portion of the spectral envelope. The magnitude of the time-varying modal blood flow velocity can be computed directly from this waveform by using Equation 1, assuming that the probe is aligned with the flow of blood. The direction of blood flow is indicated by the sign (in this case negative) of the Doppler fre-

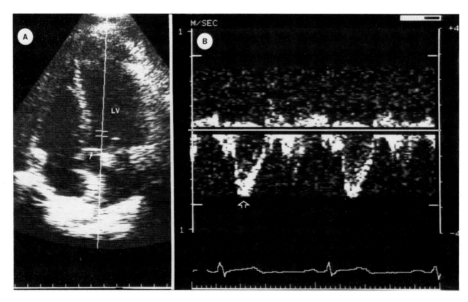

Figure 22-1 Pulsed Doppler measurement of velocities of the left ventricular outflow tract from an apical window in a normal subject. A, Stop-frame view of an apical four-chamber two-dimensional echocardiogram with anterior angulation to image the aortic valve. The Doppler cursor, represented by a line running from the apex through the left ventricular outflow tract, indicates the direction of the interrogating ultrasound beam. The double crosshatch marks delineate the position of the sample volume in the left ventricle (LV) just proximal to the aortic valve (arrow). B, Fast Fourier transform of the pulsed Doppler-shifted frequencies plotted as velocity (±1 m/s on the left) or Doppler frequency shift (± 4 kHz on the right) versus time (full sweep = 2 seconds). The electrocardiogram is presented beneath the spectrum. During systole a negative parabolic flow velocity envelope representing blood flow moving away from the transducer (open arrow) is recorded; no flow velocity signals are recorded during diastole.

quency shift, and the waveform can be timed easily relative to the simultaneous electrocardiogram.

In this example from a normal person the blood flow is well organized and reasonably uniform, so that all ultrasound reflectors are traveling at similar rates and directions, resulting in a narrow range of detectable Doppler frequency shifts. Accordingly, the width of the spectral waveform is narrow and has the appearance of a discrete envelope. By contrast, in abnormal situations that result in unsteady flow conditions, the velocity regimen may become disorganized and chaotic, resulting in an increased range of detectable Doppler frequency shifts (spectral broadening) such that the spectral envelope becomes wider and less well defined. As we shall see, these conditions will pertain when turbulent intracardiac blood flow is generated. Finally, the location of the region sampled for the Doppler measurements, ascertained from concurrent two-dimensional echocardiography, places the observed flow regime in an appropriate spatial context with respect to the valves and chambers.

In the assessment of valvular heart disease we are particularly interested in the changes in blood flow velocity that occur at localized obstructions or constrictions. At a constant volume rate of blood flow, Q, the velocity of flow, v, and the cross-sectional area of flow, A, are related.

$$Q = A \cdot v \tag{2}$$

If a narrowed or stenotic segment (with a smaller cross-sectional area) is interposed (for example, a stenotic valve) in the path of flow, the velocity of the blood flow in the stenotic segment, v_2, must increase relative to the velocity of the blood flow proximal to the stenosis, v_1, since the cross-sectional flow area and the flow velocity are inversely related (Equation 2). A pressure head is required to drive the blood flow across this obstruction such that the pressure proximal to the stenosis, P_1, is greater than the pressure distal to the stenosis, P_2. The dissipation of potential energy represented by this pressure gradient, $P_1 - P_2$, is accounted for by three factors: (1) increased kinetic energy required for the higher velocities of blood flow in this stenotic segment, (2) local inertial losses encountered during acceleration and deceleration of pulsatile flow through the stenotic segment, and (3) viscous frictional losses dissipated in the stenotic segment. These factors are summarized in the Bernoulli equation[1]:

$$P_1 - P_2 = \tfrac{1}{2}\rho\left(v_2^2 - v_1^2\right) + \rho\int_1^2 \frac{dv}{dt}ds + R(v) \tag{3}$$
$$\text{Kinetic losses} \quad \text{Inertial losses} \quad \text{Frictional losses}$$

where ρ is the mass density of blood, dv/dt is the acceleration (or deceleration) of blood through a finite displacement (ds), and $R(v)$ is frictional loss expressed as a function of blood flow velocity. In the practical application of these principles to the assessment of blood flow across stenotic valves in humans,[2-4] the frictional and inertial losses are neglected (as they are small) and v_1 is assumed to be much smaller than v_2 (which is generally true), such that $P_1 - P_2 = \tfrac{1}{2}\rho v_2^2$, or (for velocity expressed in meters per second and pressure expressed in millimeters of mercury):

$$P_1 - P_2 = 4v_2^2 \tag{4}$$

Since v_2 is measurable from the Doppler frequency shift of blood flow through the stenotic orifice, an estimate of the pressure drop across constricted segments can be obtained.

DOPPLER ECHOCARDIOGRAPHIC INSTRUMENTATION

There are two principal types of Doppler instrumentation: pulsed wave and continuous wave. By analogy to M-mode echocardiography, pulsed-wave Doppler transmits short bursts of signal at a given pulse repetition frequency and the returning Doppler-shifted reflected ultrasound signal is sampled once for each pulse transmission. Thus the location of the origin of the Doppler-shifted signal (range gate depth) is uniquely determined because it can be calculated from the speed of sound in biologic tissues and the known time elapsed between signal transmission and reception. By contrast, the continuous-wave Doppler mode employs two piezoelectric crystals, one that transmits constantly and another that receives reflected signals continuously. Under these circumstances it is not possible to determine uniquely the depth from which the Doppler-shifted signal originates because the time from transmission of a given signal to reception of the *corresponding* reflected signal is unknown. Accordingly the pulsed-wave Doppler technique affords range specificity as the sampling depth along the ultrasound beam can be controlled, whereas the continuous-wave Doppler technique is ambiguous with respect to range and consequently Doppler frequency shifts from moving blood all along the ultrasound beam are included in the composite signal. Thus pulsed-wave Doppler provides precise spatial localization of flow velocity measurements, whereas continuous-wave Doppler only localizes the measurement to the path of the Doppler beam.

There is a trade-off, however, for the range specificity inherent in the pulsed Doppler mode. Sampling from sites at increasing depths from within the thorax requires a longer delay between ultrasound signal transmission and reception (to allow for the greater distance the signal must travel) and hence pulse repetition frequency, PRF, decreases. It can be shown that the maximal Doppler-shifted frequency, f_D, which can be determined unambiguously, is equal to ½ (PRF) (referred to as the Nyquist limit).[1] Thus as pulse repetition frequency declines as a result of increasing range, the maximal unambiguously detectable Doppler-shifted frequency (and hence flow velocity) declines. Doppler-shifted frequencies exceeding the Nyquist limit result in signal aliasing such that the corresponding flow velocity cannot be measured. Unfortunately, in adults the depth of cardiac structures from the chest wall results in a Nyquist limit for pulsed-wave Doppler that does not permit quantitation of Doppler frequency shifts from high-velocity jets due to valvular obstruction because of the phenomenon of aliasing. By contrast, continuous-wave Doppler is not subject to the same limitations and in theory is not restricted as to the maximal detectable Doppler frequency shift. Thus the high blood flow velocities encountered in valvular heart disease are measurable by continuous-wave Doppler at the expense of localization of range. The two modes of operation are complementary: A complete Doppler examination requires pulsed-wave Doppler to localize a flow velocity disturbance relative to cardiac structures and continuous-wave Doppler to quantitate the high velocities in the flow disturbance.

High-pulse repetition frequency Doppler is a conceptual hybrid between pulsed-wave and continuous-wave Doppler. For sampling at a given range depth, conventional pulsed Doppler is constrained to a pulsed repetition frequency that permits sufficient transit time for the ultrasound signal to travel to the target and return to the transducer. In high-pulse repetition frequency Doppler, that pulse repetition frequency is increased above this limit to enable the measurement of peak velocities in excess of the Nyquist limit for the conventional system. If, for instance, the pulse repetition frequency is increased by a multiple, n, the maximal velocity that can be measured without signal aliasing also will be increased by n. However, as a trade-off for the ability to measure higher velocities, range ambiguity is introduced into the system as now there are n potential sample volumes from which the Doppler frequency shift may be coming. As discussed above, both the ability to localize the origin of a Doppler signal and the measurement of the corresponding peak velocity are important in evaluating valvular heart disease. Whether these goals are best met by performing separate pulsed- and continuous-wave Doppler studies on a given patient or by performing a pulsed Doppler examination enhanced with high-pulse repetition frequency capability remains to be determined.

Nonimaging, continuous-wave Doppler probes (for example, a Pedoff probe), which provide continuous-wave Doppler capability without localization of the direction of the beam relative to an echocardiographic image of the heart, rely on pattern recognition of the audible Doppler signal and Doppler spectrum to distinguish which normal or abnormal intracardiac flows are being interrogated. The primary advantage of this technology is that the transducer contact surface is considerably smaller than that found on Doppler probes that also perform two-dimensional echocardiographic imaging. Consequently, this transducer is ideally suited for several small acoustic windows (suprasternal notch, supraclavicular fossae, right parasternal window) where limited anatomic space and a high degree of transducer maneuverability are desirable. This capability is of particular value in the interrogation of the stenotic aortic valve.

The most recent technologic advance in the armamentarium of the Doppler echocardiographer is color flow Doppler imaging. Just as two-dimensional echocardiography displays a composite tomographic image obtained from multiple M-mode echocardiographic interrogations within the imaging plane, so also color flow Doppler images may be conceptualized as the composite tomographic display obtained from multiple pulsed Doppler interrogations within the plane. The analogy is not quite perfect because two-dimensional echocardiography displays three variables (two spatial coordinates and echo intensity) changing in real time, whereas color flow Doppler displays the real-time changes in four variables (two spatial coordinates, speed of blood flow, and direction of blood flow). To accommodate these variables, the direction of blood flow is represented by its color (commonly red for flow directed toward the transducer and blue for flow directed away from the transducer), and the speed of blood flow is represented by the brightness of the color flow signal. In order to present information roughly analogous to the spectral broadening that can be seen on pulsed-wave Doppler, some color flow Doppler formats include the admixture of a third color (green) to the signal as a reflection of the degree of disorder (or variance) in the Doppler-shifted frequency distribution. To achieve these features the color flow Doppler technique uses an algorithm (for example, an autocorrelation algorithm) to map spatially the color-encoded Doppler-determined flow velocities onto a nearly simultaneous two-dimensional echocardiographic image of the heart.

The complexity of the technology and the magnitude of the data acquisition and processing tasks are not without their drawbacks. In prototype models, significant two-dimensional echocardiographic image degradation (poor spatial resolution), low frame rates (poor temporal resolution), low pulse repetition frequency (signal aliasing at physiologic blood flow velocities), and relative insensitivity to low velocities are problems that are being addressed actively. In addition, it has been difficult to design Doppler echocardiographic equipment optimized to perform color flow Doppler studies as well as conventional Doppler studies and conventional echocardiographic imaging. In this chapter we will touch upon new approaches to evaluating valvular heart disease offered by the color flow Doppler technique with the expectation that improved technology, new computational algorithms, and enhanced display formats will evolve from these rudimentary beginnings.

STENOSIS AND INCOMPETENCE OF THE SEMILUNAR VALVES

Aortic and Pulmonic Stenosis

In patients with valvular aortic stenosis a high-velocity jet is created as the streamlines of blood flow converge to pass through the constricted orifice. Pulsed-wave Doppler can be used to demonstrate this marked increase in velocities as one moves the sample volume from just proximal to the valve to just distal to the valve; it may be helpful also in localizing and comparing the relative magnitude of serial obstructions to left ventricular outflow, as might be encountered in a patient with valvular as well as subvalvular (or supravalvular) obstruction. Usually the two-dimensional echocardiogram provides sufficient anatomic clues to localize the level of obstruction so that the primary purpose of the Doppler examination is to characterize the hemodynamic severity of the lesion. Pulsed Doppler mapping of the spatial extent of the systolic jet of aortic stenosis (the cross-sectional systolic flow area) has been attempted from short-axis images at the level of the aortic orifice.[5] While the approach is conceptually appealing, the technical demands of mapping the systolic flow area of an aortic stenotic jet (range 0.2 to 0.8 cm^2) with a sample volume measuring 0.4×0.2 cm are unlikely to be realized outside a dedicated research setting. Color flow Doppler might be expected to facilitate the task of defining the spatial extent and direction of the aortic stenotic jet, but in practice the technique has had limited success in imaging these relatively small, eccentrically directed, flow velocity disturbances.

In patients with hemodynamically significant aortic stenosis, the peak transvalvular velocities exceed the Nyquist limit of pulsed Doppler systems, resulting in signal aliasing and inability to measure the velocities. High-pulse repetition frequency and continuous-wave Doppler systems have been used to characterize the time-varying outflow velocities. Not only are peak flow velocities increased, but the duration of forward flow velocity (left ventricular ejection time) and the time from the onset of systole to peak flow velocity are progressively prolonged with increasing severity of aortic stenosis.[1] Downstream from the stenotic orifice as flow reconstitutes and velocity declines significant turbulence is generated, the magnitude of which can be quantitated by measuring the spectral broadening of the flow velocity profile downstream of the valve. In a model pulse duplicator system with simulated aortic stenosis, high-pulse repetition frequency Doppler has been used to permit measurement of the systolic spectral envelope area, an index of spectral turbulence, free from signal aliasing.[6] The systolic spectral area was linearly proportional to the mean systolic pressure gradient across the stenotic orifice ($r = 0.96$) and to the effective valve area ($r = 0.94$). Since this technique requires sampling downstream from the valve to identify the zone where blood flow shows the greatest turbulence, its application in vivo may be limited to situations where detailed interrogation of the proximal aortic root is feasible. Nonetheless it demonstrates the potential for extracting useful information from the spectral characteristics of the flow velocity envelope.

By far and away the most direct approach to quantitating the hemodynamic severity of aortic stenosis by Doppler techniques involves the use of continuous-wave Doppler to measure the maximal transaortic flow velocities.[7-14] From Equation 1 it is evident that computation of flow velocity from the measured Doppler frequency shift requires the knowledge of the angle, θ, between the Doppler beam and the stenotic jet. In practice this angle cannot be measured because the jets emerge from the stenotic orifice without a predictable orientation relative to such landmarks as the aortic walls or the plane of the aortic annulus. Instead, the echocardiographer must manipulate the transducer until the highest reproducibly recorded Doppler signal is found and assume that the ultrasound beam is nearly parallel to the jet such that θ is negligible (and hence cos [0°] = 1). To achieve this it is imperative that all available ultrasonic (apical, right parasternal, supraclavicular, and suprasternal) windows available for interrogation of the stenotic jet be utilized to increase the chances of sampling the maximal velocities in the jet.[10] While the continuous-wave Doppler ultrasound beam will sample Doppler shifts all along its path, it is assumed implicitly that the maximal velocities intercepted represent those emanating from the aortic valve orifice. In this regard in patients with multivalvular disease it is important to ascertain that the beam is not directed in the path of some other high-velocity systolic jet (eg, mitral regurgitation). Since there are no convenient means of validating these velocity measurements directly in vivo in humans, these velocity measurements have been related to the transaortic pressure gradient measured at the time of cardiac catheterization by using Equation 4.

Linear correlations ($r = 0.79$ to 0.94) have been demonstrated for comparisons of Doppler-derived pressure gradients with the *nonsimultaneous* measurements of transaortic pressure gradients obtained by cardiac catheterization.[8,9,12] In general the results are superior in children with congenital aortic stenosis,[8] which may relate to several factors, including (1) the geometry of the orifice and resultant jet, (2) the greater accessibility of ultrasonic windows in children, and (3) the shallower depths of cardiac structures in the smaller thorax of a child.

A detailed study of *simultaneously* obtained catheter measurements of transaortic pressure gradients and pressure gradients derived from continuous-wave Doppler data measured in 100 adult patients (50 to 89 years old) demonstrated an excellent correlation between invasively determined and Doppler-derived transaortic pressure gradients.[14] Not only was there excellent agreement between invasive and Doppler-derived gradients in comparisons of patients with varying degrees of aortic stenosis (Figure 22-2), but there was close tracking of the beat-to-beat alterations in the systolic pressure gradient in a given patient. Of particular importance is the emphasis placed in this study on the examination of *physiologically comparable* data obtained by catheterization and Doppler techniques. All too often investigators have attempted to compare the *maximal instantaneous pressure gradient* calculated from peak Doppler flow velocities with *peak-to-peak pressure gradient* measured during catheterization. Although the latter measurement is a commonly used clinical index, it is not the same as (and generally underestimates) the maximal instantaneous pressure gradient because the peak aortic pressures and left ventricular pressures used to measure the peak-to-peak gradient are not simultaneous events. When maximal instantaneous pressure gradients measured directly by catheterization and derived from Doppler data are compared, the correlation is excellent ($r = 0.92$).[14]

Perhaps of greater physiologic significance is the comparison of mean systolic pressure gradients, which are measured by the incremental application of Equation 4 at successive intervals during the systolic ejection period with subsequent averaging of the results:

$$P_1 - P_2 = \frac{4}{n} \sum_{i=1}^{n} (v_2)_i^2 \qquad (5)$$

where $P_1 - P_2$ is the mean pressure gradient, n is the number of increments chosen for the iterative calculation over the systolic ejection period, and v_2 is the instantaneous peak velocity at each increment. With these techniques the linear correlation between simultaneously determined catheter-derived (ΔP_C) and Doppler-derived (ΔP_D) mean pressure gradients was excellent ($\Delta P_C = 0.98 \Delta P_D + 5.2$, standard error of the estimate (SEE) = 10 mm Hg, $r = 0.93$) with a slope approaching unity and a y-intercept near the origin.[14]

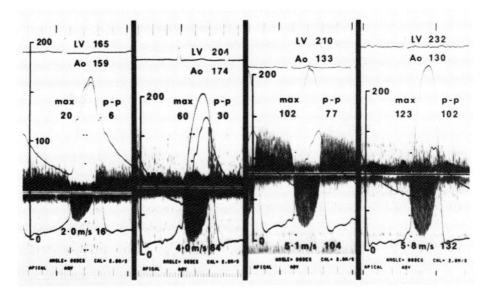

Figure 22-2 Simultaneous continuous-wave Doppler spectra of transaortic flow velocities and dual-catheter aortic pressure gradient measurements in four patients with different degrees of aortic stenosis. Tracings from top to bottom in each panel are those of the electrocardiogram, left ventricular pressure, ascending aortic pressure, and continuous-wave Doppler spectra. The upper half of each panel is annotated with the peak left ventricular pressure (LV), peak ascending aortic pressure (Ao), maximal aortic pressure gradient (max), and peak-to-peak aortic pressure gradient (p-p). The lower half of each panel is annotated with the peak transaortic flow velocity and the Doppler-derived estimate of the maximal aortic pressure gradient. The maximal Doppler-derived gradients reflect accurately the simultaneously recorded maximal catheter gradient but exceed the peak-to-peak catheter gradient. *Source:* Reprinted with permission from *Circulation* (1985;71:1162–1169), Copyright © 1985, American Heart Association.

In theory, high-pulse repetition frequency Doppler technology should be capable of producing results similar to those reported for continuous-wave Doppler. This expectation has not been fulfilled entirely, however; when compared with continuous-wave Doppler the high-pulse repetition frequency technique significantly underestimates peak velocities (particularly in patients with more severe aortic stenosis) and correlates less well with catheterization data.[15]

Quantitation of transaortic flow velocity and pressure gradients, whether by Doppler or catheterization techniques, is insufficient to characterize fully the severity of left ventricular obstruction because at any given orifice size the pressure drop across the valve and the velocity of blood flow are dependent on the volume rate of flow (that is, cardiac output). There are two potential approaches to measuring effective aortic valve area with Doppler techniques: (1) substitution of Doppler-derived mean transaortic pressure gradients into the Gorlin equation[16–20] and (2) direct application of the continuity equation.[18,20–24] The Gorlin equation,[16,17]

$$\text{Orifice area} = K \frac{\overline{Q}}{\sqrt{\overline{P_1 - P_2}}} \quad (6)$$

where K is an empiric constant, $\overline{Q}$ is mean transvalvular flow, and $\overline{P_1 - P_2}$ is mean transvalvular gradient, neglects blood flow velocity proximal to the valve and local acceleration effects—assumptions also made when $\overline{P_1 - P_2}$ is calculated from Doppler-derived velocity data. Thus by assuming a constant value for K and substituting Doppler-derived values for $\overline{P_1 - P_2}$, orifice area can be computed provided that $\overline{Q}$ is known. The latter can be measured with an independent technique (thermodilution or Fick technique) or estimated by Doppler echocardiography, since the average systolic flow rate ($\overline{Q}$) can be measured at a reference site (for example, the left ventricular outflow tract) from knowledge of the cross-sectional flow area (A), estimated anatomically from the two-dimensional echocardiogram; the systolic flow velocity integral (FVI), measured by pulsed Doppler sampling in the left ventricular outflow tract; and the systolic ejection period (SEP), measured from the FVI:

$$\overline{Q} = \frac{(A)(FVI)}{(SEP)} \quad (7)$$

By employing these principles several investigators have measured aortic and pulmonic valve area by substituting either Doppler- or catheterization-derived mean pressure gradients into the Gorlin equation and have achieved reasonably linear proportionality between these measurements ($r = 0.87$ to 0.96),[18–20] even when the systolic flow rate was measured by Doppler echocardiographic techniques.[19,20]

As an alternative to substitution of Doppler-derived mean pressure gradients into the Gorlin equation, the aortic valve area can be calculated from Doppler data by direct substitution of flow velocities measured at the aortic orifice (AO) and at a reference site (R) into the continuity equation[21]:

$$(A_{AO})(FVI_{AO}) = (A_R)(FVI_R) \quad (8)$$

where A is cross-sectional flow area and FVI is flow velocity integral. For this equation to be valid the blood flow per beat ($A \times FVI$) at the aortic orifice must be equal to that at the reference site, which means that if aortic regurgitation is present the reference site must be the left ventricular outflow tract; whereas if aortic regurgitation is absent the reference site may be at any other competent valve (mitral valve, pulmonic valve). A_{AO} represents the estimate of effective aortic orifice area, which is solved for by substituting Doppler-derived measures of FVI_{AO} and FVI_R and a two-dimensional echocardiographic anatomic estimate of A_R into the equation. Alternatively, an independent technique (thermodilution or Fick technique) could be used to measure reference flows in the continuity equation. The use of Doppler data in the continuity equation generally has resulted in better correlation with catheterization data than has the substitution of Doppler-derived pressure gradients into the Gorlin equation.[18–20] The potential for the *totally noninvasive* prediction of aortic valve area using the continuity equation (Doppler-derived flow velocities and an echocardiographically measured reference cross-sectional area) is illustrated by the excellent linear correlation between the catheterization-derived aortic valve area (AVA_c) and the Doppler-derived aortic valve area (AVA_D) when the left ventricular outflow tract is used as the reference site[24] ($r = 0.95$, $AVA_c = 1.07 [AVA_D] - 0.04$, SEE = 0.15 cm^2.)

While it is beyond the scope of this discussion to critique the limitations inherent in the invasive measurement of aortic valve area, which has served as the standard of comparison for Doppler techniques, a few comments on the advantages and disadvantages of the Doppler techniques are in order. Probably the two most important technical factors governing the accuracy of the Doppler technique are (1) the ability to align the Doppler probe parallel to flow so that maximal velocities in the stenotic jet and at the reference flow site are obtained and (2) the ability of the two-dimensional echocardiogram to provide an accurate measure of the anatomic flow area at the reference site. Implicit in the latter statement is the assumption that the anatomic flow area is a reasonable approximation of the effective hemodynamic flow area. Since in practice Doppler measurements are *not* made simultaneously at the aortic valve and at the reference site, it is important that the patient be in a reasonably steady state with a stable, regular heart rhythm.

The Doppler application of the continuity equation offers some potential advantages over either the substitution of invasive data or Doppler data into the Gorlin equation. Both the continuity equation and the Gorlin equation assume that the coefficient of velocity and the coefficient of orifice contraction at the aortic valve orifice are constant or can be neglected over a wide range of hemodynamic conditions. In both equations the coefficient of velocity is assumed to be unity, which is equivalent to saying that the velocity profile is relatively flat (with negligible boundary effects) across the cross-sectional flow area so that the maximal velocity measured is a good approximation to the spatial average velocity. The coefficient of orifice contraction relates the effective hemodynamic orifice cross-sectional area to the anatomic cross-sectional area. This coefficient is assigned an empirically determined constant value (incorporated in K) in the Gorlin equation (Equation 6) and is neglected in the continuity equation as presented above (although an empiric correction factor could be introduced). The continuity equation is independent of assumptions regarding the coefficient of nozzle discharge at the aortic valve orifice, whereas the Gorlin equation (Equation 6) makes the critical assumption that this coefficient, which accounts for energy losses affecting the measured pressure gradient (see Equation 3), is constant (incorporated in K) over a wide range of hemodynamic conditions. These considerations have been explored and validated in hydraulic model studies, which suggest that in patients with low cardiac output the Gorlin equation may be less accurate than the continuity equation, resulting in underestimation of valve area and suggesting that Doppler application of the continuity equation potentially can offer more accurate assessment of aortic valve area under these conditions.[17]

Doppler echocardiographic estimation of aortic valve area by the continuity equation in patients with mixed aortic stenosis and regurgitation offers an alternative to the measurement of left ventricular stroke output from the left ventricular angiogram by measuring the reference systolic flow per beat with Doppler techniques at the left ventricular outflow tract. The Doppler approach is critically dependent on accurate measurement of the anatomic area of the left ventricular outflow tract and assumes that the flow velocities can be sampled at a point proximal to where the velocity streamlines converge as they pass through the stenotic aortic orifice.[23,24] If these conditions are met, some of the data scatter in the correlation between catheterization-derived (Gorlin equation) and Doppler-derived (continuity equation) measures of aortic valve area ($r = 0.80$) in studies that included patients with aortic regurgitation[22] may be attributable to the inaccuracy of angiographic estimates of left ventricular stroke output.

Aortic and Pulmonic Regurgitation

In normal persons a pulsed Doppler sample volume positioned proximal to the aortic valve will detect no signal above background noise during diastole (Figure 22-1), whereas in patients with aortic regurgitation a high-velocity (usually resulting in signal aliasing) diastolic flow velocity signal distinct from the characteristic M-shaped envelope of the mitral inflow velocity profile is recorded[25–27] (Figure 22-3). With the use of supravalvular aortography as the standard for judging the presence or absence of aortic regurgitation, pulsed Doppler echocardiography is highly sensitive (96%) and specific (96%) for diagnosing aortic regurgitation and is superior to auscultation, M-mode echocardiography, and two-dimensional echocardiography.[26] Even in the presence of coexisting mitral stenosis, where auscultation may be confusing and M-mode echocardiographic evidence for anterior mitral leaflet flutter may be obscured, pulsed Doppler echocardiography successfully identifies aortic regurgitation (sensitivity 97%, specificity 90%) provided that care is taken to distinguish the abnormal diastolic signals of aortic regurgitation and mitral stenosis.[28] Although a sense of

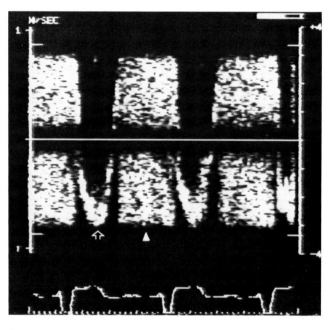

Figure 22-3 Pulsed Doppler spectra obtained by using an apical window with the sample volume located in the left ventricular outflow tract in a patient with aortic regurgitation. Calibrations and labels are the same as those in the legend to Figure 22-1. A systolic flow velocity envelope (open arrow) representing blood flow out of the aorta is followed by an abnormal diastolic aliased signal (▲) representing regurgitant blood flow.

the location of the regurgitation jet can be achieved by sampling multiple sites beneath the aortic valve with pulsed Doppler, color flow Doppler provides direct imaging of the regurgitant flow juxtaposed to the anatomic structures and hence permits the distinction of paravalvular (Figure 22-4) from valvular (Figure 22-5) aortic insufficiency (see color insert).

Measurement of the severity of aortic regurgitation is hampered by the lack of adequate reference standards for comparison, although qualitative grading by supravalvular aortography and measurement of the regurgitant fraction by catheterization techniques have provided the basis for most clinical comparisons. Three major approaches to Doppler estimation of the severity of aortic regurgitation have been evaluated (Table 22-1): (1) spatial mapping of the regurgitant flow velocity disturbance, (2) measurement of the regurgitant fraction, and (3) measurement of the rate of decline of instantaneous peak aortic regurgitant flow velocity during diastole.

Mapping of the spatial distribution of the flow velocity disturbance created by the aortic regurgitant jet is predicted on the assumption that the dissipation of kinetic energy of the regurgitant jet into the left ventricular blood pool is proportional to the spatial extent of the flow velocity disturbance and related to the hemodynamic severity of the regurgitant lesion. The flow velocity disturbance is not simply a measure of the regurgitant volume but rather is a complex interaction between the mass of regurgitant blood moving with a velocity dictated by the transvalvular pressure gradient, the nozzle characteristics of the regurgitant orifice, the passive constraints/filling properties of the left ventricular chamber, and the transmitral inflow.

Notwithstanding these complexities, investigators have demonstrated correlations between aortographic severity of aortic regurgitation and assessment by pulsed Doppler mapping of the spatial extent of the flow velocity disturbance relative to intracardiac structures ($r = 0.88$)[29,30] or the apparent volume of the flow velocity disturbance relative to the size of the left ventricular outflow tract ($r = 0.67$).[31] Alternatively, investigators have attempted to measure the cross-sectional area of the aortic regurgitant jet (as an indicator of regurgitant aortic valve area) by pulsed Doppler mapping in the short-axis plane just below the aortic valve and have demonstrated a correlation ($r = 0.88$) with angiographic grading of the severity of aortic incompetence.[32]

Color flow Doppler images directly the flow velocity disturbance of aortic regurgitation as an abnormal diastolic flow signal originating from the aortic valve, extending into the left ventricular outflow tract, and having a mosaic pattern of colors suggestive of disordered flow. In vitro assessment of regurgitant jets by color flow Doppler in a pulse duplicator system has suggested that the length and maximal width of the flow velocity disturbance created by aortic regurgitation are poor predictors of regurgitant fraction or regurgitant orifice size and are strongly influenced by the diastolic transaortic pressure gradient.[33] Only the minimal proximal width of the flow velocity disturbance (at transaortic pressure gradients greater than 80 mm Hg) was predictive of aortic regurgitant orifice size, and the same parameter correlated well with the regurgitant fraction independent of the transvalvular pressure gradient.[33] These principles have been applied to the grading of the severity of aortic regurgitation in patients by color flow Doppler. A measure of the short-axis area of the flow velocity disturbance relative to the anatomic left ventricular outflow tract area proved to be the best predictor of angiographic severity of aortic regurgitation in one study.[34] These investigators reasoned by analogy to the in vitro studies that the cross-sectional area of the flow velocity disturbance is a better predictor of the severity of regurgitation because it relates to regurgitant orifice size, whereas other parameters of the flow velocity disturbance, such as maximal length and area of the flow velocity disturbance, are subject to dependence on the kinetic energy of the regurgitant mass of blood.[34] Of note, however, in both the clinical[34] and in vitro[33] studies, imaging of the flow velocity disturbance was performed close to but not precisely at the regurgitant orifice. Under the most controlled circumstances in the in vitro experiments, the flow velocity measurement of jet width was made as far as 1.0 cm from the regurgitant orifice, and the measured

Table 22-1 Quantitation of Aortic Regurgitation

Approach	Doppler Technique
• Measurement of the spatial extent of the flow velocity disturbance created by the regurgitant jet —To estimate regurgitant orifice size —To estimate dispersion of abnormal flow velocities in the left ventricle	• Pulsed-wave Doppler mapping or color flow Doppler mapping of the flow velocity disturbance onto the two-dimensional echocardiographic image of the left ventricular outflow tract
• Measurement of the regurgitant fraction —From antegrade and retrograde flow velocities in the aorta —From antegrade transaortic flow and a reference flow site	• Pulsed-wave Doppler of flow velocity profiles combined with two-dimensional echocardiographic estimates of flow cross-sectional area
• Measurement of the time required for aortic and left ventricular pressures to approach equilibration during diastole	• Continuous-wave Doppler of the aortic regurgitant flow velocity profile as a reflection of the instantaneous aorta-to-left-ventricular diastolic pressure gradient

proximal width of the flow velocity disturbance exceeded the actual orifice diameter by two- to threefold.[33] Thus these measurements of proximal jet width and cross-sectional area may actually be indices of the proximal spread of the flow velocity disturbance and not pure reflections of regurgitant orifice size. By contrast, in an animal model of surgically created aortic regurgitation, measurement of the total area of the flow velocity disturbance from two orthogonal planes normalized for left ventricular end-diastolic cavity area bore a strong linear correlation to the regurgitant fraction ($r = 0.89$) over a wide range of hemodynamic conditions.[35] Obviously, further work is needed to define the hemodynamic determinants of the spatial distribution of the flow velocity disturbance resulting from aortic regurgitation so that those parameters best predictive of the hemodynamic severity of the regurgitant lesion can be identified.

Doppler-derived indices of the ratio of regurgitant flow to total forward flow across the aortic valve (regurgitant fraction) have been devised and compared with regurgitant fractions measured by cardiac catheterization. One means of doing this is to compare the magnitudes (peak velocity or flow velocity integral) of forward systolic flow velocity (a reflection of total left ventricular output) and reverse diastolic flow velocity (a reflection of regurgitant flow) by sampling aortic flow with pulsed Doppler distal to the aortic valve (descending aorta[36,37] or abdominal aorta[38]). In one study that used pulsed Doppler sampling of descending aortic flow and an echocardiographic correction factor for the change in aortic diameter between systole and diastole, a good correlation ($r = 0.90$) was found with the regurgitant fraction measured by cardiac catheterization.[36] Another method of estimating the regurgitant fraction by Doppler techniques involves the comparison of forward flow at the aortic valve (Q_{AO}) (representing total left ventricular output) to a reference forward flow (Q_R) at a competent valve (mitral valve[39,40] or pulmonic valve[41,42]), since the difference between the two will represent the regurgitant flow. Since transvalvular flow, Q, can be computed as

$$Q = FVI \times A \times HR \qquad (9)$$

(where Q is transvalvular flow, FVI is the flow velocity integral acquired by pulsed Doppler, A is the cross-sectional flow area approximated by an echocardiographic measurement of anatomic cross-sectional area, and HR is heart rate), we can use these measurements to compute the regurgitant fraction as ($Q_{AO} - Q_R$)/Q_{AO}. By using pulmonic blood flow as the reference, one group of investigators has demonstrated an excellent correlation between the regurgitant fraction measured by these Doppler techniques and the regurgitant fraction measured by cardiac catheterization ($r = 0.96$); less good correlation was found with qualitative angiographic grading of aortic regurgitation ($r = 0.80$), the implication being that angiographic grading may be a less accurate measure of the severity of regurgitation than either the invasive or noninvasive measures of the regurgitant fraction.[42]

Continuous-wave Doppler measurement of the peak aortic regurgitant flow velocities permits the characterization of the time course of these regurgitant flow velocities. Since the instantaneous peak velocity, v, is related to the diastolic pressure difference between the aorta, P_1, and the left ventricle, P_2 ($P_1 - P_2 = 4v^2$; see the derivation of Equation 4), the monotonic decline of the instantaneous peak aortic flow velocities (Figure 22-6) is related temporally to the decline of the transaortic valve pressure gradient as aortic pressure and left ventricular pressure seek equilibration during diastole. Indices of this decline in instantaneous regurgitant flow velocities, such as the slope of the flow velocity profile (equivalent to deceleration of the regurgitant flow[30,43]) or the time required for peak regurgitant flow velocity to fall by a given proportion,[43,44] can be used to measure the severity of aortic regurgitation because in more severe grades of regurgitation left ventricular diastolic pressure will rise sooner and aortic diastolic pressure will fall sooner, resulting in earlier equilibration of these pressures. One group of investigators has demonstrated that the time required for peak velocity, v, to fall to $v/\sqrt{2}$ correlates well ($r = 0.91$) with the time required for the transvalvular pressure gradient, ΔP, to fall to half its peak value ($\Delta P = 4v^2$, thus $[\Delta P/2] = 4[v/\sqrt{2}]^2$); measurement of this time interval from continuous-wave Doppler tracings in patients with varying degrees of aortic regurgitation was inversely correlated ($r = -0.88$) with the regurgitant fraction measured by cardiac catheterization.[44] This relatively simple Doppler index of the severity of aortic regurgitation exploits the physiologic relationship of the response of the left ventricular chamber to regurgitant volume loading. One may expect, however, that the relationship may also depend on the chronicity of the aortic regurgitation and in particular on the chamber compliance and degree of adaptive dilation of the left ventricle to the pathologic volume loading.

Pulmonary regurgitation is identified by Doppler echocardiography in a manner analogous to the detection of aortic regurgitation. Validation and quantitation of these observations are hampered somewhat by the absence of adequate diagnostic standards for comparison. In apparently normal healthy persons, a retrograde diastolic flow velocity signal can be identified frequently (in one study a prevalence of 92% was reported[45]) by pulsed Doppler with the sample volume located just proximal to the pulmonic valve. Whether

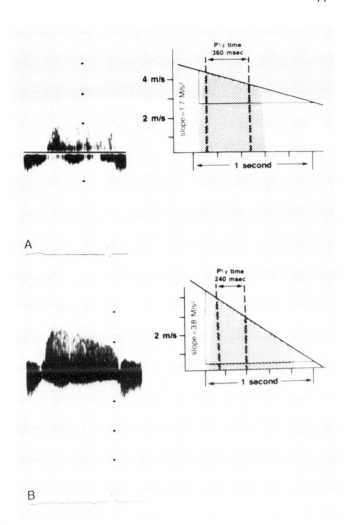

Figure 22-6 Continuous-wave Doppler spectra of aortic regurgitant flow velocities with a simultaneous electrocardiogram from a patient with mild (A) and a patient with severe (B) aortic regurgitation. The slope of the decline in flow velocities is steeper and the time required for the transvalvular pressure gradient to fall to one half its peak value ($P\frac{1}{2}$ time) is shorter in the patient with severe aortic regurgitation. Calibrations and computational techniques are illustrated in the accompanying schematics to the right of each spectrum. *Source:* Reprinted with permission from *Journal of the American College of Cardiology* (1986;1341–1347), Copyright © 1986, American College of Cardiology.

---

the majority of these observed spectra represent the detection of "physiologic" pulmonary incompetence, technical artifact, or diastolic flow inadvertently recorded from nearby structures (for example, the left main coronary artery) is unknown and subject to current investigation. In general these subpulmonic diastolic flow velocity signals are detected only within the first centimeter proximal to the pulmonic valve,[45] and this location may serve as a criterion to distinguish them from the signals seen in "pathologic" pulmonary regurgitation, which usually can be mapped to more proximal sites in the right ventricular outflow tract.

STENOSIS AND INCOMPETENCE OF THE ATRIOVENTRICULAR VALVES

Mitral and Tricuspid Stenosis

The normal transmitral flow velocity profile sampled by pulsed Doppler exhibits an M-shaped envelope during diastole that reflects two distinct peak velocities, the first corresponding to early rapid diastolic filling (v_E) and the second (v_A) corresponding to late diastolic filling due to atrial systole (Figure 22-7). In a patient with obstruction to transmitral flow, the instantaneous velocities increase, spectral broadening of the flow velocity envelope occurs, and the slope of decline of transmitral velocities following v_E decreases. By analogy to our previous discussion of the Doppler findings in aortic valve obstruction, these increased velocities, which can be measured by continuous-wave Doppler, are due to the converging streamlines of flow through the narrowed orifice and can be related to the instantaneous transvalvular pressure gradient, $P_1 - P_2$ ($P_1 - P_2 = 4v^2$; see derivation of Equation 4).[2,4] As in the case of aortic stenosis, incremental iteration of this formula throughout the diastolic filling period permits the calculation of the mean transvalvular pressure gradient in mitral stenosis (Equation 5). By application of the continuity equation (Equation 8) it is possible to estimate mitral valve area from the integral of the transmitral flow velocity profile and the noninvasive (or invasive) measurements of a reference flow. In one study using the Fick technique to measure reference flows, the Doppler estimate of mitral valve area in patients with pure mitral stenosis correlated well with catheterization-derived measurements ($r = 0.97$).[3] As has been accomplished for the noninvasive assessment of aortic valve area, it should be possible to measure reference flows noninvasively so that the mitral valve area can be determined by the continuity equation without catheterization of the right side of the heart.

Direct application of the continuity equation to solve for mitral valve area has given way in clinical practice to a simpler empiric approach based on the assessment of the rate of decline of instantaneous transmitral flow velocity following v_E. As the mitral valve becomes more severely stenotic, there is greater impedance to early diastolic filling: The early diastolic filling period is prolonged and the rate of decay of the transvalvular pressure gradient is lessened. Observations from patients with mitral stenosis, in whom the transmitral diastolic pressure gradient was measured with dual catheter systems, have demonstrated that the time interval from the peak transvalvular pressure gradient to its decay to 50% of its maximal value (the pressure half-time) is a good index of the anatomic severity of mitral stenosis.[46] In addition, in any given patient the attendant increases

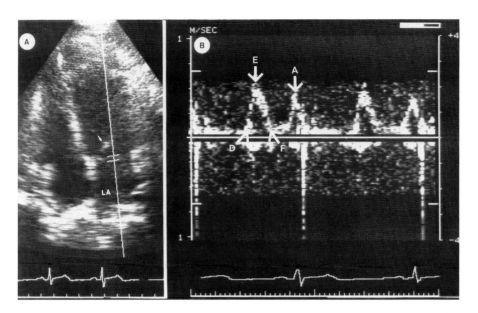

Figure 22-7 Pulsed Doppler measurement of transmitral flow velocities from an apical window in a normal subject. A, Stop-frame view of an apical four-chamber two-dimensional echocardiogram. The Doppler cursor, represented by a line running from the apex through the mitral orifice, indicates the direction of the interrogating ultrasound beam. The double crosshatch marks delineate the position of the sample volume in the left atrium (LA) just proximal to the mitral valve (arrow) at the level of the mitral annulus. B, Fast Fourier transform of the pulsed Doppler-shifted frequencies plotted as velocity ($\pm$ 1 m/s on the left) or Doppler frequency shift ($\pm$ 4 kHz on the right) versus time (full sweep = 2 seconds). The electrocardiogram is presented beneath the spectrum. During diastole, mitral inflow velocities are characterized by an early (E) peak and a later peak timing with atrial systole (A). The early diastolic filling period is defined by the interval between points D and F, and the decline of transmitral velocities following point E is relatively steep (compare with Figure 22-8). During systole no flow velocity signals are detected.

in transvalvular pressure gradient, heart rate, and cardiac output during exercise did not have a large effect on the pressure half-time.[46]

Hatle et al.[47] adapted these concepts to the evaluation of the transmitral flow velocity waveform and noted that the pressure half-time was equivalent to the time required for peak velocity during early diastole, v_E, to decline to $v_E/\sqrt{2}$ (see Equation 4). As had been previously demonstrated for pressure half-time measured directly from hemodynamic tracings, these investigators demonstrated that the analogous measurement from the Doppler flow velocity profiles was relatively unaffected by the increased transmitral flow occurring during exercise and correlated inversely ($r = -0.74$) with mitral valve area assessed by cardiac catheterization.[47] Given this linear relationship and the observation that a mitral valve area of 1.0 cm² corresponded roughly to a pressure half-time of 220 milliseconds, an empiric relationship between mitral valve area (MVA) and pressure half-time ($t\frac{1}{2}$) measured from the transmitral flow velocity profile was established[47–49] (Figure 22-8):

$$MVA \text{ cm}^2 = \frac{220}{t\frac{1}{2}\text{ms}} \quad (10)$$

Application of this equation to the estimation of mitral valve area by Doppler-derived pressure half-times has

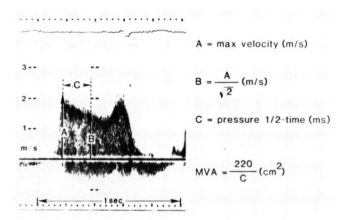

Figure 22-8 Continuous-wave Doppler spectrum of diastolic transmitral flow with a simultaneous electrocardiogram from a patient with mitral stenosis (compare with Figure 22-7). Estimation of mitral valve area (MVA) is achieved by measuring the maximal velocity during early diastole, dividing it by the square root of 2, and determining the time interval required (pressure ½-time) for early diastolic velocities to decay to that value. The mitral valve area in square centimeters is approximated by the quotient of 220 and the pressure half-time in milliseconds. *Source:* Reprinted with permission from *Circulation* (1986;73:100–107), Copyright © 1986, American Heart Association.

resulted in good correlations with measurements made during cardiac catheterization both for patients with unoperated mitral stenosis ($r = 0.85$) and for patients who have undergone mitral commissurotomy

($r = 0.90$).[48] Recent preliminary data suggest that whereas the Doppler pressure half-time technique may predict mitral valve area accurately in the unoperated patient, measurements made immediately after balloon mitral valvuloplasty correlate poorly with hemodynamically determined mitral valve area.[50,51] These observations underscore the problems associated with applying empiric formulas to new clinical situations, where baseline conditions and assumptions may not be satisfied.

Tricuspid stenosis is an elusive clinical diagnosis that can be difficult to confirm even with catheterization techniques. Two-dimensional echocardiography can demonstrate thickened tricuspid valve leaflets with restricted leaflet tip excursion and proximal leaflet doming mimicking the echocardiographic findings in mitral stenosis,[52] but a practical means of measuring anatomic orifice area has not been reported. Recent experience with Doppler assessment of transtricuspid flow velocities suggests that the qualitative shape of the flow velocity profile reflects the simultaneously measured changes in the transtricuspid valve diastolic pressure gradient.[53] It is to be hoped that future work will define an empiric relationship between the pressure half-time of the tricuspid valve diastolic pressure gradient and the tricuspid valve area.

Mitral and Tricuspid Regurgitation

During systole, a pulsed Doppler sample volume situated in the left atrium behind the mitral valve should detect no flow velocity signals above background noise in a normal person (Figure 22-7). When there is mitral regurgitation, pulsed Doppler detects a pansystolic aliased signal representing the high-velocity regurgitant jet of blood passing from the high-pressure left ventricular cavity to the low-pressure left atrial cavity (Figure 22-9). Lesser degrees of regurgitation may result in abnormal flow velocity signals that do not span all of systole.[54] Pulsed Doppler detection of mitral regurgitation is both sensitive (89% to 91%) and specific (84% to 96%) as compared with left ventricular angiography by several independent investigators.[25,54–56] Continuous-wave Doppler eliminates signal aliasing and clearly demonstrates that the regurgitant flow velocity is directed in the opposite direction to mitral inflow, is confined to systole, and attains instantaneous peak velocities proportional to the square of the instantaneous pressure difference between the left ventricle and left atrium (Figure 22-9). Color flow Doppler can image the spatial extent of the flow velocity disturbance in the left atrium created by mitral regurgitation and may provide an indication as to the direction or origin of the regurgitant jet that is dependent on the leaflet responsible for the incompetence and the nature of the pathologic process.[57] Color

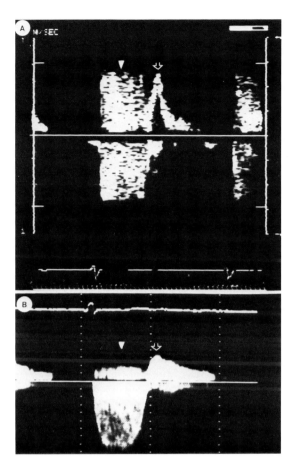

Figure 22-9 Pulsed- and continuous-wave Doppler spectra from a patient with mitral regurgitation. A, Pulsed-wave Doppler spectrum from a sample volume positioned in the left atrium just proximal to the mitral valve; an apical window was used. Calibrations and labels are the same as those in the legend to Figure 22-7. A diastolic flow velocity envelope (open arrow) representing left ventricular filling velocities is preceded by an abnormal systolic aliased signal (▼) representing regurgitant blood flow. B, Continuous-wave Doppler spectrum obtained with the transducer positioned at the apical window and the beam directed toward the mitral valve. Positive flow velocities toward the transducer (open arrow), representing transmitral diastolic filling, are recorded. Negative systolic flow velocities (▼), representing regurgitant blood flow moving away from the transducer toward the left atrium, illustrate the direction and magnitude of regurgitant velocities that could not be assessed in the aliased pulsed Doppler signal. Calibration dots are 0.5 m/s apart along the vertical axis and 0.5 second apart along the horizonal axis. The horizontal, solid baseline represents zero velocity. The simultaneous electrocardiogram is displayed above the spectrum.

flow Doppler has also been used during supine bicycle exercise to detect "functional" mitral regurgitation resulting from acute ischemic myocardial dysfunction.[58]

Quantitation of mitral regurgitation by Doppler techniques is subject to technical and theoretical considerations similar to those encountered in aortic regurgitation. Two principal approaches have been adopted (Table 22-2): (1) quantitation of the spatial extent of the left atrial flow velocity disturbance and

Table 22-2 Quantitation of Mitral Regurgitation

Approach	Doppler Technique
• Measurement of the spatial extent of the flow velocity disturbance created by the regurgitation jet —To estimate dispersion of abnormal flow velocities in the left atrium	• Pulsed-wave Doppler mapping or color flow Doppler mapping of the flow velocity disturbance onto the two-dimensional echocardiographic image of the left atrium
• Measurement of regurgitant fraction —From antegrade transmitral flow and a reference flow site	• Pulsed-wave or continuous-wave Doppler of flow velocity profiles combined with two-dimensional echocardiographic estimates of flow cross-sectional area

(2) measurement of the regurgitant fraction. The spatial extent of the left atrial flow velocity disturbance can be measured relative to left atrial size, providing rough correlations with the severity of mitral regurgitation as judged by qualitative angiographic grading.[55,56,59–61] In one study, pulsed Doppler mapping of the regurgitant flow velocity disturbance in the left atrium from two orthogonal planes (to define the length, width, and height of the region of disturbed flow) resulted in a linear correlation ($r = 0.70$) and a better exponential correlation ($r = 0.88$) with a three-point qualitative angiographic grading scale.[56] Color flow Doppler also has been used to map the distribution of the mitral regurgitant flow disturbance in the left atrium.[62,63] When the maximal regurgitant flow area obtained from multiple imaging windows was expressed as a percentage of left atrial area (obtained from the same plane as the maximal regurgitant area), the correlation with the regurgitant fraction measured by catheterization was better ($r = 0.78$) than the correlation obtained when the transverse dimension or length of the flow disturbance zone was used.[63] When this color Doppler index of the severity of mitral regurgitation was compared with a three-point qualitative angiographic grading scale, the mean values of the index were significantly different for each of the groups.[63]

Just as the severity of aortic insufficiency can be measured by Doppler quantitation of transaortic blood flow and a reference flow at a nonregurgitant valve, in similar fashion mitral regurgitant flow can be calculated as the difference between Doppler-derived transmitral flow and transaortic flow (serving as the reference site) (see Equation 9). In a canine model of mitral regurgitation where the flow rate of simulated mitral regurgitation could be measured accurately by an electromagnetic flow probe mounted on a left-ventricle to left-atrium conduit, pulsed Doppler quantitation of aortic and mitral transvalvular flows resulted in good predictions of regurgitant flow ($r = 0.84$) and regurgitant fraction ($r = 0.83$).[64] This approach has been adapted to measure the regurgitant fraction in patients with mitral regurgitation and has resulted in a similar degree of accuracy in predicting the regurgitant fraction determined by catheterization techniques.[40,65]

The diagnosis of tricuspid regurgitation is of clinical importance in the assessment of dysfunction of the tricuspid valve (for example, due to right ventricular dilation) as well as intrinsic tricuspid valve disease (Ebstein's anomaly, rheumatic disease, endocarditis), yet a definitive diagnostic standard remains elusive. Pulsed Doppler interrogation of superior vena caval flow to compare the ratio of maximal systolic flow velocity to maximal diastolic velocity has been used to detect tricuspid regurgitation; ratios less than 1.3 are indicative of moderate to severe tricuspid regurgitation.[66] The identification of abnormally high-velocity (usually aliased) systolic flow signals localized proximal to the tricuspid valve by pulsed Doppler has been taken as evidence for the presence of tricuspid regurgitation.[45] Such signals, which generally are of limited duration (not reproducibly holosystolic) and spacially confined to the portion of the right atrium just proximal to the tricuspid valve, are frequently encountered (44% of persons) in apparently healthy normal volunteers,[45] raising the suspicion that such Doppler findings either are artifactual or reflect trivial degrees of "physiologic" regurgitant flow that are of little clinical relevance.

In general, holosystolic signals of hemodynamically important tricuspid regurgitation are detectable more than 1.0 cm proximal to the tricuspid valve. The spatial extent of the right atrial flow velocity disturbance due to tricuspid regurgitation can be measured by pulsed Doppler mapping of its depth of penetration into the right atrium or the area subtended by the flow disturbance.[67] These Doppler techniques of detecting tricuspid regurgitation and quantitating its severity have been compared with saline contrast echocardiography (see also Chapter 21) using right ventricular angiography as the diagnostic standard. Notwithstanding the limitations of radiographic contrast angiography in defining the presence and severity of tricuspid regurgitation, both saline contrast echocardiography (sensitivity 82%; specificity 100%; correlation with angiographic grading of regurgitant severity, $r = 0.84$) and pulsed-Doppler echocardiography (sensitivity 91%; specificity 86%; correlation with angiographic grading of regurgitant severity, $r = 0.82$) provided reasonably good predictions of the angiographic assessment.[68] It is reassuring that there is general agreement among these three methodologies; the relative ease of performance of the Doppler examination and its lack of a requirement for an intravascular access make it a particularly attractive technique as compared with either saline contrast echocardiographic or radiographic contrast angiography.

Continuous-wave Doppler measurement of the instantaneous transvalvular velocities of tricuspid regurgitation provides a method of noninvasively assessing the right-ventricular to right-atrial systolic pressure gradient (see Equation 4).[69,70] By assuming an arbitrary value for right atrial pressure or by measuring central venous pressure from the jugular venous pulse, right ventricular systolic pressure and pulmonary artery pressure (in the absence of obstruction of the right ventricular outflow tract) can be estimated as the sum of right atrial pressure and the right-ventricular to right-atrial pressure gradient (measured by continuous-wave Doppler). Application of these techniques in patients has resulted in good linear correlations ($r = 0.93$ to 0.97) with systolic pressures in the right side of the heart measured by catheterization techniques, providing a reliable means of assessing pulmonary hypertension secondary to valvular lesions in the left side of the heart.[69,70] Since tricuspid regurgitation frequently is detected by Doppler in patients with pulmonary hypertension (80% of patients with a pulmonary systolic pressure greater than 35 mm Hg), the technique is applicable in the majority of patients with clinically relevant right ventricular systolic hypertension.[70]

DOPPLER ASSESSMENT OF PROSTHETIC HEART VALVES

A detailed discussion of the Doppler assessment of the postoperative results of valve replacement or repair is beyond the scope of this review. Nonetheless, with the improved valvular prostheses and the intense interest in valvular and annular reconstruction, there has been renewed interest in the use of Doppler echocardiographic techniques to assess postoperative valvular function noninvasively. Prior to the widespread application of Doppler techniques, echocardiographic evaluation of prosthetic valves was limited to the relatively gross assessment of valve structure (for example, for thrombi or vegetations), valve ring motion relative to cardiac structures (for example, exaggerated motion with valvular dehiscence), occluder motion (for example, ball or disk excursion) relative to the prosthetic sewing ring, and the timing of these events.[71–74] One significant limitation to ultrasonic imaging of mechanical prosthetic valves is that the structural materials used in these valves are highly reflective of ultrasound and produce intense signals with considerable generation of artifacts; thus detailed structural evaluation is problematic. To some extent this is less a problem with bioprosthetic valves, and under ideal circumstances two-dimensional echocardiography can image the prolapsed, fractured, torn, or pathologically altered leaflets of porcine heterograft prostheses.[75]

Doppler echocardiography provides the potential for assessing prosthetic valve *function* to complement the *structural* information provided by two-dimensional echocardiography. Since all prosthetic valves result in a small degree of obstruction to flow, it is important to establish the "normal" ranges for Doppler-derived transprosthetic valve pressure gradients[76–80] and to demonstrate the reproducibility of these measurements.[81] As can be anticipated from analogous studies using invasive measurement techniques,[82] these values will vary with the type of prosthesis, the size of the prosthesis, and the position of implantation. One study of simultaneously measured transmitral and transtricuspid pressure gradients derived by continuous-wave Doppler and catheterization techniques in patients with prosthetic atrioventricular valves demonstrated an excellent correlation between noninvasive and invasive determinations of transprosthetic valve pressure gradients ($r = 0.96$), and this close agreement was obtained for both mechanical and biologic prostheses.[80] Ideally, one would seek to identify the effective hemodynamic area of the orifice of the prosthetic valve rather than simply measure the transvalvular gradient, as the latter will be highly dependent on cardiac output.

Although one might be tempted to use these Doppler-derived transvalvular pressure gradients and flow velocities in the equations used for measurement of the area of a native valve, the unusual geometry of the orifice of a prosthetic valve (particularly of a mechanical valve) and the increased inertia of a mechanical occluder make it likely that the coefficients of orifice contraction and the coefficients of nozzle discharge for these valves are different from those of native valves. This may explain in part the poor correlation ($r = 0.14$) found in one study between orifice area as determined by the substitution of derived Doppler velocities into the continuity equation and orifice area as measured hemodynamically with the Gorlin equation.[80] Similarly, the correlation between catheterization-derived prosthetic valve area and valve area assessed by the Doppler pressure half-time technique[80] was poor. While it is certainly possible that some of the discrepancies may be related to inaccuracies in the catheterization technique, further studies are required to assess the limitations and applicability of Doppler techniques toward assessing prosthetic valvular area.

Doppler detection of prosthetic valve or periprosthetic valve regurgitation[83,84] and evaluation of the efficacy of various reconstructive procedures aimed at reducing atrioventricular valve regurgitation[85,86] are particularly attractive applications of this noninvasive technique. Interpretation of the presence or absence of a regurgitant flow velocity signal proximal to a prosthetic valve is not as straightforward as one might anticipate. For instance, as many as 42% of patients with

Bjork-Shiley valves in the aortic position may have aortic regurgitation detected by pulsed Doppler despite the absence of clinical evidence for prosthetic valve dysfunction; a smaller number (11%) of such patients with mitral prostheses will have mitral regurgitation detected.[87] The corresponding percentages for bioprosthetic valves (Hancock and Carpentier-Edwards porcine heterografts) in the aortic (20%) and mitral (19%) positions are lower but still raise important questions about the definition of "normal" prosthetic valve function as detected by Doppler echocardiography.[87]

As was discussed previously, there are many limitations to the semiquantitative pulsed Doppler mapping of the spatial extent of a disturbance in regurgitant flow velocity. Nonetheless, this technique is of some value in distinguishing clinically important prosthetic valve dysfunction from trivial (presumably "normal") prosthetic regurgitation, where the spatial distribution of the abnormal flow velocity signals generally is limited. Special care must be taken to interrogate multiple sites proximal to the prosthetic valve (both central and peripheral to the presumed axis of flow), as the direction of the regurgitant jet can be unpredictable and can be missed easily during a casual survey.[88] In addition, attention must be directed to using multiple echocardiographic windows (for example, the parasternal window for the aortic and mitral valves, where the ultrasound beam does not pass through the prosthesis) to avoid "acoustic shadowing" by the prosthesis that may result in ultrasound energy insufficient to reach the area to be interrogated.[88,89]

Color flow Doppler offers great potential for detecting these eccentric jets of regurgitant flow, which might be overlooked by even a thorough and systematic pulsed Doppler examination. Unfortunately, the technique is not immune from the artifacts and limitations that pulsed- and continuous-wave Doppler echocardiography encounters with prosthetic valves. Confusing artifactual color signals, referred to as "ghosting" and probably representing Doppler shifts created by moving cardiac or prosthetic valve structures, can create signals that can be confused with prosthetic valve regurgitation. In addition, as has been demonstrated in model studies, the central poppet of ball-and-cage valves or the leaflets of the tilting disk and bileaflet valves may attenuate sufficiently or reflect the ultrasound signal such that regurgitant flow "masking" occurs—a phenomenon related to acoustic shadowing.[89] Nonetheless, with an understanding of these potential artifacts and limitations of the Doppler evaluation of prosthetic valves, useful clinical noninvasive assessment can be achieved based on serial assessment of a given patient (using that patient's baseline Doppler measurements as a control) and comparison of Doppler measurements to a reference library of "normal" Doppler parameters for a given prosthetic valve at a given position.

SUMMARY

The noninvasive assessment of valvular heart disease has been extended from a structural description of pathologic valvular anatomy and chamber architecture by two-dimensional echocardiography to a functional assessment of the resulting disordered blood flow by Doppler echocardiography. The physiologic parameters of transvalvular pressure gradients and effective orifice size are now measurable by these noninvasive Doppler techniques. The convergence of the physiologic and morphologic assessment of valvular disease in the integrated Doppler and two-dimensional echocardiographic examination is an extremely powerful clinical tool for the diagnosis and assessment of valvular heart disease.

REFERENCES

1. Hatle L, Angelsen B. *Doppler Ultrasound in Cardiology*. Philadelphia, Pa: Lea & Febiger; 1985:22–26, 37–40.

2. Holen J, Aaslid R, Landmark K, et al. Determination of pressure gradient in mitral stenosis with a non-invasive ultrasound Doppler technique. *Acta Med Scand*. 1976;199:455–460.

3. Holen J, Aaslid R, Landmark K, et al. Determination of effective orifice area in mitral stenosis from non-invasive Doppler data and mitral flow rate. *Acta Med Scand*. 1977;201:83–88.

4. Hatle L, Brubakk AD, Tromsdal A, et al. Non-invasive assessment of pressure drop in mitral stenosis by Doppler ultrasound. *Br Heart J*. 1978;40:131–140.

5. Veyrat C, Gourtchiglouian C, Dumora P, et al. A new non-invasive estimation of stenotic aortic valve area by pulsed Doppler mapping. *Br Heart J*. 1987;57:44–50.

6. Cannon SR, Richards KL, Morgann RG. Comparison of Doppler echocardiographic peak frequency and turbulence parameters in the quantification of aortic stenosis in a pulsatile flow model. *Circulation*. 1985;71:129–135.

7. Hatle L, Angelsen BA, Tromsdal A. Non-invasive assessment of aortic stenosis by Doppler ultrasound. *Br Heart J*. 1980;43:284–292.

8. Lima CO, Sahn DJ, Valdes Cruz LM, et al. Prediction of the severity of left ventricular outflow tract obstruction by quantitative 2-dimensional echocardiographic Doppler studies. *Circulation*. 1983;68:348–354.

9. Berger M, Berdoff RL, Gallerstein PE, et al. Evaluation of aortic stenosis by continuous wave Doppler ultrasound. *J Am Coll Cardiol*. 1984;3:150–156.

10. Williams GA, Labovitz AJ, Nelson JG, et al. Value of multiple echocardiographic views in the evaluation of aortic stenosis in adults by continuous wave Doppler. *Am J Cardiol*. 1985;55:445–449.

11. Simpson IA, Houston AB, Sheldon CD, et al. Clinical value of Doppler echocardiography in the assessment of adults with aortic stenosis. *Br Heart J*. 1985;53:636–639.

12. Hegrenaes L, Hatle L. Aortic stenosis in adults: Non-invasive estimation of pressure differences by continuous wave Doppler echocardiography. *Br Heart J*. 1985;54:396–404.

13. Agatston AS, Chengot M, Rao A, et al. Doppler diagnosis of valvular aortic stenosis in patients over 60 years of age. *Am J Cardiol*. 1985;56:106–109.

14. Currie PJ, Seward JB, Reeder GS, et al. Continuous wave Doppler echocardiographic assessment of the severity of calcific aortic stenosis: A simultaneous Doppler catheter correlative study in 100 adult patients. *Circulation*. 1985;71:1162–1169.

15. Stewart WT, Galvin KA, Gillam LD, et al. Comparison of high pulse repetition frequency and continuous wave Doppler echocardiography in the assessment of high flow velocity in patients with valvular stenosis and regurgitation. *J Am Coll Cardiol*. 1985;6:565–571.

16. Gorlin R, Gorlin SG. Hydraulic formula for calculation of the area of the stenotic mitral valve, other cardiac valves and central circulatory shunts. *Am Heart J*. 1951;41:1–29.

17. Segal J, Lerner DJ, Miller C, et al. When should Doppler-determined valve area be better than the Gorlin formula?: Variation in hydraulic constants in low flow states. *J Am Coll Cardiol*. 1987;9:1294–1305.

18. Warth DC, Stewart WJ, Block PC, et al. A new method to calculate aortic valve area without left heart catheterization. *Circulation*. 1984;70:978–983.

19. Kosturakis D, Allen HD, Goldberg SJ, et al. Non-invasive quantitation of stenotic semilunar valve areas by Doppler echocardiography. *J Am Coll Cardiol*. 1984;3:1256–1262.

20. Teirstein P, Yeager M, Yock PG, et al. Doppler echocardiographic measurement of aortic valve area in aortic stenosis: A noninvasive application of the Gorlin formula. *J Am Coll Cardiol*. 1986;8:1059–1065.

21. Richards KL, Cannon SR, Miller JR, et al. Calculation of aortic valve area by Doppler echocardiography: A direct application of the continuity equation. *Circulation*. 1986;73:964–969.

22. Skjaerpe T, Hegrenaes L, Hatle L. Non-invasive estimation of valve area in patients with aortic stenosis by Doppler ultrasound and 2-dimensional echocardiography. *Circulation*. 1985;72:810–818.

23. Otto CM, Pearlman AS, Comess KA, et al. Determination of the stenotic aortic valve area in adults using Doppler echocardiography. *J Am Coll Cardiol*. 1986;7:509–517.

24. Zoghbi WA, Farmer KL, Soto JG, et al. Accurate non-invasive quantification of stenotic aortic valve area by Doppler echocardiography. *Circulation*. 1986;73:452–459.

25. Quinones MA, Young JB, Waggoner AD, et al. Assessment of pulsed Doppler echocardiography in detection and quantification of aortic and mitral regurgitation. *Br Heart J*. 1980;44:612–620.

26. Grayburn PA, Smith MD, Handshoe R, et al. Detection of aortic insufficiency by standard echocardiography, pulsed Doppler echocardiography and auscultation. *Ann Intern Med*. 1986;104:599–605.

27. Louie EK, Mason TJ, Shah R, et al. Determinants of anterior mitral leaflet fluttering in pure aortic regurgitation from pulsed Doppler study of the early diastolic interaction between the regurgitant jet and mitral inflow. *Am J Cardiol*. 1988;61:1085–1091.

28. Saal AK, Gross BW, Franklin DW, et al. Non-invasive detection of aortic insufficiency in patients with mitral stenosis by pulsed Doppler echocardiography. *J Am Coll Cardiol*. 1985;5:176–181.

29. Ciobanu, Abbasi A, Allen M, et al. Pulsed Doppler echocardiography in the diagnosis and estimation of severity of aortic insufficiency. *Am J Cardiol*. 1982;49:339–343.

30. Labovitz AJ, Ferrara RP, Kern MJ, et al. Quantitative evaluation of aortic insufficiency by continuous wave Doppler echocardiography. *J Am Coll Cardiol*. 1986;8:1341–1347.

31. Veyrat C, Ameur A, Gourtchiglouian C, et al. Calculation of pulsed Doppler left ventricular outflow tract regurgitant index for grading the severity of aortic regurgitation. *Am Heart J*. 1984;108:507–515.

32. Veyrat C, Lessana A, Abitbol C, et al. New indexes for assessing aortic regurgitation with 2-dimensional Doppler echocardiographic measurement of the regurgitant aortic valvular area. *Circulation*. 1983;68:998–1005.

33. Switzer DF, Yoganathan AP, Nanda NC, et al. Calibration of color Doppler flow mapping during extreme hemodynamic conditions in vitro: A foundation for a reliable quantitative grading system for aortic incompetence. *Circulation*. 1987;75:837–846.

34. Perry GJ, Helmcke F, Nanda NC, et al. Evaluation of aortic insufficiency by Doppler color flow mapping. *J Am Coll Cardiol*. 1987;9:952–959.

35. Louie EK, Krukenkamp I, Hariman RJ, et al. Quantitative assessment of aortic regurgitation by color flow Doppler in an open chest canine model. *Cardiovasc Res*. 1989;22:145–151.

36. Touche T, Prasquier R, Nitenberg A, et al. Assessment and follow-up of patients with aortic regurgitation by an updated Doppler echocardiographic measurement of the regurgitant fraction in the aortic arch. *Circulation*. 1985;72:819–824.

37. Diebold B, Peronneau P, Blanchard D, et al. Non-invasive quantification of aortic regurgitation by Doppler echocardiography. *Br Heart J*. 1983;49:167–173.

38. Takenaka K, Dabestani A, Gardin JM, et al. A simple Doppler echocardiographic method for estimating severity of aortic regurgitation. *Am J Cardiol*. 1986;57:1340–1343.

39. Zhang Y, Nitter-Hauge S, Ihlen H, et al. Measurement of aortic regurgitation by Doppler echocardiography. *Br Heart J*. 1986;55:32–38.

40. Rokey R, Sterling LL, Zoghbi WA, et al. Determination of regurgitant fraction in isolated mitral or aortic regurgitation by pulsed Doppler two-dimensional echocardiography. *J Am Coll Cardiol*. 1986;7:1273–1278.

41. Goldberg SJ, Allen HD. Quantitative asessment by Doppler echocardiography of pulmonary or aortic regurgitation. *Am J Cardiol*. 1985;56:131–135.

42. Kitabatake A, Ito H, Inoue M, et al. A new approach to non-invasive evaluation of aortic regurgitant fraction by two-dimensional Doppler echocardiography. *Circulation*. 1985;72:523–529.

43. Masuyama T, Kodama K, Kitabatake A, et al. Non-invasive evaluation of aortic regurgitation by continuous wave Doppler echocardiography. *Circulation*. 1986;73:460–466.

44. Teague SM, Heinsimer JA, Anderson JL, et al. Quantification of aortic regurgitation utilizing continuous wave Doppler ultrasound. *J Am Coll Cardiol*. 1986;8:592–599.

45. Kostucki W, Vandenbossche J, Friart A, et al. Pulsed Doppler regurgitant flow patterns of normal valves. *Am J Cardiol*. 1986;58:309–313.

46. Libanoff AJ, Rodbard S. Atrioventricular pressure half-time: Measurement of mitral valve orifice area. *Circulation*. 1968;38:144–150.

47. Hatle L, Angelsen B, Tromsdal A. Non-invasive assessment of atrioventricular pressure half-time by Doppler ultrasound. *Circulation*. 1979;60:1096–1104.

48. Smith HD, Handshoe R, Handshoe S, et al. Comparative accuracy of two-dimensional echocardiography and Doppler pressure half-time methods in assessing severity of mitral stenosis in patients with and without prior commissurotomy. *Circulation*. 1986;73:100–107.

49. Pearlman JD, Gibson RS. Doppler measurement of left atrial depressurization and mitral valve area in patients with suspected mitral stenosis: Validation of a new method. *Am Heart J*. 1987;113:868–873.

50. Wilkins G, Thomas J, Abascal V, et al. Failure of the Doppler pressure half-time to accurately demonstrate change in mitral valve area following percutaneous mitral valvotomy. *J Am Coll Cardiol*. 1987;9:219A. Abstract.

51. Reid C, McKay C, Chandraratna P, et al. Mechanism of increase in mitral valve area by double balloon catheter balloon valvuloplasty in adults with mitral stenosis: echocardiographic-Doppler correlation. *J Am Coll Cardiol.* 1987;9:217A. Abstract.

52. Guyer DE, Gillam LD, Foale RA, et al. Comparison of the echocardiographic and hemodynamic diagnosis of rheumatic tricuspid stenosis. *J Am Coll Cardiol.* 1984;3:1135–1144.

53. Pierez JE, Ludbrook PA, Ahumada GG. Usefulness of Doppler echocardiography in detecting tricuspid valve stenosis. *Am J Cardiol.* 1985;55:601–603.

54. Pons-llado G, Carreras-Costa F, Ballester-Rodes M, et al. Pulsed Doppler patterns of left atrial flow in mitral regurgitation. *Am J Cardiol.* 1986;57:806–810.

55. Abbasi AS, Allen MW, De Cristofaro D, et al. Detection and estimation of the degree of mitral regurgitation by Doppler echocardiography. *Circulation.* 1980;61:143–147.

56. Veyrat C, Ameur A, Bas S, et al. Pulsed Doppler echocardiographic indices for assessing mitral regurgitation. *Br Heart J.* 1984;51:130–138.

57. Miyatake K, Yamamoto K, Park YD, et al. Diagnosis of mitral valve perforation by real-time two-dimensional Doppler flow imaging technique. *J Am Coll Cardiol.* 1986;8:1235–1239.

58. Zachariah ZP, Hsiung MC, Nanda NC, et al. Color Doppler assessment of mitral regurgitation induced by supine exercise in patients with coronary artery disease. *Am J Cardiol.* 1987;59:1266–1270.

59. Miyatake K, Kinoshita N, Nagata S, et al. Intracardiac flow pattern in mitral regurgitation studied with combined use of the ultrasonic pulsed Doppler technique and cross-sectional echocardiography. *Am J Cardiol.* 1980;45:155–162.

60. Blanchard D, Diebold B, Peronneau P, et al. Non-invasive diagnosis of mitral regurgitation by Doppler echocardiography. *Br Heart J.* 1981;45:589–593.

61. Panidis IP, McAllister M, Ross J, et al. Prevalence and severity of mitral regurgitation in the mitral valve prolapse syndrome: A Doppler echocardiographic study of 80 patients. *J Am Coll Cardiol.* 1986;7:975–981.

62. Liu MW, Louie EK. Independent pulsed Doppler mapping techniques. Limitations in the prediction of the angiographic severity of mitral regurgitation. *Chest.* 1989;96:1263–1267.

63. Helmcke F, Nanda NC, Hsiung MC, et al. Color Doppler assessment of mitral regurgitation with orthogonal planes. *Circulation.* 1987;75:175–183.

64. Ascah KJ, Stewart WJ, Jiang L, et al. A Doppler-two-dimensional echocardiographic method for quantitation of mitral regurgitation. *Circulation.* 1985;72:377–383.

65. Zhang Y, Ihlen H, Myhre E, et al. Measurement of mitral regurgitation by Doppler echocardiography. *Br Heart J.* 1985;54:384–391.

66. Garcia-Dorado D, Falzgraf S, Almazan A, et al. Diagnosis of functional tricuspid insufficiency by pulsed-wave Doppler ultrasound. *Circulation.* 1982;66:1315–1321.

67. Miyatake K, Okamoto M, Kinoshita N, et al. Evaluation of tricuspid regurgitation by pulsed Doppler and 2-dimensional echocardiography. *Circulation.* 1982;66:777–784.

68. Curtius JM, Thyssen M, Bruwer HM, et al. Doppler versus contrast echocardiography for diagnosis of tricuspid regurgitation. *Am J Cardiol.* 1985;56:333–336.

69. Yock P, Popp R. Non-invasive measurement of right ventricular systolic pressure by Doppler ultrasound in patients with tricuspid regurgitation. *Circulation.* 1984;70:657–662.

70. Berger M, Haimowitz A, Van Tosh A, et al. Quantitative assessment of pulmonary hypertension in patients with tricuspid regurgitation using continuous wave Doppler ultrasound. *J Am Coll Cardiol.* 1985;6:359–365.

71. Cunha CLP, Giulian ER, Callahan JA, et al. Echophonocardiographic findings in patients with prosthetic heart valve malfunction. *Mayo Clin Proc.* 1980;55:231–242.

72. Mintz GS, Carlson EB, Kotler MN. Comparison of non-invasive techniques in evaluation of the non tissue cardiac valve prosthesis. *Am J Cardiol.* 1982;49:39–44.

73. Kotler MN, Mintz GS, Panidis I, et al. Non-invasive evaluation of normal and abnormal prosthetic valve function. *J Am Coll Cardiol.* 1983;2:151–173.

74. Amann FW, Burckhardt D, Jenzer H, et al. Echocardiographic findings in prosthetic mitral valve dysfunction. *Am Heart J.* 1984;108:1573–1577.

75. Forman MB, Phelan BK, Robertson RM, et al. Correlation of two-dimensional echocardiography and pathologic findings in porcine valve dysfunction. *J Am Coll Cardiol.* 1985;5:224–230.

76. Holen J, Simonsen S, Froysaker T. An ultrasound Doppler technique for the noninvasive determination of the pressure gradient in the Bjork-Shiley mitral valve. *Circulation.* 1979;59:436–442.

77. Weinstein I, Marberge J, Perez J. Ultrasonic assessment of the St. Jude prosthetic valve: M-mode, 2-dimensional, and Doppler echocardiography. *Circulation.* 1983;68:897–905.

78. Panidis JP, Rose J, Mintz GS: Normal and abnormal prosthetic valve function as assessed by Doppler echocardiography. *J Am Coll Cardiol.* 1986;8:317–326.

79. Sagar KB, Wann LS, Paulsen WHJ, et al. Doppler echocardiographic evaluation of Hancock and Bjork-Shiley prosthetic valves. *J Am Coll Cardiol.* 1986;7:681–687.

80. Wilkins GT, Giliam LD, Kritzer GL, et al. Validation of continuous wave Doppler echocardiographic measurements of mitral and tricuspid prosthetic valve gradients: A simultaneous Doppler-catheter study. *Circulation.* 1986;74:786–795.

81. Ramirez ML, Wong M. Reproducibility of stand-alone continuous wave Doppler recordings of aortic flow velocity across bioprosthetic valves. *Am J Cardiol.* 1985;55:1197–1199.

82. Rashtian MY, Stevenson DM, Allen DT, et al. Flow characteristics of four commonly used mechanical valves. *Am J Cardiol.* 1986;58:743–752.

83. Veyrat C, Witchitz S, Lessana A, et al. Valvar prosthetic dysfunction: Localization and evaluation of the dysfunction using the Doppler technique. *Br Heart J.* 1985;54:273–284.

84. Ferrara RP, Labovitz AJ, Wiens RD, et al. Prosthetic mitral regurgitation detected by Doppler echocardiography. *Am J Cardiol.* 1985;55:229–230.

85. Kronzon I, Mercurio P, Winer HE, et al. Echocardiographic evaluation of Carpentier mitral valvuloplasty. *Am Heart J.* 1983;106:362–368.

86. Kenny J, Cohn L, Shemin R, et al. Doppler echocardiographic evaluation of ring mitral valvuloplasty for pure mitral regurgitation. *Am J Cardiol.* 1987;59:341–345.

87. Williams GA, Labovitz AJ. Doppler hemodynamic evaluation of prosthetic (Starr-Edwards and Bjork-Shiley) and bioprosthetic (Hancock and Carpentier-Edwards) cardiac valves. *Am J Cardiol.* 1985;56:325–332.

88. Come PC. Pitfalls in the diagnosis of periprosthetic valvular regurgitation by pulsed Doppler echocardiography. *J Am Coll Cardiol.* 1987;9:1176–1179.

89. Sprecher DL, Adamick R, Adams D, et al. In vitro color flow, pulsed and continuous wave Doppler ultrasound masking of flow by prosthetic valves. *J Am Coll Cardiol.* 1987;9:1306–1310.

Chapter 23

The Value of Radioisotope Blood Pool Imaging for Evaluation of Valvular Heart Disease

Robert O. Bonow, MD

During the past decade, major advances in cardiac valve replacement surgery have resulted in reduced operative mortality and improved long-term postoperative survival rates of patients undergoing aortic or mitral valve replacement. However, despite major recent improvements in operative techniques and prosthetic valve design, valve replacement continues to entail both immediate and long-term risks. These risks clearly are not justified in all patients with valvular heart disease. In general, aortic or mitral valve replacement is indicated when the patient first develops significant cardiac symptoms. Recent data have also identified subgroups of asymptomatic patients who may benefit from early operation before the onset of symptoms. This has been especially the case in valvular regurgitation. This realization has prompted a movement toward early operation in patients with minimal or no symptoms who manifest evidence of left ventricular dysfunction. Noninvasive imaging techniques are critical in making this evaluation and have had a major impact in patient management.

The imaging modality that appears most useful to date in evaluating patients with valvular heart disease is echocardiography, as this method is readily available and may be used to characterize valve morphology accurately as well as left ventricular wall thickness, internal dimensions, and wall stress. In addition, Doppler echocardiography may be used to visualize abnormal flow patterns across stenotic or regurgitant valves, to quantitate valve gradients, and to evaluate the severity of regurgitant lesions on a semiquantitative basis. The applications of radionuclide angiography in the evaluation and management of patients with valvular heart disease are more limited than are those of echocardiography. Nonetheless, nuclear cardiologic methods provide important information regarding global left ventricular and right ventricular function; in subsets of patients, especially those with aortic or mitral regurgitation, these methods may provide information that is critically important in patient management.

In contrast to the primary role of nuclear imaging methods in the assessment of left and right ventricular function, other applications in valvular heart disease, such as attempts to exclude coexistent coronary artery disease or to quantitate valvular regurgitation, are limited. Neither thallium 201 scintigraphy nor cardiac blood pool imaging has a satisfactory predictive accuracy in detecting the presence of concomitant coronary artery disease in patients with left ventricular pressure or volume overload. Exercise-induced regional thallium perfusion defects are common and nonspecific in patients with left ventricular hypertrophy in these conditions, and thallium scintigraphy cannot assess accurately the likelihood of underlying coronary lesions.[1] In addition, abnormal ejection fraction responses during exercise are prevalent in aortic stenosis, aortic regurgitation, and mitral regurgitation in the absence of coronary artery disease,[2-4] and such nonspecific responses limit the use of exercise radionuclide angiography to evaluate associated coronary artery disease. Similarly, regional wall motion abnormalities are common, both at rest and during exercise, in the volume-overloaded left ventricle.[5,6]

In attempting to measure valvular insufficiency, semiquantitative information can be derived from radionuclide angiography by computing the ratio of left ventricular stroke volume to right ventricular stroke volume,[7-9] from which both the regurgitant fraction and the regurgitant volume to end-diastolic volume ratio may be calculated. Such methods are imperfect for several reasons. Imprecision in defining right ventricular regions of interest results in an underestimation of right ventricular stroke volume relative to left ventricular stroke volume in many patients, such that the left-right stroke volume ratio is often greater than unity even in normal subjects. This method is also invalid if associated right-sided regurgitation is present. Finally, this method yields only the net extent of left-sided regurgitation and in the setting of combined aortic and mitral insufficiency is unable to grade the severity of each valvular leak independently. As the severity of valvular regurgitation may be assessed well by Doppler echocardiography and by magnetic resonance imaging or ultrafast computed tomographic scanning methods, the practical utility of radionuclide methods to gauge the severity of regurgitation is limited. On the other hand, radionuclide angiography is ideally suited for assessing and quantitating left and right ventricular systolic performance, and hence this technique has important applications in patients with valvular heart disease.

MITRAL STENOSIS

Symptoms in patients with pure mitral stenosis reflect reduced cardiac output, pulmonary venous congestion, pulmonary artery hypertension, and ultimately right ventricular dysfunction. In such patients, long-term survival may be limited by chronic right ventricular failure (although well-designed studies that demonstrate the effect of right ventricular function on survival are lacking). Hence, mitral valve replacement should be performed in patients who develop severe symptoms that cannot be managed medically and in patients who manifest evidence of right ventricular dysfunction. In this regard, radionuclide angiographic evaluation of right ventricular function may be valuable in managing patients with mitral stenosis.

Although left ventricular systolic performance is often preserved in pure mitral stenosis, in many patients the left ventricular ejection fraction is subnormal.[10-14] This appears to reflect the effects of chronic underfilling of the left ventricle,[11] immobilization of the posterobasal left ventricular myocardium by a rigid mitral complex,[10] or, in some patients, foci of myocardial fibrosis representing the residual effect of previous episodes of rheumatic carditis. The reduction in ejection fraction at rest is usually minor in mitral stenosis and of little clinical importance in relation to the other hemodynamic abnormalities, although occasionally left ventricular function may be markedly depressed. Recent studies using radionuclide angiography in patients with severe mitral stenosis[14] have demonstrated that left ventricular ejection fractions almost invariably increase during exercise as compared with resting values, even in patients with subnormal ejection fractions at rest. This "normal" ejection fraction response, however, belies profound hemodynamic abnormalities induced by exercise. Reduced diastolic filling periods during exercise tachycardia cause a significant decrease in left ventricular end-diastolic volume, and hence a decrease in left ventricular stroke volume. As a result, the increase in cardiac output during exercise is greatly reduced in patients with mitral stenosis compared with that in normal subjects, and in a subset of patients cardiac output actually decreases during exercise.[14] This abnormal hemodynamic response to exercise undoubtedly contributes to the symptomatic limitation experienced by such patients, and these alterations in cardiac performance may be studied readily by radionuclide angiographic techniques.

AORTIC STENOSIS

Recent reports of operation for symptomatic patients with aortic stenosis have related excellent results, with 5-year postoperative survival rates in excess of 85%. Chronic left ventricular failure is a rare cause of late death after operation, and no preoperative hemodynamic, angiographic, or echocardiographic variables are helpful in identifying patients at risk of postoperative death. Moreover, numerous studies have indicated that preoperative left ventricular dysfunction is reversible after operation in the majority of patients with aortic stenosis,[15-18] even in patients with advanced preoperative symptoms of congestive heart failure. Thus depressed left ventricular systolic function in aortic stenosis usually represents the effect of afterload excess, such that successful reduction in afterload by valve replacement or valvuloplasty results in a substantial increase in the ejection fraction after operation (Figure 23-1). Another subset of patients with aortic stenosis is characterized by normal or even supernormal left ventricular systolic function despite symptoms of dyspnea. Such patients have elevated left ventricular filling pressures related to left ventricular hypertrophy and increased afterload. Radionuclide angiography is valuable in symptomatic patients with aortic stenosis by characterizing left ventricular systolic function and

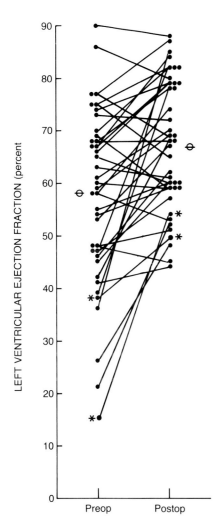

Figure 23-1 Aortic stenosis. Left ventricular ejection fractions by radionuclide angiography before (Preop) and 6 months after (Postop) aortic valve replacement in 37 symptomatic patients with critical aortic stenosis undergoing operation at the National Heart, Lung, and Blood Institute. Two other patients (asterisks) were treated by percutaneous aortic balloon valvuloplasty. Open circles with horizontal bars indicate mean values.

thereby identifying subsets with preserved versus depressed systolic function. In the subgroup with systolic dysfunction, serial postoperative studies are useful in individual patients in documenting and measuring the improvement in ventricular performance resulting from operation.

AORTIC REGURGITATION

In contrast to aortic stenosis, left ventricular systolic function is an important determinant of postoperative prognosis in patients with aortic regurgitation.[19-25] While the ejection fraction increases after operation almost uniformly in patients with left ventricular dysfunction related to aortic stenosis, improvement in the ejection fraction is less consistent in patients with ventricular dysfunction stemming from aortic regurgitation (Figure 23-2). In a subset of patients, impaired preoperative left ventricular performance reflects irreversible myocardial dysfunction rather than the reversible effects of altered loading conditions, and such patients are at risk of congestive heart failure and death after aortic valve replacement. As left ventricular systolic function is readily and accurately measured by radionuclide angiography, this technique has important

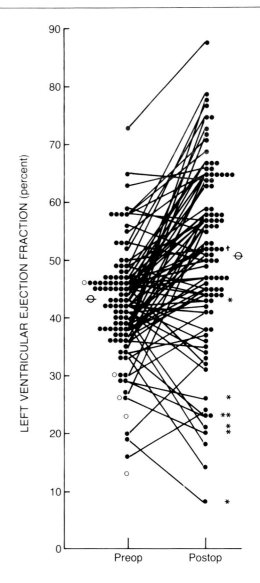

Figure 23-2 Aortic regurgitation. Left ventricular ejection fractions by radionuclide angiography before (Preop) and 6 months after (Postop) aortic valve replacement in 93 consecutive patients with chronic severe aortic regurgitation undergoing operation at the National Heart, Lung, and Blood Institute. Open circles indicate patients who died before the postoperative evaluation. Asterisks indicate patients who died of congestive heart failure, and the cross indicates a patient who died suddenly after the 6-month postoperative evaluation. Open circles with horizontal bars indicate mean values.

applications in the serial evaluation and management of patients with aortic regurgitation.

Several recent studies have addressed the influence of preoperative left ventricular systolic function on postoperative survival rates and have investigated the factors determining the reversibility of left ventricular dysfunction. These studies answer many of the critical questions regarding the timing of operation for aortic regurgitation. First, in *symptomatic* patients undergoing operation, preoperative measures of left ventricular "pump" function at rest are the most sensitive in identifying patients at risk of postoperative left ventricular dysfunction and congestive heart failure.[21–25] These indices, including the left ventricular ejection fraction measured by contrast angiography or radionuclide angiography and the left ventricular fractional shortening measured by echocardiography, continue to provide important prognostic information even in the current surgical era, in which the overall survival rate after aortic valve replacement is significantly enhanced compared with earlier results.[24] Symptomatic patients with ejection fraction and fractional shortening below the normal range comprise a high-risk group with reduced postoperative survival rates, whereas those with normal indices of left ventricular pump function have an excellent prognosis (Figure 23-3). Second, in *asymptomatic* patients the time course between the development of left ventricular dysfunction at rest and the onset of symptoms is relatively short (Figure 23-4): Two thirds or more of asymptomatic patients who manifest evidence of left ventricular dysfunction develop symptoms requiring operation within 2 to 3 years.[26]

Third, long-term postoperative prognosis and improvement in left ventricular function are enhanced in asymptomatic patients or mildly symptomatic patients with left ventricular dysfunction, compared with more severely symptomatic patients.[23–25,27] Fourth, reversal of left ventricular dysfunction after operation is also dependent upon the duration of preoperative left ventricular dysfunction[25,28]: Patients with a brief duration of preoperative ventricular dysfunction demonstrate a significantly greater decrease in left ventricular dilatation and an increase in the ejection fraction after operation compared with patients in whom the duration of preoperative left ventricular dysfunction is more prolonged (Figure 23-5). These data support the concept that postoperative survival and postoperative left ventricular function will be enhanced if asymptomatic or mildly symptomatic patients with left ventricular dysfunction undergo operation without waiting for the development of more significant symptoms. Thus, the serial monitoring of left ventricular systolic performance in aortic regurgitation is of critical importance.

Although the concern that some asymptomatic patients with aortic regurgitation may develop irreversible left ventricular dysfunction is a valid one, and although some asymptomatic patients may benefit from early operation, it must be understood that this concern pertains only to a small minority of asymptomatic patients with this disease. The vast majority of asymptomatic patients who are encountered in clinical practice maintain normal left ventricular contractile function at rest for many years and usually develop symptoms before, or coincident with, the onset

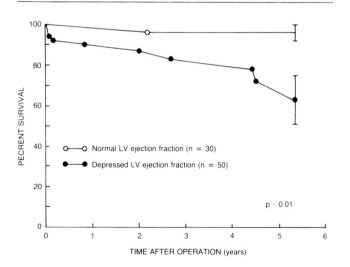

Figure 23-3 Influence of the preoperative left ventricular (LV) ejection fraction on the postoperative survival rate in 80 consecutive patients with aortic regurgitation undergoing operation from 1976 to 1983 at the National Heart, Lung, and Blood Institute. *Source:* Adapted with permission from *Circulation* (1985;72:1244), Copyright © 1985, American Heart Association.

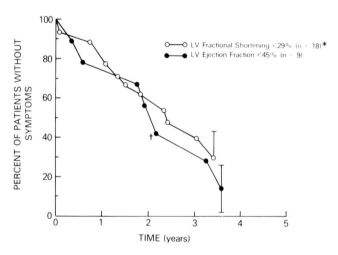

Figure 23-4 Temporal relationship in asymptomatic patients with aortic regurgitation between left ventricular (LV) dysfunction and the onset of symptoms, based on combined echocardiographic data (asterisk) from the National Heart, Lung, and Blood Institute (NHLBI) and the University of Melbourne series[26] and on radionuclide angiographic data from the NHLBI. The cross indicates a patient who died suddenly during the follow-up period.

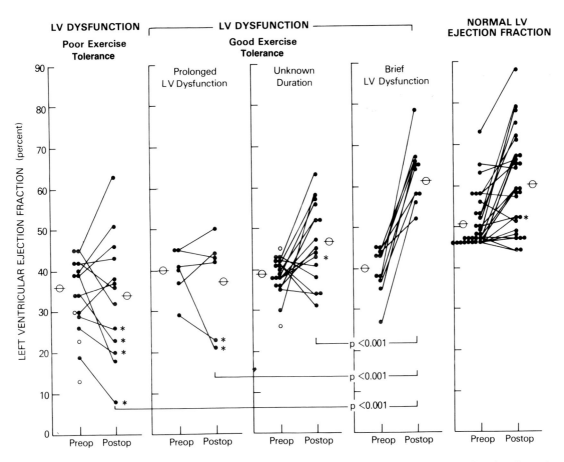

Figure 23-5 Radionuclide angiographic left ventricular (LV) ejection fractions at rest before (Preop) and 6 months after (Postop) operation in subgroups of patients with aortic regurgitation and normal versus subnormal preoperative ejection fractions. Among patients with preoperative left ventricular dysfunction, those with preserved preoperative exercise tolerance and only a brief duration of left ventricular dysfunction had significantly greater reversal of left ventricular function after operation compared with patients with prolonged preoperative left ventricular dysfunction or impaired exercise tolerance. Symbols are defined in the legend to Figure 23-2. *Source:* Adapted with permission from *Circulation* (1985;72:1244), Copyright © 1985, American Heart Association.

of depressed contractile function at rest.[29–31] Asymptomatic patients with normal left ventricular systolic function (normal ejection fraction and fractional shortening) have an excellent prognosis with only a gradual rate of deterioration during conservative, nonoperative management. The long-term follow-up experience with such patients indicates that death is rare and that fewer than 4% per year require aortic valve replacement because symptoms or left ventricular dysfunction at rest develop (Figure 23-6). Patients likely to require operation over a 10-year period because symptoms or left ventricular dysfunction develops can be identified on the basis of severity of left ventricular dilatation by echocardiography, magnitude of the decrease in the ejection fraction during exercise, or, importantly, progressive changes in left ventricular dimensions or in the resting ejection fraction during the course of serial follow-up studies.[29–31] If asymptomatic patients are followed carefully and undergo operation only after the onset of symptoms or the detection of a depressed ejection fraction at rest, the operative mortality is very

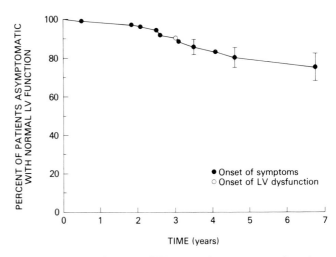

Figure 23-6 Clinical course of 77 consecutive asymptomatic patients with chronic aortic regurgitation and normal left ventricular (LV) function (normal echocardiographic fractional shortening and radionuclide angiographic ejection fractions at rest). At 7 years, 75% of patients remained asymptomatic with normal left ventricular function. *Source:* Reprinted with permission from *Circulation* (1983;68:509), Copyright © 1983, American Heart Association.

low, the long-term postoperative survival rate is excellent, and left ventricular function after operation improves in virtually every patient. Hence, although asymptomatic patients with *depressed* left ventricular contractile function at rest should undergo operation before the onset of symptoms, asymptomatic patients with *normal* left ventricular contractile function at rest do not require "prophylactic" aortic valve replacement to preserve left ventricular function.

Radionuclide angiography also permits evaluation of the ejection fraction response to exercise in aortic regurgitation, although the clinical value of this measurement remains unclear at present. Numerous studies have indicated that abnormal ejection fraction responses develop during maximal exercise in many asymptomatic patients.[2,25,32–36] Although this abnormal response may represent in some patients early evidence of true myocardial dysfunction,[35,36] such abnormal responses are very nonspecific, as the ejection fraction may be influenced dramatically by the sudden changes in loading conditions developing during exercise.[37] The likelihood of an ejection fraction decrease with exercise and the magnitude of the decrease itself are related directly to the severity of left ventricular dilatation, wall stress, and regurgitant volume.[8,9,25,34,38] Thus patients with the most severe degree of left ventricular volume overload are those most likely to have marked reductions in the ejection fraction with exercise. Such patients are also subject to the greatest alteration in loading conditions with exercise, as the regurgitant volume decreases with exercise in response to peripheral vasodilatation and shortened diastolic filling periods.[8,9]

Further evidence of the load dependence of the ejection fraction response to exercise in aortic regurgitation is the difference in ejection fraction responses when exercise is performed in the supine position compared with the upright position. In studies employing supine exercise[2,9,32–36] a much higher prevalence of abnormal ejection fraction responses was found than in those employing upright exercise,[38,39] and in a study in which both forms of exercise were compared directly a higher prevalence of abnormal responses was found in the supine position.[40] Finally, the exercise ejection fraction does not appear to be useful in clinical decision making, as the preoperative exercise ejection fraction response does not predict postoperative survival or postoperative improvement in resting left ventricular function.[25,41] Moreover, in asymptomatic patients, the exercise ejection fraction response does not predict the likelihood of subsequent development of symptoms or impaired resting ventricular function, once the patient's age and degree of ventricular dilatation at rest are accounted for.[31] Therefore, in the evaluation and management of patients with aortic regurgitation, the resting ejection fraction provides extremely valuable clinical information, whereas the exercise ejection fraction is of questionable value.

MITRAL REGURGITATION

Patients with mitral regurgitation, like patients with aortic regurgitation, may develop irreversible left ventricular dysfunction while asymptomatic. Survival after mitral valve replacement is influenced significantly by the preoperative left ventricular ejection fraction.[4,42] Determining the optimal timing of operation for mitral regurgitation, however, has been elusive, since deteriorating left ventricular function is often masked by the ability of the left ventricle to eject into a low-impedance left atrium. Hence the ejection fraction and other ejection phase indices consistently overestimate true left ventricular function.[43] This limits the practical usefulness of radionuclide angiography or of other techniques to measure the ejection fraction in the evaluation and management of patients with mitral regurgitation. While the left ventricular ejection fraction usually increases after operation for aortic regurgitation, the ejection fraction almost uniformly decreases after valve replacement for mitral regurgitation[16] (Figure 23-7).

Angiographic and echocardiographic studies suggest that patients at risk for persistent postoperative left ventricular dysfunction after mitral valve replacement can be identified by preoperative diastolic dimensions greater than 70 mm, systolic dimensions greater than 45 mm, or elevated indices of left ventricular wall stress.[44–46] However, the prognostic implications of these findings are unclear, since the available studies included only a small number of patients.[44–46] Moreover, all of the patients in these studies were severely symptomatic (and some had associated coronary artery disease), and so the extrapolation of these findings to the management of asymptomatic patients with pure mitral regurgitation is not possible. In addition, a negative impact of preoperative left ventricular size, function, or wall stress on postoperative survival has not been demonstrated, and long-term serial postoperative studies have demonstrated that the left ventricular ejection fraction progressively and consistently increases in the majority of patients over the course of 5 years after mitral valve replacement.[47] Finally, it has become apparent that mitral valve repair results in substantially improved left ventricular function as compared with mitral valve replacement.[48–50] This will lead to changes in the threshold to perform surgery as more centers become expert in valvuloplasty procedures.

Although definitive prognostic studies are lacking, it appears reasonable to recommend operation to patients with substantial symptoms and also in asymptomatic patients in whom the ejection fraction is below normal

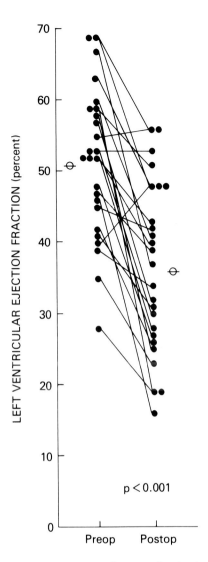

Figure 23-7 Mitral regurgitation. Left ventricular ejection fraction by radionuclide angiography before (Preop) and 6 months after (Postop) mitral valve replacement in 25 symptomatic patients with chronic mitral regurgitation undergoing operation at the National Heart, Lung, and Blood Institute. Open circles with horizontal bars indicate mean values.

or in whom the ejection fraction decreases significantly during serial studies into the low-normal range. Thus, although ejection fraction measurements do not assess accurately the true left ventricular systolic function in mitral regurgitation, serial evaluations by radionuclide angiography are useful in individual patients in helping to determine the optimal timing of valve replacement or repair.

Right ventricular dysfunction may develop during the natural history of mitral regurgitation and, as in patients with mitral stenosis, has important prognostic implications. The right ventricular ejection fraction measured by radionuclide angiography correlates with symptom severity and exercise tolerance in patients with mitral regurgitation, apparently because of the significant relationship between the right ventricular ejection fraction and pulmonary arterial and wedge pressures.[3] In addition, postoperative survival rates tend to be reduced in patients with impaired function compared with those with normal right ventricular ejection fractions.[3] These data indicate another clinically relevant application of radionuclide angiography in the evaluation of mitral regurgitation.

SUMMARY

Valve replacement or repair should be performed once significant symptoms develop in patients with valvular heart disease. Lacking important symptoms, operation should also be performed in patients with aortic regurgitation or mitral regurgitation who manifest consistent and reproducible evidence of left or right ventricular contractile dysfunction at rest. Noninvasive imaging techniques should play a major role in this evaluation. As indices of left and right ventricular pump function are the most critical clinical parameters in determining long-term prognosis, and since accurate measurement of left and right ventricular ejection fractions is readily accomplished by radionuclide angiography, this method has an important role in the serial evaluation and decision-making process for many patients. An important clinical decision, such as recommending aortic or mitral valve replacement in the asymptomatic patient, should not be based on a single echocardiographic or radionuclide angiographic measurement alone. However, when these data consistently indicate impaired contractile function at rest on repeated measurements, operation is indicated even in asymptomatic patients. This strategy should reduce the likelihood of irreversible left ventricular dysfunction in these patients and improve long-term postoperative survival rates.

REFERENCES

1. Bailey IK, Come PC, Kelley DT, et al. Thallium-201 perfusion imaging in aortic valve stenosis. *Am J Cardiol*. 1977;40:889–899.

2. Borer JS, Bacharach SL, Green MV, et al. Exercise-induced left ventricular dysfunction in symptomatic and asymptomatic patients with aortic regurgitation: Assessment with radionuclide cineangiography. *Am J Cardiol*. 1978;42:351–357.

3. Hochreiter C, Niles W, Devereux RB, et al. Mitral regurgitation: Relationship of noninvasive descriptors of right and left ventricular performance to clinical and hemodynamic findings and to prognosis in medically and surgically treated patients. *Circulation*. 1986;73:900–912.

4. Milanes JC, Paldi J, Romero M, et al. Detection of coronary artery disease in aortic stenosis by exercise gated nuclear angiography. *Am J Cardiol*. 1984;54:787–791.

5. Hecht HS, Hopkins JM. Exercise-induced regional wall motion abnormalities on radionuclide angiography: Lack of reliability for detection of coronary artery disease in the presence of valvular heart disease. *Am J Cardiol*. 1981;47:861–865.

6. Osbakken MD, Bove AA, Spann JF. Left ventricular regional wall motion and velocity of shortening in chronic mitral and aortic regurgitation. *Am J Cardiol*. 1981;47:1005–1009.

7. Urquhart J, Patterson RM, Packer M, et al. Quantification of valve regurgitation by radionuclide angiography before and after valve replacement surgery. *Am J Cardiol*. 1981;47:287–291.

8. Steingart RM, Yee C, Weinstein L, et al. Radionuclide ventriculographic study of adaptations to exercise in aortic regurgitation. *Am J Cardiol*. 1983;51:483–488.

9. Gerson MC, Engel PJ, Mantil JC, et al. Effects of dynamic and isometric exercise on the radionuclide-determined regurgitant fraction in aortic insufficiency. *J Am Coll Cardiol*. 1984;1:98–106.

10. Heller SJ, Carleton RA. Abnormal left ventricular contraction in patients with mitral stenosis. *Circulation*. 1970;42:1099–1110.

11. Dodge HT, Kennedy JW, Petersen JL. Quantitative angiocardiographic methods in the evaluation of valvular heart disease. *Prog Cardiovasc Dis*. 1973;16:1–23.

12. Bolen JL, Lopes MG, Harrison DG, et al. Analysis of left ventricular function in response to afterload changes in patients with mitral stenosis. *Circulation*. 1975;52:894–900.

13. Newman GE, Rerych SK, Bounous PE, et al. Noninvasive assessment of hemodynamic effects of mitral valve commissurotomy during rest and exercise in patients with mitral stenosis. *J Thorac Cardiovasc Surg*. 1979;78:750–756.

14. Choi BW, Barbour DJ, Leon MB, et al. Left ventricular systolic function and diastolic filling characteristics in patients with severe mitral stenosis. *J Am Coll Cardiol*. 1988;11:90A. Abstract.

15. Smith N, McAnulty JH, Rahimtoola SH. Severe aortic stenosis with impaired left ventricular function and clinical heart failure: Results of valve replacement. *Circulation*. 1978;58:255–264.

16. Thompson R, Yacoub M, Ahmed M, et al. Influence of preoperative left ventricular function on results of homograft replacement of the aortic valve for aortic stenosis. *Am J Cardiol*. 1979;43:929–938.

17. Carabello BA, Williams H, Gash AK, et al. Hemodynamic predictors of outcome in patients undergoing valve replacement. *Circulation*. 1986;74:1309–1316.

18. Redicker DE, Boucher CA, Block PC, et al. Degree of reversibility of left ventricular systolic dysfunction after aortic valve replacement for isolated aortic stenosis. *Am J Cardiol*. 1987;60:112–118.

19. Cohn PF, Gorlin R, Cohn LH, et al. Left ventricular ejection fraction as a prognostic guide in surgical treatment of coronary and valvular heart disease. *Am J Cardiol*. 1974;34:136–141.

20. Copeland JG, Griepp RB, Stinson EB, et al. Long-term follow-up after isolated aortic valve replacement. *J Thorac Cardiovasc Surg*. 1977;74:875–889.

21. Forman R, Firth BG, Barnard MS. Prognostic significance of preoperative left ventricular ejection fraction and valve lesion in patients with aortic valve replacement. *Am J Cardiol*. 1980;45:1120–1125.

22. Henry WL, Bonow RO, Borer JS, et al. Observations on the optimal time for operative intervention for aortic regurgitation, I: Evaluation of the results of aortic valve replacement in symptomatic patients. *Circulation*. 1980;61:471–483.

23. Cunha CLP, Guiliani ER, Fuster V, et al. Preoperative M-mode echocardiography as a predictor of surgical results in chronic aortic insufficiency. *J Thorac Cardiovasc Surg*. 1980;79:256–265.

24. Greves J, Rahimtoola SH, McAnulty JH, et al. Preoperative criteria predictive of late survival following valve replacement for severe aortic regurgitation. *Am Heart J*. 1981;101:300–308.

25. Bonow RO, Picone AL, McIntosh CL, et al. Survival and functional results after valve replacement for aortic regurgitation from 1976 to 1983: Impact of preoperative left ventricular function. *Circulation*. 1985;72:1244–1256.

26. Bonow RO, Rosing DR, Kent KM, et al. Timing of operation for chronic aortic regurgitation. *Am J Cardiol*. 1982;50:325–336.

27. Bonow RO, Borer JS, Rosing DR, et al. Preoperative exercise capacity in symptomatic patients with aortic regurgitation as a predictor of postoperative left ventricular function and long-term prognosis. *Circulation*. 1980;62:1280–1290.

28. Bonow RO, Rosing DR, Maron BJ, et al. Reversal of left ventricular dysfunction after valve replacement for chronic aortic regurgitation: Influence of duration of preoperative left ventricular dysfunction. *Circulation*. 1984;70:570–579.

29. Bonow RO, Rosing DR, McIntosh CL, et al. The natural history of asymptomatic patients with aortic regurgitation and normal left ventricular function. *Circulation*. 1983;68:509–517.

30. Siemienczuk D, Greenberg B, Morris C, et al. Chronic aortic insufficiency: Factors associated with progression to aortic valve replacement. *Ann Intern Med*. 1989;110:587–592.

31. Bonow RO, Lakatos E, Maron BJ, et al. The natural history of asymptomatic patients with chronic aortic regurgitation: Serial long-term changes in left ventricular function. *J Am Coll Cardiol*. 1988;11:160A. Abstract.

32. Boucher CA, Wilson RA, Kanarek DJ, et al. Exercise testing in asymptomatic or minimally symptomatic aortic regurgitation: Relationship of left ventricular ejection fraction to left ventricular filling pressure during exercise. *Circulation*. 1983;67:1091–1100.

33. Huxley RL, Gaffney A, Corbett JR, et al. Early detection of left ventricular dysfunction in chronic aortic regurgitation as assessed by contrast angiography, echocardiography, and rest and exercise scintigraphy. *Am J Cardiol*. 1983;51:1542–1550.

34. Goldman ME, Packer M, Horowitz SF, et al. Relation between exercise-induced changes in ejection fraction and systolic loading conditions at rest in aortic regurgitation. *J Am Coll Cardiol*. 1984;3:924–929.

35. Shen WF, Roubin GS, Choong CYP, et al. Evaluation of relationship between myocardial contractile state and left ventricular function in patients with aortic regurgitation. *Circulation*. 1985;71:31–38.

36. Massie BM, Kramer BL, Loge D, et al. Ejection fraction response to supine exercise in asymptomatic aortic regurgitation: Relation to simultaneous hemodynamic measurements. *J Am Coll Cardiol*. 1985;5:847–855.

37. Kawanishi DT, McKay CR, Chandraratna AN, et al. Cardiovascular response to dynamic exercise in patients with chronic symptomatic mild-to-moderate and severe aortic regurgitation. *Circulation*. 1986;73:62–72.

38. Lewis SM, Riba AL, Berger HJ, et al. Radionuclide angiographic exercise left ventricular performance in chronic aortic regurgitation: Relationship to resting echocardiographic ventricular dimensions and systolic wall stress index. *Am Heart J*. 1982;103:498–504.

39. Iskandrian AS, Hakki AH, Manno B, et al. Left ventricular function in chronic aortic regurgitation. *J Am Coll Cardiol*. 1983;1:1374–1380.

40. Shen WF, Roubin GS, Fletcher PJ, et al. Effects of upright and supine position on cardiac rest and exercise response in aortic regurgitation. *Am J Cardiol*. 1985;55:428–431.

41. Gee DS, Juni JE, Santinga JT, et al. Prognostic significance of exercise-induced left ventricular dysfunction in chronic aortic regurgitation. *Am J Cardiol*. 1985;56:605–609.

42. Phillips HR, Levine FH, Cartes JE, et al. Mitral valve replacement for isolated mitral regurgitation: Analysis of clinical course and late postoperative left ventricular ejection fraction. *Am J Cardiol*. 1981;48:647–654.

43. Berko B, Gaasch WH, Tanigawa N, et al. Disparity between ejection and end-systolic indexes of left ventricular contractility in mitral regurgitation. *Circulation.* 1987;75:1310–1319.

44. Schuler G, Peterson KL, Johnson A, et al. Temporal response of left ventricular performance to mitral valve surgery. *Circulation.* 1979; 59:1218–1231.

45. Carabello BA, Nolan SP, McGuire LB. Assessment of preoperative left ventricular function in patients with mitral regurgitation: Value of the end-systolic wall stress end-systolic volume ratio. *Circulation.* 1981;64:1212–1217.

46. Zile MR, Gaasch WH, Carroll JD, et al. Chronic mitral regurgitation: Predictive value of preoperative echocardiographic indexes of left ventricular function and wall stress. *J Am Coll Cardiol.* 1984; 3:235–242.

47. Urquhart J, Bonow RO, Maron BJ, et al. Long-term significance of the left ventricular dysfunction that occurs immediately after mitral valve replacement for mitral regurgitation. *J Am Coll Cardiol.* 1984; 3:559. Abstract.

48. David TE, Uden DE, Strauss HD. The importance of the mitral apparatus in left ventricular function after correction of mitral regurgitation. *Circulation.* 1983;68(suppl 2):II-76–II-82.

49. Goldman ME, Mora F, Guarino T, et al. Mitral valvuloplasty is superior to valve replacement for preservation of left ventricular function: An intraoperative two-dimensional echocardiographic study. *J Am Coll Cardiol.* 1987;10:568–575.

50. Hennein HA, Swain JA, McIntosh CL, et al. Comparative assessment of chordal-sparing versus chordal-resecting mitral valve replacement. *J Thorac Cardiovasc Surg.* (In press.)

Chapter 24

Evaluation of Valvular Heart Disease by Ultrafast Computed Tomography

Eulalia Roig, MD, Eva V. Chomka, MD, and Bruce H. Brundage, MD

The diagnosis of valvular heart disease usually can be accomplished at the bedside with the history and physical examination. Noninvasive techniques are useful in confirming the diagnosis and in providing more accurate information about valve anatomy and function; in selected cases they may prevent the need for cardiac catheterization.[1]

Since the first utilization of echocardiography in 1953, it has emerged as the most commonly used noninvasive technique for the diagnosis of valvular heart disease, particularly since the development of two-dimensional and Doppler technology.[2-5] Its excellent ability to image valve anatomy and function coupled with its totally noninvasive nature makes echocardiography ideal for serial evaluation of patients with valvular heart disease. However, there is considerable patient-to-patient variation in the quality of the images obtained, and the technique is highly dependent on operator skill.

Scintigraphic techniques also have been used in the evaluation of patients with valvular heart disease. Left ventricular function and regurgitant fraction can be evaluated with radionuclide angiography.[6-10] These measurements can be done at rest and during exercise.[11-13] However, the low spatial resolution of radioisotopic methods and the superimposition of the cardiac chambers are limiting factors in the evaluation of valvular disease.

Recently the first use of ultrafast computed tomography (CT) for the evaluation of valvular heart disease was reported.[14,15] Ultrafast CT is a minimally invasive method with a fast acquisition time. The technique provides high-resolution, real-time images of valvular anatomy and can assess ventricular function. This chapter reviews the current use of ultrafast CT for the evaluation of valvular heart disease.

ULTRAFAST CT

The ultrafast CT scanner (see Chapter 3) utilizes a powerful electron beam generated from a cathode positioned above the patient's head. The beam is magnetically deflected to strike four tungsten target-ring anodes that encircle the patient's thorax and generate a fan of x-rays. Two rings of stationary detectors are placed above the patient and opposite the tungsten targets. The x-rays traverse the patient's thorax; the resulting attenuation, determined by the detectors, is digitized and displayed on a television console for analysis as cross-sectional views of the thorax. Each target ring creates two adjacent 1-cm thick slices of transverse anatomy; eight adjacent slices may be obtained by cascading the beam across the four rings. After administration of a bolus of 25 to 35 mL of contrast medium, two different scanning protocols can be employed: flow mode or cine-mode. The flow mode gates scanning to the electrocardiogram. Each scan is obtained at the same point of the cardiac cycle for every heartbeat or for whatever cycle interval is desired for up to 20 scans (every other heartbeat, every third heartbeat, and so forth). The contrast medium is usually injected at a rate of 10 mL/s. The flow mode is used to evaluate the transit time of a bolus of contrast medium through the heart. By analyzing the time-density curves created, cardiac output, intracardiac shunts, and myocardial perfusion can be evaluated. Cardiac output is measured by indicator dilution analysis of the iodinated contrast medium bolus time-density curve. A good correlation with the thermodilution determination of cardiac output has been reported.[16]

In the cine-mode, one target ring is swept repetitively every 58 milliseconds for one cardiac cycle, which yields a rate of 17 scans per second. The initiation of scanning is triggered by the electrocardiogram, which provides real-time cross-sectional images of the beating heart. With the use of the four target rings in sequence, eight cross-sectional cine-views can be obtained in seven consecutive heartbeats. Contrast medium is usually injected at a rate of 8 to 10 mL/s. The cine-mode is useful for evaluation of valve motion and ventricular function and for determination of the regurgitant fraction.[17–21]

The scanner couch has the capability of a 25° horizontal slant and a 25° caudal axial tilt; thus short- or long-axis views of the heart can be obtained.[22] The long-axis view provides the best visualization of the atrioventricular valves and is obtained by tilting the scanner couch 20° feet down and slanting it 15° to the patient's left. Semilunar valves are ideally assessed in the short-axis view, which is achieved by changing the slant 13° to the right and maintaining the 20° axial tilt.

Complete evaluation of all the cardiac valves requires two or three cine-mode studies and at least one flow-mode study to analyze valve anatomy and function, cardiac output, atrial and ventricular dimensions, and left and right ventricular function.

VENTRICULAR FUNCTION

A small amount of contrast medium is required for the visualization of both ventricles, which permits the evaluation of cross-sectional images of both ventricles at each scanning level. Analyses of scans obtained at end-diastole and end-systole permit quantitative assessment of end-diastolic volume, stroke volume, ejection fraction, myocardial mass, and regional thickening and contraction. Adding the volumes indicated by each cross-sectional image of the ventricle provides an accurate measurement of the global ejection fraction, ventricular volume, and myocardial mass.[17–21,23–28] Given the stroke volume and the heart rate, cardiac output also can be determined from this functional analysis. Ventricular function can be evaluated at rest, during exercise, or with pharmacologic interventions,[29–33] making ultrafast CT an ideal technique for determining cardiac reserve.

ASSESSMENT OF REGURGITANT FRACTIONS

If only one regurgitant valvular lesion is present and there are no intracardiac shunts, the regurgitant fraction can be measured by performing both flow-mode and cine-mode studies[15]; however, the heart rate and blood pressure must be the same during both studies. Thus, analogous to angiographic determination of the regurgitant fraction, the forward stroke volume (SV) is obtained from the indicator dilution as calculated from the ultrafast CT (UFCT) time-density curve and then compared with the left ventricular stroke volume as measured by assessment of dynamic left ventricular function:

$$\text{Regurgitation fraction} = \frac{SV(\text{dynamic } UFCT) - SV(\text{time-density curve})}{SV(\text{dynamic } UFCT)} \times 100$$

Right and left ventricular volumes have been shown to be equal in normal patients as assessed by ultrafast CT. Therefore the regurgitant fraction in valvular heart disease also can be calculated by comparing the stroke volumes of both ventricles,[20,21] provided that only one ventricle has a regurgitant lesion.

A wide interobserver variability may result from the qualitative measurement of valvular regurgitation by contrast angiography. Quantitative assessment of the regurgitant fraction by conventional ventriculography also has the potential for significant error because precise measurement of the stroke volume by contrast angiography is not always possible and measurement of the forward stroke volume by an independent method[34] at another time may introduce further error.

Radionuclide angiography also has been used for evaluation of the regurgitant fraction,[7–10] but the low spatial resolution and the overlying cardiac chambers are significant limitations to the accurate measurement of ventricular volumes or ratios. Also, mild degrees of valvular regurgitation may not be detected by this technique.

Recently, measurement of the regurgitant fraction has been performed by computer analysis of digital subtraction angiography[35] and by cardiac cine-magnetic resonance imaging,[36] but these methods need further validation.

MITRAL VALVE

The leaflets of the normal mitral valve usually can be identified in the cine-mode (Figure 24-1) as thin and smooth mobile structures at the level of the atrioventricular groove, attached to the papillary muscles by chordae tendineae. Leaflets can be seen opening during diastole and closing during ventricular systole. The late diastolic reopening of the valve, due to atrial contrac-

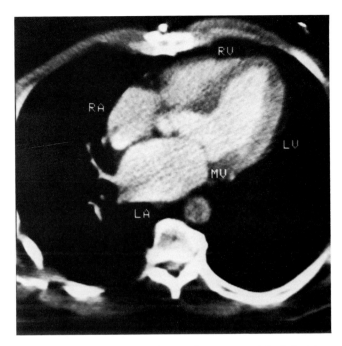

Figure 24-1 With contrast medium enhancement of the left atrium (LA) and left ventricle (LV), the normally thin leaflets of the mitral valve (MV) are visualized during ventricular systole. RA indicates the right atrium; RV, the right ventricle.

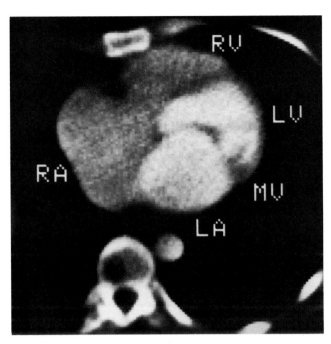

Figure 24-2 Enhancement of the left atrium (LA) and left ventricle (LV) at end-diastole demonstrates thickening of the mitral leaflets in a patient with mitral stenosis. MV indicates the mitral valve; RV, the right ventricle; RA, the right atrium.

tion, is also recognizable. The abnormal mitral valve is often even easier to visualize by ultrafast CT because of the usual increase in leaflet thickness.

The common features of the stenotic mitral valve imaged by ultrafast CT (Figure 24-2) are thickened leaflets, reduced leaflet motion, fused and sometimes shortened chordae tendineae, and calcium deposits. The motion of the valve is best evaluated in the long-axis view by the cine-mode. Elongated chordae tendineae, prolapsing leaflets, and incomplete leaflet coaptation during ventricular systole are features of regurgitant lesions. Mitral valve prolapse is diagnosed by ultrafast CT when a leaflet protrudes across the plane of the mitral valve ring into the left atrium during ventricular systole (Figure 24-3).

Usually, all four cardiac chambers can be visualized in 12 cross-sectional images of the heart by performing two six-level flow-mode or cine-mode studies. Enlargement of the left atrium and appendage is frequent in both stenotic and regurgitant lesions. The combination of enlargement of the left atrium, pulmonary artery, and right ventricle is a reliable indicator of pulmonary hypertension and may help to differentiate acute and chronic mitral regurgitation. Left ventricular enlargement is also a common finding in mitral regurgitation.

Left ventricular function is usually normal in young patients with isolated mitral stenosis. However, it has a wide spectrum in patients with mitral regurgitation. A normal or high ejection fraction has been reported in mild to severe cases of mitral regurgitation. In patients

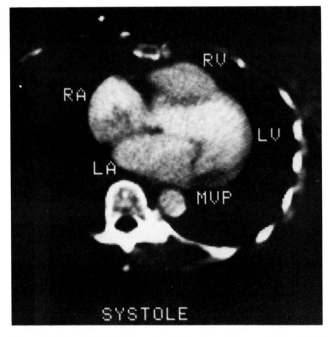

Figure 24-3 A slightly thickened mitral valve with a posterior leaflet prolapsing (MVP) into the left atrium (LA) is detected during ventricular systole. LV indicates the left ventricle; RV, the right ventricle; RA, the right atrium.

with a large amount of mitral regurgitation, a "normal" ejection fraction indicates some degree of left ventricular dysfunction. If the ejection fraction is moderately reduced (40% to 50%), severe impairment of ventricular function may be present. A low ejection fraction (less than 30%) before surgery has been associated with a poor prognosis after valve replacement.[37-40] Therefore, sequential evaluation of the left ventricular ejection fraction is important in mildly symptomatic patients in order to time valve replacement properly. End-systolic and end-diastolic volume indices also have been described as predictors of a poor postoperative prognosis.[41] Exercise ultrafast CT to assess functional capacity and left ventricular functional reserve can be performed safely in these patients. Thus ultrafast CT is useful in the evaluation of left ventricular function in patients with mitral regurgitation at rest, during exercise, or with pharmacologic interventions and is a promising technique for the serial evaluation of these patients.

The concomitance of valvular disease and coronary artery disease is not uncommon. Circulation to the papillary muscles can be affected, resulting in abnormal mitral valve function. Infarction of the papillary muscle with rupture is usually a fatal event; however, some degree of necrosis can result in abnormal contraction or retraction of the muscle or rupture of a few chordae at the tip of the papillary muscle so that mitral regurgitation results. Ultrafast CT permits visualization of both papillary muscles at several cross-sectional levels of the left ventricle and differentiation of normal and reduced myocardial blood flow[42-44]; thus an ischemic or infarcted papillary muscle may be identified. Coronary artery disease also can cause left ventricular wall motion abnormalities that alter the normal alignment of the papillary muscles in relation to the mitral leaflets. This phenomenon can result in mitral regurgitation with otherwise normally functioning papillary muscles. Finally, a "normal" papillary muscle may function normally at rest but abnormally during exercise. The analysis of wall motion by ultrafast CT at rest and during exercise in the evaluation of patients with coronary artery disease recently has been reported.[31] This technique permits the visual assessment of left ventricular wall motion in the cine-mode and objective quantification of regional abnormalities.

Ultrafast CT is able to measure wall thickness and ventricular mass accurately[27,28] and has been demonstrated to be useful for the diagnosis of hypertrophic obstructive cardiomyopathy.[45] The systolic anterior motion of the anterior leaflet of the mitral valve coapting with the septum during ventricular systole can be visualized by ultrafast CT.

Atrial thrombi are especially frequent in patients with mitral stenosis and atrial fibrillation. Ultrafast CT is useful in the diagnosis of atrial thrombi, especially those located in the left atrial appendage,[46] which usually are very difficult to diagnose by other available noninvasive techniques. Preoperative diagnosis is important to prevent intraoperative embolization.

Ultrafast CT facilitates the diagnosis of valvular masses, and it can be used to identify mitral valve vegetations or periannular abscesses in patients with bacterial endocarditis. Figure 24-4 shows a mass measuring 9 × 4 mm attached to the anterior leaflet of the mitral valve; at surgery it was found to be a mitral valve papilloma.

Mitral annular calcification is common in the elderly and is correctly diagnosed by ultrafast CT when high-density structures are visualized adjacent to the mitral valve. This entity is associated with retinal emboli and bundle-branch block.[15]

Congenital mitral valve anomalies can be evaluated by ultrafast CT.[47] Eldredge and Flicker[48] reported a study of more than 200 patients with congenital heart disease. In their series, two cases of mitral atresia, one of common atrium, and eight of complete atrioventricular canal were correctly diagnosed by ultrafast CT.

AORTIC VALVE

The normal aortic valve can be visualized in the short-axis view as a thin, three-leaflet structure that opens and closes during the cardiac cycle. Abnormal valves usually have thicker leaflets that are delineated easily by ultrafast CT.

The thickened leaflets of the stenotic aortic valve have a reduced and asymmetric valve aperture and often have calcium deposits (Figure 24-5). Differentiation

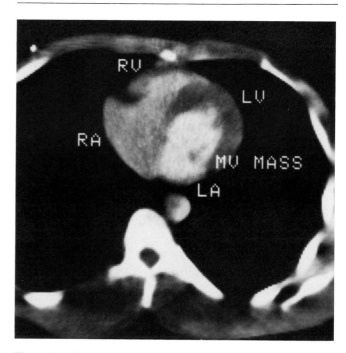

Figure 24-4 Contrast medium enhancement of the left atrium (LA) and left ventricle (LV) outlines a mitral valve mass (MV MASS), found to be a papilloma at surgery. RV indicates the right ventricle; RA, the right atrium.

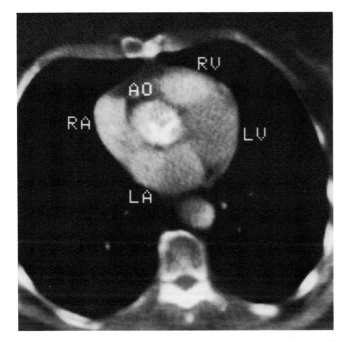

Figure 24-5 A calcified trileaflet stenotic aortic valve (AO) is well delineated in this scan. LA indicates the left atrium; LV, the left ventricle; RA, the right atrium; RV, the right ventricle.

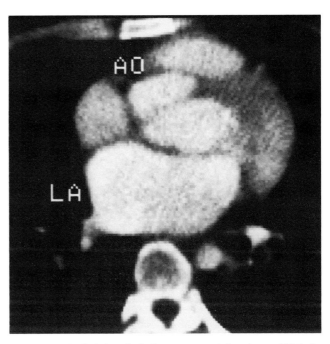

Figure 24-6 A slightly calcified aneurysm of the sinus of Valsalva involving the left coronary cusp is seen in a patient with aortic regurgitation. AO indicates the aortic valve; LA, the left atrium.

between two- and three-leaflet valves helps to determine the etiology of the stenosis.

The most common features of aortic valve regurgitation found by ultrafast CT are dilation of the aortic ring, deformity and slight thickening of the valve leaflets, elongation of one or more cusps, and eccentricity of the valve aperture. Less frequently seen but easily recognized are aneurysms of the sinus of Valsalva (Figure 24-6).

Bacterial endocarditis is a common cause of aortic regurgitation. Frequently a bicuspid aortic valve is the underlying abnormality. Vegetations can be visualized by ultrafast CT as masses on the aortic leaflets, and perforation of one or more cusps sometimes can be demonstrated. The infection may extend beyond the valve, resulting in paravalvular or myocardial wall abscesses, which can be detected by ultrafast CT. Should fistulas develop between the pulmonary and systemic circulation resulting in a left-to-right shunt, they can be diagnosed by analysis of the time-density curve obtained by scanning in the flow mode after a bolus injection of contrast medium. Measurement by ultrafast CT of cardiac output, left ventricular mass, and ejection fraction in patients with aortic stenosis may be a useful adjunct to patient management. However, assessment of left ventricular function in patients with regurgitant aortic lesions is crucial. Bonow et al[49,50] have suggested that valve replacement may be delayed in such patients until left ventricular dysfunction, with or without symptoms, develops. Others[51] have pointed out that an end-systolic volume greater than 60 mL/m^2

is associated with a higher risk for perioperative cardiac death. Recently radionuclide angiography has been used[7–10] to evaluate the regurgitant fraction and left ventricular function at rest and after exercise in patients with aortic regurgitation. Furthermore, several studies have indicated[11–13,52,53] that an abnormal exercise ejection fraction response may occur in asymptomatic patients with aortic regurgitation, indicating an abnormal reserve function, and that this exercise abnormality may precede the resting dysfunction. Current recommendations are that these patients be followed closely and valve replacement be performed when symptoms appear or resting left ventricular function deteriorates.

Ultrafast CT is a minimally invasive method that can accurately evaluate cardiac output, ejection fraction, ventricular volumes, cardiac mass, and regurgitant fraction at rest and during exercise or other forms of stress; therefore, the tool seems ideal for the serial evaluation of patients with aortic valve disease even after valve replacement. Such a detailed assessment is not possible with any other single-imaging technique.

Valvular aortic lesions frequently are associated with disease of the aorta. Ultrafast CT has been reported[54,55] to be helpful in the assessment of Marfan syndrome, aneurysms of the ascending and descending aorta, coarctation of the aorta, supravalvular aortic stenosis, and aortic dissection. Dissection of the aorta can involve the aortic root and cause acute aortic regurgitation. In a recent study by Chomka et al[54] of 40 patients with pathologic lesions of the aorta, aortic dissection was correctly diagnosed by ultrafast CT in 9 patients. In 1 of those patients rupture of the sinus of Valsalva into the pericardium was also diagnosed; in 2 patients with aortic regurgitation, previously unsuspected dissections were identified. One of these patients was taken directly to surgery without further diagnostic studies.

TRICUSPID VALVE

Assessment of the tricuspid valve by ultrafast CT is feasible, as has been reported previously.[15] The motion of the leaflets is best delineated in the long-axis view during contrast medium enhancement of the right-sided cavities.

Tricuspid stenosis is an uncommon condition, usually associated with multivalvular rheumatic heart disease or congenital anomalies. Thickening and reduced motion of the leaflets can be observed by ultrafast CT. Two cases of severe tricuspid stenosis with hypoplastic valve annulus and hypoplastic right ventricle correctly diagnosed by ultrafast CT have been reported.[48] Tricuspid regurgitation is a frequent clinical entity most often due to dilation of the tricuspid annulus secondary to enlargement of the right ventricle. Slightly thickened leaflets, right atrial and ventricular enlargement, and abnormal ventricular septal motion caused by volume overload can be observed with ultrafast CT. Less frequent causes of tricuspid regurgitation are right-sided bacterial endocarditis, tricuspid valve prolapse, Ebstein's anomaly, and carcinoid syndrome. In right-sided bacterial endocarditis, vegetation attached to the tricuspid leaflets can be visualized by ultrafast CT (Figure 24-7). Tricuspid valve prolapse can be diagnosed by ultrafast CT when a prolapsing leaflet is visualized above the tricuspid valvular plane during ventricular systole. The ultrafast CT features in Ebstein's anomaly are a downward displacement of the tricuspid septal leaflet into the right ventricle, scalloping of the valve, and marked enlargement of the right atrium. Thickened and fused tricuspid leaflets can be observed in the carcinoid syndrome.

Retrograde enhancement of the hepatic veins can be observed by ultrafast CT[15] in severe tricuspid regurgitation, as has been found by contrast echocardiography.[56] Right ventricular ejection fractions and volumes can be assessed accurately by ultrafast CT[57] because the tomographic format eliminates right atrial and ventricular overlap as occurs with radionuclide angiography and other planar imaging techniques.

PULMONIC VALVE

Pulmonic valve disease commonly is of congenital etiology. Thus pulmonic stenosis or atresia can be associated with ventricular or atrial septal defects and with a

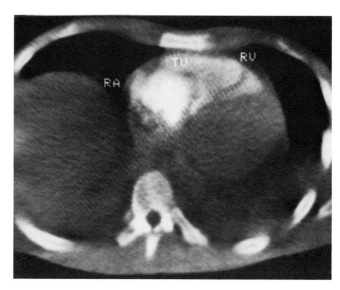

Figure 24-7 Contrast medium enhancement of the right atrium (RA) and right ventricle (RV) outlines a thickened tricuspid leaflet in a patient with right-sided bacterial endocarditis. TV indicates the tricuspid valve.

wide spectrum of other congenital anomalies. Narrowing of the outflow tract, thickening of the right ventricular wall, paradoxical septal motion, and right atrial and ventricular enlargement are the ultrafast CT features associated with pulmonic valve stenosis (Figure 24-8). Because of the 8-mm thickness of the ultrafast CT slices and the 2-mm interslice gap, right ventricular

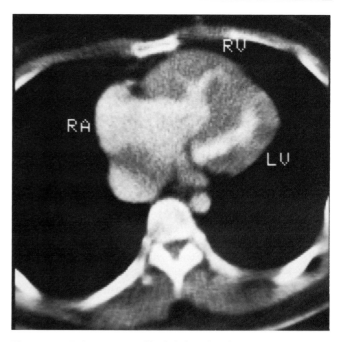

Figure 24-8 Enhancement of both left and right ventricles (LV, RV) and the right atrium (RA) identifies a markedly thickened right ventricular wall as well as an abnormal bulging of the septum into the left ventricle in a patient with pulmonic stenosis and severe right ventricular hypertension.

outflow tract obstruction and pulmonic valve stenosis may be missed in small children; also, the best view for imaging the right ventricular outflow tract has not been established. Eldredge and Flicker[48] reported that the cranial-caudal view may be best for visualization of these structures. As previously mentioned, intracardiac shunts can be diagnosed and measured and anatomic structures correctly identified by allowing a complete assessment of congenital heart disease.[47,48]

Pulmonic regurgitation is an uncommon feature that can be found in patients with severe primary pulmonary hypertension; in these patients an enlarged pulmonary artery and right-sided cavities can be identified by ultrafast CT. Right-sided bacterial endocarditis rarely may involve the pulmonic valve, but ultrafast CT probably would detect signs of volume overload such as abnormal septal motion and right ventricular enlargement.

In summary, ultrafast CT is a promising technique for the evaluation of valvular heart disease. Ultrafast CT provides real-time images of the heart, allowing the assessment of valve anatomy and function. Ventricular volumes, regurgitant fraction, ejection fraction, and myocardial mass can be measured accurately by ultrafast CT. The minimally invasive nature of this technique makes it ideal for the serial evaluation of patients.

REFERENCES

1. St John Sutton MG, St John Sutton M, Oldershaw P, et al. Valve replacement without preoperative cardiac catheterization. *N Engl J Med*. 1981;305:1233–1238.

2. Salerni R, Shaver JA. Noninvasive graphic evaluation: phonocardiography and echocardiography. In: Frankl WS, Brest AN, eds. *Valvular Heart Disease: Comprehensive Evaluation and Management*. Philadelphia, Pa: FA Davis Co; 1986:173.

3. Feigenbaum H. *Echocardiography*. 3rd ed. Philadelphia, Pa: Lea & Febiger; 1982.

4. Hatle L, Angelsen B, eds. *Doppler Ultrasound in Cardiology: Physical Principles and Clinical Applications*. Philadelphia, Pa: Lea & Febiger; 1985.

5. Nanda NC. *Doppler Echocardiography*. Tokyo, Japan: Igaku-Shoin Ltd; 1985.

6. Iskandrian AS, Hakki AH. Scintigraphic evaluation of patients with valvular heart disease. In: Frankl WS, Brest AN, eds. *Valvular Heart Disease: Comprehensive Evaluation and Management*. Philadelphia, Pa: FA Davis Co; 1986:211.

7. Thompson R, Ross I, Elmes R. Quantification of valvular regurgitation by cardiac gated pool imaging. *Br Heart J*. 1981;46:629–635.

8. Bough EW, Gandsman BJ, North DL, et al. Gated radionuclide angiographic evaluation of valve regurgitation. *Am J Cardiol*. 1980;46:423–428.

9. Sorensen SG, O'Rourke RA, Chaudhuri TK. Noninvasive quantitation of valvular regurgitation by gated equilibrium radionuclide angiography. *Circulation*. 1980;62:1089–1098.

10. Urquhart J, Patterson RE, Packer M, et al. Quantification of valve regurgitation by radionuclide angiography before and after valve replacement surgery. *Am J Cardiol*. 1981;47:287–291.

11. Borer JS, Bacharach SL, Green MV. Exercise-induced left ventricular dysfunction in symptomatic and asymptomatic patients with aortic regurgitation: Assessment with radionuclide cineangiography. *Am J Cardiol*. 1978;42:351–357.

12. Gerson MC, Engel PJ, Mantil JC, et al. Effects of dynamic and isometric exercise on the radionuclide-determined regurgitant fraction in aortic insufficiency. *J Am Coll Cardiol*. 1984;3:98–106.

13. Boucher CA, Wilson RA, Kanarek DJ, et al. Exercise testing in asymptomatic or minimally symptomatic aortic regurgitation: Relationship of left ventricular ejection fraction to left ventricular filling pressure during exercise. *Circulation*. 1983;67:1091–1100.

14. Brundage BH, Chomka EV. Ultrafast CT: A new approach to cardiac diagnosis. *Cardiac Impulse*. 1986;7:1–6.

15. Roig E, Chomka EV, Rich S, et al. Evaluation of valvular heart disease by ultrafast computed tomography. *Dynam Cardiovasc Imaging*. 1987;1:62–69.

16. Wolfkiel CJ, Ferguson JL, Chomka EV, et al. Determination of cardiac output by ultrafast computed tomography. *Am J Physiol Imaging*. 1986;1:117–123.

17. Rich S, Chomka EV, Stagl R, et al. Ultrafast computed tomography to determine left ventricular ejection fraction. *Am Heart J*. 1986;112:392–396.

18. Rumberger JA, Feiring AJ, Rees MR, et al. Quantitation of left ventricular mass and volumes in normal patients using cine computed tomography. *J Am Coll Cardiol*. 1986;7:173A. Abstract.

19. Bateman T, Whiting J, Pfaff M, et al. Cine-CT: an accurate, precise and rapid method for LVEF determination. *Circulation*. 1985;72(suppl 3):717.

20. Reiter SJ, Rumberger JA, Feiring AJ, et al. Measurement of aortic regurgitation with cine CT. *J Am Coll Cardiol*. 1986;7:154A. Abstract.

21. Stark CA, Rumberger JA, Reiter SJ, et al. Use of cine CT in assessing the severity of aortic regurgitation in patients. *Circulation*. 1986;74(suppl 2):16.

22. Rees MR, Feiring AJ, Rumberger JA, et al. Heart evaluation by cine CT: use of two new oblique views. *Radiology*. 1986;159:804–806.

23. Reiter SJ, Rumberger JA, Feiring AJ, et al. Precise determination of left and right ventricular stroke volume with cine computed tomography. *Circulation*. 1985;72(suppl 3):716.

24. Feiring AJ, Rumberger LA, Collins SM, et al. Regional ventricular function with cine CT. *J Am Coll Cardiol*. 1986;7:44A. Abstract.

25. McMillan RM, Rees MR, Weiner R, et al. Cine CT assessment of global and regional left ventricular function in ischemic heart disease. *Circulation*. 1986;74(suppl 2):484.

26. Merz R, Garrett J, Schiller N, et al. Quantitation of wall thickness by cine computed tomography. *Circulation*. 1985;72(suppl 3):722.

27. Bateman TM, Whiting JS, Forrester JS, et al. Left ventricular mass by multi-slice cardiac cine-CT: development of methods, experimental validation, and clinical application. *J Am Coll Cardiol*. 1986;7:468S.

28. Feiring AJ, Rumberger JA, Reiter SJ, et al. Determination of left ventricular mass in dogs with rapid-acquisition cardiac computed tomographic scanning. *Circulation*. 1985;72:1355–1364.

29. Chomka EV, Fletcher M, Stein M, et al. Ultrafast computed tomography during exercise bicycle ergometry. *J Am Coll Cardiol*. 1986;7:154A. Abstract.

30. Caputo GR, Dery R, Diethelm L, et al. Cine-CT evaluation of exercise induced ischemia. *Circulation*. 1986;74(suppl 2):486.

31. Roig E, Castaner A, Chomka EV, et al. Exercise ultrafast CT: A sensitive and specific test for coronary artery disease. *Clin Res*. 1987;35:320A. Abstract.

32. Stark CA, Rumberger JA, Stanford W, et al. Dobutamine stress cine CT. *Circulation*. 1986;74(suppl 2):488.

33. Chomka EV, Ferguson JL, Wolfkiel CJ, et al. The effect of contrast media on myocardial blood flow: Implications for measurements by ultrafast CT. *Circulation*. 1985;72(suppl 3):723.

34. Grossman W, Dexter L. Cardiac catheterization and angiography. In: Grossman W, ed. *Profiles in Valvular Heart Disease*. Philadelphia, Pa: Lea & Febiger; 1980:359–381.

35. Grayburn PA, Nissen SE, Elion JL, et al. Quantitation of aortic regurgitation by computer analysis of digital subtraction angiography. *J Am Coll Cardiol*. 1986;7:154A. Abstract.

36. Aurigemma G, Reichek N, Axel L, et al. Cardiac cine magnetic resonance imaging: Detection of mitral and aortic regurgitation. *J Am Coll Cardiol*. 1987;9:159A. Abstract.

37. Phillips HR, Levine FH, Carter JE, et al. Mitral valve replacement for isolated mitral regurgitation: Analysis of clinical course and late postoperative left ventricular ejection fraction. *Am J Cardiol*. 1981;48:647–654.

38. Schneider RM, Helfant RH. Timing of surgery in chronic mitral and aortic regurgitation. In: Frankl WS, Brest AN, eds. *Valvular Heart Disease: Comprehensive Evaluation and Management*. Philadelphia, Pa: FA Davis Co; 1986:361.

39. Hochreiter C, Niles N, Devereux RB, et al. Mitral regurgitation: Relationship of noninvasive descriptors of right and left ventricular performance to clinical and hemodynamic findings and to prognosis in medically and surgically treated patients. *Circulation*. 1986;73:900–912.

40. Schuler G, Peterson KL, Johnson A, et al. Temporal response of left ventricular performance to mitral valve surgery. *Circulation*. 1979;59:1218–1231.

41. Saltissi S, Crowther A, Byrne C, et al. Assessment of prognostic factors in patients undergoing surgery for non-rheumatic mitral regurgitation. *Br Heart J*. 1980;44:369–380.

42. Rumberger JA, Feiring AJ, Lipton MJ, et al. Use of ultrafast computed tomography to quantitate regional myocardial perfusion: A preliminary report. *J Am Coll Cardiol*. 1987;9:59–69.

43. Wolfkiel CJ, Ferguson JL, Chomka EV, et al. Myocardial blood flow determined by ultrafast computed tomography. *Circulation*. 1986;74(suppl 2):485.

44. Wolfkiel CJ, Ferguson JL, Chomka EV, et al. Measurement of myocardial blood flow by ultrafast computed tomography. *Circulation*. 1987;76:1262–1273.

45. Chomka EV, Wolfkiel CJ, Rich S, et al. Ultrafast computed tomography: a new method for the evaluation of hypertrophic cardiomyopathy. *Am J Noninvas Cardiol*. 1987;1:140–151.

46. Roig E, Chomka EV, Helgason CM, et al. Ultrafast computed tomography in the diagnosis of cardiac thrombus in patients with systemic embolization. *Clin Res*. 1987;35:320A. Abstract.

47. Bali C, Chomka EV, Fisher EA, et al. Ultrafast computed tomography in congenital heart disease. *Circulation*. 1987;72(suppl 3):109.

48. Eldredge WJ, Flicker S. Evaluation of congenital heart disease using cine-CT. *Am J Card Imaging*. 1987;1:38.

49. Bonow RO, Rosing DR, Kent KM, et al. Timing of operation for chronic aortic regurgitation. *Am J Cardiol*. 1982;50:325–336.

50. Bonow RO, Rosing DR, McIntosh CL, et al. The natural history of asymptomatic patients with aortic regurgitation and normal left ventricular function. *Circulation*. 1983;68:509–517.

51. Borow KM, Green LH, Mann T, et al. End-systolic volume as a predictor of postoperative left ventricular performance in volume overload from valvular regurgitation. *Am J Med*. 1980;68:655–663.

52. Samuels DA, Curfman GD, Friedlich AL, et al. Valve replacement for aortic regurgitation: Long-term follow-up with factors influencing the results. *Circulation*. 1979;60:647–654.

53. Iskandrian AS, Hakki AH, Manno B, et al. Left ventricular function in chronic aortic regurgitation. *J Am Coll Cardiol*. 1983;1:1374–1380.

54. Chomka EV, Roig E, Rich S, et al. Assessment of aortic disease by ultrafast computed tomography. *J Am Coll Cardiol*. 1987;9:159A. Abstract.

55. Brundage BH, Rich S, Spigos D. Computed tomography of the heart and great vessels: present and future. *Ann Intern Med*. 1984;101:801–809.

56. Tei C, Shah PM, Ormiston JA. Assessment of tricuspid regurgitation by directional analysis of right atrial systolic linear reflux echoes with contrast M-mode echocardiography. *Am Heart J*. 1982;103:1025–1030.

57. Mahoney LT, Smith W, Noel MP, et al. Measurement of right ventricular volume using cine computed tomography. *Circulation*. 1985;72(suppl 3):110.

Chapter 25

Magnetic Resonance Imaging of Cardiac Valvular Disease

Joseph Utz, MD

Disorders of the cardiac valves can be subdivided into two broad categories. The first category includes disorders that are characterized by gross abnormalities of valvular anatomy with subsequent redirection or misdirection of blood flow. These disorders usually are congenital in nature and frequently are accompanied by other associated anatomic abnormalities of the heart. The determination of overall cardiac anatomy is central to the examination of and the subsequent surgical treatment of these patients. Electrocardiogram (ECG)-gated spin-echo magnetic resonance imaging provides dramatic contrast between flowing blood and the myocardial wall and has been used to define cardiac anatomy in this subgroup of congenital valvular abnormalities.[1] The second group of valvular disorders is characterized by a more subtle abnormality of valvular anatomy with subsequent valvular dysfunction (stenosis or regurgitation). These disorders usually are acquired; and although they are frequently multivalvular, they result more often in hemodynamic dysfunction (such as resistance to blood flow or volume overload) than in disturbances of the path of blood flow from right atrium to right ventricle, pulmonary artery, left atrium, and left ventricle. The dynamic imaging of valvular motion and blood flow is more important than anatomic definition in this group of disorders. As a result of the lack of signal from flowing blood as well as the limited temporal resolution provided, spin-echo magnetic resonance imaging has been less helpful in these diseases. Dynamic magnetic resonance imaging with gradient-refocused echoes recently has been introduced.[2–5] This technique provides increased temporal resolution and high signal intensity from flowing blood and recently has been applied to this subgroup of cardiac valvular diseases.[6]

SPIN-ECHO MAGNETIC RESONANCE IMAGING IN CONGENITAL VALVULAR DISORDERS

The basic principles of spin-echo (cardiac-gated) magnetic resonance imaging have been described previously. The lack of signal from flowing blood coupled with cardiac gating provides dramatic contrast between intraluminal blood and the myocardial chamber wall or valvular leaflets (Figure 25-1). Two aspects of cardiac-gated spin-echo magnetic resonance imaging, however, should be emphasized. First, it is the absolute motion of the moving blood that results in the signal loss, not the direction of the flow. Thus for most practical purposes, spin-echo magnetic resonance imaging is not used to determine the velocity and direction of flow. The presence or absence of regurgitation and the direction of flow within or between cardiac chambers is difficult to ascertain with spin-echo imaging. Second, with cardiac-gated magnetic resonance imaging the images are in reality synchronized with the ECG rather than gated with the ECG such as with nuclear medicine examinations.[7] As a result of this, when a multislice

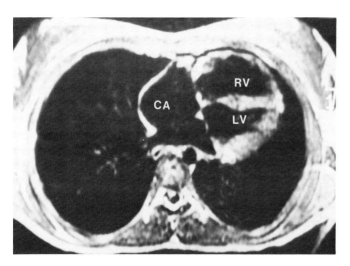

Figure 25-1 Cardiac-gated spin-echo magnetic resonance imaging in a patient with an atrial septal defect demonstrates a common atrium (CA) as well as the right ventricle (RV), the left ventricle (LV), and normal mitral and tricuspid valves.

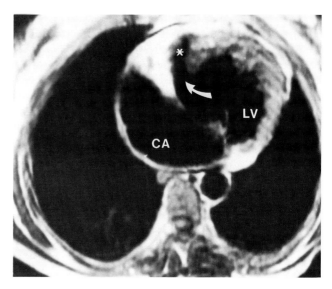

Figure 25-2 Gated spin-echo images in a patient with tricuspid atresia demonstrate a common atrium (CA), an atretic tricuspid valve, a patent mitral valve, a ventriculoseptal defect (curved arrow), the left ventricle (LV), and a hypoplastic right ventricle (*).

technique is used, each image is obtained at a slightly different phase of the cardiac cycle than the anatomically adjacent image. For example, the image obtained at the aortic outflow tract may be at end-systole, while the image obtained at the mitral and tricuspid valve plane might be at end-diastole. Dynamic information is thus difficult to obtain from spin-echo magnetic resonance images.

Multiple sequences may be used through the technique of rotated gated acquisition[8] to obtain limited dynamic information; however, the temporal resolution of the technique is limited and the imaging time is prolonged. Despite these two limitations, the application of spin-echo magnetic resonance to valvular heart disease has been described. The multiplanar capacity of magnetic resonance permits an accurate anatomic description of complex congenital heart disease without the need for exposure to ionizing radiation or intravenous contrast material. From this anatomic information, the presence of valvular disease often can be inferred. For example, the combination of a right-sided aortic arch, hypertrophied right ventricle, overriding aorta, and a ventricular septal defect defines a tetralogy of Fallot, and the presence of pulmonary stenosis can be inferred. The hypoplastic right heart syndrome recently has been examined,[9] and patients with tricuspid atresia could be separated from those with pulmonic atresia (Figure 25-2). Overall, however, with the exception of the hypoplastic right heart syndrome, valvular dysfunction is a minor part of congenital heart disease. The role of magnetic resonance imaging in these disorders is directed to the overall definition of cardiac anatomy rather than the depiction of specific valvular abnormalities.

DYNAMIC MAGNETIC RESONANCE IMAGING IN ACQUIRED VALVULAR DISEASE

Spin-echo magnetic resonance images have been applied to acquired valvular disease,[10,11] but again the evaluation of the valvular abnormality is indirect. The presence of mitral stenosis has been suggested by the combination of a dilated left atrium, normal-sized left ventricle, and increased signal within the left atrium during diastole (presumably due to slow flow). Aortic valvular disease has been inferred by the combination of ascending aortic dilatation and left ventricular hypertrophy and dilatation in a patient with a coarctation of the aorta. The evaluation of acquired abnormalities of the tricuspid and pulmonic valves with magnetic resonance imaging has yet to be described.

Although the presence or absence of valvular disease can be only suggested by spin-echo magnetic resonance images, the severity of known abnormalities such as regurgitation can be evaluated directly.[12] This may be of greater significance than the detection of the disorder. In one study of 28 regurgitant aortic or mitral valves, spin-echo images were obtained during both systole and diastole to assess chamber size and left-ventricle to right-ventricle stroke volume ratios. The stroke volume ratio increased with increasing degrees of valvular regurgitation in a linear manner and correlated well with the findings of radionuclide ventriculography. As with congenital valvular disorders, however, the evaluation of acquired valvular disease with spin-echo magnetic resonance imaging is based on

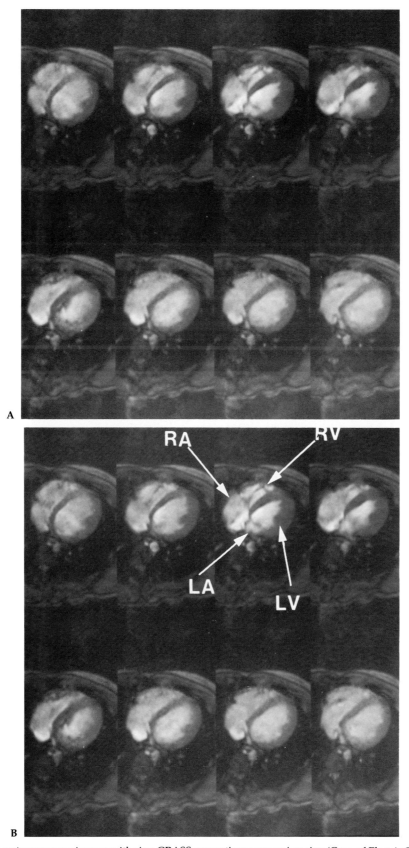

Figure 25-3 Dynamic magnetic resonance images with cine-GRASS magnetic resonance imaging (General Electric Corporation, Milwaukee, Wis). Eight alternate images from a 16-image sequence are shown at the level of the mitral valve, and the cardiac anatomy is accurately displayed. There is increased signal from flowing blood. The dynamic display of this sequence of 16 images demonstrates valvular and myocardial motion. RA indicates right atrium; RV, right ventricle; LV, left ventricle; LA, left atrium.

overall cardiac anatomy rather than the imaging of the cardiac valves or blood flow.

Acquired valvular dysfunction is one of the most common causes of nonischemic cardiac disease in the United States. As noted above, the evaluation of acquired valvular disease with spin-echo magnetic resonance is limited by the lack of signal from flowing blood and the limited temporal resolution of gated magnetic resonance images even when the rotated gated acquisition technique is utilized. A new technique of rapid dynamic magnetic resonance imaging of the heart has been introduced recently. This technique uses radiofrequency pulses of less than 90° and gradient-refocused echoes. When coupled with cardiac gating, this technique provides tomographic images of the heart with high spatial resolution, high temporal resolution of up to 32 time frames per cardiac cycle, and increased rather than decreased signal from flowing blood. An example of such a sequence is given in Figure 25-3, which shows a series of eight alternate images from a 16-image sequence at the level of the mitral valve. When displayed in a dynamic mode, wall motion and valvular dysfunction can be displayed. The images illustrate increased rather than decreased signal from intraluminal blood; more important, the signal intensity appears to be affected by the presence or absence of turbulence. This ability to detect turbulence has been applied to detection of valvular dysfunction.[6] A series of 27 patients with suspected valvular disease were examined with dynamic magnetic resonance imaging and the results were compared with those obtained by Doppler echocardiography and contrast ventriculography. Valvular incompetence was defined on magnetic resonance images as a focal triangular area of decreased signal intensity within the regurgitant chamber that extended from the valve plane and corresponded temporally to a ventricular systole for mitral regurgitation or tricuspid regurgitation and with ventricular diastole for aortic insufficiency (Figures 25-4 and 25-5). Of a total of 56 valves studied, 20 incompetent valves were detected with both magnetic resonance imaging and either echocardiography or cardiac ventriculography. There were nine false-positive examinations and no false-negative examinations. Overall, a sensitivity of 100% and a specificity of 67% were obtained in this small series.

These preliminary results have been confirmed by other investigators. Specifically, Aurigemma et al[13] evaluated the aortic valve in 46 patients and compared cine-magnetic resonance results with pulsed Doppler and color flow Doppler examinations. They noted a sensitivity and specificity of 100% for the detection of aortic regurgitation as compared with pulsed Doppler, and a sensitivity of 92% and a specificity of 50% as compared with color flow Doppler. In a similar manner,[14] the mitral valve was also evaluated in 50 patients, result-

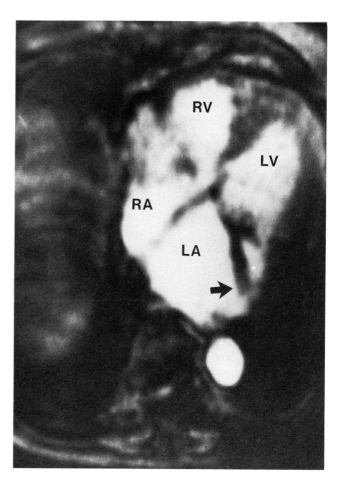

Figure 25-4 Mitral regurgitation: dynamic magnetic resonance image at the level of the mitral valve demonstrates a triangular area of decreased signal intensity (arrow) within the left atrium (LA) that corresponds with ventricular systole in a patient with mitral regurgitation. RA indicates right atrium; RV, right ventricle; LV, left ventricle.

ing in a sensitivity of 94% and a specificity of 100% for the detection of mitral regurgitation as compared with both pulsed Doppler and color flow Doppler techniques.

Even more important than the detection of regurgitation is the quantitation of regurgitation, and attempts have been made recently to measure such flow. The loss of signal on cine-nuclear magnetic resonance examinations is thought to be secondary to turbulent regurgitant flow. Using a phantom, Cook et al[15] determined that the jet area is directly related to the flow rate and inversely related to the stenotic orifice size, while the jet signal intensity is directly related to the orifice size and inversely related to the flow rate, presumably due to decreased signal intensity with increased turbulence. In clinical studies,[16] both the extent and degree of signal loss[17] have been measured and found to correlate with the severity of mitral or aortic regurgitation.

The preliminary results suggest that cine-nuclear magnetic resonance may become a powerful new tool in the detection and evaluation of valvular disease.

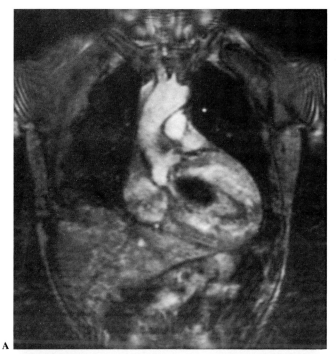

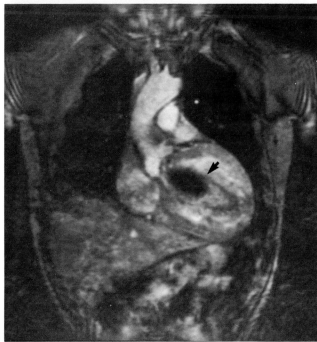

Figure 25-5 Aortic insufficiency. Dynamic magnetic resonance images in the coronal plane in a patient with aortic insufficiency demonstrate a triangular area of decreased signal intensity (arrow) within the left ventricle that corresponds with ventricular diastole, indicating regurgitant flow and aortic insufficiency.

SUMMARY

The application of spin-echo magnetic resonance imaging to valvular disorders has to date been limited by coarse temporal resolution and the overall lack of signal from flowing blood. Complex congenital heart disease can be evaluated and the known association with valvular pathology can be inferred. The evaluation of acquired valvular disease, valvular motion, and valvular incompetence with dynamic magnetic resonance imaging has yet to be explored fully.

REFERENCES

1. Fletcher BD, Jacobstein MD, Nelson AD, et al. Gated magnetic resonance imaging of congenital cardiac malformations. *Radiology*. 1984;150:137–140.
2. Herfkens RJ, Utz JA, Glover G, et al. Rapid dynamic NMR imaging of the heart: Initial experience. In: Program and abstracts of the fifth annual meeting of the Society of Magnetic Resonance in Medicine; August 19–22, 1986; Montreal, Canada. Abstracts: 247.
3. Utz J, Herfkens RJ, Glover G, et al. Rapid dynamic NMR imaging of the heart. In: Program and abstracts of the fifth annual meeting of the Society of Magnetic Resonance in Medicine; August 19–22, 1986; Montreal, Canada. Abstracts: 930.
4. Pettigrew RI, Daniels W, Churchwell A, et al. Initial clinical experience with a fast multiplane MRI technique in the study of cardiovascular flow and dynamics. In: Program and abstracts of the fifth annual meeting of the Society of Magnetic Resonance in Medicine; August 19–22, 1986; Montreal, Canada. Abstracts: 1101.
5. Sechtem U, Pflugfelder PW, White RD, et al. Cine MR imaging: Potential for the evaluation of cardiovascular function. *AJR*. 1987; 148:239–246.
6. Utz JA, Herfkens RJ, Heinsimer J, et al. Rapid dynamic MR imaging of the heart in the evaluation of valvular function. *Radiology*. 1986;161(P):198. Abstract.
7. Hedrick WR, DiSimone RN, Keen RL. NMR "gating" really means "synchronization." *J Nucl Med*. 1987;28:545–546.
8. Crooks LE, Barker B, Chang H, et al. Magnetic resonance imaging strategies for heart studies. *Radiology*. 1984;153:459–465.
9. Jacobstein MD, Fletcher BD, Goldstein S, et al. Magnetic resonance imaging in patients with hypoplastic right heart syndrome. *Am Heart J*. 1985;110:154–155.
10. Hill JA, Akins EW, Fitzsimmons JR, et al. Mitral stenosis: Imaging by nuclear magnetic resonance. *Am J Cardiol*. 1986;57:352–354.
11. Rehr RB, Filipchuk NG, Malloy C, et al. Magnetic resonance imaging in aortic valve, ascending aortic and isthmic aortic disease. *Am J Cardiol*. 1985;55:1243–1244.
12. Underwood SR, Klipstein RH, Firmin DN, et al. Magnetic resonance assessment of aortic and mitral regurgitation. *Br Heart J*. 1986;56:455–462.
13. Aurigemma G, Reichek N, Schiebler M, et al. Evaluation of aortic regurgitation by cine magnetic resonance imaging. *J Am Coll Cardiol*. 1988;11:155A. Abstract.
14. Aurigemma G, Reichek N, Schiebler M, et al. Evaluation of mitral regurgitation by cardiac cine magnetic resonance imaging. *Circulation*. 1987;76:IV-31. Abstract.
15. Cook SL, Maurer G, Berman DS, et al. Effect of flow rate and orifice size on flow jets visualized by fast NMR imaging: A phantom study. *J Am Coll Cardiol*. 1988;11:156A. Abstract.
16. Pflugfelder PW, Landzberg JS, Cassidy MM, et al. Comparison of cine magnetic resonance imaging with Doppler echocardiography for the evaluation of aortic regurgitation. *Circulation*. 1987;76:IV-31. Abstract.
17. Pflugfelder PW, Sechtem U, White RD, et al. Non-invasive evaluation of mitral regurgitation by analysis of systolic left atrial signal loss in cine magnetic resonance images. *Circulation*. 1987;76:IV-89.

Part IV

Primary Myocardial Disease

Chapter 26

Evaluation of Hypertrophic and Dilated Cardiomyopathy by Cardiac Catheterization and Angiography

Jeffrey Shanes, MD

The term "cardiomyopathy" refers to primary heart muscle disease for which no underlying cause such as atherosclerosis, hypertension, or congenital or valvular heart disease can be found.[1] The cardiomyopathies can be divided into three general categories: hypertrophic, dilated, and restrictive.[2] This chapter focuses on the angiographic and hemodynamic aspects of hypertrophic and dilated cardiomyopathy.

From the outset, it must be recognized that newer imaging techniques such as echocardiography, magnetic resonance imaging, and ultrafast computed tomography have advantages over standard angiographic techniques for evaluating patients with both dilated and (particularly) hypertrophic cardiomyopathy. These techniques allow for the actual visualization of cardiac muscle, the primary culprit in cardiomyopathy, as opposed to standard angiography, where only a silhouette of the contractile muscle is seen. Nevertheless, the angiographic features of hypertrophic and dilated cardiomyopathy continue to be important for several reasons.

First, the original descriptions of these entities were based on their angiographic appearance, since echocardiography and other newer imaging modalities were not yet available.[3,4] Thus the angiographic appearance of dilated and hypertrophic cardiomyopathy formed the basis for all subsequent descriptive work. Second, patients may have cardiac catheterization as their initial diagnostic test, and unless the patterns of these cardiomyopathies are recognized the correct diagnosis may be missed. Finally, ancillary imaging and hemodynamic data obtained at the time of catheterization, such as the detection of pressure gradients, valvular lesions, and coronary artery disease, may be important in strengthening or weakening the diagnosis of primary cardiomyopathy as well as in planning treatment strategies. Despite the potential importance of cardiac catheterization in the diagnosis of these cardiomyopathies, one must be cautious never to make a diagnosis of a dilated or hypertrophic cardiomyopathy based solely on the angiographic appearance of the heart. Any of the features that these cardiomyopathies may exhibit at catheterization can also occur as a result of other underlying diseases such as coronary, valvular, or congenital heart disease as well as hypertension. The final diagnosis of a hypertrophic or dilated cardiomyopathy must always be made only after all the clinical, hemodynamic, and other imaging data have been obtained and reviewed.

HYPERTROPHIC CARDIOMYOPATHY

The term "hypertrophic cardiomyopathy" really represents a spectrum of diseases that have one feature in common: myocardial hypertrophy of unknown cause.[5] Classically and perhaps most commonly, this hypertrophy involves disproportionate thickening of the upper septum with little or no involvement of the remaining left ventricle; this type of involvement commonly is associated with a significant systolic pressure gradient across the left ventricular outflow tract.[3] The significance and mechanism of this gradient are controversial, but it is associated with and may be explained by the narrowing of the outflow tract of the left ventricle that occurs when the anterior leaflet of the mitral valve moves anteriorly rather than posteriorly during midsystole and touches or abuts against the thickened upper septum.[6]

Numerous other patterns of hypertrophy have been described, but some of the more common types include diffuse hypertrophy of the entire ventricle,[7] disproportionate apical hypertrophy,[8,9] and finally midventricular hypertrophy associated with measurable gradients across the midventricular site of obstruction.[10]

Thus several angiographic features are typical of hypertrophic cardiomyopathy, but they vary according to the muscular pattern found. In the frontal view a generally small diastolic volume may be noted.[3,11] The papillary muscles may be quite hypertrophied and may even appear as a filling defect.[11] Because of the marked muscular thickening the ventricle may have multiple irregular indentations as a result of a prominent trabecular pattern. Commonly, an inward concavity at the midportion of the right inferior margin is noted, due to the bulging of the septum.[3] While this typically is limited to the inferior margin of the heart, a biconcave or hourglass configuration is typical in the midventricular obstructive form of the disease.[9] In patients with apical hypertrophic cardiomyopathy, gradual distal tapering or distal superior and inferior filling defects may be seen.[12] This diastolic configuration has been said to appear as a spade on a playing card; this pattern becomes more marked during systole.[9] In the lateral projection during diastole the outflow tract is narrowed by a septal concave bulge,[3] creating the so-called diastolic cone effect,[13] with its anterior border being the hypertrophied septum and its posterior border the anterior leaflet of the mitral valve.

During systole, when the systolic anterior motion of the mitral valve is present, the angiographic appearance is classic. In the lateral projection the anterior leaflet of the mitral valve normally moves in the posterior direction, which produces complete closure of the mitral orifice and widening of the outflow tract. However, with obstructive hypertrophic cardiomyopathy, after an initial posterior motion of the anterior leaflet of the mitral valve there is a sudden movement of the leaflet toward the septum, which simultaneously protrudes even further into the outflow tract during systolic contraction. This abnormal motion of the mitral valve and the disproportionate upper septal thickening give the angiographic appearance of an hourglass deformity of the outflow tract and frequently result in at least mild to moderate mitral regurgitation, which coincides with mid- to late systole[3,11] (Figure 26-1). In the lateral view during systole, depending on the extent and degree of hypertrophy, the entire left ventricular cavity may be obliterated or may show a "tonguelike"

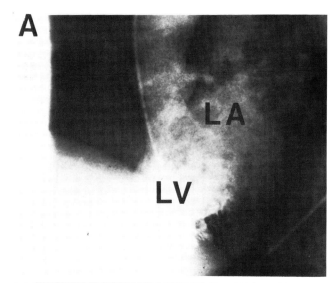

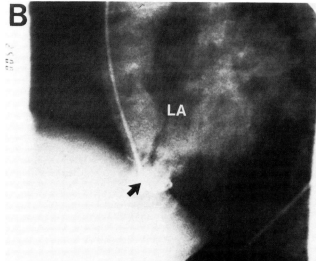

Figure 26-1 A, Left ventriculogram (left anterior oblique projection) during diastole. Note the normal left ventricular (LV) chamber size and a small amount of contrast medium in the left atrium (LA). B, During systole the left ventricular cavity is almost completely obliterated, the left ventricular outflow tract is severely narrowed (arrow), and there is moderate to severe mitral regurgitation as evidenced by contrast medium in the left atrium (LA).

appearance.[3] Attention has also been called to the fact that in hypertrophic cardiomyopathy with obstruction in the left anterior oblique projection during systole the apex of the left ventricle aligns with the left atrium rather than the aorta, as is normally the case. Some investigators believe that this abnormal alignment may contribute to the mitral regurgitation seen in patients with obstructive hypertrophic cardiomyopathy.[14]

In the frontal projection during systole once again there may be total obliteration of the left ventricular cavity with mitral regurgitation seen. An accentuation of midcavity or apical obstruction may also be apparent in the frontal projection in the presence of this type of hypertrophic pattern (Figure 26-2). In the obstructive

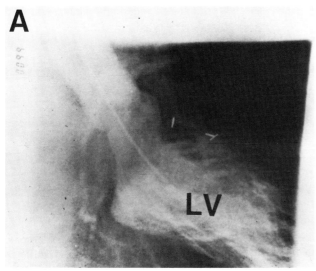

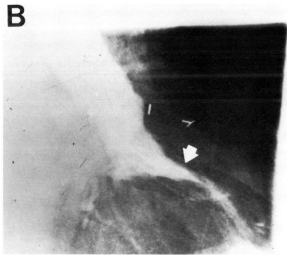

Figure 26-2 A, Left ventriculogram (right anterior oblique projection) during diastole. B, During systole the left ventricular outflow tract does not appear to be narrowed, but there is severe midcavitary obstruction (arrow). LV, left ventricle.

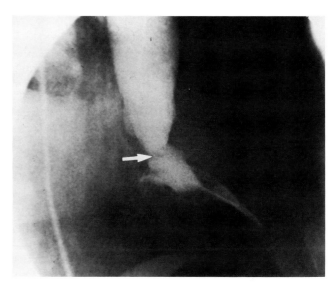

Figure 26-3 Left ventricular angiogram in the right anterior oblique projection demonstrating severe narrowing of the left ventricular (LV) outflow tract and almost complete obliteration of the left ventricular cavity during systole. Note the fine linear radiolucency (arrow), which represents contrast medium displaced by the anterior leaflet of the mitral valve moving against the septum.

form of the disease, with upper septal hypertrophy and systolic anterior motion of the mitral valve, subtle but characteristic abnormalities of the mitral valve may be seen in the frontal projection; these have been described by Simon et al.[15] Normally in this projection during systole the opposed edges of the mitral valve leaflets are not visualized, since a pool of contrast medium lies between the septum and the closed valve leaflets. However, in hypertrophic cardiomyopathy with obstruction, because of the anterior motion of the mitral valve against the septum the opposed leaflet can be seen as an irregular or linear radiolucency. Usually this radiolucency takes the shape of a V and represents the free edge of the anterior leaflet of the mitral valve, but it may take the shape of a W—a configuration that is consistent with visualization of both the anterior and posterior leaflets (Figure 26-3).

Abnormal right-sided hemodynamics are common, and right ventricular angiography is often performed in patients with hypertrophic cardiomyopathy.[3,16] Most commonly, mild to moderate pulmonary hypertension is seen as a result of reduced left ventricular compliance and hence elevated wedge pressure. A right ventricular pulmonary artery systolic pressure gradient is seen in approximately 15% of patients with hypertrophic cardiomyopathy. Usually this gradient is mild and less than that measured on the left side. However, right-sided gradients approaching 100 mm Hg have been recorded, and gradients may occur in the right ventricular outflow tract without any detectable gradient on catheterization of the left side of the heart.[17] Massive septal hypertrophy with bulging into the right ventricular outflow tract appears to cause this obstruction. Right ventricular angiography commonly reveals during diastole and even more so during systole a hypertrophied ventricular septum that encroaches upon the body and outflow tract of the right ventricle.[3,11,16,17] This bulging may be smooth but is often irregular, with masses and ridges of muscle bulging into the right ventricular cavity. These muscular bulges may cause the right ventricular apex to appear pointed or triangular.

Simultaneous biventricular angiography has been performed in an effort to characterize further the abnormal muscle patterns in hypertrophic cardiomyopathy and has also been used to differentiate secondary hypertrophy caused by aortic stenosis or hypertension as opposed to a primary hypertrophic cardiomyopathy.[18,19] In the normal or secondarily hypertrophied heart, the left ventricular endocardial surface is concave, the right and left endocardial surfaces run parallel, and the ratio between the upper and lower septal widths is approximately 1 (0.94 ± 0.06 SD). However, in patients with hypertrophic cardiomyopathy, the left ventricular endocardial surface is no longer concave but is straight or even convex, and this convexity increases during systole. However, the right ventricular border remains convex toward the right; therefore, the right ventricular to left ventricular septal endocardial surfaces no longer run parallel but are divergent. This results in a decrease of the upper-to-lower septal thickness ratio to an average of about 0.68 ± 0.01. These studies also have demonstrated that patients with the obstructive form of the disease have a significantly higher upper septal width and a higher upper-to-lower septal width ratio (0.74 ± 0.01 versus 0.66 ± 0.01) compared with patients with the nonobstructive form of the disease.

Another important aspect of hypertrophic cardiomyopathy is that the hypercontractile pattern may evolve into a dilated form of the disease with either globally or segmentally reduced left ventricular function.[20] This may occur acutely in association with a transmural myocardial infarction,[21] but more typically it occurs silently and slowly until symptoms of congestive heart failure develop.[22] The pathophysiology of this occurrence, in the absence of significant atherosclerotic disease, is unknown but has been attributed to the following potential mechanisms: (1) left ventricular dilatation as a result of chronic ischemia induced by an oxygen supply-to-demand mismatch caused by the excessive myocardial hypertrophy characteristic of these patients[23]; (2) segmental wall motion abnormalities caused by regional ischemia resulting from systolic coronary compression (bridging), a frequent finding in patients with hypertrophic cardiomyopathy[24]; (3) anterior contraction abnormalities resulting from septal myomectomy and interruption of anterior wall blood flow[25]; and (4) diffuse or regional abnormalities of wall motion contraction resulting from chronic pressure overload of the left ventricular apex, particularly in hypertrophic cardiomyopathy with midcavitary obstruction associated with high midcavity gradients. In support of the last hypothesis, in a long-term follow-up study of patients with hypertrophic cardiomyopathy, 80% of patients with midcavitary obstruction developed dilatation, whereas only 7% of patients without midcavitary obstruction developed dilatation.[25]

CORONARY ANATOMY

The majority of patients with hypertrophic cardiomyopathy have normal coronary anatomy. However, several types of abnormalities have been described. In one series of 42 consecutive patients with hypertrophic cardiomyopathy, significant atherosclerosis was present in 24% of patients aged 45 years or older, whereas no significant stenosis was found in patients aged younger than 45 years. Interestingly, the presence of angina or myocardial infarction did not correlate with the presence of obstructive coronary artery disease.[26]

Associated coronary arterial anomalies also have been described in patients with hypertrophic cardiomyopathy. These include minor variations in the take-off of the right coronary artery from the right coronary sinus, as well as more major anomalies such as the right coronary artery's arising from the left main coronary artery and the circumflex coronary artery's arising from the right coronary sinus.[26] A congenital coronary artery fistula originating from the left main coronary artery and entering into the pulmonary artery was found in one patient whom we evaluated. The presence of a coronary artery fistula in a patient with hypertrophic cardiomyopathy has not been described previously; however, it is probably rare and may be an unrelated finding.

Systolic compression of the coronary vessels has been well documented in hypertrophic cardiomyopathy. Although this angiographic finding can be seen in patients without hypertrophic cardiomyopathy, several of its features in hypertrophic cardiomyopathy distinguish it from bridging unassociated with hypertrophic cardiomyopathy. First, the degree of milking is more severe in hypertrophic cardiomyopathy and may even be associated with total occlusion during systole.[24] Second, although single-vessel involvement of the left anterior descending artery is most common, involvement of other coronary artery vessels and multivessel systolic bridging can occur, and this feature is seen almost exclusively in patients with hypertrophic cardiomyopathy.[24] Third, an association of hypertrophic cardiomyopathy with systolic compression of the septal perforators arising from either the right coronary artery or the left anterior descending coronary artery is very common.[27] Finally, when systolic bridging of the epicardial coronary vessels occurs, the angiographic appearance in hypertrophic cardiomyopathy is abrupt, long, and jagged; in patients without hypertrophic cardiomyopathy the left anterior descending coronary artery is smoothly, gradually, and progressively compressed[28] (Figure 26-4).

GRADIENTS

A gradient across the site of hypertrophy is common in obstructive cardiomyopathy. It typically occurs within the outflow tract[3] but may occur in the left ventricular midcavity[10] in this form of the disease. A significant gradient, usually considered to be greater than 20 mm Hg, is highly dependent on a number of fea-

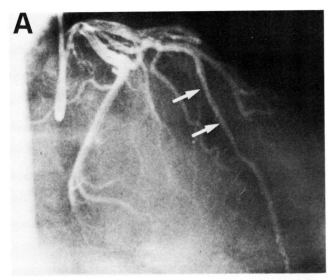
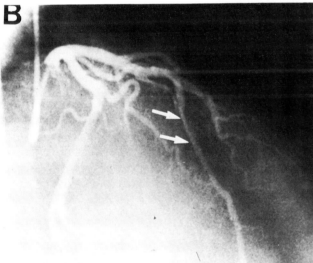

Figure 26-4 A, Coronary angiogram of the left coronary system (right anterior oblique projection) during diastole. Note the widely patent left anterior descending artery (LAD) (arrows). B, During systole the LAD becomes diffusely and irregularly narrowed (arrows), representing the bridging typically seen in hypertrophic cardiomyopathy. Source: Courtesy of R Krone and the Cardiac Catheterization Laboratory, The Jewish Hospital of St. Louis, St. Louis, MO.

tures, however, and its presence or absence neither confirms nor militates against the diagnosis of hypertrophic cardiomyopathy. This is true for several reasons. First, catheter placement is extremely important. An end-hole catheter entrapped within the trabeculae of the left ventricle commonly demonstrates a systolic gradient as measured between it and the left ventricular outflow tract or aorta.[29,30] It is important, therefore, to inject by hand a small amount of contrast medium into the catheter to establish the location of the catheter and to exclude catheter entrapment. Use of the trans-septal technique also will alleviate this potential artifact[31]; however, apical and midcavitary obstruction might be missed by this technique.

As with some of the angiographic patterns of hypertrophic cardiomyopathy, such as systolic cavitary obliteration (which can develop even in normal patients[32,33]), the presence of a significant gradient is highly dependent on several factors, including (1) left ventricular loading conditions,[34] (2) the inotropic state of the heart,[35] and (3) the thickness of the left ventricular septum and free wall (whether primary as in hypertrophic cardiomyopathy or secondary as in hypertensive heart disease)[36]; a significant gradient may even be caused by infiltrative diseases such as sarcoid.[37] Thus the interplay of these three factors may cause gradients to be present in patients without hypertrophic cardiomyopathy, gradients to be absent in patients with hypertrophic cardiomyopathy, and gradients to occur on one occasion and to be absent on another in the same patient. In general, the greater the degree of hypertrophy, the more reduced the preload and afterload and the greater the inotropic state of the heart; for example, during an infusion of isoproterenol a systolic gradient is more likely to occur even in normal patients.[35] Similarly, patients in whom an obstructive hypertrophic cardiomyopathy is present may not always demonstrate a systolic gradient, particularly under conditions of increased preload and afterload or when the inotropic state of the heart is diminished such as after β-blockade.[38] Conversely, in patients in whom the diagnosis of hypertrophic cardiomyopathy is strongly suspected but no gradients are found, provocative maneuvers designed to alter the contractile state of the heart or to change loading conditions may produce significant gradients and help to confirm the diagnosis. Such maneuvers may include the inhalation of amyl nitrite, the intravenous administration of isoproterenol, or the Valsalva maneuver. Particularly specific is the hemodynamic change induced by manipulation of a left ventricular catheter. Normally the post-premature ventricular contraction (PVC) beat will be characterized by an increase in the aortic pressure due to a larger left ventricular stroke volume that results from a longer period of left ventricular filling and an increased inotropic state of the myocardium, commonly referred to as post-extra systolic potentiation (PEP). In hypertrophic cardiomyopathy, however, the PEP results in enhanced obstruction to left ventricular outflow, and this causes an increased systolic gradient with no change or a reduction in the aortic pressure and an enhancement of the classic spike-and-dome arterial pressure tracing. This post-PVC hemodynamic phenomenon was described by Brockenbrough et al[39] and typically is referred to as the Brockenbrough effect (Figure 26-5).

DILATED CARDIOMYOPATHY

According to the World Health Organization, "cardiomyopathy" refers to "heart muscle disease of un-

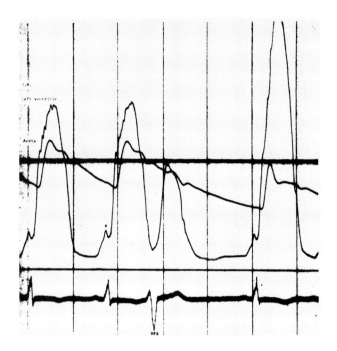

Figure 26-5 Simultaneous pressure recordings from the left ventricle and aorta. A moderate gradient is noted, but on the post-PVC beat the left ventricular to aortic gradient becomes severe. Also note the spike-and-dome character of the aortic pressure waveform. *Source:* Courtesy of A Susmano.

known cause."[2] Nevertheless, the term "dilated cardiomyopathy" commonly is used to describe patients with left ventricular dilatation and global hypokinesia regardless of etiology.[40] Therefore, "idiopathic dilated cardiomyopathy" is probably a better term to use in order to avoid confusion.

The first role of angiography in patients with suspected dilated cardiomyopathy, regardless of etiology, is to determine whether dilatation or a depression of the systolic contraction is present. The normal end-diastolic volume is 70 ± 20 mL/m². When the ventricular end-diastolic volume exceeds 100 mL/m² (greater than 2 SD above the normal average), clearly left ventricular dilatation is present.[41] An ejection fraction (EF) of less than 55% generally is considered to represent abnormal systolic function. Since minor reductions of systolic performance only rarely are thought to be clinically important,[42] some investigators have insisted that the term "dilated cardiomyopathy" be used only when the EF is less than 40%.[43] Nevertheless, in patients being evaluated by catheterization, small reductions in the EF or increases in ventricular volume may be the first indication of significant cardiac disease. Deal et al[44] found that, of 24 patients being evaluated for ventricular tachycardia with no overt clinical evidence of heart disease, abnormalities of cardiac size and function were present in 70%. Of the patients with a reduced left ventricular EF, the mean EF was only modestly reduced, with an average EF of 46%. Similarly, patients with only mild left ventricular dilatation may have severe reductions in the EF to the point of requiring cardiac transplantation because of heart failure symptoms refractory to medical therapy.[45] Thus when unexplained dilatation or reductions in EF are present a cardiomyopathy can be said to be present. However, the significance of such findings may be trivial or crucial depending on the symptomatic state of the patient and the natural history of the disease.

Besides left ventricular dilatation and reduced systolic function, diffuse rather than regional wall motion abnormalities have been thought to be characteristic of dilated cardiomyopathy, particularly of the idiopathic type.[46] In fact, recent data have demonstrated that regional wall motion abnormalities are common in patients with dilated cardiomyopathy, even in the absence of left bundle-branch block or significant coronary artery disease.[47] The reason for this finding is unknown but it appears to correlate with better hemodynamics and prognosis. Thus it is important to recognize that the presence of regional wall motion abnormalities does not necessarily indicate that coronary artery disease or other focal pathologic processes such as myocarditis or sarcoidosis are present.

Another reason for performing catheterization in patients with a dilated cardiomyopathy is to exclude potentially treatable causes of left ventricular dilatation and failure. Thus coronary angiography may reveal the presence of significant but unsuspected coronary artery disease. When large areas of reversible ischemia are present, myocardial revascularization has been shown to be effective in significantly improving left ventricular function.[48] Unfortunately, in some patients a distinction between ischemic and infarcted tissue, or ischemic and myopathic tissue in the case of left ventricular dysfunction seemingly out of proportion to the extent of coronary artery disease, may be challenging. Nevertheless, it is important to do so because revascularization of ischemic tissue may be rewarding, whereas revascularization of primarily myopathic or infarcted tissue will yield minimal results at best. Differentiation of ischemic and infarcted myocardium may be aided by measuring the EF in the basal state and once again during the post-PVC beat[49] or during a second left ventriculogram after preload reduction with nitroglycerin.[50] A 10% or greater increase in the EF is usually associated with a good outcome after revascularization. Differentiating primarily myopathic from ischemic myocardium is more difficult. Exercise testing with either thallium or radionuclide ventriculography[51] may be helpful in this regard.

Mitral regurgitation is seen frequently in patients with dilated cardiomyopathy due to distortion of the mitral apparatus resulting from concomitant left ventricular dilatation. Usually this is of only a mild degree[52] and clearly a secondary phenomenon, but on occasion it may be interpreted as the cause of the myopathy rather than the result of it. Mitral valve replacement in the case of primary mitral regurgitation may help even in patients with reduced EFs, whereas in patients with secondary mitral regurgitation due to left ventricular dilatation the results of surgery are poor. Once again differentiation may be difficult. In general, however, extremely low EFs,[52] a shorter duration of illness, a small or absent V wave on pulmonary capillary wedge tracings (particularly with exercise), and the absence of structurally apparent mitral valve disease on an echocardiogram favor a primary diagnosis of cardiomyopathy as the cause rather than the result of significant mitral regurgitation.

Left ventricular angiography in patients with cardiomyopathy occasionally may demonstrate filling defects, particularly in the apex, which usually represent thrombus formation. Although relatively specific, two-dimensional echocardiography[53] and ultrafast computed tomography[54] are probably more sensitive for the detection of left ventricular thrombi as well as for defining their geometry, which may be predictive of outcome and the need for long-term anticoagulation therapy.[55] Finally, cardiac catheterization may yield prognostic data in patients with dilated cardiomyopathy. The absence of coronary artery disease, the presence of regional wall motion abnormalities, a higher EF, a higher left ventricular systolic pressure (greater than 120 mm Hg), a lower left ventricular end-diastolic pressure (less than 13 mm Hg), and a cardiac index greater than 3 L/min/m^2 have been reported to correlate with better survival rates.[47,56–58] However, the data in this regard are controversial, and it is not advisable to perform catheterization in patients with dilated cardiomyopathy for the sole purpose of prognosis.

SUMMARY

Cardiac catheterization and angiography may supply important information in the evaluation of both hypertrophic and dilated cardiomyopathy. Although a combination of newer imaging techniques may yield similar or complementary information in a less invasive way, no single test currently available can provide the array of hemodynamic and angiographic data that can be obtained by cardiac catheterization with left ventricular and coronary angiography. It should not be considered necessary, however, to perform this test in all patients with either dilated or hypertrophic cardiomyopathy. The potential benefit of this test must be balanced against its inherent risks. Clearly, newer imaging techniques have reduced the need to perform this procedure in many patients with both dilated and hypertrophic cardiomyopathies.

REFERENCES

1. Mattingly TW. The clinical concept of pulmonary myocardial tissue. *Circulation*. 1965;32:845–851.

2. Report of the WHO/ISFC task force on the definition and classification of cardiomyopathies. *Br Heart J*. 1980;44:672–673.

3. Braunwald E, Lambrew CT, Rockoff SD, et al. Idiopathic hypertrophic subaortic stenosis, I: A description of the disease, based upon an analysis of 64 patients. *Circulation*. 1964;30(suppl 4):IV3–119.

4. Croxson RS, Raphael MJ. Angiographic assessment of congestive cardiomyopathy. *Br Heart J*. 1969;31:390–391.

5. Wynne J, Braunwald E. The cardiomyopathies and myocarditides. In: Braunwald E, ed. *Heart Disease*. Philadelphia, Pa: WB Saunders Co; 1980:1437–1498.

6. Pollick C, Morgan CD, Gilbert BW, et al. Muscular subaortic stenosis: The temporal relationship between systolic anterior motion of the anterior mitral leaflet and the pressure gradient. *Circulation*. 1982;66:1087–1094.

7. Henry WL, Clark CE, Roberts WC, et al. Differences in the distribution of myocardial abnormalities in patients with obstructive and non-obstructive asymmetric septal hypertrophy (ASH): Echocardiographic and gross anatomic findings. *Circulation*. 1974;50:447–455.

8. Keren G, Belhassen B, Sherez J, et al. Apical hypertrophic evaluation by noninvasive and invasive techniques in 23 patients. *Circulation*. 1985;71:45–56.

9. Yamaguchi H, Ishimura T, Mishiyama S, et al. Hypertrophic non-obstructive cardiomyopathy with giant negative T waves (apical hypertrophy): Ventriculographic and echocardiographic features in 30 patients. *Am J Cardiol*. 1979;44:401–412.

10. Falicov RE, Resnekov L, Bharati S, et al. Midventricular obstruction: A variant of obstructive cardiomyopathy. *Am J Cardiol*. 1976;37:432–437.

11. Cohen J, Effat H, Goodwin JF, et al. Hypertrophic obstructive cardiomyopathy. *Br Heart J*. 1964;26:16–32.

12. Louie EK, Maron BJ. Apical hypertrophic cardiomyopathy: Clinical and two-dimensional echocardiographic assessment. *Ann Intern Med*. 1987;106:663–670.

13. Braunwald E, Roberts WC, Goldblatt A, et al. Aortic stenosis: Physiological, pathological and clinical concepts. *Ann Intern Med*. 1963;58:494–521.

14. Adelman AG, McLoughlin MJ, Marquis Y, et al. Left ventricular cineangiographic observations in muscular subaortic stenosis. *Am J Cardiol*. 1969;24:689–697.

15. Simon AL, Ross J Jr, Gault JH. Angiographic anatomy of the left ventricle and mitral valve in idiopathic hypertrophic subaortic stenosis. *Circulation*. 1967;36:852–867.

16. Wigle ED, Heimbecker RO, Gunton RW. Idiopathic ventricular septal hypertrophy causing muscular subaortic stenosis. *Circulation*. 1962;26:325–340.

17. Nordenstrom B, Ovenfors CO. Low subvalvular aortic and pulmonic stenosis with hypertrophy and abnormal arrangement of the muscle bundles of the myocardium. *Acta Radiol*. 1962;57:321–339.

18. Redwood DR, Scherer JL, Epstein SE. Biventricular cineangiography in the evaluation of patients with asymmetric septal hypertrophy. *Circulation*. 1974;49:1116–1121.

19. Delius W, Wirtzfeld A, Schinz A, et al. Evaluation of the ventricular septum by biventricular cine angiography in congestive and hypertrophic cardiomyopathies. In: Kaltenbach M, Loogen F, Olsen EGJ, eds. *Cardiomyopathy and Myocardial Biopsy*. New York, NY: Springer-Verlag Inc; 1978:205–217.

20. Roberts WC, Ferrans VJ. Morphologic observations in the cardiomyopathies. In: Fowler NO, ed. *Myocardial Disease*. New York, NY: Grune & Stratton Inc; 1973:59–115.

21. Waller BF, Maron BJ, Epstein SE, et al. Transmural myocardial infarction in hypertrophic cardiomyopathy: A cause of conversion from left ventricular asymmetry to symmetry and from normal-sized to dilated left ventricular cavity. *Chest.* 1981;79:461–465.

22. ten Cate FJ, Roelandt J. Progression to left ventricular dilatation in patients with hypertrophic obstructive cardiomyopathy. *Am Heart J.* 1979;97:762–765.

23. Sutton MG, Tajik AJ, Smith HC, et al. Angina in idiopathic hypertrophic subaortic stenosis: A clinical correlate of regional left ventricular dysfunction: A videometric and echocardiographic study. *Circulation.* 1980;61:561–568.

24. Kitazume H, Kramer JR, Krauthamer D, et al. Myocardial bridges in obstructive hypertrophic cardiomyopathy. *Am Heart J.* 1983;106:131–135.

25. Fighali S, Krajcer Z, Edelman S, et al. Progression of hypertrophic cardiomyopathy into a hypokinetic left ventricle: Higher incidence in patients with midventricular obstruction. *J Am Coll Cardiol.* 1987;9:288–294.

26. Walston A II, Behar VS. Spectrum of coronary artery disease in idiopathic hypertrophic subaortic stenosis. *Am J Cardiol.* 1976;38:12–16.

27. Pichard AD, Meller J, Teichholz LE, et al. Septal perforator compression (narrowing) in idiopathic hypertrophic subaortic stenosis. *Am J Cardiol.* 1977;40:310–314.

28. Brugada P, Bar FW, de Zwaan C, et al. "Sawfish": systolic narrowing of the left anterior descending coronary artery: An angiographic sign of hypertrophic cardiomyopathy. *Circulation.* 1982;66:800–803.

29. Criley JM, Lewis KB, White RI Jr. Pressure gradients without obstruction: A new concept of "hypertrophic subaortic stenosis." *Circulation.* 1965;32:881–887.

30. Adelman AL, Wigle ED. Two types of intraventricular pressure differences in the same patient: Left ventricular catheter entrapment and right ventricular outflow tract obstruction. *Circulation.* 1968;38:649–655.

31. Peterson KL, Karliner JS, Ross J Jr. Profiles in congestive and hypertrophic cardiomyopathy. In: Grossman W, ed. *Cardiac Catheterization and Angiography.* Philadelphia, Pa: Lea & Febiger; 1980:346–357.

32. Raizner AE, Chahine RA, Ishimori T, et al. Clinical correlates of left ventricular cavity obliteration. *Am J Cardiol.* 1977;40:303–309.

33. Grosse R, Maskin C, Spindola-Franco H, et al. Production of left ventricular cavitary obliteration in normal men. *Circulation.* 1981;64:448–455.

34. Shabetai R. A new syndrome in hypovolemic shock: Systolic murmur and intraventricular pressure gradient. *Am J Cardiol.* 1969;24:404–408.

35. White RI Jr, Criley JM, Lewis KB, et al. Experimental production of intracavity pressure differences: Possible significance in the interpretation of human hemodynamic studies. *Am J Cardiol.* 1967;19:806–817.

36. Maron BJ, Gottdiener JS, Roberts WC, et al. Left ventricular outflow tract obstruction due to systolic anterior motion of the anterior mitral leaflet in patients with concentric left ventricular hypertrophy. *Circulation.* 1978;57:527–533.

37. Awdeh MR, Erwin S, Young JM, et al. Systolic anterior motion of the mitral valve caused by sarcoid involving the septum. *South Med J.* 1978;71:969–971.

38. Glancy DL, Shepherd RL, Beiser D, et al. The dynamic nature of left ventricular outflow obstruction in idiopathic hypertrophic subaortic stenosis. *Ann Intern Med.* 1971;75:589–593.

39. Brockenbrough EC, Braunwald E, Morrow AG. A hemodynamic technique for the detection of hypertrophic subaortic stenosis. *Circulation.* 1961;23:189–194.

40. Johnson RA, Palacios I. Dilated cardiomyopathies of the adult (first of two parts). *N Engl J Med.* 1982;307:1051–1058.

41. Dodge HT. Hemodynamic aspects of cardiac failure. In: Braunwald E, ed. *The Myocardium: Failure and Infarction.* New York, NY: HP Publishing Co; 1974;70–79.

42. Johnson RA. Heart failure. In: Johnson RA, Haber E, Austen WG, eds. *The Practice of Cardiology.* Boston, Mass: Little, Brown & Co; 1980:31–95.

43. Feild BJ, Baxley WA, Russell RO Jr, et al. Left ventricular function and hypertrophy in cardiomyopathy with depressed ejection fraction. *Circulation.* 1973;47:1022–1031.

44. Deal BJ, Miller SM, Scagliotti D, et al. Ventricular tachycardia in a young population without overt heart disease. *Circulation.* 1986;73:1111–1118.

45. Keren A, Billingham ME, Weintraub D, et al. Mildly dilated congestive cardiomyopathy. *Circulation.* 1985;72:302–309.

46. Goldman MR, Boucher CA. Value of radionuclide imaging techniques in assessing cardiomyopathy. *Am J Cardiol.* 1980;46:1232–1236.

47. Wallis DE, O'Connell JB, Henkin RE, et al. Segmental wall motion abnormalities in dilated cardiomyopathy: A common finding and good prognostic sign. *J Am Coll Cardiol.* 1984;4:674–679.

48. Shanes JG, Kondos GT, Levitsky S, et al. Coronary artery obstruction: A potentially reversible cause of dilated cardiomyopathy. *Am Heart J.* 1985;110:173–178.

49. Dyke SH, Cohn PF, Gorlin R, et al. Detection of residual myocardial function in coronary artery disease using post-extra systolic potentiation. *Circulation.* 1974;50:694–699.

50. Helfant RH, Pine R, Meister SG, et al. Nitroglycerin to unmask reversible asynergy: Correlation with post coronary bypass ventriculography. *Circulation.* 1974;50:108–113.

51. Okada RD, Boucher CA, Strauss HW, et al. Exercise radionuclide imaging approaches to coronary artery disease. *Am J Cardiol.* 1980;46:1188–1204.

52. Bolen JL, Alderman EL. Ventriculographic and hemodynamic features of mitral regurgitation of cardiomyopathic, rheumatic and non-rheumatic etiology. *Am J Cardiol.* 1977;39:177–183.

53. van Meurs-van Woezik H, Meltzer RS, van den Brand M, et al. Superiority of echocardiography over angiocardiography in diagnosing a left ventricular thrombus. *Chest.* 1981;80:321–323.

54. Yoshida H, Tsunoda K, Yamada Z, et al. Assessment of an intracardiac mural thrombus by contrast enhanced computed tomography. *J Cardiogr.* 1982;12:645–654.

55. Visser CA, Kan G, Meltzer RS, et al. Embolic potential of left ventricular thrombus after myocardial infarction: A two-dimensional echocardiographic study of 119 patients. *J Am Coll Cardiol.* 1985;5:1276–1280.

56. Likoff MJ, Chandler SL, Kay HR. Clinical determinants of mortality in chronic congestive heart failure secondary to idiopathic dilated or to ischemic cardiomyopathy. *Am J Cardiol.* 1987;59:634–638.

57. Schwarz F, Mall G, Zebe H, et al. Determinants of survival in patients with congestive cardiomyopathy: Quantitative morphologic findings and left ventricular hemodynamics. *Circulation.* 1984;70:923–928.

58. Franciosa JA, Wilen M, Ziesche S, et al. Survival in men with severe chronic left ventricular failure due to either coronary heart disease or idiopathic dilated cardiomyopathy. *Am J Cardiol.* 1983;51:831–836.

Chapter 27

Evaluation of Hypertrophic and Dilated Cardiomyopathy by Echocardiography

Andrew J. Buda, MD

Echocardiography (echo) has been instrumental in the detection, differentiation, and evaluation of a variety of cardiomyopathies. M-Mode echo, two-dimensional echo, pulsed-wave and continuous-wave Doppler, and, most recently, color flow Doppler imaging have provided us with important insights into the pathophysiologic mechanisms and further understanding of these cardiomyopathies. Clearly, the development of these echo-Doppler techniques has greatly expanded our ability to examine carefully and to manage patients with cardiomyopathy. In this chapter I review the present state of knowledge related to the echocardiographic assessment of hypertrophic and dilated cardiomyopathies. No attempt has been made to be entirely comprehensive or complete because of the page limitations of this chapter, and only a small number of key references have been cited. Rather, I emphasize and focus on the impact of more recent contributions of echo and Doppler studies to the characterization and understanding of pathophysiologic mechanisms of hypertrophic and dilated cardiomyopathy.

HYPERTROPHIC CARDIOMYOPATHY

Hypertrophic cardiomyopathy (HCM) may be defined as ventricular hypertrophy without identifiable cause. Echocardiography has made a particularly important contribution to the diagnosis and understanding of HCM. Although this disease has been recognized for three decades, its diagnosis prior to echo was notoriously difficult without cardiac catheterization. Echo has made the identification of HCM relatively straightfoward and has allowed further study of its genetic transmission. For the interested reader, excellent comprehensive reviews of HCM have recently been published by Wigle et al,[1] Wigle,[2] and Maron et al.[3] Each review contains a complete and extensive bibliography that cites most important contributions to the field.

Echocardiographic Features

The single most characteristic feature of HCM is asymmetric septal hypertrophy (ASH) (Figure 27-1A). When initially described, ASH was defined as a ratio of septal and posterior wall thickness at end-diastole equal to or greater than 1.3, which was believed to be specific for HCM.[4] However, on subsequent studies, this ratio was found to be relatively nonspecific, since it was found in a wide variety of persons other than those with HCM, including normal infants and young children; patients with congenital heart disease, myocarditis, type II hyperlipidemia, coronary artery disease, primary pulmonary hypertension, systemic hypertension, and other forms of valvular and subvalvular obstruction; and normal athletes. As a result of this relative

---

The assistance of Rita Pinton, MD, with the figures and the excellent secretarial assistance of Ms. JoEllen Mahs are gratefully appreciated.

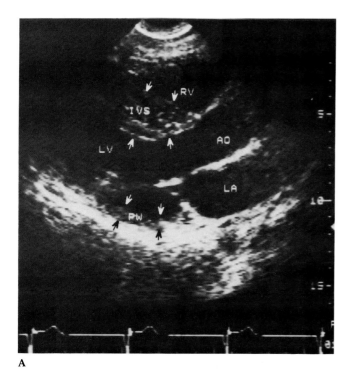

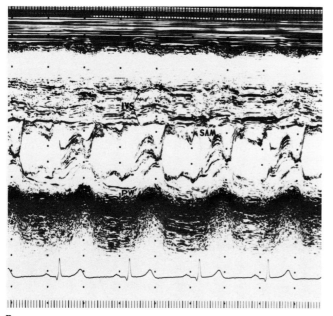

Figure 27-1 A, Parasternal long-axis projection of a two-dimensional echocardiogram from a patient with HCM. Note the marked asymmetric septal hypertrophy (ASH). AO indicates aorta; IVS, interventricular septum; LA, left atrium; LV, left ventricle; PW, posterior wall; RV, right ventricle. B, An M-mode echocardiogram from a patient with HCM to illustrate systolic anterior motion (SAM) and marked ASH.

nonspecificity, a more stringent criterion of ASH with a ratio of 1.5:1 or greater was adopted and is now generally used.

This modified criterion of ASH (ratio 1.5:1 or greater) is considered to be relatively sensitive and specific for HCM. For example, in one recent study by Doi et al,[5] this ratio of ASH was found to be 79% sensitive and 94% specific for HCM. However, other investigators have found that ASH is even more sensitive for HCM. In fact, Maron[6] reported that ASH was close to 95% sensitive for HCM. He attributed the apparent difference between his results and those of Doi et al[5] to differences in patient populations studied and to differences in methods of analysis. It should be noted, however, that ASH is not entirely sensitive or specific for HCM. A symmetric pattern of left ventricular hypertrophy may occur in HCM, but this occurs uncommonly, probably in fewer than 5% of cases, although symmetric hypertrophy has been reported more frequently by some investigators. On the other hand, asymmetric thickening of the left ventricular wall is not unique to HCM and may be present in approximately 5% to 10% of adult patients with other congenital or acquired heart diseases, particularly those with right ventricular hypertension.

A number of other echo features are characteristic of HCM (Figure 27-1B): systolic anterior motion (SAM), septal hypokinesis (interventricular septal amplitude 5 mm or less), decreased systolic mitral-septal distance (less than 25 mm), small left ventricular cavity (less than 25 mm), and mid-systolic closure of the aortic valve. None of these echo features on its own is diagnostic of HCM. Rather than any one individual echo feature, it is the constellation of echo findings that strongly suggests the diagnosis of HCM.

Systolic anterior motion was one of the earliest echo features identified and was suggested to be associated always with a left ventricular outflow tract gradient.[7] It is now well appreciated that its presence does not necessarily predict a resting gradient. Systolic anterior motion has been rarely noted in patients with complete transposition of the great arteries, aortic valve stenosis, aortic regurgitation, hypovolemic shock, coarctation of the aorta, ischemic heart disease, hypertensive heart disease, and a number of other conditions. Clearly, in most of these disease states, SAM is not associated with a left ventricular outflow tract gradient. However, in the setting of HCM, there is a definite but imprecise correlation between the degree of SAM and the presence of a left ventricular outflow tract gradient. Patients with HCM and a resting left ventricular outflow tract gradient generally have severe SAM, defined as SAM with septal contact for 30% or more of echocardiographic systole, whereas those without a resting gradient have minimal or no SAM.[8] In addition, the time of onset of SAM-septal contact further determines the left ventricular outflow tract gradient.[9] Early and prolonged SAM-septal contact is associated with a high pressure gradient, whereas late-onset and short SAM-septal contact is associated with a small pressure gradient. However, if SAM-septal contact occurs after 55%

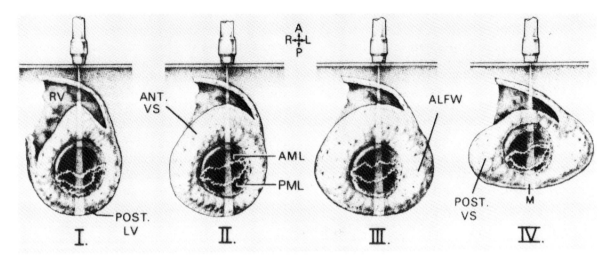

Figure 27-2 Artistic representation of morphologic variability in hypertrophic cardiomyopathy, shown here in the short-axis cross-sectional plane at the mitral valve level. The expected approximate path of the conventional M-mode echo beam (M) through the anterior aspect of the septum and posterior free wall is shown in each heart. The M-mode echocardiogram would record identical values for wall thicknesses and septum-free wall ratio (i.e., 2.0) in morphologic types I, II, and III, although these hearts actually differ considerably with regard to distribution of left ventricular (LV) hypertrophy. M-Mode echocardiography greatly overestimates the magnitude of hypertrophy in type I (which is quite localized) and underestimates the marked diffused increase in LV mass in type III; only in type II does the M-mode accurately reflect the distribution of LV hypertrophy. In type IV, the M-mode echo beam does not traverse the thickened portions of the LV wall in the posterior septum and anterolateral free wall (ALFW). AML indicates anterior mitral leaflet; A or ANT., anterior; L, left; LVFW, LV free wall; P or POST., posterior; PML, posterior mitral leaflet; R, right; RV, right ventricle; VS, ventricular septum. *Source:* Reprinted with permission from *American Journal of Cardiology* (1985;55:836), Copyright © 1985, Cahners Publishing Company Inc.

of the systolic ejection period, usually no pressure gradient is present.[1]

Extent and Distribution of Hypertrophy

Two-dimensional echo, by its ability to evaluate the extent and distribution of left ventricular hypertrophy, further defines the distribution of hypertrophy in HCM and has been particularly useful in identifying asymmetric hypertrophy of left ventricular segments not typically interrogated by the M-mode beam. Thus patients with isolated lateral wall hypertrophy, isolated midventricular hypertrophy, or isolated apical hypertrophy (all of which occur in approximately 5% of patients with HCM) can be identified readily. In addition, the full extent and distribution of the hypertrophic process can be visualized and better appreciated. In this regard, there is a substantial degree of heterogeneity in the distribution of left ventricular hypertrophy.

Four common distributions of asymmetric hypertrophy have been identified by Maron et al[6] (Figure 27-2). In 10% of cases, only the anterior aspect of the ventricular septum is thickened (type I); in 20% of cases, the anterior and posterior aspects of the ventricular septum are involved (type II); in approximately 50%, the septum and anterolateral wall are thickened (type III); and in approximately 20%, only posteroseptal, apical-septal, or anterolateral wall involvement is present (type IV) (Figure 27-3). In addition, the extent of hypertrophy along the length of the septum may be variable. For example, in approximately 25%, the hypertrophy is localized to the proximal aspect of the septum; in 25%, it extends down to the papillary mus-

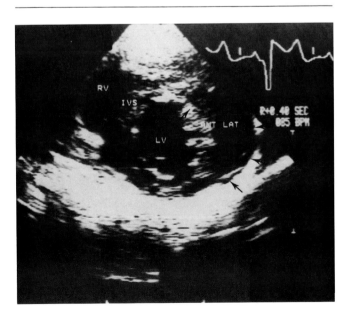

Figure 27-3 Type IV ASH involving the anterolateral (ANT LAT) region of the left ventricle (LV). Note that the septal (IVS) region has normal wall thickness and is not involved. RV indicates right ventricle.

cles; and in approximately 50%, the hypertrophy involves the whole length of the ventricular septum.[1] Thus a variety of individual variations of the hypertrophic process have been noted by two-dimensional echocardiography. In fact, 70 individual and distinctive patterns of hypertrophy in HCM now have been identified.[10]

Differences in echocardiographic features have also been observed in the hemodynamic subtypes of HCM (Figure 27-4A and B). In patients with a resting left ventricular outflow tract gradient, the septum is markedly thickened, the dimension of the left ventricular outflow tract is greatly reduced, and, as previously stated, the SAM is prominent with prolonged SAM-septal contact, exceeding 30% of echocardiographic systole.[8] Conversely, in patients with nonobstructive HCM, the septum is not as thickened, the left ventricular outflow tract is not as reduced, and there is little if any SAM. Finally, patients with latent obstruction (no resting but a provocable gradient) have intermediate values for these echo features. These morphologic characteristics are thus useful in predicting the presence or absence of a left ventricular outflow tract gradient. However, as is discussed later, Doppler echocardiography may be even more useful in further identifying and measuring the presence of an outflow gradient.

Echocardiographic Evidence Supporting Obstruction

The concept of obstruction in HCM has generated considerable controversy over the past two decades, and the actual presence of an obstructive element continues to be debated vigorously.[11] A number of investigators consider the presence of a left ventricular outflow tract gradient as indicative of obstruction, whereas others consider it an epiphenomenon related to excessively rapid early systolic ejection with resultant cavity obliteration or elimination. The opponents of an obstructive component in HCM argue that the SAM simply reflects the distortion of the left ventricular cavity associated with cavity obliteration and is of no pathophysiologic significance. However, a number of echo observations have supported the obstructive hypothesis in HCM.

Simultaneous combined echo-hemodynamic studies have better established the temporal sequence of left ventricular events during ejection in HCM.[12] From these combined studies, several observations have supported an actual obstructive component in HCM. First, the left ventricular outflow tract gradient begins almost simultaneously with the onset of SAM-septal contact. This argues for a definite association between SAM and the left ventricular outflow tract gradient and supports the notion that SAM contributes to the production of this outflow gradient. Second, the severity of the left ventricular outflow tract gradient generally correlates with the severity of the SAM, further supporting a possible cause-and-effect relationship.[9] Third, the aortic valve closes partially at the time that SAM-septal contact begins, further suggesting a hemodynamic effect produced by SAM. Fourth, the left ventricular ejection time is prolonged in patients with a left ventricular outflow tract gradient, and there exists a direct relationship between the severity of this gradient and

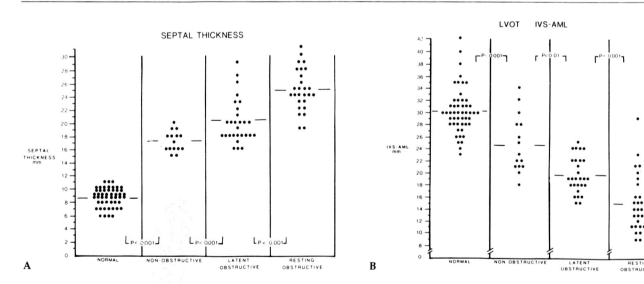

Figure 27-4 A, Individual values for septal thickness together with group means (bars) in normal subjects and in patients with no obstruction, latent obstruction, and obstruction at rest. B, Individual values for left ventricular outflow tract (LVOT) dimensions (IVS-AML) at the onset of systole with group means (bars) in normal subjects and in patients with no obstruction, latent obstruction, and obstruction at rest. IVS-AML indicates interventricular septum-anterior mitral leaflet dimensions. *Source:* Reprinted with permission from *American Journal of Cardiology* (1980;45:866–867), Copyright © 1980, Cahners Publishing Company Inc.

the left ventricular ejection time. Finally, septal myotomy-myectomy invariably relieves the left ventricular outflow tract gradient, increases the left ventricular outflow tract dimension, and improves cardiac symptoms in most patients.[13] These echo-hemodynamic observations tend to support the coexistence of an actual left ventricular outflow tract obstruction, although some investigators continue to dispute this claim.[11]

Doppler echo findings have provided additional support for the existence of an actual left ventricular outflow tract obstruction and have added further insight into its mechanisms. Although Doppler findings indicate that aortic flow persists throughout systolic ejection and that the left ventricle continues to eject at the time the gradient is present, the aortic flow decelerates rapidly in midsystole at about the time SAM-septal contact occurs. In a Doppler study by Maron et al,[14] approximately 90% of the integral of the forward flow velocity curve occurred in the first half of systole and about 50% occurred before the onset of mitral septal contact or the gradient in patients with obstructive HCM (Figure 27-5). This and other Doppler studies have supported the view originally proposed in 1971 by Wigle et al[15] that a Venturi effect may produce SAM, which in turn produces the left ventricular outflow tract gradient.

Doppler studies comparing left ventricular ejection dynamics in HCM and aortic stenosis lend further support to the existence of an actual left ventricular outflow tract obstruction.[16] Patients with obstructive HCM have significant prolongation of both the time to peak velocity and the ejection time compared with patients with nonobstructive HCM. Furthermore, there is no difference in these time intervals between patients with obstructive HCM and patients with valvular aortic stenosis, additionally suggesting the presence of a true obstruction. On the other hand, patients with nonobstructive HCM have times to peak velocity and ejection times comparable to those of a normal population.

Thus scattered evidence from a variety of echo, Doppler, and combined echo-hemodynamic studies provides support to the hypothesis that actual left ventricular outflow obstruction occurs in HCM. The mechanism for this obstruction appears to be related to the Venturi phenomenon, which produces SAM, which then further narrows the left ventricular outflow tract and creates a dynamic obstruction to left ventricular ejection.

Doppler Evaluation

A recent Doppler echo study by Yock et al[17] has characterized further the complex flow velocity patterns in patients with HCM. In their study, increased flow velocities in the left ventricular outflow tract ranged between 2.0 and 5.5 m/s (mean 4.1 m/s) and had a distinctive contour of an increasing slope (dV/dt) in midsystole (Figure 27-6). At the onset of SAM of the mitral valve, flow velocities were elevated to a mean of 1.5 m/s. The high flow velocities of the left ventricular outflow tract were not generally conducted to the aorta; the average peak outflow tract velocity was 4.1 m/s, compared with 1.4 m/s in the aorta, and the velocity contour recorded at the aortic valve level demonstrated an early peak compared with normal outflow. Yock et al[17] offered two possible explanations for this observation: (1) Since the jet is formed approximately 3 to 4 cm below the aortic leaflets, based on pulsed Doppler findings, the combined influences of sheer, friction, and boundary layer separation produced disturbed flow and dissipation of energy with progressive reduction in velocities; and (2) since the jet of the left ventricular outflow tract is directed posteriorly and laterally relative to the aortic annulus, the outflow jet is directed into the region of the posterior outflow tract below the aortic valve rather than through the leaflets. This misdirection of the jet may lead to disturbed flow and dissipation of energy and high velocities before the jet reaches the ascending aorta. The pathologic description of mural endocardial plaques located in the outflow tract would tend to support this latter hypothesis. In addition, Yock et al[17] found that these Doppler flow velocities responded in a characteristic manner to provocative

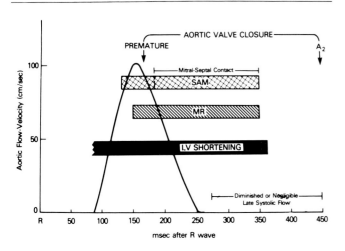

Figure 27-5 Timing relationship of aortic flow velocity to systolic anterior motion (SAM) of the mitral valve, partial midsystolic aortic valve closure, left ventricular (LV) cavity shortening, and mitral regurgitation (MR) in 20 patients with obstructive hypertrophic cardiomyopathy. Time intervals shown represent mean values for the patient group, corrected for heart rate. A bifid flow velocity pattern is depicted because it was detected in all but four patients. The onset of mitral systolic anterior motion shown here was arbitrarily taken as the time when the mitral valve came within 5 mm of the septum. A_2 indicates the aortic valve closure component of the second heart sound. Source: Reprinted with permission from Journal of the American College of Cardiology (1985;6:10), Copyright © 1985, American College of Cardiology.

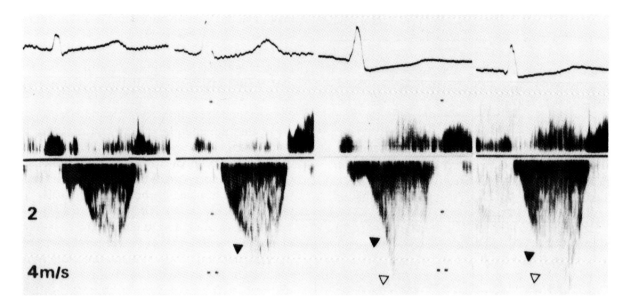

Figure 27-6 Four examples of left ventricular outflow tract jets recorded by continuous-wave Doppler ultrasound. Velocities range from 3.4 to 5.5 m/s. The increasing slope as the jets accelerate toward peak velocities (solid symbols) is typical for left ventricular outflow signals in hypertrophic cardiomyopathy. At the highest velocities, there is often some decrease in signal intensity or amplitude (open symbols). *Source:* Reprinted wth permission from *Journal of the American College of Cardiology* (1986;8:1047–1058), Copyright © 1986, American College of Cardiology.

maneuvers. With amyl nitrite inhalation, the velocity of the left ventricular outflow tract increased, indicating an increased pressure gradient. A similar response was observed immediately after a Valsalva maneuver or during postextrasystolic potentiation. These Doppler observations contribute importantly to our understanding of the complex ejection flow relationships in HCM.

There are a number of technical considerations related to Doppler studies in HCM, as emphasized by Yock et al.[17] Since the flow patterns in HCM are complex, precise transducer position and angulation are critical to obtaining optimal spectral contours in recording the outflow jet velocities. Since these high-velocity jets commonly exceed the Nyquist limit, continuous-wave Doppler is required to complement the pulse-wave mode and to ensure accurate resolution of the entire spectral signal and the peak velocities. The outflow jet is optimally recorded with the ultrasound beam midway between the axes of mitral inflow and aortic outflow. This contrasts with the normal situation and may relate to the relatively acute angle between the outflow tract and aorta in HCM. Furthermore, careful Doppler examination from a variety of locations may be necessary to localize the relatively narrow outflow jet. However, in most patients, the maximal left ventricular outflow signal will still be obtained from the apex or suprasternal notch.

The severity of the SAM appears to contribute to the mitral regurgitation, which commonly occurs in association with obstruction of the left ventricular outflow tract. Doppler echo is further useful in assessing both the temporal relationships and the severity of this mitral regurgitation, which is detected in the majority of patients with obstructive HCM. Pulsed-wave Doppler can map the mitral regurgitant jet in the left atrium and determine its severity. In addition, color flow Doppler mapping, as is discussed later, may be of even greater value in spatially localizing and in defining the severity of this coexisting mitral regurgitation. Although it is usually mild in effect, this mitral regurgitation probably contributes to the decrease in forward aortic flow during mid- and late systole and to more complete emptying of the left ventricle. On the other hand, mitral regurgitation in nonobstructive HCM appears to be more variable, occurring in only a small proportion of patients.

There may be problems in distinguishing the signal from the left ventricular outflow tract from that of mitral regurgitation, since both are high-velocity signals that may appear to be similar and may be separated by less than 1 cm at their origins. There are several specific characteristics of spectral contour and timing that allow their differentiation (Figure 27-7). First, the upstroke of the left ventricular outflow tract velocity signal is less abrupt than that of mitral regurgitation and typically demonstrates a terminal acceleration curve (increased dV/dt). Second, the peak velocity of mitral regurgitation is greater than the peak velocity of the outflow tract jet since the left ventricular-atrial pressure gradient is greater than the left ventricular-aortic pressure gradient. Third, the onset of the left ventricular outflow tract jet is often delayed relative to that of the mitral

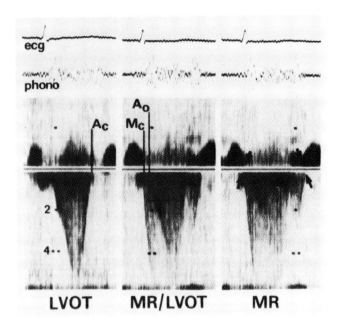

Figure 27-7 Recordings showing separation of continuous-wave Doppler signals from the left ventricular outflow tract (LVOT) jet and the mitral regurgitation (MR) jet. The panel on the left is a recording of the outflow jet alone, the velocity calibrations in meters per second. The panel on the right shows the mitral regurgitation jet, with temporal continuity between mitral inflow (above the baseline) and mitral regurgitation (below the baseline) indicated by the arrows. The middle panel shows the overlapping outflow/mitral regurgitation signal obtained with the beam in an intermediate position. A_c indicates aortic closure; A_o, aortic opening; M_c, mitral closure. *Source:* Reprinted with permission from *Journal of the American College of Cardiology* (1986;8:1047–1058), Copyright © 1986, American College of Cardiology.

regurgitation. Finally, pulsed Doppler allows further determination of whether the high-velocity jet in question is in the left atrium (indicating mitral regurgitation) or in the left ventricular outflow tract. All of these features should allow differentiation of the outflow tract jet from that of mitral regurgitation despite the fact that it is frequently possible to record both jets from the same transducer position.

It has been suggested that Doppler echocardiography may provide a means of measuring the left ventricular outflow tract gradient in patients with HCM with the use of the modified Bernoulli equation, $P = 4V^2$, where P is the pressure gradient in millimeters of mercury and V is the flow velocity in meters per second. In support of this suggestion is a recent study by Stewart et al,[18] who measured simultaneous Doppler velocity and hemodynamics intraoperatively at rest; before and after myectomy; and during interventions with isoproterenol, volume loading, and phenylephrine. They found a good correlation ($r = 0.93$) between the Doppler-derived gradient and the peak instantaneous gradient measured invasively. In addition, changes in gradient and velocity due to interventions correlated well ($r = 0.96$). However, patients with nonobstructive HCM may have peak aortic flow velocities that are markedly increased and comparable to those of patients with obstructive HCM. For example, in one recent study,[16] the average peak velocity in patients with nonobstructive HCM was 2.6 m/s, similar to the average peak velocity in patients with obstructive HCM (2.5 m/s). If the Bernoulli equation had been applied to the nonobstructed patients, left ventricular outflow tract gradients ranging between 18 mm Hg and 46 mm Hg would have been calculated in the "nonobstructive" group. Therefore, predicting left ventricular outflow tract gradients using Doppler flow velocity data may have some limitations. Further experience is necessary before the accuracy and limitations of Doppler echocardiographic evaluation of the left ventricular outflow tract gradient in HCM will be appreciated fully.

Owing to the markedly disorganized myofibril architecture, patchy fibrosis, localized hypertrophy, increased myocardial mass, and disturbances in left ventricular relaxation, diastolic dysfunction is a prominent feature of HCM. Initial echo studies using digitized M-mode tracings demonstrated abnormalities of left ventricular relaxation with prolonged left ventricular isovolumic relaxation and impaired diastolic filling in both nonobstructive and obstructive forms of HCM.[19] The mechanisms of these abnormalities in relaxation are uncertain, but they appear to be independent of the presence of a left ventricular outflow tract gradient. More recently, Doppler studies of left ventricular filling patterns have extended these M-mode echo findings further by demonstrating decreased flow velocities during the rapid-filling phase of diastole and abnormal ratios of early-to-late diastolic peak flow velocities, indicating a greater contribution of atrial systole to diastolic filling[20] (Figure 27-8). However, the presence of significant mitral regurgitation with its associated elevated early diastolic left atrial pressure increases the velocity of early transmitral flow, masking the abnormalities of diastolic function in those patients with HCM and SAM, and may tend to normalize diastolic flow velocities. Thus Doppler echocardiographic assessment of left ventricular diastolic abnormalities in HCM may be limited when significant mitral regurgitation is present.

Initial studies using color flow Doppler mapping in obstructive HCM have demonstrated acceleration of the left ventricular outflow tract jet, just proximal to mitral leaflet-septal contact and systolic narrowing of the jet at the level of mitral leaflet-septal contact.[18,21] Other color Doppler studies have demonstrated the simultaneous appearance of SAM, turbulent systolic aortic flow, and mid- and late mitral incompetence in patients with obstructive HCM. Thus experience with color Doppler studies in HCM is limited, but this technique promises to contribute further to the understanding of the abnormal flow dynamics in HCM.

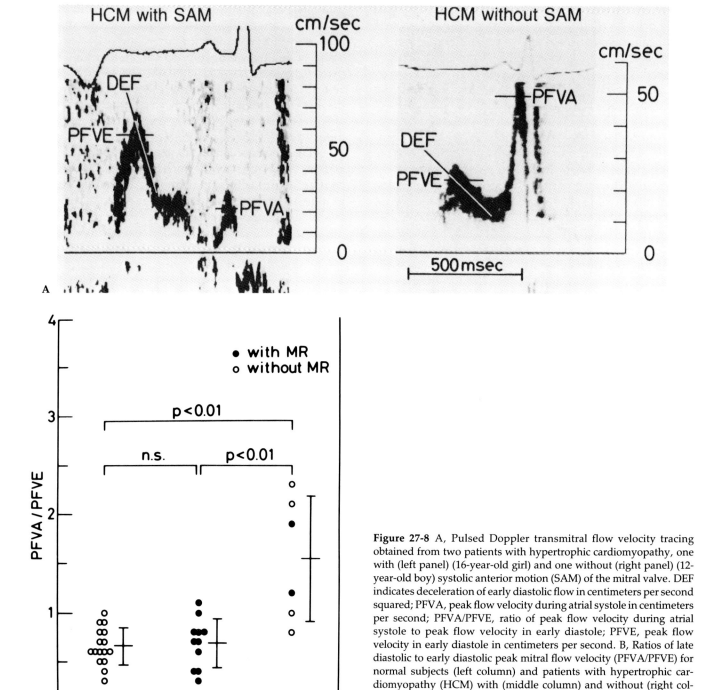

Figure 27-8 A, Pulsed Doppler transmitral flow velocity tracing obtained from two patients with hypertrophic cardiomyopathy, one with (left panel) (16-year-old girl) and one without (right panel) (12-year-old boy) systolic anterior motion (SAM) of the mitral valve. DEF indicates deceleration of early diastolic flow in centimeters per second squared; PFVA, peak flow velocity during atrial systole in centimeters per second; PFVA/PFVE, ratio of peak flow velocity during atrial systole to peak flow velocity in early diastole; PFVE, peak flow velocity in early diastole in centimeters per second. B, Ratios of late diastolic to early diastolic peak mitral flow velocity (PFVA/PFVE) for normal subjects (left column) and patients with hypertrophic cardiomyopathy (HCM) with (middle column) and without (right column) systolic anterior motion (SAM) of the mitral valve. MR indicates mitral regurgitation. *Source:* Reprinted with permission from *Journal of the American College of Cardiology* (1986;7:1266,1268), Copyright © 1986, American College of Cardiology.

Echocardiographic Assessment of Genetic Transmission and Therapeutic Results

Of considerable importance, echo has allowed careful study of the genetic transmission of HCM. This disease is often familial and transmitted in a pattern consistent with an autosomal dominant trait. It has been estimated that approximately 55% of cases are transmitted in this manner and 45% occur sporadically. Echo surveys of families have identified older relatives of patients with HCM who have mild and localized left ventricular hypertrophy. These observations suggest that the spectrum of HCM may include a subclinical form with mild morphologic traits. Although echo has been used to study families with HCM, it should be recognized that a single echo during childhood cannot entirely exclude HCM, since its morphologic expression may not be complete until adulthood and a mature body size are achieved.

Serial echocardiographic studies have followed the progression of HCM. During childhood, serial echo studies have shown that the magnitude and extent of left ventricular hypertrophy may increase substantially,[22] whereas the majority of adult patients have little change in left ventricular wall thickness and cavity dimension. However, a small but important proportion of patients develop left ventricular systolic dysfunction, progressive wall thinning (a change of thickness of 5 mm or more), and relative cavity enlargement.[23] In this latter subgroup of patients, which comprises an estimated 10% of patients with severely symptomatic HCM, these morphologic changes are associated with refractory heart failure and a poor prognosis.

A variety of echocardiographic approaches have been used to evaluate objectively the therapeutic approaches in HCM. Shapira et al[13] demonstrated M-mode and two-dimensional echo changes after septal myectomy. The subaortic left ventricular septal thickness decreased from a mean of 2.1 cm to 1.4 cm and the magnitude of SAM was significantly reduced. Two-dimensional echo further allowed evaluation of the exact subaortic area and extent of septal thinning produced by the septal myectomy. Lorell et al[24] used digitized M-mode echo to examine diastolic left ventricular relaxation in HCM. Left ventricular posterior wall endocardium and epicardium and septal wall endocardium were digitized every 5 milliseconds over one cardiac cycle to provide data on left ventricular internal dimension and posterior wall thickness. The peak rate of left ventricular diastolic posterior wall thinning (dPW/dt) and the peak rate of left ventricular early dimension change during diastole (dD_{LV}/dt) were derived by computer analysis, and the left ventricular isovolumic relaxation time was calculated. Lorell et al[24] demonstrated that nifedipine significantly improved left ventricular relaxation patterns and this correlated with symptomatic improvement. After sublingual nifedipine administration, the left ventricular isovolumic time increased, as did the peak rate of left ventricular diastolic posterior wall thinning and the rate of change in left ventricular posterior wall diastolic thickness. Also, the peak rate of the left ventricular diastolic dimension improved markedly, reflecting improved ventricular filling. This echocardiographic approach also has been used to demonstrate improvements in left ventricular diastolic function during intravenous and oral therapy with other calcium antagonists in patients with HCM.

Recently, Iwase et al[25] used exercise pulsed Doppler echocardiography to study the effects of diltiazem hydrochloride on left ventricular diastolic function in patients with HCM. In their study, resting and postexercise transmitral flow velocity was measured before and after diltiazem therapy. The peak velocity in the rapid-filling and atrial-contraction phases, the ratio of peak velocity in the atrial contraction phase to that in the rapid-filling phase, and the pressure half-time were measured to evaluate left ventricular diastolic function. Iwase et al[25] found that early left ventricular filling velocity increased and the ratio of late-to-early filling velocities decreased after diltiazem therapy. Thus both of these parameters of diastolic function tended to normalize after diltiazem therapy. In addition, the responses of these parameters to exercise were no different from those in normal subjects. Their Doppler results suggested that diltiazem decreased the left ventricular diastolic abnormality noted on dynamic exercise of mild intensity in patients with HCM.

Conclusions

It is clear from this brief overview that echocardiography has made important contributions to the characterization, understanding, and evaluation of patients with HCM. However, there remain several unanswered questions; undoubtedly, further studies with these echo techniques will continue to contribute important information about this intriguing disease over years to come. Of particular importance will be the combination of echo data with other diagnostic modalities. These combined studies should improve further our insights into the mechanisms and pathophysiology of HCM.

DILATED CARDIOMYOPATHY

Dilated cardiomyopathy (DCM), previously referred to as congestive cardiomyopathy, is a primary disorder of the myocardium resulting in left ventricular enlargement and dysfunction. A number of underlying disease processes of toxic, infectious, infiltrative, or metabolic etiology may contribute to the production of DCM. However, in the vast majority of cases, DCM is idiopathic in nature with no identifiable cause. Although echocardiography has been exceedingly useful in the identification of patients with DCM, it has been less useful than in the case of HCM in the determination of underlying etiology or in the elucidation of pathophysiologic mechanisms.

Echocardiographic Features

M-Mode echocardiography can identify easily the increased left ventricular dimensions and decreased fractional shortening that are hallmark features of DCM

(Figure 27-9). Other typical M-mode echo characteristics include posterior mitral valve displacement in the left ventricular cavity, producing a large E-point–septal separation, decreased mitral valve opening, left atrial enlargement, a poorly moving aorta, and gradual closure of the aortic valve.

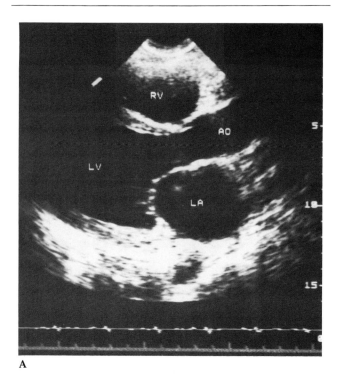

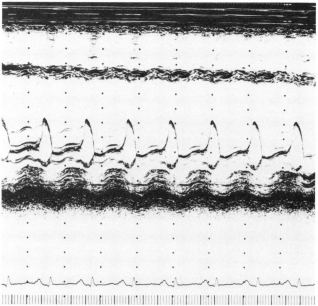

Figure 27-9 A, Parasternal long-axis two-dimensional echocardiogram near end-systole in a patient with dilated cardiomyopathy (DCM). Note the enlarged end-systolic left ventricular (LV) cavity and dilatation of the left atrium (LA). AO indicates aorta; RV, right ventricle. B, M-Mode echocardiogram from a patient with DCM. The left ventricle is dilated with a decreased percentage fractional shortening, and there is marked E-point–septal separation.

Combined hemodynamic-echo studies have allowed more comprehensive evaluation of left ventricular mechanics in DCM.[26] From these combined studies, end-systolic stress-dimension or stress-shortening relationships may be examined to delineate further the role of contractile function in DCM. Both end-systolic circumferential and meridional wall stress are elevated in DCM. However, these increased stresses do not entirely explain the decreases in ejection phase indices since contractile function, as assessed by end-systolic stress dimension or end-systolic stress shortening, is depressed.

Two-dimensional echo further allows more accurate assessment of the enlarged left ventricular volumes, the decreased global ejection fraction, and the more spheric left ventricular chamber shape, as well as the identification of discrete wall motion abnormalities that often are a clue to the presence of underlying ischemic disease as the cause of the cardiomyopathy. In addition, echo is often useful in excluding concomitant valvular or pericardial disease when these are suspected as contributing to the left ventricular dysfunction. However, in the majority of cases, clinical two-dimensional echo cannot accurately differentiate between the various potential underlying disease mechanisms that may result in DCM. The exception to this is when ultrasound tissue characterization, grossly by inspection or by more precise computer-assisted methods, may implicate a specific disease process, such as amyloidosis. In this regard, the value of echo tissue characterization in the differentiation of cardiomyopathies is reviewed in Chapter 32.

Although two-dimensional echo provides excellent diagnostic abilities, this technique has not been particularly useful in the serial evaluation of DCM.[27] Patients with DCM may improve clinically or may deteriorate with little or no change in left ventricular dimension, volume, or systolic performance. Furthermore, systolic abnormalities, once established, remain relatively fixed and change little with inotropic or vasodilator intervention.[28] As a result, M-mode and two-dimensional echo has been disappointing as a tool to follow the progression of disease in DCM or to evaluate the effect of therapeutic interventions.

Left Ventricular Mural Thrombus

In addition to evaluation of left ventricular volumes and function, two-dimensional echo also allows identification of left ventricular mural thrombus, which frequently occurs in DCM. For example, Gottdiener et al[29] visualized thrombus by two-dimensional echo in 44 (36%) of 123 patients with DCM. In their study, 11% of patients had evidence of systemic embolization. Certain features, including a protruding configuration or free mobility of the thrombus, may help to identify patients at increased risk of embolization. In these instances, two-dimensional echo may be useful in recommending anticoagulation in patients with DCM. However, despite the fact that anticoagulation is often empirically recommended when thrombus is visualized by two-dimensional echo, the precise role of anticoagulation in these patients remains ill-defined and awaits further randomized, controlled trials.

Doppler Evaluation

Doppler echo provides a new and potentially powerful tool for analyzing the dynamics of flow in DCM. Several features of DCM have now been identified by Doppler echocardiography.[30] Most notable, peak aortic flow velocity is decreased with a lower aortic flow velocity integral, and decreased aortic acceleration and ejection times (Figure 27-10). The mechanisms for these flow abnormalities are likely related to reduced left ventricular systolic performance. In addition, pulsed Doppler echo has also been useful in assessing the degree of mitral regurgitation that frequently accompanies DCM. This mitral regurgitant volume depends on the ventriculoatrial pressure gradient, the size of the regurgitant orifice, and the duration of the regurgitation. Keren et al[31] used two-dimensional echo and Doppler echo to investigate the dynamics of mitral regurgitation in patients with DCM. They found that the percentages of total regurgitant volume occurring during the pre-ejection, ejection, and postejection periods were 13%, 78%, and 8%, respectively, suggesting that the mitral regurgitation in DCM occurs predominantly during the aortic ejection period.

Measurement of mitral valve flow velocity by pulsed Doppler echo is also useful in evaluating left ventricular diastolic filling properties. The determinants of left ventricular filling are complex and involve multiple factors, including left ventricular relaxation, compliance, and filling pressure, among others. Takenaka et al[32] recently used pulsed Doppler echo to demonstrate left ventricular filling abnormalities in patients with DCM (Figure 27-11). The sample volume was positioned in the left atrium in the apical four-chamber or apical long-axis view to search carefully for holosystolic turbulent flow indicating mitral regurgitation. The sample volume then was positioned in the mitral orifice at a level close to the tips of the mitral leaflets during diastole to record transmitral flow velocity. Four mitral flow velocity indices were measured: peak flow velocity during early diastole (PFVE), peak flow velocity during atrial systole (PFVA), the ratio of PFVA to PFVE, and the deceleration half-time of early diastolic flow. Peak mitral flow velocity during early diastole was decreased and the ratio of late diastolic to early diastolic filling

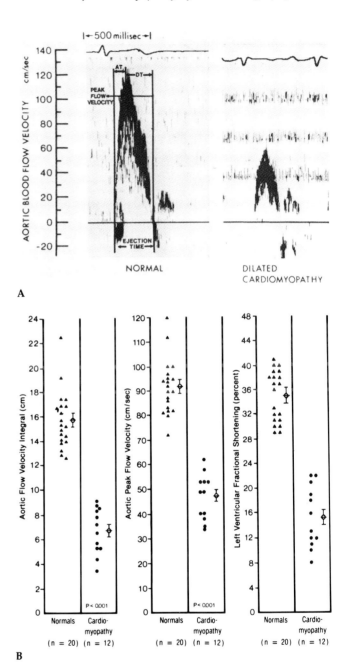

Figure 27-10 A, Doppler blood flow velocity recordings from the ascending aorta in a normal subject (left panel) and in a patient with dilated cardiomyopathy (right panel). Aortic blood flow velocity in centimeters per second is displayed on the vertical axis and time is shown on the horizontal axis. The normal Doppler acceleration time (AT), deceleration time (DT), and ejection time are indicated by arrows. Note that peak aortic flow velocity, flow velocity integral, average acceleration, and ejection time are greater in the normal subject than in the patient with cardiomyopathy. B, Aortic flow velocity integral (left panel), aortic peak flow velocity (middle panel), and left ventricular percentage fractional shortening (right panel) are compared in normal subjects and in patients with cardiomyopathy. Note that the data for all three measurements separate normal subjects from patients with cardiomyopathy to a similar degree. The shaded area represents the normal range for left ventricular percentage fractional shortening. *Source:* Reprinted with permission from *American Heart Journal* (1983;106:1057), Copyright © 1983, CV Mosby Company.

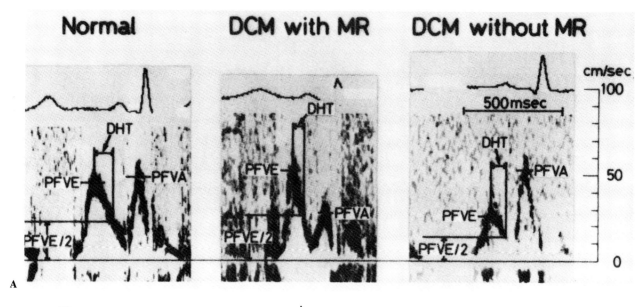

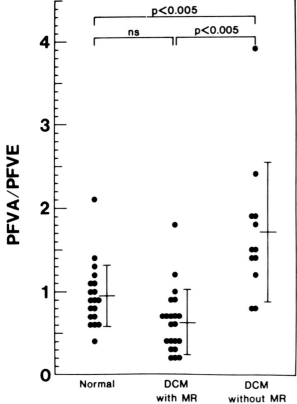

Figure 27-11 A, Pulsed Doppler mitral flow velocity tracings recorded from the apical position in a normal subject (40-year-old woman), a patient with dilated cardiomyopathy (DCM) and mitral regurgitation (MR) (58-year-old woman), and a patient with dilated cardiomyopathy without MR (63-year-old man). Three mitral flow velocity indices were measured: peak flow velocity in early diastole (PFVE, centimeters per second), peak flow velocity during atrial diastole (PFVA, centimeters per second), and deceleration half-time of early diastolic flow (DHT, milliseconds). B, Ratio of late diastolic to early diastolic peak mitral flow velocity (PFVA/PFVE) in normal subjects and patients with dilated cardiomyopathy (DCM) with and without mitral regurgitation (MR). Bracketed values represent mean ± standard deviation. *Source:* Reprinted with permission from *American Journal of Cardiology* (1986;143:144,146), Copyright © 1986, Cahners Publishing Company Inc.

(PFVA/PFVE) was increased in DCM. However, a complicating factor was the presence of significant mitral regurgitation, which enhanced early left ventricular filling by increasing left atrial pressure during early diastole. In patients with DCM and significant mitral regurgitation, both peak early diastolic filling and the ratio of early-to-late filling were normalized. In addition, the deceleration half-time of early diastolic mitral flow was shorter in patients with DCM and mitral regurgitation than it was in normal subjects. This may also be explained by the effect of mitral regurgitation, since mitral regurgitation is known to produce a decrease in deceleration time. Thus abnormalities of peak diastolic mitral flow velocity were detected in patients with DCM without mitral regurgitation but not in patients with DCM and mitral regurgitation, suggesting that mitral regurgitation, as in the case of HCM, masks left ventricular filling abnormalities in patients with DCM.

The ability to measure characteristics of ventricular ejection and filling in combination with other structural and functional characteristics suggests that combined echo-Doppler techniques may better characterize patients with DCM and may potentially improve serial follow-up and the assessment of therapeutic interventions. However, further studies are required to evaluate fully the impact of these combined approaches.

Conclusions

As in the case of HCM, echo has made important contributions to our understanding and characterization of DCM. Important issues that await resolution are the evaluation of the progression of the disease in these patients and a noninvasive approach to the evaluation of pharmacologic interventions. There is optimism that echo, particularly in combination with Doppler techniques, will improve our abilities to assess these patients. However, additional studies are necessary before the potential of these techniques can be appreciated fully in patients with DCM.

REFERENCES

1. Wigle ED, Sasson Z, Henderson MA, et al. Hypertrophic cardiomyopathy: The importance of the site and the extent of hypertrophy: A review. *Prog Cardiovasc Dis.* 1985;28:1–83.

2. Wigle ED. Hypertrophic cardiomyopathy: A 1987 viewpoint. *Circulation.* 1987;75:311–322.

3. Maron BJ, Bonow RO, Cannon RO III, et al. Hypertrophic cardiomyopathy: Interrelations of clinical manifestations, pathophysiology, and therapy. *N Engl J Med.* 1987;316:780–789, 844–852.

4. Henry WL, Clark CE, Epstein SE. Asymmetric septal hypertrophy (ASH): Echocardiographic identification of the pathognomonic anatomic abnormality of IHSS. *Circulation.* 1973;47:225–233.

5. Doi YL, McKenna WJ, Gehrke J, et al. M-mode echocardiography in hypertrophic cardiomyopathy: Diagnostic criteria and prediction of obstruction. *Am J Cardiol.* 1980;45:6–14.

6. Maron BJ. Asymmetry in hypertrophic cardiomyopathy: The septal to free wall thickness ratio revisited. *Am J Cardiol.* 1985;55:835–838. Editorial.

7. Popp R, Harrison D. Ultrasound in the diagnosis and evaluation of therapy of idiopathic hypertrophic subaortic stenosis. *Circulation.* 1969;40:905–914.

8. Gilbert BW, Pollick C, Adelman AG, et al. Hypertrophic cardiomyopathy: Subclassification by M mode echocardiography. *Am J Cardiol.* 1980;45:861–872.

9. Pollick C, Rakowski H, Wigle ED. Muscular subaortic stenosis: The quantitative relationship between systolic anterior motion and the pressure gradient. *Circulation.* 1984;69:43–49.

10. Ciro E, Nichols PF, Maron BJ. Heterogeneous morphologic expression of genetically transmitted hypertrophic cardiomyopathy: Two-dimensional echocardiographic analysis. *Circulation.* 1983;67:1227–1233.

11. Criley JM, Siegel RJ. Has "obstruction" hindered our understanding of hypertrophic cardiomyopathy? *Circulation.* 1985;72:1148–1154.

12. Pollick C, Morgan CD, Gilbert BW, et al. Muscular subaortic stenosis: The temporal relationship between systolic anterior motion of the anterior mitral leaflet and pressure gradient. *Circulation.* 1982;66:1087–1093.

13. Shapira JN, Stemple DR, Martin RP, et al. Single and two-dimensional echocardiographic visualization of the effects of septal myectomy in idiopathic hypertrophic subaortic stenosis. *Circulation.* 1978;58:850–860.

14. Maron BJ, Gottdiener JS, Arce J, et al. Dynamic subaortic obstruction in hypertrophic cardiomyopathy: Analysis by pulsed Doppler echocardiography. *J Am Coll Cardiol.* 1985;6:1–15.

15. Wigle ED, Adelman AG, Silver MD. Pathophysiological considerations in muscular subaortic stenosis. In: O'Connor M, Wolstenholme GEW, eds. *Hypertrophic Obstructive Cardiomyopathy.* London, England: Churchill Livingstone: 1971; Ciba Foundation Study Group No 47:63–70.

16. Cogswell TL, Sagar KB, Wann LS. Left ventricular ejection dynamics in hypertrophic cardiomyopathy and aortic stenosis: Comparison with the use of Doppler echocardiography. *Am Heart J.* 1987;113:110–116.

17. Yock PG, Hatle L, Popp RL. Patterns and timing of Doppler-detected intracavitary and aortic flow in hypertrophic cardiomyopathy. *J Am Coll Cardiol.* 1986;8:1047–1058.

18. Stewart WJ, Schiavone WA, Salcedo EE, et al. Intraoperative Doppler echocardiography in hypertrophic cardiomyopathy: Correlations with the obstructive gradient. *J Am Coll Cardiol.* 1987;10:327–335.

19. Sanderson JE, Traill TA, St John Sutton MG, et al. Left ventricular relaxation and filling in hypertrophic cardiomyopathy: An echocardiographic study. *Br Heart J.* 1978;40:596–601.

20. Takenaka K, Dabestani A, Gardin JM, et al. Left ventricular filling in hypertrophic cardiomyopathy: A pulsed Doppler echocardiographic study. *J Am Coll Cardiol.* 1986;7:1263–1271.

21. Tencate FJ, Mayala AP, Vletter WB, et al. Color-coded Doppler imaging of systolic flow patterns in hypertrophic cardiomyopathy. *Int J Card Imaging.* 1985;1:217–223.

22. Maron BJ, Spirito P, Wesley Y, et al. Development and progression of left ventricular hypertrophy in children with hypertrophic cardiomyopathy. *N Engl J Med.* 1987;315:610–614.

23. Spirito P, Maron BJ, Bonow RO, et al. Occurrence and significance of progressive left ventricular wall thinning and relative cavity dilatation in hypertrophic cardiomyopathy. *Am J Cardiol.* 1987;60:123–129.

24. Lorell BH, Paulus WJ, Grossman W, et al. Modification of abnormal left ventricular diastolic properties by nifedipine in patients with hypertrophic cardiomyopathy. *Circulation.* 1982;65:499–507.

25. Iwase M, Sotobata I, Takagi S, et al. Effects of diltiazem on left ventricular diastolic behavior in patients with hypertrophic cardiomyopathy: Evaluation with exercise pulsed Doppler echocardiography. *J Am Coll Cardiol.* 1987;9:1099–1105.

26. Laskey WK, St John Sutton M, Zeevi G, et al. Left ventricular mechanics in dilated cardiomyopathy. *Am J Cardiol.* 1984;54:620–625.

27. Engler R, Ray R, Higgins CB, et al. Clinical assessment and follow-up of functional capacity in patients with congestive cardiopathy. *Am J Cardiol.* 1982;49:1832–1837.

28. Haq A, Rakowski H, Baigrie R, et al. Vasodilator therapy in refractory congestive heart failure: A comparative analysis of hemodynamic and noninvasive studies. *Am J Cardiol.* 1982;49:439–444.

29. Gottdiener JS, Gay JA, VanVoorhees L, et al. Frequency and embolic potential of left ventricular thrombus in dilated cardiomyopathy: Assessment by two-dimensional echocardiography. *Am J Cardiol.* 1983;52:1281–1285.

30. Gardin JM, Iseri LT, Elkayam U, et al. Evaluation of dilated cardiomyopathy by pulsed Doppler echocardiography. *Am Heart J.* 1983;106:1057–1065.

31. Keren G, LeJemtel TH, Zelcer AA, et al. Time variation of mitral regurgitant flow in patients with dilated cardiomyopathy. *Circulation.* 1986;74:684–692.

32. Takenaka K, Dabestani A, Gardin JM, et al. Pulsed Doppler echocardiography study of left ventricular filling in dilated cardiomyopathy. *Am J Cardiol.* 1986;58:143–147.

Chapter 28

Evaluation of Cardiomyopathy by Radioisotopes

Robert A. Quaife, MD, and John B. O'Connell, MD

DILATED CARDIOMYOPATHY

Dilated cardiomyopathy (DC) is characterized by abnormal left ventricular contractility and chamber dilatation that leads to symptomatic impairment of systolic function.[1] This definition may be applied only in the absence of coronary artery, valvular, or pericardial disease. Because the clinical manifestations mimic that of end-stage ischemic heart disease, one task of noninvasive radionuclide evaluation is directed toward differentiation of DC from ischemic heart disease (IHD). Noninvasive radionuclide imaging techniques are also helpful in the evaluation of specific heart muscle diseases such as active myocarditis, sarcoidosis cordis, muscular dystrophy, and metabolic disorders. There are also radionuclide imaging techniques that assess physiologic impairment. These include analysis of left ventricular function using technetium 99m-labeled gated blood pool imaging, perfusion imaging with thallium 201 chloride, and metabolic imaging with biochemically active tracers.

Radionuclide Angiography

Gated cardiac blood pool scanning or radionuclide angiography (RNA) radiolabels intravascular blood volume to evaluate right or left ventricular ejection fraction, segmental wall motion, ventricular size, interventricular septal thickness, and left ventricular filling parameters of diastolic function.

Patients with DC have biventricular enlargement, reduced left ventricular ejection fraction (less than 0.50), and diffusely diminished wall motion by RNA (Figure 28-1). It is difficult, however, to determine whether these abnormalities are due to DC or IHD.[2,3] Greenberg and co-workers[4] evaluated segmental wall motion abnormalities and found that they were present in both DC and IHD. Wall motion abnormalities

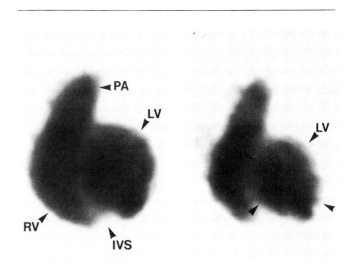

Figure 28-1 Radionuclide angiogram from a patient with DC in the 45° left anterior oblique view. The image on the left is end-diastole and the image on the right is end-systole. There is minimal change in left ventricular (LV) chamber size between the two images in this patient with a 12% ejection fraction. RV indicates right ventricle; IVS, interventricular septum; PA, pulmonary artery.

localized to the septum and apex were more notable in patients with DC. When right and left ventricular functions were compared, the right ventricular ejection fraction (RVEF) was often normal in IHD, in contrast to DC, where RVEF was frequently impaired. In the study by Wallis et al,[5] patients with DC and segmental wall motion abnormalities identified by RNA had a significantly higher left ventricular ejection fraction (LVEF) and 1-year survival rate than did those with diffuse wall motion abnormalities. Perfusion studies, however, showed that only a rough correlation ($r = 0.58$) between qualitative ^{201}TlCl uptake and RNA regional wall motion abnormalities was demonstrated.[6] Caglar and associates[7] found no significant difference between these disease states when comparing left ventricular systolic indices; however, the RVEF was decreased selectively in patients with DC versus those with IHD. Because DC represents a generalized process involving both ventricles and IHD is primarily a left ventricular disease, in the absence of right ventricular infarction these results are consistent with accepted pathophysiologic concepts. Furthermore, when previous right ventricular infarction was documented, the RVEF was always greater than 0.35, which is higher than the mean RVEF (0.32) in DC.[8] The relevance of this observation is clouded by the small patient population and the variation in RVEF within the groups, particularly in light of the fact that others have not confirmed these observations.[4]

Exercise RNA has also been suggested as a means of differentiating IHD from DC. Although the resting LVEF is not different in the two groups, the exercise ejection fraction consistently decreases in IHD and increases in DC.[9] The fall in ejection fraction in IHD is postulated to result from exercise-induced ischemic contractile abnormalities, whereas patients with DC have normal coronary arteries and an attenuated but directionally normal exercise response. The addition of exercise to RNA thus increases the specificity in the differentiation between IHD and DC. Unfortunately, there is overlap in the exercise response, and hence exercise RNA cannot be considered a definitive diagnostic test.

Recently the importance of left ventricular diastolic dysfunction in DC has been re-emphasized. Impairment of left ventricular diastolic performance leads to elevation of left ventricular filling pressure, which is transmitted to the pulmonary vascular bed and results in symptoms of dyspnea. Radionuclide ventriculography may be used to evaluate diastolic as well as systolic function of the left ventricle. When a third-degree polynomial function is fitted to the time-activity curve, the peak filling rate and the time to peak filling rate may be calculated (Figure 28-2). The peak filling rate and the time to peak filling rate reflect isovolumic relaxation. Bonow and co-workers[10] described the normal range

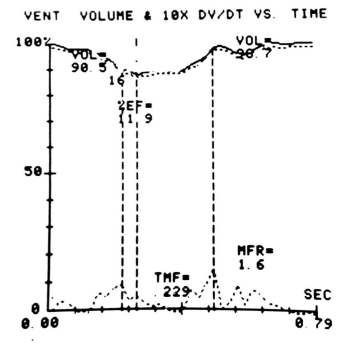

Figure 28-2 This time-activity curve from the patient in Figure 28-1 shows the parameters of diastolic function. TMF indicates the time of maximal left ventricular (LV) filling (229 milliseconds); MFR, the maximal LV filling rate expressed in EDV per second (1.6); %EF, the LV ejection fraction (11.9%); VENT., ventricular.

for peak left ventricular filling rate as 2.5 to 5.0 end-diastolic volumes (EDV) per second (mean 3.3 ± 0.6 [SD] EDV per second) and the time to peak filling rate as 90 to 180 milliseconds (mean 136 ± 22 milliseconds) in normal subjects. In normal volunteers, the peak left ventricular filling rate and the time to peak filling rate did not correlate with heart rate, age, or echocardiographic left ventricular end-diastolic dimension.

Soufer et al[11] evaluated left ventricular systolic and diastolic filling parameters in patients with congestive heart failure. After evaluation of the time-activity curves, the patients were divided according to diastolic function. One third of the patients with a peak filling rate of less than 2.5 EDV per second had pulmonary edema and a mean LVEF of 59% with a mean peak filling rate of 1.95 EDV per second. Eighty-six percent of these patients had coronary artery disease or hypertension, or both, as the cause of their heart failure. In patients with a borderline or normal peak filling rate, diagnostic heterogeneity was noted. Other authors have noted a decreased left ventricular peak filling rate and prolongation of the time to peak filling rate in patients with DC as well.[11,12] Abnormalities of diastolic function are not specific and therefore will not differentiate IHD from DC.

Diastolic function determined by Doppler echocardiography, including peak rapid-filling phase and atrial "kick" flow velocity coupled with an estimation of left ventricular volume (cross-sectional mitral valve area method), was compared with RNA time-activity curves. Echocardiography when compared with RNA correctly identified diastolic abnormalities in DC with an 86% accuracy.[13] Furthermore, a direct correlation ($r = 0.72$) was shown between echocardiographic duration of early diastolic peak flow velocity and the interval from end-systole to the end of rapid ventricular filling on radionuclide angiography.[14] The correlation improved ($r = 0.83$) when the duration of the isovolumic relaxation was separated from the time to diastolic flow peak. These studies indicate that abnormalities in diastolic function of the left ventricle may be quantitated before overt systolic dysfunction is present.

Thallium Perfusion Imaging

Thallium 201 myocardial perfusion scintigraphy has been suggested as a means of differentiating IHD from DC.[15] When both ^{201}TlCl perfusion imaging and RNA using ^{99m}Tc-labeled serum albumin were compared, areas of poor myocardial perfusion correlated to regions of segmental wall motion abnormalities, and IHD could be distinguished from DC. In patients with IHD, only defects involving more than 40% of the left ventricular circumference correlated with the segmental wall motion abnormalities noted on RNA. In contrast, DC showed either homogeneous uptake of the left ventricle or perfusion defects that were less than 20% of the circumference.[16] Saltissi et al,[17] however, showed that reversible perfusion defects were equally common (60%) in both groups. The presence of a fixed perfusion defect, more than 40% of the left ventricle, or a "reverse" redistribution defect favored IHD. These investigators postulated that thinning of the ventricular myocardium was the reason for the perfusion defect. If this theory is correct, left ventricular fibrosis may produce regional perfusion abnormalities and segmental wall motion abnormalities in DC. In summary, conventional isotopic images are useful in the evaluation (Table 28-1) and follow-up of patients with DC.

Metabolic Imaging

Another noninvasive approach to the definition of the etiology and pathophysiology of DC includes fatty acid metabolism imaging with standard radionuclide tracers and positron emission tomography. When gated imaging with 17-[^{123}I]iodoheptadecanoic acid was used to evaluate fatty acid turnover and abnormal clearance rates, DC patients showed regional abnormalities in fatty acid clearance that did not correlate with specific coronary artery distributions.[18] Patients with coronary artery disease, however, had defects of tracer accumulation that followed the distribution of stenotic coronary arteries and were exaggerated by stress. Although ischemia produced abnormalities in substrate metabolism and fatty acid clearance, the precise cause of these defects is incompletely understood. Rabinovitch and associates[19] evaluated this radionuclide in various types of cardiomyopathy and noted that only 12% had abnormal images and clearance rates. In each of these two patients, a carnitine deficiency, one associated with the Kearns-Sayre syndrome, could be implicated as a factor in the cardiomyopathy.[20]

Positron emission tomographic imaging using radiolabeled fatty acids, primarily [^{11}C]palmitate, has been applied recently to the metabolic evaluation in DC. This positron-emitting radiotracer detects shifts in cellular myocardial metabolism from normal fatty acid substrate utilization to the glycolytic pathway of glucose metabolism produced by ischemia. Thus in regions of ischemia [^{11}C]palmitate accumulates and washes out slowly.[21,22] Geltman and co-workers[23] evaluated dilated and hypertrophic cardiomyopathy using [^{11}C]palmitate and found numerous regions of noncontiguous discrete accumulation of palmitate in DC as compared with IHD or normal controls. These areas had irregular borders, classified as "moth-eaten," which did not correlate to regional wall motion abnormalities. The etiology of this uptake is postulated to represent fibrosis with abnormal or absent fatty acid metabolism. Furthermore, these abnormalities did not correlate with ^{201}TlCl defects or regions of depressed oxidative phosphorylation by glucose radiotracer imaging. The differentiation of IHD from DC could be accomplished only because the defects in DC typically were heterogeneous and multiple, and had irregular borders; in IHD the regional defects had regular borders and were more homogeneous.

Glucose metabolism was investigated by Perloff et al[24] in Duchenne muscular dystrophy using radionuclide ventriculography, ^{201}TlCl perfusion imaging, and metabolic imaging with [2-^{18}F]fluoro-deoxyglucose and [^{13}N]ammonia. Segmental wall motion abnormalities in the posterobasal and lateral free wall were identified in these subjects by RNA and correlated with autopsy distribution of fibrosis. Metabolic imaging

Table 28-1 Radionuclide Techniques Used To Differentiate Ischemic from Dilated Cardiomyopathy in Congestive Heart Failure

Evaluation	Ischemic Heart Disease	Dilated Cardiomyopathy
Left ventricular ejection fraction	↓ ↓	↓ ↓
Exercise left ventricular ejection fraction	↓	Slightly ↓
Right ventricular ejection fraction (resting)	Normal to ↓	↓
Segmental wall motion abnormality	Present	Present
Thallium perfusion defects	+ +	±

revealed abnormal [^{13}N]ammonia perfusion in segments with increased glucose utilization, as measured by fluorine (^{18}F) tomography, whereas ^{201}TlCl scintigraphy showed a reduction in perfusion to these regions. These investigators concluded that the radiotracer abnormalities indicated abnormal metabolism associated with the underlying disease process (Duchenne muscular dystrophy) and that these regions of myocardium may be genetically targeted.[25,26]

SPECIFIC HEART MUSCLE DISEASE

Anthracycline Cardiotoxicity

Isotopic imaging may be useful in specific heart muscle diseases induced by heavy metals, alcohol, or anthracyclines. Doxorubicin hydroxychloride (Adriamycin) is a potent chemotherapeutic agent that serves as the cornerstone of protocols commonly used in oncology; however, it has a dose-related cardiotoxicity that may limit its usefulness. As a result, doxorubicin produces irreversible heart failure when large cumulative doses are administered without monitoring. Before the application of endomyocardial biopsy, it was recommended that the total dose not exceed 550 mg/m^2. Gottdiener and co-workers[27] evaluated RNA as a noninvasive means of assessing left ventricular dysfunction in anthracycline cardiotoxicity. They demonstrated that the addition of graded exercise to resting RNA improved the sensitivity for detection of subclinical left ventricular dysfunction. When 32 subjects underwent rest/exercise RNA after completion of a course of "maximum" dose doxorubicin (480 to 550 mg/m^2), abnormal results were obtained in 25% when only the resting scan was analyzed but rose to 38% with exercise data acquisition. Other investigators have confirmed that sensitivity for detection of asymptomatic left ventricular dysfunction in anthracycline cardiotoxicity is improved with exercise RNA; however, the specificity decreases.[28,29] When endomyocardial biopsy and resting RNA were compared, biopsy was superior in predicting the development of clinically significant anthracycline toxicity.[29] The sensitivity of the isotopic studies, however, was 100%; therefore, this technique could be used as a screening test to select a population for biopsy.

When Fourier amplitude and phase analysis is applied to rest and exercise RNA, the sensitivity is increased.[30] In the study by Alcan and associates,[30] standard rest RNA, rest and exercise RNA, and rest and exercise RNA with Fourier analysis were compared. The sensitivity for the detection left ventricular dysfunction was evaluated, yielding the following: 12% for rest studies alone, 59% with Fourier analysis of rest studies, 95% for Fourier rest and exercise studies, and 100% sensitivity for Fourier-analyzed serial rest and exercise studies. By using the Fourier amplitude technique analysis, regional areas of low-amplitude contraction that correlate to regional decreases in stroke volume may be identified. Thus temporal Fourier analysis uniquely may provide segmental data of the regional toxic effects of anthracycline administration.

Myocarditis

The application of endomyocardial biopsy for the definition of etiology in DC has supported the hypothesis that chronic inflammation (active myocarditis) stimulated by a viral infection may result in a clinical syndrome indistinguishable from DC. Although the reported incidence varies because of differences in pathologic interpretation, active myocarditis is histologically identified in approximately 10% of patients with new-onset heart failure of undetermined etiology.[31] It has been suggested that immunosuppressive therapy may result in hemodynamic and clinical improvement in a subset of these patients based on uncontrolled preliminary studies.[32] The efficacy of immunosuppressive therapy in biopsy-proved myocarditis currently is being assessed in a multicenter study (Myocarditis Treatment Trial), which will also determine the role of endomyocardial biopsy and inflammation-avid radioisotopic imaging in evaluating patients with recent-onset heart failure.[33] If immunosuppressive therapy is shown to be efficacious in this condition, it will be recommended that all patients who have DC be screened for the presence of active myocardial inflammation. The lack of universal availability of endomyocardial biopsy may lead to the recommendation that a noninvasive-accessible technique be used to select patients for referral to biopsy centers.

The histologic diagnosis of active myocarditis includes both an inflammatory infiltrate and evidence of subsequent myocytic necrosis. Imaging techniques designed to detect each of these factors have been studied in small patient populations. The uptake of gallium 67, an inflammation-avid radioisotope, has been described in patients with symptomatic DC (Figure 28-3).[34] When patients with positive myocardial images received prednisone and azathioprine, 40% showed clinical and hemodynamic improvement. When parallel studies employing ^{67}Ga imaging and endomyocardial biopsy were performed in 67 patients with new-onset DC, histologic myocarditis was identified in 8% of the population.[35] The incidence of myocarditis on biopsy in those patients with a positive gallium scan was 36%, whereas myocarditis was detected in only 1.8% of those with a negative gallium scan. Gallium 67 imaging, therefore, improves the yield of biopsy fourfold and with a high sensitivity. Gallium 67

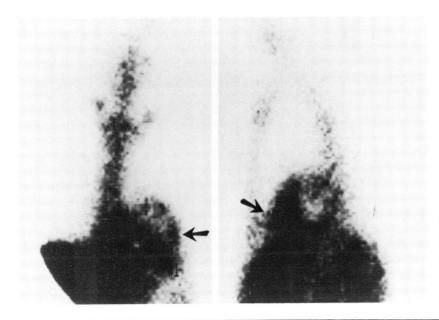

Figure 28-3 Gallium 67 imaging in a patient with acute myocarditis showing dense diffuse uptake at arrows. Left, posteroanterior projection; right, left anterior oblique view with minimal right ventricular uptake. *Source:* Adapted with permission from *Circulation* (1984;70:60), Copyright © 1984, American Heart Association.

imaging has disadvantages in that the radiation dose may be significant if multiple images are required, and there is a 72-hour delay from the initial injection to imaging, which may be unacceptable. Definitive recommendations regarding the role of ^{67}Ga imaging must await the results of the Myocarditis Treatment Trial in which paired ^{67}Ga scans and endomyocardial biopsies will be performed in a large patient population.

Radioisotopic imaging has been designed to identify myocytic necrosis in patients with DC and no clinical evidence of coronary artery disease. Technetium 99m pyrophosphate (^{99m}Tc-PYP) was the first such isotope to be investigated. In preliminary studies, some patients with acute myocarditis were noted to have uptake of ^{99m}Tc-PYP over the myocardium.[36] Because of the lack of specificity of this imaging technique, however, the interest in further study waned rapidly. More recently, monoclonal indium 111–labeled antimyosin antibody imaging has been studied in 28 patients suspected of having myocarditis.[37] Active myocarditis was identified on biopsy of 9 of these patients, all of whom had positive antimyosin antibody images. Eight of the remaining patients had positive antimyosin scans but negative biopsies. The sensitivity was therefore 100% with a specificity of 58%. In a comparison of the preliminary studies using ^{67}Ga imaging to those of ^{111}In-labeled antimyosin antibody imaging, these techniques had similar sensitivities and specificities. Perhaps some of the "false-positive" images represented "false-negative" biopsies owing to sampling error. Although it would be difficult to prove this hypothesis, investigations are in process that are designed to compare the results of immunosuppression and the untreated natural history of patients with "isotope-positive" cardiomyopathy with those of both DC (biopsy and isotope-negative) and active myocarditis. Because of the preliminary nature of these investigations, it can only be recommended that patients with suspected myocarditis undergo endomyocardial biopsy to confirm the diagnosis because these imaging techniques cannot be considered diagnostic at this time.

Sarcoidosis

Other noninfectious inflammatory reactions have been implicated as having a causal role in the congestive heart failure of DC. One such state is sarcoidosis, the hallmark of which is the presence of noncaseating granulomas involving multiple organ systems. When sarcoid involves the heart there may be an acute myocarditis that progresses rapidly to DC. Thallium 201, ^{67}Ga, and ^{99m}Tc-PYP scanning have been suggested as diagnostic procedures in the evaluation of myocardial sarcoidosis.[38–40] Gallium 67 imaging results in diffuse myocardial uptake suggesting chronic inflammation. Thallium 201 perfusion imaging shows multiple defects, most commonly in the inferior and apical regions, which may redistribute following exercise. Perfusion defects may be present even when cardiac symptoms are absent. Technetium 99m pyrophosphate imaging shows homogeneous regions of uptake that most probably correlate with areas of cell death and calcium deposition.

HYPERTROPHIC CARDIOMYOPATHY

Hypertrophic cardiomyopathy (HC) is characterized by accentuated left ventricular contractility, impaired diastolic filling, and subaortic obstruction. Both the hypercontractile state and the impaired left ventricular relaxation contribute to the production of symptoms. Radionuclide imaging has focused on the evaluation of abnormal chamber dimensions and geometry produced by septal hypertrophy, myocardial perfusion abnormalities in the interventricular septum, and abnormal diastolic function of the left ventricle. Thallium 201 imaging has been used in the assessment of wall thickness and perfusion using single-photon emission-computed tomography. Left ventricular systolic and diastolic function may be evaluated by RNA, which enables serial noninvasive evaluation of left ventricular ejection and filling.

Angina-like chest discomfort is a common finding in HC despite "normal" coronary anatomy by contrast angiography. The etiology of the ischemic symptoms is unclear but may be accounted for by increased left ventricular wall stress or perfusion abnormalities within the left ventricular myocardium.[41] The hypertrophied myocardium is both histologically and metabolically abnormal, with abnormal peak oxygen consumption, coronary blood flow, and lactate utilization. Increased lactate production suggests that there is myocardial ischemia possibly resulting in angina. Pitcher and co-workers[42] evaluated 23 patients with HC, 15 of whom had normal coronary arteries. Ten had perfusion defects that redistributed and 14 had fixed defects. These results were confirmed in a familial HC population in which 90% of asymptomatic patients had perfusion abnormalities by ^{201}TlCl imaging.[43] The reason for the defects is not completely understood, although postulates include abnormal myocytes with fibrosis-distorting perfusion and tissue uptake of thallium or myocardial ischemia due to myocytic hypertrophy and increased wall stress with poor perfusion. Necropsy studies have shown infiltration of the vasculature in the septum, which may limit tissue perfusion and result in transmural myocardial fibrosis and scarring. Thus patients with fixed thallium defects may have had a transmural infarction, while patients with reversible defects may have myocardial ischemia.[44]

Radionuclide angiography has been used to assess systolic and diastolic dysfunction (Figures 28-4 and 28-5). Bonow and co-workers[10] first reported the evaluation of RNA time-activity curves in HC. All of their patients had normal or supranormal systolic function with a mean ejection fraction of 0.75. The peak left ventricular ejection rate was also accelerated in HC. Although the peak filling rate was normal in 81% of patients, the contribution to total left ventricular filling by rapid diastolic filling was diminished, thus requiring atrial systole to supply the primary work of left ventricular filling.[45] The contribution of atrial systole to left ventricular filling was 16% in normal subjects and 31% in patients with HC by analysis of diastolic segments of the time-activity curves. In this study verapamil did not significantly change systolic function but increased the peak filling rate from 3.3 to 4.2 EDV per second and reduced the time to peak filling from 177 to 164 milliseconds, suggesting improved diastolic function. Evaluation of left ventricular isovolumic relaxation using RNA was studied by Betocchi and associates,[46] who found that this interval was prolonged in patients with HC (95 milliseconds) compared with control subjects (50 milliseconds). The greatest prolongation was seen in patients with significant outflow tract obstruction. In HC, the peak filling rate was less and the time to peak filling was delayed as a result of prolongation of the

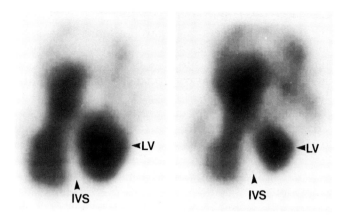

Figure 28-4 The radionuclide angiographic images in the 45° left anterior oblique projection (left, end-diastole; right, end-systole) are from a patient with hypertrophic cardiomyopathy. Note the thick interventricular septum (IVS), which further thickens during systole. A significant decrease in left ventricular (LV) volume correlates with a 0.68 ejection fraction.

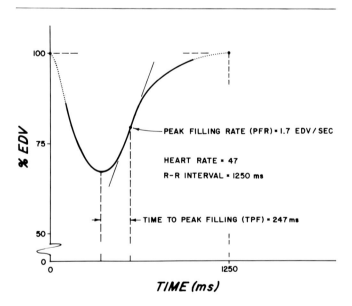

Figure 28-5 The time-activity curve from the patient in Figure 28-4. Note the prolonged time to peak filling of 274 milliseconds and the reduced peak filling rate of 1.7 EDV per second.

isovolumic relaxation. Verapamil significantly reduced the duration of isovolumic relaxation from 95 milliseconds to 80 milliseconds while it improved the peak filling rate and the time to peak filling. The analysis of time-activity curves has confirmed the observation that the primary pathophysiologic derangement in HC is an abnormality in left ventricular relaxation. Noninvasive radionuclide assessment of HC may therefore be used to assess left ventricular wall thickness, left ventricular internal dimension, myocardial perfusion abnormalities, and left ventricular diastolic properties. Serial evaluation of left ventricular systolic and diastolic function may be useful in following therapeutic interventions such as the administration of verapamil and may provide a correlation between improvement in left ventricular diastolic function and symptoms.[42–47]

RESTRICTIVE CARDIOMYOPATHY

Restrictive cardiomyopathy (RC) is a relatively rare pathophysiologic state. Etiologies for this condition include endocardial fibroelastosis, scleroderma, amyloidosis, hemochromatosis, and neoplasms. The initial clinical problem is the differentiation of RC from constrictive pericarditis. In general, radionuclide studies have not been applied for this purpose.

Amyloidosis, an uncommon disease, is the most common cause of RC. Primary amyloidosis may affect the myocardium in 90% of afflicted patients at autopsy; however, only 54% of patients with secondary amyloidosis have cardiac involvement. Most patients with amyloid heart disease have symptoms of congestive heart failure at the time of diagnosis and in general the prognosis is poor. Recently, ^{99m}Tc-PYP myocardial imaging has been shown to be diffusely positive (Figure 28-6).[48] Patients with ^{99m}Tc-PYP uptake usually have abnormal left ventricular chamber size and wall thickness on echocardiography.[49] Falk and co-workers[50] evaluated 20 consecutive patients with biopsy-proved primary amyloidosis. These patients had left ventricular dysfunction and echocardiographic studies suggestive of amyloidosis. The ^{99m}Tc-PYP images were scored as follows: 0, negative scan; 1+, equivocal uptake by the heart; 2+, definite uptake with intensity less than the ribs; 3+, uptake intensity equal to the ribs but less than the sternum; and 4+, intensity equal to or greater than the sternum. Falk et al[50] found that 82% of patients with echocardiographic criteria of cardiac amyloid had 3+ uptake diffusely over the myocardium. No strongly positive scans were found in the control group, suggesting that ^{99m}Tc-PYP imaging may be more sensitive than echocardiographic studies in amyloidosis. The mechanism of myocardial uptake is unknown, although there is a suggestion that the abnormal paraprotein binds calcium; since ^{99m}Tc-PYP uptake follows calcium deposition, the protein itself and not myocytic damage may account for the uptake. Furthermore, the sensitivity of uptake may vary depending on the characteristics of the protein deposited. In secondary amyloidosis, where the amyloid A (AA) protein rather than the amyloid light chain (AL) protein is deposited in the myocardium, there is decreased avidity accounting for the fact that only 50% of scans are positive. Technetium 99m pyrophosphate scintigraphy is a readily applicable, noninvasive means of assessing patients with possible cardiac amyloidosis.

Other examples of RC include progressive systemic sclerosis (PSS) and hemochromatosis. Perfusion defects with the use of thallium 201 imaging have been observed by Follansbee et al[51] in patients with diffuse PSS, which appear during the stress of exercise and redistribute with rest. These authors also found that during maximal exercise 23% had evidence of cardiac involvement; they concluded that PSS results in myocardial perfusion abnormalities most probably related to disturbances in the microcirculation from the underlying disease. In pathologic correlations, it is suggested that there are regions of myocardial necrosis resulting from ischemic injury possibly related to intermittent spasm of coronary arteries (myocardial Raynaud phenomenon.).[52,53]

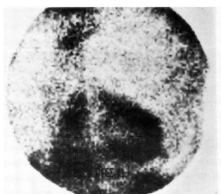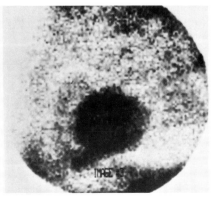

Figure 28-6 Anteroposterior (left) and left lateral (right) images exhibit technetium 99m pyrophosphate uptake in a patient with cardiac amyloidosis.

Finally, iron overload has been investigated by Leone and co-workers[54] in patients with excessive transfusion requirements due to β-thalassemia. Radionuclide angiography accompanied by exercise showed normal resting ejection fractions with a blunted exercise response. Patients who required more than 100 transfusions were more likely to have abnormalities in the exercise ejection fraction. These authors concluded that RNA was useful in the evaluation of cardiac iron overload in patients requiring regular transfusions.

SUMMARY

Radioisotopes are useful in the diagnosis of heart muscle disease, defining the extent of pathophysiologic abnormalities and the serial evaluation of cardiac function. New insights into the etiology, pathogenesis, and therapeutic interventions of the cardiomyopathies have resulted directly from the widespread application of these techniques.

REFERENCES

1. Report of the WHO/ISFC task force on the definition and classification of cardiomyopathies. Br Heart J. 1980;44:672–673.

2. Goldman MR, Boucher CA. Value of radionuclide imaging techniques in assessing cardiomyopathy. Am J Cardiol. 1980;46:1232–1236.

3. Beller GA. Nuclear cardiology: Current indications and clinical usefulness. Curr Probl Cardiol. 1985;10:1–76.

4. Greenberg JM, Murphy JH, Okada RD, et al. Value and limitations of radionuclide angiography in determining the cause of reduced left ventricular ejection fraction: Comparison of idiopathic dilated cardiomyopathy and coronary artery disease. Am J Cardiol. 1985;55:541–544.

5. Wallis DE, O'Connell JB, Henkin RE, et al. Segmental wall motion abnormalities in dilated cardiomyopathy: A common finding and good prognostic sign. J Am Coll Cardiol. 1984;4:674–679.

6. Yamaguchi S, Tsuiki K, Hayasaka M, et al. Segmental wall motion abnormalities in dilated cardiomyopathy: Hemodynamic characteristics and comparison with thallium-201 myocardial scintigraphy. Am Heart J. 1987;113:1123–1128.

7. Caglar N, Araki H, Nagata Y, et al. Evaluation of right ventricular function in patients with idiopathic dilated and ischemic cardiomyopathy by equilibrium radionuclide ventriculography. Jpn Heart J. 1986;27:1–9.

8. Araki H, Caglar N, Hisano R, et al. Right and left ventricular dysfunction in patients with dilated cardiomyopathy: A study with equilibrium radionuclide ventriculography. J Cardiogr. 1985;15:155–161.

9. Schoolmeester WL, Simpson AG, Sauerbrunn BJ, et al. Radionuclide angiographic assessment of left ventricular function during exercise in patients with a severely reduced ejection fraction. Am J Cardiol. 1981;47:804–809.

10. Bonow RO, Rosing DR, Bacharach SL, et al. Effects of verapamil on left ventricular systolic function and diastolic filling in patients with hypertrophic cardiomyopathy. Circulation. 1981;64:787–796.

11. Soufer R, Wohlgelernter D, Vita NA, et al. Intact systolic left ventricular function in clinical congestive heart failure. Am J Cardiol. 1985;55:1032–1036.

12. Lavine SJ, Krishnaswami V, Shreiner DP, et al. Left ventricular diastolic filling in patients with left ventricular dysfunction. Int J Cardiol. 1985;8:423–436.

13. Spirito P, Maron BJ, Bonow RO. Noninvasive assessment of left ventricular diastolic function: Comparative analysis of Doppler echocardiographic and radionuclide angiographic techniques. J Am Coll Cardiol. 1986;7:518–526.

14. Friedman BJ, Drinkovic N, Miles H, et al. Assessment of left ventricular diastolic function: Comparison of Doppler echocardiography and gated blood pool scintigraphy. J Am Coll Cardiol. 1986;8:1348–1354.

15. Dunn RF, Uren RF, Sadick N, et al. Comparison of thallium-201 scanning in idiopathic dilated cardiomyopathy and severe coronary artery disease. Circulation. 1982;66:804–810.

16. Bulkley BH, Hutchins GM, Bailey I, et al. Thallium-201 imaging and gated cardiac blood pool scans in patients with ischemic and idiopathic congestive cardiomyopathy: A clinical and pathological study. Circulation. 1977;55:753–760.

17. Saltissi S, Hockings B, Croft DN, et al. Thallium-201 myocardial imaging in patients with dilated and ischaemic cardiomyopathy. Br Heart J. 1981;46:290–295.

18. Hock A, Freundleib C, Vyska K, et al. Myocardial imaging and metabolic studies with 17-[^{123}I]iodoheptadecanoic acid in patients with idiopathic congestive cardiomyopathy. J Nucl Med. 1983;24:22–28.

19. Rabinovitch MA, Kalff V, Allen R, et al. Omega-^{123}I-hexadecanoic acid metabolic probe of cardiomyopathy. Eur J Nucl Med. 1985;10:222–227.

20. Tripp ME, Katcher ML, Peters HA, et al. Systemic carnitine deficiency presenting as familial endocardial fibroelastosis, a treatable cardiomyopathy. N Engl J Med. 1981;305:385–390.

21. Lerch RA, Bergmann SR, Ambos HD, et al. Effect of flow-independent reduction of metabolism on regional myocardial clearance of ^{11}C-palmitate. Circulation. 1982;65:731–738.

22. Henze E, Grossman RG, Nagafi A, et al. Measurement of C-11 palmitate kinetics after metabolic interventions in normals and patients with cardiomyopathy using positron emission computed tomography. Am J Cardiol. 1982;49:1023. Abstract.

23. Geltman EM, Smith JL, Beecher D, et al. Altered regional myocardial metabolism in congestive cardiomyopathy detected by positron tomography. Am J Med. 1983;74:773–785.

24. Perloff JK, Henze E, Schelbert HR. Alterations in regional myocardial metabolism, perfusion and wall motion in Duchenne muscular dystrophy studied by radionuclide imaging. Circulation. 1984;69:33–42.

25. Schelbert HR, Henze E, Phelps ME. Emission tomography of the heart. Semin Nucl Med. 1980;10:355–373.

26. Bergmann SR, Fox KAA, Geltman EM, et al. Positron emission tomography of the heart. Prog Cardiovasc Dis. 1985;28:165–194.

27. Gottdiener JS, Mathisen DJ, Borer JS, et al. Doxorubicin cardiotoxicity: Assessment of late left ventricular dysfunction by radionuclide cineangiography. Ann Intern Med. 1981;94:430–435.

28. Miller PP, Gill JB, Fischman AJ, et al. New radionuclides for cardiac imaging. Prog Cardiovasc Dis. 1986;28:419–434.

29. McKillop JH, Bristow MR, Goris ML, et al. Sensitivity and specificity of radionuclide ejection fraction in doxorubicin cardiotoxicity. Am Heart J. 1983;106:1048–1056.

30. Alcan KE, Robeson W, Graham MC, et al. Early detection of anthracycline-induced cardiotoxicity by stress radionuclide cine-

angiography in conjunction with Fourier amplitude and phase analysis. *Clin Nucl Med*. 1985;11:160–166.

31. O'Connell JB, Costanzo-Nordin MR, Subramanian R, Robinson JA. Dilated cardiomyopathy: Emerging role of endomyocardial biopsy. *Curr Probl Cardiol*. 1986;11:447–507.

32. O'Connell JB, Costanzo-Nordin MR, Engelmeier RS, et al. Prognosis and treatment of cardiomyopathy and myocarditis. *Heart Vessels*. 1985;1(suppl):176–179.

33. Mason JW. Endomyocardial biopsy: The balance of success and failure. *Circulation*. 1985;75:185–188.

34. O'Connell JB, Robinson JA, Henkin RE, et al. Immunosuppressive therapy in patients with congestive cardiomyopathy and myocardial uptake of gallium-67. *Circulation*. 1981;64:780–786.

35. O'Connell JB, Henkin RE, Robinson JA, et al. Gallium-67 imaging in patients with dilated cardiomyopathy and biopsy-proven myocarditis. *Circulation*. 1984;70:58–62.

36. Mitsutake A, Nakamura M, Inou T, et al. Intense persistent myocardial-avid technetium 99m pyrophosphate scintigraphy in acute myocarditis. *Am Heart J*. 1981;101:683–684.

37. Yasuda T, Palaciois IF, Dec GW, et al. Indium-111 monoclonal antimyosin antibody imaging in the diagnosis of acute myocarditis. *Circulation*. 1987;76:306–311.

38. Forman MB, Sandler MP, Sacks GA, et al. Radionuclide imaging in myocardial sarcoidosis: Demonstration of myocardial uptake of technetium pyrophosphate-99m and gallium. *Chest*. 1983;83:578–580.

39. Bulkley BH, Rouleau J, Strauss HW, et al. Sarcoid heart disease: Diagnosis by thallium-201 myocardial perfusion imaging. *Am J Cardiol*. 1976;37:125. Abstract.

40. Malkler PT, Lavine SJ, Denenberg BS, et al. Redistribution of the thallium scan in myocardial sarcoidosis. *J Nucl Med*. 1981;22:428–432. Concise Communication.

41. Cannon RO, Rosing DR, Maron BJ, et al. Myocardial ischemia in patients with hypertrophic cardiomyopathy: Contribution of inadequate vasodilator reserve and elevated left ventricular filling pressures. *Circulation*. 1985;71:234–243.

42. Pitcher D, Wainwright R, Maisey PC, et al. Assessment of chest pain in hypertrophic cardiomyopathy using exercise thallium-201 myocardial scintigraphy. *Br Heart J*. 1980;44:650–656.

43. Nagata S, Park Y, Minamikawa T, et al. Thallium perfusion and cardiac enzyme abnormalities in patients with familial hypertrophic cardiomyopathy. *Am Heart J*. 1985;109:1317–1322.

44. Rubin KA, Morrison J, Padnick MB, et al. Idiopathic hypertrophic subaortic stenosis: Evaluation of anginal symptoms with thallium-201 myocardial imaging. *Am J Cardiol*. 1987;44:1040–1045.

45. Bonow RO, Frederick TM, Bacharach SL, et al. Atrial systole and left ventricular filling in hypertrophic cardiomyopathy: Effect of verapamil. *Am J Cardiol*. 1983;51:1386–1391.

46. Betocchi S, Bonow RO, Bacharach SL, et al. Isovolumic relaxation period in hypertrophic cardiomyopathy: Assessment by radionuclide angiography. *J Am Coll Cardiol*. 1986;7:74–81.

47. Suzuki Y, Kadota K, Nohara R, et al. Recognition of regional hypertrophy in hypertrophic cardiomyopathy using thallium-201 emission computed tomography; Comparison with two-dimensional echocardiography. *Am J Cardiol*. 1984;53:1095–1102.

48. Sobol SM, Brown JM, Bunker SR, et al. Noninvasive diagnosis of cardiac amyloidosis by technetium-99m pyrophosphate myocardial scintigraphy. *Am Heart J*. 1981;103:563–565.

49. Wizenberg TA, Muz J, Sohn YH, et al. Value of positive myocardial technetium-99m pyrophosphate scintigraphy in the noninvasive diagnosis of cardiac amyloidosis. *Am Heart J*. 1982;103:468–473.

50. Falk RH, Lee VW, Rubinow A, et al. Sensitivity of technetium-99m pyrophosphate scintigraphy in diagnosing cardiac amyloidosis. *Am J Cardiol*. 1983;51:826–830.

51. Follansbee WP, Curtiss EI, Medsger TA, et al. Physiologic abnormalities of cardiac function in progressive systemic sclerosis with diffuse scleroderma. *N Engl J Med*. 1984;310:142–148.

52. Bulkley BH, Ridolfi RL, Salyer WR, et al. Myocardial lesions of progressive systemic sclerosis, a cause of cardiac dysfunction. *Circulation*. 1976;53:483–490.

53. Nitenberg A, Foult JM, Kahan A, et al. Reduced coronary flow and resistance reserve in primary scleroderma myocardial disease. *Am Heart J*. 1986;112:309–315.

54. Leone MB, Borer JS, Bacharach SL, et al. Detection of early cardiac dysfunction in patients with severe beta-thalassemia and chronic iron overload. *N Engl J Med*. 1979;301:1143–1148.

Chapter 29

Evaluation of Hypertrophic Cardiomyopathy and Dilated Cardiomyopathy by Ultrafast Computed Tomography

Eva V. Chomka, MD, and Bruce H. Brundage, MD

Cardiomyopathies constitute a heterogeneous group of abnormalities presenting with clinical features ranging from latent to advanced symptomatology. This chapter describes some of the ultrafast computed tomographic (CT) findings in persons with different cardiomyopathies along with some of the general aspects used to evaluate cardiomyopathies. According to strict definition, cardiomyopathies constitute a muscle disorder unrelated to a definable etiology such as coronary artery disease, hypertension, valve disease, or congenital or pericardial abnormalities.

The two major categories of cardiomyopathy considered are nondilated cardiomyopathy (hypertrophic cardiomyopathy [HCM] and restrictive cardiomyopathy) and dilated cardiomyopathy (DCM).[1–10] Emphasis is placed on HCM and DCM. The classic form of HCM is genetically transmitted and has been well described.[1,2,4–8,11] Variants with apical hypertrophy may occur.[9,10] There may also be acquired forms, such as in the older patient with hypertension.[12] Restrictive cardiomyopathy in the United States is most commonly the result of amyloidosis; however, it can also be due to eosinophilic disease, including Löffler endocarditis.[13,14] More common etiologies for DCM include alcohol abuse, acute inflammation, Chagas disease, anthracycline toxicity, uremia, peripartum state, and sarcoidosis.[4,5,15–22] Outside the United States, particularly in less-developed countries, idiopathic cardiomyopathy is the most common presentation of DCM, and this will serve as the classic example. Within the United States, as our capabilities to treat coronary artery disease increase and prolong life, the most common form of cardiomyopathy arises from the sequelae of coronary artery disease.[23] This deserves separate consideration, since it is outside the strict rubric of classic DCM.

The history and physical examination are still the crux of the preliminary evaluation leading to an initial diagnosis. The history, however, may not always differentiate between the two conditions. In the younger age group (less than 30 years), sudden death may be the initial manifestation of HCM.[24] Persons with either condition, however, may be asymptomatic, may experience dyspnea on exertion, or may have palpitations.

Hypertrophic cardiomyopathy may manifest as anginal pain, believed to be due to microvascular disease with normal epicardial coronary arteries.[25] Dilated cardiomyopathy may present with atypical chest discomfort. The physical finding of classic HCM is a systolic murmur that increases with standing. This murmur may be confused with that in aortic stenosis or mitral regurgitation. Dilated cardiomyopathy more commonly presents with an apical holosystolic murmur of mitral regurgitation. Distinguishing anatomically between the types of cardiomyopathy is important because of the distinct natural history of each disease process. Both HCM and DCM have been associated with increased mortality. Hypertrophic cardiomyopathy has been particularly associated with increased mortality in younger patients, whereas DCM does not appear to have such a preferential younger age predilection.[24] Now that cardiac transplantation with adequate immunosuppression has become available, the therapeutic options for cardiomyopathy have been expanded.[17,26] Ultrafast CT allows serial evaluations of a patient's clinical course with or without therapeutic interventions. Quantitative anatomic and physiologic evaluations are possible with ultrafast CT.

HYPERTROPHIC CARDIOMYOPATHY

The pathogenesis of HCM is unknown; however, there are data to support a genetic etiology in some patients.[11] The clinical history of hypertrophic cardiomyopathy may be variable. Although a great deal of emphasis has been placed on symptomatic patients, a recent study indicates that asymptomatic patients may have a more benign course, thus changing some perceptions regarding the spectrum of presentation.[27] Ultrafast CT can be used to provide both subjective and quantitative information supporting the clinical assessment of HCM. Hypertrophic cardiomyopathy is classified as a nondilated cardiomyopathy, and it may be obstructive or nonobstructive depending on the state of the left ventricular outflow tract. The left ventricular myocardial wall thickness is usually increased and the left and right ventricular cavities are nondilated; however, there are cases of subendocardial or transmural necrosis with resultant cavity dilatation. Hypertrophic cardiomyopathy may exist with concomitant coronary artery disease.

Ultrafast CT can evaluate global and regional wall thickness. Asymmetric septal hypertrophy, that is, disproportionate septal hypertrophy compared with the free wall of the left ventricle, is one of the well-known features of HCM. It is not, however, pathognomonic for HCM.[28] It is usually most pronounced at the mid-basal and mid-left ventricular level. There may be considerable variability in the septal hypertrophy. Patterns of basal and mid-left ventricular, apical, or diffusely symmetrically increased wall thickness may be identified. We have observed symmetric and basal mid-left ventricular hypertrophy.[29] Our study of patients with HCM demonstrated that regional wall involvement was septum > anterior > lateral > posterior wall. This distribution has been documented by other imaging modalities and by autopsy (Figure 29-1). Midcavity obliteration and apical hypertrophy also can be identified with ultrafast CT. In addition to the increased wall thickness, hypertrophied papillary muscle and chordae tendineae may be seen.

Mitral valve function, including systolic anterior mitral leaflet motion, can be evaluated by acquiring 50-millisecond scans continuously during one cardiac cycle. The anterior mitral valve leaflet may be thickened, and the thickness can be measured. The relationship of the mitral valve to the septum can be defined easily, and narrowing of the left ventricular outflow tract also can be identified.

Multiple anatomic and functional parameters can be measured with ultrafast CT by using the 50-millisecond

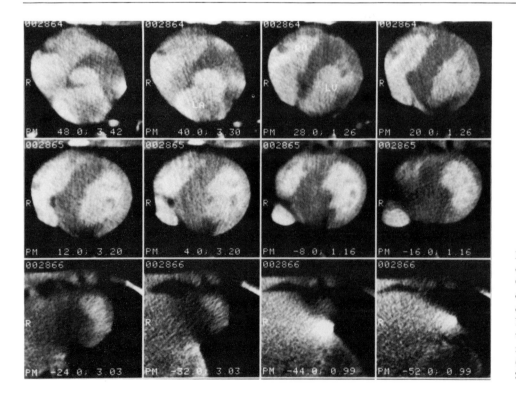

Figure 29-1 Hypertrophic cardiomyopathy. Individual ultrafast CT scans are shown of the left ventricle (LV) from base to apex at end-diastole. RV indicates right ventricle; LA, left atrium. *Source:* Reprinted with permission from *American Journal of Noninvasive Cardiology* (1987;1:143), Copyright © 1987, S Karger AG.

per scan acquisition (Table 29-1). Systolic function can be measured in a tomographic single-level evaluation or globally. The area ejection fraction (Figure 29-2) is calculated as follows:

$$\frac{\text{End-diastolic area} - \text{End-Systolic area}}{\text{End-diastolic area}} = \text{Area ejection fraction}$$

This area ejection fraction method has been used to compare the mid-left ventricular ejection fraction with the ejection fraction obtained from biplane ventriculography ($r = 0.91$, standard error of estimate 6.6; $y = 1.1 \times -8.5$).[30] Although the mid-left ventricular ejection fraction serves as an approximation of the global left ventricular function, assessment of the entire global left ventricular function is preferred.[31] The left ventricular ejection fraction typically is normal or increased, related to decreased end-diastolic volumes and increased chamber emptying in HCM. Summation of the tomographic left ventricular area ejection fractions provides the global left ventricular ejection fraction. Ultrafast CT measurements of anatomy and function in normal patients and patients with HCM are

Table 29-1 Morphologic and Quantitative Analysis of HCM and DCM

Morphology	Quantification
Wall thickness	Left ventricular ejection fraction
Global and regional	Mid-left ventricle
Patterns of mural hypertrophy	Global-left ventricle
Cavity obliteration	Left ventricular mass
Mitral valve motion	Wall thickness
Mitral valve thickening	Left ventricular aneurysm
Left ventricular outflow tract anatomy	quantification
	Diastolic filling rate (global and tomographic regional/level)
Left atrial thrombus	Left ventricular volume (end-systolic, end-diastolic, stroke volume)
Left ventricular thrombus	
Ischemia (regional wall motion abnormality)	
Left ventricular aneurysm	Left atrial dimension
Coronary artery calcification	Left atrial volume
	Regional wall motion (quantification: rest and exercise)
	Myocardial blood flow
	Regurgitant fraction (single valve)
	Cardiac output

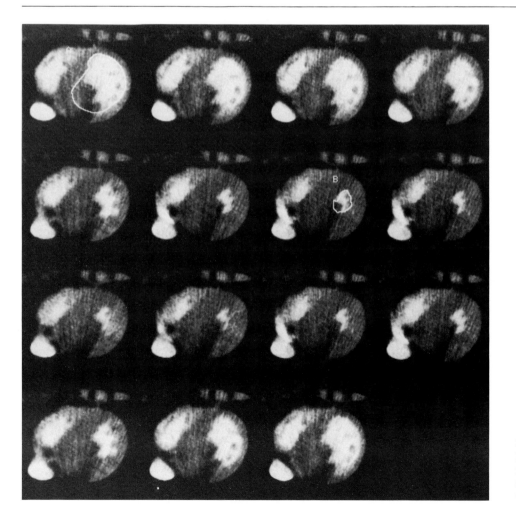

Figure 29-2 One cardiac cycle of the left ventricle at one anatomic level (8-mm thickness) with the area ejection fraction of the patient shown in Figure 29-1.

presented in Tables 29-2, 29-3, and 29-4; the measurements in patients with HCM are contrasted with those in a group of normal patients[10] in Table 29-5.

Total left ventricular volumes can be measured throughout the cardiac cycle by summing the values obtained at each level.[32,33] Patients with HCM have normal or decreased left ventricular volumes. Precise end-systolic, end-diastolic, and stroke volumes can be obtained. Ultrafast CT volume measurements have been validated in dogs and in normal humans.[32,33]

Measurement of the left ventricular mass is important in order to assess progression or regression of disease or the results of therapeutic intervention in any cardiomyopathic state. This measurement may be obtained either by subtracting endocardial surface areas from epicardial surface areas and multiplying by slice thickness or by circumscribing the left ventricular contour at each tomographic level. One then multiplies by the specific gravity of the myocardium and adds to obtain global parameters. Measurement of the left ventricular mass in dogs has been verified for ultrafast CT by comparison with autopsy weight, with a 0.99 correlation.[34] Reproducible measurement of left ventricular mass has also been demonstrated in humans,

Table 29-2 Normal Patients

Patient	Sex	Age (yr)	Scanning Angle (°axial tilt, °slant)	Mid-LV Area EF (%)	Global EF (%)	LV Mass (g)	Number of Levels— Total LV	Diastolic Filling (mL/s)	Heart Rate (beats/min)
1	M	38	10, −15	82	85	133	7	102	77
2	F	23	20, −13	57	63	154	10	111	69
3	M	27	+10, −14	63	59	141	8	142	68
4	M	37	+15, −12	67	70	94	10	73	58
5	M	60	+10, −12	61	57	170	8	56	84
6	F	54	+12, −12	71	66	126	10	56	90
7	F	75	+15, −15	68	72	155	9	72	77
8	M	63	+11, −15	72	74	190	9	117	46
9	M	23	+13, −12	57	52	140	8	68	64
10	F	61	+20, −13	67	72	131	8	116	66
11	F	48	+20, −13	61	66	129	8	134	67
Mean ± SD		46.4 ±17.8		65.2 ±8.6	66.9 ±9.2	142.1 ±25.2		95.2 ±31.4	69.6 ±12.2

*LV, Left ventricle; EF, ejection fraction.
Source: Reprinted with permission from *American Journal of Noninvasive Cardiology* (1987;1:140–151), Copyright © 1987, S Karger AG.

Table 29-3 Patients with HCM

Patient	Sex	Age (yr)	Scanning Angle (°axial tilt, °slant)	Mid-LV Area EF (%)	Global EF (%)	LV Mass (g)	Number of Levels— Total LV	Diastolic Filling (mL/s)	Heart Rate (beats/min)
1	F	38	+8, −10	65	Incomplete	Incomplete	Incomplete	145	57
2	M	61	+9, −10	57	60	342	8	84	82
3	F	55	+10, −12.5	53	57	427	10	102	58
4	M	72	+13, −13.5	70	69	284	10	108	65
5	M	60	+5, −12	70	Incomplete	Incomplete	Incomplete	31	78
6	M	28	+10, −15	85	82	293	10	89	91
7	M	64	+13, −12.5	69	70	296	11	136	64
8	F	23	+15, −12	73	68	308	11	55	54
9	F	72	+14, −12	79	73	149	10	166	68
10	F	52	+16, −12	92	87	325	9	188	64
11	M	70	+20, −13	63	60	186	8	80	76
Mean ± SD		54.1 ±17.3		70.5 ±11.5	69.6 ±10.1	290 ±82		107.6 ±47.3	68.8 ±11.6

*LV, Left ventricle; EF, ejection fraction.
Source: Reprinted with permission from *American Journal of Noninvasive Cardiology* (1987;1:140–151), Copyright © 1987, S Karger AG.

Table 29-4 Patients with HCM

Patient		Wall Thickness (mm)*				Hypertrophic Pattern	Mitral Motion	Left Atrial Dimension (mm)
		S	A	L	P			
1	B					Septum: M	Normal	Level not visualized
	M	12.2	3.7	7.2	6.5			
2	B	28.0	13.8	13.8		LVH/RVH	SAM	37
	M	22.7	14.1	10.8	10.2	Septum: B + M		
						Anterior: B + M		
3	B	31.9	19.5	7.3		Septum: B + M	SAM	55
	M	33.2	10.9	8.2	10.6	Anterior: B + M		
4	B	23.9	9.9	11.0		Septum: B + M	SAM	46
	M	17.5	10.7	9.7	7			
5	B	18.7	34.5	18.7		Concentric LVH entire septum Lateral: B	Normal	Level not visualized
	M	19	20.2	20.9	17			
6	B	20.4	28.6	17		RVH: septum B + M	SAM	42
	M	23.3	12.2	8.7	10.2	Anterior: B + M Lateral: B		
7	B	27.3	11.9	11.3		Septum B + M	SAM	50
	M	18.5	9.8	7.1	7.9			
8	B	36	38.3	37.6		Septum: B + M	SAM	32
	M	36	38.7	18.3	9.9	Anterior: B + M Lateral: B		
9	B	15.3	9.2	7.7		Septum: B + M	Normal	42
	M	14.5	8.8	5.8	8.8			
10	B	23.71	18.6	10.5		Septum: B + M	SAM	56
	M	17.6	16.6	11.8	19	Anterior: B + M		
11	B					Septum: B + M	Normal	56
	M	19.7	7.4	12.3	12.3	Lateral: M		
Mean ± SD	B	25.0 ± 6.5	20.4 ± 11.0	15.9 ± 9.2	—			46 ± 8.4
	M	21.8 ± 8.1	13.9 ± 9.4	11.0 ± 4.8	10.8 ± 3.9			

*S, septal; A, anterior; L, lateral; P, posterior; LVH, left ventricular hypertrophy; RVH, right ventricular hypertrophy; B, basal; M, mid-LV; SAM, systolic anterior motion.
Source: Reprinted with permission from *American Journal of Noninvasive Cardiology* (1987;1:140–151), Copyright © 1987, S Karger AG.

Table 29-5 Summary of Data* (Mean ± Standard Deviation)

	Normal	HCM	t	P
Age (yr)	46.4 ± 17.8	54.1 ± 17.3	1.027	
Global EF (%)	66.9 ± 9.2	69.6 ± 10.1	0.625	
LV mass (g)	142.1 ± 25.2	290 ± 82	5.642	<.05
Diastolic filling rate (mL/s)	95.2 ± 31.4	107.6 ± 47.3	0.688	
Heart rate (beats/min)	69.6 ± 12.2	68.8 ± 11.6	0.158	

Wall Thickness (mm)				
	Normal (B)	HCM (B)		
Septal	10.0 ± 1.8	25.0 ± 6.5	−6.99	<.05
Anterior	11.0 ± 2.0	20.4 ± 11.0	−2.798	<.05
Lateral	9.2 ± 2.3	15.9 ± 9.2	2.225	<.05
	Normal (M)	HCM (M)		
Septal	9.4 ± 2.4	21.8 ± 8.1	−4.807	<.05
Anterior	8.3 ± 1.8	13.9 ± 9.4	−1.958	<.05
Lateral	8.8 ± 3.2	11.0 ± 4.8	−1.269	
Posterior	8.4 ± 1.3	10.8 ± 3.9	−1.926	<.05

*t, Unpaired t test value; EF, ejection fraction; LV, left ventricle; B, basal; M, mid-LV.

with a mean variability of 0.0 ± 1.5%.[35] Both intraobserver and interobserver variabilities were excellent ($r = 0.99$ and 0.97, respectively) (unpublished data, ER, BHB, EVC, 1987). Abnormal wall thinning can be seen by ultrafast CT and may be related to small-vessel disease in HCM.[25] Subsequent to our initial report, we have scanned one patient with prominent apical thinning and abnormal wall motion who had normal epicardial coronary arteries by coronary angiography.

Abnormal diastolic filling may be the initial manifestation of any cardiomyopathy, and in fact is probably the most common functional abnormality found in HCM. The abnormal left ventricular compliance may be related to mural hypertrophy and cellular disarray. Early diastolic filling is prolonged, and the rate of filling is reduced. Abnormal left ventricular diastolic relaxation and distensibility compliance are probably related to abnormal calcium sequestration and myocardial elastic properties. The diastolic relaxation abnormality results in increased left ventricular end-diastolic pressure; thus the atrial contribution to diastolic filling becomes more important. Evaluation of diastolic filling in patients with HCM has been performed by ultrafast CT[29] and is substantially impaired compared with that in normal persons.[36]

Mitral regurgitation occurs in HCM, probably related to the abnormal left ventricular geometry and resultant abnormal tension on papillary muscles and chordae tendineae.[37] Measurement of mitral regurgitation can be performed by ultrafast CT. First, cardiac output is obtained in the flow mode.[38] A time-density curve is generated and, through knowledge of the contrast coefficient of the contrast agent used and the amount of contrast medium injected, cardiac output may be derived by measuring the gamma variate curve. This measurement is based on the fundamentals of the Stewart-Hamilton equation. By knowing the heart rate at the time of the acquisition and the cardiac output, one can then obtain forward stroke volume: cardiac output = heart rate × stroke volume. Alternatively, by using the cine-mode one can also scan the left ventricle from base to apex and obtain direct global left ventricular end-diastolic and end-systolic volume measurements. If the heart rate is the same during this data acquisition, one can compare the ventricular stroke volume (analogous to angiographic stroke volume) with the forward stroke volume obtained in the flow mode. Regurgitant volume may also be obtained by comparing the two ventricular stroke volumes (provided that there is no tricuspid regurgitation). Similar measurements can be performed in the right ventricle if there is no left-sided regurgitation.

$$\frac{\text{Left ventricular stroke volume} - \text{Right ventricular stroke volume}}{\text{Left ventricular stroke volume}}$$
= Regurgitant fraction for a left-sided cardiac valve lesion

Computation of regurgitant fraction has been validated in experimentally induced aortic regurgitation.[39]

Left ventricular thrombus can also be detected by ultrafast CT. The left atrium is well visualized and accurate volume measurements can be made. Left atrial dimensions were quantified in a recent series evaluating HCM and separated into normal size and abnormal size.[29] Patients with HCM may have an enlarged left atrium secondary to mitral regurgitation. Ultrafast CT has demonstrated potential in evaluating and identifying thrombi in the left atrium, including the left atrial appendage, and is therefore useful in both HCM and DCM. Its tomographic format allows precise localization of atrial thrombi, whereas echocardiography is sometimes limited in left atrial assessment.[40]

Ultrafast CT can be used to evaluate the results of surgery in HCM. Patients who have undergone septal myomectomy-myotomy can be assessed for septal configuration, contour, apical diverticula (Figure 29-3), mass and wall thickness, and systolic and diastolic left ventricular function. The natural history of HCM also may be defined, since ultrafast CT can document serial

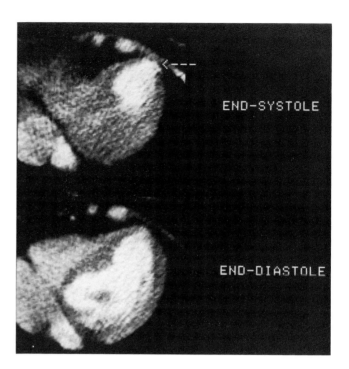

Figure 29-3 Hypertrophic cardiomyopathy. Scans of end-systole and end-diastole at the posteromedial papillary muscle (PM) level demonstrate the results of myomectomy, and an apical diverticulum presumably the result of a ventricular sump. In addition, a left-sided superior vena cava is noted. LV indicates left ventricle. *Source:* Reprinted with permission from *American Journal of Noninvasive Cardiology* (1987;1:142), Copyright © 1987, S Karger AG.

changes. A few patients with HCM may progress to DCM, and this can be documented by ultrafast CT.

Ultrafast CT can well delineate right ventricular morphology, and volume, mass, and function can be measured. Some patients with left ventricular HCM may also have associated right ventricular hypertrophy. Evaluation of the right ventricle has posed a problem for radionuclide and echocardiographic imaging techniques because of its unusual configuration. Ultrafast CT is not constrained by these limitations.

Associated congenital abnormalities such as a left-sided superior vena cava can be identified by ultrafast CT. Atrial fibrillation sometimes occurs in patients with HCM. Therefore, the ejection fraction as measured by ultrafast CT scanning will vary beat to beat compared with that measured during the sinus rhythm. This problem is true, however, of any technique that evaluates ventricular function during atrial fibrillation.

RESTRICTIVE CARDIOMYOPATHY

Restrictive cardiomyopathy may present as dyspnea at rest or during exertion. Left ventricular systolic function visually is normal or nearly normal. Diastolic function is impaired. With ultrafast CT one can differentiate between restrictive cardiomyopathy and constrictive pericarditis by evaluating the pericardial thickness. The pericardium is thickened or calcified in constrictive pericarditis, whereas it is normal in restrictive cardiomyopathy.[41] In addition, one can evaluate right ventricular dynamics along with those of the left ventricle.

DILATED CARDIOMYOPATHY

Dilated cardiomyopathy is a condition with increased left ventricular volume and decreased left ventricular systolic function. The etiologies are numerous[42] and often impossible to identify once left ventricular dilatation occurs. Recent clinical studies have demonstrated the value of afterload therapy in prolonging life.[43,44] Endomyocardial biopsy is used to evaluate myocarditis, a potentially treatable cause of DCM.[17,45] The value of endomyocardial biopsy in the evaluation of myocardial inflammation has been questioned because of significant interobserver variability in the interpretation of biopsy results.[42] Initially there was great enthusiasm for immunosuppressive therapy in myocarditis.[46] Controlled studies are under way, but currently therapeutic recommendations are considered on an individual basis. Myocardial wall thickness can be normal or thinned in this condition. Although the classic case demonstrates diffuse hypokinesia, regional wall motion abnormalities can exist in a minority of patients, making differentiation between dilated and ischemic cardiomyopathy difficult in some cases. Coronary angiography can help to resolve this dilemma. Ultrafast CT can also identify coronary artery calcification, which is almost always present if the cardiomyopathy has an ischemic basis.[47]

Left ventricular morphology can be well assessed by ultrafast CT (Figure 29-4). Whereas many patients with HCM have normal ejection fractions, patients with DCM may have borderline normal or (more commonly) decreased ejection fractions. As the heart dilates the radius of the heart increases and therefore the volumes of the heart increase. According to Laplace's law, as the heart enlarges wall stress increases. It is thus more difficult for the heart to contract. Measures to reduce wall stress, such as afterload vasodilator therapy, currently are being tried because several studies have demonstrated increased survival with this therapy.[43,44] The

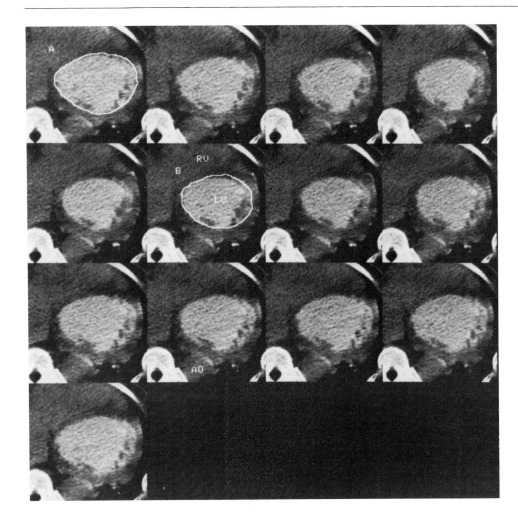

Figure 29-4 Dilated cardiomyopathy. One cardiac cycle at one ventricular level is shown with the area computed for the end-diastolic and end-systolic frames.

therapeutic effectiveness is probably related to decreasing the left ventricular volume. Left ventricular diastolic filling parameters may also be evaluated in DCM as they are in HCM.[29,36]

Mitral regurgitation in DCM results from mitral annular dilatation and left ventricular dilatation. The regurgitant fraction can be calculated from the precise measurement of right and left ventricular stroke volumes (see HCM) by ultrafast CT. Left ventricular mass can be measured in DCM as it is in HCM.

The right ventricle can be also be evaluated for volume, global ejection fraction, and the wall motion abnormality. The right ventricle may be dilated or normal in DCM.

Special consideration of ischemic cardiomyopathy (see Figure 29-5) is appropriate because of its increasing prevalence. It is a secondary form of cardiomyopathy. Cardiomyopathy related to coronary artery disease, or ischemic cardiomyopathy, has a greater preponderance of regional wall motion abnormalities than does DCM. There are situations in which left ventricular dilatation may be due to reversible ischemia, although this is a relatively infrequent phenomenon. In some such situations, revascularization may decrease left ventricular size and improve function. As patients with ischemia survive longer as a result of increasing therapeutic intervention, the left ventricle may dilate. Afterload therapy recently has been found to be beneficial in ischemic cardiomyopathy.[48]

Left ventricular thrombi are more frequent in DCM than in HCM, and ultrafast CT with its high degree of resolution can detect this complication.[3] The enlarged atria can also be evaluated for the presence of thrombi and for measurement of atrial volumes. The identification of thrombi, however, probably does not predict embolization.[3,49] The left atrium's configuration, with its four pulmonary veins, is prone to partial volume effects that must be considered in evaluating the upper portion and appendage of the left atrium, in particular. The left ventricle must also be scanned from base to apex, since many thrombi exist in the apex and can be missed if the whole left ventricle is not scanned.

Ultrafast CT has potential for measuring myocardial blood flow.[50] Patients with DCM may have diffuse decreases or regional decreases in myocardial blood flow.

Ultrafast CT can aid in assessing left and/or right ventricular function at rest and during exercise using

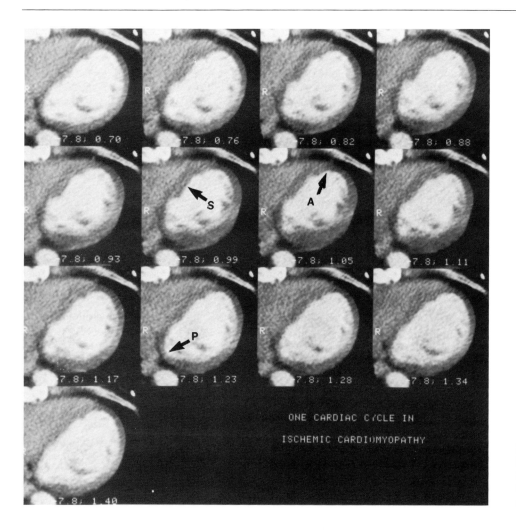

Figure 29-5 Ischemic cardiomyopathy. One cardiac cycle area is shown with septal (S), anterior (A), and posterior (P) (arrows) regional wall motion abnormalities.

bicycle ergometry, which may be helpful in evaluating dilated cardiomyopathies.[51,52] One of the advantages of ultrafast CT is that the images are acquired tomographically, allowing both regional and global evaluation. It may be possible also to evaluate left ventricular wall stress in conjunction with carotid noninvasive tracings, as has been performed by echocardiography.[53]

Dilated cardiomyopathy and (less frequently) HCM may present with recurrent ventricular tachycardia. Ultrafast CT can be used for identification of the dysrhythmia locus,[54] as in radionuclide evaluation, with superior spatial resolution.

Ultrafast CT has demonstrated potential in screening patients for coronary artery disease by detecting coronary calcification. This may be useful in both patients with DCM and patients with HCM. Preliminary data suggest a sensitivity of 91% and a specificity of 100% in patients who have undergone coronary angiography.[47]

Contrast medium is required for evaluating left ventricular function; therefore, contrast medium allergy is a relative contraindication to ultrafast CT studies, as is pregnancy. Patients typically receive 3 to 6 rad of exposure for the complete study, including evaluation of coronary artery calcification and left ventricular function.

Ultrafast CT is an evolving technology that is capable of precise, reproducible, morphologic measurements to evaluate anatomy and physiology in HCM and DCM. Its tomographic acquisition allows for left ventricular evaluation at a level of resolution not previously achieved by other imaging techniques, thus adding more pathophysiologic insight. The ability to evaluate both left ventricular function and flow offers a multifaceted approach to answering a variety of clinically relevant questions. Future technologic improvements should serve to enhance and improve the resolution of this modality.

REFERENCES

1. Wigle ED, Sasson Z, Henderson MA, et al. Hypertrophic cardiomyopathy: The importance of the site and extent of hypertrophy: A review. *Prog Cardiovasc Dis*. 1985;28:1–83.

2. Braunwald E, Lambrew CT, Rockoff SD, et al. Idiopathic hypertrophic subaortic stenosis: A description of the disease based upon analysis of 64 patients. *Circulation*. 1964;30(suppl 5):3–119.

3. Fuster V, Gersh BJ, Giulani ER, et al. The natural history of idiopathic cardiomyopathy. *Am J Cardiol*. 1981;47:525–531.

4. Johnson RA, Palacios I. Dilated cardiomyopathies of the adult. *N Engl J Med*. 1982;307:1051–1058. Part 1.

5. Johnson RA, Palacios I. Dilated cardiomyopathies of the adult. *N Engl J Med*. 1982;307:1119–1126. Part 2.

6. Criley JM, Lewis KB, White RI. Pressure gradients without obstruction: A new concept of hypertrophic subaortic stenosis. *Circulation*. 1965;32:881–885.

7. Dinsmore RE, Sanders CA, Harthorne JW. Mitral regurgitation in idiopathic hypertrophic subaortic stenosis. *N Engl J Med*. 1966;275:1225–1228.

8. Menges H Jr, Brandenburg RO, Brown AL Jr. The clinical, hemodynamic and pathologic diagnosis of muscular subvalvular aortic stenosis. *Circulation*. 1961;24:1126–1136.

9. Maron BJ, Bonow RO, Seshagiri TN, et al. Hypertrophic cardiomyopathy with ventricular septal hypertrophy localized to the apical region of the left ventricle. *Am J Cardiol*. 1982;49:1838–1848.

10. Yamaguchi H, Ischimura T, Nishiyama S, et al. Hypertrophic nonobstructive cardiomyopathy with giant negative T waves (apical hypertrophy): Ventriculographic and echocardiographic features in 30 patients. *Am J Cardiol*. 1979;44:401–412.

11. Maron BJ, Bonow RO, Cannon R, et al. Hypertrophic cardiomyopathy: Interrelations of clinical manifestations, pathophysiology and therapy. *N Engl J Med*. 1987;316:780–789, 844–852.

12. Whiting RB, Powell WJ Jr, Dinsmore RE, et al. Idiopathic hypertrophic subaortic stenosis in the elderly. *N Engl J Med*. 1971;285:196–200.

13. Meanzy E, Shabetai R, Bhargana B, et al. Cardiac amyloidosis, constrictive pericarditis and restrictive cardiomyopathy. *Am J Cardiol*. 1976;38:547–556.

14. Roberts WC, Liegler DC, Carbone PP. Endomyocardial disease and eosinophilia: a clinical and pathological spectrum. *Am J Med*. 1969;46:28–42.

15. Burch GEV, Giles TD. Alcoholic cardiomyopathy: Concept of the disease and its treatment. *Am J Med*. 1971;50:141–145.

16. Rubin E. Alcoholic myopathy in heart and skeletal muscle. *N Engl J Med*. 1979;301:28–33.

17. Mason JW, Billingham ME, Ricci DR. Treatment of acute inflammatory myocarditis assisted by endomyocardial biopsy. *Am J Cardiol*. 1980;45:1037–1044.

18. Laranja FS, Dias E, Nobrega E, et al. Chagas disease: A clinical, epidemiologic and pathologic study. *Circulation*. 1956;14:1035–1060.

19. Henderson IC, Frei E III. Adriamycin and the heart. *N Engl J Med*. 1979;300:310–312.

20. Alexander J, Dainck N, Beyer HJ, et al. Serial assessment of doxorubicin endotoxicity with quantitative radionuclide angiocardiography. *N Engl J Med*. 1979;300:278–283.

21. O'Connell JB, Costanzo-Nordin MR, Subramanian R, et al. Peripartum cardiomyopathy: Clinical, hemodynamic, histologic and prognostic characteristics. *J Am Coll Cardiol*. 1986;8:52–56.

22. Silverman KJ, Hutchins GM, Bulkley BH. Cardiac sarcoid: A clinicopathologic study of 84 unselected patients with systemic sarcoidosis. *Circulation*. 1978;58:1204–1211.

23. Yatteau RF, Peter RH, Behar VS, et al. Ischemic cardiomyopathy: the myopathy of coronary artery disease: Natural history and results of medical versus surgical treatment. *Am J Cardiol*. 1974;34:520–525.

24. Maron BJ, Roberts WC, Epstein SE. Sudden death in hypertrophic cardiomyopathy: A profile of 78 patients. *Circulation*. 1982;65:1388–1396.

25. Maron BJ, Wolfson JK, Epstein SE, et al. Intramural (small vessel) coronary artery disease in hypertrophic cardiomyopathy. *J Am Coll Cardiol*. 1986;8:545–557.

26. Oyer PE, Stinson EB, Jamieson SW, et al. Cyclosporine in cardiac transplantation: A 2½ year follow-up. *Transplant Proc*. 1983;15:2546–2552.

27. Spirito P, Chiarella F, Carratino L, et al. Clinical course and prognosis of hypertrophic cardiomyopathy in an outpatient population. *N Engl J Med*. 1989;320:749–755.

28. Epstein SE, Henry WL, Clark SE, et al. Asymmetric septal hypertrophy. *Ann Intern Med*. 1974;81:650–680.

29. Chomka EV, Wolfkiel CJ, Rich S, et al. Ultrafast computed tomography: A new method for the evaluation of hypertrophic cardiomyopathy. *Am J Noninvas Cardiol.* 1987;1:140–151.

30. Rich S, Chomka EV, Stagl R, et al. Determination of left ventricular ejection fraction using ultrafast computed tomography. *Am Heart J.* 1986;112:392–396.

31. Feiring AJ, Rumberger JA, Reiter SJ, et al. Sectional and segmental variability of left ventricular function: Experimental and clinical studies using ultrafast computed tomography. *J Am Coll Cardiol.* 1988;12:415–425.

32. Reiter SJ, Rumberger JA, Feiring AJ, et al. Precise determination and right ventricular stroke volume with cine computed tomography. *Circulation.* 1988;74:890–900.

33. Rumberger JA, Feiring AJ, Rees MR, et al. Quantitation of left ventricular mass and volumes in normal patients using cine computed tomography. *J Am Coll Cardiol.* 1986;7:173A. Abstract.

34. Higgins CB, Ell S, Marcus ML, et al. Determination of left ventricular mass in dogs with rapid-acquisition cardiac computed tomographic scanning. *Circulation.* 1985;72:1355–1364.

35. Roig E, Chomka EV, LoGalbo C, et al. Variability of left ventricular mass measurement by ultrafast computed tomography. *J Am Coll Cardiol.* 1988;11:157A. Abstract.

36. Rumberger JA, Weiss RM, Feiring AJ, et al. Patterns of regional diastolic function in the normal human left ventricle: An ultrafast computed tomographic study. *J Am Coll Cardiol.* 1989;14:119–126.

37. Roberts WC. Congenital cardiovascular abnormalities usually silent until adulthood. In: Roberts WC, ed. *Adult Congenital Heart Disease.* Philadelphia, Pa: FA Davis Co; 1987:670.

38. Wolfkiel CJ, Ferguson JL, Chomka EV, et al. Determination of cardiac output by ultrafast computed tomography. *Am J Physiol Imaging.* 1986;1:117–123.

39. Reiter SJ, Rumberger JA, Stanford W, et al. Quantitative determination of aortic regurgitant volumes in dogs by ultrafast computed tomography. *Circulation.* 1987;70:728–735.

40. Helgason C, Chomka EV, Louie E, et al. The potential role for ultrafast cardiac computed tomography in patients with stroke. *Stroke.* 1989;20:465–472.

41. Isner JM, Carter BL, Bankoff MS, et al. Differentiation of constrictive pericarditis from restrictive cardiomyopathy by computed tomographic imaging. *Am Heart J.* 1983;105:1019–1025.

42. Shanes JG, Ghali J, Billingham ME, et al. Interobserver variability in the pathologic interpretation of endomyocardial biopsy results. *Circulation.* 1987;75:401–405.

43. Cohn JN, Archibald DG, Ziesche S, et al. Effect of vasodilator therapy on mortality in chronic congestive heart failure. *N Engl J Med.* 1986;314:1547–1552.

44. CONSENSUS trial study group. Effects of enalapril on mortality in severe congestive heart failure. *N Engl J Med.* 1987;316:1429–1435.

45. Dec GW Jr, Palacios IF, Fallon JT, et al. Active myocarditis in the spectrum of acute dilated cardiomyopathies: Clinical features, histologic correlates and clinical outcome. *N Engl J Med.* 1985;312:885–880.

46. O'Connell JB, Robinson JA, Hewnkin RE, et al. Immunosuppressive therapy in patients with congestive cardiomyopathy and myocardial uptake of gallium-67. *Circulation.* 1981;64:780–786.

47. Tanenbaum SR, Kondos GT, Veselik KE, et al. Detection of calcific deposits in coronary arteries by ultrafast computed tomography and correlation with angiography. *Am J Cardiol.* 1989;63:870–872.

48. Pfeffer MA, Lamas GA, Vaughan DE, et al. Effect of captopril on progressive ventricular dilatation after anterior myocardial infarction. *N Engl J Med.* 1988;319:80–85.

49. Gottdiener JS, Gay JA, Van Voorhees L, et al. Frequency and embolic potential of left ventricular thrombus in dilated cardiomyopathy: Assessment by two-dimensional echocardiography. *Am J Cardiol.* 1983;52:1281–1285.

50. Wolfkiel CJ, Ferguson JL, Chomka EV, et al. Measurement of myocardial blood flow by ultrafast computed tomography. *Circulation.* 1987;76:1262–1273.

51. Chomka EV, Fletcher MC, Stein M, et al. Ultrafast computed tomography during exercise bicycle ergometry. *J Am Coll Cardiol.* 1986;7:154A. Abstract.

52. Roig E, Chomka EV, Castaner A, et al. Exercise ultrafast computed tomography for the detection of coronary artery disease. *J Am Coll Cardiol.* 1989;13:1073–1081.

53. Borow KM, Neumann A, Lang RM. Milrinone versus dobutamine: contribution of altered myocardial mechanics and augmented inotropic state to improving left ventricular performance. *Circulation.* 1986;73:153–161.

54. Abbott JA, Botvinick EH, Scheinman ED, et al. Noninvasive localization of accessory pathways. *J Am Coll Cardiol.* 1989;13:8A. Abstract.

Chapter 30

Evaluation of Cardiomyopathy by Magnetic Resonance Imaging

Udo Sechtem, MD, and Charles B. Higgins, MD

Assessment of cardiomyopathies by cardiac imaging is directed toward the detection and description of the underlying anatomic, functional, and metabolic abnormalities. Magnetic resonance imaging (MRI) offers noninvasiveness, three-dimensional imaging, and potential for tissue characterization. More recently, dynamic MRI has been able to provide information on cardiac function and blood flow. This chapter describes typical MRI findings in patients with various forms of cardiomyopathy.

HYPERTROPHIC CARDIOMYOPATHY

Magnetic resonance imaging is a technique that allows clear identification of endocardial and epicardial borders and is therefore useful for defining the extent, location, and severity of abnormal wall thickness in hypertrophic cardiomyopathy. The characteristic feature of the disease is the disproportionate thickening of the interventricular septum compared with the left ventricular free wall. Whereas normal subjects show similar thicknesses of the septal, inferior, lateral, and posterior walls with a ratio of septal to posterolateral wall thickness of 1.06, spin-echo MR images demonstrate a wide spectrum of distribution of abnormal wall thickness through the left ventricular myocardium in patients with hypertrophic cardiomyopathy.[1,2] Although it is limited to the outflow portion of the septum in most cases,[1] abnormal wall thickness is also seen in other parts of the septum and the lateral wall (Figure 30-1). A comparison of MRI and two-dimensional echocardiograms shows general agreement in defining the distribution of abnormal wall thickness. However, the classification by echocardiography is based mainly on the short-axis view of the left ventricle, making direct comparison with the fixed orthogonal planes used for MRI difficult. More recently, most MR imagers have become equipped with the hardware necessary to image the heart in any desired plane. Thus direct comparison with echocardiographic measurements of wall thickness is possible (Figure 30-2).

Compared with echocardiography, the advantages of MRI in assessing the extent of hypertrophy include the wider field of view, three-dimensional imaging using multislice techniques, and sharper discrimination of myocardial walls. Magnetic resonance imaging offers three-dimensional measurement of left ventricular mass,[3–5] which may be useful for documenting therapeutic interventions aimed at a reduction of muscle mass. Because of its rapidity and widespread availability, echocardiography will remain the most common and most practical technique for confirming hypertrophic cardiomyopathy. Nevertheless, MRI is possible in most patients and may be helpful in those in whom echocardiographic quality is poor.

Functional assessment of patients with hypertrophic cardiomyopathy by MRI has been limited by the long imaging times required for permutating gated spin-echo images (Figure 30-3). Recently, more rapid functional evaluation has become available with dynamic imaging.[6] With the introduction of new pulse se-

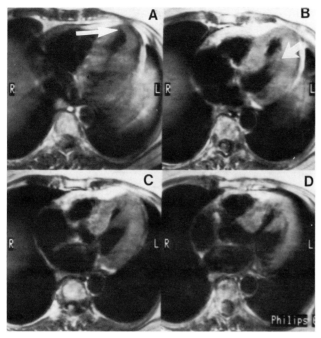

Figure 30-1 Transverse spin-echo images (echo time [TE], 30 milliseconds) in a patient with hypertrophic cardiomyopathy. A, The most apical section. The arrow indicates the left ventricular apex, which shows normal wall thickness. B, Ten millimeters more craniad, at the level of the atrioventricular valves, one head of the anterior papillary muscle is seen arising from the anterolateral wall (arrow). Septal thickening is less pronounced in the anterior portion. C, The interventricular septum extends into the left ventricular outflow tract. A second head of the anterior papillary muscle is seen arising from the anterior wall. D, The most craniad section shows the proximity of the hypertrophied outflow portion of the septum and the anterior papillary muscle.

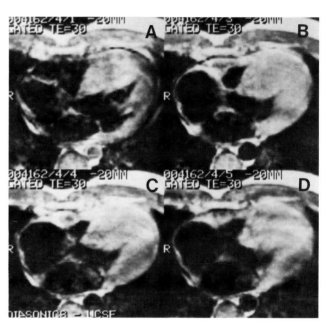

Figure 30-3 Spin-echo images of a transverse section of the heart with hypertrophic cardiomyopathy at four different times of the cardiac cycle. A, End-diastolic image. Note the massive hypertrophy of the septum and the septal portion of the anterior wall. B, Image acquired 200 milliseconds after end-diastolic image. Wall thickening of both the septum and the lateral wall is evident. C, End-systolic image. Most of the left ventricular cavity is obliterated by hypertrophied muscle. D, Image acquired 100 milliseconds after end-systolic image. *Source:* Reprinted with permission from *American Journal of Cardiology* (1987;59:145–151), Copyright © 1987, Cahners Publishing Company Inc.

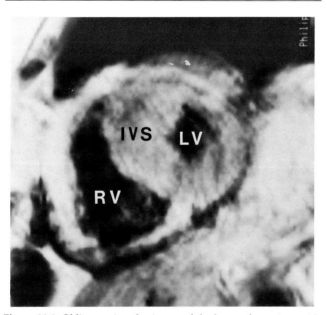

Figure 30-2 Oblique spin-echo image of the heart of a patient with hypertrophic cardiomyopathy. Comparable to echocardiographic views, this short-axis section demonstrates pronounced thickening of the interventricular septum (IVS) and normal dimensions of the lateral and inferior walls. RV indicates right ventricle; LV, left ventricle.

quences using reduced flip angles, short repetition and echo times, and gradient-refocused echoes,[7,8] dynamic MR images show normally flowing blood with higher intensity than that in the myocardium. In contrast, turbulent blood flow leads to signal loss in these images. Because of the short repetition time of 20 milliseconds, many MR images can be obtained during one cardiac cycle. Reviewing dynamic images in a closed cine-loop permits assessment of cardiac contraction and intracardiac blood flow. In patients with a resting subvalvular gradient, the characteristic narrowing of the left ventricular outflow tract causes turbulent blood flow distal to the juxtaposition of the interventricular septum and the anterior mitral leaflet and/or the anterior papillary muscles.[9,10] This results in signal loss within the left ventricular outflow tract on dynamic MR images (Figure 30-4). Measurement of flow velocity with dynamic MRI is possible[11,12] and may allow estimation of the subvalvular gradient. In these patients, turbulence also may occur within the left atrium as a result of mitral regurgitation and may lead to an area of low signal intensity on dynamic MR images. Analysis of the area of signal loss allows classification of valve incompetence, since the extension of this area

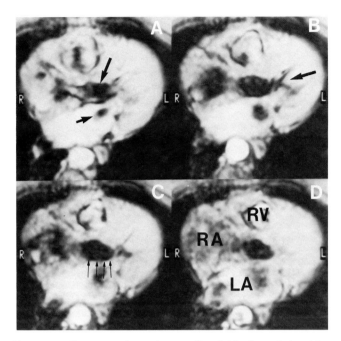

Figure 30-4 Transverse dynamic scans (fast-field echo technique) in a patient with hypertrophic cardiomyopathy. A, Early systolic image shows hypertrophy of the left ventricular myocardium; low signal intensity of blood in the outflow tract, indicating turbulent blood flow (arrow); and focal low signal intensity in the left atrium due to mitral regurgitation (short arrow). B, Fifty-seven milliseconds later. As in Figure 30-1, there is a prominent anterior papillary muscle (arrow) arising atypically from the anterior wall instead of from the anterolateral wall of the left ventricle. C, Another 57 milliseconds later, mitral regurgitation is still evident. The anterior leaflet of the mitral valve is in an abnormal position (small arrows), possibly a result of abnormal forces exerted by the papillary muscle. D, Near the end of systole, the flow signal is still present in the outflow tract. LA indicates left atrium; RA, right atrium; RV, right ventricular outflow tract.

(Figure 30-5) and the globally diminished or absent left ventricular wall thickening are clearly demonstrated by spin-echo MRI.[15] Another potential application is the detection of left ventricular thrombi in these cardiomyopathic patients.

Dynamic MRI may prove useful for discrimination between patients with idiopathic cardiomyopathy and those with severe coronary artery disease if the latter exhibit at least some region of normal wall motion. Since dynamic MRI is well suited for three-dimensional determination of ventricular volumes,[16,17] subtle changes in left and right ventricular function with therapeutic intervention may be detected. Again, the presence and extent of associated valvular regurgitation can be assessed with this technique.

Similar morphologic and functional abnormalities have been noted in patients with idiopathic congestive cardiomyopathy and those with alcoholic cardiomyopathy. Magnetic resonance spectroscopy (MRS) in hamsters with cardiomyopathy induced by prolonged ethanol administration has provided some insight into the biochemical changes associated with this disease.[18] Alcoholic hamsters had significantly higher inorganic phosphate and lower adenosine triphosphate levels, while maintaining normal intracellular pH, phosphocreatine, and creatinine. Magnetic resonance spectroscopy could also demonstrate that treatment with verapamil during long-term alcohol consumption prevented the development of these metabolic changes. In another animal model of cardiomyopathy, Syrian hamsters with a hereditary form of cardiomyopathy uniformly showed a reduction in

correlates well with the angiographic estimate of the severity of mitral regurgitation.[13]

In addition to the ability to measure the thickness of the myocardium and to assess the presence and extent of outflow tract obstruction and mitral incompetence, the inherent soft tissue contrast on MR images may permit tissue characterization in selected patients with hypertrophic cardiomyopathy. Variable myocardial signal intensity in a patient with hypertrophic cardiomyopathy has been described in a case report. At surgery, areas with low signal intensity showed myocardial fibrosis, whereas a region with high signal intensity appeared edematous by inspection.[14]

CONGESTIVE CARDIOMYOPATHY

There have been very few reports describing MR findings in patients with congestive cardiomyopathy. The typical enlargement of all four cardiac chambers

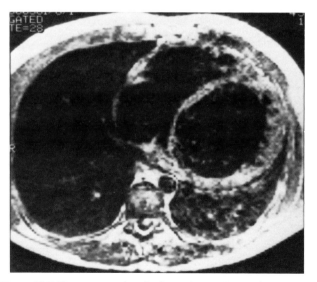

Figure 30-5 Transverse spin-echo image in a patient with congestive cardiomyopathy. Enlargement of the right-sided chambers and the left ventricle is evident. *Source:* Reprinted with permission from *American Journal of Cardiology* (1987;59:145–151), Copyright © 1987, Cahners Publishing Company Inc.

phosphocreatine with variable adenosine triphosphate levels.[19]

In humans, MRS has been limited by the resolution of surface coils requiring a sample size of less than the 10-mm thickness of the left ventricular wall. Certain favorable circumstances, however, such as increased left ventricular wall thickness combined with minimal skeletal muscle development, may permit MRS measurements of cardiac muscle.[20] In this case, both myocardium and skeletal muscle showed a phosphocreatine to inorganic phosphate ratio of half that for a normal control. Importantly, therapeutic interventions designed to improve myocardial energy production could be monitored by MRS.

DOXORUBICIN HYDROCHLORIDE CARDIOMYOPATHY

Present noninvasive techniques to detect doxorubicin-induced cardiomyopathy rely on assessment of myocardial function rather than on direct measurement of altered tissue properties. Since relaxation times depend on water, lipid, and macromolecular content, as well as physiochemical interactions and exchange processes, MR spectrometry has been used in rat models to evaluate the effect of doxorubicin-induced cardiomyopathy on myocardial relaxation times.[21,22] In excised rat hearts with histologic evidence of chronic cardiotoxicity, T_1 values were significantly higher than those in control animals.[21] This difference was most pronounced in rats with gross evidence of toxicity or heart failure. There were no changes in T_2 relaxation times or tissue water content. The alterations in T_1 were explained by membrane dysfunction leading to changes in calcium and sodium content of the myocardium.[21] However, the observed changes in T_1 were small, and the standard deviation of the group values was in the same order of magnitude.

Currently, measurements of T_1 relaxation time from MR images are limited by cardiac motion and flow artifacts as well as by restrictions of repetition times and irregularities of heartbeat intervals.[23] In our experience with a dog model of doxorubicin cardiomyopathy, MRI showed the progressive dilatation of the left ventricle associated with increasing toxicity (Figure 30-6). However, it was not possible to distinguish between normal hearts and those with mild, moderate, or severe cardiomyopathy by measuring relaxation times from in vivo MR images. This was probably due to the abovementioned difficulties of in vivo measurements and the absence of changes in water content between groups. Unexpectedly, post-mortem imaging with fixed repetition time (TR) intervals also failed to reveal differences of statistical significance.[24]

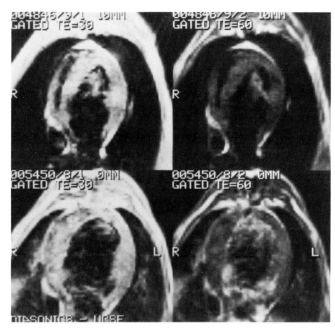

Figure 30-6 Transverse spin-echo images in a dog with doxorubicin cardiomyopathy. Top left, First echo image before treatment. Top right, Second echo image before treatment. Bottom left, First echo image after development of left ventricular failure. Note dilatation of the left ventricle. At autopsy, there was histologic evidence of severe doxorubicin cardiomyopathy. Bottom right, Second echo image after development of left ventricular failure.

TRANSPLANT REJECTION

Another type of cardiomyopathy is found in cardiac allografts suffering transplant rejection. For this problem there is an urgent need for a noninvasive technique to identify subtle changes in myocardial tissue composition indicating graft rejection. The process of cardiac allograft rejection involves generalized inflammation and tissue edema, a process for which proton MRI is known to be sensitive.[25,26] Initial work in a dog model of heterotopic cardiac transplants[27] demonstrated a significant increase in T_2 relaxation times and intensity values for the transplanted hearts compared with native hearts from 3 days to 14 weeks after transplantation. The T_1 relaxation times of native and transplanted hearts showed no significant difference on in vivo electrocardiogram-gated studies. However, T_1 values calculated from post-mortem studies were significantly longer in the transplanted hearts than in the native hearts. By extending these experiments it could be shown that immunosuppressive treatment prevents prolongation of in vivo T_2 relaxation times. Again, the extent of the increase in T_2 correlated well with the intensity of rejection in the untreated group of animals.[28] Characteristic abnormalities of rejected transplants on MR images include increased thickness and increased signal intensity on second echo images compared with those of normal control subjects.

A recent study[29] in humans with cardiac allografts found an increase in T_1 and T_2 relaxation time in hearts with late rejection events more than 25 days after surgery. This group used a short TR pulse sequence with a TR approximating the T_1 value of the tissue being imaged in order to improve the accuracy of in vivo T_1 measurements. Of 15 late rejections, 14 were correctly identified on the basis of measurements of relaxation time. Again, a significant increase in left ventricular wall thickness was noted in rejecting allografts. The diagnostic applicability of the method is limited by the fact that nonrejecting grafts also had increases in T_1 and T_2 values when imaged within 24 days of graft implantation. This is likely caused by a prolonged period of cold anoxic arrest and subsequent graft edema that seems to subside 3 to 4 weeks after implantation.

The bioenergetic processes that occur during cardiac allograft rejection can be assessed by using in vivo phosphorus 31 MRS. Ratios of phosphocreatine to inorganic phosphate or phosphocreatine to β-adenosine triphosphate were compared in allografts and isografts in a rat model 2 days after subcutaneous transplantation.[30] Neither ratio changed in isografts but decreased continuously in allografts, with the difference becoming significant by days 3 and 4, respectively. The isograft group showed no change in intracellular pH; however, the allograft group demonstrated an initial alkaline shift followed by acidosis. The multitude of abnormal MRS findings suggests that monitoring of abnormal bioenergetics associated with cardiac allograft rejection may become a useful means for following patients with cardiac transplants once ^{31}P spectroscopy has become clinically feasible.

RESTRICTIVE CARDIOMYOPATHY

Restrictive cardiomyopathy is an uncommon disease that is usually diagnosed by cardiac catheterization combined with endomyocardial biopsy. Hemodynamic changes include elevated right and left ventricular filling pressures resulting in reduced flow within the atria. Consequently, MRI reveals enlargement of both atria, enlargement of the inferior vena cava and pulmonary veins, and a prominent flow signal within the atria at multiple phases of the cardiac cycle (Figure 30-7). Normal pericardial thickness as seen by MRI allows differentiation from constrictive pericarditis.[31] The presence of amyloid heart disease is suggested by widespread thickening of all chamber walls and atrioventricular

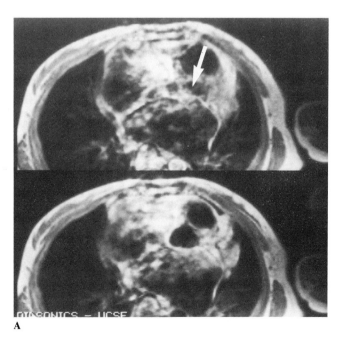
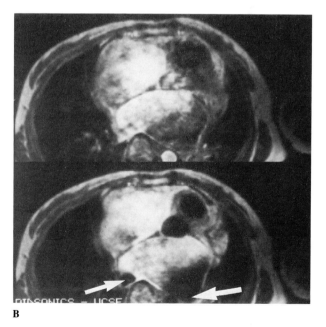

Figure 30-7 Transverse spin-echo images in a patient with restrictive cardiomyopathy of unknown cause. A, Top, End-diastolic image (echo time [TE], 30 milliseconds) obtained at the R wave of the electrocardiogram shows prominent signal in both atria due to the stasis of blood. Note the flow signal in the aortic root (arrow). Bottom, End-systolic image. The flow signal in the right atrium is more prominent. B, Top, Same section of the heart as in A but the second spin-echo image (TE, 60 milliseconds). The flow signal now fills both atria at end-diastole. Bottom, Image during systole shows the flow signal to be even more prominent than during diastole. Note the absence of signal at the entry site of pulmonary veins (arrows). *Source:* Reprinted with permission from *American Journal of Cardiology* (1987;59:480–482), Copyright © 1987, Cahners Publishing Company Inc.

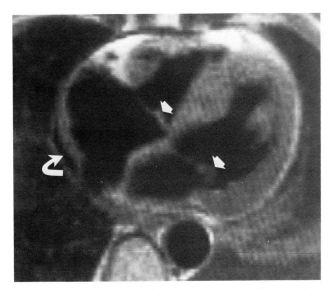

Figure 30-8 Transverse spin-echo image in a patient with amyloidosis showing thickening of the ventricular and atrial septa, right atrial wall, and atrioventricular valves (arrows). Note the small pericardial effusion over the right atrial wall (curved arrow). *Source:* Reprinted with permission from *American Journal of Cardiology* (1987;59:480–482), Copyright © 1987, Cahners Publishing Inc.

valve leaflets (Figure 30-8). However, differentiation of amyloid from normal myocardium using magnetic relaxation times calculated from MR images could not be accomplished with current techniques. Dynamic MRI is useful for assessment of left ventricular function that may be decreased or normal in restrictive cardiomyopathy and in the demonstration of associated tricuspid and mitral regurgitation.

SUMMARY

Compared with other cardiac imaging modalities, MRI offers the unique features of true three-dimensional imaging with excellent anatomic resolution and functional information, including depiction of blood flow. Moreover, in certain types of cardiomyopathy, tissue characterization by MRI and MRS opens a new approach to diagnosis and monitoring of therapeutic interventions. These unique advantages may outweigh the well-known limitations of the technique, including high cost, limited availability, and long imaging times. Presently, MRI is not the method of choice in a patient with cardiomyopathy. However, further technical and scientific developments ought to secure a clinical role of MRI for the depiction of anatomy and measurement of function in cardiomyopathic patients, in assessment of patients with drug-induced cardiomyopathy, and for the detection of rejection in recipients of cardiac transplants.

REFERENCES

1. Higgins CB, Byrd BF III, Stark D, et al. Magnetic resonance imaging in hypertrophic cardiomyopathy. *Am J Cardiol.* 1985; 55:1121–1126.

2. Been M, Kean D, Smith MA, et al. Nuclear magnetic resonance in hypertrophic cardiomyopathy. *Br Heart J.* 1985;54:48–52.

3. Caputo GR, Tscholakoff D, Sechtem U, et al. Measurement of canine left ventricular mass by using MR imaging. *Am J Roentgenol.* 1987;148:33–38.

4. Florentine MS, Grosskreutz CL, Chang W, et al. Measurement of left ventricular mass in vivo using gated nuclear magnetic resonance imaging. *J Am Coll Cardiol.* 1986;8:107–112.

5. Keller AM, Peshock RM, Malloy CR, et al. In vivo measurement of myocardial mass using nuclear magnetic resonance imaging. *J Am Coll Cardiol.* 1986;8:113–117.

6. Sechtem U, Pflugfelder PW, White RD, et al. Cine MR imaging: Potential for the evaluation of cardiovascular function. *Am J Roentgenol.* 1987;148:239–246.

7. Haase A, Frahm J, Matthei D, et al. FLASH imaging: Rapid NMR imaging using low flip-angle pulses. *J Magn Reson.* 1986; 67:258–266.

8. Wehrli FW. Fast-scan imaging: principles and contrast phenomenology. In: Higgins CB, Hricak H, eds. *Magnetic Resonance Imaging of the Body.* New York, NY: Raven Press; 1987:23–38.

9. Jiang L, Levine R, King ME, et al. An integrated mechanism for systolic anterior motion of the mitral valve in hypertrophic cardiomyopathy based on echocardiographic observations. *Am Heart J.* 1987;113:633–644.

10. Nagata S, Nimura Y, Beppu S, et al. Mechanism of systolic anterior motion of the mitral valve and site of intraventricular pressure gradient in hypertrophic obstructive cardiomyopathy. *Br Heart J.* 1983;49:234–243.

11. Firmin DN, Nayler GL, Klipstein RH, et al. In vivo validation of MR velocity imaging. *J Comput Assist Tomogr.* 1987;11:751–756.

12. Nayler GL, Firmin DN, Longmore DB: Blood flow imaging by cine magnetic resonance. *J Comput Assist Tomogr.* 1986;10:715–722.

13. Pflugfelder PW, Sechtem U, White RD. Non-invasive evaluation of mitral regurgitation by analysis of systolic left atrial signal loss in cine magnetic resonance images. *Circulation.* 1987;76(suppl 4):89.

14. Farmer D, Higgins CB, Yee E, et al. Tissue characterization by magnetic resonance imaging in hypertrophic cardiomyopathy. *Am J Cardiol.* 1985;55:230–232.

15. Sechtem U, Sommerhoff B, Marciewicz W, et al. Regional left ventricular wall thickening by magnetic resonance imaging: Evaluation in normal persons and patients with global and regional dysfunction. *Am J Cardiol.* 1987;59:145–151.

16. Sechtem U, Pflugfelder PW, Gould RG, et al. Measurement of right and left ventricular volumes in healthy individuals with cine MR imaging. *Radiology.* 1987;163:697–702.

17. Utz JA, Herfkens RJ, Heinsimer JA, et al. Cine MR determination of left ventricular ejection fraction. *Am J Roentgenol.* 1987; 148:839–843.

18. Wu S, White R, Wikman-Coffelt J, et al. The preventive effect of verapamil on ethanol-induced cardiac depression: Phosphorus-31 nuclear magnetic resonance and high-pressure liquid chromatographic studies of hamsters. *Circulation.* 1987;75:1058–1064.

19. Marciewicz W, Wu S, Parmley WW, et al. Evaluation of the hereditary Syrian hamster cardiomyopathy by P-31 nuclear magnetic resonance spectroscopy: Improvement after acute verapamil therapy. *Circ Res.* 1986;59:597–604.

20. Whitman GJR, Chance B, Bode H, et al. Diagnosis and therapeutic evaluation of a pediatric case of cardiomyopathy using phosphorus-31 nuclear magnetic resonance spectroscopy. *J Am Coll Cardiol*. 1985;5:745–749.

21. Thompson RC, Canby RC, Lojeski EW, et al. Adriamycin cardiotoxicity and proton nuclear magnetic resonance relaxation properties. *Am Heart J*. 1987;113:1144–1149.

22. Ng TC, Daugherty JP, Evanochko WT, et al. Detection of antineoplastic agent induced cardiotoxicity by P-31 NMR of perfused rat hearts. *Biochem Biophys Res Commun*. 1983;110:339–347.

23. Prato FS, Drost DJ, King M, et al. Cardiac T1 calculations from MR spin-echo images. *Magn Reson Med*. 1987;4:227–243.

24. Sechtem U, Caputo G, Finkbeiner W, et al. Evaluation of canine adriamycin cardiomyopathy by MRI. *Circulation*. 1986;74 (Suppl. III):348.

25. Johnston DL, Brady TJ, Ratner AV, et al. Assessment of myocardial ischemia with proton magnetic resonance: Effects of a 3 hr coronary occlusion with and without reperfusion. *Circulation*. 1985;71:595–601.

26. Pflugfelder PW, Wisenberg G, Prato FS, et al. Early detection of canine myocardial infarction by magnetic resonance in vivo. *Circulation*. 1985;71:587–594.

27. Tscholakoff D, Aherne T, Yee ES, et al. Cardiac transplantation in dogs: evaluation with MR. *Radiology*. 1985;157:697–702.

28. Aherne T, Tscholakoff D, Finkbeiner W, et al. Magnetic resonance imaging of cardiac transplants: the evaluation of rejection of cardiac allografts with and without immunosuppression. *Circulation*. 1986;74:145–156.

29. Wisenberg G, Pflugfelder PW, Kostuk WJ, et al. Diagnostic applicability of magnetic resonance imaging in assessing human cardiac allograft rejection. *Am J Cardiol*. 1987;60:130–136.

30. Canby RC, Evanochko WT, Barrett LV, et al. Monitoring the bioenergetics of cardiac allograft rejection using in vivo P-31 nuclear magnetic resonance spectroscopy. *J Am Coll Cardiol*. 1987;9:1067–1074.

31. Sechtem U, Higgins CB, Sommerhoff BA, et al. Magnetic resonance imaging of restrictive cardiomyopathy. *Am J Cardiol*. 1987;59:480–482.

Chapter 31

Ultrasonic Myocardial Tissue Characterization

Julio E. Pérez, MD, Steve M. Collins, PhD, James G. Miller, PhD, and David J. Skorton, MD

Recent advances in digital computer image processing and signal analysis have provided a strong impetus to the use of ultrasound for precise diagnostic applications in cardiology. For example, the use of real-time, two-dimensional Doppler echocardiography, providing color-encoded depiction of intracardiac blood flow, is the latest of such applications to become widely employed clinically. Color Doppler echocardiography has revolutionized the diagnostic capabilities of ultrasound in the setting of congenital and valvular heart disease. Conventional two-dimensional echocardiography continues to maintain an important role as a high-resolution imaging modality used to depict cardiac anatomy and the mechanics of ventricular wall motion. Its ability to define precisely the boundaries between tissues with different acoustic impedances (for example, myocardium versus blood) confers on this modality high sensitivity and specificity for the detection of segmental abnormalities of ventricular wall motion and the visualization of intracardiac mural thrombi. As opposed to these bright reflections from large, smooth interfaces, the intramural echoes that emanate from myocardial tissue as a result of interaction with the incoming ultrasonic waves have received relatively little attention as diagnostic features of cardiac pathology and physiology. Thus ultrasonic characterization of myocardial tissue has a purpose fundamentally different from that of conventional echocardiography. Tissue characterization has as its goal the description, mostly in quantitative terms indicative of ultrasonic energy, of the physical state or architecture of myocardium per se, rather than its delineation based on wall thickness, thickening or thinning patterns, and motion, which are otherwise well outlined by conventional echocardiographic modalities. Over the last 10 to 15 years there has been a growing interest in the development of this new branch of cardiac ultrasound that holds promise for the noninvasive and bedside quantitative differentiation of normal tissue from abnormal tissue, including ischemic, infarcted, contused, myopathic, and rejecting transplanted myocardium.[1,2]

For some time, there has been intriguing evidence suggesting that structural alterations in myocardium associated with specific pathologic states are represented by peculiar images even on conventional M-mode and two-dimensional echocardiograms. Thus myocardial replacement by fibrous tissue has been noted with the use of conventional ultrasound as bright intramural echoes[3] indicating an increase in echo amplitudes.[4] In addition, the intramural septal region in patients with hypertrophic cardiomyopathy has been described as having a "ground-glass" appearance, indicating a change in the spatial pattern of the echo amplitudes.[5] Similarly, the myocardium of patients with infiltrative cardiomyopathy related to amyloidosis has been

---

This work was supported in part by Grants HL17646 and HL32295 (Specialized Centers of Research in Ischemic Heart Disease at Washington University and The University of Iowa, respectively), New Investigator Research Award HL36733 (Dr. Pérez), and Research Career Development Award KO4-HL01290 (Dr. Skorton) from the National Heart, Lung, and Blood Institute, and by Grant IAG-30 from the American Heart Association—Iowa Affiliate.

described by conventional echocardiography as "shining" or "sparkling."[6] This suggests that fiber disarray in hypertrophic cardiomyopathy and amyloid deposits in infiltrative cardiomyopathy are probably responsible for the observed peculiar image pattern in each entity. However, the main difficulty with these descriptions is their qualitative nature. Thus the images subjectively may be interpreted differently by various observers and may be affected by the gain settings used by individual operators. In addition, the sensitivity and specificity for detection of disease and the histopathologic verification of the tissue determinants responsible for the ultrasonic characteristics need to be clarified further in quantitative terms.

The purpose of this chapter is to review the current status of ultrasonic characterization of myocardium as a method that attempts to define, mostly in quantitative terms, the physical state of cardiac muscle using novel ultrasonic parameters that relate to its structural components. These parameters are used to differentiate normal, viable myocardium from tissue that has undergone structural or functional alterations (or both) and, in some approaches, are used ultimately to generate computer-assisted, real-time ultrasonic images based on quantitative indices of tissue structure. We shall start with a general description of the approaches taken by groups of investigators in the field, some basic concepts concerning the interaction of ultrasound and myocardium, and the basis for instrumentation differences. This is followed by examples of reports on ultrasonic characterization of specific clinicopathologic states and a perspective on the present and future roles of this new diagnostic modality in cardiovascular medicine.

Investigators active in the field of ultrasonic characterization of myocardium have taken one of three general approaches to the problem: (1) techniques that rely directly on analysis of the radiofrequency signals that emanate from myocardium and that result from interactions between ultrasound and tissue along the ultrasonic beam, (2) methods based on qualitative description of alterations in conventional echocardiographic images, and (3) techniques that involve computer-assisted analysis of digitized data derived from conventional echocardiographic images.

BASIC CONCEPTS OF ULTRASONIC TISSUE CHARACTERIZATION

Ultrasonic waves travel through biologic tissues with a *propagation speed* that is dependent on the density and the compressibility of the medium. The ultrasonic *wavelength* is the length of space encompassing one cycle of the ultrasonic waves. The ultrasonic wavelength is determined by the ratio of the propagation speed to the ultrasonic frequency. The frequency is determined by the choice of ultrasonic transducer used to make the image. The product of propagation speed and density of the tissue yields the tissue's *acoustic impedance*. One feature of ultrasound as a diagnostic tool relates to the property of waves to be *reflected* when they reach a boundary between tissues with an *acoustic impedance mismatch*. If the wavelength is small when compared with the boundary dimensions, then by definition the reflection that occurs is called a *specular reflection* (Figure 31-1A). This is the typical reflection that occurs, for example, at the endocardium-blood interface, and that is so important for definition of edges in conventional echocardiography. On the other hand, when the boundaries between the different tissues (or between components within the tissue) are smaller than the wavelength, the incident wave will undergo *scattering* (Figure 31-1B). Scattering typically occurs in a multidirectional fashion, in particular in the case of heterogeneous media or in suspensions with particles such as the bloodstream. Those waves that are redirected back to the transmitting transducer after interacting with the inhomogeneities of the tissue are defined as being *backscattered* (Figure 31-1C).

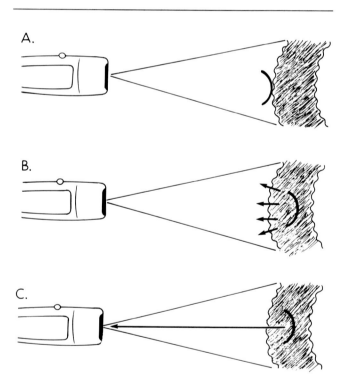

Figure 31-1 Different modes of ultrasonic reflection from tissue. A is a schematic representation of the occurrence of specular (mirrorlike) reflection; this occurs when the ultrasonic wavefront is reflected from a region of acoustic impedance mismatch (for example, tissue boundary) and the wavelength is of smaller magnitude than the boundary. In B, scattering occurs in multiple directions after the incidence of ultrasound with components of tissue smaller than the wavelength. In C, backscattering is shown as ultrasound that is redirected back to the transmitting transducer after interacting with the inhomogeneities of the tissue.

In addition, as the sound travels through tissue, the intensity and amplitude of the waves are diminished as a function of the depth into the tissue. This phenomenon is called *attenuation*. The losses resulting from attenuation can be due in part to absorption (conversion into heat) and in part to reflection and scattering, as defined above. Both ultrasonic attenuation and backscatter are quantitative tissue parameters that have been shown to be useful in characterizing myocardial pathologic states[7-9] in experimental animals. Both attenuation and backscatter depend upon the frequency used for the ultrasonic measurements (Figure 31-2). It is useful to express backscatter as a function of a measurable bandwidth of frequencies that may be employed clinically. The use of averaging over a range of frequencies tends to minimize the variability in the precise measurement of backscatter because of the phenomenon of *phase cancelation*.[10,11] The use of piezoelectric elements in diagnostic ultrasonic transducers results in phase cancelation effects. These effects are caused by distortions of the ultrasonic wavefronts due to inhomogeneity in the tissue, leading to degradations in the received signal at the transducer. In addition, reflected waves interfere with each other (in constructive and destructive interactions), also contributing to phase cancelation. Previous experiments have demon-

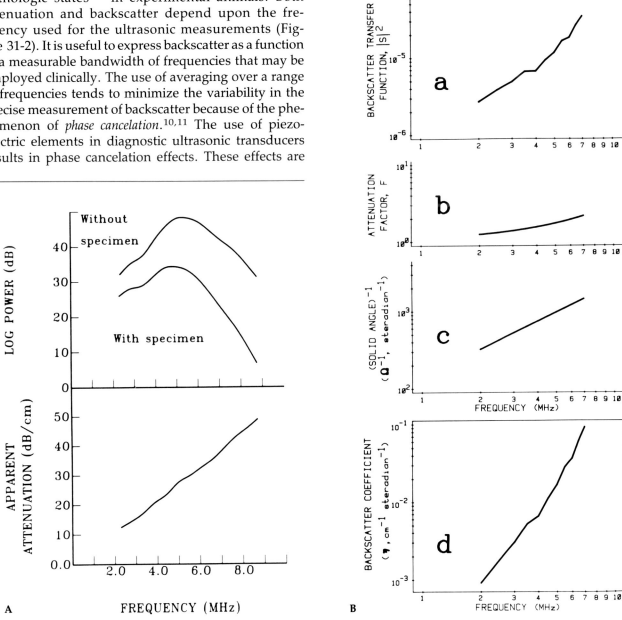

Figure 31-2 Dependence of ultrasonic attenuation and backscatter on frequency. A, Top, A plot of the ultrasonic energy transmitted in vitro with no specimen present (top curve) and with a specimen present (bottom curve). Bottom, The attenuation versus frequency plot determined by subtracting the two curves shown in the top panel and correcting for the thickness of the specimen. The linear dependence of attenuation on frequency is evident. *Source:* Reprinted from *Proceedings of the Institute of Electrical and Electronics Engineers Ultrasonics Symposium* (83CH1947-1:782–793) by JG Miller, JE Pérez, JG Mottley et al with permission of IEEE Inc, © 1983. B, Backscatter coefficient of normal dog myocardium in vivo (d) derived from the product of the backscatter, transfer function (a), an attenuation factor (b), and the inverse solid angle subtended by the scattering volume (c). This demonstrates the relative linear dependence of backscatter parameters on frequency. *Source:* Reprinted with permission from *Progress in Cardiovascular Diseases* (1985;28:85–110), Copyright © 1985, Grune & Stratton Inc.

strated the feasibility of reducing the variability in ultrasonic measurements due to phase cancelation effects either by averaging data obtained from multiple sites, as opposed to single sites, or by averaging the results obtained over a range of frequencies. The backscatter measured over a range of frequencies is defined as *integrated backscatter*, which is a spatially localized, relative measure of the scattering efficiency of a volume of tissue. By combining spatial averaging (that is, averaging ultrasonic data acquired from multiple sites) and using integrated backscatter, the variability in the measurements can be reduced even further.[12]

Instrumentation Approaches

There is little doubt about the impact that sophisticated electronic components and digital computers have had in the development of the field of myocardial tissue characterization with ultrasound. The progress made has been intimately linked to advances in electrical engineering and physics applied to improvement of imaging systems for use in diagnostic ultrasonic imaging. Different groups of investigators have undertaken different methodologic approaches to develop novel indices that have become descriptors of normal and abnormal myocardium. These different quantitative approaches are based in part on different instrumentation for data acquisition and analysis. The following discussion contrasts the two approaches used by the authors in their respective institutions.

Radiofrequency Signal Analysis

At Washington University in St Louis, the basis of the ultrasonic measurements of myocardium has been the analysis of unprocessed radiofrequency signals. This approach permits measurement of the degree of myocardial ultrasonic attenuation in transmission studies (that is, with one transducer on either side of the tissue) as well as the extent of myocardial ultrasonic backscatter, with only one transducer acting as both transmitter and receiver, as is typical of clinical echocardiography. In experiments performed in vitro for the determination of the attenuation of tissue, the signal loss resulting from transmission through a measured thickness of tissue is compared with the signal loss resulting from transmission through the same thickness of saline. By a substitution technique, the attenuation coefficient is obtained as a function of frequency used for the measurements. The resulting slope (determined by fitting a least-squares line) is expressed in decibels per centimeter of thickness of tissue per megahertz.[13] Although the attenuation characteristics are important features related to tissue structure, this transmission measurement approach does not lend itself well to applications in vivo. Thus measurements of backscatter were developed by detection of the signals emanating from a segment of myocardium using the same transducer for transmission and reception. An electronic gate, typically a few microseconds in duration, selects a volume of myocardium chosen to avoid specular reflections from epicardial and endocardial surfaces. The power spectrum obtained from measurements in tissue is then compared with a reference power spectrum obtained from the measured reflection from a standard surface such as a steel plate to obtain the *backscatter transfer function. Integrated backscatter* is the frequency-average of this transfer function over the useful bandwidth of the system.[14] As mentioned earlier, the use of integrated backscatter (as well as spatial averaging of data from multiple sites) reduces variability in measurements. One of the advantages of approaches based on quantitative assessment of myocardial characteristics using attenuation and backscatter is to minimize yet another potential source of variability: the operator dependence of methods based only on descriptions derived from conventional echocardiographic images. Furthermore, quantitative assessment based on backscatter represents a possible tool for performing serial quantitative measurements in the same segment of tissue to monitor sequential pathologic alterations in structural components.[15]

Even though these measurements provided absolute characterization of acoustic properties of tissue in animals, they originally did not lend themselves directly to the generation of cardiac images. A transitional step was the construction of an M-mode–based system to measure integrated backscatter in real time, but without the spatial orientation of two-dimensional echocardiography.[16] In more recent work performed in St Louis, these fundamental concepts derived from the radiofrequency-backscattered signals have been implemented in a commercial two-dimensional echocardiographic imaging device, modified to compute in real time the integrated backscatter along each A-line of the sector before passing the resulting signals to the scan converter to generate images based on integrated backscatter.[17,18] This is achieved with image resolution virtually identical with that of conventional echocardiography and thus represents a significant step toward obtaining quantitative tissue characterization in real time in patients. The current implementation permits the measurement of the physiologic cardiac cycle-dependent variation of myocardial integrated backscatter (see below) and relative values of the time-averaged backscatter. Additional refinements in this system in the future may permit quantitative comparisons of time-averaged backscatter from patient to patient and among patient populations with different myocardial pathologic processes.

Quantitative Analysis of Gray Level (Image) Data

Experiments performed at the University of Iowa in Iowa City have demonstrated the feasibility of myocardial characterization with ultrasound, using techniques based on computer analysis of gray level echocardiographic image data obtained by conventional imaging systems. Thus the methods employ evaluation of regional myocardial average echo amplitude and the spatial distribution (texture) of amplitudes in a quantitative, statistical manner. This approach is in contrast to the subjective visual methods of other techniques that rely on the display of acoustically abnormal regions of the heart. The approach used at Iowa City measures the overall distribution and spatial pattern of echo amplitudes[19] and hence does not derive such tissue acoustic parameters as attenuation and backscatter. An example of a technique commonly used to evaluate texture is the group of *run-length measures*. A *gray level run* refers to a set of picture elements that run in a given direction and have the same or similar gray level. A *run-length matrix* representing the number of runs of varying lengths and gray levels can then be derived. A variety of statistical parameters of image gray levels can be computed, indicative of the heterogeneity of their appearance in the image and representing the relative occurrence and size of the individual echo reflections in the tissue. Thus these and other gray level texture calculations assess the heterogeneity of distribution of echogenic reflections in the echocardiographic image (Figure 31-3).

Acoustic Properties of Normal Myocardium

Ultrasonic characterization data from normal myocardium has become available both from experimental work performed primarily in dogs and from images obtained in normal subjects. With respect to absolute acoustic measurements, experiments in dogs have determined that the slope of the ultrasonic attenuation versus frequency relationship in normal tissue has an approximate value of 0.6 dB/cm per megahertz.[20] The time-averaged integrated backscatter values (obtained at a center frequency of approximately 5 MHz) of normal myocardium in dogs approximates 52 to 55 dB below (less than) that of a perfect reflector surface.[9,21] If the integrated backscatter is measured gated to a physiologic cardiac signal such as the electrocardiogram, the ventricular pressure waveform, or the first-time derivative of the ventricular pressure, values of backscatter are found to be higher at end-diastole and lower at end-systole. This reflects the cyclic variation of integrated backscatter characteristic of normal myocardium.[22–24] The difference between end-diastolic and end-systolic

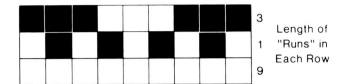

Figure 31-3 The concept of measures of gray level run-length image textures. This schematic digital region of interest contains three rows, each consisting of nine pixels. A gray level run is defined as a set of consecutive pixels having the same or similar gray level value. This length of a run is the number of pixels traversed before a significant alteration in gray level occurs. In the top row, runs are three pixels in length; in the middle row, one pixel in length; and in the bottom row, nine pixels in length.

integrated backscatter is approximately 5 dB in normal dog hearts and in human myocardium in normal subjects.[18] Additional observations will be required to determine whether the extent of cyclic variation of integrated backscatter in normal myocardium is dependent on factors such as patient age, gender, myocardial collagen content, or underlying systemic diseases that involve the heart (arterial hypertension, segmental involvement of myocardial areas such as occurs in atherosclerotic coronary artery disease, and the like). With respect to image gray level analysis, normal myocardium exhibits relatively homogeneous statistical values of gray levels, indicative of the absence of large intramural variations of scatterer structures, but with detectable variation of gray levels and of image texture as a result of cardiac systole.[25,26]

The definition of normal myocardial characteristics, in both textural and quantitative acoustic terms, has permitted recognition of abnormalities in tissue of both experimental animals and patients. The discussion that follows describes several examples of these abnormalities as detected by a variety of methods and relates the experimental observations to clinically relevant situations.

APPLICATIONS OF ULTRASONIC MYOCARDIAL TISSUE CHARACTERIZATION

Ischemic Heart Disease

Since the advent of effective methods of reestablishing myocardial perfusion during acute infarction, increased attention has turned to methods of assessing the amount of myocardium at risk of necrosis, the extent and severity of regional ischemia, the size of the ensuing myocardial infarction, and the amount of viable myocardium remaining. Standard clinical parameters, including electrocardiography, assay of myocyte enzymes, and radionuclide scintigraphy, offer only incomplete information concerning these problems. Thus improved methods of reaching these diagnostic goals in ischemic heart disease would be most welcome from the clinician's point of view. Qualitative and quantitative ultrasonic tissue characterization techniques have been applied to several subsets of ischemic heart disease, including acute ischemia, reperfusion, acute infarction, and chronic infarction resulting in regional myocardial scar.

Qualitative Observations

Simple visual inspection of conventional echocardiograms occasionally has identified regions of abnormal myocardial perfusion by a change in regional gray level. Fraker and co-workers[27] were able to identify acute increases in regional echo amplitude as soon as 15 minutes after coronary occlusion. The approximate size of the region exhibiting increased echo amplitude correlated with infarct size as determined by histochemical techniques. Werner et al[28] identified a subset of patients with acute myocardial necrosis exhibiting a *decrease* in regional echo amplitude. Patients with this finding had a high risk of severe complications of myocardial infarction, including free wall rupture and death. In the more chronic phase of myocardial infarction, as mentioned above, Rasmussen and co-workers[3] noted that, in patients with remote infarction, regions of scar could be identified by an increase in echo amplitude on M-mode echograms (Figure 31-4).

These qualitative observations helped to establish the feasibility of ultrasonic analysis of myocardial tissue in ischemic heart disease. However, the findings are subjective and therefore may lack reproducibility. Moreover, these display-based observations are greatly dependent on adjustment of overall gain, time-gain compensation, and other system parameters under operator control. In an attempt to improve the visual perception of differences in regional echo amplitude, other investigators have used color-encoding of echo amplitude information. Parisi and co-workers[29] identified increases in echo amplitude as part of the evolution of myocardial infarction based on color-encoded echocardiograms. Similarly, Logan-Sinclair et al[4] and Shaw et al[30] identified regions of fibrosis in hearts with chronic myocardial infarction. As with the gray scale qualitative techniques, color-encoded echocardiograms are dependent on instrument adjustments.

Quantitative Observations

Acute Ischemia. In an early study of acute myocardial infarction, Namery and Lele[8] noted a decrease in acoustic impedance soon after coronary occlusion in dogs. In a subsequent group of studies, investigators at Wash-

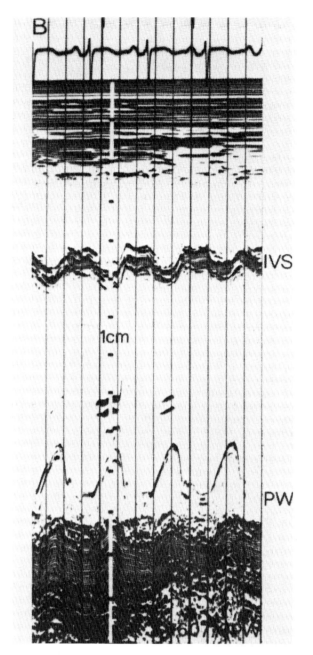

Figure 31-4 M-Mode echocardiogram of a patient with scarring of the interventricular septum (IVS) due to myocardial infarction. The echo reflections from the septum are visibly stronger (darker) than those from the posterior left ventricular wall (PW). *Source:* Reprinted with permission from *Circulation* (1978;57:230–237), Copyright © 1978, American Heart Association.

ington University in St Louis have evaluated various aspects of the acoustic characteristics of acute ischemia. In early investigations, attenuation through excised myocardial specimens was measured by using a transmission technique. With this system, Mimbs et al[31] identified significant decreases in attenuation through ischemic tissue as soon as 15 minutes after coronary occlusion in dogs. Decreased attenuation was noted up to 24 hours after occlusion, thereafter being replaced by an *increase* in attenuation (Figure 31-5). Two significant implications resulted from this early study that served to motivate further this area of research: (1) Ischemic myocardial injury could be identified as early as 15 minutes after coronary occlusion, much sooner than light microscopic evidence of ischemic damage could be expected; and (2) the early decrease, followed by a later increase, in attenuation suggested the possibility that the stage of an acute injury could be estimated by using ultrasonic methods. Earlier studies by Mimbs et al[13] of ultrasonic attenuation in excised tissue specimens had already established that the amount of change in attenuation in chronic infarction correlated with the size of the injury as assessed by creatine kinase assays.

Although promising and of theoretic interest, this two-transducer method of measuring attenuation through excised tissue specimens obviously was not clinically applicable. O'Donnell and co-workers[9] described a measure of the amount of energy returning to a single transducer (acting as both transmitter and receiver), termed ultrasonic "integrated backscatter." Ultrasonic integrated backscatter, a potentially clinically applicable variable, was shown by Mimbs et al[32] to be increased as soon as 1 hour after acute coronary occlusion. Schnittger and associates,[33] using a fundamentally different calculation technique, identified an increase in the ratio of mean regional echo amplitude to the standard deviation of echo amplitudes (the "mean to standard deviation ratio") within 30 minutes after coronary occlusion in dogs.

Most of these observations were based on the total amount of energy returning to the transducer, averaged throughout the heart cycle; the *cyclic variation* of ultrasonic backscatter with cardiac contraction is also a variable of great interest. Madaras et al.[22] first reported a cardiac cycle-dependent variation in ultrasonic backscatter, with maximal values occurring near end-diastole and minimal values near end-systole. As demonstrated by Barzilai and co-workers[23] (Figure 31-6) and confirmed by Fitzgerald et al.[34] and Sagar et al.,[35] this cardiac cycle variation is blunted by ischemia. This phenomenon is now generally accepted to be an accompaniment of the acute ischemic process. The physiologic basis of normal cardiac cycle-dependent backscatter variation is uncertain but may be related to cardiac contractile function,[24] the elasticity of myocardial tissue, or contraction-dependent variations in myocardial scatterer geometry.

Calculation of integrated backscatter requires the evaluation of the ultrasonic data in its radiofrequency form. To extend the findings just discussed, other investigators have used the image or video ultrasonic data to calculate overall echo amplitude or gray level in a region of interest (as an analog of integrated ultrasonic backscatter) and to assess the regional pattern of gray levels within a region, sometimes referred to as tissue "texture." McPherson et al[36] have used quantitative texture analysis to identify acute myocardial ischemia 2 hours after coronary occlusion in a closed-chest dog model. Chandrasekaran and co-workers[37] have confirmed these findings in another experimental preparation. The relative advantages of the video or image

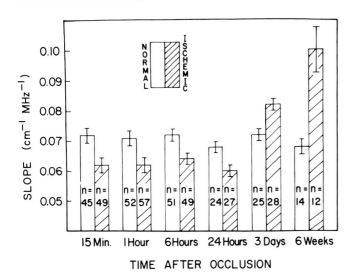

Figure 31-5 A quantitative index of ultrasonic attenuation (indicated as "slope") is plotted for zones of ischemic injury and normal zones from the same canine hearts studied in vitro at specified intervals after coronary artery occlusion. *Source:* Reprinted with permission from *Progress in Cardiovascular Diseases* (1985;28:85–110), Copyright © 1985, Grune & Stratton Inc.

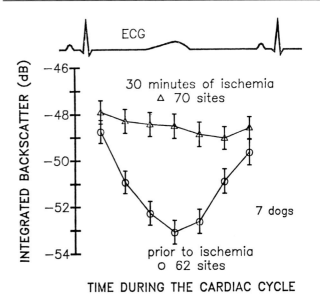

Figure 31-6 Blunting of the normal cardiac cycle-dependent cyclic variation of integrated backscatter and elevation of the time-averaged integrated backscatter after 30 minutes of myocardial ischemia in dogs. *Source:* Reprinted from *Proceedings of the Institute of Electrical and Electronics Engineers Ultrasonics Symposium* (83CH1947-1:782–793) by JG Miller, JE Pérez, JG Mottley et al with permission of IEEE Inc, © 1983.

analysis techniques in terms of more modest instrumentation and data storage requirements (compared with those of radiofrequency methods) may be offset by the fact that these techniques are dependent on echocardiographic instrument settings. Data based on radiofrequency ultrasonic analysis are somewhat less dependent on instrumentation, although methods of gain compensation for attenuation will affect any ultrasonic data acquired through the chest wall.

In summary of the work in acute ischemia, all measurements of ultrasonic attenuation, backscatter, the cyclic variation of backscatter, and regional echocardiographic gray level patterns or image texture are capable of identifying acutely ischemic tissue. The success of this broad variety of techniques in identifying ischemic tissue suggests the robustness of the ultrasonic tissue characterization method and gives the promise of clinical relevance in the near future.

Reperfusion. The ability to identify successfully reperfused tissue (that is, tissue that achieves functional recovery) and the ability to identify viable myocardium when deciding whether or not to attempt reperfusion are extremely topical in today's clinical environment. Recent echocardiographic tissue characterization data suggest that acoustic analysis techniques may aid in these determinations. For example, Rasmussen and co-workers[38] showed in an experimental model that alterations in ultrasonic backscatter that occurred after 20 minutes of coronary occlusion reversed after 30 minutes of myocardial reperfusion. In a study evaluating regional echo amplitude (gray level) data, Haendchen et al[39] evaluated the effect of 3 hours of coronary occlusion followed by 1 hour of reperfusion. Viable myocardium showed characteristic differences in mean echo amplitudes and skewness of gray level distributions compared with regions irreversibly infarcted. Glueck et al[15] have shown that the cardiac cycle-dependent variation in backscatter (which was blunted by ischemia) returns toward normal after variable periods of reperfusion. These findings were confirmed by Sagar et al[35] (Figure 31-7). Furthermore, Wickline et al[40] have demonstrated that the extent of alterations in the cyclic variation of backscatter resulting from ischemia and the magnitude of its subsequent recovery after reperfusion are not merely related to the level of segmental wall thickening. Thus preliminary observations suggest that reperfused and viable myocardium may be differentiated from necrotic tissue based on ultrasonic tissue characterization methods.

Evolving Infarction. Myocardial infarction that has been present for a few days (but not sufficiently long to result in significant deposition of collagen) also results in characteristic acoustic changes. For example, Mimbs et al,[31] as mentioned above, noted that attenuation

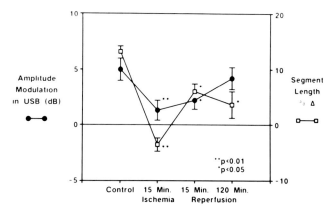

Figure 31-7 Alteration of cardiac cycle-dependent variation in ultrasonic backscatter (USB) and of regional contraction during ischemia and reperfusion. Both the amplitude modulation (cyclic variation) in ultrasonic backscatter and segmental shortening are decreased with ischemia and return toward normal with reperfusion. *Source:* Reprinted with permission from *Circulation* (1987;75:436–442), Copyright © 1987, American Heart Association.

through infarcted myocardium was increased compared with normal tissue, beginning approximately 3 days after coronary occlusion. This differed from their observations in earlier time periods after occlusion, when attenuation was decreased compared with normal. This finding suggests that the very early changes in acoustic properties may be related to changes in myocardial perfusion, alterations in the formed elements of blood within the tissue region, changes in contractility, and other acute accompaniments of ischemia, whereas alterations in acoustic properties 2 or more days after coronary occlusion may be due to early cellular infiltration and other accompaniments of the necrotic process. Parisi et al[29] demonstrated progressive increases in echo image amplitude as displayed by color-encoding techniques during evolving infarction. Skorton et al[41] demonstrated in a closed-chest dog model of 2-day-old myocardial infarction that the distribution of regional echo amplitudes differentiated infarcted from normal tissue: Regions of myocardial infarction showed a decrease in the kurtosis (peakedness) of the echo amplitude distribution compared with normal myocardium.

Chronic Myocardial Infarction with Scar Tissue. In a variety of studies performed weeks to years after myocardial infarction, the presence of scar tissue in the infarcted region has been demonstrable by ultrasonic tissue characterization techniques. In this field of endeavor one of the early observations by Rasmussen and co-workers,[3] described above, suggested that scar tissue was identifiable on M-mode echocardiograms as areas of increased echo reflection. Similar findings have been reported by Parisi and co-workers[29] and Shaw et al.[30]

Using in vitro measurements of attenuation through excised myocardial specimens, O'Donnell et al.[42] demonstrated marked increases in attenuation of the ultrasonic signal that correlated well with increased collagen content at 2 weeks, 4 weeks, or 6 weeks after an acute myocardial infarction in a dog model. Furthermore, integrated backscatter showed a substantial increase in remote experimental myocardial infarction.[21,43] Hoyt et al,[44] in studying specimens of fibrotic human myocardium, showed a linear correlation between collagen content (assessed by hydroxyproline concentration) and integrated ultrasonic backscatter (Figure 31-8).

In all of these studies it appears likely that the deposition of collagen as scar tissue was responsible for the acoustic alterations, namely an increase in attenuation and backscatter of the abnormal tissue.

Cardiomyopathies

Conventional echocardiographic analysis commonly permits differentiation of the major congestive, hypertrophic, and restrictive cardiomyopathies. Nonetheless, diagnostic ambiguity can occur, such as in the differentiation of hypertensive left ventricular hypertrophy from concentric hypertrophic cardiomyopathy. In addition to their gross morphologic features, the cardiomyopathies are also characterized by alterations in myocardial composition and structure as well as the deposition of collagen or other substances (such as amyloid in amyloid heart disease or iron in hemochromatosis). Since the acoustic properties of soft tissue are determined in part by tissue architecture and by the presence of abnormally reflecting substances, ultrasonic tissue characterization techniques have the potential for identifying and differentiating among a variety of cardiomyopathies. Although the amount of information that has been gathered in studies of cardiomyopathies is much less than that available from the study of ischemic heart disease, several intriguing early qualitative and quantitative observations suggest the clinical potential of ultrasonic tissue characterization in cardiomyopathies.

Qualitative Observations

The typical echocardiographic features of hypertrophic cardiomyopathy include left ventricular hypertrophy, often predominantly involving the ventricular septum; abnormal diastolic function; a hyperdynamic ventricle; and, in the case of obstructive hypertrophic cardiomyopathy, systolic anterior motion of the mitral valve apparatus. Investigators have also noted an unusual appearance of echoes returning from the ventricular septum in some patients with hypertrophic cardiomyopathy. Martin and co-workers,[5] for example, noted an unusual "ground-glass" texture in the ventricular septum of some patients with hypertrophic cardiomyopathy, presumably related to the myofibrillar disarray characteristic of this disorder. Bhandari and Nanda[45] systematically evaluated the clinical and histopathologic significance of abnormal image texture as displayed on conventional two-dimensional echocardiograms (Figure 31-9). The majority of cases of hypertrophic cardiomyopathy in their series displayed abnormal visual texture characterized by the presence of discrete, "highly refractile echoes." Unfortunately, this finding was somewhat nonspecific, also being noted in other cardiomyopathies.

Another example of a cardiomyopathy that may result in abnormal echocardiographic image texture is amyloid heart disease. Chiaramida et al[46] reported bright regional myocardial texture in a patient with amyloidosis. Siqueira-Filho et al[6] noted an unusual "granular sparkling texture" in 92% of their cases of amyloid heart disease. Bhandari and Nanda[45] noted that approximately half of their cases of amyloid heart disease exhibited "highly refractile echoes."

Other workers have used color encoding of echocardiographic data to identify cardiomyopathies. Davies et al[47] noted increased echo brightness in patients with endomyocardial involvement in hypereosinophilic disease.

Quantitative Observations

The few experimental and clinical investigations so far reported have suggested the presence of quantitative abnormalities of acoustic properties in car-

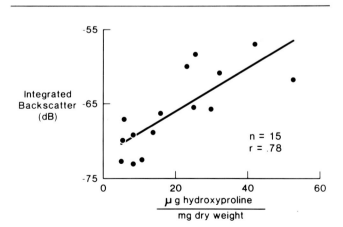

Figure 31-8 Relationship between the magnitude of integrated ultrasonic backscatter and collagen content of the myocardium as assessed by hydroxyproline assay. A linear relationship was found between backscatter and collagen content in these fibrotic specimens of human myocardium. *Source:* Reprinted with permission from *Circulation* (1985;71:740–744), Copyright © 1985, American Heart Association.

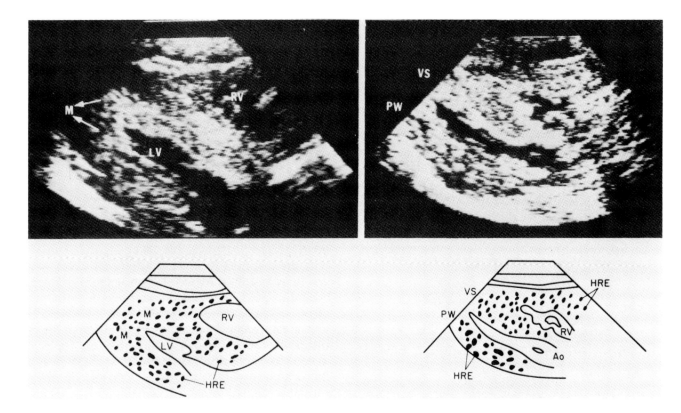

Figure 31-9 Examples of abnormal echocardiographic image texture in subcostal views of patients with cardiomyopathies. Left, An image of the heart in a patient with cardiac amyloidosis; right, an image from a patient with Pompe disease. In both cases the ventricular walls are thickened; bright, so-called "highly refractile echoes" (HRE) also appear within the myocardium. Ao indicates aorta; LV, left ventricle; M, myocardium; PW, posterior wall; RV, right ventricle; VS, ventricular septum. Source: Reprinted with permission from *American Journal of Cardiology* (1983;51:817–825), Copyright © 1983, Cahners Publishing Company Inc.

diomyopathies. In an experimental model of anthracycline-induced cardiomyopathy, Mimbs et al[48] demonstrated increases in integrated backscatter in cardiomyopathic and fibrotic tissue in rabbits administered doxorubicin. Pérez and associates[49] evaluated the acoustic properties of the myocardium in Syrian hamsters during the evolution of a spontaneous cardiomyopathy that is a well-characterized feature of the natural history of this animal model. The cardiomyopathic tissue demonstrated increased ultrasonic backscatter and less homogeneity of backscatter within regions of fibrosis (Figure 31-10). Thus quantitative abnormalities in ultrasonic backscatter appear to accompany the fibrosis and calcification characteristic of certain cardiomyopathies.

By using quantitative analysis of image or video data, Aylward et al[50] demonstrated that quantitative echocardiographic texture analysis was capable of discriminating among amyloid heart disease, hypertrophic cardiomyopathy, hypertensive left ventricular hypertrophy, and normal tissue. Angermann and co-workers[51] have used quantitative texture analysis to differentiate dilated cardiomyopathies from normal tissue.

Taken together, these few qualitative and quantitative observations suggest a place for ultrasonic tissue characterization in the diagnosis and characterization of a variety of cardiomyopathies.

Figure 31-10 Integrated ultrasonic backscatter gray scale map (left) and histologic section (right) of the same region on the left ventricle in a 6-month-old cardiomyopathic hamster. The ultrasonic map shows two zones of increased backscatter, corresponding to the two focal fibrocalcific lesions in the histologic section. von Kossa stain was used to delineate calcium. The area depicted includes the 2.5 × 2.5-mm region interrogated by ultrasonic backscatter analysis (original magnification of photomicrograph × 42, reduced by 19%). Source: Reprinted with permission from *Journal of the American College of Cardiology* (1984;4:88–95), Copyright © 1984, American College of Cardiology.

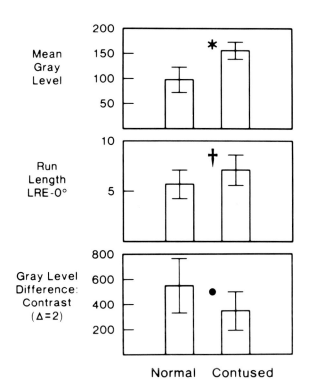

Figure 31-11 Quantitative texture analysis of normal versus contused myocardium. These data, taken from standard two-dimensional echocardiographic images, illustrate the significant differences in mean gray level and in two selected quantitative texture measures between normal and contused tissue. *P = .0001; †P = .0018; •P = .0015; LRE-0°, long-run emphasis calculated in horizontal direction (perpendicular to the ultrasonic beam); Δ = 2, gray-level difference contrast calculated for a pixel separation of two. *Source:* Reprinted with permission from *Circulation* (1983;68:217–223), Copyright © 1983, American Heart Association.

Traumatic Heart Disease

In an early study using quantitative texture analysis, Skorton and associates[19] demonstrated significant differences between the quantitative texture of contused myocardium and that of normal tissue (Figure 31-11). Quantitation of classic acoustic parameters (for example, attenuation and backscatter) has not to our knowledge been reported in acute myocardial contusion.

Investigations of Vascular Wall (Atherosclerosis)

High-frequency echocardiographic techniques have been used to investigate the acoustic properties of atherosclerosis. Picano and co-workers[52] have shown increases in ultrasonic backscatter in regions of fibrofatty infiltration and calcified, complex atherosclerotic plaque in human aorta in vitro. McPherson et al,[53] using high-frequency epicardial echocardiography in humans, have demonstrated significant differences in the average gray level of atherosclerotic wall compared with normal coronary arterial wall. Fibrous regions in human aortic atherosclerotic plaques examined in vitro by Barzilai et al[54] were also shown to be significantly different from calcified and fibrofatty regions by analysis of backscatter. These observations suggest a possible role for high-frequency echocardiographic techniques in the identification of peripheral vascular atherosclerosis and, when used in the intraoperative setting, coronary atherosclerosis.

POTENTIAL OF ULTRASONIC TISSUE CHARACTERIZATION IN CLINICAL CARDIOLOGY

Having reviewed the fundamental principles, methods, and some experimental and clinical results obtained thus far with ultrasonic tissue characterization, we now face the question of the potential place of ultrasonic tissue characterization in the clinical cardiologist's diagnostic armamentarium. What role might the techniques play in the management of patients with clinical cardiac disorders?

Diagnosis

The most obvious clinical application of ultrasonic tissue characterization techniques is in the diagnosis of a variety of cardiovascular disorders. Based on the results reviewed briefly above, it appears that the disorders most likely to be amenable to identification by ultrasonic tissue characterization techniques will be those associated with (1) alterations in myocardial water content or the content of formed elements of blood; (2) alterations in cardiac contractility; (3) myocardial necrosis and infiltration by inflammatory cells; (4) disruption of normal architecture (as in hypertrophic cardiomyopathy); (5) deposition of abnormal material (amyloid, iron); and (6) deposition of collagen. Because of the broad nature of functional and structural abnormalities that have been identified by tissue characterization techniques, it appears likely that the clinical potential of these techniques is similarly diverse and covers the gamut of ischemic, myopathic, and many other cardiac disorders. Although it must be acknowledged that many of these disorders currently are identifiable with standard echocardiographic analysis techniques, some diagnostic ambiguity exists and this ambiguity may be resolved partially or entirely by the successful application of ultrasonic tissue characterization techniques. For example, abnormal regional wall motion (defined by echocardiography or other imaging techniques) is characteristic of ischemic heart disease. However, the differentiation among acute, subacute, and chronic ischemic injury may not be possible based on wall motion analysis alone. Ultrasonic tissue characterization may permit distinction of collagen-infiltrated chronic infarction from edematous acute infarction from reperfused myocardium based on specific tissue characteristics. Similarly, various cardiomyopathies may be difficult to distinguish based on only standard echocardiographic criteria. For example, hypertrophic cardiomyopathy may simulate infiltrative cardiomyopathy,[55] hypertensive heart disease in the elderly may be difficult to distinguish from hypertrophic cardiomyopathy,[56] and amyloidosis may mimic hypertrophic cardiomyopathy.[57] Ultrasonic tissue characterization-based analysis of structural composition may help to resolve some of these ambiguities.

Staging of Cardiac Disease

Delineation of the extent of abnormal myocardium in ischemic heart disease or in the cardiomyopathies is a difficult task at present for the clinician. Particularly in ischemic syndromes, the delineation of the extent of myocardial infarction and the amount of tissue rendered irreversibly necrotic versus tissue still viable after reperfusion is an important unmet need in clinical cardiology. Because of the excellent inherent spatial resolution of ultrasonic imaging, especially with relatively high-frequency transducers, distinction of normal versus abnormal myocardium may be possible across the transmural extent of the wall as well as in different parts of the left ventricular circumference. Some investigations have already suggested that ultrasonic tissue characteristics may permit estimation of the size of myocardial ischemia or infarction[13,27]; further informative work in this area obviously would be beneficial to clinicians.

Assessment of Interventions

The final area in which tissue characterization techniques should be of great use is in the assessment of the response of the myocardium to interventions such as those used to limit the size of acute myocardial infarction. Complete assessment of the value of chemical thrombolysis or mechanical procedures (angioplasty) awaits reliable methods of defining the extent of infarcted versus viable tissue; ultrasonic tissue characterization may be one of the evolving imaging techniques to help meet this goal.

SUMMARY

Ultrasonic imaging of the heart is an important diagnostic method because of its excellent spatial and temporal resolution, safety, portability, and relatively inexpensive nature. Standard methods of echocardiographic analysis largely involve assessment of cardiac chamber and wall anatomy and function as well as the assessment of blood flow using Doppler techniques. Ultrasonic tissue characterization promises to add a unique dimension to the ultrasonic evaluation of the heart by supplying relevant information concerning tissue composition and function based on the assessment of myocardial physical characteristics. The successful completion of initial clinical trials currently being conducted may supply the clinician with a valuable new tool in the diagnosis and management of patients with cardiac disorders.

REFERENCES

1. Miller JG, Pérez JE, Sobel BE. Ultrasonic characterization of myocardium. *Prog Cardiovasc Dis*. 1985;28:85–110.

2. Aylward PE, McPherson DD, Kerber RE, et al. Ultrasound tissue characterization in ischemic heart disease. *Echocardiography*. 1986;3:385–398.

3. Rasmussen S, Corya BC, Feigenbaum H, et al. Detection of myocardial scar tissue by M-mode echocardiography. *Circulation*. 1978;57:230–237.

4. Logan-Sinclair R, Wong CM, Gibson DG. Clinical application of amplitude processing of echocardiographic images. *Br Heart J*. 1981;45:621–627.

5. Martin RP, Rakowski H, French J, et al. Idiopathic hypertrophic subaortic stenosis viewed by wide-angle, phased-array echocardiography. *Circulation*. 1979;59:1206–1217.

6. Siqueira-Filho AG, Cunha CLP, Tajik AJ, et al. M-mode and two-dimensional echocardiographic features in cardiac amyloidosis. *Circulation*. 1981;63:188–196.

7. Miller JG, Yuhas DE, Mimbs JW, et al. Ultrasonic tissue characterization: Correlation between biochemical and ultrasonic indices of myocardial injury. *Proc IEEE Ultrason Symp*. 1976;76CH1120-5SU:33–43.

8. Namery J, Lele PP. Ultrasonic detection of myocardial infarction in dog. *Proc IEEE Ultrason Symp*. 1972;72CHO708-8SU:491–494.

9. O'Donnell M, Bauwens D, Mimbs JW, et al. Broadband integrated backscatter: An approach to spatially localized tissue characterization in vivo. *Proc IEEE Ultrason Symp*. 1979;79CH1482-9:175–178.

10. Busse LJ, Miller JG, Yuhas DE, et al. Phase cancellation effects: A source of attenuation artifact eliminated by a CdS acoustoelectric receiver. In: White D, ed. *Ultrasound in Medicine*. vol. 3. New York, NY: Plenum Publishing Corp; 1977:1519–1535.

11. Busse LJ, Miller JG. Detection of spatially nonuniform ultrasonic radiation with phase sensitive (piezoelectric) and phase insensitive (acoustoelectric) receivers. *J Acoust Soc Am*. 1981;70:1377–1386.

12. Miller JG, Pérez JE, Mottley JG, et al. Myocardial tissue characterization: An approach based on quantitative backscatter and attenuation. *Proc IEEE Ultrason Symp*. 1983;83CH1947-1:782–793.

13. Mimbs JW, Yuhas DE, Miller JG, et al. Detection of myocardial infarction in vitro based on altered attenuation of ultrasound. *Circ Res*. 1977;41:192–198.

14. O'Donnell M, Miller JG. Quantitative broadband ultrasonic backscatter: An approach to nondestructive evaluation in acoustically inhomogeneous materials. *J Appl Physiol*. 1981;52:1056–1065.

15. Glueck RM, Mottley JG, Miller JG, et al. Effects of coronary artery occlusion and reperfusion on cardiac cycle-dependent variation of myocardial ultrasonic backscatter. *Circ Res.* 1985;56:683–689.

16. Thomas LJ III, Wickline SA, Pérez JE, et al. A real-time integrated backscatter measurement system for quantitative cardiac tissue characterization. *IEEE Trans Ultrason Ferroelect Freq Control.* 1986;UFFC-33:27–32.

17. Thomas LJ III, Barzilai B, Pérez JE, et al. Quantitative real-time, imaging of myocardium based on ultrasonic integrated backscatter. *IEEE Trans Ultrason Ferroelect Freq Control* 1989;36:466–470.

18. Vered Z, Barzilai B, Mohr GA, et al. Quantitative ultrasonic tissue characterization with real-time integrated backscatter imaging in normal human subjects and in patients with dilated cardiomyopathy. *Circulation.* 1987;76:1067–1073.

19. Skorton DJ, Collins SM, Nichols J, et al. Quantitative texture analysis in two-dimensional echocardiography: Application to the diagnosis of experimental myocardial contusion. *Circulation.* 1983;68:217–223.

20. O'Donnell M, Mimbs JW, Sobel BE, et al. Ultrasonic attenuation of myocardial tissue: Dependence on time after excision and on temperature. *J Acoust Soc Am.* 1977;62:1054–1057.

21. O'Donnell M, Mimbs JW, Miller JG: Relationship between collagen and ultrasonic backscatter in myocardial tissue. *J Acoust Soc Am.* 1981;69:580–588.

22. Madaras EI, Barzilai B, Pérez JE, et al. Changes in myocardial backscatter throughout the cardiac cycle. *Ultrason Imaging* 1983;5:229–239.

23. Barzilai B, Madaras EI, Sobel BE, et al. Effects of myocardial contraction on ultrasonic backscatter before and after ischemia. *Am J Physiol.* 1984;247:H478–H483.

24. Wickline SA, Thomas LJ III, Miller JG, et al. A relationship between ultrasonic integrated backscatter and myocardial contractile function. *J Clin Invest.* 1985;76:2151–2160.

25. Olshansky B, Collins SM, Skorton DJ, et al. Variation of left ventricular myocardial gray level in two-dimensional echocardiograms as a result of cardiac contraction. *Circulation.* 1984;70:972–977.

26. Collins SM, Skorton DJ, Prasad NV, et al. Quantitative echocardiographic image texture: Normal contraction-related variability. *IEEE Trans Med Imaging.* 1985;MI-4:185–192.

27. Fraker TD Jr, Nelson AD, Arthur JA, et al. Altered acoustic reflectance on two-dimensional echocardiography as an early predictor of myocardial infarct size. *Am J Cardiol.* 1984;53:1699–1702.

28. Werner JA, Speck SM, Greene HL, et al. Discrete intramural sonolucency: A new echocardiographic finding in acute myocardial infarction. *Am J Cardiol.* 1981;47:404. Abstract.

29. Parisi AF, Nieminen M, O'Boyle JE, et al. Enhanced detection of the evolution of tissue changes after acute myocardial infarction using color-encoded two-dimensional echocardiography. *Circulation.* 1982;66:764–770.

30. Shaw TRD, Logan-Sinclair RB, Surin C, et al. Relation between regional echo intensity and myocardial connective tissue in chronic left ventricular disease. *Br Heart J.* 1984;51:46–53.

31. Mimbs JW, O'Donnell M, Miller JG, et al. Changes in ultrasonic attenuation indicative of early myocardial ischemic injury. *Am J Physiol.* 1979;236:H340–H344.

32. Mimbs JW, Bauwens D, Cohen RD, et al. Effects of myocardial ischemia on quantitative ultrasonic backscatter and identification of responsible determinants. *Circ Res.* 1981;49:89–96.

33. Schnittger I, Vieli A, Heiserman JE, et al. Ultrasonic tissue characterization: Detection of acute myocardial ischemia in dogs. *Circulation.* 1985;72:193–199.

34. Fitzgerald PJ, McDaniel MM, Rolett EL, et al. Two-dimensional ultrasonic variation in myocardium throughout the cardiac cycle. *Ultrason Imaging.* 1986;8:241–251.

35. Sagar KB, Rhyne TL, Warltier DC, et al. Intramyocardial variability in integrated backscatter: Effects of coronary occlusion and reperfusion. *Circulation.* 1987;75:436–442.

36. McPherson DD, Aylward PE, Knosp BM, et al. Ultrasound characterization of acute myocardial ischemia by quantitative texture analysis. *Ultrason Imaging.* 1986;8:227–240.

37. Chandrasekaran K, Chu A, Greenleaf JF, et al. 2D echo quantitative texture analysis of acutely ischemic myocardium. *Circulation.* 1986;74(suppl 2):II-271. Abstract.

38. Rasmussen S, Lovelace DE, Knoebel SB, et al. Echocardiographic detection of ischemic and infarcted myocardium. *J Am Coll Cardiol.* 1984;3:733–743.

39. Haendchen RV, Ong K, Fishbein MC, et al. Early differentiation of infarcted and noninfarcted reperfused myocardium in dogs by quantitative analysis of regional myocardial echo amplitudes. *Circ Res.* 1985;57:718–728.

40. Wickline SA, Thomas LJ III, Miller JG, et al. Sensitive detection of the effects of reperfusion on myocardium by ultrasonic tissue characterization with integrated backscatter. *Circulation.* 1986;74:389–400.

41. Skorton DJ, Melton HE Jr, Pandian NG, et al. Detection of acute myocardial infarction in closed-chest dogs by analysis of regional two-dimensional echocardiographic gray-level distributions. *Circ Res.* 1983;52:36–44.

42. O'Donnell M, Mimbs JW, Miller JG. The relationship between collagen and ultrasonic attenuation in myocardial tissue. *J Acoust Soc Am.* 1979;65:512–517.

43. Mimbs JW, O'Donnell M, Bauwens D, et al. The dependence of ultrasonic attenuation and backscatter on collagen content in dog and rabbit hearts. *Circ Res.* 1980;47:49–58.

44. Hoyt RH, Collins SM, Skorton DJ, et al. Assessment of fibrosis in infarcted human hearts by analysis of ultrasonic backscatter. *Circulation.* 1985;71:740–744.

45. Bhandari AK, Nanda NC. Myocardial texture characterization by two-dimensional echocardiography. *Am J Cardiol.* 1983;51:817–825.

46. Chiaramida SA, Goldman MA, Zema MJ, et al. Real-time cross-sectional echocardiographic diagnosis of infiltrative cardiomyopathy due to amyloid. *J Clin Ultrasound.* 1980;8:58–62.

47. Davies J, Gibson DG, Foale R, et al. Echocardiographic features of eosinophilic endomyocardial disease. *Br Heart J.* 1982;48:434–440.

48. Mimbs JW, O'Donnell M, Miller JG, et al. Detection of cardiomyopathic changes induced by doxorubicin based on quantitative analysis of ultrasonic backscatter. *Am J Cardiol.* 1981;47:1056–1060.

49. Pérez JE, Barzilai B, Madaras EI, et al. Applicability of ultrasonic tissue characterization for longitudinal assessment and differentiation of calcification and fibrosis in cardiomyopathy. *J Am Coll Cardiol.* 1984;4:88–95.

50. Chandrasekaran K, Aylward PE, Fleagle SR, et al. Feasibility of identifying amyloid and hypertrophic cardiomyopathy with the use of computerized quantitative texture analysis of clinical echocardiographic data. *J Am Coll Cardiol.* 1989;13:832–840.

51. Angermann CE, Hart RJ, Stempfle U, et al. Frame by frame quantitation of myocardial backscatter: Analysis of standard 2D echo images and radio frequency signals. *Circulation.* 1986;74(suppl 2):II-270. Abstract.

52. Picano E, Landini L, Distante A, et al. Different degrees of atherosclerosis detected by backscattered ultrasound: An in vitro study on fixed human aortic walls. *J Clin Ultrasound.* 1983;11:375–379.

53. McPherson DD, Sirna SJ, Haugen JA, et al. Acoustic properties of normal and atherosclerotic human coronary arteries: In vitro and in vivo observations. *Circulation.* 1987;76:IV-43. Abstract.

54. Barzilai B, Saffitz JE, Miller JG, et al. Quantitative ultrasonic characterization of the nature of atherosclerotic plaques in human aorta. *Circ Res*. 1987;60:459–463.

55. Frustaci A, Loperfido F, Pennestri F. Hypertrophic cardiomyopathy simulating an infiltrative myocardial disease. *Br Heart J*. 1985;54:329–332.

56. Topol EJ, Traill TA, Fortuin NJ. Hypertensive hypertrophic cardiomyopathy of the elderly. *N Engl J Med*. 1985;312:277–283.

57. Sedlis SP, Saffitz JE, Schwob VS, et al. Cardiac amyloidosis simulating hypertrophic cardiomyopathy. *Am J Cardiol*. 1984;53:969–970.

Chapter 32

Evaluation of Myocardial Metabolism by Nuclear Magnetic Resonance Spectroscopy

Saul Schaefer, MD

The 1980s have seen the advent of nuclear magnetic resonance (NMR) as an investigative and clinical tool in the study of the cardiovascular system. Nuclear magnetic resonance spectroscopy has unique potential because it can determine the concentration and kinetics of metabolically important compounds, measure intracellular pH, and establish the fate of exogenously administered tracers. This is in contrast to conventional NMR imaging, which provides tomographic images of the body based on the density and relaxation properties of mobile protons. This chapter describes briefly the principles of NMR spectroscopy and its use in the characterization of cardiovascular physiology and disease.

THE NMR EXPERIMENT

The NMR experiment can be performed with many different nuclei, including hydrogen, phosphorus, carbon, fluorine, and sodium. The feasibility of observing any one of these nuclei depends on (1) its natural abundance in the body and its molar concentration, (2) the gyromagnetic ratio of the nucleus (a value that defines its resonant frequency in a given magnetic field), and (3) the sensitivity of the nucleus to the NMR experiment (a measure of the signal strength obtained from the nucleus relative to the hydrogen nucleus) (Table 32-1). Although theoretically any of these nuclei can be used to produce images, the hydrogen nucleus is ideal for this since it has both the highest concentration and the highest sensitivity. In addition to the hydrogen nucleus, other nuclei can be examined with NMR spectroscopy despite their less favorable NMR characteristics.

Table 32-1 Isotopes with Nuclear Spin Found in Tissue

Isotope	Molar Concentration (mol. L^{-1})	Gyromagnetic Ratio (MHz T^{-1})	Sensitivity (Relative to ^{1}H)
^{1}H	99.0	42.58	1.000
^{14}N	1.6	3.08	—
^{31}P	0.35	17.24	0.066
^{13}C	0.10	10.71	0.016
^{23}Na	0.078	11.26	0.093
^{39}K	0.045	1.99	0.0005
^{17}O	0.031	5.77	0.029
^{2}H	0.015	6.53	0.0096
^{19}F	0.0066	40.05	0.830

SPECTROSCOPY

Nuclear magnetic resonance spectroscopy uses the same experimental approach as imaging, but generally without the spatial encoding that defines small pixel volumes. Instead, spectroscopy examines relatively large volumes of tissue and obtains detailed information on the concentration and chemical environment of the chosen nucleus within the sample. The NMR experiment for spectroscopy is conceptually much simpler than the imaging experiment. Within a uniform applied magnetic field, the tissue is excited by a radiofrequency (RF) pulse near the resonant frequency of the nuclei of interest. After this excitation, the nuclei reorient themselves back into the axis of the magnetic field, giving off energy and hence emitting a signal that decreases exponentially with time, called a free induction decay (FID). After detection of this signal, a mathematical operation called a Fourier transformation is performed to define the intensity of different frequency components of the FID. If all the nuclei of interest are chemically identical, the FID will have only one frequency and the resulting Fourier transformation will have only one peak whose intensity is proportional to the strength of the signal. If, however, the nuclei are in several different chemical environments, the FID will have a number of frequency components and its Fourier transformation will have several peaks, each one proportional to the strength of the signal at that resonant frequency. This representation of intensity (on the y axis) versus frequency (on the x axis) is termed the "spectrum."

The presence of species with different resonant frequencies in the spectrum is due to a phenomenon termed "chemical shift." As noted earlier, the resonant frequency of a nucleus is proportional to the magnetic field applied to it. In addition to the main magnetic field, each nucleus is subject to a small magnetic field produced by the electrons surrounding it. This local electron-induced field changes the resonant frequency of that nucleus slightly, a change termed "chemical shift," since the electron distribution is altered by the chemical status of the nucleus. The chemical shift difference between one resonance and another is defined in hertz (frequency) and increases with increasing field strength. For uniformity, chemical shift is often expressed in parts per million, a number that is independent of the applied field strength. Chemical shift allows the identification of different peaks in a spectrum, since each peak is formed by the nucleus of interest in a slightly different chemical environment. For example, in the normal heart spectrum shown in Figure 32-1, the beta phosphate of adenosine triphosphate (ATP) has a resonant frequency that differs from the phosphate in phosphocreatine (PCr) by approximately 16 ppm. Similarly, the chemical shifts of the alpha and gamma peaks of ATP also differ from the PCr peak, but by lesser amounts.

The spectrum also provides information about concentrations of the various species, since the intensity of each of the defined peaks is proportional to the number of nuclei in that chemical environment. Although conceptually simple, this analysis may be complicated by the fact that different molecules may have exactly the same resonant frequency, as in the overlap of the resonances of adenosine diphosphate (ADP) with the alpha and gamma resonances of ATP. If a standard is available, absolute concentrations of each molecule can be

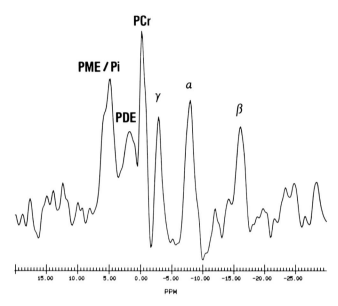

Figure 32-1 Spectrum from a normal human heart obtained at 1.5 T using the image-selected in vivo spectroscopic localization technique. The x axis denotes the chemical shift (resonant frequency) of each phosphate resonance, with the phosphocreatine resonance (PCr) set at zero ppm. The intensity (area) of each peak is proportional to the concentration of that phosphate molecule in the sample. PME/Pi indicates phosphomonoester/inorganic phosphate; PDE, phosphodiester; γ, gamma phosphate of adenosine triphosphate (ATP); α, alpha phosphate of ATP; β, beta phosphate of ATP.

calculated. However, in in vivo spectroscopy, relative concentrations are often used owing to the technical difficulties in obtaining accurate standardization.

Principles of chemical shift can also be used to measure intracellular pH noninvasively. In some molecules, the electron distribution, and therefore the resonant frequency, is sensitive to, and changes with, pH. Notably, the state of orthophosphate (HPO_4^- or H_2PO_4) is pH-dependent and its chemical shift with respect to PCr (which is not pH-dependent) is a function of intracellular pH. Thus the change in the pH of a tissue can be examined serially after interventions or during disease states.

Saturation Transfer

In addition to measuring concentrations of compounds and intracellular pH, NMR spectroscopy can determine steady state reaction rates. Using a method called saturation transfer, the magnetization state of one molecule is altered by an RF pulse at its resonant frequency until it can no longer be observed in a subsequent NMR scan. If this molecule is part of a reaction, some of its nuclei will have been transferred to another molecule whose NMR signal is now diminished. For example, the rate of the creatine kinase (CK) reaction

$$\text{PCr + ADP} \xleftrightarrow{\text{CK}} \text{ATP + Creatine}$$

can be measured by pulsing PCr at its resonant frequency. Since the phosphate in PCr is transferred to the gamma phosphate of ATP via this reaction, the intensity of the gamma ATP peak will decrease in proportion to the rate at which saturated phosphate moieties are being transferred from PCr to ATP. The reaction rate can then be determined if the kinetics are in the steady state.

Technical Aspects

Although NMR spectroscopy conceptually is a simpler experiment than imaging, important factors such as the low sensitivity of nuclei of interest and the difficulty in obtaining spectra from localized regions currently limit its use in humans. There are also stringent technical requirements for its success. The signal strength in an NMR experiment is a function of the number of visible NMR nuclei in the sample. This number is related to the sample volume, applied field strength, and temperature. Since the concentrations and NMR sensitivity of the nuclei of interest for spectroscopy are much lower than those of the water protons used in imaging, sample sizes in in vivo spectroscopy are in the order of cubic centimeters, compared with the imaging pixel volumes of cubic millimeters. Also, spectroscopy must be performed at higher field strengths (typically greater than 1.5 T) and requires greater magnetic field homogeneity than is necessary for imaging. Finally, because of the lower signal obtained, many one-pulse experiments must be averaged to achieve an adequate signal-to-noise ratio in the spectrum for most experiments.

Spatial localization of the tissue sample is crucial to the success of in vivo spectroscopy. In its simplest form, this localization is accomplished by transmitting and receiving the RF signal with a surface coil placed over the tissue of interest. The volume of tissue sampled by the surface coil can be defined by isocontour lines (Figure 32-2), which correspond to the diminishing power deposition and detection sensitivity of the coil. Thus, the tissue close to the coil experiences a large RF pulse and contributes a greater percentage of signal than does the tissue farther away from the coil. This inhomogeneity and lack of volume definition do not preclude experiments in which the coil can be placed on the tissue of interest (such as in open-chest animal experiments or peripheral muscle studies). However, in studies in which there are other structures between the coil and the tissue being examined, such as the chest wall and the human heart, other techniques are required to localize a volume of tissue for examination.

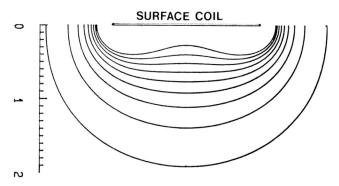

Figure 32-2 A computer-generated sensitivity isocontour plot from a surface coil. The grid on the left defines the distance from the surface coil as a function of the coil's radius. Each line corresponds to a 10% decrement in sensitivity and hence signal strength as the distance from the coil increases. Nuclei more than one radius away from the coil contribute little to the overall signal.

Localization Techniques

Localization techniques can rely on the variation of the observed pulse produced by a surface or volume coil (the B_1 field) or on changes in the field of the magnet (B_0 field) produced by applied gradients. The major techniques useful for human studies include rotating frame techniques, depth-resolved surface-coil spectroscopy (DRESS), image-selected in vivo spectroscopy (ISIS), and magnetic resonance spectroscopic imaging (MRSI).

Rotating Frame

Rotating frame, a B_1 technique, relies on the fact that a set of nuclei in a given isocontour of the coil experiences a degree of excitation dependent on its distance from the coil as well as the length of the excitation pulse.[1] Each set of nuclei at a different distance from the coil rotates at a rate that is a function of its position. If the excitation pulses are lengthened for each successive data accumulation, a series of spectra are obtained that correspond to varying depths in the sample. The resulting data set is a series of spectra corresponding to convex anatomic sections moving away from the coil (Figure 32-3). A major disadvantage of this technique is that the localization must be inferred primarily from the appearance of the spectra, since there is no reliable independent indicator of the location of the tissue providing the signal.

Depth-Resolved Surface-Coil Spectroscopy

The dress technique uses both B_1 and B_0 techniques for localization.[2] The nuclei in a plane parallel to the surface coil are selected by a B_0 gradient excitation sequence and detected by the surface coil. The depth of the region of interest is defined by the selective excitation, whereas its lateral dimensions are determined by

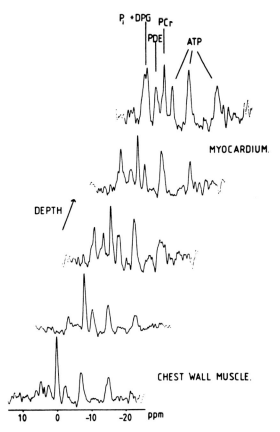

Figure 32-3 The series of spectra resulting from a human heart study using the rotating frame localization technique. These spectra are displayed as function of distance from the coil placed on the anterior chest wall. The anteriorly located spectra show a high phosphocreatine (PCr) peak, corresponding to chest wall skeletal muscle. The more posterior spectra have lower PCr intensities and are consistent with cardiac muscle spectra. ATP, adenosine triphosphate; P_i inorganic phosphate; PDE, phosphodiester; DPG, diphosphoglyceric acid. *Source:* Reprinted with permission from *Proceedings of the National Academy of Sciences* (1987;84:4283–4287), Copyright © 1987, MJ Blackledge et al.

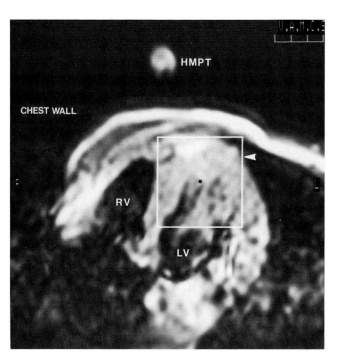

Figure 32-4 A transaxial proton nuclear magnetic resonance image of the human subject shown in Figure 32-1. The image-selected in vivo spectroscopic volume (arrow) was chosen from this image to perform the localized spectroscopic experiment. Using magnetic field gradients and special radiofrequency pulses, a phosphorus 31 spectrum is obtained from the chosen sample volume. RV, right ventricle; LV, left ventricle.

the sensitive volume of the surface coil. Reliance on the sensitivity of the surface coil for lateral definition of the sample volume, as well as the inherent inhomogeneity of the coil's RF field, creates uncertainty about exact spatial definition of the observed signal.

Image-Selected in Vivo Spectroscopy

The ISIS technique is a B_o technique that uses proton NMR imaging to allow selection of a volume of interest (VOI).[3] As is shown in Figure 32-4, a three-dimensional VOI is chosen from the image to encompass the tissue of interest. A series of gradients and 180° RF pulses are used to select a cuboidal volume that is then detected by a surface coil. The technique is attractive because a known VOI can be selected prior to the spectroscopic experiment. However, computer modeling has shown that there can be contribution from nuclei outside the VOI and that the signal intensity within the VOI is nonuniform. Finally, since ISIS sums eight separate FIDs, this technique is sensitive to motion.

Magnetic Resonance Spectroscopic Imaging

The MRSI technique (sometimes termed "chemical shift imaging") uses B_o gradients to phase-encode spins in three dimensions,[4] a process similar to that used to encode protons spatially in NMR imaging. The signal is obtained from a large region that can then be broken down into smaller VOIs after the experiment, each VOI having its own spectrum. Several VOIs can be summed to provide a spectrum from a region chosen using a corresponding proton NMR image. Unfortunately, the size of the smallest VOI obtainable with reasonable study times is limited by the sensitivity of the nucleus under study and, for phosphorus, currently is in the range of 20 to 30 cm^3. Also, the spatial definition of each VOI is inexact, having significant contributions from surrounding VOIs.

With the above background in techniques, we will now consider the use and results of spectroscopy of different nuclei in both animal and human studies.

SPECTROSCOPY OF BIOLOGICALLY IMPORTANT NUCLEI

Phosphorus

Phosphorus 31 is the most frequently studied nucleus because of its importance in biologic processes and the relative ease of its detection. As seen in Figure 32-5, the quantitation of ATP, PCr, and inorganic phosphate (P_i) can provide insight into basic energy-producing reactions of the cell. Thus many experiments have been performed to examine the changes in the concentrations of these compounds or their reaction rates with interventions, or both.

Workload

Several issues have interested scientists in characterizing the normal heart. First, the question of whether high-energy phosphates (HEP) change during the cardiac cycle along with phasic changes in cardiac work and perfusion was addressed by Fossel et al,[5] who used a glucose-perfused isolated heart preparation. They demonstrated cyclic changes in ATP and PCr, with the lowest values of each occurring during cardiac systole. These results were supported by a subsequent in vivo rat study[6] but were contradicted by other in vivo studies in rats[7] and dogs.[8] The issue of substrate availability, a potential explanation for these differences, was addressed by Zweier and Jacobus,[9] who used a perfused heart preparation. Hearts in which pyruvate was used as a substrate had higher developed pressures and increased PCr concentrations than did those in which glucose perfusion was used, indicating a more favorable metabolic status.

A second issue is the metabolic response of the heart to large changes in workload. This is an important issue since its resolution would help to determine the mechanisms by which cardiac metabolic rates are linked to work requirements. Although it is known that the level of ATP is tightly regulated over a wide range of workloads and oxygen consumptions, the changes in PCr concentration or creatine kinase (CK) reaction rates are incompletely characterized. Bittl and Ingwall[10] and Bittl et al,[11] in both perfused and open-chest rat studies, demonstrated a fall in PCr content and an increase in CK flux as the rate-pressure product was increased threefold by changing pressure and/or inotropic state. In contrast, Balaban et al[12] used a catheter coil in an in vivo dog preparation and varied the rate-pressure product fivefold by atrial pacing. Under these conditions, the PCr-ATP ratio did not change. Thus with certain stimulations the heart will increase CK flux and utilize PCr to maintain the concentration of ATP. Differences in results may be explained by differences in the protocols, especially the different methods used to alter myocardial oxygen demand.

Ischemia and Infarction

Phosphorus 31 NMR is ideally suited for studying the metabolic changes associated with myocardial ischemia and infarction. Studies of the ischemic heart have shown that oxygen deprivation causes a fairly rapid fall in PCr coupled with a rise in P_i. Although the concentration of ATP initially is maintained, it too falls with continued ischemia. Several important issues relative to ischemia and high-energy phosphates (HEP) have been addressed. These include (1) the temporal, and possibly causal, relationship between changes in HEP during ischemia and changes in myocardial function; (2) the relationship between metabolic recovery and recovery of function after reperfusion; and (3) the effect of pharmacologic interventions on HEP during and after ischemia. Finally, efforts to develop and refine ^{31}P NMR techniques for the identification of ischemia and infarction in humans have accelerated with the improvement in localization techniques.

Initial studies of global ischemia or hypoxia were performed in perfused heart preparations. These studies were instrumental in characterizing the alterations in HEP and intracellular pH during ischemia and hypoxia. Recently, ^{31}P NMR studies have progressed to more representative in vivo experiments. In an in vivo dog study, Guth et al[13] were able to correlate the reductions in transmural myocardial blood flow after acute coronary occlusion with changes in both contractile

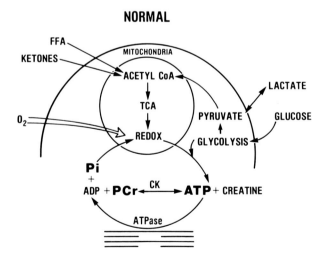

Figure 32-5 A schematic of the major energy-producing reactions of the myocardial cell. The phosphorus 31 NMR observable compounds, P_i, PCr, and ATP have a central role in these processes. The status of free fatty acids (FFA), lactate, and components of the tricarboxylic acid cycle (TCA) can be observed by NMR spectroscopy of hydrogen and carbon nuclei. ADP, adenosine diphosphate; CK, creatine kinase; redox, oxidation-reduction reactions; acetyl CoA, acetylcoenzyme A.

function and levels of P_i, PCr, and pH (Figure 32-6). Although the concentration of ATP during ischemia was not correlated with contractile dysfunction, its continued depression following reperfusion was highly correlated with the degree of functional impairment at that time. This was contrasted with the rapid return of other phosphates to normal levels with resumption of blood flow. The effect of ATP depletion in the persistence of postischemic dysfunction has been noted by others,[14] although the roles of glycolytic products (lactate, hydrogen ions, or reduced nicotinamide-adenine dinucleotide) may also be significant in the prolonged depression of left ventricular function.[15] These findings have led to pharmacologic efforts to blunt the metabolic changes due to ischemia, including the use of adenosine nucleotide precursors,[16] free radical scavengers,[17] calcium-channel blockers,[18] and β-adrenergic blockers.[19]

By using direct biochemical measurements of ATP and its breakdown products, Lange et al[20] showed that verapamil maintained endocardial ATP concentrations and total adenylate energy charge during 15 minutes of ischemia and 90 minutes of reperfusion. Although control animals in this study demonstrated the expected fall in ATP concentration during ischemia, they also had higher rate-pressure products than did the treated animals. After this experiment, Lavanchy et al[18] used a perfused rat heart model to assess the effects of diltiazem on HEP. When diltiazem was administered in the perfusate as the hearts were subjected to 24 minutes of ischemia followed by 30 minutes of reperfusion, there was a smaller increase in P_i during both periods than in control hearts. Also, PCr returned to control levels during reperfusion in treated hearts, whereas control hearts showed continued depression of PCr. Nuclear magnetic resonance measurements did not detect any differences in ATP concentration between groups, whereas biochemical measurements showed higher levels in treated hearts at the end of the reperfusion period. Treated hearts had lower rate-pressure products than did controls, and the resulting lower energy expenditure may explain some of the findings.

To determine the effects of β-adrenergic blockade, Malloy et al[19] used an in vivo rabbit preparation of regional ischemia treated with propranolol. After 30 minutes of ischemia, the intracellular concentrations of ATP were higher, and those of PCr were lower, than in untreated animals. While pH was not different between groups, the rate-pressure product was again reduced throughout the experiment.

In summary, ^{31}P NMR spectroscopy has been used to measure noninvasively both the metabolic consequences of ischemia and the role of pharmacologic therapy in altering those responses. However, many of these studies have been conducted in models that may not represent human physiology accurately and with

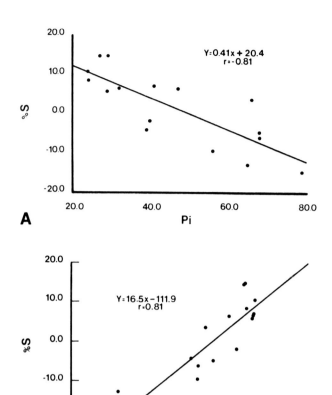

Figure 32-6 The relationship between regional systolic shortening (%S) and (A) relative inorganic phosphate concentration (Pi) and (B) myocardial pH before and during coronary artery occlusion in the dog. Measurements were obtained using a 2.5-cm surface coil sutured to the epicardium. *Source:* Reprinted with permission from *Journal of the American College of Cardiology* (1987;10:673), Copyright © 1987, American College of Cardiology.

drug regimens that have confounding hemodynamic effects.

Cardiomyopathies

Phosphorus 31 NMR spectroscopy has been used to study cardiomyopathic processes that may involve changes in the availability or utilization of HEP. These processes include congenital cardiomyopathies and those induced by exogenous substances such as doxorubicin hydrochloride and alcohol. Markiewicz et al[21] used a perfused heart preparation from cardiomyopathic Syrian hamsters to measure HEP and their changes with calcium-channel blockade. They found lower PCr and higher P_i concentrations in this model, abnormalities that were corrected partially with verapamil. Further studies in this model showed that isoproterenol improved the performance and metabolic profile of these cardiomyopathic hearts.[22]

Similar abnormalities in HEP were found by Keller et al[23] in a perfused rabbit heart model of acute and chronic doxorubicin toxicity. Although absolute concentrations were not measured, acute therapy lowered the PCr-ATP ratio, a change consistent with a fall in PCr resulting from impairment of oxidative phosphorylation. These hearts also exhibited lipid deposits on histologic examination. In contrast, chronic doxorubicin therapy did not alter the PCr-ATP ratio, and the corresponding histology demonstrated only loss of myocytes.

The metabolic profile of alcohol-induced cardiomyopathy is different from that seen in the above cardiomyopathies. In a perfused hamster heart preparation,[24] the PCr was maintained at normal levels after 6 months of 50% ethanol ingestion. This was accompanied by decreases in ATP and increases in P_i concentrations, along with functional abnormalities. Although the cause for this different response is unknown, it may be due to a unique effect of ethanol in altering oxidative phosphorylation while preserving PCr. This may occur via inhibition of Na^+-K^+-stimulated ATPase activity and depressed adenine nucleotide translocation.

Transplantation

Phosphorus 31 NMR has been proposed as a tool to provide early and sensitive indications of rejection after cardiac transplantation. In a promising study of transplanted allograft rat hearts, Canby et al[25] performed serial surface coil ^{31}P NMR for 2 to 7 days after transplantation. Compared with control isografts, the rejecting hearts showed progressive temporal decreases in the ratios of PCr to P_i and PCr to ATP, primarily due to a fall in PCr. Intracellular pH, in contrast, initially became alkylotic, followed by acidosis on day 7. Histologic examination of allografts from days 5, 6, and 7 showed active rejection, raising the question of the sensitivity and specificity of NMR in this setting. If further studies demonstrate that NMR (either proton or phosphorus) can detect and monitor episodes of rejection early in their course, this technique may become clinically useful in this setting.

Human Studies

The difficulties of spatial localization have limited the number of cardiac studies using ^{31}P NMR in humans. However, several investigators have shown the feasibility and potential of this technique in selected circumstances. The earliest demonstration of this was in an 8-month-old female child with a congenital cardiomyopathy and massive cardiomegaly.[26] When only a surface coil was used for localization, her PCr-Pi ratio was found to be 1.0, a value that was half that of a normal control subject. Furthermore, glucose or carbohydrate loading raised this ratio to 1.8, suggesting that ^{31}P NMR could guide therapy in certain cardiac disorders. This examination was possible because of the child's cardiomegaly and poorly developed skeletal musculature, thereby heavily weighting the signal with cardiac rather than skeletal muscle.

In a normal man, Bottomley[27] first demonstrated the capability of measuring cardiac PCr and ATP using the DRESS method described earlier. His calculated value for the PCr-ATP ratio was 1.3, which was in agreement with earlier values in animal hearts. Studies of canine myocardial infarction[28] were extended to clinical examinations in four patients 5 to 9 days after myocardial infarction.[29] Again by using a surface coil with DRESS, spectra were obtained that were localized spatially to 0.5-cm thick sections. Some of these showed low PCr-Pi ratios (approximately 1.1) that may have corresponded to infarcted myocardium. Phosphocreatine to adenosine triphosphate ratios ranged from 1.2 to 2.2, values that were not dissimilar from control values, and suggested contamination from skeletal muscle in some instances. Although promising, this study also pointed up the difficulties in performing truly localized cardiac spectroscopy in man.

Blackledge and colleagues[30] used the rotating frame method to perform ^{31}P NMR in six subjects. In normal subjects, the PCr-ATP ratio was 1.55 ± 0.20, whereas a patient with right ventricular hypertrophy and congestive heart failure had an abnormal ratio of 0.9.[31] In contrast, the PCr-ATP ratio was unchanged in patients with hypertrophic cardiomyopathy involving only the septum and left ventricle. This finding may indicate insensitivity of the method to possible metabolic abnormalities in this cardiomyopathy or, more likely, the inability of the rotating frame technique to localize to myocardium away from the right ventricle.

In view of the limitations of these localization methods, implementation of ISIS and MRSI for cardiac examinations was pursued. In 10 normal subjects, Schaefer et al[32] demonstrated the feasibility of performing reproducible, image-guided, localized ^{31}P NMR studies using a modified ISIS technique. In this study, the volume of interest was selected using proton NMR images to include the septum, apex, and left ventricular free wall. Spectra were collected in 35 minutes over an average sample volume of 78 cm³. In agreement with the ratios in earlier studies, the PCr-ATP ratio was calculated at 1.33 ± 0.19. While there was no evidence of significant skeletal muscle contamination, these sample volumes were rather large and constrained to the orginally chosen spatial coordinates. Figure 32-1 is an example of a human heart spectrum obtained by using this technique.

Magnetic resonance spectroscopic imaging, like ISIS, allows the choice of a volume of interest from proton NMR images. In addition, MRSI provides the flexibility of moving the sample volume *after* the study is completed. Figure 32-7 shows the original spectrum obtained from a normal subject using this technique as well as the spectrum from a 35-cm³ voxel displaced 5 mm anteriorly. While the original spectrum has a PCr-ATP ratio of 1.56 (consistent with cardiac muscle), the more anterior voxel has a higher ratio and evidence of skeletal muscle contamination. To interrogate a region, the voxel can also be moved in all three orthogonal axes and spectra processed from different regions of interest. Although this method is likely to become the preferred technique for human spectroscopy, it too is limited by the physics and biology of ^{31}P NMR. Current magnet field strengths limit minimal voxel sizes to at least 20 cm³ to obtain analyzable spectra in reasonable time periods (that is, 30 minutes). If greater signal sensitivity can be achieved using higher field strengths, smaller voxels could be sampled and then summed to determine the metabolic profile of an organ or region. While this has been demonstrated for large, stationary organs such as liver and brain, metabolic profiles of the heart have not yet been achieved.

Hydrogen

The proton nucleus of hydrogen is ideal for NMR experiments because of its high natural abundance and sensitivity advantages that have made it the choice for imaging. However, the vast majority of protons in tissues are in water, with concentrations in the order of 70 to 80 mol/L. Compounds of interest that contain protons, such as glucose, lactate, amino acids, and free fatty acids, are present in much smaller (millimol per liter) concentrations. In addition to the significantly lower amplitude of the spectral peaks generated by these compounds, there are many such individual peaks with overlapping chemical shifts. Identification of these compounds usually requires higher field strengths and special water suppression and editing techniques to eliminate the large water signal and spectral overlap. Despite these difficulties, preliminary work has been performed to address the changes in cardiac proton spectra occurring with ischemia or infarction. These studies have been based on prior observations of increased lipids or lactate with ischemia, changes that previously were measurable only with direct biochemical analysis of tissue or blood measurements.

The mobile lipid resonances in postischemic myocardium were examined in a canine model of 15 minutes of coronary occlusion followed by 3 hours of reperfusion.[33] High-resolution spectra were obtained from post-mortem tissue samples using an 8.5-T magnet and single-frequency irradiation for water suppression.

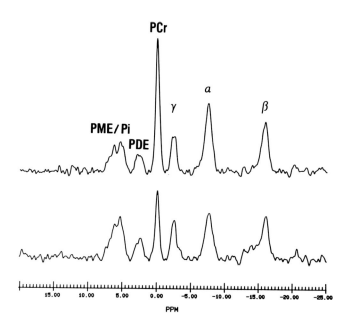

Figure 32-7 Human spectra obtained in 30 minutes using magnetic resonance spectroscopic imaging. The lower spectrum is from the voxel originally chosen from the proton NMR image (as in Figure 32-4) and shows the expected relative myocardial concentrations of PCr and ATP. The upper spectrum is from the same size voxel displaced 5 mm anteriorly and generated after the examination from the original data set. The high PCr peak indicates significant skeletal muscle contamination in this sample. For definitions of peaks see legend to Figure 32-1.

When compared with the creatine resonance, the mobile lipid peaks increased approximately 100% from control spectra, a change consistent with the accumulation of nonoxidized fatty acids. Even greater increases in mobile lipids were seen when a similar spectroscopic technique was used in a canine model of 24-hour coronary artery occlusion[34] (Figure 32-8). Although tissue samples from moderately ischemic zones showed these increases, severely ischemic tissue (1% of control blood flow) did not, despite a marked decrease in the creatine peak. These results suggested that proton spectroscopy may be capable of identifying "stunned" myocardium after brief periods of ischemia or of differentiating ischemic from normal or necrotic myocardium.

Extension of these in vitro, high-field results to another system was attempted by Miller et al,[35] who performed MRSI of ex vivo perfused dog hearts after 3 hours of coronary occlusion and 1 hour of reperfusion. The study was conducted at 1.4 T, and chemical shift images were obtained in 2 hours with an in-plane resolution of 1.5 mm. In two of the three hearts, prominent lipid peaks were seen in the border zone pixels, a finding confirmed in one animal with higher field in vitro spectroscopy. Although these preliminary studies are promising, they also serve to highlight the difficulties in

Sodium and Potassium

The electrical and osmotic gradients across cardiac cells depend, in part, on the intracellular and extracellular concentrations of sodium ($^{23}Na^+$) and potassium ($^{39}K^+$). Measurement of these ions, and the changes in their concentrations under various physiologic conditions, could lend insight into basic mechanisms governing resting and action potentials. In the NMR experiment, sodium and potassium have spins of 3/2 (termed quadrupole moments), which result in greater line broadening of peaks than with phosphorus or protons. The study of these ions has been limited by several factors in addition to the problems of line broadening, molar concentration, and sensitivity. The examination of intracellular potassium (and possibly intracellular sodium) is complicated by the fact that its major fraction is relatively immobile and its NMR is "invisible,"[40] thus leading to falsely low values of molar concentration. The examination of sodium is also difficult because of its high concentration in the extracellular compartment (150 mmol/L) compared with intracellular sodium (10 mmol/L). This large signal obscures the signal from intracellular sodium, which is the nucleus of greater interest. This problem has been addressed by using "shift reagents," aqueous compounds that alter the resonant frequency of nuclei that are in close proximity to the reagent.[41]

By using a dysprosium-based shift reagent in a perfused rat heart preparation, Elgavish et al[42] examined the response of intracellular sodium concentration to global hypoxia. They noted that a significant increase in intracellular sodium concentration after 30 minutes of hypoxia predicted the lack of functional recovery with reoxygenation 10 minutes later. Also, the rise in intracellular sodium paralleled the duration of ischemia in hearts that did not recover with reperfusion. Although the mechanisms for this effect are unknown, it is likely that the rise in intracellular sodium was a marker of injury sufficient to prevent the cell from maintaining normal ionic gradients.

In a different type of experiment, Cannon et al[43] successfully obtained sodium images of excised dog hearts after 1 hour of ischemia and reflow. These images were produced in 3 to 4 hours with the use of a 2.7-T magnet and demonstrated increased total sodium signal in the regions of myocardial damage with concurrent increases in measured concentrations of tissue sodium. While intracellular, rather than total, sodium concentration may be a more sensitive indicator of ischemia, there are even greater problems in the detection of sodium since its concentration is relatively low (10 mmol/L) compared with that of extracellular sodium (150 mmol/L). Thus the feasibility of clinical, or even experimental, sodium spectroscopy or imaging in

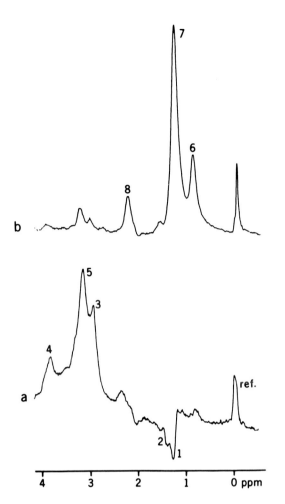

Figure 32-8 Hydrogen 1 NMR spectra from excised canine myocardial tissue after 24 hours of coronary artery occlusion: a is from control tissue; b is from tissue with a 59% reduction in myocardial blood flow. The mobile lipid peaks (6, 7, and 8) increase with ischemia, while the creatine (3) and choline/carnitine (5) peaks are diminished. *Source:* Reprinted with permission from *Magnetic Resonance Medicine* (1987;5:26), Copyright © 1987, Academic Press.

performing in vivo proton spectroscopy of the heart at current field strengths. Despite substantial (two- to five-fold) increases in mobile lipids under ischemic conditions, their concentrations remain in the millimolar range, thus limiting the successful implementation of in vivo cardiac proton spectroscopy.

Measurement of lactate with proton spectroscopy is subject to similar difficulties and has been limited to studies of perfused hearts. In these preparations, lactate concentration has increased dramatically during ischemia[36] and returned to normal after reperfusion.[37] Advances in water-suppression techniques[38] and localization schemes[39] may enable lactate detection in the heart in the future; at present, the use of NMR in this area must await further technologic developments.

in vivo cardiac studies will require significant technical advances.

Carbon

The most common naturally abundant form of carbon, ^{12}C, has no net magnetic moment and is invisible to NMR. The NMR-visible isotope ^{13}C, although naturally present in low concentrations, can be administered exogenously and used as a tracer of specific carbon-containing metabolites. The labeling of endogenous metabolites allows the noninvasive monitoring of metabolism under normal or stressed conditions. For example, the rate of myocardial glycogen synthesis was observed by Neurohr et al[44] with the use of exogenous ^{13}C (as D-[1-^{13}C]glucose). During anoxia, they observed the degradation of labeled glycogen and noted that glycogenolysis lagged behind the fall in PCr seen in this model.

Experiments with ^{13}C have also shown that metabolite levels in the perfused heart depend on the substrate used (pyruvate versus lactate) and that lactate is preferred over glucose as a substrate even in the presence of insulin.[45] Malloy et al[46] extended mathematical models of the tricarboxylic acid cycle to measure the fractional enrichment of acetylcoenzyme A entering the cycle and the relative activities of the oxidative versus anaplerotic pathways.

From these and other data, it is clear that ^{13}C NMR studies will remain a unique and valuable tool for investigation of carbon metabolism in small animal models and perfused hearts. Their clinical potential, however, is limited by the low NMR sensitivity of ^{13}C and its requirement for exogenous administration.

SUMMARY

Nuclear magnetic resonance spectroscopy has established itself clearly as an important investigative tool for the study of cardiac metabolism and energetics. In animal models, it can provide insight into basic metabolic processes in both health and disease. Its nondestructive nature and capacity for serial measurements in the same system have given scientists the ability to monitor changes in biologic systems over time, while its ability to measure different nuclei and hence different metabolic processes gives it immense flexibility.

In relative contrast to animal studies, human cardiac NMR spectroscopy is in its infancy. To date, cardiac ^{31}P spectroscopy has been implemented successfully in only a few centers. These studies have examined primarily normal subjects to establish the feasibility of such examinations with a variety of techniques. The few reports of metabolic abnormalities detected by ^{31}P NMR have been limited by problems of accurate spatial localization and large sample volumes.

The ultimate utility of human spectroscopy with phosphorus or any other nucleus will depend on the success of efforts to increase the sensitivity and spatial selectivity of the NMR experiment. These efforts include utilizing systems of higher field strength and homogeneity, as well as improving pulse sequences and radiofrequency coil designs. In addition, demonstration of some degree of sensitivity and specificity of metabolic abnormalities in disease states will be necessary before clinical use is indicated. If these requirements are met, human cardiac spectroscopy may provide useful information about cardiac metabolic processes and, combined with proton NMR imaging, may enable complete noninvasive evaluation with one technology. The rapid advances over the past few years attest to the feasibility of this goal.

REFERENCES

1. Garwood M, Schleich T, Matson GB, et al. Spatial localization of tissue metabolites by phosphorus-31 NMR rotating frame zeugmatography. *J Magn Reson*. 1984;60:268–279.

2. Bottomley PA, Foster TH, Darrow RD. Depth resolved surface coil spectroscopy (DRESS) for in vivo ^1H, ^{31}P, and ^{13}C NMR. *J Magn Reson*. 1984;59:338–342.

3. Oridge RJ, Connelly A, Lohman JAB. Image-selected in vivo spectroscopy (ISIS): A new technique for spatially selective NMR spectroscopy. *J Magn Reson*. 1985;66:283–294.

4. Brown TR, Kincaid BM, Ugurbil K. NMR chemical shift imaging in three dimensions. *Proc Natl Acad Sci U S A*. 1982;79:3523–3526.

5. Fossel ET, Morgan HE, Ingwall JS. Measurement of changes in high-energy phosphates in the cardiac cycle by using gated P-31 nuclear magnetic resonance. *Proc Natl Acad Sci U S A*. 1980;77:3654–3658.

6. Toyo-Oka T, Nagayama K, Umeda M, et al. Rhythmic change of myocardial phosphate metabolite content in cardiac cycle observed by depth-selected and EKG-gated in vivo ^{31}P-NMR spectroscopy in a whole animal. *Biochem Biophys Res Commun*. 1986;135:808–815.

7. Koretsky AP, Wang S, Murphy-Boesch J, et al. P-31 NMR spectroscopy of rat organs in situ using chronically implanted RF coils. *Proc Natl Acad Sci U S A*. 1983;80:7491–7595.

8. Kantor HL, Briggs RW, Metz KR, et al. Gated in vivo examination of cardiac metabolites with ^{31}P nuclear magnetic resonance. *Am J Physiol*. 1986;251:H171–175.

9. Zweier JL, Jacobus WE. Substrate-induced alterations of high energy phosphate metabolism and contractile function in the perfused heart. *J Biol Chem*. 1987;262:8015–8021.

10. Bittl JA, Ingwall JS. Reaction rates of creatine kinase and ATP synthesis in the isolated rat heart. *J Biol Chem*. 1985;260:3512–3517.

11. Bittl JA, Balschi JA, Ingwall JS. Effects of norepinephrine infusion on myocardial high-energy phosphate content and turnover in the living rat. *J Clin Invest*. 1987;79:1852–1859.

12. Balaban RS, Kantor HL, Katz LA, et al. Relation between work and phosphate metabolite in the in vivo paced mammalian heart. *Science*. 1986;232:1121–1123.

13. Guth BG, Martin JF, Heusch G, et al. Regional myocardial blood flow, function and metabolism using phosphorus-31 nuclear magnetic resonance spectroscopy during ischemia and reperfusion in dogs. *J Am Coll Cardiol*. 1987;10:673–681.

14. Flaherty JT, Weisfeldt M, Bulkley BH, et al. Mechanisms of ischemic myocardial cell damage assessed by phosphorus-31 nuclear magnetic resonance. *Circulation*. 1982;56:561–570.

15. Neely JR, Grotyohann LW. Role of glycolytic products in damage to ischemic myocardium. *Circ Res*. 1984;55:816–824.

16. Swain JL, Hines JJ, Sabina RL, et al. Accelerated repletion of ATP and GTP pools in postischemic canine myocardium using a precursor of purine de novo synthesis. *Circ Res*. 1982;51:102–105.

17. Przyklenk K, Kloner RA. Superoxide dismutase plus catalase improve contractile function in the canine model of "stunned myocardium." *Circ Res*. 1986;58:148–156.

18. Lavanchy N, Martin J, Rossi A. Effects of diltiazem on the energy metabolism of the isolated rat heart submitted to ischemia: A ^{31}P NMR study. *J Mol Cell Cardiol*. 1986;18:931–941.

19. Malloy CR, Matthews PM, Smith MB, et al. Influence of propranolol on acidosis and high energy phosphates in ischemic myocardium of the rabbit. *Cardiovasc Res*. 1986;20:710–720.

20. Lange R, Ingwall J, Hale SL, et al. Preservation of high-energy phosphates by verapamil in reperfused myocardium. *Circulation*. 1984;70:734–741.

21. Markiewicz W, Wu SS, Parmley WW, et al. Evaluation of the herditary Syrian hamster cardiomyopathy by ^{31}P nuclear magnetic resonance spectroscopy: Improvement after acute verapamil therapy. *Circ Res*. 1986;59:597–604.

22. Camacho SA, Wikman-Coffelt J, Wu ST, et al. Improvement in myocardial performance without a decrease in high-energy phosphate metabolites after isoproterenol in Syrian cardiomyopathic hamsters. *Circulation*. 1988;77:712–719.

23. Keller AM, Jackson JA, Peshock RM, et al. Nuclear magnetic resonance study of high-energy phosphate stores in models of Adriamycin cardiotoxicity. *Magn Reson Med*. 1986;3:834–843.

24. Wu S, White R, Wikman-Coffelt J, et al. The preventive effect of verapamil on ethanol-induced cardiac depression: Phosphorus-31 nuclear magnetic reonance and high-pressure liquid chromatographic studies of hamsters. *Circulation*. 1987;75:1058–1064.

25. Canby RC, Evanochko WT, Barrett LV, et al. Monitoring the bioenergetics of cardiac allograft rejection using in vivo P-31 nuclear magnetic resonance spectroscopy. *J Am Coll Cardiol*. 1987;9:1067–1074.

26. Whitman GJR, Chance B, Bode H, et al. Diagnosis and therapeutic evaluation of a pediatric case of cardiomyopathy using phosphorus-31 nuclear magnetic resonance spectroscopy. *J Am Coll Cardiol*. 1985;5:745–749.

27. Bottomley PA. Noninvasive study of high-energy phosphate metabolism in human heart by depth-resolved ^{31}P NMR spectroscopy. *Science*. 1985;229:769–772.

28. Bottomley PA, Herfkens RJ, Smith LS, et al. Noninvasive detection and monitoring of regional myocardial ischemia in situ using depth-resolve ^{31}P NMR spectroscopy. *Proc Natl Acad Sci U S A*. 1985;82:8747–8751.

29. Bottomley PA, Herfkens RJ, Smith LS, et al. Altered phosphate metabolism in myocardial infarction: P-31 MR spectroscopy. *Radiology*. 1987;165:703–707.

30. Blackledge MJ, Rajagopalan B, Oberhaensli RD, et al. Quantitative studies of human cardiac metabolism by ^{31}P rotating-frame NMR. *Proc Natl Acad Sci U S A*. 1987;84:4283–4287.

31. Rajagopalan B, McKenna W, Blackledge M, et al. Measurement of phosphorus metabolites in hearts of patients with hypertrophic cardiomyopathy by MRS. Presented at the Sixth Annual Meeting of the Society of Magnetic Resonance in Medicine; August 17–21, 1987. Abstract.

32. Schaefer S, Gober J, Valenza M, et al. Magnetic resonance imaging guided phosphorus-31 spectroscopy of the human heart. *J Am Coll Cardiol*. 1988;12:1449–1455.

33. Reeves RC, Evanochko WT, Canby RC, et al. H-1 NMR spectroscopic detection of increased myocardial lipids in post-ischemic "stunned" myocardium. Presented at the Fifth Annual Meeting of the Society of Magnetic Resonance in Medicine; August 19–22, 1986. Abstract.

34. Evanochko WT, Reeves RC, Sakai TT, et al. Proton NMR spectroscopy in myocardial ischemic insult. *Magn Res Med*. 1987; 5:23–31.

35. Miller DD, Rosen BR, Dragotakes D, et al. Nuclear magnetic resonance detection of increased lipid in the ischemic border of reperfused myocardial infarction by 3-dimensional chemical shift imaging. *Clin Res*. 1987;35:306A. Abstract.

36. Ugurbil K, Petein M, Maidan R, et al. High resolution proton NMR studies of perfused rat hearts. *FEBS Lett*. 1984;167:73–78.

37. Keller AM, Barr ML, Cannon PJ. Rapid lactate measurement in ischemic perfused hearts using continuous negative echo acquisition during steady-state frequency selective excitation (CASTLE). Presented at the Sixth Annual Meeting of the Society of Magnetic Resonance in Medicine; August 17–21, 1987. Abstract.

38. Hardy CJ, Dumoulin CL. POTSHOT: A new technique for lipid and water suppression in 1H spectroscopy: Towards lactate detection in humans. Presented at the Fifth Annual Meeting of the Society of Magnetic Resonance in Medicine; August 19–22, 1986. Abstract.

39. Luyten PR, den Hollander JA. ^{1}H MR observation of metabolites in human tissues in situ. Presented at the Fifth Annual Meeting of the Society of Magnetic Resonance in Medicine; August 19–22, 1986. Abstract.

40. Fossel ET, Hoefeler H. Observation of intracellular potassium and sodium in the heart by NMR: A major fraction of potassium is "invisible." *Magn Reson Med*. 1986;3:534–540.

41. Pike MM, Frazer JC, Dedrick DF, et al. ^{23}Na and ^{39}K nuclear magnetic resonance studies of perfused rat hearts: Discrimination of intra- and extracellular ions using a shift reagent. *Biophys J*. 1985;48:159–173.

42. Elgavish GA, Foster RE, Canby RC, et al. Prediction of reversibility of function in the hypoxic or ischemic isolated perfused rat heart using sodium-23 and phosphorus-31 NMR. Presented at the Sixth Annual Meeting of the Society of Magnetic Resonance in Medicine; 1987. Abstract.

43. Cannon PJ, Maudsley AA, Hilal SK, et al. Sodium nuclear magnetic resonance imaging of myocardial tissue of dogs after coronary artery occlusion and reperfusion. *J Am Coll Cardiol*. 1986; 7:573–579.

44. Neurohr KJ, Gollin G, Neurohr JM, et al. Carbon-13 nuclear magnetic resonance studies of myocardial glycogen metabolism in live guinea pigs. *Biochemistry*. 1984;23:5029–5035.

45. Sherry AD, Nunnally RL, Peshock RM. Metabolic studies of pyruvate- and lactate-perfused guinea pigs hearts by ^{13}C NMR: Determination of substrate preference by glutamate isotopomer distribution. *J Biol Chem*. 1985;260:9272–9279.

46. Malloy CR, Sherry AD, Jeffrey FM. Carbon flux through citric acid cycle pathways in perfused heart by ^{13}C NMR spectroscopy. *FEBS Lett*. 1987;212:58–62.

Part V
Pericardial Disease

Chapter 33

Evaluation of Pericardial Disease by Cardiac Catheterization, Fluoroscopy, and Angiocardiography

Dale C. Wortham, MD, and W. John Nicholas, MD

The membranous covering of the heart, the pericardium, is a structure whose function has been much debated and investigated since antiquity.[1] This chapter examines the current modalities available in the cardiac catheterization laboratory (hemodynamics, fluoroscopy, and angiocardiography) that evaluate the anatomy and, more importantly, the physiology of the pericardium in various disease states. For the reader interested in a broader examination of anatomy, function, or specific diseases of the pericardium, numerous monographs, books, and textbooks are available.[2-5]

The pericardium consists of a serous membrane covering the surface of the heart and epicardial fat (visceral pericardium), which reflects back on itself just above the origin of the great vessels to line the inner aspect of an outer fibrous sac and form the parietal pericardium.[3] In the healthy state, whether the pericardium has important functions has been a topic of discussion, particularly since early observations verified that congenital absence of the pericardium or pericardectomy had no obvious untoward effects.[6] However, recent clinical and experimental evidence has led to an extensive list of the functions of the normal pericardium. These are summarized by Spodick[7] in a review that separates the roles of the pericardium into mechanical, membranous, and ligamentous functions. Just as remarkable as this impressive list of functions of the pericardium is the frequency with which disease processes affect the pericardium, either primarily or secondarily, acutely or chronically, and the dramatic consequences caused by this pericardial involvement because of its critical position in relationship to the heart.

The more common etiologies of pericarditis are shown in Table 33-1. The major hemodynamic manifestations caused by diseased pericardium result from a change in one or both of two different pericardial volumes: the pericardial fluid volume, which is the volume of pericardial fluid (usually less than 50 mL) located between the parietal and visceral pericardium; and the total pericardial volume, which consists of the heart, the blood within the heart, and pericardial fluid. A crucial change in either of these two volumes results basically in two common hemodynamic presentations for all diseases of the pericardium. An increase in the pericardial fluid volume ultimately results in cardiac tamponade, whereas a decrease in the total pericardial volume occurs with constrictive pericarditis. Less frequently, a combination of the two can exist as effusive-constrictive pericarditis.[8,9] When the above conditions are suspected or need to be differentiated from other pathologic states, cardiac catheterization and angiography can document their presence and severity.

CARDIAC TAMPONADE

Cardiac tamponade exists when there is an increase in the volume of a substance (gas, solid, or liquid) in the pericardial sac that results in an increase in the intrapericardial pressure that is transmitted to the intrapericardial surfaces of the venae cavae, atria, ventricles,

Table 33-1 Etiologies of Pericarditis

1. Idiopathic
2. Infectious (viral, bacterial, fungal, tuberculous, parasitic)
3. Neoplastic (primary, metastatic)
4. Rheumatologic (rheumatic fever, rheumatoid arthritis, systemic lupus erythematosus, scleroderma)
5. Arteritides (Takayasu's arteritis, polyarteritis nodosa, Wegener's granulomatosis)
6. Systemic diseases (sarcoidosis, Behçet's syndrome, Whipple's disease, myxedema, hypereosinophilic syndrome)
7. Uremia
8. Trauma (penetrating and nonpenetrating, surgical)
9. Drugs (procainamide hydrochloride, hydralazine hydrochloride, penicillin, isoniazid, anticoagulants, methysergide)
10. Radiation
11. Autoimmune (Dressler's syndrome, postpericardiotomy syndrome)
12. Miscellaneous (cholesterol pericarditis, chylopericarditis, dissecting aortic aneurysm)

or pulmonary veins or arteries to a degree sufficient to produce hemodynamic abnormalities.[10] Cardiac tamponade can be produced by a gas, such as air introduced experimentally into the pericardial space; by a tension pneumopericardium; or by a solid, such as clot or tumor compressing a cardiac chamber. However, tamponade is much more frequently caused by fluid in the pericardial space. The severity of the hemodynamic abnormalities will depend on a number of factors, such as the rate at which the intrapericardial pressure increases, the stiffness of the pericardium, the size and functional status of the intrapericardial cardiac structures, the volume of the pericardial fluid, the total pericardial volume, and the total circulating blood volume. The normal pericardium is poorly distensible and acutely will accommodate only minor increases in intrapericardial volume without concomitant increases in pressure. Although theoretically any disorder that affects the pericardium can result in a pericardial effusion that may progress to cardiac tamponade, the etiologies seen in clinical practice are somewhat more restrictive than those found in Table 33-1. In medical patients, the most common causes of cardiac tamponade are neoplasms, infections, uremia, anticoagulant therapy, idiopathic pericarditis, connective tissue disorders, dissecting aortic aneurysm, and myocardial rupture.[11] In surgical patients, the most common causes of tamponade are severe chest trauma or surgery.

Pathophysiology

Recent experimental and clinical investigations have expanded greatly the understanding of the events that occur with cardiac tamponade.[2,3,10,12-14] Normally the intrapericardial pressure, which reflects intrapleural pressure, is a few millimeters of mercury lower than right ventricular diastolic pressure, which likewise is a few millimeters of mercury lower than left ventricular diastolic pressure. As the volume of pericardial fluid increases, the intrapericardial pressure corresponding will increase, resulting in a decrease in myocardial transmural filling pressure (the intracavity pressure minus the pericardial pressure). When the intrapericardial pressure increases to a level that equals right ventricular diastolic pressure, transmural filling pressure equals zero, and cardiac tamponade exists. Cardiac output and blood pressure are maintained initially by tachycardia and reflex vasoconstriction from increased autonomic tone. Further elevations of intrapericardial pressure will result in a negative transmural filling pressure throughout diastole for both ventricles. This event suggests that continued ventricular filling occurs by diastolic suction.[15] Although transmural filling pressure is an important concept, difficulties in obtaining accurate in vivo intrapericardial pressure measurements (particularly with fluid-filled catheters) lead to inaccuracies in the calculation of transmural filling pressure.[16] Further increases in intrapericardial pressure result in identical increases in right ventricular diastolic pressure until left ventricular diastolic pressure is reached. If pericardial pressure continues to rise, right and left ventricular diastolic pressures rise concomitantly so that all three pressures remain equal.

The reductions in right ventricular filling and volume are accompanied by an absolute reduction in vena caval flows and an increase in systemic venous pressure. Studies using flow probes with pressure and volume measurements by angiography or echocardiography have demonstrated that, even though the total right ventricular volume is decreased, relative expansion during inspiration is preserved.[2,3] Augmentation of right ventricular filling during inspiration is important in producing a paradoxical pulse.[17] With inspiration there is increased flow through the venae cavae into the right heart, producing an increased stroke volume both with and without cardiac tamponade.[18] Conversely, inspiration causes decreased left ventricular diastolic filling, volume, and output.[19] It is this combination of events that produces pulsus paradoxus, an exaggerated fall (more than 10 mm Hg) in systemic systolic arterial pressure during inspiration. Occasionally, patients may have cardiac tamponade without pulsus paradoxus. This can occur in patients with left ventricular failure, atrial septal defect, aortic insufficiency, or hypovolemia. Pulsus paradoxus and an elevated venous pressure are frequently lacking in acute cardiac tamponade with hypovolemia secondary to penetrating chest trauma. However, the absence of these signs in cardiac tamponade and hypovolemia of a more chronic nature was first reported by Antman and associates[20] and termed "low-pressure cardiac tamponade."

Cardiac Catheterization

Standard criteria for the diagnosis of cardiac tamponade should include the demonstration of a pericardial effusion in the presence of elevated systemic venous pressure, and usually with a paradoxical arterial pulse. The venous pressure and paradoxical pulse should diminish when the fluid is removed.[2] Although these criteria do not absolutely require cardiac catheterization, it is the single best means available to document the characteristic venous and systemic arterial pressure findings as well as the severity of the clinical situation. In addition, it can be used to direct treatment through pericardiocentesis and to confirm the relief of tamponade after fluid removal.

Pericardial fluid can be demonstrated by a number of techniques. Listed roughly in order of invasiveness, these are epicardial fat line, echocardiography, nuclear magnetic resonance, computed tomography, radioisotopic imaging, right atrial catheterization, angiocardiography, coronary angiography, pericardiocentesis, and open thoracotomy. The widespread availability and safety of echocardiography, combined with its high degree of sensitivity and specificity, have made it the principal current means of detecting pericardial fluid.[21]

Hemodynamic Studies

Right heart catheterization should be performed whenever possible to document the characteristic hemodynamic findings of cardiac tamponade as well as their resolution after treatment. Normal venous or right atrial pressure tracings display three positive waves and two negative troughs. These are the a, c, and v waves, and the x and y troughs or descents, respectively. The a and v waves reflect atrial and ventricular systole. The c wave corresponds to displacement of the tricuspid valve apparatus toward the right atrium during isovolumic contraction. The x descent is due to atrial relaxation and downward displacement of the tricuspid valve during right ventricular systole. The y descent follows tricuspid valve opening and corresponds to right ventricular diastolic filling. Abnormalities of the x and y descents are important in identifying and distinguishing the compressive pericardial disorders.

In cardiac tamponade, right atrial pressure tracings show a prominent x descent corresponding to peak flows in the superior and inferior venae cavae during ventricular systole. The prominent x descent with an abbreviated or absent y descent suggests that atrial filling occurs only during ventricular emptying. If pericardial pressure is measured, the right atrial and pericardial pressures should be equal. The absence of an elevated pericardial pressure that equals right atrial pressure with its characteristic waveform makes tamponade unlikely. Right ventricular diastolic pressure is elevated and also equals right atrial pressure and intrapericardial pressure. The rate of early right ventricular diastolic filling is decreased in tamponade; therefore, the right ventricular "dip and plateau" waveform found in constrictive pericarditis is absent. Right ventricular systolic pressure may be slightly elevated, normal, or decreased. Pulmonary arterial diastolic pressure should be very similar to right ventricular diastolic pressure and mean pulmonary capillary wedge pressure, in the absence of concurrent diseases. Left ventricular diastolic pressure is elevated and equal to pericardial pressure, right atrial pressure, and right ventricular diastolic pressure, unless there is preexisting left ventricular failure. Figure 33-1 illustrates the typical hemodynamic findings of a patient with cardiac tamponade. These simultaneous pressure recordings were made from the right atrium, right ventricle, pulmonary artery, left ventricle, and aorta, using two high-fidelity micromanometer-tipped catheters with three and two pressure sensors, respectively. With concomitant tamponade and left ventricular failure, the left ventricular diastolic pressure may exceed right atrial or pericardial pressure, but with pericardiocentesis there should be improvement in the right-sided hemodynamics and a change in the right atrial pressure waveform with a return of the y descent.

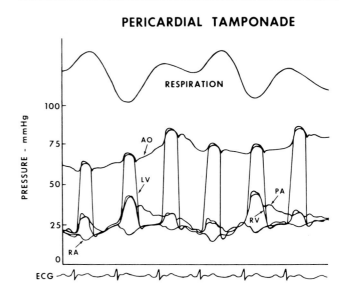

Figure 33-1 Simultaneous pressure tracings from the right atrium, right ventricle, pulmonary artery, left ventricle, and aorta made from solid-state micromanometric transducers on two multisensor catheters in a patient with cardiac tamponade. Note the elevation and equalization of diastolic pressures. The left-sided systolic arterial pressure falls with inspiration (pulsus paradoxus). The right atrial pressure waveform demonstrates a prominent x descent and an absent y descent. AO, aortic; ECG, electrocardiogram; LV, left ventricular; PA, pulmonary arterial; RV, right ventricular; RA, right atrial. Inspiration coincides with a downward deflection of respiratory tracing. *Source:* Courtesy of Steven Bailey MD, San Antonio, TX.

Cardiac Fluoroscopy and Angiocardiography

Fluoroscopic examination of the heart when cardiac tamponade is present is useful only in providing indirect evidence of a pericardial effusion. This evidence, which is separation of the outer cardiac border or silhouette from a marker of the epicardial or endocardial surface of the heart, can be inferred by several observations. Frequently with fluoroscopy the anterior epicardial fat lines (epicardial surface of the heart) and the pericardial fat line (parietal pericardium) may be visualized in the lateral projection. Separation of these two lines suggests that a pericardial effusion is present. Occasionally the presence of extensive coronary artery calcification, as may be seen in a uremic patient suspected of having a pericardial effusion, provides another marker representing the epicardial surface of the heart, whose distance from the outer cardiac silhouette may be used to estimate the size of a pericardial effusion. An additional method of detecting a pericardial effusion with fluoroscopy is accomplished at the time of right heart catheterization. The tip of the catheter while it is in the right atrium is rotated laterally until it is in contact with the right atrial endocardial surface. An increased distance from the catheter tip (endocardial surface) to the border of the cardiac silhouette (parietal pericardium) suggests that a pericardial effusion is present. Fluoroscopy of the heart in the presence of a significant pericardial effusion reveals an enlarged cardiac silhouette with greatly diminished or absent cardiac pulsations. When no pulsations can be seen along the right and left heart borders in a large globular heart, tamponade may be present. Patients with dilated cardiomyopathy obviously will have large hearts with diminished cardiac motion, but pulsations along the heart borders should not be abolished entirely, as is seen with large effusions or in tamponade.

There continues to be interest in angiocardiography in cardiac tamponade. As early as 1958, Steinberg et al[22] reported their angiographic observations on pericardial effusions; in 1968 Steinberg[23] discussed the diagnosis of effusive restrictive pericarditis. Vena caval and right atrial angiography in moderately severe tamponade is associated with straightening of the right atrial border, whereas with more severe tamponade the right atrial border is actually concave.[24] Additionally, there is usually increased separation between the opacified right atrium and the outer right atrial border owing to the presence of pericardial fluid. More recently, in an attempt to identify specific angiographic appearances in cardiac tamponade, Miller et al[25] described tapering or local lateral indentation of the intrapericardial superior vena cava and an exaggerated phasic variation in superior caval diameter at the right atrial junction. Calculated left ventricular angiographic end-diastolic and end-systolic volumes are decreased in more severe tamponade, with a normal or supranormal ejection fraction. Although no published data exist using digital angiography in cardiac tamponade, the utility of this technique with its better contrast agent resolution while allowing the use of less contrast material in patients with severe tamponade is obvious. Additionally, calculations such as ventricular volumes, ejection fraction, or those performed by Miller et al[25] can be made more easily and more accurately by using available computer software and the digitized, angiographic images.

Differential Diagnosis

Cardiac tamponade must be distinguished from other clinical conditions that may present with an elevation of venous pressure and a paradoxical pulse. These are cor pulmonale, right ventricular infarction, pulmonary embolism, left ventricular failure, restrictive cardiomyopathy, constrictive pericarditis, superior vena caval syndrome, tension pneumothorax, and extra pericardial cardiac compression by clot or tumor. Careful attention to the clinical setting, the use of noninvasive tests, and (when necessary) cardiac catheterization will differentiate these various conditions.

CONSTRICTIVE PERICARDITIS

In constrictive pericarditis the pericardium becomes thickened and noncompliant; all chambers of the heart are affected. The visceral and parietal surfaces become fused, obliterating the potential pericardial space and eliminating the lubricating function of the pericardial fluid. When compared with cardiac tamponade or effusive-constrictive pericarditis, constrictive pericarditis is an uncommon disease. As with tamponade, the compressive disorders of the pericardium inhibit cardiac diastolic function while systolic function frequently is unaltered. Systolic function, however, can be impaired if the pericardial disease process involves the mycoardium. Constrictive pericarditis may develop insidiously as a chronic disorder or subacutely within weeks after the pericardial insult.

Etiology

The changing clinical presentations of constrictive pericarditis from the classic triad of increased venous pressure, ascites, and a small quiet heart have made the recognition of constriction more interesting and challenging. Whereas tuberculosis was once the most frequent inciter, now the idiopathic forms predominate.[26] Chronic pericardial disease has become the most frequent cardiac disorder after radiotherapy.[27] As cardiothoracic surgery has become more commonplace, postthoracotomy constriction has been recognized increasingly.[28] The postoperative incidence at Emory University has been reported to be 0.2%.[29] Other common etiologies include postinfectious complications; chronic renal failure; neoplastic infiltration; collagen vascular disorders, such as progressive systemic sclerosis and rheumatoid arthritis; and the aftereffects of penetrating or blunt chest trauma.[26] Unusual causes include an association with atrial septal defect and with mulibrey nanism (a form of dwarfism seen primarily in Finland).[2]

Pathophysiology

In constrictive pericarditis, progressive pericardial fibrosis and scarring result in a stiff, noncompliant sac that restricts diastolic filling of all chambers of the heart. The common physiologic derangement of impaired diastolic filling results in similarities in the pathophysiology of constrictive pericarditis and cardiac tamponade (elevated intracardiac pressures, reduced chamber filling, and stroke volume). However, important differences in diastolic filling patterns and in the influence of negative intrathoracic pressure enable one to recognize whether the compression is caused by constriction, tamponade, or both.

Pericardial constriction causes progressive compression of the heart until the end-diastolic volume of the heart is determined by the pericardial volume. This restraint by the pericardium is exaggerated in mid-diastole and the later portions of diastole, causing early rapid diastolic ventricular filling. In cardiac tamponade the compression of the heart is continuous, preventing rapid early diastolic ventricular filling, hence the lack of a prominent y descent in the right atrial waveform and the absence of an early diastolic dip-and-plateau pattern in the right and left ventricular waveforms in tamponade. A second important difference is that in tamponade the negative intrathoracic pressure of inspiration is transmitted to the nonrigid pericardium, resulting in increased right atrial and right ventricular filling. This is important in the production of pulsus paradoxus and accounts for the usual lack of Kussmaul's sign (increased systemic venous pressure with inspiration) in cardiac tamponade.

Cardiac Catheterization

In spite of the utility of echocardiography,[30] computed tomography,[31,32] and nuclear magnetic resonance[33] in diagnosing constrictive pericarditis, the most important step in the assessment of a patient with suspected constriction is still right and left heart catheterization with angiography. This not only confirms the presence and severity of the restrictive hemodynamics but allows assessment of coexisting disease processes such as myocardial involvement, valvular disease, and coronary artery disease. In addition, cardiac catheterization will aid in the sometimes difficult differentiation between constrictive pericarditis and restrictive cardiomyopathy.

Hemodynamics

The characteristic hemodynamic findings of constrictive pericarditis are (1) equalization of elevated right and left ventricular diastolic pressures, (2) an early diastolic pressure dip followed by a plateau in both ventricles ("square root sign" or dip-and-plateau), and (3) elevated atrial pressures with prominent x and y descents.

Diastolic Equalization

The rigid pericardium constrains atrial and ventricular relaxation and filling such that at ventricular mid-diastole, during passive ventricular filling, all intracardiac pressures are nearly equal. The mean right atrial, mean left atrial (pulmonary capillary wedge), and ventricular diastolic pressures are equal or nearly equal (within 5 mm Hg or less). Although not pathognomonic of constrictive pericarditis, simultaneous mid-diastolic ventricular equalization is the most important measurement to document. This can be accomplished with fluid-filled catheters and accurately calibrated and leveled transducers, or by high-fidelity micromanometer-tipped catheters (Figure 33-2). These catheters, available with serial pressure sensors, are ideally suited for simultaneously recording accurate pressures and waveforms from multiple sites (such as right atrium, right ventricle, and pulmonary artery) with a single catheter. The right ventricular diastolic pressure is usually 12 to 30 mm Hg and is more than one third of the systolic pressure, which is usually less than 45 mm Hg. This value is in contrast to that in restrictive cardiomyopathy, where left ventricular diastolic pressure tends to exceed right ventricular diastolic pressure at rest or, if not, with exercise or saline infusion. Furthermore, there is a lack of respiratory variation of both right and left ventricular systolic pressures that is rarely absent in cardiac tamponade. Kaul et al[34] described a unique hemodynamic response in which right and left ventricular systolic and diastolic pressures in a patient with constrictive pericarditis failed to change during postextrasystolic beats. This finding was in marked contrast to the findings in normal subjects and patients with other forms of heart

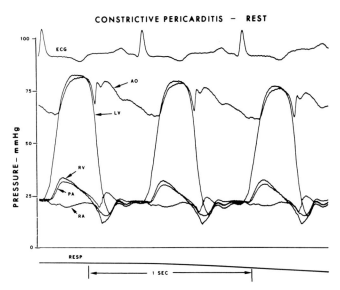

Figure 33-2 Simultaneous pressure tracings from the right atrium, right ventricle, pulmonary artery, left ventricle, and aorta made from solid-state micromanometric transducers on two multisensor catheters in a patient with constrictive pericarditis. Note equalization of elevated right and left ventricular diastolic pressures, which also equal pulmonary arterial diastolic and mean right atrial pressures. The right and left ventricular systolic pressures did not vary significantly with respiration. Abbreviations are as described in the legend to Figure 33-1. *Source:* Courtesy of S Bailey, Fort Sam Houston, TX.

disease, such as tetralogy of Fallot, valvular disease, hypertrophic cardiomyopathy, and congestive heart failure.

Ventricular Dip and Plateau

Rapid early diastolic ventricular filling results in an early dip in the pressure waveform of both ventricles. This is followed by an abrupt increase in diastolic pressure that remains relatively constant throughout later diastole. These two phenomena give rise to the characteristic dip-and-plateau pattern or square root sign seen in constrictive pericarditis (Figure 33-3). Tyberg et al[35] found that the sudden early plateau phase of diastole occurred 0.09 to 0.12 second after the aortic closing sound and coincided with the pericardial knock.

X and Y Descents

Pressure patterns from the right atrium and ventricle in constrictive pericarditis were first described in 1946 by Bloomfield et al.[36] Right atrial mean pressure is elevated, with prominent x and y descents that give the pressure waveform its characteristic M or W configuration (Figure 33-4). The prominent y descent corresponds to rapid early diastolic ventricular filling after tricuspid valve opening, whereas the steep x descent results when ventricular ejection causes displacement of the atrioventricular valves.[7,37] The nadirs of the x and y descents fail to reach baseline, reflecting decreased

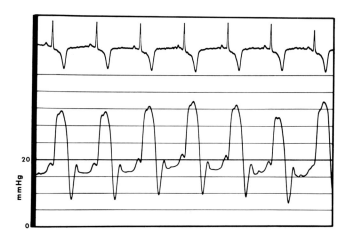

Figure 33-3 Right ventricular pressure recording from a patient with surgically confirmed constrictive pericarditis. Note the early diastolic pressure dip followed by a plateau of elevated diastolic pressure (dip-and-plateau pattern, or square root sign). This pattern was also present in the left ventricular diastolic pressure tracing.

total pericardial volume with a steep pressure-volume relationship.

Latent or Occult Constriction

Bush[38] reported a series of 19 patients with normal baseline hemodynamics who, after the rapid intravenous administration of 1000 mL of saline, developed characteristic pressures of constrictive pericarditis. These patients, who were described as having occult constrictive pericardial disease, were characterized by chronic symptoms of fatigue, dyspnea, chest pain, or a history of pericarditis.

Angiocardiography and Cardiac Fluoroscopy

Although the definitive diagnosis of constrictive pericarditis is made on the basis of hemodynamic catheterization data, cardiac fluoroscopy and angiography can provide important corroborative evidence and exclude other conditions. Fluoroscopy will demonstrate pericardial calcification in at least 50% of patients with constrictive pericarditis, although dense calcification can be present without hemodynamic findings of constriction. Conversely, pericardial calcification may be absent in severe constriction and is usually absent in effusive constrictive pericarditis. The calcifications are usually linear or plaquelike, and are located in the area of the atrioventricular groove or along the diaphragmatic surface of the heart (Figure 33-5). The heart size is usually normal or only slightly increased, with preserved cardiac pulsations, in contrast to cardiac tamponade. The distinctive location and motion of the pericardial calcifications (when present) allow differentiation from other calcified structures such as cardiac

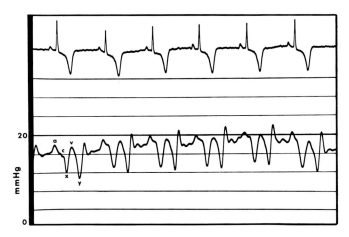

Figure 33-4 Right atrial pressure recording from the same patient shown in Figure 33-3. The mean right atrial pressure is elevated with prominent x and y descents, which give the pressure waveform its characteristic M or W configuration.

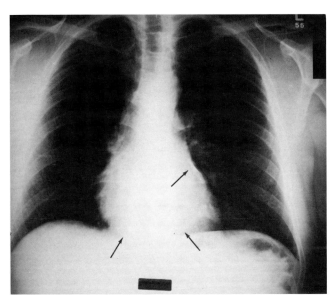

Figure 33-5 Chest roentgenogram of a patient with constrictive pericarditis demonstrating extensive pericardial calcification (arrows).

valves, coronary arteries, cardiac chambers, or extra cardiac lesions.

Angiocardiography can demonstrate superior vena caval dilatation, thickening and straightening of the right atrial border, reduced right ventricular diastolic motion, prolonged transthoracic circulation, and left atrial enlargement.[39-41] Superior vena caval angiography is useful in excluding other conditions associated with elevated venous pressure, such as superior vena caval syndrome. Left ventricular angiography shows a decreased end-diastolic volume and a small end-systolic volume, resulting in an ejection fraction that is normal to increased with a reduced stroke volume. By tracing and digitizing left ventriculograms, Tyberg and associates[42] were able to show increased early diastolic filling rates in patients with constrictive pericarditis when compared with normal subjects and patients with restrictive cardiomyopathy. This process, which required frame-by-frame tracing of the left ventricular silhouettes between end-systole and end-diastole, could now be more easily accomplished with the use of digital cineangiography.

Lesions may be produced by constrictive pericarditis that simulate pulmonic stenosis, mitral stenosis, aortic stenosis, or right ventricular outflow tract obstruction. These obstructions result from a constricting ring or band of pericardium in the atrioventricular groove, often from the remaining pericardium after partial pericardiectomy.[43-45] Cardiac angiography can be key in documenting these functionally stenotic lesions.

By using selective coronary angiography, Alexander and associates[46] described the angiographic appearance of the coronary arteries in constrictive pericarditis. Seven of eight patients with constrictive pericarditis had lack of motion in some or all of the major epicardial coronary arteries. This was in contradistinction to patients with congestive or restrictive cardiomyopathy. Pericardial stripping in five of the patients with constriction revealed epicardial involvement corresponding to the regions of absent coronary artery motion. Goldberg et al[47] reported diastolic segmental coronary artery obliteration in a patient without angina pectoris who had constrictive pericarditis. Figure 33-6 shows end-systolic (a) and end-diastolic (b) frames in the right anterior oblique projection of a left coronary artery cineangiogram in a patient with constrictive pericarditis. The second obtuse marginal branch of the circumflex coronary artery was nearly obliterated at end-diastole while being widely patent at end-systole and early diastole. At the time of pericardiectomy, this area of coronary arterial narrowing was found to be covered by a thick rim of calcium within the pericardium. The extent to which diastolic coronary arterial flow to the myocardium is restricted in constrictive pericarditis is unknown. Digital coronary angiographic techniques, such as contrast myocardial appearance times using the technique described by Vogel et al,[48] could be useful in assessing the functional significance of coronary artery abnormalities in constrictive pericarditis.

Differential Diagnosis of Constrictive Pericarditis

In addition to cardiac tamponade and restrictive cardiomyopathy as mentioned above, the hemodynamic pattern of constriction can be mimicked by several other conditions. The pericardium under normal conditions is relatively nondistensible and resists acute cardiac dilatation.[49] Acute right ventricular dilatation in asso-

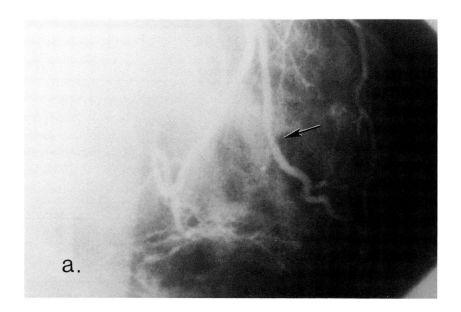

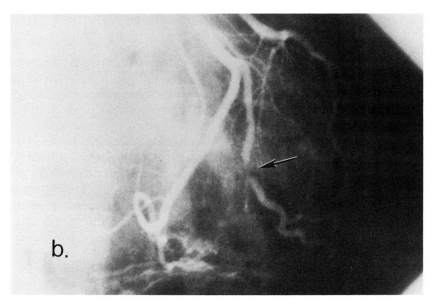

Figure 33-6 Right anterior oblique projection of a left coronary artery cineangiogram from a patient with constrictive pericarditis. The obtuse marginal branch of the circumflex coronary artery was widely patent during systole (a) but became nearly occluded at end-diastole (b).

ciation with right ventricular infarction, tricuspid insufficiency, or massive pulmonary embolism may produce a hemodynamic pattern resembling pericardial constriction.[36,50-52] Acute mitral regurgitation with associated volume overload of the left ventricle may produce diastolic pressures resembling those seen with a constrictive pericardium.[53] Finally, intrapericardial and extrapericardial compression of the heart may simulate the hemodynamic profile of constrictive pericarditis.[54,55]

EFFUSIVE-CONSTRICTIVE PERICARDITIS

An intermediate condition that combines features of cardiac tamponade with pericardial effusion and constrictive pericarditis was termed "subacute effusive-constrictive pericarditis" by Hancock.[8] The initial hemodynamic findings in effusive-constrictive pericarditis suggest cardiac tamponade; however, after removal of pericardial fluid and return of pericardial pressure to normal, the hemodynamic features of constrictive pericarditis emerge. Although the etiologies can be the same as those mentioned for constrictive pericarditis, the most common offenders are tuberculosis, neoplasms, and mediastinal irradiation.[3] Fowler[2] reported this combined entity to be more common than constrictive pericarditis alone. A transient phase of constriction has been reported in some patients with a resolving effusive process, perhaps dispelling the notion that constriction is an irreversible process.[56]

RESTRICTIVE CARDIOMYOPATHY

Restrictive cardiomyopathies are infrequent infiltrative disorders of the myocardium or endocardium associated with elevated filling pressure in the ventricles and normal or nearly normal systolic function. Etiologies include amyloid heart disease, neoplasm, fibroelastosis, endomyocardial fibrosis, Löffler hypereosinophilic syndrome, hemochromatosis, and glycogen storage diseases, although in some cases no cause can be identified. These disorders occasionally can be nearly indistinguishable from constrictive pericarditis, even after cardiac catheterization. Although diastolic equalization can be present in restrictive cardiomyopathy, most commonly the left ventricular diastolic pressure is elevated 5 to 6 mm Hg above right ventricular diastolic pressure. Maneuvers at the time of catheterization, such as rapid saline infusion or exercise, can be used to accentuate differences between right and left ventricular diastolic pressures. High-fidelity micromanometer-tipped catheters demonstrate an ambiguous right ventricular dip-and-plateau pattern in restriction.[57] Digitized left ventriculograms used to characterize diastolic filling rates vary markedly between restictive cardiomyopathy and constrictive pericarditis. Tyberg and associates[42] found that in constrictive pericarditis, 85% of left ventricular filling occurred during the first half of systole, compared with 45% in amyloid heart disease. Ventricularization of the right atrial pressure tracing may be noted in amyloidosis.[58] Myocardial biopsy can be useful in some patients by detecting the presence of amyloidosis, hemochromatosis, or myocardial fibrosis, while in others the biopsy may be normal or nonspecific.[59]

CONGENITAL ANOMALIES OF THE PERICARDIUM

Congenital anomalies of the pericardium are rare and usually come to medical attention because of an abnormality on a routine chest radiograph. Congenital pericardial abnormalities are most frequently of two types: defects of the pericardial sac and pericardial cysts. Defects of the pericardium may be either complete or partial. Complete absence of the pericardium is extremely rare and is not associated with symptoms or an impaired prognosis. Cardiac catheterization of these patients reveals normal hemodynamics.[60] Partial absence of the pericardium is more common than total absence of the pericardium; absence of the left side of the pericardium, either partial or complete, is more common than right-sided defects.

The majority of patients with partial defects likewise do not have symptoms or associated cardiac abnormalities. However, congenital anomalies of the heart and lungs occur in approximately one third of reported cases, and include tetralogy of Fallot, atrial septal defect, patent ductus arteriosus, mitral stenosis, pulmonary sequestration, and bronchogenic cysts.[60] Patients without associated congenital cardiac abnormalities may come to cardiac catheterization because of symptoms (chest pain, dyspnea, palpitation, dizziness, or syncope), an abnormal physical examination (systolic murmur, wide splitting of the second heart sound), an abnormal electrocardiogram (right-axis deviation, incomplete right bundle-branch block), or an abnormal chest radiograph. The abnormal chest radiograph is quite characteristic, with complete or significant defects of the left side of the pericardium. These abnormalities may include a leftward shift of the cardiac silhouette, with the right heart border obscured by the underlying spine; an elongated left heart border with a prominent aortic knob and pulmonary artery; radiolucent bands between either the aortic knob and main pulmonary artery segment or between the diaphragm and base of the heart, caused by interposed lung; and prominence of the left atrial appendage.

Additionally, cardiac catheterization may be performed to exclude other conditions associated with prominence of the pulmonary arteries; left or right heart borders; or hili such as found in neoplasm, mitral valve disease, pulmonary hypertension, pulmonary stenosis, aneurysm, atrial septal defect, or lymphadenopathy. Hemodynamic studies usually are normal in partial pericardial defects without associated congenital cardiac lesions. Dimond et al[61] reported the first case of a partial left-sided pericardial defect demonstrated by cineangiography that was associated with herniation of the left atrial appendage. Documentation of partial pericardial defects may be important because of reports of sudden death related to herniation and strangulation of the heart through the pericardial defect.[60] In partial left-sided defects the levo-phase of pulmonary artery cineangiography is usually sufficient to document herniation of the left atrium or the left atrial appendage through the defect. Partial right-sided pericardial defects can be demonstrated by right atrial cineangiography showing protrusion of the right atrium or right ventricle through the defect. Lajos et al[62] reported localized narrowing of the left anterior descending coronary artery shown by coronary angiography, which presumably was secondary to external compression by an anterior fibrous rim of pericardium in a partial pericardial defect; however, after surgical release of the fibrous ring, postoperative coronary angiography revealed the lesion to be unchanged.

Pericardial cysts, like pericardial defects, are detected most often because of an abnormality on a routine chest radiograph. Most patients are asymptomatic but occasionally may complain of atypical chest pain, dyspnea, or cough. The cysts, which are filled with a clear transudative fluid, are most often located at the right cardiophrenic angle. Rarely they may occur at the left cardiophrenic angle or other locations. Fluoroscopy will reveal a rounded or oval density that is contiguous with the cardiac silhouette, which may or may not pulsate with cardiac motion. Infrequently, calcification in the wall of the cyst can be detected.[2] Fluoroscopy has also been used to direct needle aspiration of larger cysts, although this is not usually indicated. Cineangiography of the appropriate cardiac chamber will demonstrate lack of communication between the cyst and the heart. Ventriculography can be useful in some patients in excluding other diagnoses such as left ventricular aneurysm when a left-sided pericardial cyst is present. Surgical confirmation of pericardial cysts may be necessary to exclude malignancy.

SUMMARY

In summary, a variety of noninvasive and invasive techniques are available to the clinician for diagnosing abnormalities of the structure and function of the pericardium. Abnormalities seen with congenital defects of the pericardium, pericardial cyst, or acute and chronic pericarditis can be defined or suggested by imaging modalities such as computed tomography, nuclear magnetic resonance, echocardiography, fluoroscopy, and angiocardiography. In constrictive pericarditis and cardiac tamponade, these techniques provide both critical anatomic information and evidence of the physiologic effects of these two disease processes. Hemodynamic evaluation, however, remains an integral part in confirming the diagnosis of constrictive pericarditis and cardiac tamponade. Generally, it is the combination of the findings of one or more imaging techniques in conjunction with the hemodynamic profile that provides the clinician with the information not only to confirm the diagnosis but to recommend appropriate therapy. The vast number of imaging modalities now available, combined with their cost and the slight risk of cardiac catheterization, challenges the clinician to understand the utility of each in diagnosing the entire spectrum of pericardial disease.

REFERENCES

1. Spodick DH. Medical history of the pericardium. *Am J Cardiol.* 1970;26:447-454.

2. Fowler NO. *The Pericardium in Health and Disease.* New York, NY: Futura Publishing Co; 1985:1-363.

3. Shabetai R. *The Pericardium.* New York, NY: Grune & Stratton Inc; 1981:1-421.

4. Hurst JW, Logue RB, Rackley CE, eds. *The Heart.* 6th ed. Philadelphia, Pa: WB Saunders Co, 1986;1363-1393.

5. Braunwald E, ed. *Heart Disease.* 3rd ed. Philadelphia, Pa: WB Saunders Co; 1987;1484-1534.

6. Moore TC, Shumacker HB Jr. Congenital and experimentally produced pericardial defects. *Angiology.* 1953;4:1-11.

7. Spodick DH. The normal and diseased pericardium. *J Am Coll Cardiol.* 1983;1:240-251.

8. Hancock EW. Subacute effusive constrictive pericarditis. *Circulation.* 1971;43:183-192.

9. Mann T, Brodie BR, Grossman W, et al. Effusive constrictive hemodynamic pattern due to neoplastic involvement of the pericardium. *Am J Cardiol.* 1978;41:781-786.

10. Fowler NO, Gabel M. Regional cardiac tamponade: A hemodynamic study. *J Am Coll Cardiol.* 1987;10:164-169.

11. Guberman BA, Fowler NO, Engel PJ, et al. Cardiac tamponade in medical patients. *Circulation.* 1981;64:633-640.

12. Refsum H, Junemann M, Lipton MJ, et al. Ventricular diastolic pressure-volume relations and the pericardium. *Circulation.* 1981;64:997-1004.

13. Shabetai R, Fowler NO, Guntheroth WG. The hemodynamics of cardiac tamponade and constrictive pericarditis. *Am J Cardiol.* 1970;26:480-489.

14. Fowler NO, Gabel M. The hemodynamic effects of cardiac tamponade: Mainly the result of atrial, not ventricular, compression. *Circulation.* 1985;71:154-157.

15. Brecher GA. Experimental evidence of ventricular diastolic suction. *Circ Res.* 1956;4:513-518.

16. Santamore WP, Constantinescu M, Little WC. Direct assessment of right ventricular transmural pressure. *Circulation.* 1987;75:744-747.

17. Shabetai R, Fowler NO, Fenton JC, et al. Pulsus paradoxus. *Clin Invest.* 1965;44:1882-1898.

18. Shabetai R, Fowler NO, Guntheroth WG. The hemodynamics of cardiac tamponade and constrictive pericarditis. *Am J Cardiol.* 1970;26:480-489.

19. Ferguson R, Bristow D, Mintz F, et al. The effects of pericardial tamponade on left ventricular volumes and function as calculated from aortic thermodilution curves. *Clin Res.* 1963;11:100. Abstract.

20. Antman EM, Cargill V, Grossman W. Low-pressure cardiac tamponade. *Ann Intern Med.* 1979;91:403-406.

21. Feigenbaum H. *Echocardiography.* Philadelphia, Pa: Lea & Febiger; 1986:558-565.

22. Steinberg I, von Gal HV, Finby N. Roentgen diagnosis of pericardial effusion: New angiographic observations. *Am J Roentgenol.* 1958;79:321-332.

23. Steinberg I. Angiocardiography in diagnosis of pericardial effusion and pulmonary stenosis in Hodgkin's disease. *Am J Roentgenol.* 1968;102:619-626.

24. Spitz HB, Holmes JC. Right atrial contour in cardiac tamponade. *Radiology.* 1972;103:69-75.

25. Miller SW, Feldman L, Palacios I, et al. Compression of the superior vena cava and right atrium in cardiac tamponade. *Am J Cardiol.* 1982;50:1287-1292.

26. Cameron J, Oesterle SN, Baldwin JC, et al. The etiologic spectrum of constrictive pericarditis. *Am Heart J.* 1987;113:354-360.

27. Applefeld MM, Wiernik PH. Cardiac disease after radiation therapy for Hodgkin's disease: Analysis of 48 patients. *Am J Cardiol.* 1983;51:1679-1681.

28. Cohen MV, Greenberg MA. Constrictive pericarditis: Early and late complication of cardiac surgery. *Am J Cardiol.* 1979;43:657-661.

29. Kutcher MA, King SB, Alimurung BN, et al. Constrictive pericarditis as a complication of cardiac surgery: Recognition of an entity. *Am J Cardiol.* 1982;50:742-748.

30. Lewis BS. Real time two dimensional echocardiography in constrictive pericarditis. *Am J Cardiol.* 1982;49:1789-1793.

31. Isner JM, Carter BL, Bankoff MS, et al. Computed tomography in the diagnosis of pericardial disease. *Ann Intern Med.* 1982;97:473–479.

32. Sutton FJ, Whitley NO, Applefeld MM. The role of echocardiography and computed tomography in the evaluation of constrictive pericarditis. *Am Heart J.* 1984;109:350–355.

33. Soulen RL, Stark DD, Higgins CB. Magnet resonance imaging of constrictive pericardial disease. *Am J Cardiol.* 1985;55:480–484.

34. Kaul U, Gupta CD, Anand IS, et al. Characteristic postextrasystolic ventricular pressure response in constrictive pericarditis. *Am Heart J.* 1981;102:461–462.

35. Tyberg TI, Goodyer AVN, Langou RA. Genesis of pericardial knock in constrictive pericarditis. *Am J Cardiol.* 1980;46:570–575.

36. Bloomfield RA, Lauson HD, Cournand A, et al. Recording of right heart pressures in normal subjects and in patients with chronic pulmonary disease and various types of cardio-circulatory disease. *J Clin Invest.* 1946;25:639–664.

37. Lorell BH, Grossman W. Profiles in constrictive pericarditis, restrictive cardiomyopathy, and cardiac tamponade. In: Grossman W, ed. *Cardiac Catheterization and Angiography.* 3rd ed. Philadelphia, Pa: Lea & Febiger; 1986:427–445.

38. Bush CA. Occult constrictive pericardial disease diagnosed by rapid volume expansion and correction by pericardiectomy. *Circulation.* 1977;56:924–930.

39. Figley MM, Bagshaw MA. Angiocardiographic aspects of constrictive pericarditis. *Radiology.* 1957;69:46–52.

40. Fowler NO. Constrictive pericarditis: New Aspects. *Am J Cardiol.* 1982;50:1014–1017.

41. Steinberg I. Roentgenography of pericardial disease. *Am J Cardiol.* 1961;7:33–47.

42. Tyberg TI, Goodyer AVN, Hurst VW III, et al. Left ventricular filling in differentiating restrictive amyloid cardiomyopathy and constrictive pericarditis. *Am J Cardiol.* 1981;47:791–796.

43. Mounsey P. Annular constrictive pericarditis: With an account of a patient with functional pulmonary, mitral, and aortic stenosis. *Br Heart J.* 1959;21:325–334.

44. Barros JL, Gomez FP. Pulmonary stenosis due to external compression by a pericardial band. *Br Heart J.* 1967;29:947–949.

45. Nishimura RA, Kazmier FJ, Smith HC, et al. Right ventricular outflow obstruction caused by constrictive pericardial disease. *Am J Cardiol.* 1985;55:1447–1448.

46. Alexander J, Kelley MJ, Cohen LS, et al. The angiographic appearance of the coronary arteries in constrictive pericarditis. *Radiology.* 1979;131:609–617.

47. Goldberg E, Stein J, Berger M, et al. Diastolic segmental coronary artery obliteration in constrictive pericarditis. *Cathet Cardiovasc Diagn.* 1981;7:197–202.

48. Vogel R, LeFree M, Bates E, et al. Application of digital techniques to selective coronary arteriography: Use of myocardial contrast appearance time to measure coronary flow reserve. *Am Heart J.* 1984;107:153–164.

49. Holt JP. The normal pericardium. *Am J Cardiol.* 1970;26:455–465.

50. Lorell B, Leinbach RC, Pohost GM, et al. Right ventricular infarction: Clinical diagnosis and differentiation from cardiac tamponade and pericardial constriction. *Am J Cardiol.* 1979;43:465–471.

51. Butman S, Olson HG, Aronow WS, et al. Remote right ventricular myocardial infarction mimicking chronic pericardial constriction. *Am Heart J.* 1982;103:912–914.

52. Jensen DP, Goolsby JP, Oliva PB. Hemodynamic pattern resembling pericardial constriction after acute inferior myocardial infarction with right ventricular infarction. *Am J Cardiol.* 1978;42:858–861.

53. Bartle SH, Hermann HJ. Acute mitral regurgitation in man: hemodynamic evidence and observations indicating an early role for the pericardium. *Circulation.* 1967;36:839–851.

54. Dell'Italia LJ, Walsh RA. Hemodynamic profile of constrictive pericarditis produced by a massive right pleural effusion. *Cathet Cardiovasc Diagn.* 1984;10:471–477.

55. Little WC, Primm RK, Karp RB, et al. Clotted hemopericardium with the hemodynamic characteristics of constrictive pericarditis. *Am J Cardiol.* 1980;45:386–388.

56. Sagrista-Sauleda J, Permanyer-Miralda G, Candell-Riera J, et al. Transient cardiac constriction: An unrecognized pattern of evolution in effusive acute idiopathic pericarditis. *Am J Cardiol.* 1987;59:961–966.

57. Hirota Y, Kohriyama T, Hayashi T, et al. Idiopathic restrictive cardiomyopathy: Differences of left ventricular relaxation and diastolic wave forms from constrictive pericarditis. *Am J Cardiol.* 1983;52:421–423.

58. Gowda S, Salem BI, Haikai M. Ventricularization of right atrial wave form in amyloid restrictive cardiomyopathy. *Cathet Cardiovasc Diagn.* 1985;11:483–491.

59. Benotti JR, Gorssman W, Cohn PF. Clinical profile of restrictive cardiomyopathy. *Circulation.* 1980;61:1206–1212.

60. Nasser WK. Congenital absence of the left pericardium. *Am J Cardiol.* 1970;26:466–470.

61. Dimond EG, Kittle CF, Voth DW. Extreme hypertrophy of the left atrial appendage: The case of the giant dog ear. *Am J Cardiol.* 1960;5:122–125.

62. Lajos TZ, Bunnell IL, Colokathis BP, et al. Coronary artery insufficiency secondary to congenital pericardial defect. *Chest.* 1970;58:73–76.

Chapter 34

Echocardiography of Pericardial Disease

Ivan A. D'Cruz, MD, FRCP

Echocardiography was introduced into clinical medicine in 1965 when Feigenbaum et al. demonstrated, experimentally and in patients, the ability of ultrasound to diagnose the presence of pericardial effusions.[1] Over the next decade, M-mode echocardiography was widely and rapidly accepted as a reliable and convenient noninvasive way to detect pericardial fluid and pericardial thickening. During the late 1970s, the two-dimensional (2-D) technique was found to enhance the echocardiographic visualization of the amount and distribution of fluid within the pericardial sac.

The echocardiographic manifestations of tamponade were first described in 1975.[2] Since then various additional ultrasound signs of tamponade have been reported; their value and limitations have gradually become better understood.

A clinician responsible for a patient with known or suspected pericardial disease should obtain the following diagnostic information:

1. Is a pericardial effusion present or not?
2. How much pericardial fluid is present?
3. Is the pericardial effusion inflammatory (pericarditis) or transudative (hydropericardium)?
4. On serial echocardiograms, is the amount of pericardial fluid increasing, decreasing, or unchanged?
5. Is the pericardial effusion loculated or freely distributed?
6. Is tamponade present?
7. Is the pericardium thickened, sclerotic, or calcified?
8. Is pericardial constriction present?
9. Is any solid material (clots, neoplastic masses, fibrin, or fibrous bands) seen within the effusion?
10. Are there any abnormalities of cardiac structure or function present, apart from the effusion?
11. What is the etiology of the effusion?
12. Is the effusion suitable for paracentesis, if this is thought indicated?

Except for the query on etiology, echocardiography, in expert hands, is capable of answering all these questions, at least in some patients with pericardial effusions.

Until the mid 1970s, only the M-mode technique was available; subsequently 2-D echocardiography has been in general use. However, M-mode echocardiography is by no means obsolete. M-mode and 2-D echocardiography should be viewed as complementing each other. For example, the M-mode technique is capable of better resolution and can therefore demonstrate the texture of thickened pericardium better than the 2-D technique. On the other hand, the distribution and relative quantity of fluid in different regions or recesses of the pericardial sac are better appreciated on the 2-D image. In large pericardial effusions with tamponade, alternation in cardiac position and respiratory variation in ventricular dimensions or mitral excursion are best recorded on the M-mode tracing. The right atrial wall

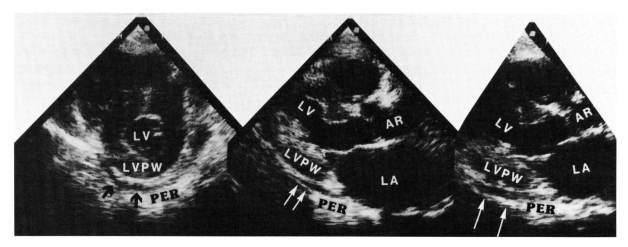

Figure 34-1 Two-dimensional echocardiogram of a patient with a small pericardial effusion. The effusion manifests as a narrow sonolucent space (arrows) between the left ventricular posterior wall (LVPW) and the pericardium (PER), in the parasternal short-axis view (left) and long-axis view (middle and right). Since the effusion is scanty in amount and localized, a slight change in transducer direction can demonstrate more pericardial fluid at right than at center. AR, aortic root; LA, left atrium; LV, left ventricle.

sign of tamponade, however, can only be visualized on the 2-D echocardiogram.

Normally the parietal and visceral pericardium (epicardium) are entirely apposed, and echocardiography demonstrates no echo-free pericardial space. Pericardial effusions manifest ultrasonically as sonolucent spaces separating the ventricular or atrial wall from the parietal pericardium. Pericardial fluid anterior and posterior to the heart can be visualized on the M-mode echocardiogram; the 2-D echocardiogram has the further advantage that the full extent and distribution of fluid within the pericardial sac can be appreciated by imaging the heart and pericardium in multiple planes, using several different views: parasternal long and short axes, apical four-chamber and long axis, and subcostal. If pericardial paracentesis is contemplated, 2-D echocardiography is valuable in ascertaining whether sufficient pericardial fluid is present at the intended site of the tap.

ECHOCARDIOGRAPHIC (M-MODE AND 2-D) ESTIMATION OF THE SIZE OF PERICARDIAL EFFUSIONS

Horowitz et al. correlated M-mode echocardiographic appearances with the amount of fluid obtained by suction from the pericardial sac in patients undergoing thoracic surgery[3] and described M-mode patterns characteristic of small, moderate, and large pericardial effusions. Since precise quantification of the quantity of pericardial fluid is not feasible, it is customary to report the size of a pericardial effusion using such descriptive terms as (in ascending scale) minimal, small, small to moderate, moderate, moderate to large, large, and very large.

In small pericardial effusions (Figure 34-1), a narrow sonolucent band or crescent less than 1 cm in width is visible, representing fluid in the posterior pericardial space. This narrow echo-free space is seen in long-axis as well as short-axis views and is more prominent in systole than in diastole. A very small rim of fluid may or may not be discerned in the pericardial plane anteriorly or laterally.

In pericardial effusions of moderate size, the width of the circumventricular sonolucent space is greater (maximum width 1 to 2 cm posteriorly, and is also obvious anteriorly), and the space persists in diastole as well as in systole. A posterior effusion of moderate size is seen in Figure 34-2.

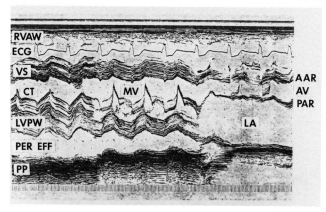

Figure 34-2 M-Mode echocardiogram of a patient with a posterior pericardial effusion, showing a scan from the left ventricle at the chordae level to the aortic root–left atrium level. The pericardial effusion (PER EFF) is seen as a space between the left ventricular posterior wall (LVPW) and the parietal pericardium (PP). Pericardial fluid is not seen anteriorly or behind the left atrium (LA). RVAW, right ventricular septum; CT, chordae tendineae; AAR, anterior aortic root; AV, aortic valve; PAR, posterior aortic root; ECG, electrocardiogram; MV, mitral valve; VS, ventricular septum.

In large pericardial effusions the sonolucent pericardial space is even larger (Figure 34-3), and undue mobility of the ventricles becomes evident. The term "swinging heart" has been applied to conspicuous pendulous or bouncing cardiac motion of this type. In some of these patients, alternation of cardiac position can be recorded on the M-mode echocardiogram, providing an explanation for electrical alternans on the electrocardiogram (ECG). In large effusions it is the rule for pericardial fluid to be seen behind the left atrium (in the oblique sinus), as well as around the right atrium. Occasionally, such juxta-atrial accumulations of pericardial fluid may be present even in small or moderate effusions. In small effusions it is not uncommon to detect fluid in the groove between the right atrium and ventricle in the subcostal four-chamber view.

Loculated Pericardial Effusions

Since echocardiography is usually done with the patient in a supine flat or partial left lateral position, pericardial fluid tends to gravitate to the posterior pericardial space, and relatively less fluid is seen in the anterior pericardium. However, in certain subacute or chronic varieties of pericarditis, intrapericardial adhesions develop that produce loculation of pericardial fluid.[4] The pericardium posterior to the left ventricle is a common site for loculation (Figure 34-4); less often, pericardial fluid is loculated anteriorly or adjacent to the right or left atrium. Common clinical settings in which such fluid pockets are seen include after cardiac surgery, chronic renal failure, systemic lupus, tuberculous pericarditis, and hemopericardium. If these pockets are sufficiently large and of high enough tension, tamponade can result (see below).

Intrapericardial Echoes Signifying Solid Material

Organizing exudate or fibrinous material within the pericardium commonly accompanies pericarditis of various etiologies. On the echocardiogram such material produces ill-defined amorphous or smudgy images that partly or completely replace the echo-free appearance of fluid. This is a very common finding in

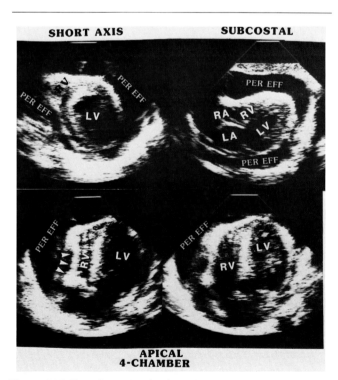

Figure 34-3 Two-dimensional echocardiogram of a patient with a large pericardial effusion (PER EFF) with tamponade. In the parasternal short-axis view and the subcostal four-chamber view (top), as well as in the apical four-chamber view (bottom), the abundant pericardial fluid surrounds all the cardiac chambers. Conspicuous "collapse" or compression (arrows) of the right ventricular wall is seen in diastole (bottom left) but not in systole (bottom right). RV, right ventricle; LV, left ventricle; RA, right atrium; LA, left atrium.

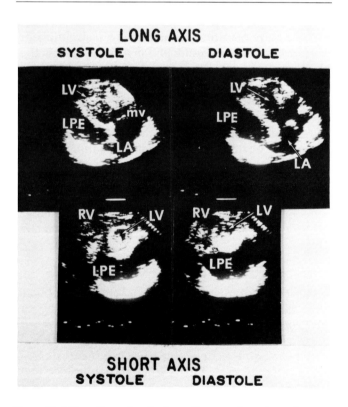

Figure 34-4 Two-dimensional echocardiogram of a patient who presented with a large loculated posterior pericardial effusion 6 weeks after coronary artery bypass surgery. The left ventricle (LV) appears compressed by the loculated pericardial effusion (LPE), more so during early diastole (right frames) than in systole (left frames). Note the deformation also of left atrial contour, in the right top frame. Hemodynamic data indicated tamponade, which was promptly relieved when the pericardial effusion was drained. LA, left atrium; MV, mitral valve; RV, right ventricle.

patients who have had cardiac surgery within the previous few days.

Less common are well-defined fibrous bands coexisting with moderate to large pericardial effusions and bridging the space between visceral and parietal pericardium.[4-5] Sometimes such fibrinous or fibrous strands are tethered at one end only, so that a ribbon-like structure floats or undulates freely within the pericardial effusion. Intrapericardial clots may be visualized on the 2-D echocardiogram.[6] Neoplastic intrapericardial masses[7] are very rare, although malignant pericardial involvement is common, the primary growth being in lung or breast. The usual echocardiographic manifestation of such involvement is pericardial thickening and a moderate to large effusion.

ECHOCARDIOGRAPHIC SIMULATORS OF PERICARDIAL EFFUSIONS

A small echo-free space anterior to the right ventricle, especially in obese individuals, is usually attributable to pericardial fat.

Massive infiltration of the pericardium by malignant neoplastic tissue or lymphoma (tumor encasement of the heart) with little or no fluid exudation can simulate a pericardial effusion inasmuch as it can present as a sonolucent space anterior and posterior to the heart. This diagnosis may be considered if a dry pericardial tap is obtained in patients with known cancer or lymphoma.

Left pleural effusions constitute the most common differential diagnosis of pericardial effusions. They present as large sonolucent spaces posterior to the heart; large pleural effusions may sometimes also extend anteriorly to the right ventricle. Their distinction from a large pericardial effusion can be made as follows: (1) Left pleural effusions are posterior to the plane of the descending thoracic aorta; pericardial effusions are anterior to the descending aorta[8]; (2) large pericardial effusions are associated with exaggerated oscillating motion of the heart, but pleural effusions are not; (3) the contours of fluid-filled pericardial and pleural spaces are characteristic and different from each other, especially in the short-axis view.

Many cardiologists forget that other serosal fluid spaces adjacent to the heart can simulate loculated pericardial effusions[9] in the subcostal view. An example is right pleural effusions that are contiguous with the right atrium. (It is worth noting that pericardial cysts may also present as a sonolucent space related to the right atrium.[10]) The uppermost part of a tense ascites is anteroinferior to the heart, separated by the diaphragm, but in the subcostal echocardiographic view would appear anterior to the right ventricle. The identification of these several different types of serosal spaces and their differentiation from pericardial effusions can be accomplished by careful subcostal 2-D scanning in all possible planes, with special attention to determining (1) the contours of the sonolucent space, (2) which cardiac chamber (or chambers) is related to the space, and (3) the anatomic relationship of the space to the liver, inferior vena cava, and hepatic veins.[11]

Anterior pericardial effusions may be simulated by various other causes of sonolucent or partly sonolucent spaces, including thymic cysts, pericardial cysts, lymphomas, dermoid cysts, diaphragmatic hernias, giant atrial diverticula, retrosternal hematomas or abscesses, apical left ventricular pseudoaneurysm, and the like. Posterior pericardial effusions may be simulated by left pleural effusions, left ventricular pseudoaneurysm, pancreatic pseudocysts, ascitic fluid, or retroventricular extension of a dilated left atrium.

CARDIAC TAMPONADE

Over the last 14 years, much progress has been made in this important diagnostic area. Several echocardiographic signs of tamponade have been described, but a definite statement in the echocardiographic report as to whether tamponade is present or not is often difficult, because the sensitivity and specificity of these signs are still controversial. Moreover, various authors who have published their experience on this topic have not all used the same clinical criteria for tamponade.

The following abnormalities have been found of value in signifying the presence of tamponade, in patients with large pericardial effusions:

1. Large phasic respiratory changes in ventricular dimensions and mitral motion are seen. Normally slight changes in left and right ventricular dimensions (on the order of 1 to 3 mm) do occur with respiration. In tamponade these changes are conspicuous and very obvious on the M-mode tracing (Figure 34-5). During inspiration, the right ventricular size increases, left ventricular size decreases, mitral diastolic excursion diminishes, mitral ejection fraction (EF) slope decreases, and sometimes aortic valve systolic opening shortens in duration.[2,12-13] In expiration all the opposite changes occur. During apnea, no phasic changes are seen. Reciprocal changes in right and left ventricular area can also be appreciated on the 2-D echocardiogram, especially in the short-axis and four-chamber views. Similar changes in ventricular dimension with respiration, of a reciprocal nature, have been described in other conditions associated with enhanced inspiratory effort and large swings in intrapleural pressure, such as pulmonary embolism and chronic obstructive lung disease.

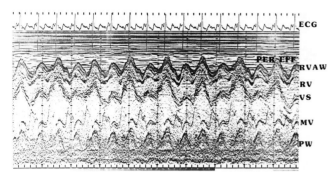

Figure 34-5 M-Mode echocardiogram of a patient with a large pericardial effusion (PER EFF), showing the ventricles at basal (mitral valve) level. Note the large respiratory fluctuations in the right ventricular dimension, with reciprocal changes in left ventricular dimension. During inspiration the right ventricular size increases, and the left ventricular size decreases, and mitral valve (MV) excursions decrease in amplitude. Note also alternation in the position of the right ventricular anterior wall (RVAW) from beat to beat. Alternation of the ECG is minimal in this lead but better seen in some other leads. VS, ventricular septum; PW, posterior cardiac wall at left atrial–left ventricular junction; RA, right atrium; RV, right ventricle; ECG, electrocardiogram.

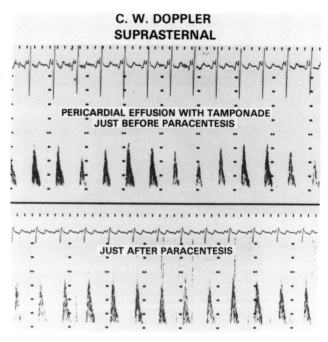

Figure 34-6 Continuous wave Doppler recording from the suprasternal notch in a patient with a large pericardial effusion and tamponade. Note the large respiratory fluctuations in peak velocity of flow in the ascending aorta just before paracentesis (top), which are virtually absent just after paracentesis (bottom).

2. Doppler recording of peak systolic flow velocity in the left ventricular outflow tract (in pulsed wave Doppler from the apex) as well as in the ascending aorta (on continuous wave Doppler from the suprasternal notch) shows marked respiratory fluctuations (inspiratory decrease, expiratory increase; Figure 34-6). This recent addition to the routine echocardiographic evaluation of patients with pericardial effusions is, of course, the Doppler equivalent of pulsus paradoxus, a cardinal physical sign of tamponade for more than a century.

It is probable that Doppler recordings of flows across the cardiac valves and venae cavae will soon become routine echocardiographic practice, since preliminary studies have indicated the technique is promising in identifying the presence of tamponade: Inspiratory increases in flow-velocity integrals of about 80% across tricuspid and pulmonic valves, and inspiratory decreases of about 35% across mitral and aortic valves, were reported by Leeman et al.[14] in patients with tamponade; in those with pericardial effusions but no tamponade, the respiratory fluctuations were a quarter of this magnitude. Appleton et al.[15] noted an expiratory decrease in diastolic forward flow in the superior vena cava in patients with tamponade, but patients with pericardial effusion not in tamponade did not manifest this Doppler finding.

3. Right ventricular compression or narrowing[16] is indicative of tamponade, as seen in Figure 34-3. On the 2-D echocardiogram the whole right ventricle may be narrowed in some cases. In others, an early diastolic inward motion ("dimpling") of the right ventricular anterior wall, especially the right ventricular outflow tract, may be discerned in diastole[17] on the 2-D echocardiogram, as well as on the M-mode tracing. It should be added that this abnormality is sometimes encountered in patients who have moderate to large pericardial effusions but no clinical evidence of tamponade at all (false-positive). It is probable that right ventricular wall hypertrophy may prevent this dimpling even when frank tamponade exists (false-negative).

4. Right atrial wall indentation (Figure 34-7) is considered by most echocardiographers to be a more sensitive and reliable sign of tamponade than right

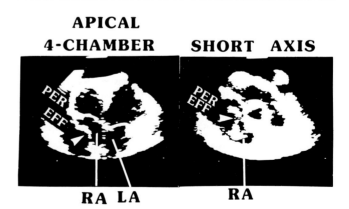

Figure 34-7 Two-dimensional echocardiogram of a patient with a large pericardial effusion (PER EFF) and tamponade. Right atrial wall indentation (arrowhead) is well seen in the apical four-chamber view (left) and also in the parasternal short-axis view at aortic root level (right). RA, right atrium; LA, left atrium.

ventricular wall dimpling.[18] The right atrial wall transiently infolds just after inscription of the P wave (i.e., late diastole and early systole). This sign should be looked for in apical four-chamber, subcostal four-chamber, and parasternal short-axis views. However, it may be *too* sensitive, being sometimes unaccompanied by clinical or hemodynamic tamponade.

5. Alternation in position of the heart, as manifested particularly on the right ventricular anterior wall echocardiogram on the M-mode tracing, is usually not quoted as a sign of tamponade. However, such an appearance, accompanied by electrical alternans on the twelve-lead electrocardiogram (ECG), is almost invariably associated with tamponade in clinical practice.

Tamponade Produced by Loculated Pericardial Effusions

It is not generally recognized that loculated pericardial effusions, if sufficiently large and under high enough pressure, can cause the typical hemodynamic changes of tamponade and be as life threatening as the usual nonloculated effusion tamponade. Several well-documented examples of such instances, where the loculated effusion was visualized by echocardiography, have been published.[19–20] The chamber selectively compressed was in some cases the right atrium, in others the right ventricle, and in yet others the left ventricle.

On the 2-D echocardiogram, compression and distortion of the affected cardiac chamber are evident, but the signs of tamponade described above for nonloculated (circumcardiac) pericardial effusions are not seen. It is worth noting, however, that deformation and narrowing of the left ventricular chamber and a remarkable paradoxical motion of the left ventricular posteroinferior wall have been demonstrated (Figure 34-4) in an occasional patient with tamponade secondary to a large, posteriorly loculated effusion following cardiac surgery.[21]

ROLE OF ECHOCARDIOGRAPHY IN PERICARDIAL PARACENTESIS

Chandaratna et al. first described the use of contrast echocardiography during pericardial tapping.[22] Saline is injected through the needle, repeatedly if necessary, to ascertain the position of the needle tip, especially if bloody fluid is being aspirated. The appearance of microbubbles within the pericardial fluid confirms the intrapericardial location of the needle, but if contrast medium appears within a cardiac chamber (and not in pericardial fluid), it means that the needle has been advanced too far. Moreover, 2-D echocardiography permits a rough estimation of the amount of effusion remaining within the pericardial sac, and this should help decide when to terminate the procedure. The routine use of 2-D echocardiography in conjunction with pericardial parencentesis has been advocated, but this recommendation has not yet been widely adopted for everyday clinical practice.

Apart from the possible use of 2-D echocardiography *during* paracentesis, it must be emphasized that echocardiography has an important role *just before* pericardial paracentesis. Too often a pericardial tap is attempted (1) on the basis of an echocardiogram taken several days before, (2) without ascertaining whether a pericardial effusion is present in sufficient quantity at the proposed site of aspiration, or (3) without any cardiac ultrasound examination at all. The unpleasant surprise of a dry pericardial tap or of aspirating blood from a cardiac chamber could be avoided by 2-D echocardiographic visualization of the size and distribution of the pericardial effusion just before the procedure.

PERICARDIAL THICKENING, SCLEROSIS, AND CALCIFICATION

Schnittger et al.[23] correlated the presence of pericardial thickening (noted at surgery) with various M-mode patterns. When a pericardial effusion coexists, thickening of the epicardium (visceral pericardium) and of the parietal pericardium manifests as two dense, well-defined layers separated by the echo-free pericardial space. During the recovery phase of acute pericarditis of any etiology, and in the week or two following cardiac surgery, it is common to observe that the pericardial space is not entirely clear but is filled with mottled or irregular, indistinct echoes that represent organizing exudate. During recovery from pyogenic pericarditis or hemopericardium, dense, stratified linear echoes may rapidly develop, which, in my experience, should alert the echocardiographer to the possible development of pericardial constriction.

In the absence of a pericardial effusion, pericardial thickening may be diagnosed if, on the M-mode tracing, a dense, well-defined layer appears contiguous with the ventricular wall and moves along with it in systole and diastole (Figure 34-8). This layer is more echo-dense than the ventricular myocardium, and its thickness varies from a few millimeters to 1 or 2 cm or more. It may be difficult or impossible to state whether "thick pericardium" as visualized on the echocardiogram represents pericardium alone or pericardium and pleura fused together.

Sometimes very thick pericardium has a stratified or layered morphology on an M-mode view (Figure 34-9). When the pericardium is calcified (calcific shell visible on the chest x-ray film), pericardial echoes are

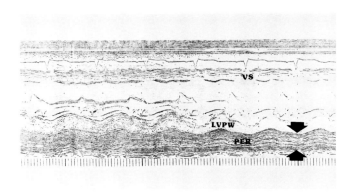

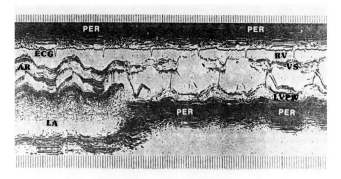

Figure 34-8 M-Mode echocardiogram of a patient with very dense and thick pericardium (arrows). At lower gain setting in the last three beats, the myocardial echoes of the ventricular septum (VS) and left ventricular posterior wall (LVPW) were suppressed, but the dense pericardial echoes persisted, indicating a high degree of pericardial sclerosis.

Figure 34-9 M-Mode echocardiographic scan from the aortic root (AR)–left atrium (LA) level to the left ventricular level, in a patient with hemodynamically proved pericardial constriction. The very thick calcified pericardium (PER) is seen as a wide, dense echo-band anterior as well as posterior to the heart. The left ventricular posterior wall (LVPW) does not show normal gradual posterior motion (ventricular expansion) during diastole, a sign of pericardial constriction. RV, right ventricle; LV, left ventricle; VS, ventricular septum; ECG, electrocardiogram.

extremely dense and thick. Occasionally pericardial calcification may penetrate into the atrial or ventricular wall and appear intramyocardial on the echocardiogram.[24]

On the 2-D echocardiogram, pericardial thickening manifests as a rim or shell of unduly dense echoes encasing all or part of the ventricular perimeter (Figure 34-10). This is one diagnostic area in which the older M-mode technique is by no means obsolete; with its greater power of resolution, it is very useful in depicting finer imaging detail, including "texture" and gradations of density of pericardial echoes (Figure 34-8).

PERICARDIAL CONSTRICTION

Constrictive pericarditis is now rare in the United States. Only a very small percentage of patients with pericardial thickening on echocardiography eventually prove to have hemodynamically important constriction.

The ultrasound diagnosis of pericardial constriction is difficult because it is a physiologic rather than an anatomic diagnosis and therefore less suited to cardiac imaging techniques, and the echocardiographic signs for pericardial constriction are nonspecific or present only in a minority of patients.

On the 2-D echocardiogram, constrictive pericarditis may be suspected if, in a patient with thick pericardium, (1) the ventricles are low normal to small in size and contract normally, whereas the atria are both dilated; (2) the inferior vena cava is dilated and does not narrow with inspiration, and the hepatic veins are also very dilated; or (3) the atrial and ventricular septa bulge toward the left side during inspiration. These features are described by Lewis in five patients with constrictive pericarditis of tuberculous etiology.[25]

Another valuable though subtle sign is rapid but abruptly limited expansion of the ventricles in early diastole. Several M-mode echocardiographic abnormalities have been described in association with pericardial constriction. Perhaps the most useful is absence of the gradual ventricular expansion that normally occurs in diastole (Figure 34-9), best appreciated at slow heart rates with long diastolic periods. In pericardial constriction, the left ventricular posterior wall is not

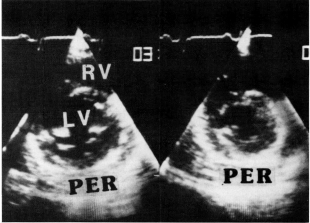

Figure 34-10 Two-dimensional echocardiogram in the parasternal short-axis view of a 10-year-old child who had been admitted for acute idiopathic pericarditis 2 months before. Note the dense posterior echoes produced by the thickened pericardium (PER). Symptoms and signs of constrictive pericarditis were absent. RV, right ventricle; LV, left ventricle.

more than 1 mm farther from the chest wall in late diastole (at onset of the atrial kick) than it is at the end of the early diastolic rapid filling phase.[26]

A second abnormality is abnormal systolic motion of the ventricular septum, to varying degrees: hypokinesis, akinesis, or paradoxical anterior motion.[27] An abnormal brief anterior motion very early in diastole has been described by Spanish echocardiographers.[28] An unduly prominent transient anterior motion (atrial kick) late in diastole has also been reported as a sign of pericardial constriction.

An abnormally large-amplitude pulmonic valve *a* wave, an unduly rapid mitral EF slope, and striking disparity between cardiac width at the ventricular and atrial levels on M-mode (Figure 34-9) and 2-D echocardiography are three more signs.

The presence of two or more of these echocardiographic signs in a patient with definite pericardial thickening (particularly if the thickening is very prominent and dense, suggesting calcification) should raise the suspicion of pericardial constriction. This suspicion is further strengthened if the jugular venous pressure is much elevated, the liver enlarged, and left ventricular systolic function undoubtedly normal (on echocardiography).

One of the most important (yet difficult) differential diagnoses in cardiology is that between pericardial constriction and restrictive cardiomyopathy (particularly amyloid). These two entities closely resemble each other in clinical presentation, physical signs, ECG, chest x-ray film appearance, and even hemodynamic findings. Echocardiography can be of real help by demonstrating abnormally thick ventricular walls with (usually) somewhat subnormal contractility in restrictive cardiomyopathy, and normal wall thickness and contractility with densely thickened pericardium in pericardial constriction. Recent work suggests that Doppler recordings of mitral flow might also show certain differences between the two entities.[29]

ABSENT PERICARDIUM

Congenital absence of the left pericardium may be associated with the following echocardiographic abnormalities: excessive motion of the left ventricular posterior wall, paradoxical systolic anterior motion of the ventricular septum, and right ventricular dilatation.[30] Similar findings are typical of right ventricular volume overload states. In the latter, the presence of an atrial septal defect or significant tricuspid regurgitation could be detected by Doppler and 2-D echocardiography.

Partial absence of the left pericardium has been reported to cause herniation of the left atrial appendage, which may dilate to aneurysmal proportions.

SUMMARY

The introduction of cardiac ultrasound 2 decades ago revolutionized the diagnosis of pericardial disease. M-mode echocardiography soon became the procedure of choice for detecting the presence and size of pericardial effusions, and even the presence of tamponade. Subsequently, the development of the 2-D technique extended the diagnostic scope of echocardiography, by providing additional information about the distribution and loculation of pericardial fluid; further 2-D echocardiographic signs of tamponade were also established.

Echocardiography can also identify pericardial thickening and intrapericardial adhesions, as well as fibrous material or neoplastic masses within the pericardium. With regard to constrictive pericarditis, echocardiographic data are suggestive rather than definitive and have to be considered along with physical signs, the chest x-ray film, and other clinical information.

Echocardiography is cost effective with respect to diagnostic yield of pericardial abnormalities in the following clinical circumstances:

1. symptoms, physical signs, or ECG suggestive of acute or subacute pericarditis
2. unexplained enlargement of the cardiac shadow on the chest x-ray film
3. unexplained congestive cardiac failure
4. unexplained atrial tachyarrhythmia in a patient with a lung or mediastinal mass
5. any cardiovascular symptom in a patient with end-stage renal disease; recent radiation to chest; recent myocardial infarction, cardiac surgery, or chest trauma; or past history of lymphoma, lung malignancy, or breast malignancy

Other cardiac imaging techniques, such as computed tomography or magnetic resonance imaging, are particularly helpful in supplementing echocardiography if the latter (and/or the chest x-ray film) suggests mediastinal pathology, or a paracardiac cystic or solid mass.

REFERENCES

1. Feigenbaum H, Waldhausen JA, Hyde LP. Ultrasound diagnosis of pericardial effusion. *JAMA.* 1965;191:107–110.

2. D'Cruz IA, Cohen HC, Prabhu R, Glick L. Diagnosis of cardiac tamponade by echocardiography. *Circulation.* 1975;52:460–464.

3. Horowitz MS, Schultz CS, Stinson EG, Harrison DG, Popp RL. Sensitivity and specificity of echocardiographic diagnosis of pericardial effusion. *Circulation.* 1974;50:239–247.

4. Martin RP, Bowden R, Filly K, Popp RL. Intrapericardial abnormalities in patients with pericardial effusion. *Circulation.* 1980;61:568–572.

5. Chiaramida SA, Goldman MA, Zema MJ, Pizarello RA, Goldberg HM. Echocardiographic identification of intrapericardial fibrous

strands in acute pericarditis with pericardial effusion. *Chest.* 1980;77:85–88.

6. Schuster AH, Nanda NC. Pericardiocentesis induced intrapericardial thrombus. *Am Heart J.* 1982;104:308.

7. Chandaratna PA, Aronow WS. Detection of pericardial metastases by cross-sectional echocardiography. *Circulation.* 1982;63:197–199.

8. Haaz WS, Mintz GS, Kotler MN, Parry W, Segal BL. Two-dimensional echocardiographic recognition of the descending thoracic aorta. Value in differentiating pericardial from pleural effusion. *Am J Cardiol.* 1980;46:739–743.

9. D'Cruz IA. Echocardiographic simulation of pericardial fluid accumulation by right pleural effusion. *Chest.* 1984;86:451–453.

10. Hynes JK, Tajik AJ, Osborn MJ, Orszulak TA, Seward JB. Two-dimensional echocardiographic diagnosis of pericardial cyst. *Mayo Clin Proc.* 1983;58:60.

11. D'Cruz IA. Echocardiographic simulation of pericardial effusion by ascites. *Chest.* 1984;85:93–95.

12. Cosio F, Martinez JP, Serrano CM. Abnormal septal motion in cardiac tamponade with pulsus paradoxus. *Chest.* 1977;71:787–788.

13. Settle HP, Adolph RJ, Fowler NO, Engel P, Agruss NS, Levenson NI. Echocardiographic study of cardiac tamponade. *Circulation.* 1977;56:951–959.

14. Leeman DE, Riley MF, Carl LV, Come PC. Doppler echocardiography in cardiac tamponade: Exaggerated respiratory variation in transvalvular blood flow velocity integrals. *J Am Coll Cardiol.* 1987;9(suppl):17A. Abstract.

15. Appleton CP, Hatle LK, Popp RL. Superior vena cava flow velocity patterns can diagnose cardiac tamponade in patients with pericardial effusions. *J Am Coll Cardiol.* 1987;9:118A. Abstract.

16. Armstrong WF, Schilt BF, Helper DJ, Dillon JC, Feigenbaum H. Diastolic collapse of the right ventricle with tamponade: An echocardiographic study. *Circulation.* 1982;65:1491–1496.

17. Williams GJ, Partridge JB. Right ventricular diastolic collapse: An echocardiographic sign of tamponade. *Br Heart J.* 1983;49:292.

18. Gillam LD, Guyer DE, Gibson TC, King ME, Marshall JE, Weyman AE. Hydrodynamic compression of the right atrium: a new echocardiographic sign of cardiac tamponade. *Circulation.* 1983;68:294.

19. Kronzon I, Cohen ML, Winer HE. Cardiac tamponade by loculated pericardial hematoma. *J Am Coll Cardiol.* 1983;1:913.

20. Friedman MJ, Sahn DJ, Haber K. Two-dimensional echocardiography and B-mode ultrasonography for the diagnosis of loculated pericardial effusion. *Circulation.* 1979;60:1644.

21. D'Cruz IA, Kensey K, Campbell C, Replogle R, Jain M. Two-dimensional echocardiography in cardiac tamponade occurring after cardiac surgery. *J Am Coll Cardiol.* 1985;5:1250–1252.

22. Chandaratna PAN, First J, Langevin E, O'Dell R. Echocardiographic contrast studies during pericardiocentesis. *Ann Intern Med.* 1977;87:199–200.

23. Schnittger I, Bowden RE, Abrams J, Popp RL. Echocardiography: Pericardial thickening and constrictive pericarditis. *Am J Cardiol.* 1978;42:388–395.

24. D'Cruz IA, Levinsky R, Anagnostopoulos C, Cohen HC. Echocardiographic diagnosis of partial pericardial constriction of the left ventricle. *Radiology.* 1978;127:755–756.

25. Lewis BS. Real time two-dimensional echocardiography in constrictive pericarditis. *Am J Cardiol.* 1982;49:1789–1793.

26. Voelkel AG, Pietro DA, Folland ED, Fisher ML, Parisi AF. Echocardiographic features of constrictive pericarditis. *Circulation.* 1978;58:871–875.

27. Pool PE, Seagren SC, Abbasi AS, Charuzi Y, Kraus R. Echocardiographic manifestations of constrictive pericarditis: Abnormal septal motion. *Chest.* 1975;68:684–688.

28. Candell-Riera J, Delcastillo G, Permanyer-Miralda G, Soler-Soler J. Echocardiographic features of the interventricular septum in chronic constrictive pericarditis. *Circulation.* 1978;57:1154–1158.

29. Hatle L, Appleton CP, Popp RL. Constrictive pericarditis and restrictive cardiomyopathy, differentiation by Doppler recording of atrioventricular flow velocities. *J Am Coll Cardiol.* 1987;9(suppl):17A. Abstract.

30. Payvandi MN, Kerber RE. Echocardiography in congenital and acquired absence of the pericardium. *Circulation.* 1976;53:86.

I am grateful to Ms. Stephanie Hopkins for her word-processing and secretarial skills, and Ms. Karen McBride for her photographic expertise.

Chapter 35

Computed Tomography in the Diagnosis of Pericardial Disease

William Stanford, MD

The exact incidence of pericardial disease is not known. The frequent reports seen in the literature stem from an increasing awareness of those disease processes involving the pericardium, coupled with better identification of the changes that occur following sternal splitting operations and drug and radiation therapies.[1] Several imaging modalities can be used to assess pericardial disease. The two more commonly used are ultrasound and computed tomography (CT); however, clinical experience with magnetic resonance imaging (MRI) is increasing. In this chapter I will discuss pericardial anatomy as it relates to CT imaging, point out some anatomic variations and pitfalls of the procedure, illustrate how various disease processes affect CT images of the pericardium, and attempt to show some of the advantages and disadvantages of CT in the evaluation of these disease processes. I will conclude by discussing how the newer imaging modality, ultrafast CT, provides additional physiological information for the evaluation of pericardial disease.

ADVANTAGES AND DISADVANTAGES OF CT AS COMPARED WITH ULTRASOUND AND MAGNETIC RESONANCE IMAGING

Ultrasound should be the initial imaging modality in evaluating pericardial disease. It has the advantages of being readily available even in the small hospital setting and of being fully transportable so that severely ill patients may be studied without the necessity of transport. As will be pointed out subsequently, it does very well in imaging pericardial effusions but less well in evaluating pericardial thickening, loculated effusions, and focal abnormalities of the pericardium. Technical problems can degrade the ultrasound images; examples are interference with transducer placement due to the ribs and sternum and interposed lung. The technique is highly operator dependent, and the images may not be reproducible.[2] Furthermore, the composition of a pericardial effusion, if present, may not be apparent from the images.

The advantages of CT are that overlying bone, lung, and bowel gas do not interfere wth image quality. Pericardial anatomy is well seen, and intrapericardial masses and loculated effusions are routinely visualized, as is calcification. The images are reproducible and easily interpretable, even by the neophyte imager. While motion artifact in conventional CT may interfere with the definition of cardiac anatomy, it does not affect the pericardium, and the pericardial anatomy is consistently visualized.[1] An additional advantage of CT is its value in outlining puncture routes to avoid lung and pleura during needle aspirations of tumors and loculated effusions.[2] The disadvantages of CT are radiation exposure and the adverse effects of the contrast medium, if one is administered. Additional disadvantages are occasional motion artifacts, which can interfere with image resolution, and surgical clips and implantable metallic devices, which may cause streak artifacts. Other shortcomings lie in the inability of the scanner to reconstruct images except in coronal or sagittal planes and the possibility of missing lesions if axial slice thicknesses are not contiguous or if the patient moves during the scan. On occasion, patients' size may preclude their fitting into the scanner gantry opening.

Magnetic resonance imaging is also very useful for evaluating pericardial disease but requires longer scan times, and gating problems may interfere with anatomic definition. Magnetic resonance imaging has the ability to reconstruct images in multiple planes and does not have to rely on a reformat sequence, as is the case with conventional CT. The disadvantages are those of inferior resolution and the inability to see thin calcifications.

NORMAL PERICARDIAL ANATOMY

The pericardium is a flasklike fibroserous sac surrounding the heart. It attaches to the great vessels cephalad to the aortic and pulmonary valves and has a wide base anchored to the central tendon of the left hemidiaphragm.[1] The parietal pericardium consists of three layers: an outer connective tissue layer, a middle fibrous layer, and an inner serous layer,[3] which is continuous with the epicardial layer of the ventricular myocardium. Under normal conditions the two serous layers are everywhere in contact[4] and are lubricated by up to 25 mL of an ultrafiltrate of serum.[5] Fibrous attachments of the pericardium to the sternum may be present, and some patients may also have attachments to the dorsal spine.[6] In most instances the pericardium is separated from the sternum and anterior thoracic wall by fat, and possibly by interposed lung and pleura. Posteriorly the pericardium is separated from the spine by the esophagus, descending aorta, and mediastinal surfaces of lung and pleura.[4]

On CT the ventral pericardium is commonly seen as a 1- to 2-mm line of soft tissue density lying between mediastinal fat ventrally and epicardial fat dorsally (Figure 35-1), except inferiorly at its insertion into the diaphragm, where there is apparent thickening to 3 to 4 mm.[2] Some aspect of the ventral pericardium is visualized in virtually 100% of patients. The dorsal pericardium is seen only 25% of the time; however, visualization may be improved by imaging perpendicularly to the long axis of the heart. At its attachment to the great vessels, there is often little to no fat so the pericardium is frequently not visualized. The thin visceral pericardium (epicardium) is likewise usually not seen. The same is true of the pericardial cavity unless it is distended by air, fluid, or a mass lesion.[2]

There are recognized pitfalls in imaging the pericardium. As previously stated, inferior thickening at the attachment to the diaphragm is a common finding and should not be confused with a pathological process. Similarly, nodular thickenings in the area of the right

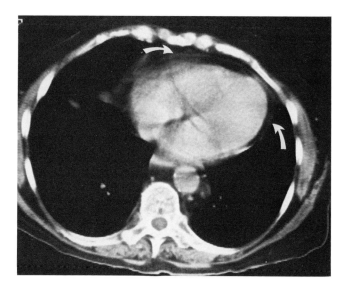

Figure 35-1 Computed tomograph at the level of midventricle, showing normal pericardium (arrows) surrounding both the right and left ventricle.

atrium and ventricle are also normal. Superoposteriorly at its attachment to the aorta, there is often a retroaortic recess that should not be mistaken for adenopathy.[7] Additional recesses can also be confusing and will be discussed in the section on effusions.

IMAGING TECHNIQUES AND PROTOCOLS

Imaging sequences vary, but one of our frequently used sequences consists of a series of axial images 10 mm in thickness, at interslice intervals of 10 mm. The images take 2 to 5 seconds to acquire, and because of this, CT is not suitable for use with agitated patients or patients who cannot hold their breath for the time required for the imaging sequence. Computed tomography is also not useful for patients who cannot lie supine or are too ill to be transported to the imaging suite.

Examinations may be done either with or without contrast agent administration. Noncontrast images show excellent pericardial anatomy, if there is a moderate amount of epicardial and mediastinal fat to accentuate the fat/pericardial interface. Pericardial calcification is readily apparent, and cysts and effusions may be easily identified. In these examinations, Hounsfield numbers are useful in defining the composition of the pericardial mass or effusion.

Enhancement has certain advantages. In our protocols up to 150 mL of Renografin 60 (or a similar contrast agent) is administered by multiple bolus or by bolus and intravenous drip infusions.[8] We find contrast agent administration to be useful for outlining cardiac chambers, for enhancing the pericardium in inflammatory states, and for differentiating between pericardial thickening and adjacent atelectatic lung. Highly vascular tumors tend to show better definition with a contrast agent, and it also helps define pericardial anatomy in patients with minimal amounts of epicardial fat. It is routinely used in ultrafast CT imaging, as will be discussed later. Contrast media are, of course, contraindicated in patients with allergic histories and renal compromise.

CONGENITAL ABSENCE OF THE PERICARDIUM

The incidence of congenital absence of the pericardium varies between 1 in 700 and 1 in 13,000, as determined from necropsy studies. The condition is three times more frequent in males,[9] and the diagnosis is often made later in life. The entire parietal pericardium may be absent or just the left hemipericardium; however, the most common presentation is a localized defect in the pericardium over the left atrial appendage.[10] An absence is suspected when the posteroanterior chest film shows an abnormal contour to the left side of the heart.[11] The CT findings of pericardial absence are a lack of pericardial continuity and a herniation of the entire left ventricle or of just the left atrial appendage through the defect. In addition, there are often changes in the positional axis of the heart and great vessels,[2] and with lateral positioning, the heart may displace into the left chest. These findings, along with the demonstration of direct contact between the heart and great vessels and lung, appear to be the most reliable of the CT findings. A localized defect is of greater clinical significance since herniation of an atrial appendage could result in strangulation or dysrhythmia. Conversely, in complete absence, there are often no symptoms. Overall, computed tomography appears to be an excellent method of detecting pericardial defects.

PERICARDIAL CYST

Cysts of the pericardium are relatively uncommon and are seen in 1 in 100,000 people (Figure 35-2).[12–13] They present as 3- to 8-cm mass lesions located in the right or left pericardiophrenic angles and often occur in asymptomatic patients. More commonly they are on the right. The CT appearance is distinctive. They lie in close approximation to but usually do not communicate with the pericardium and are filled with a clear, low-density fluid. They have a wall thin enough not to be visible on CT.[2] They are near water density, and the Hounsfield numbers cluster around zero but can go as high as 20 to 40.[14] A pericardial diverticulum has a similar CT appearance, with the only difference being that the diverticulum communicates directly with the pericardial cavity.

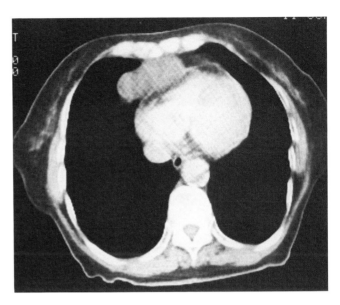

Figure 35-2 Pericardial cyst presenting as a low-density loculated collection in the right pericardiophrenic space. The cyst density is between that of fat and soft tissue. *Source:* Courtesy of G DeMarino MD, Pittsburgh, PA.

Cysts need to be differentiated from mediastinal fat, solid tumors, and ventricular aneurysms. The CT appearance is usually distinctive and the diagnosis readily apparent. Whereas previously the patient required exploration to determine the diagnosis, with CT the present management is to forgo operation and follow the patient.

PERICARDIAL EFFUSION

The pericardium can react to injury by fluid exudation, fibrin production, and cellular proliferation[15]; all can occur either independently or concomitantly. Fluid initially accumulates in the most caudal portion of the pericardial sac, and small effusions are best seen dorsolaterally to the left ventricle. The fluid is seen on CT as a thin, elliptical density, dorsal to the left ventricular myocardium. As the effusion increases, the fluid extends and is then seen ventral to the right atrium and ventricle (Figure 35-3). Massive pericardial effusions surround the heart and can compress the diaphragm and change the shape and axis of the heart. In these instances, the extent of the axis distortion is often proportional to the volume of the effusion. With massive effusions, the heart may actually "float" within the fluid and have its apex tilted cephalad.[1] At times the fluid may extend far enough superiorly to surround the origins of the great vessels (Figure 35-4).[16] In addition to these locations, there are recesses within the pericardium where fluid can accumulate. These include the transverse sinus behind the ascending aorta and pulmonary trunk, the oblique sinus behind the left atrium, and the pulmonic recess between the left pulmonary

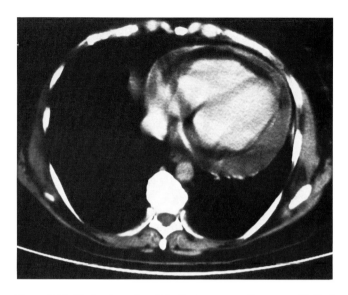

Figure 35-3 Moderate pericardial effusion, presenting as an elliptical density dorsal to the left ventricle. There is a small rim of fluid anterior to the right ventricle.

artery and left superior pulmonary vein.[17] Fluid in these areas is usually readily identifiable by CT.

The amount of fluid and its physiological effect can vary. The normal pericardium may become enormously distended with a chronic effusion that produces little or no hemodynamic alteration. Conversely, a small effusion can lead to cardiac tamponade if the fluid accumulates rapidly or if it accumulates in an area where a thickened pericardium limits distensibility (Figure 35-5).

Fluid composition can also vary. Effusions may be serous or chylous and have CT numbers from 0 to 40 HU. Exudates and blood generally have CT numbers that approach those of soft tissue, and in these cases it may be hard to distinguish between fluid and a thickened pericardium. The use of a contrast agent and the manipulation of window and level settings are helpful in trying to resolve this issue.

Effusions can be free-flowing or present as loculations contained in spaces formed by pericardial adhe-

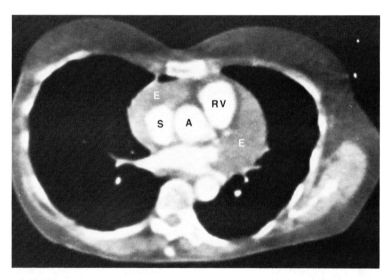

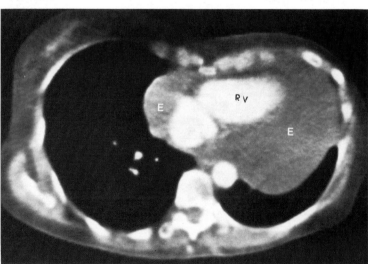

Figure 35-4 Top, Large pericardial effusion (E), surrounding the superior vena cava (S), aorta (A), and right ventricular outflow tract (RV). Bottom, Inferiorly, the right ventricle (RV) appears to be "floating" in the effusion (E).

CT in the Diagnosis of Pericardial Disease 455

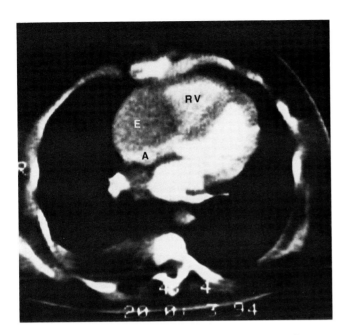

Figure 35-5 Loculated pericardial effusion (E) in a postoperative coronary artery bypass graft patient. The fluid was compressing the right atrium (A) and producing tamponade. It did not encroach on the right ventricle (RV). Following the scan, she was taken to an operating room where the fluid was drained. There was immediate resolution of her symptoms.

phy, are reproducible, and may accurately assess ventricular compression. Often CT will be able to define the composition of the effusion.

PERICARDIAL THICKENING

Pericardial thickening results from trauma and inflammation, from primary or metastatic tumor infiltration, and as a sequela of median sternotomy incisions and radiation injury. The thickening can be localized or generalized and can range from a few millimeters to more than 5 cm. It may involve just the parietal layer or both parietal and visceral layers and at times may infiltrate the myocardium and obliterate the pericardial cavity. Often the greatest thickening occurs ventral to the right atrium and ventricle.

The organisms responsible for pericardial thickening are changing. Staphylococcal and tuberculosis organisms are less often seen and are being replaced by coxsackie, echo, and polio viruses. Noninfectious inflammations arising from radiation and dialysis injury are also being increasingly identified.[1-2]

Calcification occurs as the sequel of end stage tuberculous pericarditis (Figure 35-7). Other etiologies are bacterial and viral infections, hemorrhage, and rheumatoid involvement.[15] The calcification may be present in patches or be continuous and up to 2 cm in thickness. It often occurs in the atrioventricular or interventricular grooves and at the crux of the heart and does not necessarily result in constrictive physiology. It is exquisitely imaged on the nonenhanced CT.

sions (Figure 35-6). Positioning the patient for prone or lateral decubitus views may help define the borders of these collections.

While most physicians feel echocardiography should remain the primary imaging modality in diagnosing pericardial effusions, many feel CT to be an important adjuvant. Computerized tomographic images show loculations and thickening better than echocardiogra-

Postsurgical scans are increasingly being used to assess for effusion or bleeding as an etiology of low

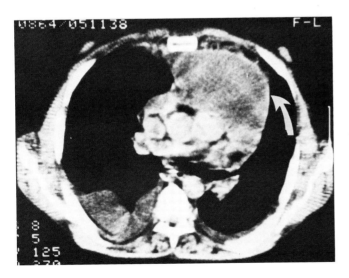 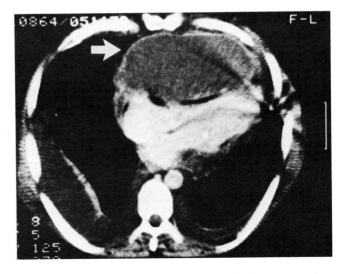

Figure 35-6 Large loculated pericardial effusion anterior to and compressing the right ventricle (straight arrow). Superiorly, the effusion extends to the origin of the great vessels (curved arrow). *Source:* Courtesy of G Rienmuller, Muenchen, West Germany.

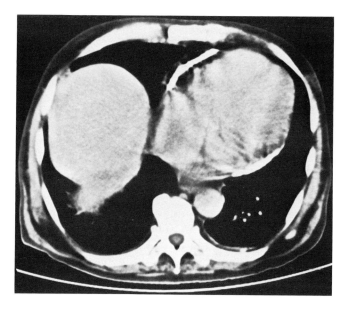

Figure 35-7 Calcification of the pericardium, anterior to the right ventricle and posterolateral to the left ventricle. The calcification is readily seen on CT.

cardiac output. The scans can also assess the extent of a pericardial resection if one has been carried out.

At times a high-density pericardial effusion may mimic thickening, and in these instances changing the position of the patient may change the effusion and hence the pericardial contour. The change does not occur in fibrous thickening, and this difference may be used to help differentiate these lesions. Another source of confusion is adjacent atelectatic lung. Lung often enhances with the use of a contrast agent and can be differentiated from pericardial thickening per se.

BENIGN TUMORS

Benign tumors of the pericardium are infrequent. Teratomas, bronchogenic cysts, leiomyomas, hemangiomas, and lipomas are the primary lesions.[6] Intrapericardial lipomas are readily identifiable on CT because they have densities ranging between −55 to −120 HU. If the patient is asymptomatic, these are usually not removed, and the patient is followed. Hemangiomas often enhance significantly with contrast agent use and are also identifiable. The CT numbers of bronchogenic cysts may vary depending upon the composition of the cyst contents, and this may make interpretation difficult. The same is true with leiomyomas. In both, the CT numbers are those of soft tissue, and malignancy cannot be excluded. In imaging teratomas, CT has an advantage over ultrasound in that it can readily identify both the fat and tooth calcifications that are often present.

MALIGNANT DISEASE

Metastases from breast and lung are the most common malignant tumors of the pericardium (Figure 35-8). Of the primary tumors involving the pericardium, malignant mesothelioma (Figure 35-9) appears to be the most frequently seen. The latter tumor may present as a solitary mass or be multicentric in origin, or it may form diffuse plaques that encase the visceral and parietal pericardium. Any area of the pericardium can be involved, but the pericardium over the right or left ventricle is the most often affected. Nodular involvement is seen, and thickenings can be up to 3 cm. The thickening can result in constrictive physiology, but often there is little to no impairment of ventricular function.

Lymphoma is a primary malignancy that also involves the pericardium. These tumors commonly present with a diffuse infiltration that at times causes constrictive physiology by virtue of total encasement.

CONSTRICTIVE PERICARDITIS

Many disease processes (tuberculosis, mediastinal fibrosis, tumor, infection) can result in constrictive physiology. Tissue planes between the parietal and visceral pericardium are lost, and fibrosis may prevent diastolic filling and result in low-output states. The extent, location, and thickness of the pericardial involvement are important, especially if pericardiectomy is contemplated. At times the fibrotic process extends into the myocardium, which makes pericardiectomy extremely difficult. Conversely, if there is a thickened parietal pericardium in juxtaposition to a thin visceral pericardium, or if there is an adequate tissue plane between the parietal and visceral layers, then stripping of the pericardium is a relatively easy operative procedure. The same is true if the parietal layer is separated from the epicardial layer by pockets of fluid or gelatinous material. The axial slice images of CT define pericardial thickness, demonstrate obliterations of the pericardial cavity, and show separations of the pericardial layers by fluid or gelatinous material. The images also show the extent of involvement and whether a single chamber or multiple chambers are affected.

It is also important to differentiate restrictive cardiomyopathy from constrictive pericarditis. Ultrafast CT is extremely useful in this differentiation because in cardiomyopathy the pericardium is of normal thickness, whereas in constrictive states the pericardium is thick and noncompliant.

One disadvantage is the inability of conventional CT to evaluate ventricular function; however, this is now possible with ultrafast CT.

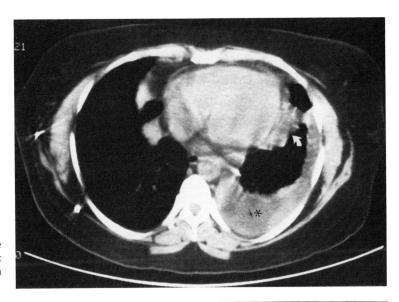

Figure 35-8 Metastatic deposits from breast carcinoma are producing thickening of the pericardium lateral to the left ventricle (arrow). Posteriorly, there is a large pleural effusion layering out in the left chest (asterisk).

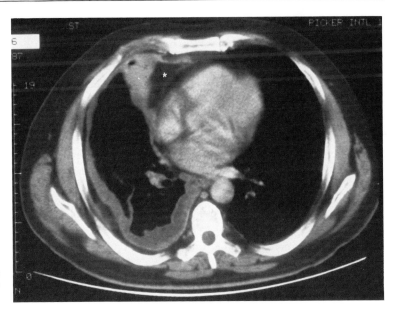

Figure 35-9 Mesothelioma encasing the right hemithorax and involving the pericardium over the right atrium. Anteriorly, there is a loculated fluid collection (asterisk).

ASSOCIATED LESIONS

Foramen of Morgagni hernias and eventrations of the diaphragm can present as parasternal and retrosternal masses, and these need to be differentiated from pericardial disease. Computed tomography can identify these entities and may be the modality of choice for differentiating them.

ULTRAFAST CT IN PERICARDIAL DISEASE

Ultrafast CT, by virtue of its imaging speed, adds an additional dimension to the CT evaluation of pericardial disease. It retains the advantages of conventional CT in showing pericardial anatomy; however, in addition it images ventricular contractions at up to 17 images per second and plays them back in a closed loop movie mode, thus allowing the observer to see the actual contraction pattern of the ventricle. Conventional CT, because of the relatively slow scanning times of 1.5 to 5 seconds per image, is too slow to image the beating heart. It is also possible to calculate from ultrafast images left and right ventricular end-diastolic and end-systolic volumes, stroke volume, ejection fraction, left ventricular mass, and cardiac output. This is important information in those patients suspected of having constrictive physiology. My current practice is to image these patients with echocardiography and then to do an ultrafast CT scan to evaluate pericardial thickening and wall motion abnormalities and to determine the physiologic parameters mentioned above. This sequence allows a comprehensive evaluation of both pericardial anatomy and ventricular function.

SUMMARY

Conventional CT and ultrafast CT are complementary to echocardiography in evaluating pericardial disease. Currently, conventional CT is the imaging modality of choice for evaluating small loculated effusions, localized pericardial thickening, and calcification, while ultrafast CT is the modality of choice in evaluating pericardial constriction and ventricular function.

REFERENCES

1. Naidich DP, Zerhouni EA, Siegelman SS. *Computerized Tomography of the Thorax.* New York: Raven Press; 1984:269–289.
2. Moncada R, Demos TC, Posniak HV, Hammer R. Computed tomography of pericardial heart disease. In: Taveras JM, Ferrucci JT, eds. *Radiology: Diagnosis, Imaging, Intervention.* Vol 2. Philadelphia: JB Lippincott Co, 1986:1–19.
3. Ishihara T, Ferrans VJ, Jones M, et al. Histologic and ultrastructural features of normal parietal pericardium. *Am J Cardiol.* 1980;46:744–753.
4. Clemente CD, ed. *Gray's Anatomy.* 30th ed. Philadelphia: Lea & Febiger; 1985:622.
5. Holt JP. The normal pericardium. *Am J Cardiol.* 1970;26:455–465.
6. Lipton MJ, Herfkens RJ, Gamsu G. Computed tomography of the heart and pericardium. In: Moss AA, Gamsu G, Gerot HK, eds. *Computed Tomography of the Body.* Philadelphia: WB Saunders Co; 1983:414–423.
7. Aronberg DJ, Peterson RR, Glazer HS, et al. The superior sinus of the pericardium: CT appearance. *Radiology.* 1984;153:489–492.
8. Chiu L, Stanford W, Yiu-Chiu V. *Computed Tomographic Angiography of the Mediastinum.* St. Louis: WH Green; 1986:148–182.
9. Moore RL. Congenital deficiencies of the pericardium. *Arch Surg.* 1925;11:765–777.
10. Nasser WK. Congenital absence of the left pericardium. *Am J Cardiol.* 1970;26:466–470.
11. Fraser RG, Pare JAP. *Diagnosis of Diseases of the Chest.* 2nd ed. Philadelphia: WB Saunders Co; 1977:656.
12. LeRoux BT. Pericardial coelomic cysts. *Thorax.* 1959;14:27–35.
13. Engle DE, Tresch DD, Boncek LI, et al. Misdiagnosis of pericardial cyst by echocardiography and computed tomography scanning. *Arch Intern Med.* 1983;143:351–352.
14. Brunner DR, Whitley NO. A pericardial cyst with high CT numbers. *AJR.* 1984;142:279–280.
15. Roberts WC, Spray TL. Pericardial heart disease: A study of its causes, consequences and morphologic features. *Cardiovasc Clin.* 1976;7:11–65.
16. Lee JK, Sagel SS, Stanley RJ. *Computed Body Tomography.* New York: Raven Press; 1983:122–126.
17. Levy-Ravetch M, Auh YH, Rubenstein WA, Whalen JP, Zazam E. CT of pericardial recesses. *AJR.* 1985;144:707–714.

Chapter 36

Evaluation of Pericardial Disease by Magnetic Resonance Imaging

Madeleine R. Fisher, MD

Since the introduction of magnetic resonance imaging (MRI), multiple clinical applications for the technique have been discovered. Magnetic resonance imaging has been found to be particularly useful for demonstration of cardiovascular abnormalities because of the inherent contrast provided between flowing blood and the cardiac chambers and blood vessel walls.[1-2] As a consequence of the unique ability of MRI to allow delineation of the cardiovascular structures and surrounding anatomy, investigation into its uses for pericardial disease has been ongoing. Traditionally, echocardiography and angiography have been used to demonstrate pericardial disease. However, certain pericardial diseases are not readily demonstrated by these techniques because of the inability to visualize the pericardium directly. With the advent of computed tomography (CT), the pericardium began to be better depicted.[3-4] After the introduction of MRI, it was clear this technique could not only delineate the pericardium well but also delineate the underlying cardiovascular structures and depict surrounding pathology, due to its large field of view. MRI, when compared to echocardiography, allows a larger field of view, is not dependent upon operator expertise, and is not limited as a consequence of difficulties with a patient's body habitus.

This chapter will discuss the normal anatomy of the pericardium, technical approaches to evaluation of the pericardium by MRI, the normal MRI appearance of the pericardium, congenital anomalies, and a variety of pathologic processes.

ANATOMY

The pericardium is a fibroserous sac that completely surrounds the heart and extends for a short distance along the aorta, the pulmonary trunk, the venae cavae, and the pulmonary veins. The outer layer or pericardium is mainly fibrous tissue and is continuous with the inner layer or epicardium, which is closely applied to the heart surface.[5-6] The apposing surfaces of these two layers are lined with a smooth serous membrane, between which is the pericardial cavity. It normally contains a small amount of fluid.

The superior part of the epicardium surrounds the ascending aorta and pulmonary artery trunk. Anteriorly, the epicardium lines the groove between the ascending aorta and the pulmonary artery to form a preaortic compartment or recess. Behind the ascending aorta, immediately above the level of the right pulmonary artery, the epicardium reflects posteriorly to become the parietal pericardium, forming a retroaortic recess.[7] These are termed superior pericardial recesses. The pericardium covers the ascending aorta for approximately one half the distance from the aortic valve to the origin of the innominate artery (2 to 3 cm).[5] The pulmonary trunk is covered to a point just below its bifurcation. The investments of the venae cavae (particularly the inferior vena cava) and pulmonary veins are relatively short. The fibrous coat of the pericardium is continuous with the external coat of the great vessels and the mediastinal pleura. The lower portion, or base, is in close contact with the diaphragm.

TECHNIQUE

Several choices are available for cardiac imaging and specifically for imaging the pericardium. It is essential that electrocardiogram (ECG) gating be employed in order to adequately visualize the cardiovascular structures. Due to the nature of MRI, cardiac synchronization does not increase the image acquisition time. Magnetic resonance imaging uses short data acquisition periods (approximately 20 to 75 milliseconds), with long intervals between successive acquisitions.[8] The earliest technique employed with cardiovascular MRI used multislice ECG gating that resulted in several adjacent cardiac levels being acquired at different phases in the cardiac cycle. This type of technique may cause some uncertainty as to the actual pericardial thickness in relation to the phase of the cardiac cycle. In this type of multisection imaging, as there is advancement from section to section, the time interval between the ECG trigger and data collection changes. For example, the first section is synchronous with the trigger signal; the second is also synchronous but delayed by 100 milliseconds, and so on to the last section, which is delayed by 400 milliseconds. Thus, although several sections are sampled in a span of 3 to 10 minutes, each is sampled at a different but fixed point in the cardiac cycle. As mentioned, this can result in some uncertainty as to the actual pericardial thickness in relation to the phase of the cardiac cycle. This has also been seen in CT scans of the pericardium where images are obtained over several cardiac cycles. The reported minimum thickness of the pericardium on CT scans is more than twice the dimensions measured at gross pathologic examination.[3,6]

Another imaging strategy consists of using a cycled multisection imaging format. Again, an ECG-triggered spin echo pulse sequence is utilized; however, this technique allows imaging of each section at different phases in the cardiac cycle. Many terms have been used to describe this technique, including rotating gated acquisition,[9] cycled acquisition,[8] and permutation acquisition.[10] An example of this technique consists of the first section's being obtained after a 5- to 10-millisecond delay with respect to the trigger; each subsequent section has an incremental 100-millisecond delay. After the first acquisition, acquired in the format of section 1,2,3,4 . . ., the next acquisition rotates (cycles, permutes) the acquisition format so that the second acquisition acquires images at 2,3,4,5,1. The next acquisition acquires images at 3,4,5,1,2, and so forth until each of the sections is sampled at different times in the cardiac cycle. The software is programmed to reorder the sections such that they are displayed at one anatomic level throughout the cardiac cycle.

Gradient echo imaging is an MRI pulse sequence that allows rapid acquisition of images of the heart obtained at one anatomic level with 16 to 32 phases of the cardiac cycle. Contrary to the spin echo technique, where the blood has a low signal due to normal velocities, on the gradient echo technique blood has a high signal intensity. The gradient echo technique utilizes a variable flip angle to acquire images. The echo is formed by inversion of the measurement gradient, rather than by the 180° pulse used with spin echo imaging. Gradient echo imaging has allowed accurate demonstration of the cardiovascular structures, has improved delineation of valvular pathology and segmental myocardial wall motion, and has allowed precise demonstration of the pericardium.[11]

Each of the above-described cardiac MRI techniques has a role in the evaluation of pericardial disease.

MRI APPEARANCE OF THE NORMAL PERICARDIUM

The first report on the MRI appearance of the pericardium showed the pericardium as a dark line between the epipericardial and epicardial fat layers that was uniformly visible over the right anterior aspect of the heart on transverse images (Figure 36-1). This low-intensity line was attributed to the pericardium and possibly to some normal pericardial fluid.[12] The low-intensity MRI signal of the pericardium is postulated to be secondary to the relative paucity of mobile hydrogen nuclei in dense fibrous structures and the short T2 and fairly long T1 relaxation times. It is also postulated that physiologic pericardial fluid contributes to the thickness of the low-intensity line. The serous fluid within the pericardial sac remains of low intensity due to nonlaminar flow. The motion of the heart causes the nonlaminar motion of this fluid confined to the pericardial sac. The presence of nonlaminar flow suggests there are components of accelerated flow and multiple flow vectors within the imaged plane, leading to signal loss on first and second echo images because of the effects caused by changes in the spin phase.[13] A recent study of the MRI appearance of normal pericardium[14] determined the optimal phase of the cardiac cycle for visualizing the pericardium, identified the optimal transverse level of the heart for visualizing the pericardium, and established the dimensions of the pericardium at various phases of the cardiac cycle and at various anatomic levels.

The contrast provided on magnetic resonance images by the epicardial and epipericardial fat delineates the pericardium better along the anterior aspect of the right ventricle. The pericardial region located inferior and posterior to the left ventricle is less often visible because low-intensity lung tissue is adjacent to the pericardium, rather than fat. The pericardium is better visualized during systole. The pericardial thickness is seen to

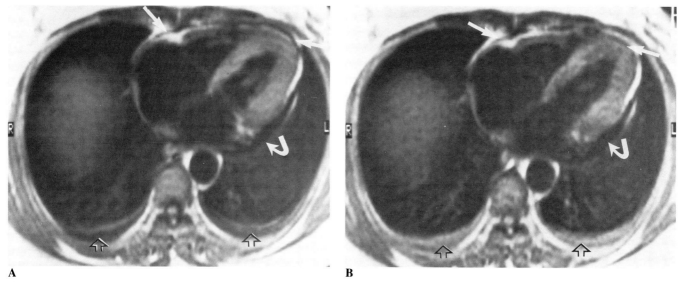

Figure 36-1 Transverse gated magnetic resonance images demonstrate the normal pericardium, seen as a dark line (straight arrows) along the right and anterior aspect of the heart. There is a posterior pericardial effusion (curved arrow), seen as low intensity on both the first (left) and second (right) spin echo images. In contrast, there are bilateral pleural effusions (open arrows) that are low intensity on the first spin echo image (left) and increase in intensity on the second spin echo image (right). The nonlaminar motion of pericardial fluid results in the signal loss observed with simple pericardial effusions.

increase significantly during the systolic phase. The precise reason why the pericardium appears thicker on systole is not known; however, there are several theories. It is hypothesized that there is improved systolic delineation of the pericardium on magnetic resonance images along the anterior aspect of the right ventricle due to the inability of the parietal pericardium to move as much as the visceral pericardium during cardiac ejection, due to its anchorage to the sternum anteriorly.[14] As a consequence, this difference in movement maximizes the pericardial space during systole. Another hypothesis is that there is a systolic increase in fluid anterior to the heart. This is seen in patients with pericardial effusions. It is postulated this may also be seen in patients who have physiologic pericardial fluid.

The optimal transverse level of the heart for visualizing the pericardium is the more caudal sections of the heart.[14] The thickness of the pericardial line is considerably increased on these sections and could measure up to 7 mm. It has been hypothesized that this thickened appearance, in a normal patient, results from three factors. First, the pericardium tends to be thicker in this region because of its diaphragmatic insertion. Second, since the inferior wall of the heart and the covering pericardium are almost parallel to the axial MRI sections used, this tangential sectioning may contribute to the increased thickness. Third, the normal pericardial fluid preferentially gathers in this region. Because there has been demonstration of variations in pericardial thickness depending upon the plane of the section, the most reliable section for measurement of pericardial thickness is felt to be at an anatomic level that displays the right atrium, right ventricle, and left ventricle.[14]

The average width of the pericardium in magnetic resonance images is 1.9 ± 0.6 mm in systole; in comparison, anatomic studies measure the parietal pericardium's width at 0.4 to 1 mm. It is felt that a small amount of physiologic pericardial fluid accounts for this apparent increased width. Although the width of the concomitant layer of pericardial fluid is found to be less than the thickness of the parietal pericardium, this fluid could contribute at least half of the thickness of the pericardial line seen on MRI. Additionally, the pericardial fluid need not be equally distributed around the heart. This would account for the pericardium's being of minimal thickness in some regions and of more than (presumed) normal thickness in others. The volume of physiologic pericardial fluid varies among normal subjects, and this may also lead to different thicknesses of the fluid layer. Therefore, it is likely that the pericardial width observed in magnetic resonance images is due to both the pericardium and the physiologic pericardial fluid.

In summary, the normal pericardium is optimally delineated along the anterior aspect of the right ventricle. The low-intensity pericardial line is better seen on MRI when there is a large amount of epicardial and epipericardial fat. Since there is less fat located behind the posterior and lateral walls of the left ventricle and around the right atrium, the pericardium may only be demonstrated in short distances and is not readily distinguishable from the surrounding low-intensity lung fields. The length of the visible pericardial line differs depending on the anatomic section being studied. The section that includes the four chambers shows the smallest amount of pericardium, and the section that includes the right atrium, left ventricle, and right ventricle shows the most.[14] The systolic phase of imaging also allows better delineation of the pericardium, especially when it measures less than 2 to 4 mm.

CONGENITAL ANOMALIES

A congenital defect of the pericardium is a rare anomaly. The most common site of involvement is the left side. Partial absence of the left side of the pericardium is found in approximately 1 of every 10,000 autopsies.[15] The chest x-ray film findings are usually fairly characteristic of congenital absence of the left side of the pericardium. On chest x-ray films there is usually leftware displacement of the heart, with prominence in the pulmonary outflow tract. Recently MRI has been shown to demonstrate absence of the pericardium,[16] and it can provide a definitive diagnosis in confusing or atypical cases by allowing direct visualization of the presence or absence of the pericardial reflection.

The coronal plane of imaging in MRI is similar to that seen on the posteroanterior x-ray film and allows for direct correlation between the two imaging studies. Magnetic resonance imaging will demonstrate the structure causing the bulge seen on the chest x-ray film and may show a lack of pericardium at the site of absence.[16]

MRI APPEARANCE OF DISEASED PERICARDIUM

Pericardium

As mentioned in the above section, the normal pericardium is seen as a thin, low-intensity line of 2 mm width, highlighted by the epipericardial fat (Figure 36-1). Thickening of the pericardium may be seen as a low signal intensity, medium signal intensity, and/or high signal intensity. In patients whose pericardium appeared thickened and of low intensity in MRI studies, pathologic examination revealed a thickened, fibrous pericardium with varying degrees of calcification.[17] In patients whose MRI studies demonstrated a medium signal intensity, pathologic examination showed a fibrous peel (Figure 36-2). Pathologic examination of patients with high–signal intensity thickening of the pericardium demonstrated a dense, fibrinous exudate involving the pericardium. This is the type of image seen in patients with uremia. Two features allow differentiation of the low–signal intensity fluid from the low–signal intensity fibrous or calcific pericardium: (1) Calcification produces irregular borders that extend into the right atrioventricular groove, displacing the fat that usually fills this space, and (2) the typical distribution of pericardial fluid is not seen when thickened fibrous tissue or calcifications exist.

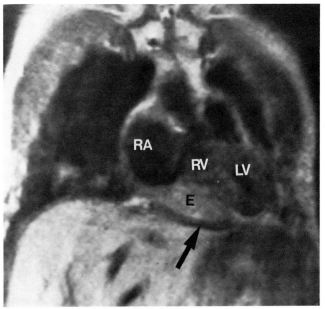

A

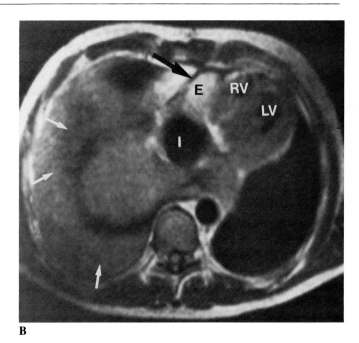

B

Figure 36-2 Thickened pericardium and a loculated pericardial effusion, resulting in effusive constrictive pericarditis. Left, Coronal gated magnetic resonance image demonstrates marked thickening of the pericardium along the inferior heart surface (black arrow). The loculated effusion (E) is seen as intermediate intensity on the first spin echo image because of the lack of nonlaminar flow. Right, Transverse gated magnetic resonance image shows right ventricle (RV) compression secondary to the loculated effusion. Also, the inferior vena cava (I) is dilated. Note on the second spin echo image the relative increase in signal intensity of the effusion. Redemonstrated is marked thickening of the pericardium along the right side of the heart. There is a large right pleural effusion (small arrows). RA, right atrium; LV, left ventricle. *Source:* Reprinted with permission from *Diagnostic Imaging* (1988:130–133), Copyright © 1988, Miller Freeman Publications Inc.

Usually an enlarged anterosuperior pericardial recess is indicative of a pericardial effusion and is not observed in cases of pericardial thickening.[17] A technique that allows visualization of one anatomic level throughout the cardiac cycle may aid in differentiating pericardial thickening from pericardial effusion. Usually the thickened pericardium has a constant appearance as opposed to an observed change in the thickness of the pericardial fluid layer during the cardiac cycle.

Patients who have undergone cardiac surgery may, even after 10 months, show pericardial thickening and yet not have features of constrictive pericarditis. This usually occurs when a median sternotomy surgical approach has been used and when no attempt is made to close the pericardium. Thus pericardial thickening alone is not diagnostic of pericardial constriction, as demonstrated by these postoperative patients who have pericardial thickening and no clinical symptoms.

Differences in signal intensity of the pericardium are found between patients who have recently undergone cardiac surgery and those with chronic pericarditis. In patients with chronic constrictive pericarditis, the pericardium is of low signal intensity due to thick fibrous tissue or calcification.[17] In contrast, a medium signal intensity surrounds the heart following cardiac surgery, which is felt to represent an organizing inflammatory reaction within and around the pericardium.

When there is a question of constrictive pericarditis versus restrictive cardiomyopathy, MRI has been particularly useful in excluding the former. Often echocardiography and conventional CT show equivocal findings.[17] Magnetic resonance imaging may also allow differentiation between these two pathologies when cardiac catheterization findings are equivocal.

Pericardial Cavity

Pericardial Effusions

Magnetic resonance imaging has been shown to outline the distribution of pericardial fluid and allow an estimation of its volume; these volume estimates correlate with those of echocardiography.[17] Additionally, MRI may demonstrate pericardial effusions not found on echocardiographic examination. Often these collections are located posterolateral to the right atrium, which is not optimally seen by echocardiographic study.

Simple Effusion. Pericardial effusions are seen as low signal intensity on the first echo of a spin echo image. These simple pericardial effusions usually move freely within the pericardial sac. As mentioned above, motion of the heart results in nonlaminar motion of the fluid confined within the pericardial sac. Nonlaminar flow results in signal loss on both the first and second echo images (Figure 36-1) because of effects caused by changes in the spin phase.[13] Therefore, the amount of fluid seen within the pericardial sac on MRI is based upon the observed thickness of the curvilinear low-intensity line around the heart, which is a combined effect of imaging the properties of the fibrous tissue of the pericardium and the nonlaminar motion of serous fluid. When there is thickened pericardium, as mentioned in the above section, and there is a question of differentiating fluid from thickening, defining the location, borders, and appearance of the superior pericardial recess may allow correct identification. In addition, when the pericardium is thickened and there are clinical symptoms suggestive of constrictive pericarditis, analysis of the effect on the underlying cardiac chambers may aid in the differentiation of a simple pericardial effusion from thickened pericardium.

The distribution of pericardial fluid is not entirely even.[17] Pericardial fluid collections have been observed mainly dorsolateral to the left ventricle. Additionally, many patients show a considerable amount of fluid dorsolateral to the right atrium. The anterosuperior pericardial recess, normally a small triangle between the aorta and the pulmonary artery, may be enlarged in patients with pericardial effusions. A fluid layer greater than 5 mm anterior to the right ventricle is considered indicative of a moderate effusion. As a result of the uneven distribution of fluid around the heart, there is no direct correlation between the thickness of the fluid layer in front of the right ventricle and the fluid volume.

Loculated Pericardial Effusion. In patients with loculated pericardial effusions, the signal intensity of the fluid may be intermediate on the first echo of the spin echo image and show a relative increase on the second echo (Figure 36-2). This is seen because of the nonlaminar motion of the fluid confined within the pericardial sac. The loculated pericardial effusion may result in compression of the underlying cardiac chambers and present a clinical picture of constrictive pericarditis. The thickened pericardium often is seen best on the second echo images as a low–signal intensity line defining the high–signal intensity loculated pericardial effusion (Figure 36-2).

Another cause of complex pericardial effusions is uremia. In these patients a medium to high signal intensity from the pericardial fluid may be seen, in addition to visible adhesions between the visceral and parietal pericardium.[17]

Infected Pericardial Effusion. Collections within the pericardial sac with an intermediate signal intensity that becomes relatively increased on the second spin echo image (but not to the intensity seen with loculated simple pericardial effusions) have occurred in effusions with an infectious etiology. In two recent cases at our institution, both patients presented a similar MRI appearance. There was a large amount of medium-intensity material surrounded by a thickened pericardium that modestly increased in signal on the second echo. In one patient, air was visualized within the fluid collection, even though there had been no previous surgical intervention or pericardiocentesis. In each of these cases the organism responsible was *Staphylococcus aureus*; the identification was made by surgical evacuation and subsequent microbiologic culture. In the literature, tuberculous pericarditis has also produced an increased signal intensity in the pericardial sac.[17]

Intrapericardial Hematoma

Magnetic resonance imaging has been found to be superior to echocardiography in defining the nature and localization of processes within the pericardial sac.[17] It allows differentiation of clotted hemopericardium from simple pericardial effusion (Figure 36-3). There have been reports of false-negative echocardiograms in the presence of pericardial hematomas and clots,[18] while MRI has allowed this differentiation.[17] In MRI the appearance of blood within the pericardial sac is quite different from that of simple pericardial fluid. Gating to every heartbeat results in a relatively T1-weighted sequence for imaging the heart.

Hematomas are seen as regions of high signal intensity that may have medium signal intensity within them. The high signal intensity corresponds to the serous blood components, while the medium intensity represents the clotted portion of the hematoma. Additionally, there may be pericardial thickening associated with intrapericardial hematomas. The pericardial thickening may be seen as medium intensity.[17] In patients with hemopericardium, echocardiography may demonstrate only pericardial thickening or a mass surrounding the heart, rather than the precise etiology.[17-18]

Intrapericardial Masses

The demonstration of a mass within the pericardial cavity is based upon visualization of a mass that is surrounded on its lateral border by pericardium. Pericardium is usually highlighted by the surrounding epipericardial fat. An intrapericardial pheochromocytoma has been diagnosed by MRI, where a CT study was negative. A metaiodobenzylguanidine (MIBG) iodine 131 scan localized the mass, and MRI subsequently provided precise anatomic localization of the mass.[19] This case demonstrates that while MRI is not a screening modality, it may be used as a complementary technique in conjunction with the MIBG study. Magnetic resonance imaging of an intrapericardial mass also allows identification of the cardiac chambers and adjacent vasculature due to the signal void provided by

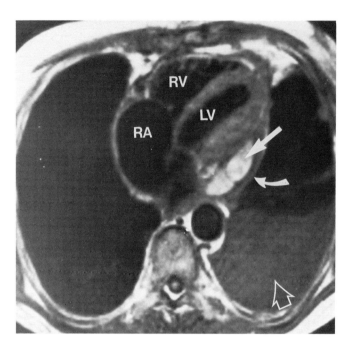

Figure 36-3 Transverse gated magnetic resonance image demonstrating on this relatively T1-weighted sequence a high-intensity intrapericardial collection (straight arrow) typical of subacute blood. This hematoma occurred after coronary artery bypass graft surgery. Note the thickened, medium-intensity pericardium (curved arrow). The hematoma is causing marked compression of the left ventricle (LV). Additionally, there is a large left pleural effusion (open arrow). RV, right ventricle; RA, right atrium.

rapidly flowing blood. Therefore, the bolus of contrast medium that often is required for CT scanning in order to accurately identify a mass is unnecessary in MRI. The primary difficulty with MRI remains determining whether the mass is outside the pericardial space, within the pericardial space, or within a cardiac chamber. When the pericardium is visualized surrounding the mass, highlighted by fat, an intrapericardial location is confirmed.

One type of intrapericardial mass is pericardial cysts.[17] On MRI pericardial cysts are seen as masses with fairly long T1 and T2 relaxation values, indicative of fluid-filled structures. The lesions have low signal intensity on the first echo of a spin echo image and a relative increase in signal intensity on the second echo. Additionally, a line of low signal intensity, corresponding to the parietal pericardium, is visualized surrounding the cysts.[17]

Paracardiac Masses

Magnetic resonance imaging allows demonstration of paracardiac masses by showing the low-intensity line

Evaluation of Pericardial Disease by MRI 465

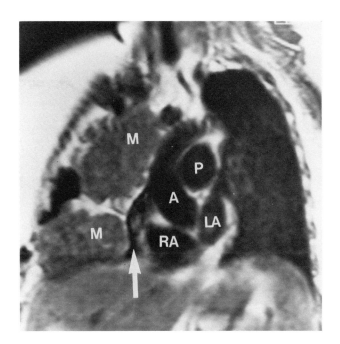

Figure 36-4 Oblique coronal gated magnetic resonance image demonstrates a large mediastinal mass (M) abutting the pericardium (arrow) along the right side of the heart. There is no invasion into the pericardium. RA, right atrium; LA, left atrium; A, aorta; P, pulmonary artery.

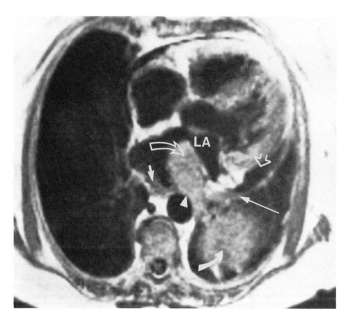

Figure 36-5 Transverse gated magnetic resonance image demonstrates a left lung mass (curved white arrow) invading the pericardium along the left posterior lateral aspect of the heart (white straight arrow). There is a simple pericardial effusion posteriorly (open straight arrow). Note metastatic invasion (open curved arrow) into the left atrium (LA) along the posterior wall of the left atrium (thin arrow) and invasion into the fat anterior to the aorta (arrowhead).
Source: Reprinted with permission from *Diagnostic Imaging* (1988; 130–133), Copyright © 1988, Miller Freeman Publications Inc.

of the pericardium intervening between the mass and the heart. A variety of paracardiac masses have been demonstrated with MRI,[20] and all have a variable signal intensity that is predominantly low signal intensity on the first spin echo image.[20] Magnetic resonance imaging allows depiction of both the size and position of the mass relative to the pericardium and cardiac chamber walls (Figure 36-4). However, specific information in regard to the pathologic nature of the mass cannot be obtained by this method. Similar signal intensity has been seen with teratoma, mesothelioma, rhabdomyosarcoma, uterine sarcomas, and plasmocytomas.[20]

Magnetic resonance imaging has been particularly useful for determining invasion of the pericardium secondary to paracardiac masses (Figure 36-5). Since MRI allows delineation of the pericardial line, it can determine whether mediastinal masses extend into the paracardial sac. In patients with known lung or mediastinal masses who develop dysrhythmias, MRI may be used to determine if invasion of the underlying cardiac walls is causing the dysrhythmias.

Computed tomography and MRI may have equivalent efficacy for diagnosis of pulmonary and mediastinal masses. However, gated MRI has been found to be superior for evaluating masses in the vicinity of central cardiovascular structures and for determining whether there is tumor-associated vascular compromise.[21]

Constrictive Pericarditis versus Restrictive Cardiomyopathy

Often the differentiation of constrictive pericarditis from restrictive cardiomyopathy, based on clinical symptoms or from the findings at heart catheterization, is difficult. Loculated effusion resulting in an effusive constrictive pericarditis (Figure 36-2), hemopericardium associated with pericardial thickening (Figure 36-3), and complex collections within the pericardial sac, such as loculated infected pericardial fluid each result in a symptom complex that raises the question of constrictive pericarditis versus restrictive cardiomyopathy. In this type of patient MRI can provide the diagnosis that may not be obtainable by echocardiography. Studies investigating the use of MRI for defining constrictive pericardial disease have shown that pericardial thickness is 5 mm or greater in patients with constrictive pericarditis (Figures 36-2 and 36-3). Additional findings consisted of dilatation of the right atrium, inferior vena cava, and hepatic veins and straightening of the ventricular septum.[22] An important finding is a small, tube-shaped right ventricle, which is the most important morphologic sign of pericardial constriction. Analysis of the various associated findings of MRI regarding pericardial constriction revealed that the presence of an enlarged right atrium and inferior vena cava was not specific for pericardial constriction, since this enlargement may also be seen in patients with restrictive cardiomyopathy. However, a small tube-shaped right ventricle and the straightening of the ventricular septum are indicative of pericardial constriction.[17,22] Thus thickening of the pericardium as detected by MRI is highly indicative of pericardial constriction in patients with typical clinical symptoms and a compressed right ventricle. As mentioned previously, pericardial thickening may be seen in postoperative patients who have no clinical symptoms of constriction.

Magnetic resonance imaging has also been able to define a normal pericardium in patients with restrictive cardiomyopathy, thereby differentiating them from patients with constrictive pericarditis. Endomyocardial biopsies subsequently taken from these patients confirmed restrictive cardiomyopathy.[17] These patients have the MRI features of dilatation of the right atrium and inferior vena cava. Magnetic resonance imaging is better able to exclude pericardial thickening in patients with a restrictive cardiomyopathy than echocardiography and conventional CT, which often show equivocal findings.

Magnetic resonance imaging has been found to be superior to echocardiography in defining the nature and localization of processes causing the clinical picture of constrictive pericarditis, such as hemopericardium. Clotted hemopericardium has been reported to mimic constrictive pericarditis.[22-25] While echocardiography has yielded false-negative results in the presence of intrapericardial clots,[18] MRI can determine if an intrapericardial collection is a hematoma, due to the intrinsic soft tissue contrast. Echocardiography has had difficulty in identifying the nature of such an intrapericardial collection, or in identifying an intrapericardial collection at all.[18] An intrapericardial hematoma is seen as a region of high signal intensity on gated cardiac spin echo images that are relatively T1 weighted.

SUMMARY

The noninvasive diagnosis of pericardial disease is usually accomplished by echocardiography. However, there are certain limitations on the usefulness of echocardiography: studies of good diagnostic quality cannot be achieved in all patients; the anterior portion of the pericardium may be difficult to visualize secondary to focusing problems in the near field of the sonographic beam, and making a diagnosis of pericardial constriction is particularly problematic.[26] Echocardiography has also not been particularly helpful for the evaluation for paracardiac masses and their relation to the pericardium. Magnetic resonance imaging, in contrast, has been shown to combine the advantages of excellent delineation of intracardiac and pericardial structures with the ability to demonstrate additional tissue characteristics. Both the appearance and thickness of the pericardial structures can be accurately evaluated during the cardiac cycle with MRI. A variety of pericardial abnormalities, including congenital anomalies, pericardial thickening, pericardial effusions (both simple and complex), hemopericardium, and intrapericardial and paracardiac masses, have been well defined by MRI. Additionally, MRI has been particularly useful in the differentiation of constrictive pericarditis from restrictive cardiomyopathy.

REFERENCES

1. Herfkens R, Higgins CB, Hricak H, et al. Nuclear magnetic resonance imaging of the cardiovascular system: Normal and pathologic findings. *Radiology*. 1983;147:749–759.

2. Higgins CB, Byrd B, McNamara M. Magnetic resonance imaging of the heart: A review of the experience in 172 patients. *Radiology*. 1985;155:671–679.

3. Silverman PM, Harell GS. Computed tomography of the normal pericardium. *Invest Radiol*. 1983;18:141–144.

4. Moncada R, Baker M, Salinas M, et al. Diagnostic role of computed tomography in pericardial heart disease. Congenital defects, thickening, neoplasms, and effusions. *Am Heart J*. 1982;103:263–282.

5. Cooley RN, Schrieber MH, eds. *Radiology of the Heart and Great Vessels*. 3rd ed. Baltimore: Williams & Wilkins; 1987:578–579.

6. Ferrans VJ, Ishihara T, Roberts WC. Anatomy of the pericardium. In: Reddy PS, Leon DF, Shaver JA, eds. *Pericardial Disease*. New York: Raven Press; 1982:77–92.

7. McMurdo KK, Webb WR, Von Schulthess GK, Gamsu G. MR of superior pericardial recesses. *AJR*. 1985;145:985–988.

8. Crooks LE, Barker B, Chang H, et al. Magnetic resonance imaging strategies for heart studies. *Radiology*. 1984;153:459–465.

9. Fisher MR, Von Schulthess GK, Higgins CB. Multiphasic cardiac magnetic resonance imaging. Normal regional left ventricular wall thickening. *AJR*. 1985;145:27–30.

10. Von Schulthess GK, Fisher MR, Crooks LE, Higgins CB. The nature of intracardiac signal on gated MR images in normals and patients with abnormal left ventricular function. *Radiology*. 1985; 156:125–132.

11. Fisher MR, Steinberg F, Rogers LF. MR assessment of wall motion in MI patients utilizing gradient echo imaging. Presented at Radiographic Society of North America. November, 1986; Chicago, Ill. Abstract.

12. Stark D, Higgins CB, Lanzer P. Magnetic resonance imaging of the pericardium: Normal and pathologic findings. *Radiology*. 1984; 150:469–474.

13. Von Schulthess G, Higgins CB. Blood flow imaging with MR: Spin-phase phenomenon. *Radiology*. 1985;157:687–695.

14. Sechtem U, Tscholakoff D, Higgins CB. MRI of the normal pericardium. *AJR*. 1986;147:239–244.

15. Ellis K, Leeds NE, Himmelstein A. Congenital deficiencies in the parietal pericardium. *AJR*. 1959;82:125–127.

16. Guiterrez FR, Shackelford GD, McKnight RC, Levitt RG, Hartmann A. Diagnosis of congenital absence of left pericardium by MR imaging. *J Comput Assist Tomogr*. 1985;9:551–553.

17. Sechtem V, Tscholakoff D, Higgins CB. MRI of abnormal pericardium. *AJR*. 1986;147:245–256.

18. Kerber R, Payvandi M. Echocardiography in acute hemopericardium: Production of false negative echocardiograms by pericardial clots. *Circulation*. 1977;55:111–124.

19. Fisher MR, Higgins CB, Andereck W. MR imaging of an intrapericardial pheochromocytoma. *J Comput Assist Tomogr*. 1985;9:1103–1105.

20. Amparo EG, Higgins CB, Farmer D, Gamsu G, McNamara M. Gated MRI of cardiac and paracardiac masses: Initial experience. *AJR*. 1984;143(6):1151–1156.

21. Von Schulthess GK, McMurdo K, et al. MRI of mediastinal masses. *Radiology*. 1986;158(2):289–296.

22. Soulen RL, Stark D, Higgins CB. Magnetic resonance imaging of constrictive pericardial disease. *Am J Cardiol*. 1985;55:480–484.

23. Little WC, Prim RK, Karp RB, Hood WP. Clotted hemopericardium with hemodynamic characteristics of constrictive pericarditis. *Am J Cardiol*. 1980;45:386–388.

24. Dunlap TE, Sorkin RP, Mori KW, Popat KD. Massive organized intrapericardial hematoma mimicking constrictive pericarditis. *Am Heart J*. 1982;104:1373–1375.

25. Low RI, Arthur A, Kelly PB, Takeda PA. Clotted hemopericardium post myocardial infarction presenting as effusive constrictive pericarditis. *Am Heart J*. 1985;109:905–908.

26. Sutton FJ, Whitley NO, Applefield MM. The role of echocardiography and computed tomography in the evaluation of constrictive pericarditis. *Am Heart J*. 1985;109:350–355.

Part VI
Congenital Heart Disease

Chapter 37

The Evaluation of Congenital Heart Defects by Echocardiography and Doppler

Elizabeth A. Fisher, MD

Diagnosis of congenital heart defects requires a systematic examination for anatomic and hemodynamic evidence of septal defects, intra- and extracardiac shunts, inflow and outflow tract obstructions, valvular insufficiency, cardiac chamber and great vessel positions, and venoatrial, atrioventricular, and ventriculoarterial connections. Echocardiography can provide detailed anatomic information in real time, at the bedside, during cardiac surgery,[1] and even in utero.[2] It is done without the use of ionizing radiation or iodinated contrast medium. Coupled with Doppler ultrasound data on the direction, velocity, timing, and character of blood flow, it permits calculation of hemodynamics. With echocardiography and Doppler, it is possible to noninvasively assess systolic and diastolic ventricular function,[3-4] ventricular volumes,[5] systemic and pulmonary blood flow and pulmonary-to-systemic blood flow ratio,[6] pressure gradients,[7] valve areas,[8] and ventricular and pulmonary artery pressures.[9] Furthermore, no extensive computer facilities are required for analysis.

There are no contraindications to the use of echocardiography in the frequencies employed for clinical studies. The method is limited only by factors that prevent proper alignment of signals. These include interposition of bone or air or an uncooperative patient. Echocardiography and Doppler studies are replacing cardiac catheterizations in many clinical settings, including preoperative evaluation.[10-11] It is therefore essential that these studies be performed under optimum conditions, which include sedation of restless or uncooperative patients. Chloral hydrate, 50 to 100 mg/kg, works well in children.

A full two-dimensional echocardiogram should be done first, using the highest frequency transducer possible for the patient (5 to 7.5 MHz for infants, 1.9 to 3.5 MHz for older children and adults). It may be necessary to change transducers several times during the study to obtain optimal images from different transducer positions. Take advantage of all acoustic windows—parasternal, apical, subcostal, and suprasternal. Routine pulsed wave Doppler interrogation of all valves, septa, and great vessels, followed by continuous wave examination of areas of high-velocity flow exceeding pulsed wave measurement capacity, should then be done.

Two-dimensional imaging with continuous wave Doppler is especially important with congenital heart defects in which there may be more than one source of high-velocity flow. High-velocity signals should be recorded from several transducer positions in order to obtain the maximum velocity. Color flow mapping can greatly expedite location of areas of abnormal flow, which should be further investigated with pulsed or continuous wave Doppler.

In the following sections, the use of echocardiographic and Doppler studies to evaluate various types of congenital heart defects will be illustrated.

SEPTAL DEFECTS/LEFT-TO-RIGHT SHUNTS

Direct visualization of a defect in the ventricular (Figure 37-1) or atrial (Figure 37-2) septum by two-dimen-

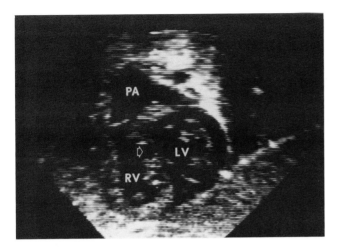

Figure 37-1 Two-dimensional echocardiogram, subcostal long-axis view (inverted), right ventricular outflow tract. A large defect (arrowhead) is present in the perimembranous ventricular septum. LV, left ventricle; PA, pulmonary artery; RV, right ventricle.

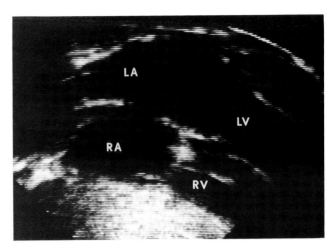

Figure 37-2 Two-dimensional echocardiogram, subcostal four-chamber view (inverted). A large defect is present in the midatrial septum. Absence of right ventricular dilatation is due to an associated large patent ductus arteriosus. RV, right ventricle; RA, right atrium; LV, left ventricle; LA, left atrium.

sional echocardiography is generally diagnostic if the defect is of significant size.[11–12] Imaging of a patent ductus arteriosus is adequate in some instances[13] but unreliable in others. In patients with normal pulmonary vascular resistance and no associated defects modifying the usual hemodynamics, a left-to-right shunt results in a volume overload pattern appropriate to the lesion. Dilatation of the right side of the heart and reversed systolic motion of the ventricular septum are typical of atrial septal defect, while dilatation of the left side of the heart is typical of ventricular septal defect and patent ductus arteriosus.

Shunt flow across an atrial or ventricular septal defect can be demonstrated by color flow mapping (Figures 37-3 and 37-4; see color insert) or pulsed (Figures 37-5 and 37-6) or continuous wave Doppler.[11,14] This may be particularly helpful if the defect is small or cannot be satisfactorily imaged. In patent ductus arteriosus (Figures 37-7 and 37-8), the flow pattern in the pulmonary artery root may be normal in systole but show swirling, antegrade flow continuing through diastole or retrograde diastolic flow, depending upon the sampling site.[15] Typically, ductal flow is directed along the left lateral and superior

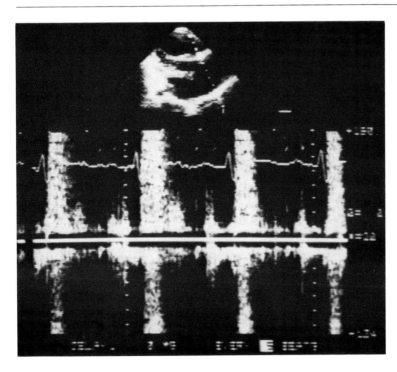

Figure 37-5 Pulsed wave Doppler study, same patient as in Figure 37-3. A jet of disturbed high-velocity (at least 2.94 m/s) systolic flow toward the transducer is recorded at the right septal surface. The study confirms the presence of a left-to-right shunt across a restrictive subaortic ventricular septal defect. Since the maximum systolic velocity of the jet exceeds the measurement capacity of pulsed wave, calculation of the pressure gradient across the defect requires continuous wave study.

Evaluation of Congenital Heart Defects by Echocardiography and Doppler   471

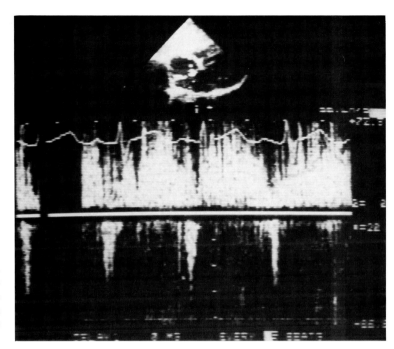

Figure 37-6 Pulsed wave Doppler, same patient as in Figures 37-2 and 37-4. Disturbed systolic and diastolic flow across the midatrial septum from left to right confirms the presence of a left-to-right shunt at atrial level. The low peak velocity (0.7 m/s) is consistent with a large, unrestrictive defect.

wall of the main pulmonary artery and can be shown to originate in the proximal descending aorta.

The most important hemodynamic factors in left-to-right shunting lesions are the size of the shunt and the pulmonary artery pressure and resistance. The size of a left-to-right shunt, expressed as the ratio of pulmonary-to-systemic blood flow or QP/QS, can be determined by two-dimensional echocardiography and Doppler.[6] Pulmonary and systemic flows (Q) are calculated from the mean Doppler systolic flow velocity (Vm) and the two-dimensional echocardiographic diameter (D) of the region in which flow is measured, as

$$Q = Vm \times \pi (D/2)^2$$

In atrial and ventricular septal defects, pulmonary artery root velocity and diameter are used to calculate pulmonary flow, and aortic root velocity and diameter are used to calculate systemic flow. In patent ductus arteriosus, aortic root velocity and diameter are used to

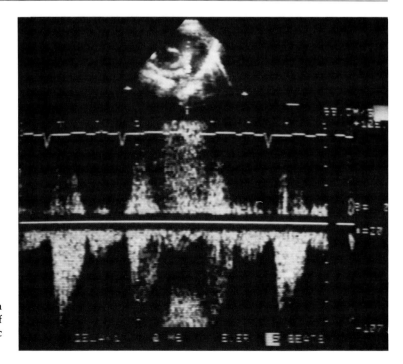

Figure 37-8 Pulsed wave Doppler study, same patient as in Figure 37-7. With the sample volume moving in and out of the retrograde jet, both antegrade and retrograde diastolic flow are recorded. Systolic antegrade flow is normal.

calculate pulmonary flow. The turbulent flow created by the shunt makes pulmonary artery root measurements invalid for systemic flow calculations in patent ductus, however. Some investigators have used right ventricular outflow tract measurements instead. Flows cannot be calculated in the presence of semilunar valve stenosis; the greatest source of error is in measurement of vessel diameter.

Pulmonary artery pressure can be calculated in some patients by the simplified Bernoulli equation, which allows calculation of the pressure drop across a restrictive orifice. This equation is

$$P_1 - P_2 = 4(V_2^2 - V_1^2)$$

As V_1 is usually low and can be ignored, the equation can be further simplified to

$$P_1 - P_2 = 4V_2^2$$

This equation can be used to calculate pulmonary artery systolic pressure in some patients with direct systemic-pulmonary connections,[16] such as a Waterston shunt, by measuring the peak systolic velocity of shunt flow in the pulmonary artery and subtracting the calculated pressure gradient from systemic blood pressure measured by cuff. This method is not reliable when applied to tube grafts or Blalock-Taussig shunts.

Right ventricular systolic pressure can also be calculated from the peak systolic velocity of the shunt flow across a restrictive ventricular septal defect[14] or from the regurgitant jet of tricuspid regurgitation.[9] In the former, the calculated gradient is subtracted from cuff systolic blood pressure, which equals left ventricular systolic pressure if there is no left ventricular outflow tract stenosis. In the latter, the calculated gradient across the tricuspid valve is added to the central venous pressure. If venous pressure is unknown, a value of 10 mm Hg is assumed. If there is no right ventricular outflow tract stenosis, pulmonary artery systolic pressure equals right ventricular systolic pressure.

In the absence of high-velocity flow from tricuspid regurgitation, a restrictive ventricular septal defect, or a direct aortopulmonary connection, the best Doppler index for separating patients with elevated pulmonary artery pressure from those with normal pulmonary artery pressure appears to be the time to peak velocity (TPV) of pulmonary artery flow, also termed acceleration time (AT); this is the time from onset of systolic flow to peak flow in the pulmonary artery root.[17]

OUTFLOW OBSTRUCTION/STENOSIS

Congenital obstruction of the left or right ventricular outflow tract occurs most commonly at a valvular level. In semilunar valve stenosis, the valve leaflets are thickened, and their systolic motion is jerky and restricted

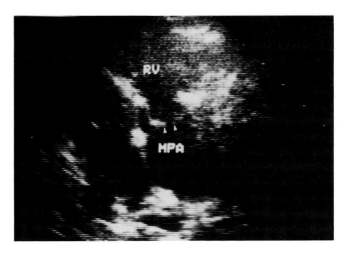

Figure 37-9 Two-dimensional echocardiogram, parasternal long-axis view, right ventricular outflow. In this systolic frame, separation of the stenotic pulmonary valve leaflets at their free edges is very restricted. The main pulmonary artery (MPA) is dilated. RV, right ventricle.

(Figure 37-9). At peak systole, separation of the leaflets at their free edges is much less than at their annular attachments.[18–19] If the stenosis is significant, there will be increased wall thickness of the ventricle proximal to the obstruction and poststenotic dilatation of the arterial root.

The severity of stenosis is assessed by measurement of the pressure gradient across the valve or by calculation of valve area, both measurements classically requiring cardiac catheterization. The maximum Doppler flow velocity (V_2) across the valve (Figure 37-10) can be used to calculate the peak instantaneous systolic pressure gradient between ventricle and arterial root by the simplified Bernoulli equation.[7,20] The calculated gradient correlates well with that measured at cardiac catheterization if measurements are made under similar conditions and if great care is taken to record the highest Doppler velocity. Color flow mapping is useful to help locate the direction of jet flow across the valve so that the Doppler signal can be properly positioned to record the highest velocity. Left ventricular systolic pressure can be calculated by adding the calculated gradient across a stenotic aortic valve to the cuff systolic blood pressure.

Since pressure gradients are dependent upon flow, knowledge of aortic valve area may be of critical importance in judging the severity of aortic stenosis. Two methods have been used for this calculation from echocardiography/Doppler data, the Gorlin equation and the continuity equation.[8] The Gorlin equation is

$$AVA = \frac{CO/SEP}{44.5 \sqrt{\text{mean gradient}}}$$

where AVA is aortic valve area, CO is cardiac output, and SEP is systolic ejection period. Cardiac output is

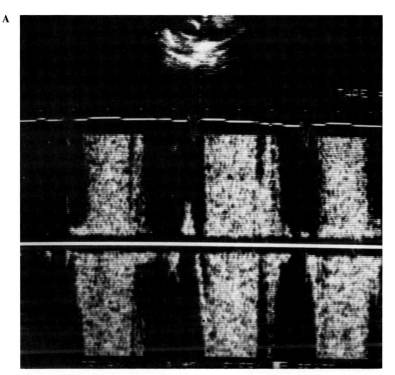

Figure 37-10 Doppler study, same patient as in Figure 37-9. A, The pulsed wave sample volume is in the main pulmonary artery just above the valve. Disturbed, very high velocity systolic flow is recorded, confirming the presence of valvular pulmonary stenosis. B, Imaging continuous wave recording across the right ventricular outflow shows a peak systolic velocity of 5.2m/s. The calculated peak instantaneous systolic gradient across the valve of at least 108 mm Hg indicated severe stenosis.

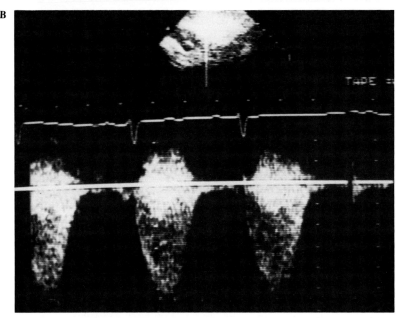

calculated as the mean systolic velocity in the left ventricular outflow tract times the cross-sectional area of the outflow tract, calculated from the diameter measured by two-dimensional echocardiography. Systolic ejection period is calculated as aortic valve ejection time, as measured from the aortic valve Doppler tracing, divided by the RR interval, times 60 seconds. The mean transvalvular gradient is calculated by applying the simplified Bernoulli equation point by point to the Doppler tracing and averaging the valves. The continuity equation is

$$A_1 \times Vm_1 = A_2 \times Vm_2$$

where A_1 is the area of the left ventricular outflow tract, calculated from diameter measured by two-dimensional echocardiography; Vm_1 is the mean systolic velocity in the left ventricular outflow tract; A_2 is the aortic valve area; and Vm_2 is the mean systolic velocity across the aortic valve. Solving for aortic valve area, we obtain

$$A_2 = \frac{A_1 \times Vm_2}{Vm_2}$$

Aortic coarctation most commonly occurs as a discrete narrowing of the proximal descending aorta near

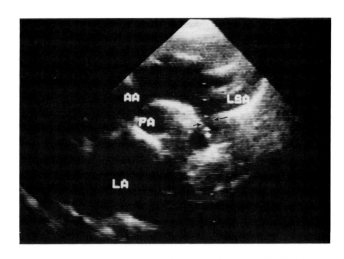

Figure 37-11 Two-dimensional echocardiogram, suprasternal notch view, aortic arch. Discrete coarctation of the aorta just distal to the left subclavian artery (LSA). AA, ascending aorta; PA, pulmonary artery; LA, left atrium.

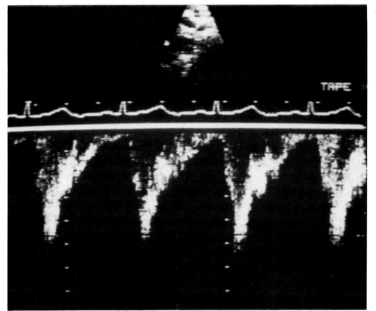

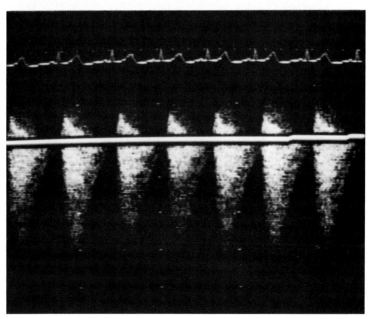

Figure 37-12 Doppler study, same patient as in Figure 37-11. Both pulsed wave (A) and continuous wave (B) studies show continuous flow with velocity peaking in systole. The pulsed wave velocity grossly underestimated the pressure gradient across the coarctation. With a continuous wave peak velocity of 3 m/s, a peak instantaneous systolic gradient of at least 36 mm Hg can be calculated.

the origin of the left subclavian artery. In some patients the anatomy of the aortic arch is very well demonstrated by two-dimensional echocardiography from the suprasternal notch (Figure 37-11), but in other patients coarctation may not be obvious on two-dimensional examination.[21] Doppler ultrasound examination of the area of suspected coarctation will confirm the presence of significant obstruction with disturbed, continuous antegrade flow peaking in late systole,[22] frequently with very high systolic velocity reflecting the systolic pressure gradient across the coarctation (Figure 37-12). The presence of continuous flow is more significant than demonstration of a high velocity in systole. The highest velocity jet may be very difficult to locate due to its eccentric position. Continuous antegrade flow, however, is easily demonstrated and indicates the presence of a significant pressure gradient across the obstruction throughout the cardiac cycle. If the high-velocity jet can be recorded, the pressure gradient across a coarctation may be calculated from the peak velocities distal and proximal to the obstruction, by the simplified Bernoulli equation. V_1 can be ignored if it is low. Of course, the gradient can also be calculated noninvasively by measurement of cuff blood pressures in upper and lower extremities.

MALALIGNMENT

In certain congenital heart defects, there is loss of normal continuity or alignment of the atrial and ventricular septa or, more commonly, portions of the ventricular septum. A ventricular septal defect is always present, and an atrioventricular or semilunar valve over-rides the defect. A large defect in the perimembranous or infundibular septum with malalignment of the remainder of the ventricular septum occurs in the most common example of this type of defect, tetralogy of Fallot (Figure 37-13).[23] Right ventricular outflow tract obstruction is always present at the subvalvular level and may also occur at valvular and supravalvular levels (Figure 37-14). Other examples of malalignment defects include double outlet right ventricle, truncus arteriosus, and over-riding tricuspid or mitral valve.

In tetralogy, color flow mapping is useful to document systolic flow from the right ventricle into the aortic root (Figure 37-15; see color insert). As the velocity of flow across the large defect is not high, conventional pulsed and continuous wave Doppler contribute little to examination of the ventricular septum. Pulsed wave Doppler is very useful to assess the site of right ventricular outflow tract obstruction, however. Disturbed high-velocity antegrade systolic flow is found in the right ventricular outflow tract and may be present in the pulmonary artery root as well. Because of the high velocity, assess-

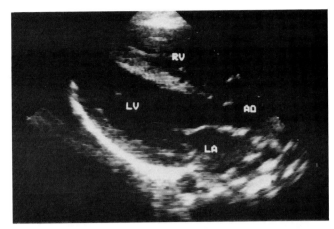

Figure 37-13 Two-dimensional echocardiogram, parasternal long-axis view. The large, malaligned, perimembranous infundibular ventricular septal defect with an overriding aortic root is consistent with tetralogy of Fallot. AO, aorta; other abbreviations are explained in the legend for Figure 37-2.

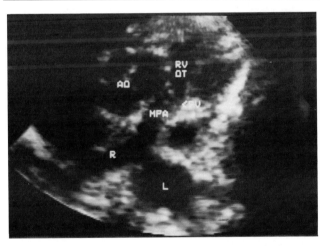

Figure 37-14 Two-dimensional echocardiogram, parasternal short-axis view, great arteries. The aortic root (AO) is dilated. Severe right ventricular outflow tract (RVOT) and pulmonary valvular (PV) and supravalvular stenosis are present in this patient with tetralogy of Fallot. R, right pulmonary artery; L, left pulmonary artery; MPA, main pulmonary artery.

ment of the severity of stenosis requires continuous wave Doppler.[20] In patients with pulmonary atresia, no antegrade systolic flow is detected in the outflow tract. In those patients, flow in the pulmonary arteries is dependent upon a patent ductus arteriosus, bronchial collateral arteries, or a surgical systemic-pulmonary anastomosis. The flow is therefore continuous and disturbed, and the velocity peaks in systole (Figure 37-16).[16] Despite the presence of a long segment of right ventricular outflow tract stenosis, the maximum velocity of flow across the right ventricular outflow tract can be used to estimate the peak instantaneous systolic pressure gradient between the right ventricle and the pulmonary artery by the simplified Bernoulli equation.[20]

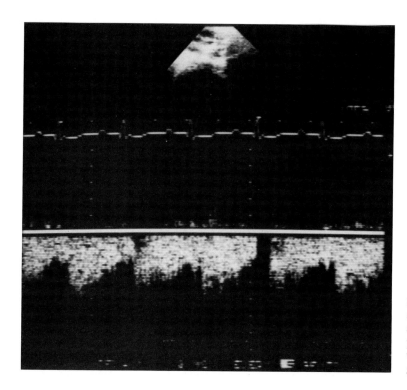

Figure 37-16 Pulsed wave Doppler study, left pulmonary artery, pulmonary atresia with modified left Blalock-Taussig anastomosis. Continuous flow peaking in systole confirms patency of the surgical connection, but the peak systolic velocity of only 0.8 m/s grossly underestimates the pressure gradient between the subclavian and pulmonary arteries.

GREAT ARTERY MALPOSITION

An abnormal relationship of the great artery roots is an obvious finding in several congenital heart defects. The most common defect with great artery malposition is complete transposition of the great arteries. In this defect, the aortic root is anterior to the pulmonary artery root.[24] It is generally also to the right, occasionally directly anterior, or even slightly to the left of the pulmonary artery root (Figure 37-17). Great artery malposition is also found in corrected transposition of the great arteries, double outlet right ventricle, univentricular heart, and other complex congenital heart defects. Once the presence of arterial malposition is recognized, care must be taken to establish the ventriculoarterial and atrioventricular connections and to search for associated abnormalities of septation, outflow or inflow obstructions, and venoatrial connections. In transposition of the great arteries, the ventriculoarterial connections are by definition abnormal or discordant. The aorta arises from the right ventricle and the pulmonary artery from the left ventricle (Figure 37-18).[24] In complete transposition, the atrioventricular connections are concordant. In corrected transposition, they are discordant. Doppler studies are valuable to establish the presence of septal defects, inflow or outflow obstructions, or valvular insufficiency.

UNIVENTRICULAR HEART/VENTRICULAR HYPOPLASIA

Several congenital heart defects are characterized by the presence of a large ventricular chamber with or without a second, markedly hypoplastic ventricle. Key points to consider in reaching a precise anatomic diagnosis are the anatomy and connections of the atrioventricular valves and the position and connections of the great arteries. Two examples of this type of defect are tricuspid atresia and single ventricle. In tricuspid atresia[25] there is an absence of the right atrioventricular connection. The only outlet of the right atrium is at atrial level. A single, left-sided mitral valve connects the left atrium and left ventricle (Figure 37-19). Great artery positions and connections are usually normal, although a small percentage of patients have complete transposition of the great arteries. The ventricular septal defect, if present, is usually small and restrictive. The right ventricle is nearly always very hypoplastic and may not be identifiable in some patients, giving the appearance of a single ventricle. Associated valvular or subvalvular pulmonary stenosis or atresia is common if the great arteries are normally positioned. In those patients, pulmonary blood flow may depend on a patent ductus arteriosus or a surgical systemic-pulmonary anastomosis.

If both atria connect to a single large ventricular chamber, the defect is termed single ventricle.[26] The

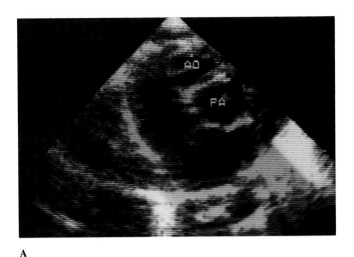
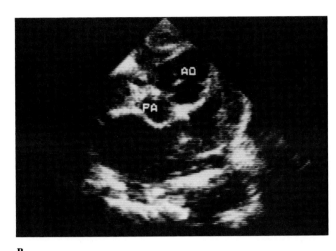

Figure 37-17 Two-dimensional echocardiograms, parasternal short-axis views, great arteries. A, The anterior great artery is to the right, an obvious malposition typical of complete transposition of the great arteries. B, The anterior great artery is to the left. This position appears normal until the posterior, right-sided great artery is identified by its branching patter (C) to be the pulmonary artery (PA). This malposition is typical of corrected transposition but is found in some patients with complete transposition as well. Other abbreviations are explained in the legend for Figure 37-14.

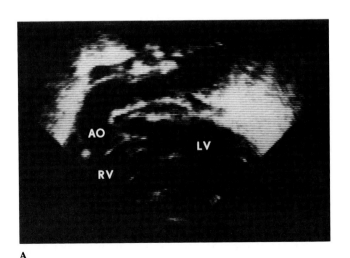
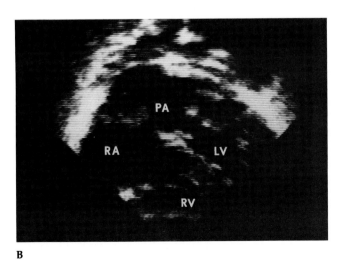

Figure 37-18 Two-dimensional echocardiogram, subcostal long-axis view (inverted), same patient as in B and C of Figure 37-17. Here, the aorta (AO) is seen to arise from the right ventricle (RV) in the left panel and the pulmonary artery (PA) from the left ventricle (LV) in the right panel. The discordant ventriculoarterial connections are typical of transposition of the great arteries. Since the atrioventricular connections were shown in other views to be concordant (right atrium [RA] to right ventricle, left atrium to left ventricle), the patient has complete transposition of the great arteries despite the arterial malposition suggestive of corrected transposition.

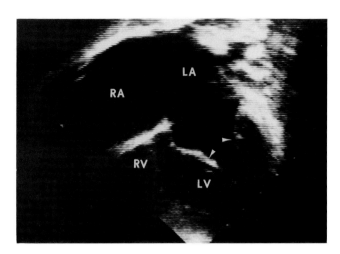

Figure 37-19 Two-dimensional echocardiogram, apical four-chamber view (inverted). A dense band of nonmoving echoes separates the right atrium (RA) and very small right ventricle (RV). The mitral valve (arrowheads) connects the left atrium (LA) and left ventricle (LV). The atrial septum is not well seen in this view. Absence of the right atrioventricular connection is diagnostic of tricuspid atresia.

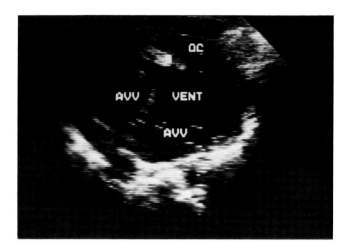

Figure 37-20 Two-dimensional echocardiogram, parasternal view between long and short axes. Both atrioventricular valves (AVV) enter a single large ventricular chamber (VENT). A small outflow chamber (OC) is situated superiorly and to the left and gives rise to the aorta (not shown). This appearance is typical of single ventricle. S, septum.

connection may be via two separate atrioventricular valves or a common atrioventricular valve. Some form of great artery malposition is very common, either complete transposition or corrected transposition of the great arteries. A small outlet chamber, situated superiorly, anteriorly, left, or right and having no atrioventricular connection, usually gives rise to the aorta (Figure 37-20). The pulmonary artery arises from the large ventricle, which receives the atrioventricular valves. Pulmonary stenosis is frequent. A ventricular septal defect connects the single ventricle and the small outlet chamber. This defect may become restrictive, resulting in significant subvalvular aortic stenosis.

Doppler is useful to assess the atrioventricular and ventriculoarterial connections and to document the presence of an obligatory right-to-left shunt at the atrial level in tricuspid atresia. The presence and severity of inflow or outflow tract obstructions and valvular insufficiency should also be determined. Continuous flow will be found in the pulmonary arteries if pulmonary blood flow is ductal or shunt dependent.

SUMMARY

Echocardiography and Doppler are sensitive, safe techniques for detailed noninvasive evaluation of the anatomy and hemodynamics of congenital heart defects. In many cases cardiac catheterization is unnecessary. In others, catheterization is greatly facilitated by information gained from prior echocardiography and/or Doppler studies. The instrument can be taken to the bedside, so studies can be done without the risks of transporting sick patients. The ability to antenatally diagnose and plan therapy for infants with congenital heart defects or dysrhythmias adds an additional dimension to a technology that is already indispensable to the evaluation of congenital heart defects.

REFERENCES

1. Gussenhoven EJ, van Herwerden LA, Roelandt J, Ligtvoet KM, Bos E, Witsenberg M. Intraoperative two-dimensional echocardiography in congenital heart disease. *J Am Coll Cardiol*. 1987; 9:565–572.

2. Silverman NH, Golbus MS. Echocardiographic techniques for assessing normal and abnormal fetal cardiac anatomy. *J Am Coll Cardiol*. 1985;5(suppl):20S–29S.

3. Douglas PS, Reichek N, Plappert T, Muhammad A, St. John Sutton MG. Comparison of echocardiographic methods for assessment of left ventricular shortening and wall stress. *J Am Coll Cardiol*. 1987;9:945–951.

4. DeMaria AN, Wisenbaugh T. Identification and treatment of diastolic dysfunction: Role of transmitral Doppler recordings. *J Am Coll Cardiol*. 1987;9:1106–1107.

5. Tortoledo FA, Quinones MA, Fernandez GC, Waggoner AD, Winters WL Jr. Quantification of left ventricular volumes by two-dimensional echocardiography: A simplified and accurate approach. *Circulation*. 1983;67:579–584.

6. Sanders SP, Yeager S, Williams RG. Measurement of systemic and pulmonary blood flow and QP/QS ratio using Doppler and two-dimensional echocardiography. *Am J Cardiol*. 1983;51:952–956.

7. Hatle L. Noninvasive assessment and differentiation of left ventricular outflow obstruction with Doppler ultrasound. *Circulation*. 1981;64:381–387.

8. Teirstein P, Yeager M, Yock PG, Popp RL. Doppler echocardiographic measurement of aortic valve area in aortic stenosis: A noninvasive application of the Gorlin formula. *J Am Coll Cardiol*. 1986;8:1059–1065.

9. Chan KL, Currie PJ, Seward JB, Hagler DJ, Mair DD, Tajik AJ. Comparison of three Doppler ultrasound methods in the prediction of pulmonary artery pressure. *J Am Coll Cardiol.* 1987;9:549–554.

10. Bush SE, Huhta JC, Vick GW, Gutgesell HP, Ott DA. Hypoplastic left heart syndrome: Is echocardiography accurate enough to guide surgical palliation? *J Am Coll Cardiol.* 1986;7:610–616.

11. Callahan MJ, Seward JB, Tajik AJ. Two-dimensional and Doppler echocardiography in atrial septal defect: Operation without catheterization. *Echocardiography.* 1984;1:521–526.

12. Canale JM, Sahn DJ, Allen HD, Goldberg SJ, Valdes-Cruz LM, Ovitt TW. Factors affecting realtime cross-sectional echocardiographic imaging of perimembranous ventricular septal defects. *Circulation.* 1981;63:689–697.

13. Sahn DJ, Allen HD. Real-time, cross-sectional echocardiographic imaging and measurement of the patent ductus arteriosus in infants and children. *Circulation.* 1978;48:343–354.

14. Halte L. Noninvasive diagnosis and assessment of ventricular septal defect by Doppler ultrasound. *Acta Med Scand.* 1981;645 (suppl):47–56.

15. Swensson RE, Valdes-Cruz LM, Sahn DJ, et al. Real-time Doppler color flow mapping for detection of patent ductus arteriosus. *J Am Coll Cardiol.* 1986;8:1105–1112.

16. Marx GR, Allen HD, Goldberg SJ. Doppler echocardiographic estimation of systolic pulmonary artery pressure in patients with aorto-pulmonary shunts. *J Am Coll Cardiol.* 1986;7:880–885.

17. Kosturakis D, Goldberg SJ, Allen HD, Loeber C. Doppler echocardiographic prediction of pulmonary arterial hypertension in congenital heart disease. *Am J Cardiol.* 1984;53:1110–1115.

18. DeMaria AN, Bommer W, Joye J, Lee G, Bonteller J, Mason DT. Value and limitations of cross-sectional echocardiography of the aortic valve in the diagnosis and quantification of valvular aortic stenosis. *Circulation.* 1980;62:304–312.

19. Weyman AE, Hurwitz RA, Girod DA, Dillin JC, Feigenbaum H, Green D. Cross-sectional echocardiographic visualization of the stenotic pulmonary valve. *Circulation.* 1977;56:769–774.

20. Johnson GL, Kwan OL, Handshoe S, Noonan JA, DeMarie AN. Accuracy of combined two-dimensional echocardiography and continuous wave Doppler recordings in the estimation of pressure gradient in right ventricular outlet obstruction. *J Am Coll Cardiol.* 1984; 3:1013–1018.

21. Sahn DJ, Allen HD, McDonald G, Goldberg SJ. Real-time cross-sectional echocardiographic diagnosis of coarctation of the aorta. *Circulation.* 1977;56:762–769.

22. Wyse RKH, Robinson PJ, Deanfield JE, Tunstall Pedo DS, Macartney FJ. Use of continuous wave Doppler ultrasound velocimetry to assess the severity of coarctation of the aorta by measurement of aortic flow velocities. *Br Heart J.* 1984;52:278–283.

23. Caldwell RL, Wyman AE, Hurwitz RA, Girod DA, Feigenbaum H. Right ventricular outflow tract assessment by cross-sectional echocardiography in tetralogy of Fallot. *Circulation.* 1979;59:395–402.

24. Bierman FZ, Williams RG. Prospective diagnosis of D-transposition of the great arteries in neonates by subxiphoid, two-dimensional echocardiography. *Circulation.* 1979;60:1496–1502.

25. Beppu S, Nimura Y. Tamai M, et al. Two-dimensional echocardiography in diagnosing tricuspid atresia: Differentiation from other hypoplastic right heart syndromes and common atrioventricular canal. *Br Heart J.* 1978;40:1174–1183.

26. Silverman NH, Schiller NB. Apex echocardiography. A two-dimensional technique for evaluating congenital heart disease. *Circulation.* 1978;57:503–511.

Chapter 38

Evaluation of Cardiac Anatomy and Function Using Ultrafast Computed Tomography in Patients with Congenital Heart Disease

W. Jay Eldredge, MD

INTRODUCTION

Over the past 20 years many significant improvements in cardiac imaging modalities have occurred. These changes are partially due to rapid improvements in technology, as well as to the more recent application of computer science to medical imaging systems. However, there is a continued search for the ideal technique for imaging the complex three-dimensional problems encountered in patients with congenital heart disease. Definition of the complicated spatial relationships found in these patients would be best accomplished using a true three-dimensional imaging technique.

Computed tomography (CT) has many characteristics that would appear to make it an ideal imaging modality for use in patients with congenital heart disease. Cross-sectional tomographic scanning utilizing multiple contiguous slices represents a true volume or three-dimensional imaging technique. When applied to cardiac diagnosis, the entire heart and great vessels may be captured within a three-dimensional cubic matrix of CT data, which then is available for later display and analysis in a variety of different formats. A wide field of view also permits the cardiac structures to be seen within the entire thorax, which facilitates the definition of their spatial relationships (Figure 38-1). In addition, CT offers both high spatial and high density resolution. This unique approach also overcomes one of the single greatest problems of the more commonly used projection imaging techniques—the loss of crucial anatomic information secondary to the obscuring of important regions of interest by overlapping structures (Figure 38-2). Conventional CT scanners have had, at best, data acquisition times in the range of 2 seconds per single tomographic slice, as well as prolonged interscan delay times. Such slow scanning times resulted in images that contained serious motion artifacts, rendering them completely unacceptable for most cardiac applications. It is because of this low temporal resolu-

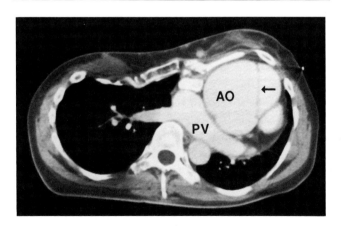

Figure 38-1 A single frame from a multilevel study in a patient with Marfan's syndrome and dissection of the aorta. The wide field of view seen with CT permits precise localization of the markedly displaced cardiac structures within the deformed thoracic cavity. The severely dilated ascending aorta (AO) is displaced leftward. A dissection of the aorta (arrow), which had not been visualized during routine aortography, is beautifully demonstrated. The pulmonary veins (PV) are displaced rightward and posteriorly.

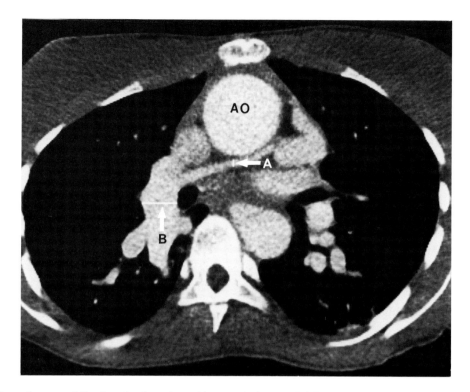

Figure 38-2 A single frame from a multilevel study of a patient with a ventricular septal defect and pulmonary atresia. The ascending aorta (AO) is dilated and displaced anteriorly. The portion of the right pulmonary artery posterior to the aorta is severely hypoplastic (point-A). This portion of the pulmonary artery is frequently abnormal in patients with various forms of congenital heart disease and is often poorly visualized due to superimposition of the overlapping, contrast material–filled ascending aorta. More distally, the artery is larger (point-B). Measurements of 3 mm at point A and 19.5 mm at point B were obtained. Again, the wide CT field of view facilitates definition of the spatial relationships of the various cardiac structures imaged.

tion that only cursory attempts to develop conventional CT as a cardiac diagnostic technique have been made.

A new multislice rapid acquisition CT scanner utilizing a magnetically deflected electron beam is now available and has been in clinical use for over 3 years. (See Chapter 3.) The details of this unique scanner have been previously described.[1-3] When the electron beam strikes one of the four tungsten target rings present within the scanner, two side-by-side 8-mm tomographic slices are obtained in 50 milliseconds. If each of the four target rings is serially scanned, eight side-by-side tomographic slices are produced in 224 milliseconds. Rapid scan times combined with multislice capability now make it possible to use the acknowledged benefits of CT for the anatomic diagnosis of congenital heart disease, the quantitation of blood flow parameters, and the analysis of cardiac function.

Data may be acquired through three modes. Each mode provides its own unique anatomic and functional information, depending upon the manner in which the CT data are acquired. The following discussion will review each of these three modes, emphasizing the individual characteristics of each. Our experience with the first 400 cases of congenital heart disease evaluated with ultrafast CT (Table 38-1) will form the basis of the discussion.

Table 38-1 First 400 Patients with Congenital Heart Disease Evaluated by Ultrafast CT

Diagnosis	Preoperative Study	Postoperative Study
Aortic stenosis	10	
Atrial septal defects		
Secundum	30	5
Primum	9	5
Sinus venosus	5	
Cardiomyopathies	18	
Coarctation of aorta	13	5
Complete atrioventricular canal	10	2
Complete transposition	14	5
Coronary artery anomalies	3	1
"Corrected" transposition	9	
Double outlet right ventricle	7	1
Ebstein's anomaly	4	
Hypoplastic right ventricle	4	
Marfan's syndrome	5	1
Mitral insufficiency	3	
Patent ductus arteriosus	6	
Pulmonary atresia	17	
Pulmonary valve stenosis	11	
Single ventricle complexes	28	
Tetralogy of Fallot	28	14
Tricuspid atresia	6	3
Truncus arteriosus	2	
Vascular ring	2	
Ventricular septal defect	110	4

VOLUME MODE

Scanning in the volume mode will be readily recognized by the radiologist since it is similar to multislice scanning with table movement using a conventional CT scanner. The major advantage that ultrafast CT offers in this mode is that the speed of data acquisition usually necessitates only a single injection of contrast material, even when covering a large area of interest. The patient is placed in the transaxial position (0° tilt, 0° slew). During the injection of iodinated contrast material, continuous 100 millisecond scans are obtained during table incrementation. Sufficient images are acquired to cover the entire region of interest. The total amount of contrast material used varies according to the patient's size and heart rate and the total distance to be covered. A dose of 1 to 2 mL/kg injected at a rate of 1 to 2 mL/s is generally adequate. Scanning should begin approximately 10 to 20 seconds after the onset of the injection and should continue until just before the termination of image acquisition. These images may then be viewed as individual cross-sectional slices or reconstructed in any operator-determined axis.

The volume mode has proved most useful in studying abnormalities of the aorta, pulmonary arteries, and pericardium. Unoperated cases of coarctation of the aorta have proved to be particularly well suited for accurate imaging using ultrafast CT. In each case the anatomy of the coarctation has been well defined, thereby permitting surgical repair without resorting to invasive angiography (Figure 38-3). Volume scanning of the aorta has also been an effective tool for evaluating the surgical results of coarctation repair and has become a useful, easily repeatable method of serially evaluating such patients. A flexible "off-axis reconstruction" program is included in the system's software. The operator can determine the plane of off-axis reconstruction in an infinite number of angles. After the initial off-axis reconstruction has occurred, multiple reconstructions of the previously reconstructed image are also possible (Figure 38-4). Using this technique, it is possible to interrogate the entire aorta in a tomographic manner in any desired direction (Figure 38-5). This capability has greatly improved the technique's accuracy in evaluating aortic coarctation. Off-axis reconstruction of images obtained in the flow or cine mode is also feasible. Reviewing images in multiple off-axis planes that are reconstructed after the patient has left the imaging suite has proved very useful in describing anatomic relationships in complex congenital lesions.

Patients with Marfan's syndrome have also been evaluated using the volume mode protocol. Dilatation of the aortic root as well as the presence or absence of aortic dissection has been evaluated in each case (Figure 38-1). A computer-assisted "measure distance" program permits rapid determination of the diameter of the aorta at multiple sites. If the volume mode study is combined with cine studies in the same patient, the presence of mitral and aortic valve abnormalities can be determined, and regurgitant volume can be calculated.

Precise measurement of pulmonary artery size has also become an invaluable use of ultrafast CT. Using the

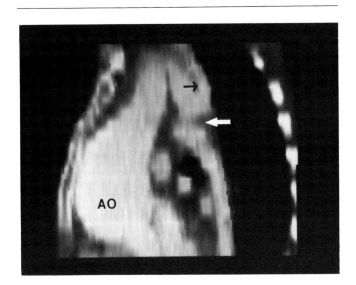

Figure 38-3 An off-axis reconstruction from a multilevel volume study in a patient with coarctation of the aorta. The ascending aorta (AO) is markedly dilated. The site of the coarctation (arrow) is beautifully demonstrated just below the origin of the left subclavian artery. The descending aorta, below the level of the coarctation, resumes a normal size.

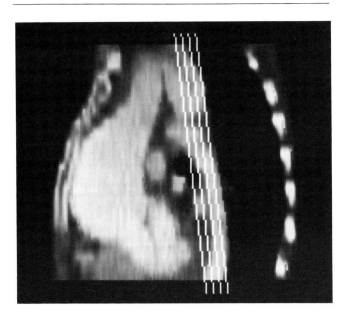

Figure 38-4 Four operator-determined planes with 4.1-mm spacing are superimposed upon the previously reconstructed image of aortic coarctation. Further reconstructions of this previously reconstructed image are then possible, giving an elegant method for comprehensive tomographic evaluation of the entire aorta in an infinite variety of views.

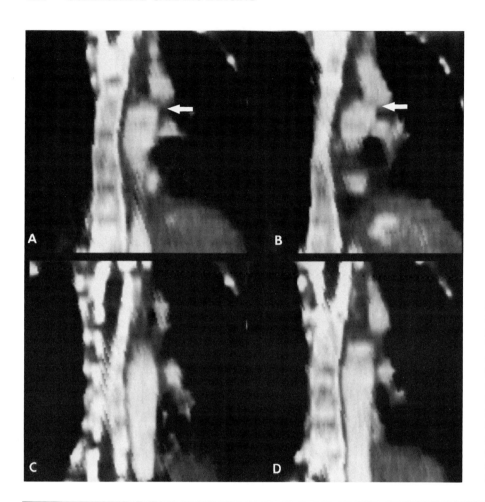

Figure 38-5 Four reconstructed images derived from the operator-determined reconstruction planes shown in Figure 38-4. The descending aorta in the area of the coarctation is seen in these four tomographic slices, progressing from anterior to posterior. In frames A and B the coarctation site is again demonstrated (arrow) in this different plane. Note the unusual curve of the descending aorta around the area of coarctation. In frames C and D the descending aorta below the coarctation comes into view within the reconstruction plane.

"measure distance" program, extremely accurate measurements of pulmonary artery size are easily obtained (Figure 38-2). The values thus obtained may then be compared to the aortic diameter for use as an index of potential operability. These measurements do not require correction for magnification, as is required with conventional angiography. Measurements made from cineangiography can also be inaccurate because of inadequate filling of the pulmonary arteries with contrast material. The extremely high density resolution of ultrafast CT obviates this problem.

FLOW MODE

Data acquisition using the flow mode is triggered by the patient's electrocardiogram (ECG). Continuous contrast medium–enhanced, ECG-triggered scans are obtained at two, four, six, or eight levels in the same phase of the cardiac cycle. The scanner may be triggered in systole, diastole, or at any percentage of the RR interval. A sufficient number of images is obtained in time to allow visualization of the wash-in and washout of the contrast material through the various structures within the scan plane (Figure 38-6).

During early trials the flow mode was used as the primary mode of data acquisition, since it is useful not only for defining cardiac anatomy but also for calculating cardiac output and pulmonary-to-systemic flow ratios. A maximum of 80 images may be obtained during one flow study at two to eight contiguous levels. Sufficient levels should be chosen to cover the entire region of interest. When the flow study is to be used for calculation of such flow parameters as the pulmonary-to-systemic flow ratio, only a four-level study should be performed. Twenty points on the computer-generated flow curve are thus obtained. This permits a better gamma variate fit to the resulting time-density curves, permitting more accurate calculations. Following an eight-level localization scan to ensure proper patient position, a bolus of iodinated contrast material is injected using a power injector. The injection of 0.3 to 0.5 mL/kg of Renografin 76 at a flow rate of 5 to 10 mL/s through a peripheral intravenous site has resulted in a tight bolus with excellent opacification of the cardiac blood pool.

Proper positioning of the patient within the scanner has proved to be extremely important. During early experience, all patients were scanned in the usual CT

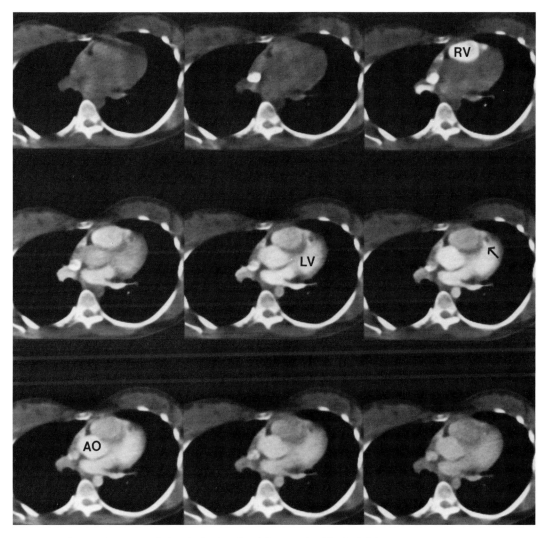

Figure 38-6 Nine selected frames from a single level of a four-level flow study. The initial frame contains no contrast material. The arrival of contrast medium in the superior vena cava is seen in the second image. With each subsequent image, the wash-in and washout of contrast material through the various cardiac structures within the imaging plane can be seen. During opacification of the left ventricle (LV), a small muscular ventricular septal defect (arrow) is imaged near the apex of the ventricle.

transaxial position (0° slew, 0° tilt). This still remains the position of choice for defining complex anatomic relationships. A long-axis position that resembles its echocardiographic counterpart is achieved by slewing the table 20° to 25° counterclockwise with no table tilt. In this position the atrial and ventricular septa as well as the left ventricular outflow tract are best visualized, and it is now the position of choice for imaging these areas. A CT short-axis position is achieved by slewing the table 20° to 25° clockwise and tilting the foot of the table 25° downward. This is the position of choice for all studies used to calculate ventricular volumes, ejection fractions, or ventricular mass. The aortic valve is also best evaluated in this position. The pulmonary valve is most consistently imaged using a cranial-caudal position, which is achieved by tilting the table 25° downward with no slew. The right ventricular outflow tract is also seen well in this position.

As can be seen from Table 38-1, a large variety of congenital cardiac defects, both isolated and in combination, have been imaged using ultrafast CT. More complete details of our initial experience have been reported elsewhere and will not be reiterated here.[4-6] Many simple cardiac defects have been imaged with a high degree of accuracy. Since the atrial septum can be scanned in its entirety, atrial septal defects have been consistently identified. Not only can the presence of a defect be determined, but its location can be precisely defined (Figure 38-7). The more common secundum variety can be readily distinguished from those located within the septum primum. It has also been possible to correctly define defects in the sinus venosus region and to determine the presence of partial anomalous pulmonary venous drainage in these cases. Patients now may have surgical closure of atrial septal defects without undergoing additional invasive studies.

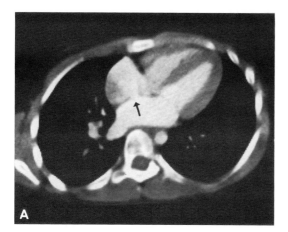
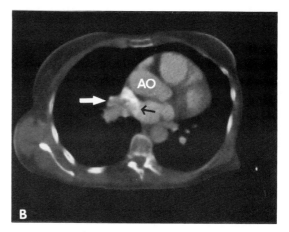

Figure 38-7 Selected images from the flow studies of patients with two different types of atrial septal defects. In frame A a secundum defect is seen (arrow) in the midportion of the atrial septum. The mitral valve can be seen at this level. In frame B a sinus venosus type of defect is imaged. The level of the tomographic slice is much higher than in frame A since the aortic root (AO) is visualized in the same imaging plane. A small amount of contrast material can be seen passing from the superior vena cava near its junction with the right atrium through the sinus venosus defect and into the high left atrium (black arrow). An anomalous right pulmonary vein (white arrow) is seen draining into the superior vena cava at the same level.

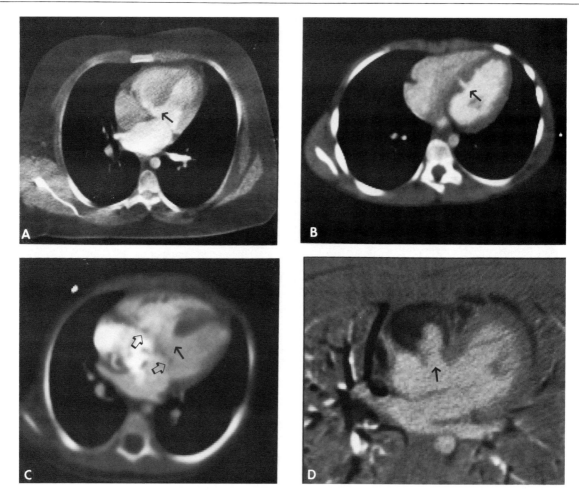

Figure 38-8 Single selected frames from the flow studies of four patients with varying types of ventricular septal defects. In frame A, the more common type of perimembranous defect (arrow) is well visualized. Frame B shows a small defect (arrow) in the midportion of the muscular septum. The defect is low, near the floor of the left ventricle. Frame C is from a patient with a complete type of atrioventricular canal. A large ventricular septal defect of the atrioventricular canal type (arrow) is demonstrated. A common atrioventricular valve (open arrows) is seen crossing through the defect. Frame D is a subtraction image processed from one level of a flow study. The left-sided cardiac structures contain contrast medium. A ventricular septal defect that is associated with a large aneurysm of the membranous septum can be seen (arrow). The aneurysm is protruding into the right ventricular cavity.

Similarly, the entire ventricular septum may be scanned superiorly from the outlet septum inferiorly through the entire muscular septum.[7] Using this technique, it has been possible to define ventricular septal defects of the supracristal (subarterial), perimembranous, muscular, and atrioventricular canal types (Figure 38-8). With experience it has also been possible to predict extension of ventricular septal defects into adjacent areas of the septum. For example, the ability to accurately determine the presence of extension of a supracristal defect inferiorly into the perimembranous septum has proved to be of great value in preoperative planning of the surgical approach to such patients.

The great advantages of cross-sectional tomographic imaging have become most obvious in patients with complex congenital cardiac malformations. Since the cross-sectional images are obtained in an orderly fashion, progressing from the great vessels caudally to the base of the heart, it is possible to utilize an orderly, segmental approach to anatomic diagnosis (Figures 38-9 and 38-10). In the most superior slices, the great vessels will be seen, and their precise spatial relationship to one another can be defined. By progressing inferiorly in a slice-by-slice fashion, the arterial-ventricular connections may then be determined, as well as the relationships between the great vessels. More inferiorly, the atrioventricular valves are imaged, and the atrioventricular connections may be defined. The systemic venous connections are also readily identifiable. Accurate definition of these complex spatial relationships is readily accomplished without encountering the problems common to other imaging techniques, such as patient position with respect to the projection image, an abnormal position of the cardiac structures within the thorax, and, most important, loss of information due to overlapping structures obscuring the region of interest.

One of the most exciting uses of ultrafast CT in patients with congenital heart disease is the ability to quantitate blood flow utilizing the same flow study that was obtained for anatomic definition. By using standard indicator-dilution theory, the mean CT number of the injected contrast material is plotted against time within any operator-determined region of interest. The computer-generated points are then fitted to a gamma variate curve, and cardiac output is determined using the Stewart-Hamilton equation. Cardiac output studies comparing values obtained from ultrafast CT to those obtained using simultaneous thermodilution measurements have shown excellent correlation.[8] The feasibility of using ultrafast CT to quantitate intracardiac shunts has been demonstrated using flow phantoms.[9] In patients with left-to-right shunts, a bimodal shunt curve may be generated by placing a region-of-interest marker over any chamber or vessel involved in the shunt (Figure 38-11). Using an algorithm previously

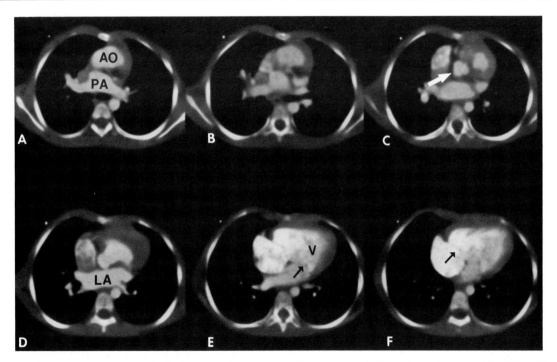

Figure 38-9 Frames from six contiguous levels of a flow study in a patient with a single ventricle. In frame A the spatial relationships of the great vessels can be determined. The aorta (AO) is anterior and to the left of the posterior, rightward pulmonary artery (PA) in the "L loop" position. Progressing caudally, a tricuspid pulmonic valve (white arrow) is visualized in frame C. The subpulmonic area is imaged in frame D in its position anterior to the left atrium (LA). In frames E and F the single ventricle (V) is demonstrated. In frame E the left (arrow) and in frame F the right atrioventricular valves (arrow) are shown both emptying into the single ventricle.

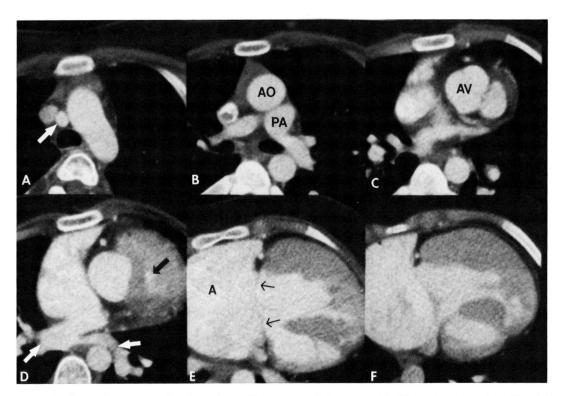

Figure 38-10 Six selected frames from a second patient with a different form of single ventricle. The patient has a right Blalock-Taussig shunt, which is seen in frame A (arrow). In frame B the aorta (AO) and pulmonary arteries (PA) are imaged. The aorta is somewhat anterior but to the right of the pulmonary artery. In frame C the rightward aortic valve (AO) is imaged nearly side by side with the main pulmonary artery. An extremely narrowed subpulmonary area (black arrow) can be seen in frame D. This proved to be the site of obstruction to the patient's pulmonary blood flow. The pulmonary veins (white arrows) are also well demonstrated. A huge common atrium (A), which connects to the single ventricle across a large common atrioventricular valve (arrows), is imaged in frames E and F.

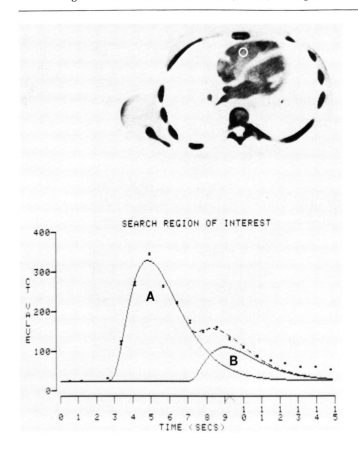

Figure 38-11 A region-of-interest marker is placed over the right ventricle in a flow study from a patient with a small- to moderate-sized ventricular septal defect. The computer generates a bimodal shunt curve by plotting the CT value of the injected contrast medium against time. The areas under the primary (A) and secondary (B) portions of the contrast agent clearance curve are calculated following a gamma variate fit to each portion. The pulmonary-to-systemic blood flow ratio is calculated using the formula QP:QS = area A/(area A − area B). The QP:QS in this patient was 1.6:1.

validated by nuclear medicine, a gamma variate curve is fitted to the primary and secondary portions of the resulting curve.[10] The pulmonary-to-systemic flow ratio may then be calculated following measurement of the area under both curves. Retrospective review of our own clinical data shows excellent correlation between the intracardiac shunt values obtained using ultrafast CT and those obtained during cardiac catheterization for pulmonary-to-systemic blood flow ratios between 1.3 and 2.5 to 1.

CINE MODE

Using the cine mode, sufficient continuous 50-millisecond scans can be obtained to cover one entire heartbeat (Figure 38-12). From two to eight levels may be obtained, depending upon the patient's heart rate. Contrast material may be injected in one of several ways, depending upon the number of levels to be scanned, to cover the particular area of interest. When two to four levels are to be scanned, a bolus injection similar to that performed in the flow mode is most often used. The delay between the injection and the onset of the scan may be determined from a previous flow study or by any other method of determining circulation time. When more than four levels are to be scanned, a slower or "strung-out" injection is used. Depending upon the patient size and heart rate, 1 to 1.5 mL/kg of contrast material is injected at a flow rate of 1 to 2 mL/s. Scanning then begins shortly before the end of the injection. By using this technique, contrast material should be well distributed throughout the cardiac structures.

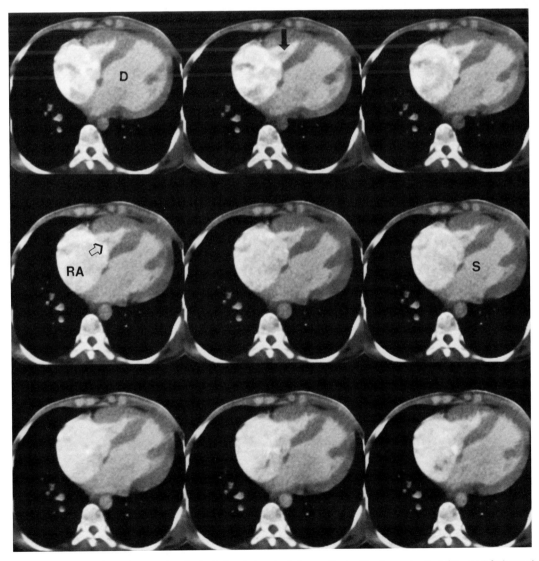

Figure 38-12 Nine selected frames from the same level of a six-level cine study in a patient with a hypoplastic right ventricle (arrow). The first frame shows the left ventricle in diastole (D). Each successive frame demonstrates progressive systolic contraction until end-systole is reached (S). A hypoplastic tricuspid valve (open arrow) and enlarged right atrium (RA) can be seen.

The cine mode is predominantly used for analysis of cardiac function. Occasionally, however, we have also used it for anatomic definition. All studies performed for the purpose of calculating ventricular volumes, ejection fractions, or ventricular mass are performed with the patient in the short-axis position.

Currently measurements of ventricular volume are done using cineangiography or two-dimensional echocardiography. Since both methods require geometric assumptions for the calculation of volume, they are subject to great error in patients with congenital heart disease who may have abnormally shaped or abnormally positioned chambers. Accurate calculation of right ventricular volume in patients with congenital heart disease is of great value, particularly when evaluating long-term postoperative results. Current methods of obtaining this information have proved to be particularly difficult and cumbersome. The ability of CT to accurately measure ventricular volumes despite abnormal shape or chamber orientation has been previously shown.[11-12] Using ultrafast CT, the entire heart may be imaged in a dynamic way from apex to base, thereby including the entire volume of the ventricles in all phases of the cardiac cycle within the cubic matrix of CT data (Figure 38-13). Using a bicycle ergometer placed on the patient couch, ventricular function data may be accurately obtained during exercise as well as at rest, adding a whole new dimension to the evaluation of both pre- and postoperative patients with congenital heart disease. Calculation of the ventricular volume then becomes merely a matter of planimetering the area of the ventricular cavity in each tomographic slice. Computer software then calculates the total volume. The extreme accuracy of these measurements has been validated, and normal values for the left ventricle established.[13-16]

If, in addition to tracing the endocardial borders for the measurement of left ventricular volume, the epicardial borders are also outlined, left ventricular mass may be calculated with amazing accuracy.[17-18] Having completed these tracings for each level from apex to base, the computer generates not only left ventricular mass but also stroke volume, end-diastolic volume, and ejection fraction for each level as well as the total. This permits evaluation of global as well as regional cardiac function. Calculation of left ventricular mass has become routine in all our patients with hypertrophic cardiomyopathies. Serial scanning of these patients with the calculation of left ventricular mass should prove an extremely sensitive method of follow-up.

Regurgitant volume may be measured following the calculation of stroke volume for each ventricle.[19-20] By

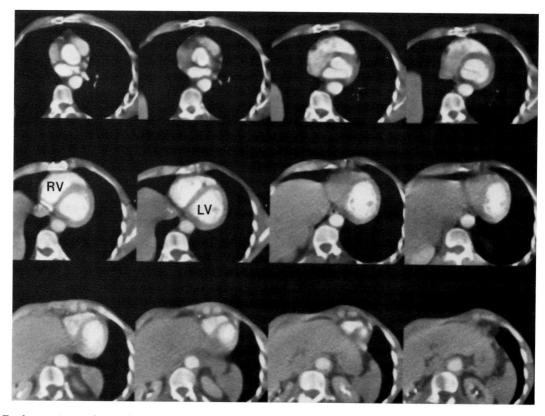

Figure 38-13 Twelve contiguous frames from a cine study are shown to demonstrate ultrafast CT's ability to encompass the entire volume of both ventricles. Calculation of both global and regional parameters of ventricular function as well as ventricular mass are readily feasible using such a technique. RV, right ventricle; LV, left ventricle.

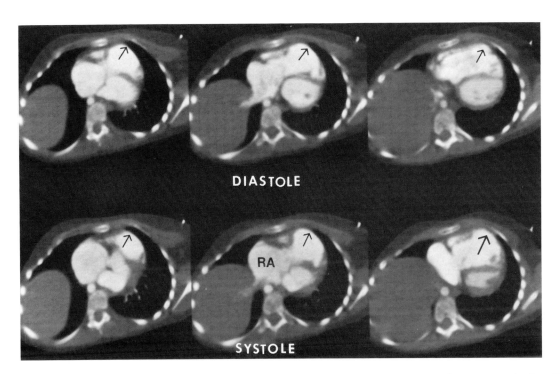

Figure 38-14 Systolic and diastolic frames from a single cine study at three different levels are shown to demonstrate ultrafast CT's ability to evaluate postoperative right ventricular function. The patient had previously undergone repair of tetralogy of Fallot incorporating a patch in the right ventricular outflow tract. A very large akinetic area of the right ventricle (arrows) is well shown using this technique. No change in wall thickness or cavity configuration can be seen between the systolic and diastolic images.

subtracting the stroke volume of one ventricle from that of the other, the regurgitant volume is obtained. In patients with congenital heart disease, the ability to assess right ventricular function as well as to quantitate pulmonary valve insufficiency both at rest and with exercise is one of the most exciting features of ultrafast CT. The long-term effects on cardiac function of surgical intervention for correction or palliation of congenital cardiac defects needs to be determined. A simple, nearly noninvasive method is now available, which should enable us to obtain this information in an accurate, quantitative way (Figure 38-14).

SUMMARY

The single ideal imaging modality for use in patients with congenital heart disease remains to be defined. Our initial experience with 400 patients with a wide variety of defects suggests that ultrafast CT offers many advantages over currently available imaging techniques. The major strength of the technique is its ability to provide accurate three-dimensional information regarding cardiac anatomy, function, and blood flow relatively noninvasively, with high spatial, density, and temporal resolution. Another major benefit is that these data may be acquired either in the resting state or with exercise or drug intervention, adding a further important dimension to cardiac diagnosis and follow-up. The exact role that ultrafast CT will eventually play in the overall cardiac imaging picture remains to be completely defined. Our initial experience, however, suggests that this modality can contribute much important diagnostic information that is not currently readily available by other techniques.

REFERENCES

1. Boyd DP. Computerized transmission tomography of the heart using scanning electron beams. In: Higgins CB, ed. *CT of the Heart and Great Vessels: Experimental Evaluation and Clinical Evaluation.* Mt. Kisko, NY: Futura Publishing; 1983:46–60.

2. Boyd DP, Lipton MJ. Cardiac computed tomography. *IEEE Proc.* 1983;7:289–307.

3. Boyd DP, Couch JL, Napel SA. Ultra cine-CT for cardiac imaging: Where have we been? What lies ahead? *Am J Card Imaging.* 1987;1:175–185.

4. Eldredge WJ, Flicker S. Evaluation of congenital heart disease using cine-CT. *Am J Card Imaging.* 1987;1:38–50.

5. Eldredge WJ, Bharati S, Flicker S, et al. Cine CT scanning in the diagnosis of congenital heart disease: Analysis of the first 42 cases. In: Doyle EF, et al., eds. *Pediatric Cardiology.* New York: Springer-Verlag; 1985:404–405.

6. Eldredge WJ, Flicker S, Steiner RM. Cine-CT in the anatomical evaluation of congenital heart disease. In: Pohost GM, et al., eds. *New Concepts in Cardiac Imaging 1987.* Chicago: Year Book Medical Publishers; 1987:256–285.

7. Eldredge WJ, Rees MR, FLicker S. Cine-CT scanning of the ventricular septum for the diagnosis of ventricular septal defect. *Circulation.* 1985;72(suppl III):27.

8. Garrett JS, Lanzer P, Jaschke W, et al. Noninvasive measurement of cardiac output by cine-CT. *Am J Cardiol.* 1985;56:657–661.

9. Garrett JS, Jaschke W, Aherne T, et al. Quantitation of intracardiac shunts by cine-CT. *J Comput Assist Tomogr.* 1988;12:82–87.

10. Treves S, Kurac A. Radionuclide evaluation of circulatory shunts. *Cardiol Clin.* 1983;1:427–439.

11. Lipton MJ, Hayashi TT, Boyd DP, et al. Measurement of left ventricular cast volume by computed tomography. *Radiology.* 1978;127:419–423.

12. Lipton MJ, Hayashi TT, Davis PL, et al. The effects of orientation on volume measurements of human left ventricular casts. *Invest Radiol.* 1980;15:469–474.

13. Reiter SJ, Rumberger JA, Stanford W, et al. Precise stroke volume measurements by cine-CT in the presence of abnormal left ventricular shape and size. *Circulation.* 1986;74(suppl II):122. Abstract.

14. Reiter SJ, Rumberger JA, Feiring AJ, et al. Precision of measurements of right and left ventricular volume by cine computed tomography. *Circulation.* 1986;74:890–900.

15. Reiter SJ, Rumberger JA, Feiring AJ, et al. Precise determination of left and right ventricular stroke volume with cine computed tomography. *Circulation.* 1985;72(suppl III):179. Abstract.

16. Mahoney LT, Smith W, Noel MP, et al. Measurement of right ventricular volume using cine computed tomography. *Circulation.* 1985;72(suppl III):28. Abstract.

17. Feiring AJ, Rumberger JA, Higgins CB, et al. Determination of left ventricular mass by rapid acquisition computed tomography. *Circulation.* 1984;70(suppl II):250. Abstract.

18. Feiring AJ, Rumberger JA, Reiter SJ, et al. Determination of left ventricular mass in dogs with rapid-acquisition cardiac computed tomographic scanning. *Circulation.* 1985;72:1355–1364.

19. Reiter SJ, Rumberger JA, Feiring AJ, et al. Measurement of aortic regurgitation with cine-CT. *J Am Coll Cardiol.* 1986;7:154A. Abstract.

20. Reiter SJ, Rumberger JA, Stanford W, et al. Quantitative determination of aortic regurgitant volume in dogs by ultrafast computed tomography. *Circulation.* 1987;76:728–735.

Chapter 39

Magnetic Resonance Imaging of Congenital Heart Disease

Barbara Kersting-Sommerhoff, MD, and Charles B. Higgins, MD

Improvements of cardiovascular surgical technique have made it possible to repair congenital heart lesions formerly considered inoperable because of the complexity of the lesion or the age or size of the patient. Such effective treatment demands a precise determination of intracardiac and great vessel anatomy. Complex intracardiac pathoanatomy in congenital heart disease is hard to define with a single imaging technique; even axial angiography, currently considered the gold standard, is not always able to evaluate complex cardiac malformations due to inadequate opacification of vessels or superimposition of cardiac structures on a projected image.

Gated magnetic resonance imaging (MRI) has been shown to be effective in the evaluation of cardiovascular pathoanatomy; it has now been used for the diagnosis of most congenital heart anomalies.[1] Congenital malformations of the great arteries, such as coarctation,[2-3] aortic arch anomalies,[4] or pulmonary arteries in conotruncal malformation,[5] have been demonstrated. Abnormalities of the heart, such as atrial level lesions (atrial septal defects and anomalous pulmonary venous connections),[6] complex ventricular lesions (Dery R, Diethelm L, Stager P, et al., unpublished data), pulmonary atresia,[7] and ventricular septal defects,[8] have been depicted by MRI.

The unique capability of MRI to differentiate flowing blood from intracardiac structures or vessel walls without the need for contrast media or exposure to the potentially harmful effects of ionizing radiation makes this technique ideally suited for the cardiovascular evaluation of children. Imaging in several orthogonal and nonorthogonal planes allows the demonstration of cardiovascular anatomy in the optimal plane for the various anomalies.

There are now several techniques for MRI of the heart. The electrocardiographically (ECG)-gated spin echo technique can be used for evaluating anatomy; this technique provides high-resolution images with a good signal-to-noise ratio. The new technique of cine MRI can be used for dynamic imaging of the heart if functional information is required.

IMAGING TECHNIQUES

Electrocardiogram-gated spin echo images are done using echo times (TE) of 30 to 40 and 60 to 80 milliseconds. If images at each anatomic level are acquired at only a single TE (usually 30 milliseconds), 10 to 12 anatomic levels can be encompassed in a multislice sequence. With a slice thickness of 10 mm, a 120-mm volume can be imaged during a single sequence. A proportionately smaller volume is used for studying infants, since the slice thickness usually must be reduced to 3 to 5 mm. Each imaging sequence can be accomplished in 6 to 10 minutes; the exact time is determined by the length of the RR interval of the ECG, which is equivalent to the repetition time (TR) for ECG-gated acquisitions. Two multislice series will encompass the entire heart and the great vessels, including the aortic arch, even in adults and require less than 20 minutes of imaging time. However, more than one imaging plane may be necessary for the complete evaluation of some lesions.

The transverse is the standard plane; images in this plane are done in all patients. For some purposes coronal and sagittal planes are added. A plane similar to the left anterior oblique view in cineangiography can be obtained by elevating the patient's right shoulder 30° to 40° and imaging in the sagittal plane. This additional plane is used for depicting aortic abnormalities since it usually displays most of the thoracic aorta on a single image. For specific purposes images can also be done perpendicularly (short axis) or parallel (long axis) to the major axis of the left ventricle. Indeed, images in any desired plane can be obtained by nonorthogonal orientation of the slice-selective gradient.

Cine MRI is a new technique used specifically to evaluate cardiac function and abnormal flow patterns. This technique of gradient-recalled acquisition in the steady state (GRASS) uses low flip angles (usually 30°) and gradient-refocused echoes. For cine magnetic resonance images the TE is 5 to 12 milliseconds and the TR is 20 to 30 milliseconds. With these parameters the cardiac cycle can be divided into 48 time frames. A total of up to 48 images is available for each acquisition period. These 48 images can be used at one anatomic level, or, if two levels are studied during an acquisition, 24 time frames are available at each level. The technical details of this technique have been described previously.[9]

The blood pool appears differently on magnetic resonance images depending on the imaging technique that is applied. If the spin echo technique is used, rapidly flowing blood generates minimal or no signal, resulting in a marked natural contrast between blood (signal void) and vessel walls or intracardiac structures. The rate of blood flow, however, differs according to the phase of the cardiac cycle; blood flowing at low velocities during diastole may produce considerable magnetic resonance signal. In cine MRI, blood has higher signal intensity than myocardium and appears white; the contrast between blood and myocardium is slightly less than in conventional magnetic resonance images. Turbulent blood flow, as seen in valvular incompetence or shunt lesions, can be detected as an area of low signal intensity or even signal loss emanating from the insufficient valve.

To obtain MRI studies of good image quality without motion artifacts, sedation is needed in infants or children less than 6 or 7 years of age; these patients are given approximately 100 mg/kg chloral hydrate (maximum dose, 2 g) orally 30 minutes prior to the MRI study.

NORMAL ANATOMY/SITUS DETERMINATION

Due to the natural contrast between flowing blood and vessel walls or cardiac structures, ECG-gated MRI clearly depicts cardiovascular anatomy (Figure 39-1). Standard transverse sections covering the heart from the top of the aortic arch to the cranial portion of the liver allow identification of all important cardiac structures. Additional images in the sagittal or coronal plane are helpful for the determination of complex cardiac relationships such as the visceroatrial situs.

Transverse images at the level of the base of the heart depict the position and relationship of the great vessels. In normal subjects the aorta lies posteriorly and to the right of the pulmonary artery. If the top of the aortic arch is imaged, the branching neck vessels are easily identified. The sizes of the aorta and the main pulmonary artery at the base of the heart are very similar, and their internal diameter can be measured. Sagittal images show the origin of the vessels from their respective ventricles. The aortic and pulmonary valves and their semilunar cusps can sometimes be identified with conventional MRI, but their visualization is not reliable and accurate enough to allow a functional evaluation. The new technique of cine MRI frequently overcomes this limitation.

At the level of the atria and the ventricles, the interatrial as well as the interventricular septa and the ventricle walls are demonstrated. Gradual thinning of the interatrial septum at the region of the fossa ovalis may result in signal loss or even regional signal drop-out; this finding may be problematic for the diagnosis of an atrial septal defect. The membranous portion of the ventricular septum can be differentiated from the muscular portion by its position and lower signal intensity. The right and left ventricular walls are sharply delineated by the natural contrast between blood and endocardium and the epicardial surface of the heart and the lung. Measurements of ventricular wall thickness and chamber size are possible on transverse or short-axis images. Pulmonary veins draining into the left atrium can be identified on adjacent levels of tranverse sections. Both lower lobe veins and the right upper lobe vein are usually easily recognized; the left superior pulmonary vein sometimes may be difficult to differentiate from the left atrial appendage. Tricuspid and mitral valves are often well visualized on transverse images; however, as with all cardiac valves, functional evaluation is limited with conventional MRI.

Since transverse tomograms depict segmental anatomy at the three important levels, complex cardiac relationships can be determined to identify great vessel relationships, the ventricular loop, and the visceroatrial situs.[10] Images through the middle of the right and left ventricles permit determination of the ventricular loop.

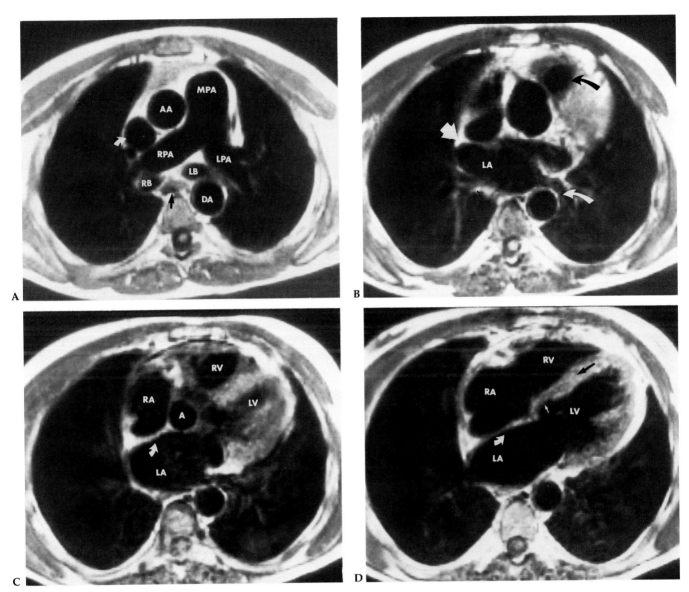

Figure 39-1 Tranverse images of the heart at four levels, demonstrating normal cardiac anatomy (10-mm slices). The image at the top left, taken just beneath the tracheal bifurcation, depicts the relationship of the great vessels. The ascending aorta (AA) lies posteriorly and to the right of the main pulmonary artery (MPA). The right (RPA) and left (LPA) central pulmonary arteries course anteriorly to the right main bronchus (RB) and over the left main bronchus (LB). The descending aorta (DA) is adjacent to the esophagus (black arrow). The curved arrow points to the superior vena cava. On the top right an image taken 30 mm caudad shows the right ventricular outflow tract (black arrow). The left lower pulmonary vein (curved white arrow) is seen entering the left atrium (LA). Note the sinus venosus portion of the atrial septum (straight white arrow). The right (RA) and left atria and right (RV) and left (LV) ventricles are demonstrated in the image at the bottom left, at the level of the aortic root (A). The curved white arrow points to the atrial septum. Ten mm lower, the image on the bottom right offers a four-chamber view of the right and left atria and ventricles. Note the exquisite delineation of the atrial (curved white arrow) and ventricular (black arrow) septa and the thinning of the perimembranous portion (straight white arrow) of the ventricular septum.

The anatomic right ventricle is characterized by its triangular configuration, its trabecular pattern, the muscular infundibular outflow tract, and the supraventricular crest between the tricuspid and semilunar valves. The morphologic left ventricle appears elliptic and smooth and does not have a muscular infundibulum, and its atrioventricular valve is farther from the apex. Coronal images help in determining the thoracic situs by showing the tracheal bifurcation into the mainstem bronchi and their relationship to the pulmonary arteries. The left central pulmonary artery is recognized by its course over the left bronchus. Visceroatrial situs solitus is present if a short right main bronchus, a right-sided inferior vena cava, a coronary sinus draining into the right atrium, and a right-sided liver are present.

ANATOMIC EVALUATION OF CONGENITAL ANOMALIES

Thoracic Aortic Anomalies

Positional as well as structural anomalies of the great vessels are clearly demonstrated in various planes by ECG-gated spin echo MRI, which can supply most if not all the information for which angiography has been traditionally used.[2-3,5] Aortic arch anomalies (Figure 39-2), such as right aortic arch with mirror image branching, left aortic arch with aberrant right subclavian artery, or double aortic arch, are easily identified in transverse sections encompassing the area at and above the aortic arch.[4,11] Aberrant vessels appear as an outpouching of the ascending or descending aorta, and their retroesophageal course can be followed. Magnetic resonance imaging provides complete evaluation of aortic arch anomalies, showing the airway compromise as well as the actual compression by the vascular anomaly. Symptomatic lesions have been associated with airway compression on magnetic resonance images.

Coarctation of the aorta is shown best on sagittal and left anterior oblique images. The site and extent of the aortic narrowing can be precisely defined, and supra- or infraductal position can be determined. Focal narrowing and poststenotic dilatation as well as involvement of arch vessels may also be seen on transverse images, depending on the extent of the coarctation and the slice thickness of the MRI study. Since restenosis or complications such as postoperative aneurysm or perianastomosal hematoma occur after surgical repair of coarctations, MRI is useful in follow-up studies for the noninvasive re-evaluation of the aorta. However, conventional MRI is not able to provide precise information about the severity of the lesion, as a pressure gradient in the aorta cannot be determined. The assessment of aortic valve competence is also limited with the spin echo technique.

Dilatation of the aortic root is a characteristic sign of Marfan's syndrome and may be differentiated from other forms of ascending aortic enlargement by frequently being limited to the proximal ascending aorta.[12] The diameters of the aortic root and the ascending aorta can be measured to establish a baseline if MRI is used as a screening and follow-up technique to monitor the aortic dilatation prior to surgery and after graft implantation.

Aortopulmonary shunts, such as a patent ductus arteriosus (Figure 39-3) or surgical shunts, can be evaluated noninvasively with MRI. A patent ductus arteriosus may occur as an isolated anomaly or in combination with a variety of congenital heart defects. Transverse scans at the level of the tracheal bifurcation clearly depict the connection between the descending aorta and the left pulmonary artery. Surgical shunts are usually best seen on transverse and coronal images; the presence of a flow void in shunts that can be followed through at least two imaging levels has been shown to be a reliable criterion for patency.[13]

Transposition or Malposition

Such positional abnormalities as transposition or malposition of the great arteries are identified with spin

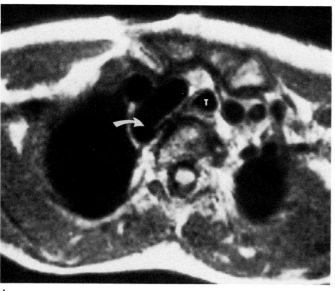 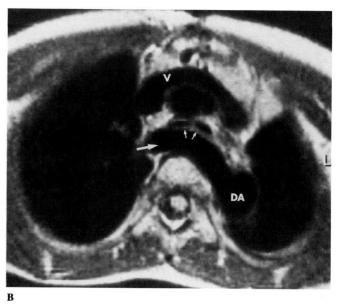

A B

Figure 39-2 Transverse images of a high right aortic arch with retrotracheal arch component. The left image depicts the right aortic arch (curved arrow). The image on the right, 10 mm more caudad at level of the tracheal bifurcation (small arrows), shows the retrotracheal aortic arch component (large arrow) and descending aorta (DA) on the left side. Note the compression of the proximal portion of the bronchi (small arrows). V, brachiocephalic vein; T, trachea.

MRI of Congenital Heart Disease   497

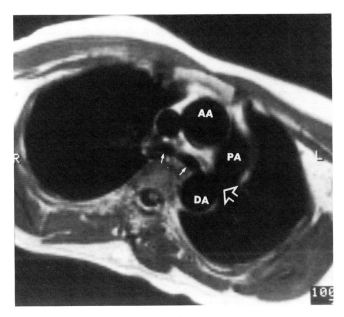

Figure 39-3 Transverse image of a patient with patent ductus arteriosus. A scan at the level of the tracheal bifurcation (small arrows) shows wide open communication (open arrow) between the pulmonary artery (PA) and the descending aorta (DA). AA, ascending aorta.

echo MRI.[10] On transverse images the aorta is visualized anterior and rightward of the main pulmonary artery in D-transposition and anterior and leftward of the pulmonary artery in L-transposition. Sagittal images show the origin of the anterior aorta from the morphological right ventricle and the posterior pulmonary artery arising from the morphological left ventricle.

Pulmonary Artery Anomalies

Pulmonary artery anomalies are often better demonstrated with MRI than with angiography, which may be compromised by opacification difficulties in pulmonary atresia or stenotic central pulmonary arteries.[5,7] In pulmonary atresia the right ventricular outflow tract ends blindly in muscle or fat tissue (Figure 39-4). Instead of a continuous signal void in the region of the main pulmonary artery, medium (muscle) or high (fat) intensity tissue intercepts the signal void and separates the ventricular chamber from the pulmonary artery. The proximal main pulmonary artery is atretic, but a distal main pulmonary artery or a confluence may be visualized on coronal or transverse scans. The central right or left pulmonary arteries may be absent; if present, they are usually hypoplastic, supplied with blood by bronchial collaterals or a patent ductus arteriosus. Collateral arteries are usually found on the side of absent or very hypoplastic central pulmonary arteries (Figure 39-5). The sizes of the pulmonary arteries are important deter-

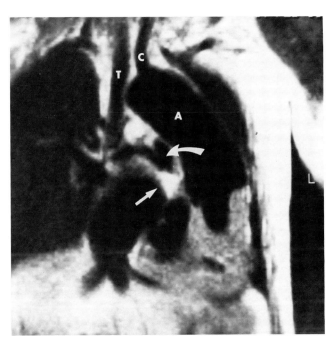

Figure 39-4 Coronal image of a patient with pulmonary atresia, a functional single ventricle, and levotransposition of the aorta. Pulmonary atresia is present as fat tissue (straight arrow) that lies between the right ventricular outflow tract and the hypoplastic pulmonary arteries. No main pulmonary artery is visualized, but a confluence (curved arrow) of the right and left pulmonary arteries is demonstrated. Note the levotransposition of the aorta (A). C, left common carotid artery; T, trachea.

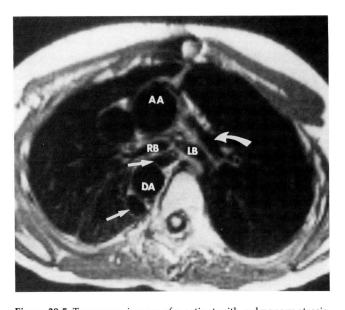

Figure 39-5 Transverse images of a patient with pulmonary atresia. The hypoplastic left pulmonary artery (curved arrow) courses over the left main bronchus (LB). Note the collateral arteries (straight arrows) on the right side, posterior to the bronchus and posterior to the descending aorta (DA). The right central pulmonary artery is not visualized in its expected position between the right bronchus (RB) and the ascending aorta (AA). Also note the absence of right hilar vessels.

minants of the likelihood of adequate surgical correction. MRI has been shown to demonstrate hypoplastic central pulmonary arteries that were missed by angiography due to difficulties in completely opacifying these vessels.

Pulmonary artery stenoses after surgical shunt procedures (Figure 39-6) and patent aortopulmonary shunts can also be demonstrated noninvasively. Patent Blalock-Taussig shunts between the subclavian and pulmonary arteries and Waterston or Potts shunts between the aorta and the pulmonary artery can usually be visualized on more than one level or in different imaging planes. A flow void in the shunt vessels is a reliable indicator of patency of these shunts.

Atrial-Level Lesions

Atrial-level lesions are best seen on transverse images. Atrial septal defects (ASDs; Figure 39-7) can be defined and classified at a high level of specificity.[6] A sinus venosus defect in the superior portion of the atrial septum near its junction with the superior vena cava can be demonstrated, as well as a septum primum defect in the lower portion of the septum near the

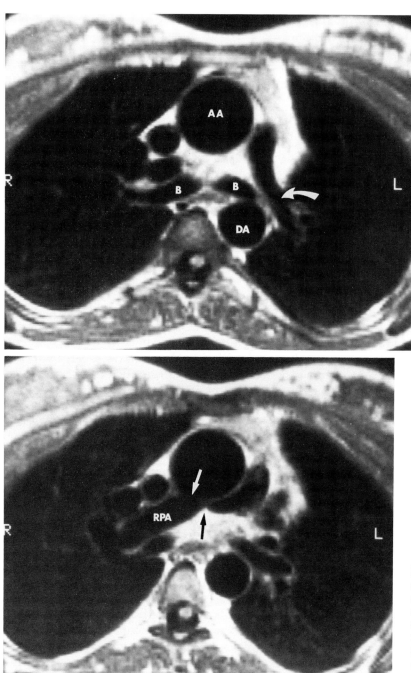

Figure 39-6 Transverse scans of a patient with tricuspid and pulmonary atresia and a stenosed right pulmonary artery after Waterston shunt implantation. The image on the top demonstrates a hypoplastic left pulmonary artery (curved arrow). The image on the bottom, taken 10 mm more craniad, clearly depicts a severe stenosis (black arrow) of the right central pulmonary artery (RPA) near the site of the Waterston shunt (white arrow). B, main bronchi; AA, ascending aorta; DA, descending aorta.

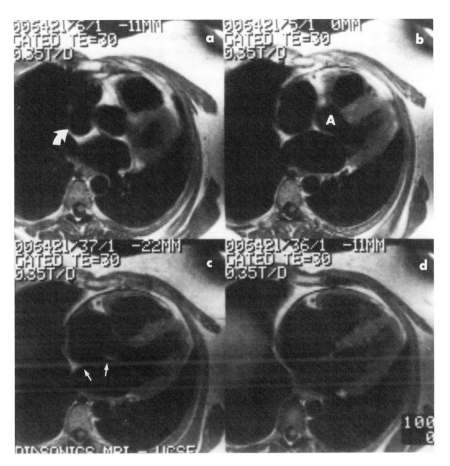

Figure 39-7 Transverse scans of a secundum atrial septum defect. Images A and B depict the intact sinus venosus portion (curved arrow in A) of the atrial septum. Images C and D demonstrate the atrial septal defect as an area of signal loss in the secundum portion of the atrial septum, with characteristically thickened residual septum at the edges of the defect (straight arrows in C). A, aortic root.

atrioventricular valves. A potential pitfall is the differentiation of a secundum ASD from signal drop-out in the region of the fossa ovalis. In septum secundum defects the thickness of the residual septum at the edges of the defect is increased, whereas there is gradual thinning of the septum and occasional signal drop-out of the septum at the fossa ovalis. In common atrium the primum and secundum portions of the atrial septum are completely absent.

Incomplete or complete endocardial cushion defects (Figure 39-8) can be identified by the association of an atrial septum primum defect with a truncated ventricular inlet septum and a common atrioventricular valve or a defect in the anterior leaflet of the mitral valve. Transverse magnetic resonance images of the incomplete endocardial cushion defect show the attachment of the mitral leaflet to the crest of the foreshortened inflow portion of the ventricular septum.

The evaluation of the drainage pattern of the pulmonary veins is important in patients with ASD, as it is sometimes anomalous. Total as well as partial anomalous pulmonary venous return, with the pulmonary veins draining into an enlarged coronary sinus or into the superior vena cava, can be demonstrated on transverse magnetic resonance images. The left upper pulmonary vein may sometimes be difficult to identify, but the other three pulmonary veins are usually observed.

Ventricular-Level Lesions

Ventricular septal defects (VSDs; Figure 39-9) are reliably demonstrated by MRI on transverse, sagittal, or coronal images, and their extent as well as their location can be defined. Even small defects (less than 5 mm) can be identified as an abrupt truncation of the ventricular septum.[8] Since the insertion of the atrioventricular valves is depicted on transverse images, the differentiation of perimembranous and muscular septal defects is possible. The inflow (posterior) and outflow (anterior) portions of the septum can also be identified; in complete endocardial cushion defect, the inflow VSD, primum ASD, and common atrioventricular valve are demonstrated on transverse or coronal images. Outflow septal defects are found in tetralogy of Fallot, truncus arteriosus, and double outlet right ventricle.

Complex Cyanotic Lesions

The unique advantage of MRI in assessing complex ventricular anomalies is the composite view of chambers and great vessels that it enables. In tetralogy of Fallot the right ventricular wall hypertrophy, infundibular hypoplasia, and over-riding aorta are well demonstrated on sagittal images. The ascending aorta over-

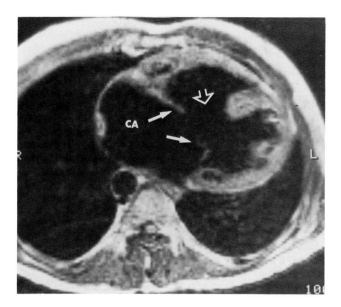
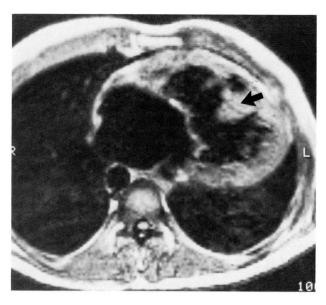

Figure 39-8 Transverse scans of a patient with complete atrioventricular canal and infundibular pulmonic stenosis. The image on the left depicts a common atrium (CA), a common atrioventricular valve (solid arrows), and an inlet ventricular septal defect (open arrow). The image on the right is 10 mm craniad and shows the remainder of the large single atrioventricular valve, which is unattached to the foreshortened ventricular septum (black arrow) and spans the two ventricles.

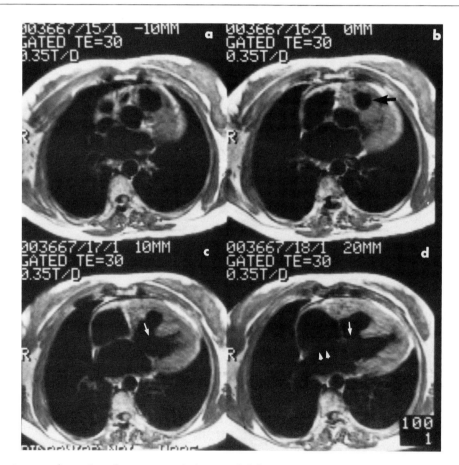

Figure 39-9 Transverse images of a perimembranous ventricular septal defect in a patient with infundibular pulmonic stenosis and right ventricular hypertrophy. Images A and B depict a narrowed right ventricular outflow tract with infundibular pulmonic stenosis (black arrow) and hypertrophied right ventricular walls. On images C and D the area of signal loss in the perimembranous part of the septum indicates the ventricular septal defect (white arrow). Note the signal drop-out in the region of the fossa ovalis (arrowheads), which is due to thinning of the atrial septum, not to an atrial septal defect.

riding a perimembranous VSD is enlarged, and the central pulmonary arteries are small.

Ebstein's anomaly is characterized by the displacement of the tricuspid valve leaflets toward the apex and atrialization of the right ventricular inflow tract. Transverse and coronal images are helpful in identifying this condition; besides a large right atrium and a small muscularized portion of the right ventricle, the displaced tricuspid leaflets can often be seen in their abnormal location.

Magnetic resonance images of patients with tricuspid atresia show that the tricuspid valve leaflets are replaced by muscle or fat extending from the right atrioventricular groove toward the center of the heart. An enlarged right atrium, an atrial septal defect, and a hypoplastic right ventricle can be visualized on transverse scans; the pulmonary arteries are usually hypoplastic or absent.

Single ventricle and common ventricle can be differentiated on coronal images. In patients with a single ventricle a bulboventricular foramen between the rudimentary and the dominant chamber can be discerned, whereas in patients with common ventricle, two equally sized chambers communicate through a large VSD. The rudimentary septum may be connected with the tricuspid and mitral valves or a common atrioventricular valve.

FUNCTIONAL EVALUATION OF CONGENITAL ANOMALIES WITH CINE MRI

The evaluation of cardiac function has been limited by the long imaging times and low temporal resolution of conventional spin echo MRI. Cine MRI overcomes this limitation; the improved time resolution facilitates the analysis of ventricular function. Since systole and diastole are readily identified, systolic wall thickening as well as wall motion are visualized easily. Measurements of systolic and diastolic wall thickness should be more precise than with conventional MRI, and the evaluation of ventricular function more reliable (Higgins CB, Holt W, Pflugfelder P, Sechtem U, unpublished data).[9]

Valves can be identified with cine MRI in systole and diastole, and the opening of the aortic valve and the motion of the pulmonic valve are visualized. Turbulent blood flow encountered in various valvular lesions results in areas of distinct signal loss in cine MRI. Thus the alteration in signal intensity of turbulent blood is useful for assessing the functional integrity of cardiac valves. In valvular regurgitation a regurgitant jet appears as an area of low signal intensity, extending from the incompetent valve into the recipient chamber. Aortic insufficiency (Figure 39-10) is characterized by a

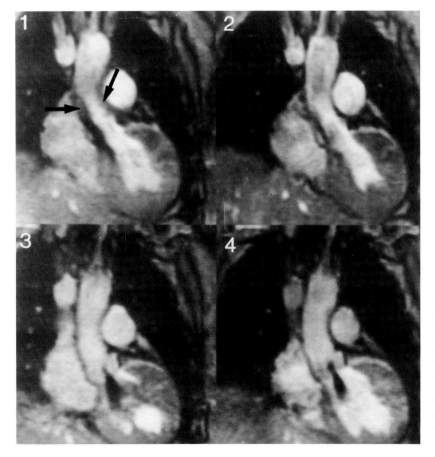

Figure 39-10 Coronal cine magnetic resonance images in a patient with moderate aortic regurgitation. Image 1, obtained in early systole, shows increased thickness of the left ventricular wall and areas of signal loss at the lateral walls of the ascending aorta (arrows). Blood flowing centrally through the aortic valve has high signal intensity. Image 2 was obtained at end-systole. At early diastole (image 3), a fan-shaped area of signal loss originating between the aortic valve cusps extends into the left ventricle, indicating aortic regurgitation. In late diastole (image 4), regurgitant turbulent blood flow is still visible.

region of signal void at several transverse levels extending from the aortic valve into the depth of the left ventricular chamber. Pulmonic regurgitation is recognized by an area of signal loss in the right ventricular outflow tract during diastole. Assessment of the severity of the lesion and quantification of the regurgitant volume is possible by comparing right and left ventricular stroke volumes, which are equivalent in normal subjects but different in valvular incompetence (Sechtem U, Pflugfelder P, Cassidy MC, White RD, Schiller ND, unpublished data).

Valvular stenosis can also be identified with cine MRI. The signal void is observed at the immobile cusps in valvular stenosis and below the valves in subvalvular aortic stenosis (Figure 39-11).

Intracardiac shunts occur frequently in congenital heart disease. A primum atrial septal defect appears as a fan of low signal intensity at the level of the attachments of the atrioventricular valves. Ventricular septal defects (Figure 39-12) also show a fan-shaped area of low signal intensity directed toward the right ventricular chamber in left-to-right shunts.[14]

Since cine MRI acquires three-dimensional information encompassing the entire heart, volume calculation for each heart chamber is possible without making geometrical assumptions. Thus the quantification of shunt volumes may become feasible.

SUMMARY

Magnetic resonance imaging can be used not only to anatomically define anomalies of the heart and the great arteries but also to evaluate function using gradient-refocused techniques. These techniques can define the abnormal flow patterns caused by stenotic and regurgitant valvular lesions and the flow across intracardiac septal defects. The completely noninvasive nature of MRI provides an attractive technique for monitoring the severity of cardiovascular anomalies.

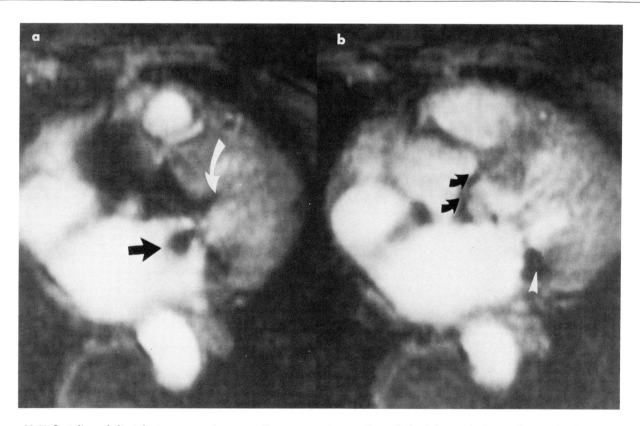

Figure 39-11 Systolic and diastolic transverse cine magnetic resonance images through the left ventricular outflow region in a patient with subaortic and aortic valve stenosis and accompanying mitral incompetence. A, In systole, an area of signal loss is visible within the left ventricular outflow tract distal to the site of the outflow obstruction (curved white arrow). There is narrowing of the area of turbulent flow at the level of the aortic valve cusps. Signal loss is again evident distal to the aortic valve. Note the fan-shaped area of signal loss within the left atrium due to mitral regurgitation (straight black arrow). B, In diastole, the aortic cusps appear thickened and have low signal intensity because of calcification (curved black arrows). Calcification is also seen adjacent to the posterior mitral valve leaflet (white arrowhead).

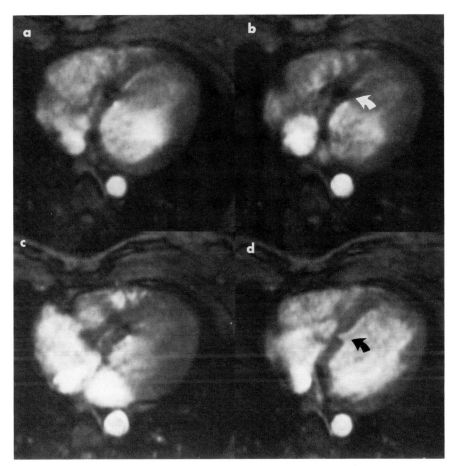

Figure 39-12 Transverse cine magnetic resonance images of a patient with a small, muscular ventricular defect. A to D show four frames taken throughout the cardiac cycle. B, In early systole, signal loss is seen within the ventricular septal defect, resulting from turbulent flow (shunt jet) through the narrow orifice (white arrow). D, The end-diastolic image shows the defect (black arrow) in the inflow portion of the ventricular septum. Note the high signal intensity of the blood relative to the myocardium.

REFERENCES

1. Didier D, Higgins CB, Fisher MR, Osaki L, Silverman NH, Cheitlin MD. Congenital heart disease: Gated MR imaging in 72 patients. *Radiology*. 1986;158:227–235.

2. Von Schulthess GK, Higashino SM, Higgins SS, Didier D, Fisher MR, Higgins CB. Coarctation of the aorta: MR imaging. *Radiology*. 1986;158:469–474.

3. Fletcher BD, Jacobstein MD. MRI of congenital abnormalities of the great arteries. *AJR*. 1986;146:941–948.

4. Kersting-Sommerhoff BA, Sechtem UP, Fisher MR, Higgins CB. MR imaging of congenital anomalies of the aortic arch. *Am J Roentgenol*. 1987;149:9–13.

5. Formanek AG, Witcofski RL, D'Souza VJ, Link KM, Karstaedt N. MR imaging of the central pulmonary arterial tree in conotruncal malformation. *AJR*. 1986;147:1127–1131.

6. Diethelm L, Dery R, Lipton MJ, Higgins CB. Atrial-level shunts: Sensitivity and specificity of MR in diagnosis. *Radiology*. 1987;162:181–186.

7. Kersting-Sommerhoff BA, Sechtem UP, Higgins CB. Evaluation of pulmonary blood supply by nuclear magnetic resonance in patients with pulmonary atresia. *J Am Coll Cardiol*. In press.

8. Didier D, Higgins CB. Identification and localization of ventricular septal defects by gated magnetic resonance imaging. *Am J Cardiol*. 1986;57:1363–1368.

9. Sechtem U, Pflugfelder PW, White RD, et al. Cine MR imaging: Potential for the evaluation of cardiovascular function. *AJR*. 1987;148:239–246.

10. Guit GL, Bluemm R, Rohmer J, et al. Levotransposition of the aorta: Identification of segmental cardiac anatomy using MR imaging. *Radiology*. 1986;161:673–679.

11. Bisset GS, Strife GL, Kirks DR, Bailey WW. Vascular rings: MR imaging. *AJR*. 1987;149:251–256.

12. Kersting-Sommerhoff BA, Sechtem UP, Schiller NB, Lipton MJ, Higgins CB. MRI of the thoracic aorta in Marfan patients. *J Comput Assist Tomogr*. 1987;11:633–639.

13. Jacobstein MD, Fletcher BD, Nelson AD, Clampitt M, Alfidi RJ, Riemenschneider TA. Magnetic resonance imaging: Evaluation of palliative systemic-pulmonary shunts. *Circulation*. 1984;70:650–656.

14. Sechtem U, Pflugfelder P, Cassidy MC, Holt W, Wolfe C, Higgins CB. Ventricular septal defect: Visualization of shunt flow and determination of shunt size by cine magnetic resonance imaging. *AJR*. 1987;149:689–692.

Part VII
Tumors of the Heart

Chapter 40

Imaging of Cardiac Tumors

Mary Jo Bertsch, MD, Eva V. Chomka, MD, and Bruce H. Brundage, MD

The clinical diagnosis of a cardiac tumor creates a significant challenge for the physician because the symptoms are variable or even absent, the physical findings may mimic many other systemic and cardiac disorders, and the tumors are rare. Regardless of their malignant potential, just by virtue of their location, cardiac tumors may be fatal if undiagnosed. Improved treatment modalities, including the surgical resection of primary cardiac neoplasms, has placed increased significance on their early diagnosis. Twenty years ago the antemortem diagnosis of a cardiac tumor was infrequent. Since then, however, there has been rapid development of several reliable cardiovascular imaging modalities that have provided an effective means of premortem diagnosis.

Tumors of the heart may be divided into primary neoplasms, benign or malignant, and secondary or metastatic neoplasms. They can be further classified by their location within the heart: intracavity, myocardial, or pericardial. Primary benign tumors are generally intracavitary pedunculated tumors. The myxoma is the most common primary tumor of the heart, and 75% occur in the left atrium. However, myxomas and other benign cardiac tumors may occur in any cardiac chamber or on valvular structures, and they may be multiple. Therefore it is important to visualize all chambers of the heart as well as intracardiac structures when evaluating a cardiac tumor. Virtually all malignant primary tumors of the heart are sarcomas. These generally occur as intracavity masses; they are characterized by rapid tumor growth, local extension into pericardium and mediastinum, and a propensity to occur in the right-sided chambers.[1]

A wide variety of neoplasms may involve the heart secondarily by metastasis or by direct extension of the primary extracardiac neoplasm. Secondary involvement of endocardium, myocardium, or pericardium have all been observed, but only rarely does a secondary neoplasm present as an intracavitary mass like a primary cardiac tumor. Most secondary neoplasms are asymptomatic and only discovered at autopsy.[1]

When a cardiac tumor is suspected clinically or discovered incidentally, the choice of imaging technique for further evaluation should be guided by the individual characteristics of the suspected tumor type, its location, the extent of cardiac involvement expected based on clinical and physical manifestations, and surgical considerations. Important points of information to obtain when imaging a cardiac tumor are (1) location of the tumor within the cardiac chambers, myocardium, or pericardium; (2) site of tumor attachment; (3) size of the tumor; (4) definition of tumor extension; (5) tissue characterization; (6) motion of the tumor within the cardiac chambers; and (7) hemodynamic consequences of the tumor. No imaging modality can provide all of this information, but many are complementary, and when used in combination one can usually obtain the necessary diagnostic information to proceed with appropriate therapy. Current imaging techniques useful in the evaluation of cardiac tumors include cardiac catheterization with angiography, digital subtraction angiography, echocardiography, nuclear imaging, x-ray computed tomography, and magnetic resonance imaging.

ANGIOGRAPHY

The first reported diagnosis of an intracardiac tumor during life was in 1952 by Goldberg with the use of angiography.[2] Cardiac catheterization with angiography offers good spatial and temporal resolution of intracavitary tumors and provides information about the hemodynamic consequences of the tumor. Hemodynamic alterations may reflect the size and position of the tumor, which may manifest itself through inflow or outflow obstruction or tamponade. Useful angiographic findings include (1) intracavitary filling defects, (2) presence of pericardial effusion, (3) distortion of the cardiac chambers, and (4) regional wall motion abnormalities, which may represent myocardial tumor infiltration.[3] The most common angiographic finding is an intracavitary defect, which may be fixed or mobile (Figures 40-1 and 40-2). A regional increase in myocardial wall thickness, particularly if associated with a pericardial effusion, is suggestive of a malignant infiltrating tumor, although definition of wall thickness is frequently difficult with cineangiography.

In most cases, right heart catheterization with contrast medium injection proximal to the suspected location of the tumor yields the necessary information. For example, if a tumor is suspected in the right atrium or right ventricle, contrast medium injection can be made in the superior vena cava or right atrium, respectively. Left heart tumors can be visualized in the levophase after contrast medium injection into the pulmonary artery. These techniques are recommended in order to prevent tumor embolization due to catheter manipulation.[4] Tumor vascularity can also be demonstrated by coronary angiography and may be noted as a "tumor blush."[5–7] The presence of a "tumor blush," however, is not specific for tumor, as there have been reports of neovascularization of cardiac thrombi.[8] The differentiation of a pericardial tumor from a myocardial tumor may also be made with coronary angiography because coronary vessels do not supply pericardial neoplasms. In addition to identifying the cardiac neoplasm, cardiac catheterization provides information regarding coexistent coronary artery and/or valvular disease, which could conceivably alter the surgical approach to a resectable cardiac tumor. Therefore, in some instances, despite adequate definition of a tumor by other imaging modalities, cardiac catheterization may still be warranted because of the likelihood of these coexisting cardiac lesions.

Although cardiac angiography has been widely used to diagnose cardiac tumors, many false-positive and false-negative studies have been observed.[9–11] Small pedunculated tumors and those that are primarily intramyocardial are most likely to be missed by angiography. False-positive results can result from streaming of nonopacified venous blood, thrombi, septal hematomas, or myocardial aneurysms. Furthermore, angiography is an invasive procedure with associated risks of ionizing radiation, catheter injury, contrast medium reactions, and tumor embolization. Therefore, angiography should be recommended only if less invasive techniques cannot provide adequate tumor definition or other concomitant heart disease is suspected, which might alter the surgical approach.

DIGITAL SUBTRACTION ANGIOGRAPHY

Digital subtraction angiography (DSA) has provided a means for making contrast angiography less invasive and has lowered the required concentration of contrast medium. The technique utilizes subtraction of unnecessary background information from a sequence of images after peripheral venous injection of contrast medium. Despite the low dye concentration, DSA generates images whose spatial resolution approaches that of conventional angiography. Although the use of DSA for evaluating cardiac tumors has not been fully assessed, there have been reports of its use in evaluating intracavitary tumors, including left atrial myxoma, right atrial rhabdomyoma, and metastatic tumor with involvement of the inferior vena cava, right atrium, and right ventricle.[12–14] Its potential advantages for assessing cardiac tumors lie in its ability to visualize all four cardiac chambers and the great vessels, with minimal invasive risk.

ECHOCARDIOGRAPHY

Echocardiography revolutionized the detection of intracardiac tumors. Its distinctive advantages are its totally noninvasive nature and its relatively low cost. Because surgical and autopsy correlations are lacking, the exact sensitivity and specificity of echocardiography in detecting cardiac tumors is not known. However, since its introduction the average number of cardiac tumors diagnosed antemortem and preoperatively has increased significantly.[15]

M-Mode echocardiography has been used extensively for the diagnosis of cardiac tumors.[16–17] This technique is most useful for pedunculated left atrial tumors (i.e., myxomas) and is considered much less sensitive for intramural tumors and tumors occurring in the ventricular cavities. The typical feature of a pedunculated left atrial tumor is the presence of a mass of echoes behind the anterior mitral valve leaflet during diastole. The EF slope is generally reduced, depending

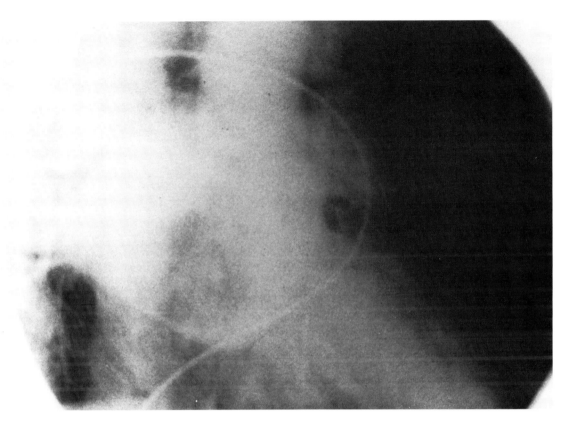

Figure 40-1 Cineangiogram of a left atrial myxoma visualized in systole during the levophase after injection of contrast medium into the pulmonary artery. The left atrial myxoma is visualized as an intracavitary defect in the left atrium.

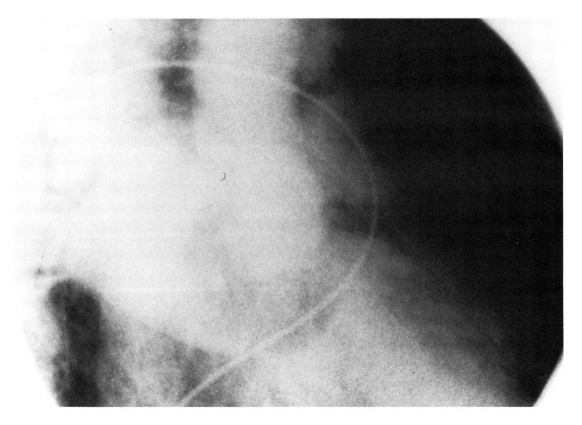

Figure 40-2 Cineangiogram of a left atrial myxoma during diastole, demonstrating prolapse of the tumor into the left ventricle.

on the degree of obstruction to ventricular filling.[16–17] Right atrial tumors can also be detected by M-mode echocardiography and appear as a mass of echoes behind the tricuspid valve in diastole.

Two-dimensional echocardiography has proved superior to M-mode techniques for the evaluation of cardiac tumors. When compared to other imaging techniques, it provides a relatively easy, inexpensive, and sensitive technique, especially for detecting intracavitary pedunculated tumors. Two-dimensional echocardiography is able to provide real-time imaging with good spatial orientation in most patients, thereby providing an accurate assessment of tumor size, location, site of attachment, and mobility (Figures 40-3 and 40-4).[16–18] These features make it particularly useful for evaluating intracavitary pedunculated tumors or tumors involving the cardiac valves, such as fibroelastomas, and may provide intraoperative guidance in their surgical removal.[19–20] Perhaps one of the greatest advantages of two-dimensional echocardiography is its ability to visualize all four cardiac chambers adequately in a good-quality study. Such thorough assessment is important because tumors, particularly myxomas, may occur in any cardiac chamber and may be present in multiple sites concomitantly.[21–23] Two-dimensional echocardiography has been used to evaluate atrial myxomas with such accuracy and completeness regarding diagnosis, size, and tumor attachment that further invasive evaluation prior to surgery was deemed unnecessary in some cases.[15] Two-dimensional echocardiography may also detect smaller and sessile tumors, which are generally not detected by M-mode echocardiography or angiography.

A potential problem when imaging an intracardiac mass is the differentiation of tumor from thrombus. Echocardiographic phased-array techniques may facilitate this differentiation. Thrombi frequently produce a layered or laminated appearance, have a broad-based attachment to the intracardiac wall, and are often associated with other cardiac abnormalities, such as a ventricular wall motion abnormality, left atrial enlargement, or a left ventricular aneurysm. Tumors, on the other hand, are mottled in appearance, pedunculated, and mobile.[16] Also, some myxomas may display areas of echolucency in the tumor mass when visualized by two-dimensional echocardiography, which correspond to areas of tumor hemorrhage.[24] Further tissue characterization has been attempted by utilizing digital analysis of the gray levels reflected from intracardiac masses. Color-coded two-dimensional echocardiography can also highlight the differences in gray scale levels within a cardiac mass.[25–26] These techniques may be helpful in determining the composition of a cardiac mass.

Secondary metastatic tumors of the heart occur much more frequently than primary cardiac neoplasms. Metastatic tumors often involve the pericardium and infiltrate the myocardium but only infrequently produce the intracavitary space-occupying lesion that is so amenable to imaging with two-dimensional echocardiography.[27] Although mild degrees of pericardial thickening and small homogeneous tumor masses that

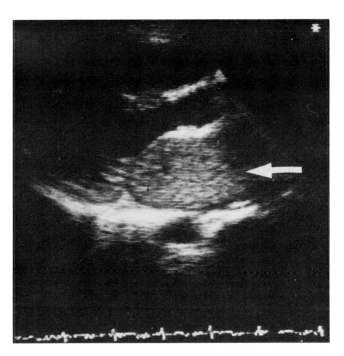

Figure 40-3 Two-dimensional echocardiogram in the parasternal long-axis view, illustrating a left atrial myxoma during systole. The myxoma appears to fill the entire left atrium and at surgery was found to have a broad attachment to the left atrial septum.

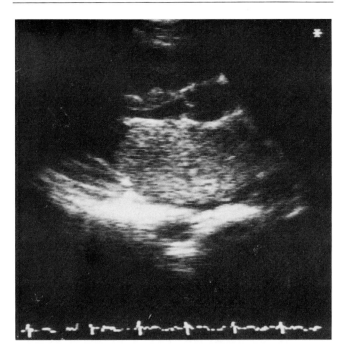

Figure 40-4 Same tumor as pictured in Figure 40-3, shown here prolapsing across the mitral valve during diastole.

do not distort the contour of the cardiac walls cannot be easily detected by two-dimensional echocardiography, frequently metastatic involvement is detected by two-dimensional echocardiography by recognition of diffuse pericardial thickening with or without effusion or the presence of extracardiac masses causing compression of the pericardium or great vessels. In addition, although intramural tumor involvement is difficult to detect by two-dimensional echocardiography, such involvement can change the acoustic and functional properties of the involved cardiac wall and produce an increase in wall thickness, cause regional wall motion abnormalities, give the appearance of a sessile mass, or produce a different acoustic pattern or echogenicity in an area adjacent to normal myocardium. The two-dimensional echocardiographic detection of intramural tumors, both primary and metastatic, has been reported.[28-31]

The detection of tumor extension into vascular structures around the heart is also possible with two-dimensional echocardiography. Nephroblastomas, renal cell carcinomas, and leiomyomas may extend into the right heart via the inferior vena cava.[32-35] The extension of left atrial sarcomas into the pulmonary veins has also been reported.[27] The ability of two-dimensional echocardiography to visualize the surrounding vascular structures of the heart provides a means of distinguishing extracardiac tumors with extension into the heart via blood vessels from primary cardiac tumors. This is an important advantage, also provided by computed tomography and magnetic resonance imaging but not always by angiography.

Although most cardiac tumors have a high acoustic reflectance and are readily detectable by two-dimensional echocardiography, highly vascular tumors that have acoustic properties similar to blood may escape detection.[6] Other drawbacks of two-dimensional echocardiography include the differentiation of intracardiac masses from artifact. False-positive diagnosis of intracardiac masses secondary to artifacts may be minimized by careful gain adjustment, appropriate transducer positioning to decrease lateral reverberation, and obtaining views in multiple planes. However, in spite of optimal technique, it is difficult at times to distinguish tumor from clot, calcium, vegetations, fibrosis, diaphragmatic hernias, atrial septal aneurysms, subepicardial hematoma, and lipomatous hypertrophy of the intra-atrial septum, all of which have been confused with intracardiac tumors.[35-36] Also, patient factors such as an abnormal chest configuration and lung disease may preclude an adequate acoustic window, in which case another imaging modality must be selected for diagnosis.

Doppler echocardiography provides a noninvasive method to evaluate the hemodynamic consequence of atrial myxomas, which can cause transvalvular flow

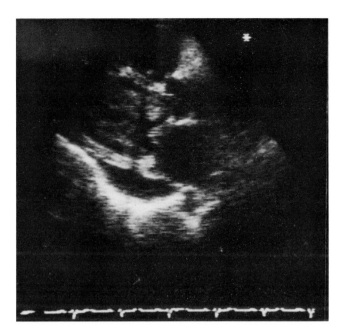

Figure 40-5 Two-dimensional echocardiogram in the parasternal long-axis view, demonstrating a large left ventricular mass. The mass was found to be a diffuse infiltrating primary liposarcoma with involvement of the left ventricular cavity and wall and had extension into the noncoronary cusp of the aortic valve. A Doppler study revealed aortic insufficiency.

disturbances and mimic tricuspid or mitral stenosis.[37-39] The hemodynamic data obtained with Doppler echocardiography may therefore complement that obtained with M-mode and two-dimensional studies (Figure 40-5).

Thus, despite the cited limitations, echocardiography remains overall an easy, safe, reliable method for the detection and serial examination of most intracardiac tumors.

NUCLEAR IMAGING TECHNIQUES

Radionuclide cardiac blood pool imaging can be utilized for the diagnosis of intracardiac tumors. Myocardial or intracavitary tumors may appear as filling defects in any cardiac chamber.[40-41] Because of poor spatial resolution, radionuclide ventriculography is much less sensitive than two-dimensional echocardiography and contrast angiography for detecting cardiac tumors. However, nuclear imaging may provide an additional noninvasive diagnostic method when two-dimensional echocardiography is technically limited.

Infiltration of the myocardium by tumor may be detected by thallium 201 myocardial perfusion scintigraphy.[42-43] Tumor invasion causes a focal defect in the perfusion scan; however, this is a nonspecific finding, and this technique is not recommended for the detection of cardiac tumors.

COMPUTED TOMOGRAPHY

The use of computed x-ray transmission tomography (CT) in the assessment of cardiac neoplasms has developed over the last 8 to 10 years. Computed tomography diagnosis of cardiac masses has lagged behind that for tumors elsewhere in the body because of slow scanning times, which create artifact due to cardiac motion. However, with the development of shorter scanning times, culminating in the recently developed ultrafast CT, the problem of motion artifact has been overcome, and CT has proved useful in the evaluation of cardiac tumors. Computed tomography was first reported to be useful in the evaluation of atrial myxoma and metastatic pericardial lesions.[44-45] Since then, its continued use in the diagnosis of many cardiac tumors, benign and malignant, has been reported.[46-47] Recently the use of ultrafast CT in the evaluation of left atrial myxomas has demonstrated this modality's ability to visualize with high spatial resolution not only the anatomy of the tumor and surrounding structures but also tumor motion throughout the cardiac cycle.[48]

Although not yet fully evaluated for the imaging of cardiac tumors, certain advantages of computed tomography have already emerged. These include its ability to produce high-quality images of all cardiac chambers, the pericardium, and the surrounding great vessels, which are unaffected by the chest configuration or the lungs. In some cases CT has proved superior to echocardiography in diagnosing intracardiac tumors in the right atrium, right ventricle, and left ventricle (Figures 40-6 and 40-7).[47,49] Computed tomography is also able to visualize the pericardium and detect intrapericardial masses more reliably than echocardiography or angiography.[50] Two-dimensional echocardiography is useful for identifying pericardial effusions but less adequate for actual visualization of the pericardium itself and masses within the pericardium. Angiography relies on indirect evidence of pericardial disease such as increased heart size or chamber distortion for diagnosis.

Perhaps the greatest advantage of computed tomography in the evaluation of cardiac tumors is its ability to assess tumor extension and involvement of the great vessels and adjacent extracardiac structures, which is an important consideration in the evaluation of any tumor. Such detailed anatomic information may not only provide characteristics of the tumor that will help differentiate benign from malignant ones but also help in planning surgical treatment (Figure 40-8).

Computed tomography's ability to discriminate tissue differences is better than that of angiography or two-dimensional echocardiography.[45] The gray scale of CT can distinguish fat from water because of the marked density differences. The low radiodensity of fatty tissue also helps in distinguishing it from the more radiodense myxomas, clots, and other solid tumors. For example, a lipoma or lipomatous hypertrophy of the intra-atrial septum is easily distinguished from most other solid tumors. Computed tomography also allows for better characterization of primary myocardial tumors, which usually present as diffuse myocardial infiltration. These tumors are often difficult to identify on two-dimensional echocardiography (Figure 40-9).[51]

Presently the use of rapid CT scanners in the diagnosis of cardiac tumors is still evolving. The primary limitation of computed tomography for cardiac diagnosis had been, in the past, slow scan times and thus poor resolution secondary to motion artifact. This, however, has been overcome with the new-generation

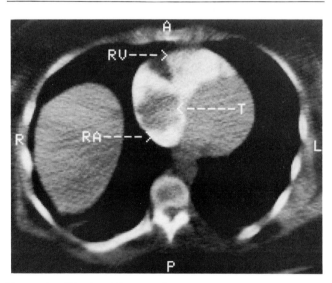

Figure 40-6 Ultrafast CT image, demonstrating a tumor mass in the right atrium. The tumor was found to be a myxoma with attachment to the atrial septum. RA, right atrium; RV, right ventricle; T, tumor.

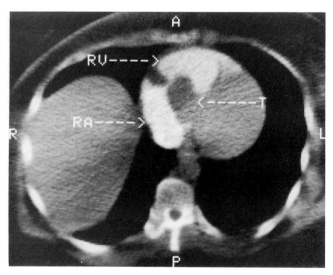

Figure 40-7 Tomographic image of the tumor pictured in Figure 40-6, taken at the same level, demonstrating prolapse of the myxoma into the right ventricle. RA, right atrium; RV, right ventricle; T, tumor.

ultrafast CT scanner. But these scanners are expensive, limited in availability, and still have the disadvantages of exposing the patient to radiation and requiring the use of contrast medium.

MAGNETIC RESONANCE IMAGING

Magnetic resonance imaging (MRI) is a noninvasive tool that can provide tomographic images of the heart in planes oriented along intrinsic cardiac axes. This technique provides the best method of tissue discrimination because of wide variations in proton densities in various tissues and their characteristics when placed in a magnetic field. These different characteristics, called relaxation times, are altered in neoplastic tissue, thus allowing for differentiation of tumor from normal tissue. Because of low signal-to-noise ratios, the images must be obtained by gating to the cardiac cycle and acquiring the images over many minutes.

Imaging of cardiac tumors with MRI is in its infancy. However, several reports of its use in evaluating right and left atrial myxomas have been published.[52–54] In these reports MRI was comparable to if not better than two-dimensional echocardiography in defining the atrial mass in terms of size, shape, tissue characteristics, and location. Furthermore, the multiplane imaging capability of MRI assists in determining the relationship of the mass to other cardiac structures.

A major advantage of MRI is its ability, like that of computed tomography, to image all cardiac chambers and the myocardium, pericardium, and great vessels. The ability of MRI to discriminate pericardial masses has also been demonstrated.[55]

The full potential of MRI for tissue discrimination has not yet been realized. The potential exists for providing data leading to specific pathologic diagnosis of cardiac masses. It would be particularly useful to be able to accurately distinguish tumor from thrombus, a frequent source of confusion with other imaging techniques. Currently MRI can readily define intramural tumor involvement directly.[55] In addition, MRI may be helpful in determining the vascularity of a cardiac mass. It has been suggested that cardiac paragangliomas, which are highly vascular tumors, may be better characterized using MRI than CT or echocardiography.[56] Magnetic resonance imaging is also very useful for surgical assessment of cardiac tumors.

Magnetic resonance imaging has the advantage of imaging without the use of ionizing radiation or contrast agents, and although its potential for the imaging of cardiac tumors has not been fully realized, certain advantages are emerging.

SUMMARY

In summary, multiple imaging techniques can be used in the evaluation of cardiac tumors. It is most important to initiate screening for a suspected tumor with a noninvasive tool. Echocardiography provides an excellent noninvasive screening tool for suspected intracardiac tumors. Computed tomography and mag-

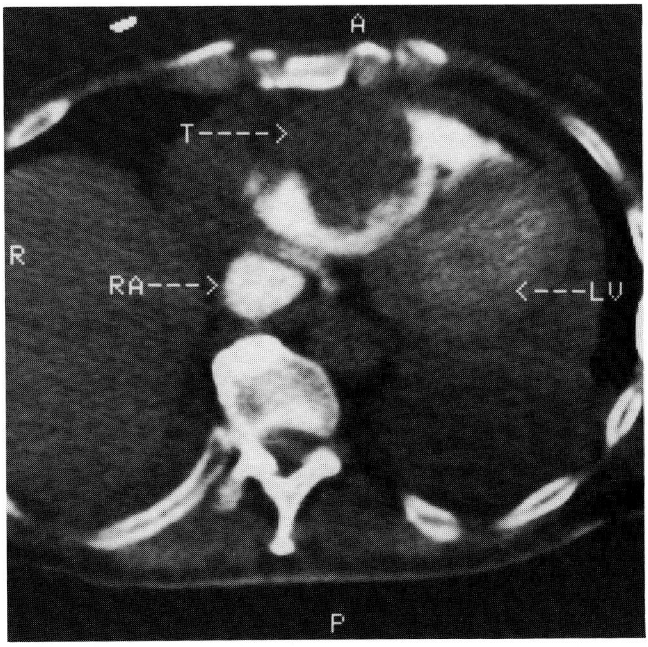

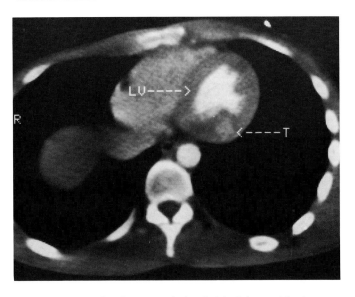

Figure 40-8 Ultrafast CT scan of an infiltrating malignant tumor in the right ventricle. RA, right atrium; LV, left ventricle; T, tumor. *Source:* Courtesy of James Talano MD, Chicago, IL.

Figure 40-9 Ultrafast CT scan at the level of the left ventricle, demonstrating an intramyocardial tumor in the wall of the left ventricle. The tumor enhances with contrast medium, which indicates it has a more vascular composition than normal myocardium. LV, left ventricle; T, tumor. *Source:* Courtesy of James Talano MD, Chicago, IL.

netic resonance imaging can provide additional anatomic information regarding involvement of surrounding extracardiac structures and intramural tumor involvement. Cardiac catheterization is reserved for patients in whom noninvasive techniques have not provided adequate tumor definition or those in whom coexisting organic heart disease is suspected. In addition, it is often necessary to use these modalities in combination to achieve the degree of tumor definition needed to plan the appropriate therapy.

REFERENCES

1. Becker RC, Loeffler JS, Leopold KA, et al. Primary tumors of the heart: A review with emphasis on diagnosis and potential treatment modalities. *Sem Surg Onc*. 1985;1:161–170.

2. Goldberg HP, Glenn F, Dotter CT, Steinberg I. Myxoma of the left atrium: Diagnosis made during life with operative and postmortem findings. *Circulation*. 1952;6:762–767.

3. Steinberg I, Miscall L, Redo SF, Goldberg H. Angiocardiography in diagnosis of cardiac tumors. *Am J Roentgenol*. 1964;91:364–376.

4. Pindyck F, Pierce EC, Baron MG, et al. Embolization of left atrial myxoma after transeptal cardiac catheterization. *Am J Cardiol*. 1972;30:569–571.

5. Marshall WH, Steiner RM, Wexler L. "Tumor Vascularity" in left atrial myxoma demonstrated by selective coronary arteriography. *Radiology*. 1969;93:815–816.

6. Stewart JA, Warnica JW, Kirk ME, et al. Left atrial myxoma: False negative echocardiographic findings in a tumor demonstrated by coronary arteriography. *Am Heart J*. 1979;98:228–232.

7. Shapiro MR, Cohen MV, Grose R, et al. Diagnosis of left atrial myxoma by coronary angiography eight years following open mitral commissurotomy. *Am Heart J*. 1983;105:325–327.

8. Standen JR. "Tumor Vascularity" in left atrial thrombus demonstrated by selective coronary arteriography. *Radiology*. 1975;116:549–550.

9. Miller JI, Mankin HT, Broadbent JC, et al. Primary cardiac tumors: Surgical considerations and results of operation. *Circulation*. 1972;45(Suppl):134–138.

10. Goodwin JF. Diagnosis of the left atrial myxoma. *Lancet*. 1963;1:464–486.

11. McGarry K, Jugdutt BI, Rossall RE. The modern diagnosis of cardiac myxoma: Role of two-dimensional echocardiography. *Clin Cardiol*. 1983;6:511–518.

12. Yiannikas J, Zaidi AR, Moodie DS. Evaluation of intracardiac masses by digital subtraction angiography. *Am Heart J*. 1984;108:600–603.

13. Detrano R, Salcedo EE, Simpfendorfer C, et al. Digital subtraction angiography in the evaluation of right heart tumors. *Am Heart J*. 1985;109:366–368.

14. Tamari I, Goldberg HL, Moses JW, et al. Left atrial myxoma: Diagnosis by digital subtraction angiography. *Cath Cardiovasc Diagn*. 1986;12:26–29.

15. Fyke FE, Seward JB, Edwards WD, et al. Primary cardiac tumors: Experience with 30 consecutive patients since the introduction of two-dimensional echocardiography. *J Am Coll Cardiol*. 1985;5:1465–1473.

16. Feigenbaum H, ed. *Echocardiography*. Philadelphia: Lea & Febiger; 1986:579–620.

17. Come AC, ed. *Diagnostic Cardiology: Noninvasive Imaging Techniques*. Philadelphia: JB Lippincott Co; 1985:509–525.

18. DePace NL, Soulen RK, Kotler MN, Mintz GS. Two-dimensional echocardiographic detection of intraatrial masses. *Am J Cardiol*. 1981;48:954–960.

19. Frumin H, O'Donnell L, Kerin NZ, et al. Two-dimensional echocardiographic detection and diagnostic features of tricuspid papillary fibroelastoma. *J Am Coll Cardiol*. 1983;2:1016–1018.

20. Topol E, Biern RO, Reitz BA. Cardiac papillary fibroelastoma and stroke. Echocardiographic diagnosis and guide to excision. *Am J Med*. 1986;80:129–132.

21. Tway KP, Shah AA, Rahimtoola SH. Multiple biatrial myxomas demonstrated by two-dimensional echocardiography. *Am J Med*. 1981;71:896–899.

22. Abramowitz R, Mojdan JR, Plzak LF, Berger BG. Two-dimensional echocardiographic diagnosis of separate myxomas of both left atrium and left ventricle. *Am J Cardiol*. 1984;53:379–382.

23. Gibbs J. The heart and tuberous sclerosis: An echocardiographic and electrocardiographic study. *Br Heart J*. 1985;54:596–599.

24. Rahilly GT, Nanda NC. Two-dimensional echocardiographic identification of tumor hemorrhages in atrial myxomas. *Am Heart J*. 1981;101:237–239.

25. Green SE, Joynt LF, Fitzgerald PJ, et al. In vivo ultrasonic tissue characterization of human intracardiac masses. *Am J Cardiol*. 1983;51:231–236.

26. Allan LD, Joseph MC, Tynan M. Clinical value of echocardiographic colour image processing in two cases of primary cardiac tumor. *Br Heart J*. 1983;49:154–156.

27. Mich RJ, Gillam LD, Weyman AE. Osteogenic sarcomas mimicking left atrial myxomas: Clinical and two-dimensional echocardiographic features. *J Am Coll Cardiol*. 1985;6:1422–1427.

28. Kutalek SP, Panidis IP, Kotler MN, et al. Metastatic tumors of the heart detected by two-dimensional echocardiography. *Am Heart J*. 1985;109:343–349.

29. Lestuzzi C, Biasi S, Nicolosi GL, et al. Secondary neoplastic infiltration of the myocardium diagnosed by two-dimensional echocardiography in seven cases with anatomic confirmation. *J Am Coll Cardiol*. 1987;9:439–445.

30. Armstrong WF, Buck JD, Hoffman R, Waller BF. Cardiac involvement by lymphoma: Detection and follow-up by two-dimensional echocardiography. *Am Heart J*. 1986;112:627–631.

31. Charuzi Y, Mills H, Buchbinder NA, Marshall LA. Primary intramural cardiac tumor: Long-term follow-up. *Am Heart J*. 1983;106:414–418.

32. Riggs T, Paul MH, DeLeon S, Ilbawi M. Two-dimensional echocardiography in evaluation of right atrial masses: five cases in pediatric patients. *Am J Cardiol*. 1981;48:961–967.

33. Politzer F, Kronzon I, Wieczorek R, et al. Intracardiac leiomyomatosis: Diagnosis and treatment. *J Am Coll Cardiol*. 1984;4:629–634.

34. Goldman A, Parmeswaran R, Kotler MN, et al. Renal cell carcinoma and right atrial tumor diagnosed by echocardiography. *Am Heart J*. 1986;110:183–186.

35. Panidis IP, Kotler MN, Mintz GS, Ross J. Clinical and echocardiographic features of right atrial masses. *Am Heart J*. 1984;107:745–758.

36. Nishimura RA, Tajik AJ, Schattenberg TT, Seward JB. Diaphragmatic hernia mimicking an atrial septal mass: A two-dimensional echocardiographic pitfall. *J Am Coll Cardiol*. 1985;5:992–995.

37. Vargas-Barron J, Lacy-Niebla MC, Keirns C, et al. Pulsed-Doppler echocardiographic analysis of atrioventricular flow changes in patients with atrial myxomas. *Am Heart J*. 1986;112:850–854.

38. Panidis IP, Mintz GS, McAllister M. Hemodynamic consequences of left atrial myxomas as assessed by Doppler ultrasound. *Am Heart J*. 1986;111:927–931.

39. Goli VD, Thadani U, Thomas SR, et al. Doppler echocardiographic profiles in obstructive right and left atrial myxomas. *J Am Coll Cardiol*. 1987;9:701–703.

40. Pohost GM, Pastore JO, McKusick KA, et al. Detection of left atrial myxoma by gated radionuclide cardiac imaging. *Circulation*. 1977;55:88–92.

41. Pitcher D, Wainwright R, Brennand-Roper D, et al. Cardiac tumours: Noninvasive detection and assessment by gated blood pool radionuclide imaging. *Br Heart J*. 1980;44:143–149.

42. Blumhardt R, Telepak RJ, Hartshorne MF, et al. Thallium imaging of benign cardiac tumor. *Clin Nucl Med*. 1983;8:297–298.

43. Helmer S, Abghari R, Stone AJ, Lee CC. Detection of benign cardiac fibroma on thallium-201 imaging in an adult. *Clin Nucl Med*. 1987;12:365–367.

44. Huggins TJ, Huggins MJ, Schnapf DJ, et al. Left atrial myxoma: Computed tomography as a diagnostic modality. *J Comput Assist Tomogr*. 1980;4:253–255.

45. Godwin JD, Axel L, Adams JR, et al. Computed tomography: A new method for diagnosing tumor of the heart. *Circulation*. 1981;63:448–451.

46. Zingas AP, Carrera JD, Murray CA, et al. Case report: Lipoma of the myocardium. *J Comput Assist Tomogr*. 1983;7:1098–1100.

47. Chaloupka JC, Fishman EK, Siegleman SS. Use of CT in the evaluation of primary cardiac tumors. *Cardiovasc Intervent Radiol*. 1986;9:132–135.

48. Bateman TM, Sethna DH, Whiting JS, et al. Comprehensive noninvasive evaluation of left atrial myxomas using cardiac cine-computed tomography. *J Am Coll Cardiol*. 1987;9:1180–1183.

49. Niehues B, Heuser L, Jansen W, et al. Noninvasive detection of intracardiac tumors by ultrasound and computed tomography. *Cardiovasc Intervent Cardiol*. 1983;6:30–36.

50. Gross BH, Glazer GM, Francis IR. CT of intracardiac and intrapericardial masses. *AJR*. 1983;140:903–907.

51. Isner JM, Falcone MW, Virmani R. Cardiac sarcoma causing "ASH" and simulating coronary artery disease. *Am J Med*. 1979;66:1025–1030.

52. Pflugfelder PW, Wisenberg G, Boughner DR. Detection of atrial myxoma by magnetic resonance imaging. *Am J Cardiol*. 1985;55:242–243.

53. Conces DJ, Vix VA, Klatte EC. Gated MR imaging of left atrial myxomas. *Radiology*. 1985;156:445–447.

54. Go RT, O'Donnell JK, Underwood DA, et al. Comparison of gated cardiac MRI and 2 D echocardiography of intracardiac neoplasms. *AJR*. 1985;145:21–25.

55. Amparo EG, Higgins CB, Farmer D, et al. Gated MRI of cardiac and pericardial masses: Initial experience. *AJR*. 1984;743:1151–1156.

56. Conti VR, Saydjari R, Amparo EG. Paraganglioma of the heart. The values of magnetic resonance imaging in the preoperative evaluation. *Chest*. 1986;90:604–606.

Part VIII

The Great Vessels

The first direct aortogram in human beings was performed by Nuvoli in 1936 by introducing a large-bore needle directly into the ascending aorta through the sternum.[1] Since then, arteriography and, more recently, alternate imaging modalities have played an ever-increasing role in the diagnosis of and therapeutic planning for diseases of the thoracic aorta and its branches. Though the physical exam and plain chest roentgenogram can be quite helpful, more precise anatomic information is often required.

Patients are referred to the Department of Radiology for evaluation of the thoracic aorta and its branches for a variety of suspected conditions. The bulk of these referrals usually fall into two categories: trauma to the chest and suspected arteriosclerotic disease. Occasionally patients may also be evaluated for suspected congenital anomalies, as well as neoplastic and inflammatory conditions.

RADIOLOGICAL MODALITIES FOR EVALUATING GREAT VESSEL DISEASE

Radiologists today have a variety of modalities at their disposal to help define vascular abnormalities. In addition to well-accepted direct intra-arterial angiography, digital subtraction angiography (Figures 41-1 and 41-2), ultrasound (Figure 41-3), computed tomography, and magnetic resonance imaging can all provide diagnostic information in selected cases.

Duplex ultrasonography, a purely noninvasive modality, has proved its utility in evaluation of the carotid bifurcation. In particular, it has been shown to be a sensitive screening method, guiding further workup with either direct intra-arterial angiography or digital subtraction angiography.[2-4] In the series reported by Glover et al.,[3] duplex ultrasonography had a tendency to mistake severely stenotic lesions for complete occlusions.[3,5] This limitation, however, does not in general significantly detract from ultrasound's role in screening.

Depending upon the clinical setting, available equipment, and experience and attitudes of both the referring physician and the radiologist, digital subtraction angiography (DSA) has a varying role to play in the evaluation of the aorta and its branches. The technology can be used in a variety of ways, ranging from peripheral venous injection of contrast material to central venous injection to direct intra-arterial injection.

There are several theoretical and practical advantages of DSA over conventional arteriography. Intravenous DSA is less invasive, not requiring an arterial puncture and thus facilitating outpatient examination. As no intra-arterial catheter manipulation is performed, the risk of stroke due to peripheral embolization or vessel dissection is greatly reduced. However, intravenous DSA usually does not reduce and in fact may increase the amount of iodinated contrast medium required, compared to conventional arteriography.

Intra-arterial DSA reduces the iodinated contrast medium requirement but, of course, requires arterial

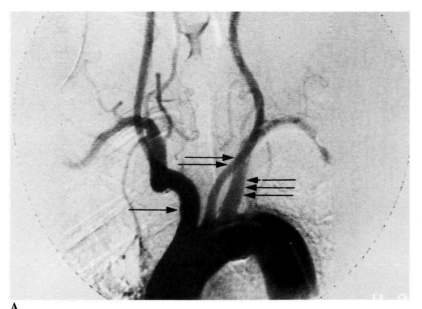

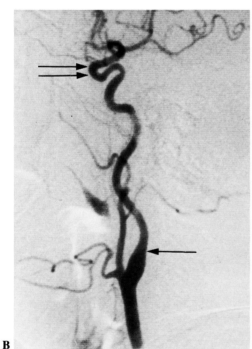
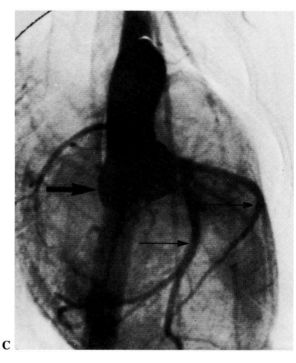

Figure 41-1 A, Normal intra-arterial digital subtraction angiogram of the aortic arch, revealing the normal branches of the aortic arch. Innominate artery, single arrow; left common carotid artery, double arrows; left subclavian artery, triple arrows. B, Normal intra-arterial digital subtraction angiogram of the left common carotid artery and its branches. Note the internal carotid artery (single arrow) and the carotid siphon (double arrows). C, Intra-arterial digital subtraction angiogram revealing the ascending aorta (thick arrow) and coronary arteries (thin arrows) of a dog.

puncture. There is, however, much less catheter manipulation required than for selective carotid and vertebral studies, thus again reducing risk of stroke.

Regardless of injection site, DSA, like all digital imaging modalities, allows computer manipulation of exam data for optimal contrast (Figure 41-2)—a clear advantage over conventional radiographic techniques. Also, primarily because of greatly reduced film requirements, cost per examination is significantly decreased.

On the other hand, like all radiographic subtraction techniques, DSA is very sensitive to patient motion. The technique may be quick and easy, or it may, in cases of tortuous vasculature, require multiple injections in multiple planes to separate confusing overlying shadows. In general, the least invasive of these options (peripheral venous injection) involves a large contrast medium requirement and produces the poorest spatial resolution.

Peripheral venous DSA may be adequate, however, for defining most abnormalities of the thoracic aorta.[6-8] Grossman et al.[6] reported that in 92% of 43 cases with a variety of aortic abnormalities, peripheral venous angiography could have replaced the conventional aortogram. Inability to adequately visualize the coronary arteries, however, rendered the modality insufficient for the evaluation of type I aortic dissection.

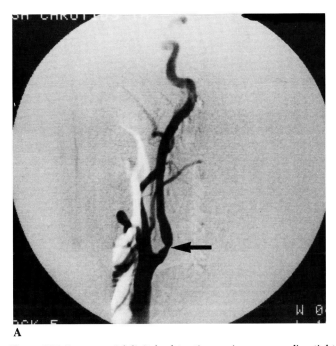
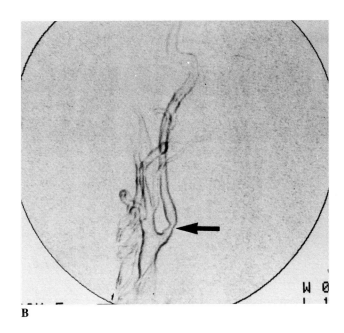

Figure 41-2 Intra-arterial digital subtraction angiogram revealing tight stenosis (arrows) of the internal carotid artery. Digital imaging allows data manipulation, such as gradient filtering (seen on the right).

Central venous injection has a contrast medium requirement similar to that of peripheral injection but improves visualization, particularly of the intracranial vessels.[9] However, depending on several factors, including patient population, central venous injection may increase the morbidity of the examination.[10–12] Aaron et al. reported that central venous DSA appeared to have greater morbidity than was reported for conventional arteriography.[10]

Direct intra-arterial DSA, while necessitating arterial puncture, brings with it a much smaller contrast medium requirement than that of either intravenous DSA or conventional arteriography. While the spatial

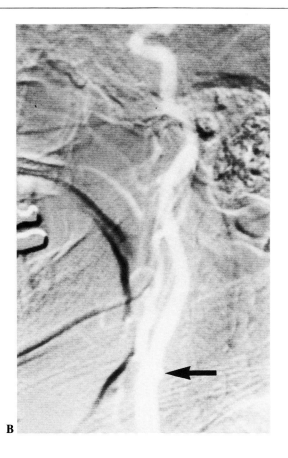

Figure 41-3 Duplex ultrasound examination (A) revealing apparent plaque in left common carotid artery (arrow), which is confirmed on the corresponding intra-arterial digital subtraction angiogram (B).

resolution is not as great as in conventional arteriography, it is adequate for diagnosis in the great majority of cases (Figures 41-1, 41-2B, 41-3B, and 41-6.)[13–15]

Computed tomography (CT) is another modality that can aid in the evaluation of the thoracic aorta and its branches. Contrast medium–infused CT can clearly define the presence and extent of aneurysms[16–17] and evaluate the carotid bifurcation.[18–19] In particular, CT can reveal information about the atheromatous wall of diseased vessels, to which conventional arteriography is blind.[20]

Magnetic resonance imaging (MRI) is a relatively new noninvasive modality with three clear advantages: no need for iodinated contrast material, ease of multiplanar image acquisition, and no ionizing radiation. Early work has demonstrated its great potential in a variety of pathologic conditions of the aorta,[21] including congenital anomalies[22] and aneurysms.[23–24] In selected cases MRI may be considered as the initial diagnostic study in evaluation of the thoracic aorta.[21,23]

Despite the recent inroads made by all these competing modalities, direct arteriography remains the gold standard for evaluation of the thoracic aorta and its branches. Although prevailing policies differ considerably from institution to institution and physician to physician, direct intra-arterial angiography often remains the procedure of choice for definitive diagnosis, particularly when surgical intervention is contemplated.

NORMAL ANATOMY

The aorta arises from the left ventricle, carrying freshly oxygenated blood directly or indirectly to all the systemic arteries. Most anatomic descriptions divide the thoracic aorta into three portions.[25–27] The proximal portion, termed the ascending aorta, stretches superiorly, anteriorly, and to the right for a distance of approximately 5 cm from the cusps of the sinuses of Valsalva to the origin of the innominate or right brachiocephalic artery. The second portion of the thoracic aorta, termed the aortic arch, begins at the origin of the right innominate artery. The arch courses superiorly, posteriorly, and to the left. Distally it is continuous with the descending, third portion of the thoracic aorta. The division between the arch and descending portion, which continues into the abdomen, is usually defined as the level of the lower border of the fourth thoracic vertebra.[25] This point corresponds approximately with the aortic isthmus, the point of fetal union between the fourth left brachial arch artery and the aorta. This is not usually seen on aortogram, but a 1- to 2-mm indentation or bulge of the anterior aortic wall at the point of the aortic isthmus is sometimes appreciated. This irregularity represents a remnant of the infundibulum of the ductus arteriosus.[26]

The first branches of the aorta are the coronary arteries, right and left, arising respectively from the right (anterior) and left (anterior) sinuses of Valsalva. The proximal portions of the coronary arteries are routinely visualized with direct aortic arteriography.

After giving off the coronary arteries, the next aortic branch is the brachiocephalic or innominate artery, which after a course of 3.75 to 5 cm divides at the level of the sternoclavicular junction into the right common carotid and right subclavian arteries.[27] The common carotid travels superiorly beneath the sternocleidomastoid muscle to divide at the approximate level of the thyroid notch into the internal carotid posteriorly and the external carotid anteriorly. The subclavian artery[27] is commonly described as having three segments, based upon the relationship to the anterior scalene muscle. The portion of the artery from its origin to the medial border of the scalene muscle is the first part, that behind the scalene the second part, and that distal to it the third. The first portion gives rise to the important branches. The vertebral artery travels superiorly to the foramen magnum, the thyrocervical trunk, and the long thoracic artery.[28]

The next branch of the aortic arch is the left common carotid, which like the right travels beneath the sternocleidomastoid and bifurcates into internal and external branches at the approximate level of the thyroid notch. The left subclavian is the next branch; it arises toward the ventral aspect of the arch. The course is similar to that of the right subclavian, with the first portion again giving rise to the vertebral and long thoracic arteries, as well as the thyrocervical trunk.[26] The left vertebral is usually larger than the right and of the two is generally the easier to selectively catheterize.

The branches of the aortic arch, though usually following this described pattern, are subject to considerable variation. A detailed description of these variations is beyond the scope of this chapter; however, a brief discussion is appropriate as they may occasionally have pathologic significance. Certainly they can often complicate attempts at catheterization and cause diagnostic confusion.

There are multiple variations of the aortic arch, ranging from six separate vessels from the arch[29] to a solitary brachiocephalic vessel subsequently giving rise to all branches to the head and neck.[30] Between these two extremes there may rarely be bi-innominate arteries with two vessels arising from the aorta, each giving rise to a subclavian and a common carotid. The innominate usually gives rise to only two branches but may give off a small branch to the thyroid, termed the thyroid ima. There may be no innominate artery as such, and the right subclavian artery may be aberrant, arising distal to the origin of the left subclavian and crossing behind the

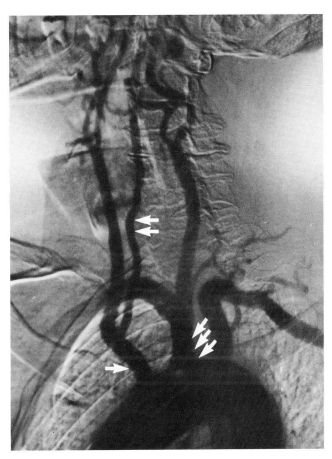

Figure 41-4 Aortogram reveals the right common carotid artery (single arrow) to be a direct branch of the aortic arch, and the right vertebral artery (double arrows) a branch of the right common carotid artery. The right subclavian artery (triple arrows) arises as the last branch of the aortic arch.

esophagus to reach the right neck (Figure 41-4). The right subclavian may arise directly from the aortic arch separate from a right common carotid.

The right vertebral, rather than originating from the right subclavian, may branch directly off the aortic arch or arise from the right common carotid (Figure 41-4). Other variations include origin from the thyrocervical trunk. The origin may also be bifid, with one segment branching from the subclavian and the other from the innominate artery, aortic arch, or thyrocervical trunk.[30–31]

The carotid arteries, particularly when the right subclavian is aberrant, may arise as a common trunk with the left carotid.[30,32] Both carotids usually bifurcate at the approximate level of the thyroid notch but may be as low as the second thoracic vertebra within the chest or as high as the first cervical vertebra. These variations, when they occur, are usually bilateral.[30,33]

The left common carotid and left innominate may arise as a common trunk, and this is one of the frequently encountered (13% to 40% incidence[30]) variations that may complicate attempts at selective catheterization. The left vertebral artery, usually the first branch of the left subclavian, originates directly from the aortic arch between the left common carotid and left subclavian arteries or, less frequently, distal to the left subclavian artery. Like the right vertebral, the left may be bifid.[30,34–35]

AORTIC TRAUMA

The aorta, deep within the thoracic cage, is well shielded from direct insult. It is, however, susceptible to injury by nonpenetrating blunt trauma, particularly due to rapid deceleration, which in modern society is most often the result of automobile accidents. The most common site of aortic rupture (50% to 85%[36–38]) is the descending portion just distal to the left subclavian at the insertion of the ligamentum arteriosum—the aortic isthmus.[39–41] Several mechanisms account for the prevalence of injury at this location.

The arch is anchored by the great vessels and ligamentum arteriosum proximally, and the distal thoracic aorta by the diaphragmatic crus. The mid-descending aorta is less well fixed and can swing anteriorly with rapid deceleration, which may result in shearing at the isthmus. Other contributory factors include local differences in the tensile strength of the aortic wall and the effect of rapid increases in intraluminal aortic pressure created by thoracic and/or abdominal compression. The relative importance of these factors is not fully understood either in aortic isthmus rupture or rupture at other, less common sites, such as at the root of the ascending aorta and at the diaphragmatic hiatus.[38,42]

Regardless of the mechanism of injury, immediate emergency evaluation and treatment of these patients are essential. It is probable that at least 80% of patients with acute rupture of the aortic isthmus die before they reach the hospital.[36,40] The outlook for patients with ascending aortic rupture is even bleaker, largely because of the high incidence of associated cardiac injuries, which are almost always immediately fatal.[38] Patients surviving acute aortic rupture are saved by mediastinal tamponade and false aneurysm formation, with containment within the vascular adventitia and/or the connective tissues of the mediastinum.[43–44]

The small percentage of patients who survive aortic rupture and reach medical attention must be evaluated and treated immediately, as without surgical intervention 49% will die in the first 24 hours after injury.[45] Prompt diagnosis and surgical repair can be lifesaving in at least 70% to 86% of cases.[45–48]

The first radiographic study of patients with suspected aortic injury is invariably the chest roentgenogram, usually obtained supine or semiupright due to the condition of the patient. While occasionally gross roentgenographic evidence of mediastinal hemorrhage

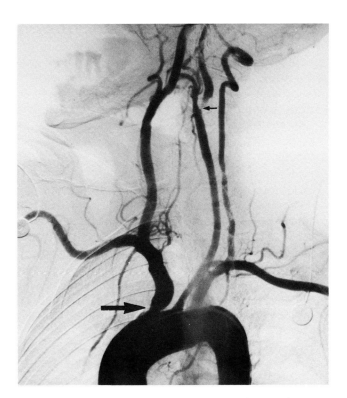

Figure 41-5 Aortogram revealing arteriosclerotic plaque at the origin of the innominate artery (large arrow) and tight stenosis at the origin of the left internal carotid artery (small arrow).

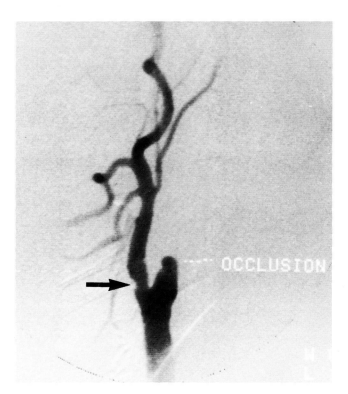

Figure 41-6 Intra-arterial digital subtraction angiogram revealing occlusion of the left internal carotid artery and stenosis of the external carotid artery (large arrow).

may be seen, the exclusion of aortic injury by a plain chest roentgenogram is extremely difficult, usually impossible.

Many plain roentgenographic signs of mediastinal hemorrhage have been described. As reported in the elegant work of Woodring et al, the most reliable are abnormal aortic contour, transverse mediastinal width of 8 cm or greater just above the aortic knob, the presence of an apical cap, 5 mm or more widening of the right paratracheal stripe, and deviation of an indwelling nasogastric tube to the right of the fourth thoracic vertebra.[45,49] Another reliable sign is deviation of the left mainstem bronchus below 40° from the horizontal.[50] Lack of nasogastric tube deviation and a paratracheal stripe thickness of less than 5 mm may together be very reliable indicators of lack of aortic rupture (98%).[45]

In summary, though the plain roentgenogram findings of aortic rupture may be obvious, they are usually nonspecific or may be absent. Thus it is generally impossible to exclude aortic rupture on the basis of roentgenogram interpretation.

In cases where there is appropriate clinical suspicion, even if based only upon a vague history suggesting the possibility of deceleration injury, arteriography must be performed. Though some have suggested that intra-arterial DSA may be as accurate as conventional arteriography,[15] in most institutions direct arteriography remains the procedure of choice for evaluation of aortic rupture. Standard technique usually involves obtaining arterial access through a femoral approach, or axillary approach if necessary, and injection of contrast medium through a catheter placed in the ascending aorta.

Positive findings are based on direct visualization of the post-traumatic pseudoaneurysm. Frank contrast medium extravasation into the periaortic soft tissues is rarely seen, as these patients will not survive to aortography. The most common finding is an irregular bulbous widening in the region of the ligamentum arteriosum (Figure 41-7). This pseudoaneurysm may appear fusiform or saccular.[51] The tear itself may not be clearly seen, but there may be a lucent line representing the intimal flap. Visualization of complete aortic transection with clearly defined dislocation of proximal and distal segments is uncommon.[52]

As stated earlier, traumatic rupture or laceration of the ascending aorta is rarely seen clinically because of its almost universal early mortality. Aortographic findings described include pseudoaneurysm formation along the proximal ascending aorta, particularly its concave surface,[38,53] and a filling defect along the wall, representing a thrombus associated with the intimal tear.[53]

Another potential site of rupture and tear of the thoracic aorta is at the level of the diaphragmatic hiatus.

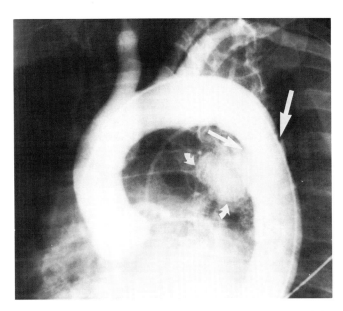

Figure 41-7 Aortogram of 60-year-old male pedestrian struck by automobile. The descending aorta is ruptured (large arrow), and there is a large pseudoaneurysm (small arrows).

innominate artery, in about 4% of cases. While injury to a major aortic branch may be isolated, it is approximately 10 times more likely to be accompanied by aortic injury.[55] Isolated injuries are most likely to be to the innominate artery, less frequently to the left subclavian artery, and least commonly to the carotid arteries.[55]

Above the thoracic inlet the carotid arteries are susceptible to insult by rotation at the level of the atlantoaxial articulation, blunt trauma, and direct laceration. Resultant injuries may include intimal tear with subsequent thrombus or dissecting aneurysm, pseudoaneurysm, and complete transection (Figure 41-8).[56] Arteriographically, with visualization achieved either through aortic arch injection or selective catheterization, intimal tears may be demonstrated, but more commonly the earliest radiographic finding is arterial narrowing due to subintimal dissection and, later, a convex filling defect. Alternatively, the internal carotid may be completely occluded. The intimal tear may lead to dissection of the internal carotid, which often ends at the entrance of the carotid into the carotid canal of the petrous bone.[56]

A dissecting carotid aneurysm may be accompanied by pseudoaneurysm formation, or pseudoaneurysms may occur alone subsequent to arterial rupture and containment within the adventitia or periarterial fascia. These lesions may bulge medially and present as a mass lesion on the posterior pharyngeal wall.[57]

These injuries are rarely encountered clinically. Stark found only three angiographically demonstrated cases reported in the literature.[52,54]

In a clinical setting, aortic laceration is accompanied by injury to the aortic branches, most commonly the

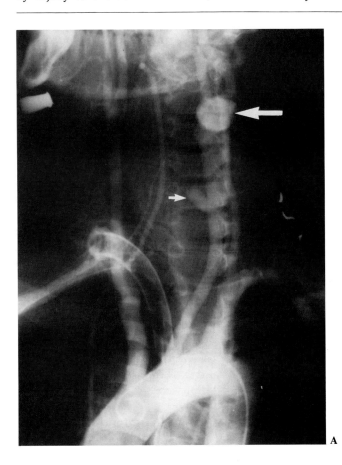

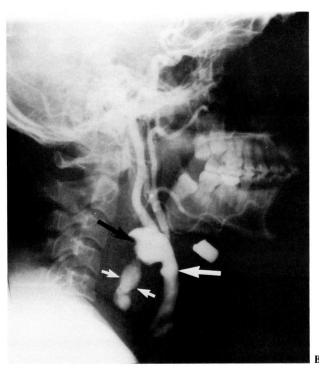

Figure 41-8 Posteroanterior (left) and lateral (right) aortogram of a patient with a gunshot wound to the neck, revealing a traumatic pseudoaneurysm of the carotid artery (large arrow) with extravasation into the soft tissues of the neck (small arrows).

Patients with direct carotid laceration or transection rarely come to angiographic examination because of the mortality of the lesions and the need for immediate intervention. However, arteriovenous fistula may result and be demonstrated radiographically.

Like the carotid artery, the vertebral artery is susceptible to rotational injury at the level of the atlantoaxial articulation and blunt or penetrating injury that leads to intimal tears, dissection, false aneurysm formation, transection, and arteriovenous fistula formation.[56,58]

Approximately 2% of patients with traumatic aortic rupture and pseudoaneurysm formation will survive without treatment and develop chronic post-traumatic aneurysms.[59] It has been shown that surgical intervention will improve the long-term survival of these patients, and elective repair is usually recommended.[60–61] As in the acute condition, plain film diagnosis is difficult,[17] and aortography generally establishes the diagnosis. Usually the false aneurysm will be identified in the region of the aortic isthmus.[17] Because this location is near the left main bronchus and recurrent laryngeal nerve, patients may present with dyspnea and hoarseness; up to 36%, however, are asymptomatic.[60]

ARTERIOSCLEROSIS AND ARTERIOSCLEROTIC ANEURYSMS

Though thoracic aneurysms may be post-traumatic, they are far more likely to be encountered clinically as one of the sequelae of arteriosclerosis. Although patient populations vary from institution to institution, in many hospitals the suspicion of arteriosclerosis is the major cause for patient referral for angiographic evaluation of the thoracic aorta and its branches. The peak incidence of atherosclerosis is in the 5th to 7th decades. Its etiology is believed to be due to a complex interplay of both reversible and nonreversible predisposing factors. Among the irreversible factors are aging and genetic predisposition; among the reversibles are cigarette smoking, hypertension, and obesity. Other factors include hyperlipidemia, diabetes mellitus and low levels of high-density lipoproteins (HDLs).[62,63]

In the affected patient, with the passage of time occlusive disease due to plaques, sometimes ulcerated, develops and causes arterial stenoses in the aorta and its branches (Figure 41-5). Usually stenoses are localized at the take-off of the brachiocephalic vessels and the bifurcation of these vessels. Other points with a propensity to form stenoses are sites of acute bending such as the level of the carotid siphon. Stenoses can lead to slow flow and thus complicate arteriosclerotic lesions with thrombosis. Thrombosis and ulcerated arteriosclerotic plaques can lead to distal emboli and, most important, to stroke. With progressive arteriosclerotic disease as well as with normal aging, the aorta and its branches become progressively more tortuous.

The combination of tortuous vasculature and the presence of arteriosclerotic plaques and thrombi combine to increase the difficulty of catheterization and increase the risk of peripheral embolization due to the procedure.

Arteriosclerotic plaques most commonly affect the common carotid bifurcations and the internal carotids at their origins (Figures 41-6 and 41-9). The vertebral and subclavian arteries are also frequently involved. In these patients, in whom arteriography is most difficult and most dangerous, it is desirable to use the less invasive modalities of duplex ultrasonography and DSA.[3,13]

Computed tomography, particularly with rapid intravenous bolus infusion of contrast medium, has also been used to visualize the carotid bifurcation. Computed tomography can evaluate disturbances in flow (Figure 41-9) and define atheromatous plaques.[19,64–65]

Arteriosclerosis is an important cause of aneurysm formation. Approximately 70% of thoracic aortic aneurysms are due to arteriosclerotic disease.[51,66–67] While post-traumatic aneurysms are generally pseudoaneurysms, those due to arteriosclerosis are almost exclusively true aneurysms. That is, the wall contains all three arterial wall layers (intima, media, and adventitia).

The pathogenesis of arteriosclerotic aneurysms and arteriosclerotic disease in general involves subintimal cholesterol deposits that grow and eventually rupture into the vascular lumen. Calcium is deposited at the resultant ulcer site, while deeper portions of the lesion disrupt the media and weaken the vessel wall, leading to progressive dilatation.[68–69] Fibrosis develops in the adventitia in an attempted healing process; however, liquefaction necrosis may occur, further weakening the vessel wall. Arteriosclerotic aneurysms of the thoracic aorta are, like traumatic aneurysms, most common distal to the left subclavian artery[68,70–72] and uncommon in the ascending aorta.

Diagnosis of aneurysms, though often suggested by plain roentgenograms, generally requires aortography to differentiate suspected aneurysm from mediastinal masses and tortuous vasculature.[68,73] Accurate diagnosis and prompt surgical intervention are essential, as the 5-year survival of patients with untreated thoracic aneurysms is approximately 20%, with the greatest risk of rupture being in aneurysms 10 cm or greater in diameter.[74] Aortic aneurysms may be accompanied by aneurysms of the aorta's major intrathoracic branches, or the branches may be individually affected.[75] Arteriosclerotic aneurysms of the thoracic aorta may be multiple and associated with aneurysms of the abdominal aorta and its branches.[76] Also, there may be a high association between aneurysm formation and obstructive lesions.[77] Arteria magna is an entity in which multiple systemic arteries, including the thoracic aorta, may be affected by multiple aneurysms. In the past this

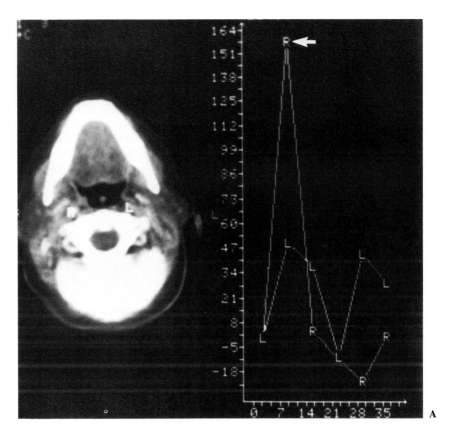

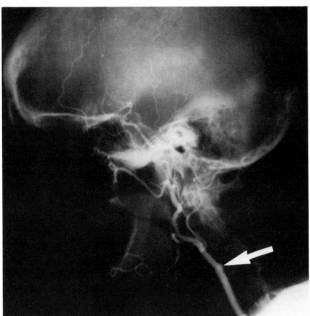

Figure 41-9 A, Dynamic CT of the internal carotid arteries. At time zero a bolus of 20 mL of 60% iodinated contrast medium was given, and six successive CT sections were obtained. Following the procedure, regions of interest within the sections were selected and the density of the selected regions plotted over time. In this case the regions of interest are the right (R) and left (L) internal carotid arteries. The graph reveals a sharp intensity peak (small arrow) at approximately 9 seconds and a sharp drop-off, suggesting prompt normal flow through the right internal carotid artery. There is a much lower peak with less clear drop-off in the region of the left internal carotid artery, suggesting reduced flow. B, These findings are confirmed on the corresponding lateral arteriogram view, revealing occlusion of the internal carotid artery near its origin (large arrow).

disease was felt to be an unusual presentation of arteriosclerotic disease. More recently, however, it has been shown that the disease is more likely due to a primary loss of elasticity in the media.[78] Because of the propensity for thoracic aorta aneurysms to be accompanied by aneurysmal or occlusive lesions elsewhere in the arterial tree, thorough physical and arteriographic examination is suggested prior to surgical intervention.[77]

AORTIC DISSECTION

Another entity that may be superimposed upon arteriosclerotic changes is aortic dissection. Though the precise etiology of aortic dissection remains the subject of some controversy, the primary predisposing factor appears to be a weakening of the arterial media, either congenital or acquired.[79] Conditions causing weakening of the media include Marfan syndrome, Ehlers-

Danlos syndrome, Turner's syndrome, Erheim's syndrome, syphilitic aortitis, giant cell aortitis, relapsing polychondritis, and arteriosclerosis.[66,80–81] Idiopathic cystic medial necrosis has been felt to be a predisposing factor, though this may be only a result of the normal aging process.[79,82–83] Hypertension superimposed upon the existing weakening of the media probably also plays a role.[79]

Regardless of the predisposing factors, aortic dissection usually involves hematoma formation in the diseased vascular media separating the intima and adventitia. The dissection begins either in the ascending portion of the thoracic aorta just above the level of the aortic valve or in the descending aorta at the level of the ligamentum arteriosum. Several classification systems have been proposed to describe aortic dissections. The most widely used is the DeBakey system, which divides these lesions into three types (Figure 41-10). Type I dissection involves an intimal tear in the ascending aorta with extension of the dissection into the descending aorta beyond the ligamentum arteriosum. Type II dissections are confined to the ascending aorta. Type III dissections initiate at the ligamentum arteriosum and usually are confined to the descending aorta, though they may extend retrograde into the arch.[84] Type I is the most common (51%), type III the next most common (42%), and type II the least common (7%). Types I and II may also be referred to as ascending dissections and type III as descending dissections.

A more recent classification system proposed by Daily divides dissections simply based upon whether or not they involve the ascending aorta. In this system all dissections involving the ascending aorta are referred to as type A and all others as type B.[85] This classification more clearly reflects the distinct differences in terms of prognosis and therapeutic planning between patients with and patients without involvement of the ascending aorta. Urgent surgical correction is recommended for ascending aorta dissections because these lesions usually rupture into the pericardium if only medical therapy is employed.[85]

As in other pathologic conditions of the aorta, plain roentgenographic examination may suggest aortic dissection but is rarely diagnostic. Aortography is usually the procedure of choice for confirming the presence of aortic dissection, identifying the site of intimal tear, and determining the extent of the aneurysm. In particular, it is important to define whether or not the ascending aorta is involved and whether there is dissection into other brachiocephalic or distal aortic branches.

If contrast medium is injected proximal to the intimal tear, there may be opacifications of both the true and false lumens, and the intimal flap may be seen between them.[86] Sometimes ultrasound will identify the intimal flap.[87] In some cases the false lumen may not fill but will still be appreciated by its distortion of the contrast medium–filled true lumen and by an increase in thickness of the aortic wall caused by the shadow of the nonopacified false lumen.[51,88] This false lumen is usually seen on the outer curve of the aorta—in the ascending aorta anterior and to the right, in the arch superior and posterior, and in the descending posterior and to the left.[88–89]

If both the true and false lumen do opacify with contrast medium injection, it may be very difficult to determine which is the true lumen.[90–91] Though there is generally faster flow within the true lumen, and the aortic branches more commonly opacify from the true lumen, these findings are not reliable.[89,91] In the proximal ascending aorta the aortic valve is continuous with the true lumen,[92–93] but even this finding may be difficult to judge. The coronary arteries also rarely fill via the false lumen.[90] When the dissection involves the ascending aorta, aortic insufficiency is common.[89]

Though arteriography can be diagnostic of aortic dissection in up to 95% to 99% of cases,[89] CT has been proposed as a reliable alternate modality[94] with an accuracy comparable to that of aortography.[95] High-resolution CT scanning can clearly differentiate true from false lumen and may provide all the preoperative imaging necessary, particularly in type III dissection. Aortography may still be necessary to quantify aortic insufficiency and coronary artery patency and to determine if vital structures are perfused solely by the false channel.[96]

Recently MRI has been shown to be another accurate modality for the assessment of aortic dissection. Early work has demonstrated its potential and suggests it may become the recommended initial modality in stable cases of suspected aortic dissection.[23]

INFECTIOUS AND AUTOIMMUNE CONDITIONS

The thoracic aorta and its branches, in addition to being susceptible to trauma, dissection, and arteriosclerosis, are less commonly affected by autoimmune or infectious diseases.[97] Among these entities is syphilis, once a common cause of thoracic aortic aneurysms.[68] Syphilitic aortitis arises 10 to 30 years after the primary infection,[47] producing an obliterative endarteritis of the vasavasorum, yielding fragmentation and dystrophic calcification of the media. These changes result most commonly in aortic regurgitation and in true aortic aneurysm formation. Syphilitic aneurysms occur most commonly in the ascending aorta and arch and less so in the descending aorta.[68,98]

Infectious agents other than syphilis may also lead to aortic aneurysms. Such aneurysms, referred to as mycotic regardless of the microbial agent involved, are usually due to bacterial infections.[99] Generally mycotic aneurysms do not occur in normal vessels and are only

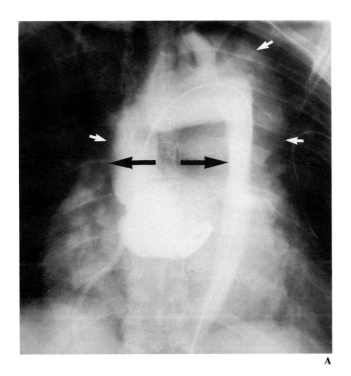

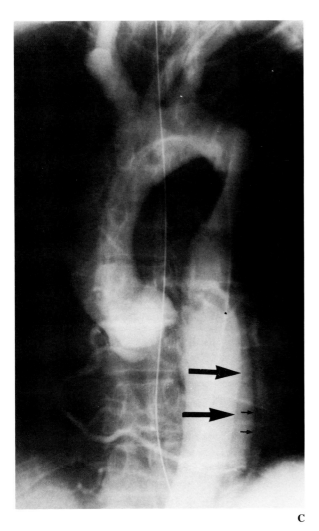
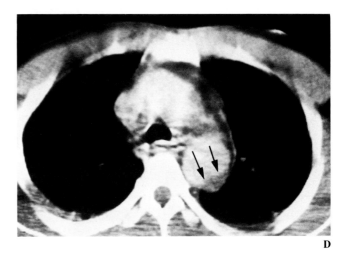

Figure 41-10 Aortic dissections. A, DeBakey type I (Dailey type A) dissection, beginning in the ascending aorta and continuing to involve the descending aorta. False lumen marked by small arrows, true lumen by large arrows. Note prosthetic aortic valve. B, DeBakey type II (Dailey type A) dissection, confined to the ascending aorta. False lumen marked by small arrows, true lumen by large arrows. C, DeBakey type III (Dailey type B) dissection, confined to the descending aorta. False lumen marked by small arrows, true lumen by large arrows. D, Infusion CT examination of case. Arrows mark false lumen. E and F, Early and late films revealing ascending and descending aortic dissections. Arrows mark intimal flaps. Long white arrows mark contrast medium reflux into the left ventricle, indicating aortic insufficiency.

Figure 41-10 continued

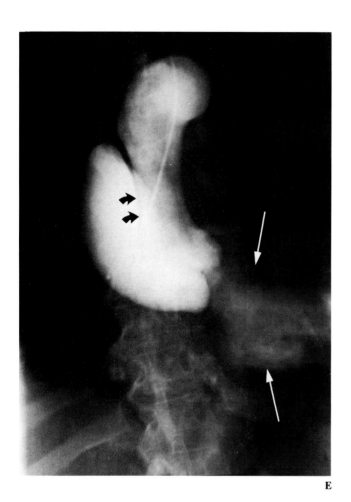

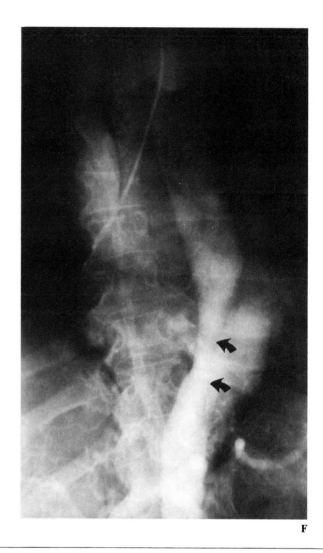

found in arteries previously insulted by processes such as arteriosclerosis, trauma, congenital defects, or syphilitic aortitis.[100] Primary lesions have a distant or unknown source, while secondary lesions have an easily identifiable intravascular source, such as subacute bacterial endocarditis.[16] Mycotic aneurysms weaken the vascular wall, predisposing it to rupture, and can provide an unsuspected focus of continued infection. Most mycotic aneurysms are of bacterial origin and are true aneurysms. False aneurysms may occur, however, particularly in tuberculosis.[36,100–101] Arteriography is the diagnostic modality of choice, and there are no particular specific radiographic findings.[100] In the thoracic aorta these aneurysms are most frequently found in the ascending portion or in the sinuses of Valsalva. Subacute bacterial endocarditis, in addition to predisposing to the development of mycotic aneurysm, may also lead to development of frank abscess of the aortic root. Angiographic diagnosis is made by demonstration of the characteristic round extraluminal collection of contrast material that has extravasated into the mediastinum. Identification of the abscesses may have significant preoperative implications.[102]

Other inflammatory conditions for which there is no known pathologic agent may also affect the aorta. Takayasu's arteritis, or "pulseless disease," is a condition of marked fibrosis and degeneration of elastic fibers in the media, together with adventitial and intimal thickening. The result is hardening and thickening of the vascular wall, which may lead to occlusion, stenosis, or aneurysmal dilatation.[103–104] The disease is felt to be autoimmune in origin and has been most commonly recognized in young Oriental females, but it occurs in all races.[105] Diagnosis is made angiographically through visualization of variable stenosis of the aorta and its branches, which may be accompanied by aneurysmal dilatation. The ascending aorta is usually dilated, while stenoses commonly occur in the subclavian and carotid arteries and the aorta distal to the origin of the left subclavian artery.

Giant cell aortitis is another inflammatory condition that may have an autoimmune etiology. The pathologic lesion is a granulomatous infiltration of the vascular media. The disease usually affects elderly females, and aneurysm or dissection may occur.[106]

SUMMARY

The radiologist plays a critical role in the prompt, accurate diagnosis of diseases of the thoracic aorta and its branches. Everyday diagnosis and therapy continues to rely heavily upon plain chest roentgenograms and direct arterial arteriography. In recent years several alternate imaging modalities, less invasive than direct arteriography, have been developed. As continued experience is gained and technologic development of these modalities progresses, the work-up of disorders of the thoracic aorta continues to evolve. The optimal use of these new technologies as well as continued improvement of our understanding of the underlying disease processes will, as always, require close cooperation between the radiologist and the referring physician.

REFERENCES

1. Cooley RN, Schreiber MH. *Radiology of the Heart and Great Vessels*. 3rd ed. Baltimore: Williams & Wilkins Co; 1978:24–27.

2. Wolverson MK, Heiberg E, Sundaram M, et al. Carotid atherosclerosis: High-resolution real-time sonography correlated with angiography. *AJR*. 1983;140:355–361.

3. Glover JL, Bendick PJ, Jackson VP, et al. Duplex ultrasonography, digital subtraction angiography, and conventional angiography in assessing carotid atherosclerosis. *Arch Surg*. 1984;119:664–669.

4. Fischer GG, Anderson DC, Farber R, et al. Prediction of carotid disease by ultrasound and digital subtraction angiography. *Arch Neurol*. 1985;42:224–227.

5. Breslau PJ, Fell G, Phillips DJ, et al. Evaluation of common carotid artery velocity patterns. *Arch Surg*. 1982;117:58–60.

6. Grossman LB, Buonocore E, Modic MT, et al. Digital subtraction angiography of the thoracic aorta. *Radiology*. 1984;150:323–325.

7. Detrano R, Moodie DS, Gill CC, et al. Intravenous digital subtraction aortography in the preoperative and postoperative evaluation of Marfan's aortic disease. *Chest*. 1985;88:249–253.

8. Chernin MM, Pond GD, Sahn DJ. Digital subtraction angiography of the aortic arch. *Cardiovasc Intervent Radiol*. 1984;7:196–203.

9. Modic MT, Weinstein MA, Paulicek W, et al. Intravenous digital subtraction angiography: Peripheral versus central injection of contrast material. *Radiology*. 1983;147:711–715.

10. Aaron JO, Hesselink JR, Oot R, et al. Complications of intravenous DSA performed for carotid artery disease: A prospective study. *Radiology*. 1984;153:675–678.

11. Pinto RS, Manuell M, Kricheff II. Complications of digital intravenous angiography: Experience in 2488 cervicocranial examinations. *AJR*. 1984;143:1295–1299.

12. Chilcote WA, Modie MT, Paulicek MS, et al. Digital subtraction angiography of the carotid arteries: A comparative study in 100 patients. *Radiology*. 1981;139:287–295.

13. Weinstein MA, Paulicek WA, Modic MT, et al. Intra-arterial digital subtraction angiography on the head and neck. *Radiology*. 1983;147:717–724.

14. Brant-Zawadzki M, Gould R, Norman D, et al. Digital subtraction cerebral angiography by intraarterial injection: Comparison with conventional angiography. *AJR*. 1983;140:347–353.

15. Mirvis SE, Pais SO, Gens DR. Thoracic aortic rupture: Advantage of intraarterial digital subtraction angiography. *AJR*. 1986;146:987–991.

16. Moore EH, Farmer DW, Geller SC, et al. Computed tomography in the diagnosis of iatrogenic false aneurysms of the ascending aorta. *AJR*. 1984;142:117–118.

17. Heystraten FM, Rosenbosch G, Kingma LM, et al. Chronic post-traumatic aneurysm of the thoracic aorta: Surgically correctable occult threat. *AJR*. 1986;146:303–308.

18. Brant-Zawadzki M, Jeffry RB. CT image reformation for non-invasive screening of the carotid bifurcation: Early experience. *AJR*. 1982;3:395–400.

19. Heinz ER, Pizer SM, Fuchs H, et al. Examination of the extracranial bifurcation by thin-section dynamic CT: Direct visualization of intimal atheroma in man (part I). *AJR*. 1984;5:355–359.

20. Leeson MD, Cacayorin ED, Iliya AR, et al. Atheromatous extracranial carotid arteries: CT evaluation correlated with arteriography and pathologic examination. *Radiology*. 1985;156:397–402.

21. Glazer HS, Gutierrez FR, Levitt AG, et al. The thoracic aorta studied by MR imaging. *Radiology*. 1985;157:149–155.

22. Fletcher BD, Jacobstein MD. MRI of congenital abnormalities of the great arteries. *AJR*. 1986;146:941–946.

23. Amparo EG, Higgins CB, Hricak H, et al. Aortic dissection: Magnetic resonance imaging. *Radiology*. 1985; 155:399–406.

24. Dinsmore RE, Liberthson RR, Wismer GI, et al. Magnetic resonance imaging of thoracic aortic aneurysms: Comparison with other imaging methods. *AJR*. 1986;146:309–314.

25. Walls EW. The blood vascular and lymphatic systems. In: Romanes GJ, ed. *Cunningham's Textbook of Anatomy*. 12th ed. Oxford: Oxford University Press; 1981:871–1005.

26. Abrams HL, Jonsson G. The normal thoracic aorta. In: Abrams HL, ed. *Abrams Angiography Vascular and Interventional Radiology*. 3rd ed, Vol. 1. Boston: Little, Brown & Co; 1983:353–366.

27. Dawson HL. *Basic Human Anatomy*. 2nd ed. New York: Appleton-Century-Crofts; 1974:148–156.

28. Hollingshead WH, Rosse C. *Textbook of Anatomy*. 3rd ed. Philadelphia: Harper & Row; 1985:84–86.

29. Johnstrude IS, Jackson DC. *A Practical Approach to Angiography*. Boston: Little, Brown & Co; 1979:385–448.

30. Haughton JM, Rosebaum AE. The normal and anomalous aortic arch and brachiocephalic arteries. In: Newton TH, Potts DG, eds. *Radiology of the Skull and Brain*. Vol 2, Book 2. St. Louis: CV Mosby; 1971;1145–1163.

31. Kiss J. Bifid origin of the right vertebral artery. *Radiology*. 1968;91:931.

32. Klinkhammer AC. *Esophagography in Anomalies of the Aortic Arch System*. Baltimore: Williams & Wilkins Co; 1969:16–30.

33. Vitek JJ, Reaves P. Thoracic bifurcation of the common carotid artery. *Neuroradiology*. 1973;5:133–139.

34. Eisenberg RA, Vines FS, Taylor SB. Bifid origin of the left vertebral artery. *Radiology*. 1986;159:429–430.

35. Suzukis S, Kuwabara Y, Matano R, et al. Duplicate origin of the left vertebral artery. *Neuroradiology*. 1978;15:27–29.

36. Parmley LF, Mattingly TW, Manion WC, et al. Non-penetrating traumatic injury of the aorta. *Circulation*. 1958;17:1086–1101.

37. Strassman G. Traumatic rupture of the aorta. *Am Heart J*. 1947;33:508–515.

38. Lundell CJ, Quinn MF, Finck EJ. Traumatic laceration of the ascending aorta: Angiographic assessment. *AJR*. 1985;145:715–719.

39. Greendyke RM. Traumatic rupture of the aorta. *JAMA*. 1966;195:527–530.

40. Harris JH, Harris WH. *The Radiology of Emergency Medicine*. 2nd ed. Baltimore: Williams & Wilkins Co; 1981:333–341.

41. Marsh CL, Moore RC. Deceleration trauma. *Am J Surg*. 1957;93:623–631.

42. Lunderall J. The mechanism of traumatic rupture of the aorta. *Acta Pathol Microbiol Scand*. 1964;62:34–46.

43. Molnar W, Pace WG. Traumatic rupture of the thoracic aorta. *Radiol Clin North Am*. 1966;4:403–414.

44. Sanborn JC, Heitzman ER, Markarian B. Traumatic rupture of the thoracic aorta: Roentgen pathologic correlations. *Radiology*. 1970;95:293–298.

45. Woodring H, Loh K, Kryscio J. Mediastinal hemorrhage: An evaluation of radiographic manifestations. *Radiology*. 1984;151:15–21.

46. Kirsh MM, Behendt DM, Orringer MB, et al. The treatment of acute traumatic rupture of the aorta: A 10 year experience. *Ann Surg*. 1976;184:308–316.

47. Turney SZ, Attar S, Ayella R, et al. Traumatic rupture of the aorta: A five-year experience. *J Thorac Cardiovasc Surg*. 1976;72:727–734.

48. McClenathan JE, Brettschneider L. Traumatic thoracic aortic aneurysms. *J Thorac Cardiovasc Surg*. 1965;50:74–82.

49. Woodring JH, Pulmano CM, Stevens RK. The right paratracheal stripe in blunt chest trauma. *Radiology*. 1982;143:605–608.

50. Marnocha KE, Maglinte DDT. Plain-film criteria in blunt chest trauma. *AJR*. 1985;144:19–21.

51. Steiner RM, Weschler RJ, Grainger RG. The thoracic aorta. In: Grainger RG, Allison DJ, eds. *Diagnostic Radiology—An Anglo-American Textbook of Imaging*. Vol 1, Section 3. Edinburgh: Churchill Livingstone Inc; 1986:704–715.

52. Stark P. Traumatic rupture of the thoracic aorta: A review. *CRC Crit Rev Diagn Imaging*. 1984;21:229–255.

53. Daniels DL, Maddison FE. Ascending aortic injury: An angiographic diagnosis. *AJR*. 1981;136:812–813.

54. Hirsch JH, Carter SJ, Chikos PM. Traumatic pseudoaneurysms of the thoracic aorta: Two unusual cases. *AJR*. 1978;130:157–160.

55. Fisher RG, Hadlock F, Ben-Manachem Y. Laceration of the thoracic aorta and brachiocephalic arteries by blunt trauma. *Radiol Clin North Am*. 1981;19:91–110.

56. Davis JM, Zimmerman RA. Injury of the carotid and vertebral arteries. *Neuroradiology*. 1983;25:55–69.

57. Silcox LE, Updegrove RA. Extracranial aneurysm of the internal carotid artery. *Arch Otolaryngol*. 1959;69:329–333.

58. Schechter MM, Gutstein RA. Aneurysms and arteriovenous fistulas of the superficial temporal vessels. *Radiology*. 1970;97:549–557.

59. Bennett DE, Cherry JK. The natural history of traumatic aneurysms of the aorta. *Surgery*. 1967;61:516–523.

60. Finkelmeier BA, Mentzer RM, Kaiser DL, et al. Chronic traumatic thoracic aneurysm. *J Thorac Cardiovasc Surg*. 1982;84:257–266.

61. DeBakey ME, Cooley DA, Crawford ES, et al. Aneurysms of the thoracic aorta. *J Thorac Surg*. 1958;36:393–420.

62. Bierman EL. Atherosclerosis and other forms of arteriosclerosis. In: Isselbacher KJ, Adams RD, Braunwald E, et al., eds. *Harrison's Principles of Internal Medicine*. 9th ed. New York: McGraw-Hill Book Co; 1980:1014–1023.

63. Fogelman AM, Edwards PA, Haberland ME. Atherosclerosis: pathology, pathogenesis, and medical management. In: Moore WS, ed. *Vascular Surgery—A Comprehensive Review*. Orlando: Grune & Stratton; 1986:45–55.

64. Tress BM, Davis S, Lavain J, et al. Incremental dynamic computed tomography: Practical method of imaging the carotid bifurcation. *AJR*. 1986;146:465–470.

65. Heinz ER, Fuchs J, Osborne D, et al. Examination of the extracranial carotid bifurcation by thin-section dynamic CT: Direct visualization of intimal atheroma in man (part 2). *AJR*. 1984;5:361–366.

66. Lang EK. The arterial system—part I. In: Teplick G, Haskins M, eds. *Surgical Radiology*. Vol 2. Philadelphia: WB Saunders Co; 1981:1443–1471.

67. DeBakey ME, McCollum CH, Graham JM. Surgical treatment of aneurysms of the descending thoracic aorta: Long term results in 500 patients. *J Cardiovasc Surg*. 1978;19:571–576.

68. Randall PA, Jarmolowski CR. Aneurysm of the thoracic aorta. In: Abrams AL, ed. *Abrams Angiography Vascular and Interventional Radiology*. 3rd ed. Boston: Little, Brown & Co; 1983:417–441.

69. Stover J, Husni EA, Aseem W. Management of thoracic aneurysms. *Int Surg*. 1967;47:344–355.

70. McNamara JJ, Pressler VM. Natural history of arteriosclerotic thoracic aortic aneurysms. *Ann Thorac Surg*. 1978;26:468–473.

71. DeBakey ME, Noon GP. Aneurysms of the thoracic aorta. *Mod Concepts Cardiovasc Surg*. 1975;44:53–58.

72. Steinberg I. The arteriosclerotic thoracic aorta: Clinical and roentgen observations. *Angiology*. 1956;7:405–418.

73. Sprayregen S, Jacobson HG. Angiographic differentiation of thoracic aneurysms and neoplasms. *Vasc Surg*. 1976;10:200–213.

74. Joyce JW, Fairbairn JF, Kincaid OW, et al. Aneurysms of the thoracic aorta: A clinical study with special reference to prognosis. *Circulation*. 1964;29:176–181.

75. Thomas TV. Intrathoracic aneurysms of the innominate and subclavian arteries. *J Thorac Cardiovasc Surg*. 1972;63:461–471.

76. Sprayregen S. Radiologic spectrum of arteriosclerotic aneurysms of the aortic arch. *NY State J Med*. 1978;78:2198–2204.

77. Ching CC, Hughes RK. Arteriosclerotic aneurysms of the thoracic aorta: Late stage of a diffuse disease. *Am J Surg*. 1967;114:853–855.

78. Randall PA, Omar MM, Rohner R, et al. Arteria magna revisited. *Radiology*. 1979;132:295–300.

79. Wheat MW. Acute dissecting aneurysms of the aorta: diagnosis and treatment—1979. *Am Heart J*. 1980;99:373–387.

80. Gore I, Hirst AE. Dissecting aneurysm of the aorta. *Prog Cardiovasc Dis*. 1973;16:103–111.

81. Strauss RG, McAdams AJ. Dissecting aneurysm in childhood. *J Pediatr*. 1970;76:578–584.

82. Schlatmann TJ, Becker AE. Pathogenesis of dissecting aneurysm of the aorta: Implications for dissecting aortic aneurysm. *Am J Cardiol*. 1977;39:21–26.

83. Schlatmann TJ, Becker AE. Histologic changes in the normal aging aorta: Implications for dissecting aortic aneurysm. *Am J Cardiol*. 1977;39:13–20.

84. DeBakey ME, Henly WS, Cooley DA, et al. Surgical management of dissecting aneurysm of the aorta. *J Thorac Cardiovasc Surg*. 1965;49:130–149.

85. Daily PO, Trueblood HW, Stinson EB, et al. Management of acute aortic dissections. *Ann Thorac Surg*. 1970;10:237–247.

86. Earnest F, Muhm JR, Sheedy PF. Roentgenographic findings in thoracic aortic dissection. *Mayo Clin Proc*. 1979;54:43–50.

87. Nicholson WJ, Cobbs BW. Echocardiographic oscillating flap in aortic root dissecting aneurysm. *Chest*. 1976;70:305–307.

88. Itzchak Y, Rosenthal T, Adar R, et al. Dissecting aneurysm of the thoracic aorta: Reappraisal of radiologic diagnosis. *AJR*. 1975;125:559–570.

89. Smith DC, Jang GC. Radiologic diagnosis of aortic dissection. In: Doroghazi PM, Slater EE, eds. *Aortic Dissection*. New York: McGraw-Hill Book Co; 1983:71–132.

90. Hayashi K, Meancy TF, Zelch JV, et al. Aortographic analysis of aortic dissection. *AJR*. 1974;122:769–782.

91. Ambos MA, Rothberg M, Lefleur RS, et al. Unsuspected aortic dissection: The chronic "healed" dissection. *AJR*. 1979;132:221–225.

92. Beachley MC, Ranniger K, Roth F. Roentgenographic evaluation of dissecting aneurysms of the aorta. *AJR*. 1974;121:617–625.

93. Soto B, Harman MA, Ceballo SR, et al. Angiographic diagnosis of dissecting aneurysm of the aorta. *AJR*. 1972;116:146–154.

94. Vasile N, Mathia D, Keita K, et al. CT of thoracic aortic dissection: Accuracy and pitfalls. *J Comput Assist Tomogr*. 1986;10:211–215.

95. Ovdkerk M, Overbosch E, Dee P. CT recognition of acute aortic dissection. *AJR*. 1983;141:671–676.

96. Thorsen MK, San Dretto MA, Lawson TL, et al. Dissecting aortic aneurysms: Accuracy of computed tomographic diagnosis. *Radiology*. 1983;148:773–777.

97. Kampmeier RH. The late manifestations of syphilis. *Med Clin North Am*. 1964;48:667–697.

98. Kampmeier RH. Saccular aneurysm of the thoracic aorta: A clinical study of 633 cases. *Ann Intern Med*. 1938;12:624–651.

99. Parkhurst GF, Decker JP. Bacterial aortitis and mycotic aneurysm of the aorta—A report of 12 cases. *Am J Pathol*. 1955;31:821–833.

100. Weintraub RA, Abrams HL. Mycotic aneurysms. *AJR*. 1968;102:354–362.

101. Felson B, Akers PV, Hall GS, et al. Mycotic tuberculous aneurysm of the thoracic aorta. *JAMA*. 1977;237:1104–1108.

102. Miller SW, Dinsmore RE. Aortic root abscess resulting from endocarditis: Spectrum of angiographic findings. *Radiology*. 1984;153:357–361.

103. Slater EE, DeSanetis RW. Disease of the aorta. In: Braunwald E, ed. *Heart Disease—A Textbook of Cardiovascular Medicine*. Vol 2. Philadelphia: WB Saunders; 1984:1546–1576.

104. Lande A, Bard A, Bole P, et al. Aortic arch syndrome (Takayasu's arteritis) arteriographic and surgical considerations. *J Cardiovasc Surg*. 1978;19:507–513.

105. Lupi-Herrera E, Sanchez-Torres G, Marcushamer J, et al. Takayasu's Arteritis—Clinical study of 107 cases. *Am Heart J*. 1977;93:94–103.

106. Ghose MK, Shensa S, Lerner PI. Arteritis of the aged (giant cell arteritis) and fever of unexplained origin. *Am J Med*. 1976;60:429–436.

Chapter 42

Evaluation of the Great Vessels by Echocardiography and Doppler

Stuart Rich, MD

Echocardiography and Doppler techniques have been used to describe the great vessels. Real-time images of the ascending and descending aorta, aortic arch, and proximal pulmonary arteries are easily obtainable in most patients. In addition, normal blood flow velocity profiles have been established using pulsed and continuous wave Doppler.[1-3] Consequently, anatomic and physiologic abnormalities of the great vessels can often be recognized by currently available echo-Doppler equipment. The major limitation to obtaining quality images is the distance between the ultrasonic probe and the structure in question, which in the case of the great vessels in adults can often be large. For that reason the quality of the information obtained is often questionable, to the point that much uncertainty may surround the diagnosis, and other confirmatory imaging is necessary. On the other hand, oftentimes echo-Doppler can make a confirmatory diagnosis in the patient in whom other imaging modalities were also equivocal. This chapter will review the application of echo-Doppler in acquired diseases of the aorta and pulmonary artery and some congenital diseases that may be seen in the adult.

AORTIC ANEURYSMS

Aneurysmal dilatation of the ascending and thoracic aorta can be reliably detected with two-dimensional echocardiography.[4] Normal size estimates for adults have been established for the aortic annulus, sinuses of Valsalva, and proximal tubular portion of the ascending aorta.[5-6] Aneurysmal dilatation of the aorta at each of these levels has been recorded by two-dimensional echocardiographic imaging.

Annuloaortic ectasia, or idiopathic dilatation of the proximal aorta and aortic annulus, is a condition most often seen in patients with Marfan syndrome.[7] Marked dilatation of the aortic root is also commonly associated with pure aortic regurgitation,[8] which is easily detectable by echocardiographic findings (such as diastolic fluttering of the mitral valve)[9] and Doppler recordings at the left ventricular outflow tract.[10] Echo-Doppler is very helpful in the management of the majority of patients with annuloaortic ectasia as it provides a noninvasive means to monitor the aortic root diameter, which will often dictate the need for surgical intervention. Because of the high morbidity of aortic dissection in patients with Marfan syndrome and annuloaortic ectasia, it has been recommended that surgery be performed even in the asymptomatic patient if the aortic root diameter reaches 6 cm by M-mode echocardiography.[11]

Aneurysms of the sinuses of Valsalva arise from congenital abnormalities of the media in the proximal aorta and most commonly involve the right coronary sinus (69%), or noncoronary sinus (26%), and rarely the left coronary sinus (5%).[12,13] As they expand, they can produce obstruction to blood flow and eventually rupture. Aneurysms of the right coronary sinus may protrude into the right ventricle and produce right ventricular outflow tract obstruction or compression of the right atrium.[14-15] Rupture usually produces a large left-to-right shunt.[16] The coronary sinuses are easily visualized in the parasternal long- and short-axis views by two-dimensional echocardiography, and assessment of rupture can be made echocardiographically, by determination of changes in volume of the cardiac chambers, and confirmed by Doppler. Thus an aortic-to-right-atrial rupture will produce right atrial and right ventricular enlargement with increased pulmonic flow velocity, and also often a detectable regurgitant continuous or diastolic jet.[17] An aortic-to-right-ventricular rupture will produce a similar constellation of signs but with the absence of right atrial enlargement, and aortic-to-left-atrial (or left-ventricular) rupture produces volume overload of the left side with regurgitant diastolic jets.[18] The echo-Doppler documentation of a left sinus of Valsalva aneurysm with an aortopulmonary tunnel has also been recently reported.[19]

AORTIC DISSECTION

Dissections of the ascending aorta are characterized echocardiographically by dilatation of the aortic root and visualization of a false channel, which appears as a parallel widening of the aortic wall, or a prominent aortic flap, which may have an undulating appearance (Figure 42-1).[20–21] Pulsed Doppler techniques have been reported to be helpful in confirming the echocardiographic appearance when the diagnosis is questioned, by revealing different flow patterns within the true and false lumens.[22–23]

Depending on image quality, echocardiography may successfully distinguish saccular from fusiform aneurysms and type A from type B dissections (Figure 42-2).[20] However, because the aortic arch and descending segments of the thoracic aorta are often poorly visualized, confirmation of a suspected dissection requires additional imaging modalities, as false-positive and false-negative diagnoses occur. Aortography has been most commonly employed, but computed tomography, especially ultrafast computed tomography, which allows the assessment of aortic flow in both the true and false channels, and magnetic resonance imaging may also provide reliable confirmatory images noninvasively.[24]

In patients with acute aortic dissection, echo-Doppler may be particularly helpful in supplying useful information about left ventricular function and the severity of aortic regurgitation. As hypertension is common in patients with aortic dissection, the status of the left ventricle is an important consideration in planning surgical treatment. In addition, documenting the presence and severity of aortic regurgitation is important for the surgeon in planning the administration of a cardioplegic solution and preparing for possible aortic valve

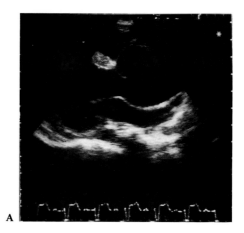
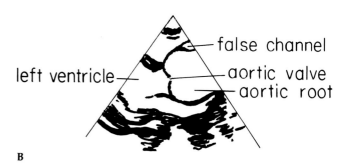

Figure 42-1 A two-dimensional echocardiographic view of a large aortic aneurysm with dissection in a patient with Marfan syndrome, made from the parasternal long-axis view. In addition to the obvious aneurysmal dilatation, an aortic flap can be seen, which during real time could be seen to be undulating in the characteristic manner of dissections.

replacement during the time of surgical repair of the aorta.

Pseudoaneurysms of the ascending aorta are uncommon and characteristically difficult to diagnose. The application of echo-Doppler in this condition has been reported as being of value by detecting an outpouching of the ascending aorta in association with abnormal flow velocities within a fistula between the aortic root and right atrium.[25] A high index of clinical suspicion, in association with high-quality ultrasonic images, is essential in making these types of diagnoses noninvasively.

COARCTATION OF THE AORTA

Coarctation of the aorta is characterized by a localized deformity that has the appearance of an indentation or localized concavity. Although coarctation may occur at any level of the thoracic aorta, it is most commonly found beyond the origin of the left subclavian artery distal to the insertion of the ligamentum arteriosum. In adults the most common type of coarctation noted is the

AORTIC TRAUMA

Although there are no published reports, echo-Doppler studies may be helpful in patients who have been subjected to blunt trauma of the chest because of its portability and real-time imaging capabilities. Most severe aortic injuries are due to motor vehicle accidents. While the aorta may be torn anywhere along its length, the most frequent point of rupture is the aortic isthmus, the site of the insertion of the ligamentum arteriosum. The injuries may vary between a small break in the aortic wall to a complete circumferential transection. If the patient survives the initial insult, a localized saccular aneurysm or pseudoaneurysm may subsequently develop at the sight of the tear. A high index of suspicion is a requisite to making these diagnoses, and confirmation by computed tomography or aortic angiography is important before surgical intervention is considered.

PATENT DUCTUS ARTERIOSUS

The main pulmonary artery and proximal left and right pulmonary arteries after the bifurcation are usually well visualized in most adults. Because the ductus arteriosus arises near the bifurcation, it is possible, although difficult, to visualize it with two-dimensional echocardiography (see Figure 37-7 in Chapter 37). Although a patent ductus arteriosus is not common in adults, its presence can be confirmed with echo-Doppler studies. The echocardiographic manifestations include evidence of left ventricular volume overload from the left-to-right shunt and changes consistent with right ventricular pressure overload if substantial pulmonary hypertension is coexistent.[32] Echo-Doppler is useful in evaluating the presence of left-to-right cardiac shunts by performing Doppler studies of the aorta and pulmonary artery, as the mean velocity-to-time intervals will be proportional to the pulmonary-to-systemic flow ratios (see Chapter 37).[33] Detection of flow in the ductus itself is also possible. The patent ductus with predominant left-to-right flow from the aorta into the pulmonary artery will reveal abnormally high flow in late systole and early diastole at the left pulmonary artery and back proximally toward the pulmonic valve.[34-35] Reduced flow in a constricted ductus and even reverse flow have been detected utilizing real-time Doppler color flow mapping.[36]

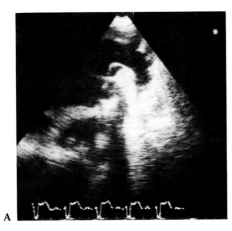
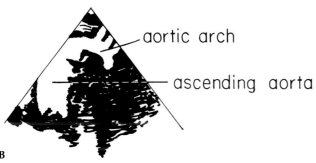

Figure 42-2 A suprasternal notch view of the ascending aorta and aortic arch of the same patient as in Figure 42-1. Two-dimensional echocardiography allowed the assessment that this dissection was limited to the ascending aorta by noting the dilatation of that segment, which ends abruptly at the aortic arch.

postductal type, which is usually very localized, appearing almost as a diaphragm within the aorta, with poststenotic dilatation immediately distal to the constriction. Direct echocardiographic visualization of coarctation has been reported by several groups, with extremely high success rates.[26-27] Echocardiographically one can see a discrete reduction in the diameter of the aortic lumen, and the area of coarctation may appear more reflective than the surrounding vascular wall because of the focal thickening in the affected region (see Figure 37-11 in Chapter 37). The severity of the coarctation can often be determined by the use of pulsed and continuous wave Doppler ultrasound.[28-29] Temporal changes in the blood flow velocity profile in the descending aorta will be noted, as compared to the profiles obtained proximal to the level of the coarctation. The peak systolic Doppler frequency shift correlates well with the gradient measured at catheterization, and the signal recorded throughout diastole reflects pan-diastolic flow across the coarctation.[29] Changes in the peak rate of acceleration, antegrade flow time, peak flow velocity, and peak rate of deceleration of aortic flow across the coarctation have been described,[30] and following surgery a return to normal flow velocity patterns has been seen.[31]

IDIOPATHIC DILATATION OF THE PULMONARY ARTERY

An uncommon entity, idiopathic dilatation of the pulmonary artery can be detected using echocardiographic techniques. Upper limits of normal size have been established for the pulmonary artery in adults,[2-3] and idiopathic dilatation of the main pulmonary artery would be suggested in a patient who has an enlarged pulmonary artery in the absence of other manifestations of pulmonary hypertension (right ventricular pressure and volume load, or abnormal Doppler flow velocity patterns consistent with pulmonary artery hypertension).

PULMONARY EMBOLISM

The majority of pulmonary emboli occur at the third-order or smaller branches of the pulmonary vascular bed and consequently are not visualized by any echocardiographic techniques. However, occasionally a patient will develop an occlusion at the main pulmonary artery or the bifurcation to the left and right arteries. Reports exist of the visualization of acute and chronic pulmonary thromboemboli in patients with proximal pulmonary thromboemboli and pulmonary hypertension.[37-38] One group was able to detect proximal thromboemboli in 10% of patients and the associated changes of right ventricular enlargement in 75%[37]; this has not been the experience of others, however.[39] The lack of echogenicity of acute thrombi, and the infrequent occurrence of associated acute pulmonary hypertension should make echocardiographic diagnoses uncommon. It should be emphasized, however, that failure to detect a pulmonary embolus by ultrasound in a patient with either acute or chronic proximal pulmonary thromboemboli should not be used as evidence against its existence.

ESTIMATION OF PULMONARY ARTERY PRESSURE BY DOPPLER TECHNIQUES

Noninvasive estimation of pulmonary artery hypertension has been performed by several different types of analyses of echo-Doppler findings (see Chapter 37). Four basic strategies have evolved, all of which have shown some promise in select groups of patients. They involve the utilization of systolic time intervals based on right-sided valve opening and closure from M-mode and Doppler recordings,[40-41] analysis of the ejection velocity profile across the pulmonic valve in the main pulmonary artery,[42-46] analysis of the regurgitant jet in patients who have tricuspid regurgitation from pulmonary hypertension,[47] and analysis of the regurgitant jet in patients who have pulmonic insufficiency and pulmonary hypertension.[48]

Right ventricular systolic time intervals are relatively easy to obtain using M-mode echocardiography. By measuring the right ventricular pre-ejection period and the right ventricular ejection time, one can assess changes in the ratio of these parameters, which appear to parallel the severity of pulmonary artery hypertension.[40-41] However, the utilization of systolic time intervals in the right ventricle is subject to the same problems regarding specificity and sensitivity that these measurements have when applied to the left ventricle (Figure 42-3).

When one looks at the systolic pulmonary artery flow velocity profile using pulsed or continuous Doppler techniques, it is apparent that patients with pulmonary hypertension develop a rapid acceleration and early deceleration in comparison to the normal pattern, which shows gradual acceleration and deceleration. The acceleration time index, defined as the ratio of the time interval from the beginning to the peak of ejection, has been shown to be inversely correlated with the mean pulmonary artery pressure. Combining conventional systolic time intervals with measurements of the systolic flow velocity profile has allowed better estimates of pulmonary artery pressure, predominantly in patients with congenital heart disease.[42-44] In these studies the ratio of the time to peak velocity to the right ventricular ejection time shows the best correlation with pulmonary artery systolic pressure (Figure 42-3).

Tricuspid regurgitation is common in patients with pulmonary hypertension; the velocity of the tricuspid

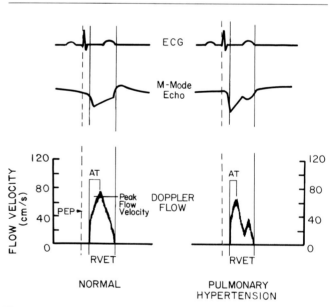

Figure 42-3 A graphic illustration of simultaneous electrocardiographic, M-mode echocardiographic, and Doppler flow tracings. The M-mode image shows pulmonic valve motion, and the Doppler demonstrates systolic flow across the pulmonic valve. Measurements of right ventricular ejection time (RVET) and the pre-ejection period (PEP) can be made from either modality. The systolic Doppler flow profile also allows measurement of peak flow velocity and the acceleration time (AT). Note that retrograde midsystolic pulmonic flow is apparent on the M-mode echocardiogram and Doppler in this patient with pulmonary hypertension. ECG, electrocardiogram.

systolic regurgitant jet is proportional to the systolic pressure within the right ventricle.[47] Therefore, continuous wave Doppler ultrasound recordings are made, and the velocity of the regurgitant jet is converted to a peak pressure gradient by applying the Bernoulli equation. The result is a pressure gradient between the right ventricle and right atrium. Estimating the right atrial pressure clinically and adding it to the Doppler-determined right-ventricular-to-right-atrial pressure gradient will provide an estimate of the actual systolic pulmonary artery pressure. This technique is particularly applicable to patients with more severe levels of pulmonary artery hypertension (Figure 42-4).

Continuous wave Doppler echocardiography has also been used to estimate pulmonary artery pressures by measuring pulmonary regurgitant flow velocity in patients with pulmonary hypertension.[48] In the face of pulmonary hypertension, the pulmonary regurgitant flow velocity pattern is characterized by a rapid rise in flow velocity immediately after closure of the pulmonic valve and a gradual deceleration until the next systole. As the level of pulmonary artery pressure increases, the pulmonary regurgitant flow velocity becomes higher, and estimates of the pulmonary-artery-to-right-ventricular pressure gradients can be made, again by applying the Bernoulli equation. In one study the Doppler-determined pressure gradient in end-diastole correlated well with measurements of the pulmonary diastolic pressure recorded at the same time (Figure 42-4).[48]

One limitation on the use of these techniques to diagnose pulmonary hypertension is that the sensitivity of the technique falls off with lower levels of pulmonary artery pressure.[46] Thus only the more severe cases are easily detected, and mild pulmonary hypertension, which is more treatable than severe cases, can easily be missed. In addition, it has not yet been established that the techniques possess enough specificity for small changes in pulmonary artery pressure to be reliably detected and to allow for following patients serially over time.

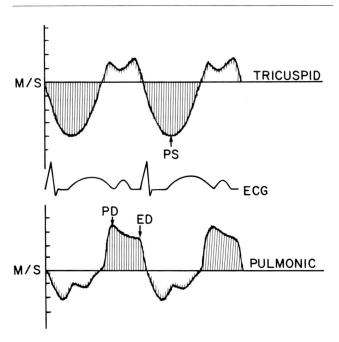

Figure 42-4 A graphic illustration of continuous wave Doppler flow velocity profiles obtained just proximal to the tricuspid and pulmonic valves in a patient with pulmonary hypertension. The peak systolic (PS) velocity of the tricuspid regurgitant jet is indicated, which can then be converted to a systolic pressure gradient between the right ventricle and right atrium. The peak diastolic (PD) and end-diastolic (ED) velocities of the pulmonary regurgitant jet are also indicated. These can be converted to pressure gradients between the pulmonary artery and right ventricle, with the peak diastolic gradient correlating with the mean pulmonary artery pressure and the end-diastolic gradient with pulmonary end-diastolic pressure. ECG, electrocardiogram.

REFERENCES

1. Gardin JM, Burn CS, Childs WJ, Henry WL. Evaluation of blood flow velocity in the ascending aorta and main pulmonary artery of normal subjects by Doppler echocardiography. *Am Heart J*. 1984; 107:310–319.

2. Wilson N, Goldberg ST, Dickinson DF, Scott O. Normal intracardiac and great artery blood velocity measurements by pulsed Doppler echocardiography. *Br Heart J*. 1985;53:451–458.

3. Gardin JM, Davidson DM, Rohan MK, et al. Relationship between age, body size, gender, and blood pressure and Doppler flow measurements in the aorta and pulmonary artery. *Am Heart J*. 1987;113:101–119.

4. DeMaria AN, Bommer W, Neumann BS, et al. Identification and localization of aneurysms of the ascending aorta by cross-sectional echocardiography. *Circulation*. 1979;59:755–761.

5. Weyman AE, Caldwell RL, Hurwitz RA, et al. Cross-sectional echocardiographic characterization of aortic obstruction, I: Supravalvular aortic stenosis and aortic hypoplasia. *Circulation*. 1978;57: 491–497.

6. Gardin JM, Henry WL, Savage DD, et al. Echocardiographic measurements in normal subjects: Evaluation of an adult population without clinically apparent heart disease. *J Clin Ultrasound*. 1979; 7:439–447.

7. McKusick VA. The cardiovascular aspects of Marfan's syndrome: A heritable disorder of connective tissue. *Circulation*. 1955;11: 321–342.

8. Roberts WC. Congenital cardiovascular abnormalities usually "silent" until adulthood. In: Roberts WC, ed. *Congenital Heart Disease in Adults*. Philadelphia: FA Davis;1979:405–453.

9. D'Cruz I, Cohen HC, Prabhu R, et al. Flutter of left ventricular structures in patients with pure aortic regurgitation, with special reference to patients with associated mitral stenosis. *Am Heart J*. 1976; 9:684–688.

10. Ciobanu M, Abbasi AS, Allen M, et al. Pulsed Doppler echocardiography in the diagnosis and estimation of severity of aortic insufficiency. *Am J Cardiol*. 1982;49:339–343.

11. Gott VL, Pyeritz RE, Magovern GJ Jr, et al. Surgical treatment of aneurysms of the ascending aorta in the Marfan syndrome. Results of composite graft repair in 50 patients. *N Engl J Med*. 1986;314: 1070–1074.

12. Sawyers JL, Adams JE, Scott HW. Surgical treatment for aneurysms of the aortic sinuses with aorticoatrial fistula. *Surgery.* 1957;41:26–32.

13. Boutefeu JM, Moret PR, Hahn C, et al. Aneurysms of the sinus of Valsalva. Report of seven cases and review of the literature. *Am J Med.* 1983;65:18–24.

14. Bulkley BH, Hutchins GM, Ross RS. Aortic sinus of Valsalva aneurysms simulating primary right-sided valvular heart disease. *Circulation.* 1975;52:696–699.

15. Kerber RE, Ridges JD, Kris JP, et al. Unruptured aneurysm of the sinus of Valsalva producing right ventricular outflow tract obstruction. *Am J Med.* 1972;53:775–783.

16. Onat A, Ersanli O, Kanui A, et al. Congenital aortic sinus aneurysms with particular reference to dissection of the interventricular septum. *Am Heart J.* 1966;72:158–163.

17. Yokoi K, Kambe T, Ichimiya S, et al. Ruptured aneurysm of the right sinus of Valsalva: Two pulsed Doppler echocardiographic studies. *J Clin Ultrasound.* 1981;9:505–510.

18. Haraoka S, Ueda M, Saito D, et al. Echocardiographic findings of a case of sinus of Valsalva aneurysm ruptured into left ventricle: Abnormal echoes in the left ventricular outflow tract. *J Cardiogr.* 1978;8:293–297.

19. Scagliotti D, Fisher EA, Deal BJ, et al. Congenital aneurysm of the left sinus of Valsalva with an aortopulmonary tunnel. *J Am Coll Cardiol.* 1986;7:443–445.

20. Mathew T, Nanda NC. Two-dimensional and Doppler echocardiographic evaluation of aortic aneurysm and dissection. *Am J Cardiol.* 1984;54:379–385.

21. Todini AR, Antignani PL. Doppler ultrasonic diagnosis of dissecting aneurysms of the aortic and great vessels. *Angiology.* 1985;36:884–888.

22. Mohri M, Nagata Y, Hisano R, et al. Detection of different blood flow patterns in the true and false lumina with aortic root dissection by pulsed Doppler echocardiography. *Clin Cardiol.* 1985;8:225–227.

23. Okamoto M, Kinoshita N, Miyatake K, et al. Detection and analysis of blood flow in aortic dissection with two-dimensional echo Doppler technique. *Ultrasound Med Biol.* 1983;2:331–335.

24. Goldman AP, Kotler MN, Scanlon MH, et al. The complementary role of magnetic resonance imaging, Doppler echocardiography, and computed tomography in the diagnosis of dissecting thoracic aneurysms. *Am Heart J.* 1986;111:970–981.

25. Wendel CH, Cornman CR, Dianzumba SB. Diagnosis of pseudoaneurysms of the ascending aorta by pulsed Doppler cross-sectional echocardiography. *Br Heart J.* 1985;53:567–570.

26. Shan DJ, Allen HD, McDonald G, et al. Real-time cross-sectional echocardiographic diagnosis of coarctation of the aorta: A prospective study of echocardiographic angiographic correlations. *Circulation.* 1977;56:762–767.

27. Weyman AE, Caldwell RL, Hurwitz RA, et al. Cross-sectional echocardiographic detection of aortic obstruction, II: Coarctation of the aorta. *Circulation.* 1978;57:498–502.

28. Wyse RK, Robinson PJ, Deanfield JE, et al. Use of continuous wave Doppler ultrasound velocimetry to assess the severity of coarctation of the aorta by measurement of aortic flow velocities. *Br Heart J.* 1984;52:278–283.

29. Houdley SD, Duster MC, Miller JF, et al. Pulsed Doppler study of a case of coarctation of the aorta: Demonstration of a continuous Doppler frequency shift. *Pediatr Cardiol.* 1986;6:275–277.

30. Shaddy RE, Snider AR, Silverman NH, et al. Pulsed Doppler findings in patients with coarctation of the aorta. *Circulation.* 1986;73:82–88.

31. Sanders SP, MacPherson D, Yeager SB. Temporal flow velocity profile in the descending aorta in coarctation. *J Am Coll Cardiol.* 1986;7:603–609.

32. Sahn DJ, Allen HD. Real-time cross-sectional echocardiographic imaging and measurement of the patent ductus arteriosus in infants and children. *Circulation.* 1978;58:343–349.

33. Cacciapuoti F, Varricchio M, D'Avino M, et al. Noninvasive evaluation of left-to-right shunts by pulsed Doppler echocardiography. *Int J Cardiol.* 1986;13:57–67.

34. Wilson N, Dickinson DR, Goldberg SJ, et al. Pulmonary artery velocity patterns in ductus arteriosus. *Br Heart J.* 1984;52:462–464.

35. Perez JE, Nordlicht SM, Geltman EM. Patent ductus arteriosus in adults: Diagnosis by suprasternal and parasternal pulsed Doppler echocardiography. *Am J Cardiol.* 1984;53:1473–1475.

36. Swensson RE, Valdes-Cruz LM, Sahn DJ, et al. Real-time Doppler color flow mapping for detection of patent ductus arteriosus. *J Am Coll Cardiol.* 1986;8:1105–1112.

37. Kasper W, Meinertz T, Henkel B, et al. Echocardiographic findings in patients with proved pulmonary embolism. *Am Heart J.* 1986;112:1284–1290.

38. Di Carlo LA, Schiller NB, Herfkens RL, et al. Noninvasive detection of proximal pulmonary artery thrombosis by two-dimensional echocardiography and computed tomography. *Am Heart J.* 1983;106:367–373.

39. Vardan S, Mookherjee S, Smulyan HS, Obeid AI. Echocardiography in pulmonary embolism. *Jpn Heart J.* 1983;24:67–70.

40. Stevenson JG, Kawabori I, Guntheroth WG. Noninvasive estimation of peak pulmonary artery pressure by M-mode echocardiography. *J Am Coll Cardiol.* 1984;5:1021–1027.

41. Riggs T, Hirschfield S, Burkat G, et al. Assessment of the pulmonary vascular bed by echocardiographic right ventricular systolic time intervals. *Circulation.* 1978;57:939–947.

42. Dabestani A, Mahan G, Gardin JM, et al. Evaluation of pulmonary artery pressure and resistance by pulsed Doppler echocardiography. *Am J Cardiol.* 1987;59:662–668.

43. Martin-Duran R, Larman M, Trugeda A, et al. Comparison of Doppler-determined elevated pulmonary arterial pressure with pressure measured at cardiac catheterization. *Am J Cardiol.* 1986;57:859–863.

44. Kitabatake A, Inoue M, Asao M, et al. Noninvasive evaluation of pulmonary hypertension by a pulsed Doppler technique. *Circulation.* 1983;68:302–309.

45. Hecht SR, Berger M, Berdoff RL, et al. Use of continuous-wave Doppler to evaluate and manage primary pulmonary hypertension. *Chest.* 1986;90:781–783.

46. Serwer GA, Cougle AG, Eckerd BM, et al. Factors affecting use of the Doppler-determined time from flow onset to maximal pulmonary artery velocity for measurement of pulmonary artery pressure in children. *Am J Cardiol.* 1986;58:352–356.

47. Berger M, Haimowitz A, Van Tosh A, et al. Quantitative assessment of pulmonary hypertension in patients with tricuspid regurgitation using continuous wave Doppler ultrasound. *J Am Coll Cardiol.* 1985;6:359–365.

48. Masuyama T, Kodama K, Kitabatake A, et al. Continuous-wave Doppler echocardiographic detection of pulmonary regurgitation and its application to noninvasive estimation of pulmonary artery pressure. *Circulation.* 1986;74:484–492.

Chapter 43

Evaluation of Diseases of the Great Vessels by Computed Tomography

Martin J. Lipton, MD, and Heber MacMahon, MD

Many forms of congenital and acquired diseases affect the great vessels, presenting a wide clinical spectrum for diagnosis. At one end of this spectrum, in adults, lie acute catastrophes, such as aortic dissection or pulmonary embolism, while at the opposite end are those asymptomatic cases with either normal or subtle findings. The chest x-ray film is usually the first noninvasive laboratory procedure obtained,[1–7] and it often provides information that prompts further studies. Disease of the mediastinum and lungs may encroach upon the great vessels, either compressing or invading them. The diagnostic issues in these patients include determining whether the disease process is primary or secondary in nature. Echocardiography is one of the most readily available noninvasive cardiac imaging techniques; however, its results in evaluating many patients in this particular category have been disappointing, mainly because the mediastinum is poorly echogenic. The majority of patients, therefore, are referred for angiography, which has proved to be the most definitive and reliable method for evaluating the aorta and the pulmonary arteries. Invasive angiography, however, is not without risk, is expensive, and usually requires patient hospitalization. Computed tomography (CT), on the other hand, can provide unique diagnostic information in both the adult and pediatric population, yet relatively few centers have experience in performing adequate numbers of cardiovascular CT studies.[8–11] Most hospital centers have a state-of-the-art conventional CT scanner, but only a few of these centers have been successful in combining radiologists, cardiologists, and physicists in a team approach to CT, which is so essential for obtaining optimal cardiovascular CT studies. This is unfortunate, as CT may preclude the need for invasive angiography. With the advent of faster CT machines that expand the applications of the technique, a closer look needs to be taken at how practical and successful this modality may be as a screening test. This chapter compares CT findings with those of other imaging modalities in patients with various diseases of the great vessels described in other chapters.

ADVANTAGES OF CT

Computed tomography is a significantly less invasive diagnostic procedure than angiography. Progressive refinements and advances in CT scanners have occurred since their introduction in 1974, with remarkable improvement in image resolution and display. Computed tomography can now acquire axial images and reconstruct from them image planes that resemble those of arteriographic projections, and this technique

---

Our gratitude is due to numerous colleagues in radiology and cardiology who made significant contributions to the CT program at University of California at San Francisco (UCSF), particularly Douglas P. Boyd, Ph.D., Professor of Radiology (Physics) at UCSF, and Robert Gould, Ph.D., Associate Professor of Radiology (Physics) at UCSF, for technical support and analytical expertise over many years. Some of the patients shown were studied in collaboration with Bruce Brundage, M.D.; Robert Herfkens, M.D.; and Charles Higgins, M.D.

can provide images comparable in diagnostic quality to those of angiography in many patients. Good results require attention to detail and CT should be thought of as a modified, minimally invasive angiographic procedure.

TECHNICAL CONSIDERATIONS AND ANATOMIC LOCALIZATION OF SCAN PLANES

Computed tomography involves the rotation of an x-ray source around the subject so that a thin fan beam of radiation is transmitted through the patient at precise anatomic levels, producing cross-sectional CT images, or CT scans. The transmitted radiation is detected as it leaves the patient by means of electronic detectors. This radiation is then converted from its analog form into digitized data. The CT computer processes and displays the data as a digital image on a cathode ray oscilloscope, using a very wide gray scale reference system. The exposure time of modern conventional whole body scanners is presently approximately 2 seconds, and this is sufficiently rapid to obtain good-quality diagnostic images of the great vessels and mediastinum. The contrast medium resolution of CT is of an order of magnitude greater than that of either fluoroscopy or x-ray film. This capability, combined with the unique cross-sectional display of CT, makes possible the imaging of vascular structures with low concentrations of contrast medium in the blood pool, achieved with relatively small peripheral intravenous injections of contrast medium.

The selection and identification of the correct anatomic levels for scanning is critically important in any cross-sectional imaging modality. An initial computed x-ray film or "scout view" overcomes this problem by registering each CT scan level with respect to the position of the CT table; an example of this localization technique is illustrated in Figure 43-1. This scan is usually obtained at the beginning of the procedure. The x-ray tube is kept stationary during this imaging time, while the patient is transported by the movement of the CT table through the pulsed fan beam of radiation. A computed digital x-ray film, illustrated in Figure 43-1, is thereby obtained in either the frontal, the lateral, or any oblique projection selected by the operator. Computed x-ray images have a wide dynamic range, which can be selected and manipulated by the operator and which allows any region to be interrogated by adjusting the CT level and window on the display monitor. A calibration system based upon the Hounsfield unit scale provides a precise density value for each pixel; the resulting image closely resembles a conventional chest x-ray film. The CT table position is then guided by computer so that each scan is registered precisely in memory at a given

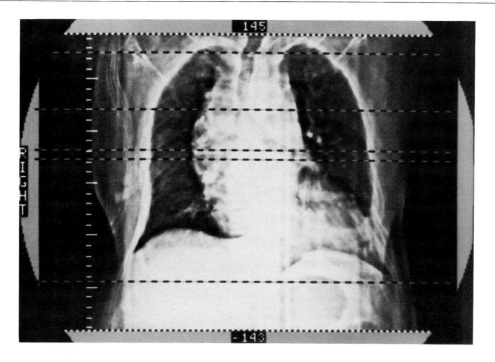

Figure 43-1 Computed chest x-ray film of a patient with a wide mediastinum. This image was obtained by advancing the table through the pulsed x-ray fan beam while the x-ray tube was kept stationary. Each pulse generates one line of the image. The digital processor constructs the image by the juxtaposition of lines. The technique ensures accurate CT slice registration for selecting the desired anatomic levels. Note the adjacent linear scale. The horizontal dotted lines, which are selected and can be varied by the operator, correspond to the scan levels to be exposed. This computed x-ray image has a wide dynamic range. Bones and soft tissues can be seen by changing the CT window and level at the console. *Source:* Reprinted from *Pediatric Cardiac Imaging* by WF Friedman and CB Higgins (Eds) with permission of WB Saunders Company, © 1984.

anatomic level and can be easily repeated as needed or used as a landmark to determine subsequent scan levels.

CONTRAST-ENHANCED STUDIES OF THE GREAT VESSELS

The usual scanning protocol for studying the great vessels is given in Table 43-1. An 18-gauge intracatheter (Deseret Corporation, Pittsburgh) is placed in an antecubital vein, and contrast medium is injected just prior to CT scanning. Contrast agents with high iodine content, such as Renografin 76, should be used for cardiac imaging, as they provide maximum opacification. Usually, a rapid bolus of approximately 0.3 to 0.5 mL/kg body weight is injected at a rate of 4 to 7 mL/s. A mechanical flow-controlled injector (Medrad Corp, Pittsburgh) is used in our laboratories to regulate the flow. This is recommended over manual injections as it provides precise contrast medium delivery, which not only permits reproducible studies for future reference but also standardizes the technique and in general greatly improves the utility of quantitative CT studies.

Following a bolus injection of contrast medium, CT scans are usually obtained in a rapid dynamic sequence. This type of scan acquisition monitors the flow of the contrast medium through the vascular structures, as shown in Figure 43-2. Quantitative data can be extracted from such CT images; this is an important and unique aspect of dynamic CT and is discussed below. In addition to this type of scan acquisition, volume imaging—the acquisition of several contiguous slices simultaneously without table movement—is also performed in most patients.[11]

Table 43-1 CT Scanning Protocol

1. Place 18-gauge intracath in either an antecubital vein or, occasionally, an external jugular vein.
2. Obtain computed x-ray image (scout view) for localization. (Figure 43-2.)
3. Use scout view to select anatomic levels through the aortic root and aortic arch and perform dynamic CT with bolus injections. Contrast agent containing high concentrations of iodine, such as Renografin 76, is injected at 5 mL/s (0.3 to 0.5 mL/kg body weight of contrast medium).
4. Next, drip infuse 50 to 80 mL of contrast medium and perform rapid contiguous scans of the chest, from lung apices to diaphragm—usually 20 to 30 scans, each 1 cm thick, are required. These images can be acquired during two or three breath holdings in most patients with conventional CT scanners.

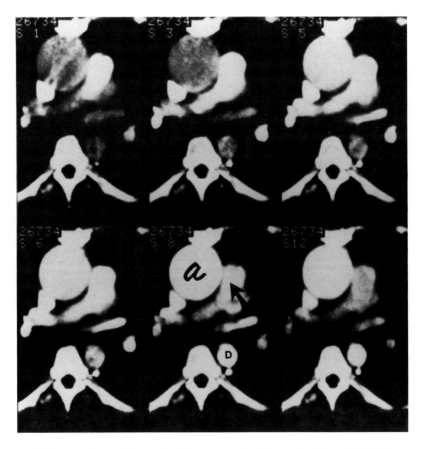

Figure 43-2 Dynamic sequence of CT scans through the aorta of the same patient shown in Figure 43-1, following a bolus injection of contrast medium into a peripheral vein. Note the progressive enhancement of first the right-sided structures and then the left. The ascending aorta is sharply demarcated and is grossly dilated. The descending aorta looks normal in size.

A CT study of the aorta can be completed in 20 to 30 minutes with most conventional scanners, using approximately 120 mL of contrast agent and acquiring 25 scans in 1-cm increments. A variety of lesions can be elegantly displayed using a combination of these dynamic CT and volume-imaging techniques. (Table 43-2 provides useful classifications of common lesions involving the aorta and pulmonary artery.) The acquisition of scans at adjacent anatomic levels not only provides a complete cross-sectional survey throughout the length of the aorta studied but also permits the operator to select any plane across any one of these axial images (as shown in Figure 43-3), which rapidly yields the reconstruction illustrated in Figure 43-4. This resembles projection radiographs routinely obtained at angiography.

Detection of Calcium

Aortic dissections can be displayed elegantly by contrast medium–enhanced CT, which characteristically defines the intimal flap, as shown in Figure 43-5. This dynamic sequence of scans at the same level illustrates the slower rate at which contrast medium–enhanced blood flows through the larger false lumen.[12–14] Time-density CT analysis can be performed with such sequences. The operator selects regions of interest over the structures of interest, as shown in Figure 43-6. The CT computer then computes and displays the data in graphic form as shown. When calcification is present in the region of the intimal flap, it can be seen inside the aortic lumen itself and is a diagnostic sign of dissection.[15–16]

Chronic rupture of the aorta may present with a localized area of curvilinear calcification, as shown in Figure 43-7A. Computed tomography scanning shows a horseshoe-shaped area in the aorta at the classic site of the ligamentum arteriosum, as shown in Figure 43-7B.

Table 43-2

Aortic lesions	
Congenital disease	coarctation and hypoplasia
	transposition
	double aortic arch
	intracardiac shunts
Acquired lesions	true aneurysms
	dissections
	rupture
Pulmonary artery lesions	atresia and hypoplasia
	pulmonary embolism
	compression and invasion
Mediastinal disease	masses (tumors, hematomas, cysts)

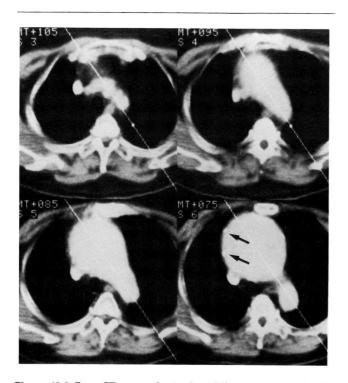

Figure 43-3 Four CT scans obtained at different anatomic levels through the aorta in the same patient as shown in Figures 43-1 and 43-2. Note the peripherally situated patchy calcifications (arrows) around the markedly dilated lumen of the ascending aorta. Note also that the whole aorta is opacified as one homogeneous structure. The oblique line indicates the plane selected for oblique reconstructions. *Source:* Reprinted from *Pediatric Cardiac Imaging* by WF Friedman and CB Higgins (Eds) with permission of WB Saunders Company, © 1984.

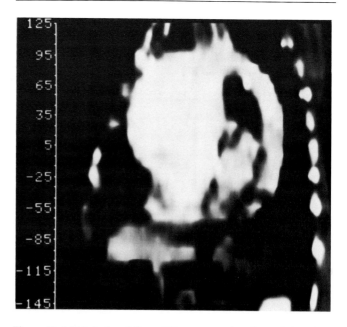

Figure 43-4 This is the oblique CT reconstruction through the plane indicated in Figure 43-3. This image illustrates the excellent demonstration of anatomic structures made possible by CT. Note that the grossly dilated ascending aorta resumes a normal diameter at the origin of the great vessels and appears to maintain its normal caliber beyond them. This patient had a true aneurysm of the aorta. *Source:* Reprinted from *Pediatric Cardiac Imaging* by WF Friedman and CB Higgins (Eds) with permission of WB Saunders Company, © 1984.

Evaluation of the Great Vessels by CT   541

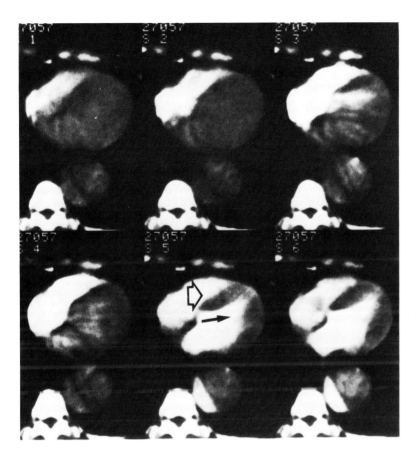

Figure 43-5 Dynamic CT sequence of scans at mid-ventricular level in a patient with hypertrophic cardiomyopathy. Note the thickened left ventricular (LV) free wall and septum, as well as the narrow and angular LV (solid arrow) chamber cavity. In addition, there is an intimal flap seen in the descending aorta in the last two lower right images, separating the compressed medial true lumen, which is well enhanced from the larger, slower filling, less-enhanced false lumen of a dissecting aneurysm. Open arrow indicates left ventricle.

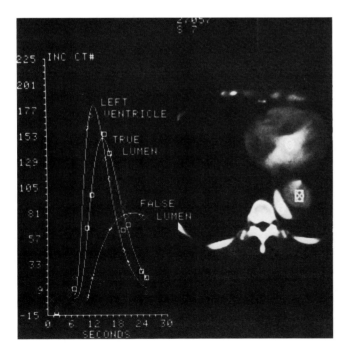

Figure 43-6 Three time-density curves for regions of interest selected by the operator over the left ventricular cavity, the true lumen, and the false lumen. Note that the false lumen curve peaks late and is of lower intensity than either of the others. This technique confirms the slow-flowing false lumen and the diagnosis of aortic dissection.

The lesion is best demonstrated in a coronal (Figure 43-7C) or sagittal (Figure 43-7D) reconstruction. Chronic vascular lesions often demonstrate regions of calcification at autopsy. Computed tomography is the best method for depicting calcification in the clinical setting, being vastly superior to both x-ray fluoroscopy and x-ray film screening combinations for this purpose. It is also superior to magnetic resonance imaging in this particular respect. This is because calcified plaques contain so few hydrogen protons that the magnetic resonance signal is weak; hence calcium looks black on spin echo magnetic resonance imaging and is more difficult to identify (depending on the weighting of the relaxation factors) since it lies adjacent to the signal void of the aortic lumen, which also appears black.

True aortic aneurysms can routinely be distinguished from dissection and chronic rupture of the aorta with contrast enhanced CT.[14] Calcification is typically curvilinear and localized to the borders of a true aneurysm, as illustrated in Figure 43-3.

QUANTITATIVE CT TECHNIQUES

The remarkably high spatial and density resolution of CT allows precise linear, area, and volume measure-

542 COMPARATIVE CARDIAC IMAGING

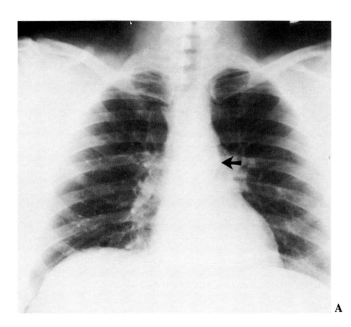

Figure 43-7 A, Chest x-ray film showing an area of curvilinear calcification (arrow) in the left mediastinum, possibly vascular in the aorta or a pulmonary artery. B, Series of contiguous 1 cm thick contrast medium–enhanced CT scans through the thorax of the patient shown at the top left showing the aortic arch above and below the aortic root and the left atrium. Note the dense, curvilinear, horseshoe-shaped anterior ring of calcification (arrows) in scans 8 through 11. C, Coronal reconstruction obtained from axial scans seen in B, showing a dense, localized area of calcification (arrowheads) in the aorta. D, Sagittal reconstruction through the aorta in same patient as in C. Note that the calcification (arrowheads) lies at the level of the ligamentum arteriosum, and there is a localized dilation of the vessel. The remainder of the aorta looks normal. This represents a chronic, calcified aortic rupture dating back to an automobile injury 18 years earlier.

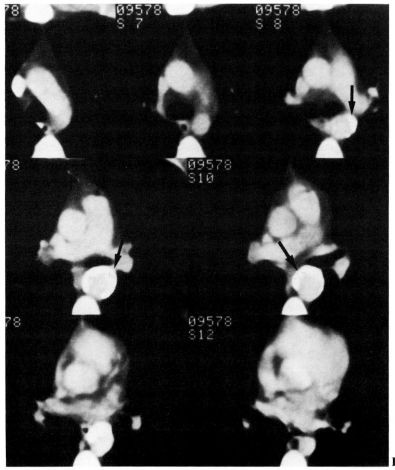

ment of vascular structures using operator-interactive computer software programs. Besides measuring dimensions, time-density measurements can also be obtained from contrast medium–enhanced diagnostic CT scanning, as illustrated in Figure 43-6. Computed tomography, therefore, not only can display the morphology of a lesion but also can evaluate the relative blood flow from routine dynamic CT studies. A combination of Doppler echocardiography, which provides the hemodynamic data, and CT may eliminate the need for many other procedures, both noninvasive and invasive.[17]

Figure 43-7 continued

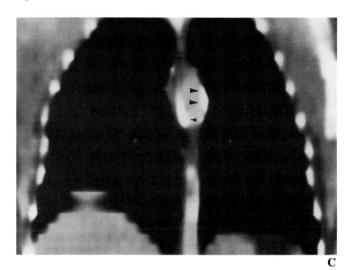

C

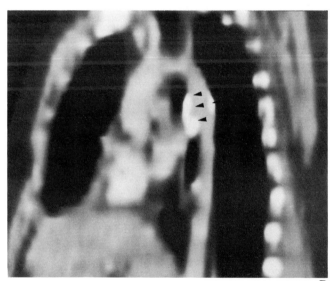

D

HIGH-SPEED CINE-CT SCANNING

The preliminary experience with conventional as well as electrocardiogram (ECG)-gated CT provided useful diagnostic images of the heart and great vessels and provided the stimulus to develop faster (millisecond) CT scanners. The Mayo clinic pioneered this early development with the dynamic spatial reconstructor (DSR).[17-18] This scanner electronically fires numerous x-ray tubes arranged around the patient to provide high-speed sequential exposures. This approach overcomes one of the major problems associated with conventional CT scanners, namely, the inertia associated with rotating a heavy x-ray tube.

Another more recent solution to this fundamental problem, inherent in all conventional CT scanners, is the incorporation of electron beam technology into the C-100 Ultrafast CT Scanner (Imatron, Inc., South San Francisco). The characteristics of this millisecond multi-

Table 43-3 Characteristics of the Imatron C-100 Cine-CT Scanner

1. Fifty meters per second exposure time
2. Multilevel scanning in pairs at two, four, six, or eight levels
3. Flexible repetition rate, up to 17 scans per second at each of two contiguous levels (total 34 scans)
4. Slice thickness options: 0.8, 0.5, or 0.3 mm
5. High-resolution mode, 100 m/s capability
6. Operator-interactive software employing a trackball-guided cursor and analysis programs for measuring distances, areas, and volumes, as well as time-density analyses for estimating blood flow
7. Multiformat options for obtaining projection images of cross-sectional scan data
8. New applications software for digital computed CT angiographic subtraction

level CT scanner are given in Table 43-3. The exposure time is 50 milliseconds for two simultaneous levels, and this scanner can acquire eight levels almost simultaneously, without moving the table. This not only reduces patient procedure time but also makes CT scan registration, patient breath holding, and cardiac motion far less troublesome. Figures 43-8 and 43-9 illustrate images obtained with this scanner.

Ultrafast CT, unlike conventional CT, can be used as a screening tool for cardiovascular studies and is also practical for acute disorders because the total scanning time is on the order of a few seconds and the total volume of contrast medium necessary is significantly reduced. Ultrafast CT provides excellent anatomic detail of the aortic and pulmonary valves and the coronary arteries, as shown in Figure 43-8. The ability to scan rapidly (in 10 to 25 seconds) the whole chest routinely in every patient also makes ultrafast CT a powerful screening and diagnostic tool for pulmonary as well as cardiac disease (Table 43-2). Furthermore, the modular design of this machine permits the addition of progressive improvements in image quality to be installed rapidly and at moderate cost.

Other important applications of high-speed scanning are with pediatric and trauma patients, where speed is crucial.[19] Fast exposures not only reduce overall procedure time, which often obviates the need for sedation in infants and avoids image degradation due to motion, but also facilitate improved diagnostic studies in restless and even uncooperative adults while minimizing the volume of contrast agent required. Such a case is illustrated in Figure 43-9. This patient was thought to have either some form of tracheal obstruction or a vascular ring. Contrast medium–enhanced high-speed multilevel CT scanning was obtained through the neck and mediastinum. The total procedure time was approximately 5 minutes, and no sedation was required (total CT scan acquisition time was actually less than 5 seconds). The CT images reveal normal anatomy of the aortic arch but also tracheostenosis. CT "movies" can also be acquired during respiration to determine whether a tracheal obstruction is of the fixed or dynamic

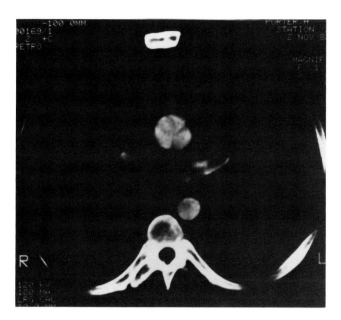

Figure 43-8 This is a conventional CT image obtained in 1.5 seconds. This scan is windowed to demonstrate the aortic valve. Note that all three leaflets can be identified and look symmetrical and normal. The contrast medium–enhanced descending aorta is also seen. Excellent images of the aortic valve are difficult to obtain with conventional nongated CT but are routinely obtained with ultrasfast cine-CT.

type. This latter application of CT (to visualize the airways) does not require contrast medium administration and is the only technique that can quantitatively evaluate the airways and also provide this degree of anatomic and functional detail.[19]

The ability of ultrafast CT to perform studies so rapidly that no form of ECG gating is necessary is a major advantage over radionuclide and magnetic resonance imaging techniques. Because scan acquisition is so fast, CT interventional studies with pharmacological agents and exercise are now possible. As ultrafast CT becomes more widely accepted and available, this capability should expand and should be more widely recommended and utilized.

SPECIFIC APPLICATIONS OF CT TO THE DIAGNOSIS OF SELECTED PULMONARY ARTERY PATHOLOGIES AND AORTIC COARCTATION

Pulmonary Embolism

A major diagnostic problem frequently encountered in adults is the diagnosis of pulmonary artery embo-

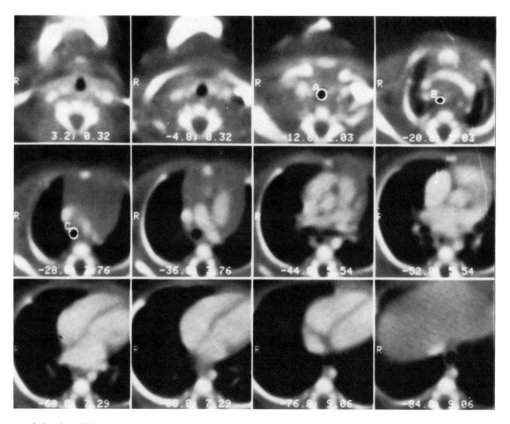

Figure 43-9 Sequence of ultrafast CT images in an infant with stridor. The clinical differential diagnosis was between a vascular ring or some form of airway obstruction. This contrast medium–enhanced CT study took only 5 minutes, and the actual scanning time was 5 seconds. The vascular anatomy, including the aortic arch, is seen to be normal. The trachea is well seen and has been outlined by the operator using an interactive computer software program at three levels. The CT computer also measures the area of the airway in the CT images, which are labeled A, B, and C. At level B the airway is significantly narrowed, measuring only 14 mm². This procedure does not require sedation because the exposures are so rapid. Movies at the same levels determine whether the tracheal stenosis is fixed or dynamic; this one was fixed.

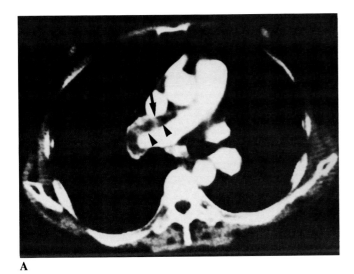

Figure 43-10 A, computed tomography scan through the level of the proximal right and main pulmonary arteries, showing an extensive filling defect (arrows) obstructing the right main pulmonary artery. Contrast medium–enhanced blood can be seen flowing anteriorly and laterally as a thin white line. Note that the left main pulmonary artery lies at a more cephalad level and cannot be evaluated on this image. B, Selective right pulmonary arteriogram confirming the extensive embolic thrombus (arrows) in the distal main and right lower lobe arteries.

lism. Large embolic obstructions can be readily seen with CT as filling defects in the main and proximal pulmonary vessels.[20] Thus CT scanning can be useful for the initial diagnosis of this entity and also for following the progress of treatment. The extent of utilization of the technique reflects preferences and also experience on the part of both the referring physician and the radiologist. Figure 43-10 illustrates the CT appearance of major chronic pulmonary embolism. Injection of contrast medium into a peripheral vein by bolus or rapid drip infusion provides a remarkably safe screening procedure for detecting this disorder. Additionally, the cardiac as well as other mediastinal and thoracic structures are automatically surveyed. In a similar way, invasion of the pulmonary artery by a mass lesion extending from either the lungs or the mediastinum can be detected and evaluated. As the quality of CT scanners continues to improve, it should be possible to identify smaller and more peripheral emboli.

Pulmonary Atresia and Intracardiac Shunts

One of the most important advantages of the use of CT to diagnose and treat congenital heart disease is its ability to demonstrate hypoplasia or atresia of the pulmonary arteries. An example of this is illustrated in Figure 43-11. Utilizing the CT data from each slice level, as described for acquired aortic lesions described above,

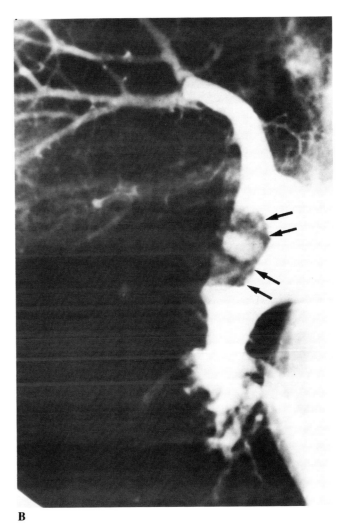

sagittal, oblique, or coronal images can be reconstructed by selecting the desired plane from any cross-sectional image (Figure 43-12). This approach displays the size and location of the great vessels and cardiac chambers as well as the abdominal situs; all these views are necessary for evaluating patients with congenital heart disease. The pulmonary arteries are typically affected in many left-to-right shunts. The size of these vessels can easily be measured from the CT images, and at the same time both the anatomic site and the shunt fraction may be estimated.[21] The advent of high-speed CT allows more scanning levels to be acquired during a given injection (40 levels are now possible within 20 seconds), making such studies safer, faster, and usually definitive.

Coarctation

An example of aortic coarctation is given in Figure 43-13. The need for initial evaluation by echocardiography in such patients may not diminish, but the excellent quality images of CT should reduce the need to resort to pre- and postoperative cardiac catheteriza-

546 COMPARATIVE CARDIAC IMAGING

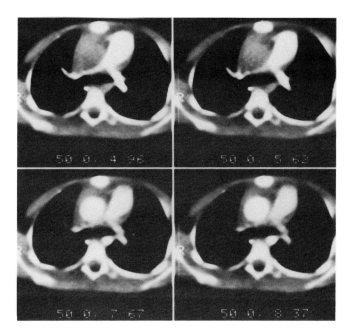

Figure 43-11 Four selected ultrafast (50-millisecond exposures) CT scans are illustrated from a dynamic series through the level of the pulmonary trunk. In the upper panels contrast medium enhancement is limited to the right side. In the lower panels the aorta is also opacified. Note the hypoplastic main pulmonary arteries and large ascending aorta and the small diameter of the descending aorta. Fast CT provides an excellent display of the anatomy in congenital heart disease; this patient had tetralogy of Fallot. *Source:* Courtesy of WJ Eldredge, Deborah Heart and Lung Center, Browns Mills, NJ.

tion procedures in the young and old. Computed tomography provides images of the pulmonic and aortic valves and may be useful in evaluating valvular stenosis. This, along with Doppler studies, is an excellent method to monitor the results of balloon valvuloplasty.

SUMMARY

Computed tomography scanning has yet to reach its full potential and acceptance as a tool for diagnosing cardiovascular disease. It provides simultaneous morphologic and quantitative data in a wide variety of diseases involving the great vessels, and can be used for initial diagnostic screening as well as interval follow-up with a greater degree of safety than invasive angiography.[10] Since multiple problems are often present in the same patient and since experience suggests that using one diagnostic modality, even echocardiography or angiography, seldom provides all the necessary cardiac information, CT scanning should be considered a useful complementary approach. It will in time be utilized with greater frequency and confidence by internists, cardiologists, and surgeons. As more high-speed CT scanners are installed, the opportunities to gain experience with this exciting new tool will expand.

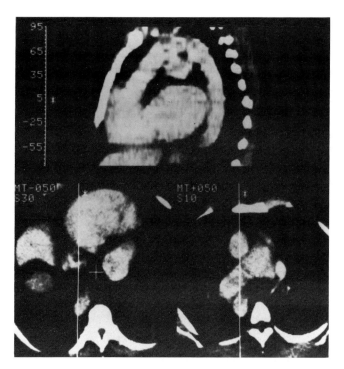

Figure 43-12 Transposition of the great vessels showing a parasagittal CT reconstruction with the aorta arising from the anterior ventricle and the aortic arch being typically unfolded, but otherwise normal. The vertical lines seen across the lower panels indicate the image plane selected from the axial images, from which the reconstructed image (above) was obtained. *Source:* Reprinted with permission from "Computed Tomography in Congenital Heart Disease" by DW Farmer, MJ Lipton, WR Webb et al in *Journal of Computer Assisted Tomography* (1984;8[4]:677–686), Copyright © 1984, Raven Press Ltd.

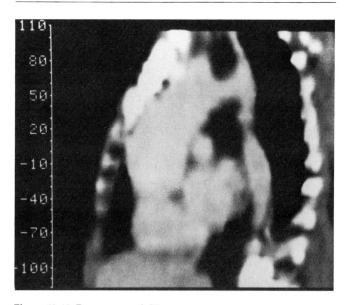

Figure 43-13 Reconstructed CT image in a sagittal plane demonstrating a coarctation of the aorta. This scan demonstrates the site of origin of the left subclavian artery, well above the coarctation. The size of the large ascending and smaller descending aorta is apparent, and precise measurements can be obtained in all planes of the vascular structures. *Source:* Reprinted with permission from "Computed Tomography in Congenital Heart Disease" by DW Farmer, MJ Lipton, WR Webb et al in *Journal of Computer Assisted Tomography* (1984;8[4]:677–686), Copyright © 1984, Raven Press Ltd.

REFERENCES

1. Woodring JH, Pulmano CM, Stevens RK. The right paratracheal stripe in blunt chest trauma. *Radiology*. 1982;143:605–608.

2. Sefczek DM, Sefczek RJ, Deeb ZL. Radiographic signs of acute traumatic rupture of the thoracic aorta. *AJR*. 1983;41:1259–1262.

3. Peters DR, Gamsu G. Displacement of the right paraspinous interface: Radiographic sign of acute traumatic rupture of the thoracic aorta. *Radiology*. 1980;134:599–603.

4. Tisnado J, Tsai FY, Als A, et al. A new radiographic sign of acute traumatic rupture of the thoracic aorta: Displacement of the nasogastric tube to the right. *Radiology*. 1977;125:603–608.

5. Gerlock AJ, Muhletaler CA, Coulam CM, et al. Traumatic aortic aneurysm: Validity of esophageal tube displacement sign. *AJR*. 1980;135:713–718.

6. Simeone JF, Minagi H, Putnam CE. Traumatic disruption of the thoracic aorta: Significance of the left apical extrapleural cap. *Radiology*. 1975;117:265–268.

7. Simeone JF, Deren MM, Cagle F. The value of the left apical cap in the diagnosis of aortic rupture. *Radiology*. 1981;139:35–37.

8. Lipton MJ, Boyd DP. Contrast media in dynamic computed tomography of the heart and great vessels. Proceedings of the CT International Workshop, Berlin. *Excerpta Medica*. 1981;204–213.

9. Lipton MJ, Higgins CB. Computed tomography: The technique and its use for the evaluation of cardiocirculatory anatomy and function. *Cardiol Clin*. 1983;1:457–471.

10. Farmer DW, Lipton MJ, Webb WR, et al. Computed tomography in congenital heart disease. *J Comput Assist Tomogr*. 1984;8:677–687.

11. Lipton MJ, Brundage BH, Higgins CB, et al. Clinical applications of dynamic computed tomography. *Prog Cardiovasc Dis*. 1986;28:349–366.

12. Godwin JD, Herfkens RJ, Skioldebrand CG, et al. Evaluation of dissections and aneurysms of the thoracic aorta by conventional and dynamic CT scanning. *Radiology*. 1980;136:125–133.

13. Turley K, Ullyot DJ, Godwin JD, et al. Repair of dissection of the thoracic aorta. Evaluation of false lumen utilizing computed tomography. *J Thorac Cardiovasc Surg*. 1981;81:61–68.

14. Skioldebrand CG, Brasch RC, Lipton MJ. Utility of computed tomography in the diagnosis of thoracic aneurysm in childhood. *Cardiovasc Intervent Radiol*. 1981;4:30–32.

15. Godwin JD, Turley K, Herfkens RJ, et al. Computed tomography for follow-up of chronic aortic dissections. *Radiology*. 1981;139:655–660.

16. White RD, Lipton MJ, Higgins CB, et al. Noninvasive evaluation of suspected thoracic aortic disease by contrast-enhanced computed tomography. *Am J Cardiol*. 1986;57:282–290.

17. Lipton MJ. Quantitation of cardiac function by cine-CT. *Radiol Clin North Am*. 1985;23:613–626.

18. Ritman EL, Robb RA, Johnson SA, et al. Quantitative imaging of the structure and function of the heart, lungs and circulation. *Mayo Clin Proc*. 1978;53:3–11.

19. Brasch RC, Gould RG, Gooding CA, et al. Upper airway obstruction in infants and children: evaluation with ultrafast computed tomography. *Radiology*. 1987;165:459–466.

20. Kereiakis DJ, Herfkens RJ, Brundage BH, et al. Computerized tomography in chronic thromboembolic pulmonary hypertension. *Am Heart J*. 1983;106:1432–1436.

21. Garrett JS, Jaschke W, Aherne T, et al. Quantitation of intracardiac shunts by cine-CT. *J Comput Assist Tomogr*. 1988;12:82–87.

Chapter 44

Evaluation of the Great Vessels by Magnetic Resonance Imaging

Joseph A. Utz, MD

Evaluation of the great vessels is an important part of cardiovascular imaging, to which magnetic resonance imaging has recently been applied. The lack of signal from flowing blood provides dramatic contrast between intraluminal blood and soft tissue, while the noninvasive nature of the examination and the ability to image in multiple planes provide added advantages over ultrasound, computed tomography (CT), and angiography. In this chapter, the application of magnetic resonance imaging to the evaluation of the aorta, pulmonary artery, and venae cavae will be described. The basic principles of vascular imaging with magnetic resonance imaging will be discussed, and examples of applications to the great vessels will be provided.

VASCULAR IMAGING WITH MAGNETIC RESONANCE

The physics of magnetic resonance has been well described and is beyond the scope of this chapter. A basic knowledge of the spin echo sequence as applied to multislice imaging is, however, important to understanding both the lack of signal often seen with flowing blood and the occasionally encountered flow artifact, which must be recognized and interpreted as flow-induced signal rather than as an anatomic abnormality. When the spin echo sequence is applied to the imaging of a stationary object such as the liver or a vessel wall, a slice-selective, 90° pulse is followed at the time TE(echo time)/2 by a 180° pulse that is also limited to the same anatomic position as the 90° pulse. At a time TE/2 later, the magnetic resonance signal is received from this same slice of tissue, and an image of that slice is reconstructed. With vascular imaging, however, while the vessel wall is stationary, the intravascular blood is in motion, and the volume of blood that receives the 90° pulse has often moved out of the imaging plane before the 180° pulse-signal reception sequence can be completed. Thus no signal is received from the blood, and a flow void is produced within the vascular lumen. This lack of signal is related to the velocity of blood flow and has been termed high-velocity signal loss. This is the most frequently encountered appearance of blood vessels on magnetic resonance images. Under conditions of slow laminar flow, however, the blood that flows out of the imaging plane is replaced by blood that has not yet been exposed to either 90° or 180° pulses. The unsaturated protons in this blood will emit a stronger signal than the partially saturated protons in adjacent stationary tissue. This will result in increased rather than decreased signal within the vessel lumen and has been termed flow-related enhancement. This is maximized when the velocity of flow is equal to the section thickness divided by the repetition rate. These two phenomena are illustrated in Figure 44-1. From the above discussion, it should be apparent that if high flow rates can cause signal loss and slow laminar flow can cause signal enhancement, intermediate flow patterns will cause intermediate levels of signal within the vascular lumen.

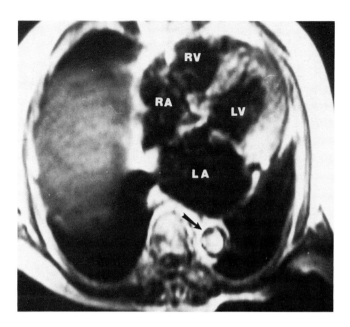

Figure 44-1 Gated spin echo magnetic resonance image at the level of the mitral valve demonstrates absence of signal due to flowing blood within the cardiac chambers (RA, RV, LV, LA). A marked increase in signal intensity (arrow) is noted in the descending thoracic aorta due to flow of unsaturated protons into the imaging plane. RA, right atrium; RV, right ventricle; LV, left ventricle; LA, left atrium.

In addition to the effects of blood velocity on the magnetic resonance signal intensity, the following must also be considered when evaluating vessels in magnetic resonance images: (1) the effects of turbulence, which causes signal loss; (2) phase shifts, which may cause alternate levels of decreasing and increasing signal in a multiecho sequence (even echo rephasing); (3) cardiac gating, which may cause an anatomic slice to be consistently imaged during diastole, resulting in relatively stagnant flow and thus increased signal; (4) the inter-relationship of the patient's heart rate and the repetition time of an ungated sequence, which, if multiples of each other, create an effect similar to cardiac gating and thus relatively increase signal on cardiac images obtained during relative diastole (diastolic pseudogating); and (5) the relationship of the direction of flow to the imaging plane and the superimposed gradients, which cause both phase and frequency shifts and thus spatial signal displacement. It thus becomes apparent that while, in the simplest case, flowing blood will produce a relative lack of signal and thus a flow void, multiple other factors will affect the appearance of blood on the image, and multiple combinations of signal intensity are encountered in clinical magnetic resonance imaging. An understanding of these factors is integral to the application of magnetic resonance imaging to the vascular system, and the interested reader is referred to several excellent review articles by Waluch and Bradley,[1] Bradley and colleagues,[2-3] Axel,[4] and Valk and co-workers.[5]

In addition to an understanding of flow signal, two practical techniques can be applied to magnetic resonance imaging of flow. The first of these is the use of a multiecho sequence, and the second is the application of multiple imaging planes. Since flow signal is an inter-relationship of flow velocity and the imaging sequence, the use of multiple echo sequences may result in flow signal that varies not only in intensity but also in distribution on separate images. Signals that vary in distribution from one sequence to the next are related to flow since an anatomic cause for signal changes should be constant in spatial distribution regardless of the pulse sequence utilized. This is illustrated in Figure 44-2, which demonstrates a patient with chronic pulmonary emboli and pulmonary hypertension. The first spin echo image demonstrates marked signal in the right pulmonary artery. The second spin echo image demonstrates persistent signal along the pulmonary artery wall due to adherent emboli, but a relative absence of signal within the vessel lumen. The lumen signal noted on the first spin echo image was due to slow flow.

The second technique that is frequently helpful is the application of multiple imaging planes in the evaluation of a vascular problem. Flow-related signal intensities are frequently affected by the relationship of flow to the imaging gradients and the direction of the magnetic field. By altering these inter-relationships by imaging a vessel in different planes (axial, sagittal, or coronal), different flow signals are often obtained, which cannot be explained by constant anatomic structures and must thus be related to flow.

The use of these two techniques, coupled with the information from the multiple characteristic signals from flowing blood, provides a foundation for vascular imaging with magnetic resonance. The remainder of this chapter will be devoted to the clinical application of magnetic resonance imaging to the aorta, pulmonary artery, and venae cavae in both congenital and acquired disorders.

AORTA

The evaluation of diseases of the aorta is frequently initiated by abnormalities seen on plain x-ray films of the chest. The initial clinical question often regards the presence of a vascular abnormality or a mediastinal mass and its relationships to the vessels. Magnetic resonance imaging, with its inherent soft tissue contrast and ability to visualize vessels in multiple planes, is often applied in this setting and will not be further discussed. The remainder of this section will deal with acquired and congenital diseases of the aorta.

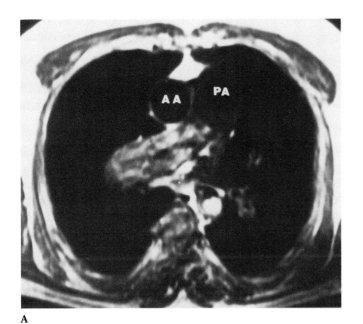 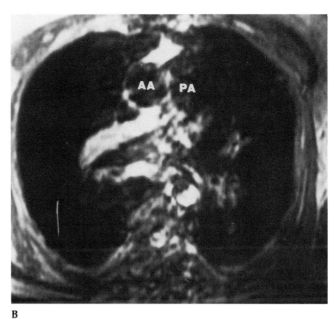

Figure 44-2 Gated spin echo images in a patient with chronic pulmonary emboli and pulmonary hypertension. The first spin echo image (left) demonstrates increased signal within the right pulmonary artery (arrow). The second spin echo image (right) demonstrates persistent signal along the vessel wall (arrow) due to adherent pulmonary emboli. There is a relative absence of signal in the vessel lumen. The signal in the vessel lumen on the first spin echo image is due to blood flow. AA, ascending aorta; PA, main pulmonary artery.

Aortic Dissection

The radiologic evaluation of suspected aortic dissection is directed first toward the demonstration of the suspected dissection and second to the definition of the proximal and distal extent of the false lumen. Both the absence of signal from flowing blood and the differences in signal intensities often obtained from blood flowing at different velocities assist in the detection of the intimal flap and false lumen that characterize aortic dissection (Figure 44-3). Contrast medium–enhanced CT studies have been used to evaluate possible aortic dissection. A bolus injection and dynamic scanning are utilized to detect differences in the arrival of contrast medium at a given location as a result of differences in the rate of blood flow between the true and false lumens. A disadvantage of this technique is the limited number of anatomic areas that can be examined as a result of the necessity of coordinating the anatomic image with the arrival of the contrast material. In contrast to bolus-enhanced CT examinations, the magnetic resonance study can produce good contrast between the intimal flap and the false lumen at all levels since timing the image to the arrival of the contrast medium bolus is not necessary.

A total of 31 patients with aortic dissection have been discussed in three recent articles evaluating magnetic resonance images and thoracic aortic disease.[6-8] In each of the three studies, the intimal flap was evaluated as well (or better) with magnetic resonance imaging as with contrast medium–enhanced CT examinations, and the proximal and distal extent of the dissection was accurately assessed when compared to CT or angiography. The classification of the dissection as either surgical (Stanford type A) or medical (Stanford type B) could thus be made, and extension of the dissection into the branch vessels could be detected. Advantages of magnetic resonance imaging included multiplanar presentations, which may assist in surgical planning, and lack of need for contrast material, which is especially important in patients with hypertension and impaired renal function. Disadvantages when compared to CT included the inability to image intimal calcifications, the displacement of which often assists in the detection of a dissection, and the occasional confusion of signal from intraluminal thrombus with signal from slowly flowing blood within the false lumen. A third consideration is the effect of the magnetic field on mechanical devices, which precludes evaluation of patients with cardiac pacemakers or patients in need of hemodynamic monitoring or respiratory assistance. When compared to angiography, a disadvantage shared by CT is the inability to assess the status of the aortic valve and coronary arteries, which is important in surgical planning for type A dissection repair.

Overall, magnetic resonance imaging is as sensitive as contrast medium–enhanced CT in evaluating patients with aortic dissection. The ability to image

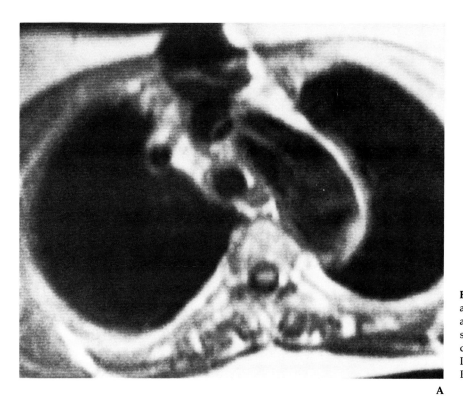

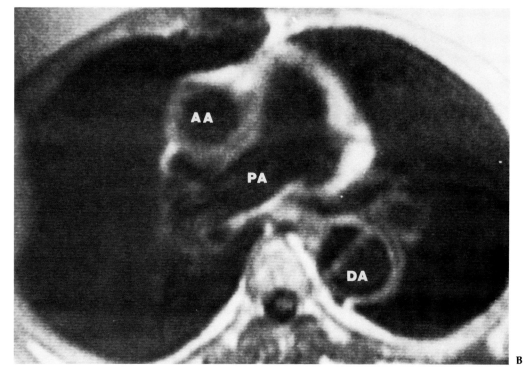

Figure 44-3 Magnetic resonance images at the aortic arch level (A) and at level of the ascending and descending aorta (B) demonstrate aortic dissection with involvement of the aortic arch and descending aorta (arrows). AA, ascending aorta; I, isthmus of aortic area; DA, descending aorta; PA, pulmonary artery.

without using contrast medium is advantageous in patients with impaired renal function and in those with type A dissection who may need an arteriogram to assess the aortic valve and coronary arteries preoperatively. Magnetic resonance imaging is less applicable to critically ill patients in need of life-support apparatus, who are better evaluated with contrast medium–enhanced CT.

Aortic Aneurysms

True[9] and false[10] aneurysms of the thoracic and abdominal aorta have been evaluated noninvasively with magnetic resonance imaging. The advantages include the noninvasive nature of the examination and the multiplanar imaging that it makes possible, which may better assess the relationship of the aneurysm to branch vessels such as the subclavian or renal arteries. Disadvantages again include the inability to image calcification and occasional confusion between signal from intraluminal thrombus and blood flow. Aneurysm size can be accurately quantitated. Like CT, the ability to accurately assess wall thickness as well as the constituents of the thickened wall offers advantages over angiography.

As mentioned above, confusion occasionally occurs in differentiating increased signal due to flowing blood from signal due to an acute thrombus; however, as previously discussed, the use of multiple imaging planes and two image sequences frequently assists in this differentiation. Organized thrombus is easily differentiated from intraluminal flow by the decrease in relative intensity on the second echo of the dual spin echo sequence, which is most often seen and which is felt to be due to fibrosis in the organizing thrombus.

A potential but unproved use of magnetic resonance imaging in the evaluation of aortic aneurysm is in the noninvasive evaluation of aortic wall constituents. The Dixon fat/water pulse sequence[11] has been utilized on an excised aorta.[12] The advantage of this sequence is its ability to separately image the fat and water contents of an aortic structure. In this manner, the aortic wall lipid deposition of early atherosclerotic plaque might be noninvasively imaged. Such images might then be used not only to follow the natural history of atherosclerosis but also to directly monitor the effects of medical or dietary therapy on atherosclerotic plaque formation. At the present time, magnetic resonance imaging is a useful noninvasive tool to evaluate aortic aneurysms with sensitivity and specificity equivalent to that of contrast medium–enhanced CT. The potential application of magnetic resonance imaging to the biochemistry of atherosclerosis has yet to be fully explored.

Developmental Abnormalities of the Aorta

The evaluation of developmental abnormalities of the aorta with magnetic resonance imaging has been described. The contrast induced by the signal void from flowing blood, coupled with the use of multiple imaging planes, has proved beneficial in the evaluation of aberrant subclavian arteries, right-sided aortic arch, and positional abnormalities of the great vessels such as transposition of the great vessels, truncus arteriosus, and corrected transposition of the great vessels.[13] In addition, the lack of ionizing radiation and the ability to conduct the study without contrast medium administration are especially beneficial in pediatric patients.

Coarctation of the aorta (Figure 44-4)[14] has been depicted with magnetic resonance imaging. The site of

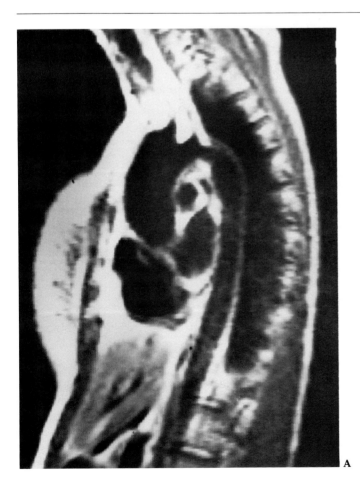

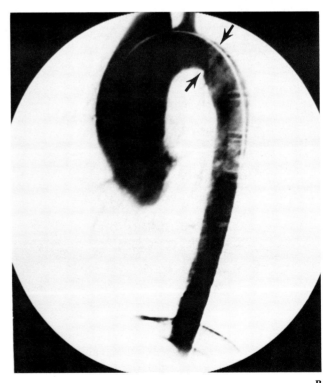

Figure 44-4 Coarctation of the aorta is demonstrated on a parasagittal magnetic resonance image in the left anterior oblique (LAO) projection (arrow) (A) and correlated with a digital angiogram (B, arrows).

the coarctation is best displayed with parasagittal images in the left anterior oblique projection. While the presence or absence of aortic narrowing can be determined accurately, differentiation between coarctation and pseudocoarctation is based on pressure differences and requires an invasive evaluation.

Arteritis

The previous discussions have described magnetic resonance imaging in the evaluation of congenital and acquired abnormalities of the aorta, and results have been favorably compared to more invasive imaging modalities such as contrast medium–enhanced CT and angiography. In inflammatory disorders of the aorta, however, such as Takayasu's arteritis, evaluation with magnetic resonance imaging has been less successful. In a study of 10 patients with Takayasu's arteritis, magnetic resonance imaging was compared to aortography.[15] False-positive results were obtained in three patients, and false-negative results in five. The majority of the areas of discrepancy occurred in the branch vessels such as the subclavian and carotid arteries and not in the aorta itself. The reason for this inadequacy may be the limited spatial resolution of magnetic resonance imaging compared to that of aortography, as well as the tortuous paths followed by these branch vessels, which complicates the tomographic imaging in any plane. Thus while magnetic resonance imaging is extremely useful in imaging the aorta itself, imaging of tortuous arteries may be less successful.

PULMONARY ARTERIES

The noninvasive evaluation of the pulmonary arterial circulation is extremely difficult. The proximity of the bones, sternum, ribs, and adjacent lung may compromise the ultrasonic evaluation, while the complex anatomy and similarity in densities between the pulmonary artery and mediastinal soft tissues often compromise the transaxial contrast medium–enhanced CT study. Gated spin echo magnetic resonance images have certain advantages in this evaluation. As noted previously, the lack of signal from flowing blood facilitates anatomic definition of pulmonary artery size, shape, and relationship to other structures, while the ability to image in a sagittal plane assists in direct evaluation of the pulmonic outflow tract. Complex congenital heart disease involving the pulmonary artery, such as transposition of the great vessels, tetralogy of Fallot, and truncus arteriosus, has been evaluated with magnetic resonance imaging to define the relationships of the pulmonary artery to the aorta and venous structures.[16] Congenital anomalies such as pulmonary artery atresia and stenosis have also been evaluated and the results obtained compared to those of CT. In a recent study of six patients with pulmonary artery hypoplasia or stenosis, magnetic resonance imaging was noted to be superior to echocardiography and angiography in the detection of pulmonary artery pathology.[17] In addition to the initial diagnosis, the postoperative evaluation in surgical interventions in the pulmonary artery is facilitated by the noninvasive nature of this examination (Figure 44-5).

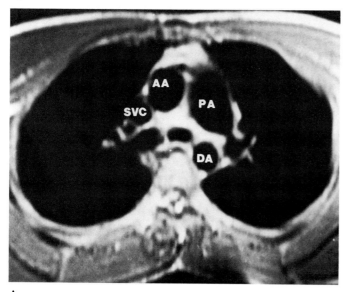
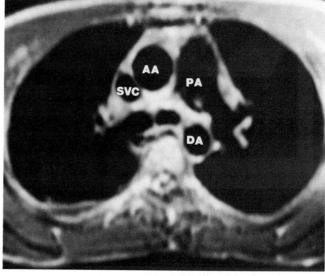

A B

Figure 44-5 Twenty-one-year-old male with atresia of the right pulmonary artery and stenosis of the left pulmonary artery (arrow), demonstrated on preoperative magnetic resonance examination (A). Postoperative examination (B) demonstrates decreased severity of arterial narrowing after surgery. AA, ascending aorta; PA, main pulmonary artery; DA, descending aorta; SVC, superior vena cava.

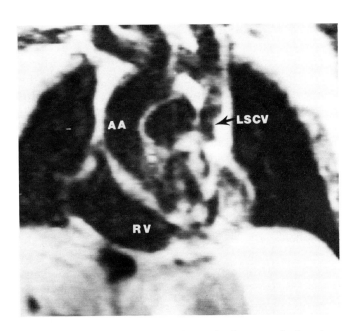

Figure 44-6 Magnetic resonance image in the coronal plane in a patient with a widened mediastinum demonstrates the persistence of a left-sided superior vena cava (LSVC). RV, right ventricle; AA, ascending aorta.

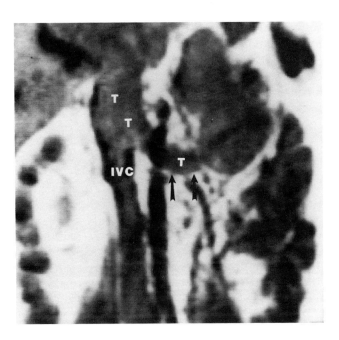

Figure 44-7 Coronal image in a patient with a large renal tumor demonstrates the extension of a tumor thrombus (T) into the left renal vein (arrow) and inferior vena cava (IVC).

In many of these patients, surgical intervention is dependent upon adequate evaluation of the distal pulmonary artery branches. As a result of stenosis or shunt lesion, adequate opacification of these vessels by cardiac catheterization may be impossible. Magnetic resonance imaging has potential application in these regions provided the stenosis is in the proximal pulmonary artery. Respiratory motion degrades evaluation of the intraparenchymal portions of the pulmonary artery, and the lack of signal from intrapulmonary air limits contrast due to the flow void within the vessel lumen. When the lesion is proximal, however, magnetic resonance images are often superior to those of cardiac catheterization and contrast medium–enhanced CT.

Postoperative evaluation with magnetic resonance imaging has been especially useful in patients with cyanotic congenital heart disease treated with palliative shunt procedures.[18] More important, perhaps, than the anatomic evaluation of the shunt is the potential use of magnetic resonance imaging to evaluate pulmonary vascular resistance. In one study of 15 patients with left-to-right shunts, the signal intensity within the pulmonary artery during systole was noted to increase in a linear fashion with pulmonary vascular resistance and with the ratio of pulmonary vascular resistance to systemic resistance.[19] This may potentially provide noninvasively obtained insight into the pathophysiologic consequences of pulmonary vascular disorders.

While the application of magnetic resonance imaging to the detection of pulmonary embolism in humans has been reported,[20] it has yet to be fully evaluated. A recent report of the detection of experimental pulmonary emboli in dogs,[21-22] however, suggests that magnetic resonance imaging may be of value in this area.

VENAE CAVAE

Congenitally acquired disorders of the venae cavae have been evaluated successfully with magnetic resonance imaging.

Congenital anomalies, such as persistent left superior vena cava (Figure 44-6),[13] azygous continuation of the inferior vena cava, duplicated inferior vena cava,[23] and renal vein anomalies, have been described. More significant, magnetic resonance has proved useful in the evaluation of vena caval obstruction[24] due to extrinsic mass or intrinsic tumor (Figure 44-7).[25-26]

SUMMARY

Magnetic resonance imaging is a useful, noninvasive modality for evaluation of the aorta, pulmonary artery, and venae cavae. High contrast is provided by flowing blood, and overall vascular assessment is facilitated by multiplanar imaging sequences. Intravascular signal, however, is complex, and signal intensity is related to turbulence and the interaction of the flow with the magnetic field. An understanding of this interaction, coupled with multiecho imaging, permits highly accurate evaluation of congenital and acquired vascular disorders without exposure to ionizing radiation or iodinated contrast material. Magnetic resonance imaging might be used as an initial screening modality when vascular abnormalities are suspected.

REFERENCES

1. Waluch V, Bradley WG. NMR even echo rephasing in slow laminar flow. *J Comput Assist Tomogr.* 1984;8:594–598.

2. Bradley WG, Waluch V. Blood flow: magnetic resonance imaging. *Radiology.* 1985;154:443–450.

3. Bradley WG, Waluch V, Lai KS, et al. The appearance of rapidly flowing blood on magnetic resonance images. *AJR.* 1984;143:1167–1174.

4. Axel L. Blood flow effects in magnetic resonance imaging. *AJR.* 1984;143:1157–1166.

5. Valk PE, Hale JD, Crooks LE, et al. MRI of blood flow: Correlation of image appearance with spin echo phase shift and signal intensity. *AJR.* 1986;146:931–939.

6. Amparo EG, Higgins CB, Hricak H, et al. Aortic dissection: Magnetic resonance imaging. *Radiology.* 1985;155:399–406.

7. Geisinger MA, Risius B, O'Donnel JA, et al. Thoracic aortic dissections: Magnetic resonance imaging. *Radiology.* 1985;155:407–412.

8. Glazer HS, Gutierrez FR, Levitt RG, et al. The thoracic aorta studied by MR imaging. *Radiology.* 1985;157:149–155.

9. Lee JKT, Ling D, Heiken JP, et al. Magnetic resonance imaging of abdominal aortic aneurysms. *AJR.* 1984;143:1197–1202.

10. Moore EH, Webb WR, Verrier ED, et al. MRI of chronic post-traumatic false aneurysms of the thoracic aorta. *AJR.* 1984;143:1195–1196.

11. Dixon WT. Simple proton spectroscopic imaging. *Radiology.* 1984;153:189–194.

12. Wesbey GE, Higgins CB, Hale JD, et al. Magnetic resonance applications in atherosclerotic vascular disease. *Cardiovasc Intervent Radiol.* 1986;8:342–350.

13. Higgins CB, Byrd BF, McNamara MT, et al. Magnetic resonance imaging of the heart: A review of the experience in 172 subjects. *Radiology.* 1985;155:671–679.

14. Ampara EG, Higgins CB, Shafton EP. Demonstration of coarctation of the aorta by magnetic resonance imaging. *AJR.* 1984;143:1192–1194.

15. Miller DL, Reinig WJ, Volkman DJ. Vascular imaging with MRI: Inadequacy of Takayasu's arteritis compared with angiography. *AJR.* 1986;146:949–954.

16. Formanek AG, Witcofski RL, D'Souza VJ, et al. MR imaging of the central pulmonary arterial tree in conotruncal malformation. *AJR.* 1986;147:1127–1131.

17. Fletcher BD, Jacobstein MD. MRI of congenital abnormalities of the great arteries. *AJR.* 1986;146:941–948.

18. Soulen RL, Donner RM. Magnetic resonance imaging of rerouted pulmonary blood flow. *Radiol Clin North Am.* 1985;23:737–744.

19. Didier D, Higgins CB. Estimation of pulmonary vascular resistance by MRI in patients with congenital cardiovascular shunt lesions. *AJR.* 1986;146:919–924.

20. Moore EH, Gamsu G, Webb WR, et al. Pulmonary embolus: Detection and follow-up using magnetic resonance. *Radiology.* 1984;153:471–472.

21. Stein MG, Crues JV, Bradley WG, et al. MR imaging of pulmonary emboli: An experimental study in dogs. *AJR.* 1986;147:1133–1137.

22. Gamsu G, Hirji M, Moore EH, et al. Experimental pulmonary emboli detected using magnetic resonance. *Radiology.* 1984;153:467–470.

23. Ellis JH, Denham JS, Bies JR, et al. Magnetic resonance imaging of systemic venous anomalies. *Comput Radiol.* 1986;10:15–22.

24. Weinreb JC, Mootz A, Cohen JM. MRI evaluation of mediastinal and thoracic inlet venous obstruction. *AJR.* 1986;146:679–684.

25. Bretan PN, Williams RD, Hricak H. Preoperative assessment of retroperitoneal pathology by magnetic resonance imaging. *Urology.* 1986;28:251–255.

26. Hricak H, Amparo E, Fisher MR, et al. Abdominal venous system: Assessment using MR. *Radiology.* 1985;156:415–422.

Part IX

Miscellaneous

Chapter 45

Comparative Safety of Cardiac Imaging Techniques

Christopher J. Wolfkiel, PhD

The physical basis of all imaging techniques is the quantization of radiation that is attenuated, reflected, or emitted by an organ. The principle of cardiac imaging by visualizing the spatial distribution of the attenuation of x-rays is the basis for cardiac angiography, digital subtraction angiography,[1] and computed tomography.[2] Reflection of ultrasonic radiation by the myocardium is the basis of echocardiography.[3] The measurement of radiation emitted from radioisotopes concentrated by the heart is the basis of nuclear cardiology planar imaging techniques, such as thallium 201 and technetium 99 imaging[4] and positron emission tomography imaging.[5] Magnetic resonance imaging techniques measure the concentration of a molecule by recording the radiofrequency signal obtained from the interaction of a nuclear magnetic resonance signal with a large, static magnetic field, smaller-gradient magnetic fields, and a radiofrequency electromagnetic field.[6] Thus the type and amount of radiation exposure are major considerations in safety comparisons of cardiac imaging techniques. This chapter will describe the types of radiation used in cardiac imaging techniques and identify the possible effects of patient exposure as a basis for safety comparisons.

Simply defined, radiation is the emission and propagation of energy through a medium. The types of radiation used in cardiac imaging are x-ray and gamma electromagnetic energy, ultrasonic energy, charged particles emitted from radionuclides, and low-frequency and static electromagnetic energy. Radiation can be described as ionizing or nonionizing based on its tissue effects. The interaction of radiation with tissue results in absorption of energy by the tissue. If enough energy is absorbed by tissue molecules, ionization and chemical damage can occur. The units of ionizing radiation dosage are rads (1 rad = 100 erg/g). Charged particles emitted during decay of radionuclides will also ionize surrounding tissue molecules and can be measured as an equivalent to a dose of x-ray radiation. However, radionuclides that emit charged particles have a greater probability of damaging tissue because of the increased mass of emitted particles (as compared to uncharged x-ray photons) and because such emissions continue as long as described by the half-life of the radionuclide. Radiobiologists characterize these differences by qualifying the linear energy transfer rate (LET) as low (x-rays) or high (beta particle emission).[7]

There are no general units of nonionizing radiation. Their effects are expressed in terms of power deposition in tissue, characterized by heating of tissue. Low-frequency electromagnetic radiation energy, such as radio waves, is part of the same spectrum as x-rays but cannot impart enough energy to ionize tissue molecules. Ultrasonic radiation is not part of the electromagnetic spectrum; it is a high-frequency, low-intensity, pressure-velocity wave. At high intensities, ultrasonic radiation can also result in tissue heating.

The biologic effect of radiation exposure is a function of the type and amount of exposure to the organism. The types of radiation used in cardiac imaging systems are summarized in Table 45-1. This review will describe the risks associated with radiation exposure from car-

Table 45-1 Types of Radiation Used in Cardiac Imaging

	Type	Modality
Ionizing	x-ray	angiography, digital subtraction angiography, computed tomography
	gamma ray	nuclear medicine
	radionuclide	nuclear medicine
Nonionizing	ultrasound	echocardiography
	magnetic fields	magnetic resonance imaging
	radio waves	magnetic resonance imaging

diac imaging systems as the main patient safety consideration. Procedural risks (i.e., of catheterization and contrast media) are discussed elsewhere in this book. Similarly, physician and technician radiation exposure are also important issues but will not be covered here.

IONIZING RADIATION IMAGING SYSTEMS

Ionization of a molecule can result from a transfer of energy during a collision with radiated particles such as x-ray photons, electrons, or positrons. Ionization can affect molecules, cells, organs, and organisms. These effects are described as either direct or indirect. Direct effects of ionizing radiation exposure are due to high-LET radiations that impart enough energy for chemical alterations of important organic molecules. Indirect effects of radiation exposure are due to the chain of events that can occur from either low- or high-LET radiation. For example, electron energization from a photon can lead to formation of a free radical molecule that can break other chemical bonds and disrupt cellular processes.

The exact mechanisms that lead from chemical damage to cellular and organ damage are not known. Because the most important effect of low-level ionizing radiation exposure is carcinogenesis,[8] the most dangerous chemical changes probably occur in the cell nucleus. Applying radiation to only the cell nucleus has been shown to break and malform chromosomes.[9] It is theorized that the newly formed "ends" of the broken chromosomes have a great affinity for one another. This can lead to three possible outcomes:

1. Repair: The ends of the broken chromosomes rejoin correctly, and no damage is detectable.
2. Rejoining: Incorrect ends rejoin, resulting in nonfunctional chromosomes.
3. No repair: The broken chromosomes never rejoin, resulting in a complete loss of genetic information.

The probabilities of the above occurrences in relation to the amount of incident radiation are not known. Thus the outcome of exposure to low-level radiation is a complete series of probabilities relating to chromosomal damage and repair, the number of cells affected as a function of the amount of incident radiation, and organ radiation sensitivity.

The effects of low-level irradiation have been described by retroactive epidemiological studies of cancer death estimates in selected populations, such as atomic bomb survivors and early radiation industry workers.[10] A precise relation between low-level ionizing radiation exposure and cancer occurrence has not been determined by these methods, for three reasons.[11] First, the cancer effect due to radiation exposure cannot be separated from natural and environmental cancer rates without very large, unbiased control groups. Second, the exact radiation exposure to these populations cannot be accurately measured. Third, there is a large latent period between exposure and cancer appearance. There is also controversy as to whether there is an age-sensitive component to low-level radiation cancer induction.[12]

In general, most estimates of cancer rates due to radiation exposure are from models that extrapolate effects from higher exposures than those encountered in imaging techniques.[13] As tabulated by the Biological Effects of Ionizing Radiation (BEIR) report,[10] estimates of the additional cancer deaths due to 10 rad of whole body exposure could vary from 0.5% to 3.1% over an expected rate of 163,800 deaths per million. An exact determination of the additional cancer rates from cardiac imaging systems is difficult to assess because the exact whole body radiation dose cannot be estimated and used as a comparison to the controversial epidemiological data.

Comparison of radiation exposure between imaging technologies is dependent on many factors: volume of tissue irradiated, duration of exposure, and type of radiation. Examples of radiation exposure by cardiac imaging modalities utilizing ionizing radiation follow.

Angiography

Radiation exposure during an angiographic procedure occurs during cine film exposure and catheter placement during fluoroscopy. The field of exposure is severely shielded, limiting exposure to a volume of space that includes the chest area. Thus whole body doses are usually not calculated and extrapolation of skin entrance doses to whole body doses is imprecise. In a study of routine angiographies, the whole body dose was calculated as 0.9 rad for a skin entrance dose of 36.1 rad.[14] Other studies have estimated the skin entrance dose from 58[12] to 28 rad.[15] More sophisticated procedures such as angioplasty can deliver four times the diagnostic catheterization exposure due to the increased fluoroscopy time.[16] Improved technology and digital subtraction angiography can reduce patient dosage. Reduction in the fluoroscopy rate from 60 to 30 pulses per second has been shown to reduce the dosage by 50%.[17] Digital subtraction angiography has been reported to markedly reduce the cine exposure of pediatric catheterization exams.[18]

Computed Tomography

Computed tomography confines radiation exposure to a series of cross-sections defined by patient position in the scanner. The amount of tissue receiving radiation is greater than the volume of tissue imaged due to scatter. Thus for scanning of adjacent levels, the smaller the distance between levels, the greater the cumulative dose.[19] The dose from a single 50-millisecond exposure of an Imatron C-100 ultrafast computed tomography scanner is 0.35 rad.[20] As an example, the radiation distribution of an eight-level series from four scans by the Imatron C-100 scanner is given (Figure 45-1). Time-repeated doses sum the effective dose. Table 45-2 summarizes the estimated entrance doses from cardiac computed tomography procedures using an Imatron C-100 ultrafast scanner.

Nuclear Cardiology

The radiation source used in nuclear cardiology procedures is a minute amount of a radiopharmaceutical chosen for its gamma energy emission or myocardial uptake properties.[21] A gamma camera counts the emitted photons produced during this agent's radioactive decay (or photons produced by interaction of tissue and emitted particles) and forms an image representing the distribution of the isotope. The patient radiation dosage is proportional to the activity given, the half-life of the isotope, and the rate of isotope expulsion from the body.[22] The organ dose is calculated from the fraction of the isotope in the organ. The estimated whole body dose is 0.22 rad/mCi for thallium 201 and 0.13 rad/mCi for technetium 99m.[23] Positron emission studies with ultrashort half-life isotopes, such as rubidium 82, impart a much lower dose, 0.0016 rad/mCi.[24]

NONIONIZING RADIATION IMAGING SYSTEMS

Two major types of nonionizing radiation used in cardiac imaging will be discussed: ultrasound and low-frequency electromagnetic radiation. These radiations are not generally associated with the popular definition of radiation. However, there are safety considerations associated with them. The two cardiac imaging modalities using these radiations are echocardiography and cardiac magnetic resonance imaging.

Echocardiography

Ultrasonic images of the heart are produced by an array of piezoelectric crystals that emit and receive reflected high-frequency sound waves. The emitted sound waves are propagated through local oscillations of velocity and pressure. These effects can have a negative effect on tissue. However, it is well established for imaging frequencies (0.1 to 5 MHz); the intensity used is lower, by a factor of 100, for any known health effects.[25] There are three effects of exposure to high-intensity ultrasonic radiation: thermal, cavitation, and direct. Thermal heating is produced by the transformation of sound waves energy into molecular vibration energy. The temperature increase is a complex function of applied power, duration, flow rates, and thermal conductivity. Cavitation, the second effect, is a process of microbubble formation; the size of the bubbles is a function of frequency and duration of exposure. Because of the pulsatile nature of echocardiography, it is assumed that microbubble formation is unlikely. The third possible effect of high-intensity ultrasonic radiation is similar to the direct effects of high-LET electromagnetic radiation. Theoretically, mechanical disrup-

Table 45-2 Computed Tomography Dosage

Procedure	No. Levels	No. Times	Exposure (rad)
Flow study	8	10	10
Wall motion (2 × 6 level)	12	12	18

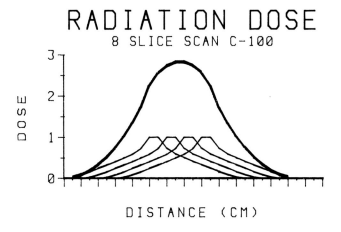

Figure 45-1 Radiation dose profile of an eight-slice computed tomography scan. The dose units are relative to the maximum skin dose of a single exposure. The radiation dose of a single exposure is a function of the electron gun current. An approximate dose for a single exposure is 0.35 rad. Thus for an eight-slice scan the maximum delivered dose is estimated at 1 rad.

tion of chemical bonds can cause the same effect as ionizing radiations (i.e., breakage, misrepair). However, no evidence has been produced to show that any of these effects are produced by cardiac imaging procedures.

Magnetic Resonance Imaging

Magnetic resonance images are produced by the interaction of tissue with a large, static magnetic field, a time-varying magnetic field, and a pulsed radiofrequency field. The health effects of these fields are currently under study; they are summarized by Budinger.[26] Large, static magnetic fields have been associated with electrocardiogram changes due to induced electromotive forces on flowing blood. However, no dysrhythmias or heart rate changes have been produced with imaging systems and hearts without pacemakers. Magnetic resonance scanning of patients with implanted pacemakers is potentially very dangerous due to pacemaker inactivation or rate induction by the magnetic field.[27] Changes in nerve tissue conductivity have been theorized to occur in very large fields (24 T). Rapidly changing magnetic fields can cause induction of internal electric currents. The current density induced is a complex function of waveform amplitude and frequency. Possible physiologic current induction effects include membrane permeability changes, visual phosphene production (light flashes), and, at very high induced currents, heart fibrillation.

The known biologic effect of radiofrequency electromagnetic radiation is tissue heating, which is proportional to incident power density, frequency, and size of the subject. A proposed safety limit of theoretical heating is the basal metabolic heat rate (1.5 W/kg) for exposures greater than 20 minutes. Magnetic resonance imaging of prosthetic heart valves has been reported not to produce significant heating.[28]

SUMMARY

All imaging technologies have the theoretical potential to cause biologic harm. Because these effects are extremely unlikely at the radiation dosages used, accurate probabilistic models are not available for comparison within a given imaging modality or between modalities. Ultrasonic radiation exposure is generally assumed to be the safest cardiac imaging technique. Thus repeated exposure to ultrasonic radiation offers little potential harm. Repeated exposure to x-ray and gamma radiation increases the cumulative risk of carcinogenesis and thus should be minimized, especially in pediatric patients. The long-term effects of exposure to electromagnetic fields during magnetic resonance imaging have not been established. As in all diagnostic tests, the risk-to-benefit ratio of a given procedure should be properly evaluated.

REFERENCES

1. Kruger RA, Rieder SJ. *Basic Concepts of Digital Subtraction Angiography*. Boston: GK Hall; 1984:13–39.

2. Zatz LM. Basic principles of computed tomography scanning. In: Newton TH, Potts DG, eds. *Radiology of the Skull and Brain: Technical Aspects of Computed Tomography*. St Louis: CV Mosby; 1981:3853–3876.

3. Geiser EA, Oliver LH. Echocardiography: Physics and instrumentation. In: Collins SM, Skorton DJ, eds. *Cardiac Imaging and Image Processing*. New York: McGraw-Hill Book Co; 1986:3–23.

4. Budinger TF, Rollo FD. Physics and instrumentation. *Prog Cardiovasc Dis*. 1972;20:19–53.

5. Correia JA, Alpert NM. Positron emission tomography in cardiology. *Radiol Clin North Am*. 1985;23:783–793.

6. Lauterbur PC. Image formation by induced local interactions: Examples employing nuclear magnetic resonance. *Nature*. 1973;242:190–191.

7. Hall EJ. *Radiobiology for the Radiobiologist*. Hagerstown, MD: Harper & Row; 1978:93–110.

8. Upton AC. The biologic effects of low-level radiation. *Sci Am*. 1982;246:41–49.

9. Munro TR. The relative radiosensitivity of the nucleus and cytoplasm of Chinese hamster fibroblasts. *Radiat Res*. 1970;42:451–470.

10. Report of the Committee on the Biological Effects of Ionizing Radiations (BEIR III). The effects of population exposure to low-levels of ionizing radiation. Washington, D.C.: National Academy Press; 1980:150–157.

11. Fabrikant JI. Estimation of risk of cancer induction in populations exposed to low-level radiation. *Invest Radiol*. 1982;17:342–349.

12. Gofman JW, O'Connor E. *X-Rays: Health Effects of Common Exams*. San Francisco: Sierra Club Books; 1985:25.

13. Whalen JP, Balter S. *Radiation Risks in Medical Imaging*. Chicago: Year Book Medical Publishers; 1984:33–53.

14. Vogel H, Westerhold R, Lohr H, et al. Radiation exposure and radiation risk in angiocardiography in adults. *Herz*. 1984;9:313–318.

15. Reuter FG. Physician and patient exposure during cardiac catheterization. *Circulation*. 1978;58:134–139.

16. Finci L, Meier B, Steffenino G, Roy P, Rutishauser W. Radiation exposure during diagnostic catheterization and single and double vessel percutaneous transluminal coronary angioplasty. *Am J Cardiol*. 1987;60:1401–1403.

17. Wondrow MA, Bove AA, Holmes DR, et al. Technical considerations for a new x-ray progressive scanning system for cardiac catheterization. *Cathet Cardiovasc Diagn*. 1988;14:126–134.

18. Levin AR, Goldberg HL, Borer JS, et al. Digital angiography in the pediatric patient with congenital heart disease: Comparison with standard methods. *Circulation*. 1983;68:374–384.

19. Pentlow KS. Dosimetry in computed tomography. In: Newton TH, Potts DG, eds. *Radiology of the Skull and Brain Technical Description of Computed Tomography*. St Louis: CV Mosby; 1982:4228–4258.

20. Chang W, Franken EA. Performance evaluation of the Imatron Cine-CT. *Radiology*. 1986;157:117.

21. Miller DD, Elmaleh DR, McKusick KA, et al. Radiopharmaceuticals for cardiac imaging. *Radiol Clin North Am*. 1985;23:765–781.

22. Roedler HD. Radiation dosimetry. In: Kristensen K, Norbygaard E, eds. *Safety and Efficacy of Radiopharmaceuticals*. Martin Nuhoff Publishers; 1984:158–178.

23. Syed IB, Flowers N, Granlith D, et al. Radiation exposure in nuclear cardiology studies. *Health Physics*. 1982;42:159–163.

24. Kearbott J. Radiation absorbed dose estimation for positron emission tomography: K-38, Rb-81, Rb-82, Cs-130. *J Nucl Med*. 1982;23:1128–1132.

25. Baker ML, Dabnymple GV. The biological effects of diagnostic ultrasound: A review. *Radiology*. 1978;126:479–483.

26. Budinger TF. Nuclear magnetic resonance (NMR) in vivo studies: Known thresholds for health effects. *J Comput Assist Tomogr*. 1981;5:800–811.

27. Erlebacher JA, Cahill PT, Pannizzo F, Knowles RJ. Effect of magnetic resonance imaging on DDD pacemakers. *Am J Cardiol*. 1986;57:437–440.

28. Hassler M, Le Bas JF, Wolf JE, Contamin C, Waksmann B, Coulomb M. Effects of the magnetic field in magnetic resonance imaging on 15 tested cardiac valve prostheses. *J Radiol*. 1986;67:661–666.

Index

A

Absent pericardium, 437
  computed tomography, 453
  echocardiography, 448
  nuclear magnetic resonance, 462
Acoustic impedance, in ultrasonic tissue characterization, 404
  mismatch, 404 (Fig. 31-1)
Acoustic properties, normal myocardium, 407
Acquired pulmonic regurgitation, 289
Acquired pulmonic stenosis, 289
Acquired valvular disease, nuclear magnetic resonance, 348–351
Acute myocardial infarction
  anterior wall, 224 (Fig. 14-1)
  computed tomography, 223–228
  echocardiography, 227–228
  enzyme method, 227
  infarct avid scintigraphy, 167–176
  nuclear magnetic resonance, 228
  positron emission tomography (PET), 227
  posterolateral wall infarct, 226
  prognosis after, 219
  radioisotope imaging, 227
Acute myocardial infarctions, clinical sequelae, 264–265
Algorithms, planimetric assessment of topographic images, 139 (Fig. 10-21)
Amplatz coronary catheters, 75, 76 (Fig. 6-19)
Amyloid heart disease, 155
Amyloidosis
  primary, 173
  restrictive cardiomyopathy and, 381, 411

Analog-to-digital conversion, 50 (Fig. 5-3)
Anaphylactoid reactions, contrast media, 97–98
Aneurysm, 149 (Fig. 10-40)
  aortic trauma and, 521
  arteriosclerotic aneurysms, 522–523
  mutuality with thrombus, 251–252
  and myocardial infarction, 149
  See also Aortic aneurysms.
Aneurysmectomy, and coronary artery disease, 112
Angina, thallium 201 and, 187, 196–197
Angiography
  cardiac tumors, 506
  constrictive pericarditis, 435
  safety aspects, 558–559
  See also Digital angiography.
Anteroseptal infarction, 148 (Fig. 10-36)
Anthracycline cardiotoxicity, imaging techniques, 378
Antimyosin antibody, 25
Aortic aneurysms
  angiography, 522–523
  computed tomography, 541
  Doppler echocardiography, 531–532
  echocardiography, 531
  magnetic resonance imaging, 552–553
Aortic arch, anatomy of, 518
Aortic dissection, 523–524
  angiography, 523–524
  aortography, 524
  classification system, 524
  computed tomography, 540
  Doppler echocardiography, 532
  echocardiography, 532
  etiology of, 523–524
  magnetic resonance imaging, 551–552

  pathogenesis, 524
Aortic insufficiency, MRI, 351 (Fig. 25-5)
Aortic lesions, ultrafast computed tomography, 343–344
Aortic regurgitation, 297, 298, 299, 331 (Fig. 23-2), 332 (Figs. 23-3, 23-4)
  aortograms, 284, 286
  bacterial endocarditis and, 343
  causes of, 285
  continuous wave Doppler, 321 (Fig. 22-6)
  diseases related to, 285
  Doppler echocardiography, 318–320
  fluoroscopy, 285
  quantitation of, 319 (Table 22-1)
  radionuclide angiography, 331–334
  ultrafast computed tomography, 343
  ventriculograms, 285–286
Aortic root, functions of, 145
Aortic root digital subtraction angiography,
  coronary artery bypass graft evaluation, 272
Aortic root injection, coronary angiography, 61
Aortic root motion, M-Mode image, 138 (Figs. 10-18, 10-19), 146 (Fig. 10-32)
Aortic stenosis, 331 (Fig. 23-1)
  aortograms, 285
  Doppler echocardiography, 315–318
  fluoroscopy, 284
  radionuclide angiography, 330–331
  ultrafast computed tomography, 342–343
  ventriculograms, 284–285

563

Aortic trauma, 519–522
  aneurysms and pseudoaneurysms, 521
  aortic rupture, common site for, 519, 520
  Doppler echocardiography, 533
  radiographic studies, 519–520
  survival of, 519, 522
Aortograms
  aortic regurgitation, 284, 286
  aortic stenosis, 285
  qualitative interpretation, 284
  quantitative interpretation, 284
Aortopulmonary shunts, 496
Apical aneurysm, 148 (Fig. 10-35)
Arteriography, great vessel disease, 518
  See also Coronary arteriography.
Arteriosclerosis, 522
  age and incidence of, 522
  areas commonly affected, 522
  stenoses, formation of, 522
Arteriosclerotic aneurysms, 522–523
  diagnosis of, 522
  etiology of, 522–523
  pathogenesis of, 522
Arteritis, magnetic resonance imaging, 554
Artifacts
  shading artifacts, 35
  thallium 201, 191 (Table 12-6)
  ultrafast CT scanner, 34–35
Asymmetric hearts, 143
Asymmetric septal hypertrophy, 361–362
Atherosclerosis, ultrasonic tissue characterization, 413
Atrial-level lesions, nuclear magnetic resonance, 498–499
Atrial pacing, 240
  left ventricular assessment studies, 55–57
Attentuation, in ultrasonic tissue characterization, 405
Autoimmune disease, great vessel disease, 526–527
Avid agents. See Infarct avid scintigraphy

B

Background subtraction, 47, 207–208 (Fig. 12-A2)
Backscattering, in ultrasonic tissue characterization, 404 (Fig. 31-1)
Bacterial endocarditis, 343, 344
Bacterial infection, aortic aneurysms and, 524–525, 526

Bayes theorem, and radioisotope imaging, 214
Benign tumors, pericardium, computed tomography, 456
Bequerels, 17
Beta-thalassemia, 382
Bicycle stress test, left ventricular assessment, 55
Bone-imaging agents, infarct avid scintigraphy, 168
Brachial artery approach
  coronary arteriography, 78–80
    bypass graft, 79
    internal mammary artery cannulation, 79
    left coronary artery cannulation, 79
    post-procedure, 80
    right coronary artery cannulation, 79
Brachial artery closure techniques, 80 (Fig. 6-25)
Brachial cutdown technique, 78 (Fig. 6-23)
Bypass surgery
  and coronary artery disease, 112
  discrepancies in study, 114
  See also Coronary artery bypass grafts.

C

Calcification
  in great vessels, computed tomography, 540–541
  pericardium, 456 (Fig. 35-7)
Carbon 11-labeled compounds, 25
Carbon, nuclear magnetic resonance spectroscopy, 426
Carcinoid syndrome, 344
Cardiac catheterization, 6
  cardiac tamponade, 431
  catheterization laboratory, 49 (Fig. 5-2)
  congenital pericardial abnormalities, 437
  constrictive pericarditis, 433
  patient selection, 215
Cardiac output
  measurement of, 231–233
  Stewart-Hamilton equation, 231
Cardiac tamponade, 429–432
  angiocardiography, 432
  cardiac catheterization, 431
  causes of, 429–430
  differential diagnosis, 432
  echocardiography, 444–446
  fluoroscopy, 432
  hemodynamic studies, 431

pathophysiology, 430–431
Cardiac tumors
  angiography, 506
  cineangiogram, 507 (Figs. 40-1, 40-2)
  classification of, 505
  computed tomography, 510–511
  digital subtraction angiography, 506
  echocardiography, 506–509
  imaging guidelines, 505
  imaging problem, 508, 509
  magnetic resonance imaging, 509, 511
  secondary metastatic tumors, 508
  ultrafast computed tomography, 510 (Fig. 40-6)
Cardiomyopathies
  angiography, 353–359
  diagnosis of, 151–156
  dilated cardiomyopathy, 150
  Doppler echocardiography, 151
  hypertrophic cardiomyopathy, 150–151, 153–154
  M-Mode echocardiography, 151 (Fig. 10-43)
  nuclear magnetic resonance imaging, 153
  phosphorus NMR spectroscopy, 422–423
  restrictive cardiomyopathy, 151, 155–156
  ultrafast computed tomography, 385–393
  ultrasonic tissue characterization, 411–413
Cardiomyopathy, 150–156
  dilated cardiomyopathy, 357–359
  doxorubicin-induced, 398
  hypertrophic cardiomyopathy, 353–357
  imaging and, 353
  use of term, 353, 357–358
  See also Dilated cardiomyopathy; Hypertrophic cardiomyopathy.
Cardiovascular effects, contrast media, 92–96
Carotid arteries, anatomy of, 519
Catheterization. See Cardiac catheterization
Centerline method, wall motion analysis, 106 (Fig. 8-3)
Chagas' disease, 153, 154 (Fig. 10-47)
Chronic ischemia, 199
Cineangiography
  aortic valve, 283 (Fig. 20-2)
  cardiac tumors, 507 (Figs. 40-1, 40-2)
  left ventricle, 281–282 (Figs. 20-1A, 20-1B)
Cine mode
  ultrafast CT scanner, 32–33

magnetic resonance imaging, 501
(Fig. 39-10), 502 (Fig. 39-11)
nuclear magnetic resonance, 160,
174–178, 494, 501–502
Cine trace, 108
Coarctation of aorta, 496
computed tomography, 545–546
Doppler echocardiography, 533
echocardiography, 532–533
magnetic resonance imaging, 553–554
Coincidence detection, 23 (Fig. 2-7)
Cold pressor, 240
Collaterals, 71, 73
functions of, 71
intercoronary collaterals, 71
intracoronary collaterals, 71
visualization of, 71, 73
Collimator, in nuclear imaging, 17–18
Color flow Doppler, 314, 319
Computed tomography (CT)
absent pericardium, 453
animal studies, 223–224
benign tumors, pericardium, 456
cardiac tumors, 510–511
constrictive pericarditis, 456
Dynamic Spatial Reconstructor
(DSR), 29
great vessels, 518
advantages of, 537–538
calcium detection, 540–541
coarctation of aorta, 545–546
contrast-enhanced studies, 539–540
intracardiac shunts, 545
pulmonary atresia, 545
pulmonary embolism, 544–545
technical aspects, 538
ultrafast computed tomography,
543–544
human studies, 224
malignant disease, pericardium,
456, 457 (Figs. 35-8, 35-9)
pericardial effusions, 453–455
pericardial cysts, 453
pericardial thickening, 455–456
pericardium, normal, 452 (Fig. 35-1)
safety aspects, 559
scanning motion, 29, 30 (Fig. 3-1)
speed limitations, 29
See also Ultrafast computed
tomography.
Computer
in nuclear imaging, 19–20
thallium 201 analysis, 207–209
Computer Acquisition of
Echocardiographic Data (CAED),
system for, 8–11
Congenital pericardial abnormalities
cardiac catheterization, 437
echocardiography, 448

nuclear magnetic resonance, 462
types of, 437–438
Congenital pulmonic regurgitation,
289
Congenital pulmonic stenosis, 289
Congenital heart defects
echocardiography, 469–478
nuclear magnetic resonance,
493–502
ultrafast computed tomography,
481–491
Congestive cardiomyopathy, 154
(Fig. 10-47) *See also* Restrictive
cardiomyopathy.
Constrictive pericarditis, 432–436
angiocardiography, 435
cardiac catheterization, 433
computed tomography, 456
differential diagnosis, 435–436
echocardiography, 447–448
etiology, 432–433
fluoroscopy, 434
hemodynamics, 433–434
nuclear magnetic resonance, 465
pathophysiology, 433
Continuous wave Doppler, 13, 14
(Fig. 1-13), 158 (Fig. 10-53), 313,
314
aortic regurgitation, 321 (Fig. 22-6)
diastolic transmitral flow, 322
(Fig. 22-8)
mitral regurgitation, 323 (Fig. 22-9)
pericardial effusion, 445 (Fig. 34-6)
Contrast echocardiography, 4–6, 8
valvular disease, 306
Contrast media
anaphylactoid reactions, 97–98
cardiovascular effects, 92–96
animal studies, 93–96
depressed myocardial
contractility, 92
vasodilation of blood vessels, 93
chemical aspects, 91–92
cost aspects, 98
diabetes, 97
hemostatic effects, 96
renal failure, 96–97
risk factors, 97
Contrast ventriculography, technical
aspects, 108–109
Coordinate system models, wall
motion analysis, 105 (Fig. 8-2), 106
Coronal spin echo image, 44 (Fig. 4-3)
Coronary anatomy (normal)
coronary artery collaterals, 71
coronary artery dominance, 71
inguinal area, 74 (Fig. 6-15)
left coronary artery, 68–71 (Figs. 6-7,
6-8, 6-9)

right antecubital fossa, 78 (Fig. 6-22)
right coronary artery, 66–68
(Figs. 6-1, 6-2, 6-4, 6-5)
Coronary arteriography
Amplatz coronary catheters, 75,
76 (Fig. 6-19)
brachial artery approach, 78–80
bypass graft, 79
internal mammary artery
cannulation, 79
left coronary artery cannulation,
79
post-procedure, 80
right coronary artery cannulation,
79
complications
major complications, 82–85
minor complications, 85–86
coronary anatomy evaluation, 80–82
coronary artery spasm, 80–81
guidewires, 74 (Fig. 6-16)
indications for, 65–66
interpretation pitfalls
catheter-induced coronary spasm,
87
inadequate x-ray penetration, 87
incomplete study, 86
myocardial bridges, 87, 88
(Fig. 6-32)
poor contrast medium injection,
86
selective cannulation of
coronary vessels, 86–87
totally occluded coronary
artery, 87–88 (Fig. 6-33)
Judkins coronary catheters, 75
(Fig. 6-17), 76 (Fig. 6-18)
limitations of, 88–89
percutaneous femoral approach,
74–78
bypass graft, 76–77
internal mammary artery
cannulation, 77–78
left coronary artery cannulation,
75
right coronary artery cannulation,
76
procedure preparation, 73
Sones coronary catheters, 79
(Fig. 6-24)
Coronary artery bypass grafts
evaluation of
aortic root digital subtraction
angiography, 272
intravenous DSA, 273
nuclear magnetic resonance,
276–278
radionuclide ventriculography,
270, 271

selective graft angiography, 271–272
thallium 201, 270, 271
ultrafast computed tomography, 273–276
weaknesses of scintigraphic methods, 271
X-ray CT, 273
success of, 269
Coronary artery collaterals, 71
Coronary artery disease
aneurysmectomy, 112
bypass surgery, 112
left ventricular angiograms, 101–110
left ventricular evaluation, post-therapy, 112–114
nuclear magnetic resonance, 257–266
patient assessment
inspection of ventriculogram, 110
left ventricular function and, 110
myocardial infarction, 110–111
preoperative evaluation, 111–112
patient selection for catheterization, 215
pretest and post-test referral bias, 215
prevalence of, 214–215
prognosis, 215–219
predictive aspects, 114
wall motion analysis, 107–108
Coronary artery dominance, 71
Coronary artery occlusion, and left ventricular dysfunction, 101, 110
Coronary artery stenosis, digital analysis of, 58–60
Coronary Artery Surgery Study (CASS) system, 66, 112, 215, 218
Coronary spasm, catheter-induced, 87
Count rates, 20
Count recovery, 20
Critical aortic stenosis, 297 (Fig. 21-3)
Crystal scintillation detector, in nuclear imaging, 18
Cyanotic lesions, nuclear magnetic resonance, 499, 501

D

Data acquisition and processing, tomographic imaging, 21–22
Death, cardiac catheterization and, 84–85
DeBakey classification, 524, 525 (Fig. 41-10)
Decay constant, 17
Dephasing effects, 44

Depth-resolved surface-coil spectroscopy, nuclear magnetic resonance spectroscopy, 419–420
Detector fan beam response, 23 (Fig. 2-7)
Diabetes
contrast media, 97
restrictive state, 155
Diastolic function assessment, echocardiography, 145
Diffuse myocardial uptake, 171
Digital angiography
advances in area, 124–125
analysis of coronary artery stenosis, 58–60
angiography with aortic root injection, 61
benefits of, 47–48, 124
cardiac imaging with, 52
contrast medium administration, 120–121
digital roadmapping, 60–61
digitization process, 49–50
direct intraventricular studies, 52–53
dual energy subtraction, 52
edge enhancement, 58 (Fig. 5-7), 59 (Fig. 5-8)
instrumentation, 119
left ventricular assessment, 121
atrial pacing studies, 55–57
bicycle stress test, 55
mask mode subtraction, 50–52, 120
storage of angiograms, 57–58
time interval differencing, 120, 121 (Fig. 9-2)
ventriculography, 121–125
videodensitometry, 53–55
Digital roadmapping, 60–61
Digital subtraction angiography
cardiac tumors, 506
great vessel disease, 515–518
Digital ventriculograms, 55–56
Dilated cardiomyopathy, 150, 255, 357–359
angiography in, 358, 359
endocardiography, 369–373
characteristic features, 369–370
Doppler echocardiography, 371–373
left ventricular mural thrombus, 370–371
fatty acid metabolism imaging, 377
mitral regurgitation, 358–359
morphology of, 387 (Table 29-1)
positron emission tomography, 377
radionuclide angiography, 375–377
thallium 201, 277

ultrafast computed tomography, 391–393
Dilation of aortic root, 496
Dipyridamole thallium 201 imaging
and acute myocardial infarction, 202
administration of, 200–201
mechanism of action, 200
in patients with physical limitations, 201
pre-major vascular surgery, 201–202
side effects, 201
Discrete myocardial uptake, 172–173
Doppler blood flow velocity, dilated cardiomyopathy, 371 (Fig. 27-10)
Doppler echocardiography, 2, 131–133
aortic aneurysms, 531–532
aortic dissection, 532
aortic regurgitation, 318–320
aortic stenosis, 315–318
aortic trauma, 533
coarctation of aorta, 533
color flow Doppler, 314, 319
continuous wave Doppler, 13, 313, 314
dilated cardiomyopathy, 371–373
Doppler principle, 11
great artery malposition, 476
high-pulse repetition frequency Doppler, 314, 315, 317
hypertrophic cardiomyopathy, 365–367
idiopathic dilation of pulmonary artery, 534
instrumentation, 313–315
laminar flow, 13–14
left ventricular assessment, 140
malalignment, 475
mitral regurgitation, 323–324
mitral valve stenosis, 321–323
multigate Doppler, 13
nonlaminar flow, 14
patent ductus arteriosus, 533
prosthetic heart valves, assessment of, 325–326
pulmonary artery pressure, 534–535
pulmonary embolism, 534
pulmonary regurgitation, 320–321
pulsed doppler, 12–13
pulsed wave Doppler, 313
special role of, 14
spectral analysis, 12
tricuspid regurgitation, 324–325
tricuspid valve stenosis, 323
univentricular heart, 478
valvular stenosis, 472–475
velocity ranges for children/adults, 133 (Table 10-1)

Doppler equation, 131, 133
Doppler frequency shift, 146 (Fig. 10-31)
  blood flow velocity measurement, 311–313
Doppler signal, 132 (Fig. 10-4)
Doppler transducer, 312
Doxorubican hydrochloride cardiomyopathy, nuclear magnetic resonance, 398
Doxorubicin hydroxychloride, cardiotoxicity, 378
Dressler's syndrome, 149
Dual energy subtraction, 52
Duplex ultrasonography, great vessel disease, 515
Dynamic mask subtraction, subtraction process, 51–52
Dynamic Spatial Reconstructor, 33
Dynamic subaortic outflow tract obstruction, 158 (Fig. 10-54)
Dysrhythmias, cardiac catheterization and, 86

E

Ear indocyanine green densitometry, 241
Ebstein's anomaly, 344
Echocardiography
  absent pericardium, 448
  acute myocardial infarction, 117–118
  aortic aneurysms, 531
  aortic dissection, 532
  cardiac tamponade, 444–446
  cardiac tumors, 506–509
  cardiomyopathy, 150–156
  coarctation of aorta, 532–533
  congenital pericardial abnormalities, 448
  constrictive pericarditis, 447–448
  contrast echocardiography, 4–6
  Doppler echocardiography, 131–133
  instrumentation, 130 (Fig. 10-1)
  left-to-right shunts, 470–472
  left ventricular assessment, 134–146
    diastolic function assessment, 145
    difficulties of, 141–142
    Doppler echocardiography, 140
    M-Mode echocardiography, 134–138
    two-dimensional echocardiography, 138–140
    ventricular dimensions assessment, 142–145
  limitations of, 133–134
  M-Mode echocardiography, 129–130
  myocardial infarction, 147–150
  pericardial effusions, 442–444
  pericardial paracentesis, 446
  pericardial thickening, 446–447
  safety aspects, 559–560
  septal defects, 469
  stress echocardiography, 156–158
  transesophageal echocardiography, 158–159
  two-dimensional echocardiography, 130–131
  valvular disease
    contrast echocardiography, 306
    information given by, 294
    intraoperative echocardiography, 306
    M-Mode echocardiography, 293, 295, 297, 305–306
    two-dimensional echocardiography, 296, 297, 300–301
    valvular inflow tract, 300–306
    ventricular outflow tract, 294–299
  valvular stenosis, 472
Echo-planar imaging, MRI, 45, 46 (Fig. 4-6)
Edge finding accuracy, ultrafast CT scanner, 36–37
Effusive-constrictive pericarditis, 436–437
Ehlers-Danlos syndrome, 523–524
Electronic components, in nuclear imaging, 18–19
E-point septal separation (EPSS), 136–137 (Fig. 10-15)
Ergonovine, 80, 81 (Fig. 6-26)
Erheim's syndrome, 524
Esophageal ultrasound, 1
Eventrations of the diaphragm, 457
Exercise
  blood pool imaging, 213–214
  cardiovascular response to, 239–240
  echocardiography, 240
  normal response to, 214
  techniques
    evaluation of, 248–249
    thallium, 184–185, 187–195
  ultrafast computed tomography, 239, 240–249

F

Fast raw averaging (FRA) board, ultrafast CT scanner, 30
Fatty acid metabolism imaging, dilated cardiomyopathy, 377
Fick technique, 321
Filtered back projection, 21
First-pass radionuclide angiocardiography, 213
Flail valve, 304 (Fig. 21-8)
FLASH (Fast Low Angle Shot), MRI, 45
Flow effects, MRI, 44
Flow mode, ultrafast CT scanner, 32
Fluoroscopy
  aortic regurgitation, 285
  aortic stenosis, 284
  aortic valve, 283
  cardiac tamponade, 432
  constrictive pericarditis, 434
  left ventricle, 281–282
  pericardial cysts, 438
Foramen of Morgagni hernias, 457
Full width at half maximum (FWHM), 19

G

Gallium 67
  myocarditis, 378–379
  sarcoidosis, 379
Gallium citrate, infarct avid scintigraphy, 167
Gamma camera, 20, 21
Gamma rays, 17, 18 (Fig. 2-1)
Gated multislice spin echo, MRI, 44–45
Gaussian curve, 19
Gaussian distribution, 19
Generalized myocardial uptake, 171
Giant cell aortitis, 526–527
Gorlin equation, 290, 296, 297, 317, 318, 472
Gradients, hypertrophic cardiomyopathy, 356–357
GRASS (Gradient Recall Acquisition of Steady State), 45, 46 (Fig. 4-5), 277, 349 (Fig. 25-3)
Great artery malposition, Doppler echocardiography, 476
Great vessel disease
  aortic dissection, 523–524
  arteriography, 518
  arteriosclerosis, 522
  arteriosclerotic aneurysms, 522–523
  autoimmune disease, 526–527
  computed tomography, 518
  digital subtraction angiography, 515–518
  duplex ultrasonography, 515
  infectious disease, 524, 526
  magnetic resonance imaging, 518
  See also specific disorders.
Great vessels, normal anatomy, 518–519
Guidewires, coronary arteriography, 74 (Fig. 6-16)

H

Half-life of radionuclide, 17
Hand-grip, 240
Heart muscle disease
  anthracycline cardiotoxicity, 378
  myocarditis, 378–379
  sarcoidosis, 379
  See also Cardiomyopathy.
Hemochromatosis, 381
Hemostatic effects, contrast media, 96
"Hibernating myocardium," 199
High-pulse repetition frequency Doppler, 314, 315, 317
High-velocity signal loss, 44
Hounsfield unit, 232 (Fig. 15-1)
Hydrogen, nuclear magnetic resonance spectroscopy, 424–425
Hypertrophic cardiomyopathy, 150–151, 153–154, 353–357
  characteristics of, 353–354, 379–380
  coronary anatomy, 356
  endocardiography, 361–369
    characteristic feature, 361–363
    distribution of disease, 363–364
    Doppler echocardiography, 365–367
    genetic assessment, 368–369
    obstructive element, 364–365
    therapeutic results, 369
  gradients, 356–357
  main feature of, 353
  morphology of, 387 (Table 29-1)
  nuclear magnetic resonance, 395–397
  pathophysiology of, 356
  progression of, 356
  radionuclide angiography, 380–381
  ultrafast computed tomography, 386–390
Hypokinesis, 111 (Fig. 8-7)
  factors related to, 111
  severe, 110 (Fig. 8-6)

I

Idiopathic dilation of pulmonary artery, Doppler echocardiography, 534
Image selected in vivo technique, nuclear magnetic resonance spectroscopy, 420
Imaging techniques, safety aspects, 557–560
Imatron C-100XL, 30
Imidodiphosphonate, infarct avid scintigraphy, 168
Incomplete mitral leaflet closure, 304

Indium 111, 25, 253
  infarct avid scintigraphy, 175–176 (Fig. 11-10)
  myocarditis, 379
Infarct avid scintigraphy
  applications
    evaluation after cardioversion, 174
    infarct localization, 174
    preoperative infarction diagnosis, 174
    prognostic value, 175–176
  avid agents
    bone-imaging agents, 168
    gallium citrate, 167
    imidodiphosphonate, 168
    indium 111 antimyosin antibody, 175–176 (Fig. 11-10)
    localization and cellular events, 168
    pyrophosphate, 168
    radionuclide behavior, 168
    technetium 99m pyrophosphate, 168, 169 (Fig. 11-1), 170, 171 (Fig. 11-4)
    technetium 99m, 167–168
  diagnostic accuracy, 172–173
  false positive scintigrams, 173
  image interpretation, 170–172
  compared to other methods, 173
  scintigraphic method, 170
  stress perfusion scintigraphy, 176
Infarct size
  estimation of, 111
  limitations in sizing techniques, 226–228
Infectious disease, great vessel disease, 524, 526
Inferior basal aneurysm, 149 (Fig. 10-39)
Inferior myocardial infarction, 148 (Fig. 10-38)
Inferoposterior infarction, 148 (Fig. 10-37)
Infundibular pulmonic stenosis, 500 (Fig. 39-8)
Inotropic stimulation, left ventricular response to, 113
Integrated backscattering, in ultrasonic tissue characterization, 406, 409
Intercoronary collaterals, 71
Intracardiac shunts, computed tomography, 545
Intracoronary collaterals, 71
Intraoperative echocardiography, valvular disease, 306
Intravenous DSA, coronary artery bypass graft evaluation, 273

Ischemic heart disease, ultrasonic tissue characterization, 407–411
Isonitriles, 25

J

Judkins coronary catheters, 75 (Fig. 6-17), 76 (Fig. 6-18)

L

Laminar flow
  Doppler techniques, 13–14
  types of, 14
Left coronary artery
  anatomy of, 68–71 (Figs. 6-7, 6-8, 6-9)
  obstruction of, 72 (Figs. 6-11, 6-12), 72, 73 (Fig. 6-13)
Left ventricular aneurysm, 216 (Fig. 13-5)
Left ventricular aneurysm and thrombus, ultrafast computed tomography, 253–255
Left ventricular angiograms
  applications, normal values, 109
  computer calculations, 109
  objectives of, 101
  tracing ventricular contour, 108–109
  ventricular volume
    calculation of, 101–104
    calculation variability, 103–104 (Table 8-2)
    regression equations, 103 (Table 8-1)
  ventriculography, 108
  wall motion analysis, 104–108
Left ventricular echocardiography
  diastolic function assessment, 145
  difficulties of, 141–142
  Doppler echocardiography, 140
  images of, 141 (Fig. 10-23)
  M-Mode echocardiography, 134–138
  transesophageal echocardiography, 158–159
  two-dimensional echocardiography, 138–140
  ventricular dimensions assessment, 142–145
Left ventricular ejection fractions
  exercise studies, 242–243
  at rest/exercise, 212 (Fig. 13-2)
  segmental EF, 243, 247 (Fig. 16-12)
Left ventricular outflow tract, 157 (Fig. 10-52)
  radioisotope imaging, 211–220
Linear backprojection, 21 (Fig. 2-5)
Lung-heart ratio, thallium 201, 196, 209

M

Magnetic resonance imaging (MRI)
  cardiac tumors, 511
  contrast mechanisms, 43
  echo-planar imaging, 45, 46
    (Fig. 4-6)
  FLASH (Fast LowAngle Shot), 45
  flow effects, 44
  gated multislice spin echo, 44–45
  GRASS (Gradient Recall Acquisition
    of Steady State), 45, 46 (Fig. 4-5)
  great vessels, 518
    aortic aneurysms, 552–553
    aortic dissection, 551–552
    arteritis, 554
    coarctation of aorta, 553–554
    postoperative evaluation, 555
    pulmonary atresia, 554
    pulmonary stenosis, 554–555
    technical aspects, 549–550
      vena cava, congenital
        abnormalities, 555
  machine, components of, 42–43
  multislice, multiphase technique,
    45
  principles of, 41–43
  safety aspects, 560
Magnetohydrodynamic effect,
  257–258
Malalignment, Doppler
  echocardiography, 475
Malignant disease, pericardium,
  computed tomography, 456, 457
  (Figs. 35-8, 35-9)
Malposition, nuclear magnetic
  resonance, 496–497
Marfan's syndrome, 483, 496, 523, 531
Mask mode subtraction, 50–52, 59
  (Fig. 5-9), 120
Methoxyisobutyl isonitrile (MIBI), 25
Misregistration artifact, 51
Mitral calcification, 155
  ultrafast computed tomography, 342
Mitral regurgitation, 304
  dilated cardiomyopathy, 358–359
  Doppler echocardiography, 323–324
  fluoroscopy, 288
  MRI, 350 (Fig. 25-4)
  quantitation of, 324 (Table 22-2)
  radionuclide angiography, 334–335
  ultrafast computed tomography,
    340–341
  ventriculograms, 288–289
Mitral stenosis
  Doppler echocardiography, 321–323
  fluoroscopy, 286
  radionuclide angiography, 330
  ultrafast computed tomography, 340
  ventriculograms, 287–288

Mitral valve, 157 (Fig. 10-52)
Mixed valve disease
  angiographic stroke volume, 290
  ventriculograms, 289–290
M-Mode echocardiography, 1–2,
  129–130
  advantages of, 129
  cardiac tumors, 506, 508
  cardiomyopathies, 151 (Fig. 10-43)
  left ventricular assessment, 134–138
  pericardial effusion, 442 (Fig. 34-2),
    445 (Fig. 34-5)
  pericardial restriction, 447
    (Fig. 34-9)
  representation of, 131 (Fig. 10-2)
  rheumatic mitral stenosis, 302
    (Fig. 21-6)
  thick pericardium, 447 (Fig. 34-8)
  tracings, 135 (Fig. 10-8), 156
    (Fig. 10-50)
  valvular disease, 293, 295, 297,
    300, 305–306
Moderate aortic stenosis, 297
  (Fig. 21-3)
Multicrystal imaging cameras, 219
Multigated equilibrium acquisition,
  213–214
Multigate Doppler, 13
Multislice, multiphase technique,
  MRI, 45
Myocardial blood flow, ultrafast
  computed tomography, 231–236
Myocardial infarction, 147–150
  aneurysm/thrombus after, 251–252
  complications of, 148–150
  and coronary artery disease, 110–111
  ultrasonic tissue characterization,
    408–411
  See also Acute myocardial infarction.
Myocardial perfusion, 5
  cardiac function and, 211–213
  thallium 201 assessment, 181–202
Myocardial rupture, 149
Myocarditis, imaging techniques,
  378–379

N

Noise
  noise artifactual, 34–35
  ultrafast CT scanner, 34–35
Nonlaminar flow, Doppler techniques,
  14
Nuclear cardiology procedures, safety
  aspects, 559
Nuclear imaging
  detection of abnormalities, 20
  equipment
    collimator, 17–18

    computer, 19–20
    crystal scintillation detector, 18
    electronic components, 18–19
  image quality, 20
  physics of, 17
  planar imaging, 20
  positron emission tomography,
    22–23
  quality control, 22
  quantification, 20
  tomographic imaging, 20–22
Nuclear magnetic resonance (NMR)
  absent pericardium, 462
  acute myocardial infarction, 118
  animal studies, 260–263
    coronary occlusion by NMR,
      260–261
    coronary occlusion and
      reperfusion, 261
    myocardial ischemia with
      gadolinium, 261–263
  cardiac tumors, 509
  cine NMR, 260, 277–278
  congenital defects
    atrial-level lesions, 498–499
    cine MRI, 494, 501–502
    cyanotic lesions, 499, 501
    malposition, 496–497
    pericardial abnormalities, 462
    pulmonary artery abnormalities,
      497–498
    thoracic aortic abnormalities, 496
    ventricular-level lesions, 499
  constrictive pericarditis,
    465
  coronary artery bypass graft
    evaluation, 276–278
  doxorubican hydrochloride
    cardiomyopathy, 398
  functional evaluation of coronary
    artery disease, 260
  gated NMR, 257–258, 259, 260
    (Fig. 18-3), 263 (Fig. 18-4), 264
    (Figs. 18-5, 18- 6), 265, 265
    (Fig. 18-7)
  human studies, 263–264
  hypertrophic cardiomyopathy,
    395–397
  NMR coronary angiograms,
    259–260
  normal heart, 258–259
  pericardial effusions, 463
  pericardial hematoma, 464
  pericardial masses, 464–465
  pericardial thickening, 462–463
  pericardium, normal, 460–462
  principle of, 257
  rapid dynamic MRI, 350, 351
    (Fig. 25-5)

restrictive cardiomyopathy, 397–398, 399–400, 466
spin-echo technique, 277
techniques of, 493–494
transplant rejection, 398–399
transverse images, 494–495 (Fig. 39-1)
valvular disease, 347–351
  acquired valvular disease, 348–351
  spin-echo NMR, 347–348
Nuclear magnetic resonance spectroscopy
  localization methods
    depth-resolved surface-coil spectroscopy, 419–420
    image selected in vivo technique, 420
    rotating frame, 419
  nuclei studied
    carbon, 426
    hydrogen, 424–425
    phosphorus, 421–424
    potassium, 425
    sodium, 425
  principles of, 417–418
  saturation transfer, 418–419
  technical requirements, 419
Nuclear medicine
  antimyosin antibody, 25
  indium 111, 25
  platelets, 25
  positron-emitting isotopes, 25
  rubidium 82, 25
  technetium 99m methoxyisobutyl isonitrile, 25
  technetium 99m, 24–25
  technetium-labeled blood flow tracers, 25
  technetium pyrophosphate, 25
  thallium 201, 23–24
Nuclear spins, 41–42
Nyquist limit, 314, 315

P

Parametric flow reserve imaging, 124–125
Partial volume effect, 20
Patent ductus arteriosus, 496, 497 (Fig. 39-3)
  Doppler echocardiography, 533
Percent resolution, 19
Percutaneous femoral approach
  coronary arteriography, 74–78
    bypass graft, 76–77
    internal mammary artery cannulation, 77–78
    left coronary artery cannulation, 75
    right coronary artery cannulation, 76
Pericardial cysts, 437–438, 453 (Fig. 35-2)
  computed tomography, 453
  fluoroscopy, 438
Pericardial disease
  absent pericardium, 448
  cardiac tamponade, 429–432
  computed tomography, 451–458
  congenital abnormalities, 437–438
  constrictive pericarditis, 432–436
  diagnostic information related to, 441
  echocardiography, 442–448
  effusive-constrictive pericarditis, 436–437
  etiologies of, 430 (Table 33-1)
  nuclear magnetic resonance, 459–466
  pericardial cysts, 437–438
  restrictive cardiomyopathy, 437
  ultrafast computed tomography, 457
  See also individual disorders.
Pericardial effusions, 454 (Figs. 35-3, 35-4), 455 (Figs. 35-3, 35-6)
  computed tomography, 453–455
  echocardiography, 442–444
  nuclear magnetic resonance, 463
Pericardial hematoma, nuclear magnetic resonance, 464
Pericardial masses, nuclear magnetic resonance, 464–465
Pericardial paracentesis, echocardiography, 446
Pericardial thickening
  computed tomography, 455–456
  echocardiography, 446–447
  nuclear magnetic resonance, 462–463
Pericardium, anatomical aspects, 429, 452, 459
Perimembranous ventricular septal defect, 500 (Fig. 39-9)
Persistent myocardial defect, 182, 183 (Figs. 12-3, 12-4)
Phase cancellation, in ultrasonic tissue characterization, 405
Phosphorus, nuclear magnetic resonance spectroscopy, 421–424
Photomultiplier tube, 18 (Fig. 2-1), 19
Photopeak, 19, (Fig. 2-3)
Picker/Imatron FASTRAC, 30
"Pixelization" of the image, 49
Planar imaging, 20
Platelets, 25

Positron emission tomography (PET), 22–23, 257
  acute myocardial infarction, 117
  advantages of, 22–23
  dilated cardiomyopathy, 377
Positron-emitting isotopes, 25
Post-test referral bias, 215
Potassium, nuclear magnetic resonance spectroscopy, 425
Pretest referral bias, 215
Progressive systemic sclerosis, 381
Propagation speed, in ultrasonic tissue characterization, 404
Prosthetic valve disease
  diagnosis of, 290
  regurgitation and, 290
Prosthetic valves, assessment of, Doppler echocardiography, 325–326
Provocable outflow tract obstruction, 157 (Fig. 10-51)
Pseudoaneurysm, 149 (Fig. 10-40)
  aortic trauma and, 521
  cardiac catheterization and, 86
Pulmonary artery abnormalities, nuclear magnetic resonance, 497–498
Pulmonary artery pressure, Doppler echocardiography, 534–535
Pulmonary atresia, 497 (Figs. 39-4, 39-5)
  computed tomography, 545
  magnetic resonance imaging, 554
Pulmonary embolism
  computed tomography, 544–545
  Doppler echocardiography, 534
Pulmonary regurgitation
  Doppler echocardiography, 320–321
  ultrafast computed tomography, 345
Pulmonary stenosis
  magnetic resonance imaging, 554–555
  ultrafast computed tomography, 344–345
Pulmonic valve disease, ventriculograms, 289
Pulsed Doppler, 12–13, 14 (Fig. 1-13), 158 (Fig. 10-54), 313
  dilated cardiomyopathy, 372 (Fig. 27-11)
  hypertrophic cardiomyopathy, 368 (Fig. 27-8)
  mitral regurgitation, 323 (Fig. 22-9)
  transmitral flow velocities, 322 (Fig. 22-7)
"Pulseless disease," 526
Pyrophosphate, infarct avid scintigraphy, 168

Q

QRS scoring system, 227
Quality control, nuclear imaging, 22
Quantitative analysis, gray level data, ultrasonic tissue characterization, 407
Q-wave infarction, thallium 201, 199

R

Radio frequency signal analysis, ultrasonic tissue characterization, 406
Radioisotope imaging, 211–220
  acute myocardial infarction, 227
  advances in, 219–220
  Bayes theorem and, 214
  first-pass radionuclide angiocardiography, 213
  general uses of, 211
  multigated equilibrium acquisition, 213–214
Radionuclide angiography, 21 (Fig. 2-4)
  aortic regurgitation, 331–334
  aortic stenosis, 330–331
  hypertrophic cardiomyopathy, 380–381
  mitral regurgitation, 334–335
  mitral stenosis, 330
  restrictive cardiomyopathy, 381–382
  systolic anterior motion, 362, 364–365
Radionuclide ventriculography, coronary artery bypass graft evaluation, 270, 271
Radiopharmaceuticals
  antimyosin antibody, 25
  indium 111, 25
  platelets, 25
  positron-emitting isotopes, 25
  rubidium 82, 25
  technetium 99m methoxyisobutyl isonitrile, 25
  technetium 99m, 24–25
  technetium-labeled blood flow tracers, 25
  technetium pyrophosphate, 25
  thallium 201, 23–24
Redistribution, 181, 183 (Fig. 12-2)
  complete, 183 (Fig. 12-2)
  computer analysis, 208
  partial, 183 (Fig. 12-2)
  reverse distribution, 182, 184 (Fig. 12-5), 198
Referral bias, thallium 201, 191
Region-of-interest marker, 488
Regurgitant fractions assessment, ultrafast computed tomography, 340

Renal failure, contrast media, 96–97
Reperfusion therapy, chest pain after, 186–187
Restrictive cardiomyopathy, 151, 155–156, 437
  etiologies of, 381, 437
  nuclear magnetic resonance, 397–398, 399–400, 466
  radionuclide angiography, 381–382
  ultrafast computed tomography, 390–391
Revascularization, thallium 201 assessment and, 199–200
Reverse distribution, 182, 184 (Fig. 12-5)
Rheumatic mitral stenosis, 302
  echocardiograms of, 302 (Fig. 21-6)
Right coronary artery, anatomy of, 66–68 (Figs. 6-1, 6-2, 6-4, 6-5)
Right ventricular infarction, 150 (Fig. 10-41)
Roadmapping, digital, 60–61
Rotating frame, nuclear magnetic resonance spectroscopy, 419
Rubidium 82, 25

S

Safety of imaging techniques
  angiography, 558–559
  computed tomography, 559
  echocardiography, 559–560
  magnetic resonance imaging, 560
  nuclear cardiology procedures, 559
Salvaged myocardium, 199
Sapirstein principle, 234
Sarcoidosis, imaging techniques, 379
Scattering, in ultrasonic tissue characterization, 404 (Fig. 31-1)
Scintigraphy. See Infarct avid scintigraphy
Scintillation camera, 18 (Fig. 2-1)
Sector ultrasound imaging, 2
Secundum atrial septum defect, 499 (Fig. 39-7)
Segmental modeling, 143
Selective cannulation of the coronary vessels, 86–87
Selective graft angiography, coronary artery bypass graft evaluation, 271–272
Sensitivity, 18
Septal penetration, 18, 19 (Fig. 2-2)
Septal rupture, 149
Septum, 18
Signal-to-noise ratio, ultrafast CT scanner, 34
Single photon emission computer tomography (SPECT), 20

Slant-hole tomography, 172 (Fig. 11-5)
Sodium, nuclear magnetic resonance spectroscopy, 425
Sonication techniques, 5–6
Spatial resolution, ultrafast CT scanner, 33–34
Spectral analysis, Doppler techniques, 12
Specular reflection, in ultrasonic tissue characterization, 404
"Sphericity" index, 143
Spin dephasing, 44, (Fig. 4-3)
Spin-echo NMR, 277
  acquired valvular disease, 347–348
  gated images, 348 (Figs. 25-1, 25-2)
  valvular disease, 347–348
Stewart-Hamilton equation, 231, 390, 487
Sones coronary catheters, 79 (Fig. 6-24)
Stress echocardiography, 156–158
Stress electrocardiography, 239
Stress ventriculography, 124
Strip chart recorder, 108
"Stunned myocardium," 199
Subtraction process, 48, 49 (Fig. 5-1)
  background subtraction, 207–208 (Fig. 12-A2)
  contrast enhancement, 53 (Fig. 5-5)
  dual energy subtraction, 52
  mask mode subtraction, 50–52, 120
Syphilitic aortitis, 524
Systemic amyloidosis, 303–304
Systolic anterior motion, hypertrophic cardiomyopathy, 362, 364–365

T

Technetium 99m, 19 (Fig. 2-3), 24–25
  infarct avid scintigraphy, 167–168
Technetium 99m methoxyisobutyl isonitrile, 25
Technetium 99m pertechnetate, 213
Technetium 99m pyrophosphate, 25
  in amyloidosis, 173 (Fig. 11-6)
  restrictive cardiomyopathy, 381
  right ventricular infarction, 174 (Fig. 11-7)
  sarcoidosis, 379
  slant-hole tomography, 172 (Fig. 11-5)
Technetium-labeled blood flow tracers, 25
Thallium 201, 21, 23–24, 25
  accessing uptake, 24

computer image analysis, 207–209
coronary artery bypass graft
  evaluation, 270, 271
dilated cardiomyopathy, 277
dipyridamole thallium 201
  imaging, 200–202
distribution patterns, 24 (Fig. 2-8)
imaging during exercise, 187–191
  for ambulatory patients with
    chest pain, 198–199
  for assessment prior to
    revascularization, 200
  factors affecting clearance, 193
    (Table 12-7)
  factors affecting test sensitivity,
    189–191
  factors affecting test specificity,
    191
  normal image, 188–189
  optimization of detection, 194–195
  patients not on thrombolytic
    therapy, 197–198
  patients with stable angina,
    195–197
  patients with thrombolytic
    therapy, 198
  post uncomplicated myocardial
    infarction, 199
  qualitative versus quantitative
    assessment, 187
  thallium clearance problems,
    191–194
imaging at rest, 185–187
  chest pain after recuperative
    therapy, 186–187
  detection of myocardial
    infarction, 185
  indications for, 185 (Table 12-1)
  left ventricular dysfunction
    exceeding infarct size, 186
  Prinzmetal angina, 187
  unstable angina, 187
kinetics of, 181
lung uptake, 184–185
optimizing detection of CAD,
  194–195
persistent defect, 182, 183
  (Figs. 12-3, 12-4)
problems in assessment of
  clearance, 191, 193–194
redistribution, 181, 183 (Fig. 12-2)
reverse distribution, 182, 184
  (Fig. 12-5)
sarcoidosis, 379
specificity of test, 191
  artifacts, 191
  false positives, 191
  referral bias, 191
superiority of, 199 (Table 12-12)

washout, 23
Thallium-activated sodium iodide,
  18, 19
Thallium scintigraphy, 240
  exercise, 248
Thoracic aortic abnormalities,
  nuclear magnetic resonance, 496
Thrombolytic therapy, evaluation of
  response to, 112–113
Thrombosis
  cardiac catheterization and, 85
  time to treatment factors, 113
Thrombus
  dilated cardiomyopathy and,
    370–371
  mutuality with left ventricular
    aneurysm, 251–252
  tumor imaging and, 508
  ultrafast computed tomography,
    342
Time-density curve, 232 (Fig. 15-1),
  232 (Fig. 15-2), 233 (Fig. 15-3)
Time-density measurements, ultrafast
  CT scanner, 36
Time-of-flight effects, 44
Tomographic imaging, 20–22
  data acquisition, 21–22
  data processing, 22
  technique in, 20–21
Transesophageal echocardiography,
  158–159
Transesophageal imaging, 3
Transplantation
  phosphorus NMR spectroscopy,
    423
  transplant rejection, nuclear
    magnetic resonance, 398–399
Traumatic heart disease, ultrasonic
  tissue characterization, 413
Tricuspid atresia, 476
Tricuspid prolapse, ultrafast
  computed tomography, 344
Tricuspid regurgitation, 344
  Doppler echocardiography, 324–325
Tricuspid stenosis
  Doppler echocardiography, 323
  ultrafast computed tomography, 344
Tricuspid valve disease
  causes of, 289
  ventriculograms, 289
Tricuspid valve endocarditis, 303
  (Fig. 21-7)
Trypanosoma cruzi, 153
"Tumor blush," 506
Tumors. *See* Cardiac tumors
Turner's syndrome, 524
Two-dimensional echocardiogram
  acute idiopathic pericarditis, 447
    (Fig. 34-10)

pericardial effusion, 442 (Fig. 34-1),
  443 (Figs. 34-3, 34-4), 445
  (Fig. 34-7)
Two-dimensional echocardiography,
  2, 3, 4, 130–131, 252, 296 (Fig. 21-2)
  cardiac tumors, 508
  left ventricular assessment, 138–140
  rheumatic mitral stenosis, 302
    (Fig. 21-6)
  valvular disease, 296, 297, 300, 301

U

Ultrafast computed tomography
  acute myocardial infarction
    benefits of, 226, 253–254, 255
    infarct size analysis, 225
    limitations of, 226
    views of anterior wall infarct, 225
      (Fig. 14-2)
  artifacts, 34–35
  cardiac tumors, 510 (Fig. 40-6)
  cine mode, 32–33
  computed tomography, 543–544
  congenital heart defects
    blood flow quantification, 487
    cine mode, 489–491
    flow mode, 484–489
    preoperative/postoperative
      studies, 482 (Table 38-1)
    volume mode, 483–484
  coronary artery bypass graft
    evaluation, 273–276
  dilated cardiomyopathy, 391–393
  edge finding accuracy, 36–37
  fast raw averaging (FRA) board, 30
  flow mode, 32
  future developments, 37
  hospital installation, 31 (Fig. 3-3),
    32
  hypertrophic cardiomyopathy,
    386–390
  left ventricular aneurysm/thrombus,
    253–255
  limitations of, 36
  modulation transfer function (MTF),
    33–34 (Fig. 3-4), 35 (Fig. 3-5)
  myocardial blood flow evaluation,
    231–236
  noise, 34–35
  pericardial disease, 457
  Picker/Imatron FASTRAC, 30
  principle of, 30–31 (Fig. 3-2)
  restrictive cardiomyopathy, 390–391
  scanning rates, 36
  signal-to-noise ratio, 34
  spatial resolution, 33–34
  time-density measurements, 36
  valvular disease, 339–345

aortic valve, 342–344
mitral valve, 340–342
pulmonic valve, 344–345
regurgitant fractions assessment, 340
tricuspid valve, 344
ventricular function, 340
volume imaging, 33
Ultrafast CT bicycle ergometry
clinical experience, 244–248
data analysis, 241–244
left ventricular ejection fractions, 242–243
quantitative analysis, 243–244
rest/exercise ejection fractions, 243
wall motion analysis, 242, 243
development of, 240
results in coronary artery patients, 245 (Table 16-2)
technique in, 240–241
Ultrasonic tissue characterization
atherosclerosis, 413
cardiomyopathies, 411–413
diagnostic applications, 413–415
interview assessment, 414
ischemic heart disease, 407–411
principles in, 404–406
quantitative analysis, gray level data, 407
radio frequency signal analysis, 406
staging applications, 414
traumatic heart disease, 413
Ultrasound
advantages of, 1
esophageal ultrasound, 1
future view, 11
general uses of, 1
principles of, 6–8
role in cardiology, 1–4
system for Computer Acquisition of Echocardiographic Data (CAED), 8–11
Univentricular heart, 476, 478
Doppler echocardiography, 478
User-programmable offline workstation (UPOW) network, 37–38

V

Valvular anatomy, 294–295, 300
Valvular heart disease
aortic stenosis, 297–298, 299
calcification of aortic valve, 295, 296
cardiac catheterization and, 85
contrast angiography, 281–290
Doppler echocardiography, 311–326
echocardiographic evaluation, 293–306
mimicking diseases, 294 (Table 21-2)
nuclear magnetic resonance, 347–351
pulmonic stenosis, 298, 299
radionuclide angiography, 329–335
ultrafast computed tomography, 339–345
vegetation in valve, 296, 344
Valvular stenosis
Doppler echocardiography, 472–475
echocardiography, 472
Vascular surgery, dipyridamole thallium 201 imaging, 201–202
Vena cava, congenital abnormalities, 555
Ventricular-level lesions, nuclear magnetic resonance, 499
Ventricular septal defect, 150 (Fig. 10-41)
Ventricular volume
calculation of, 101–104
calculation variability, 103–104 (Table 8-2)
regression equations, 103 (Table 8-1)
Ventriculograms
aortic regurgitation, 285–286
clinical studies, 122
digital, 55–56
experimental studies, 121–122
left ventricle
aortic stenosis, 284–285
improving quality of, 281–282
qualitative interpretation, 282–283
quantitative interpretation, 283
low-dose, 56
low-contrast medium dose direct approach, 122
parametric flow reserve imaging, 124–125
regional ventricular function, 123
stress ventriculography, 124
technical aspects, 108–109
ventriculographic videodensitometry, 122–123
Ventriculographic videodensitometry, 122–123
Videodensitometry, 53–55
advantages of, 54
applications of, 54–55
principle of, 54
ventriculographic, 122–123
Virtual imaging, 38
VME random access memory, 36
Voxel, 44

W

Wall motion analysis
accuracy of measures, 104–105
centerline method, 106 (Fig. 8-3)
coordinate system models, 105 (Fig. 8-2), 106
coronary artery disease, 107–108
digital imaging assessment, 55–56
exercise methods, 248
graphic presentation of abnormality, 110 (Fig. 8-5), 111 (Figs. 8-5, 8-7)
left ventricular angiograms, 104–108
normal limits of wall motion, 106–107
ultrafast CT bicycle ergometry, 242, 243
ultrafast CT scanner, 32–33, 36–37
variability of measures, 107
Waterston shunt implantation, 498 (Fig. 39-6)

X

X-ray CT, coronary artery bypass graft evaluation, 273